DYNAMIC FACTOR MODELS

ADVANCES IN ECONOMETRICS

Series Editors: Thomas B. Fomby, R. Carter Hill, Ivan Jeliazkov, Juan Carlos Escanciano, Eric Hillebrand, David Jacho-Chávez, Daniel L. Millimet and Rodney Strachan

Recent Volumes:

Volume 27A:	Missing Data Methods: Cross-Sectional Methods and Applications – Edited by David M. Drukker
Volume 27B:	Missing Data Methods: Time-Series Methods and Applications – Edited by David M. Drukker
Volume 28:	DSGE Models in Macroeconomics: Estimation, Evaluation and New Developments – Edited by Nathan Balke, Fabio Canova, Fabio Milani and Mark Wynne
Volume 29:	Essays in Honor of Jerry Hausman – Edited by Badi H. Baltagi, Whitney Newey, Hal White and R. Carter Hill
Volume 30:	30th Anniversary Edition – Edited by Dek Terrell and Daniel Millmet
Volume 31:	Structural Econometric Models – Edited by Eugene Choo and Matthew Shum
Volume 32:	VAR Models in Macroeconomics — New Developments and Applications: Essays in Honor of Christopher A. Sims – Edited by Thomas B. Fomby, Lutz Kilian and Anthony Murphy
Volume 33:	Essays in Honor of Peter C. B. Phillips – Edited by Thomas B. Fomby, Yoosoon Chang and Joon Y. Park
Volume 34:	Bayesian Model Comparison – Edited by Ivan Jeliazkov and Dale J. Poirier

ADVANCES IN ECONOMETRICS VOLUME 35

DYNAMIC FACTOR MODELS

EDITED BY

ERIC HILLEBRAND
Department of Economics and Business Economics and CREATES, Aarhus University, Aarhus, Denmark

SIEM JAN KOOPMAN
Department of Econometrics, Vrije Universiteit Amsterdam, The Netherlands, Tinbergen Institute and CREATES

United Kingdom – North America – Japan
India – Malaysia – China

Emerald Group Publishing Limited
Howard House, Wagon Lane, Bingley BD16 1WA, UK

First edition 2016

Reprints and permissions service
Contact: permissions@emeraldinsight.com

British Library Cataloguing in Publication Data
A catalogue record for this book is available from the British Library

ISBN: 978-1-78560-353-2
ISSN: 0731-9053 (Series)

CONTENTS

PART II
FACTOR STRUCTURE AND SPECIFICATION

PART III
INSTABILITY

PART IV
NOWCASTING AND FORECASTING

LIST OF CONTRIBUTORS

Anindya Banerjee	Department of Economics, University of Birmingham, Edgbaston, Birmingham, UK
Martin Belvisi	KNG Securities, London, UK
Alexander Braumann	Institute of Statistics and Mathematical Methods in Economics, Vienna University of Technology, Vienna, Austria
Jörg Breitung	Institute of Econometrics and Statistics, University of Cologne, Cologne, Germany; Deutsche Bundesbank, Frankfurt, Germany
Laurent Callot	Department of Econometrics, VU University Amsterdam, Amsterdam, The Netherlands; Tinbergen Institute, Amsterdam, The Netherlands; CREATES, Aarhus University, Aarhus, Denmark
Maximo Camacho	University of Murcia, Murcia, Spain
Jens H. E. Christensen	Federal Reserve Bank of San Francisco, San Francisco, CA, USA
Antonello D'Agostino	European Stability Mechanism, Luxemburg, Luxemburg
Manfred Deistler	Institute of Statistics and Mathematical Methods in Economics, Vienna University of Technology, Vienna, Austria; Institute for Advanced Studies, Vienna, Austria
Davide Delle Monache	Bank of Italy, Rome, Italy
Catherine Doz	Paris School of Economics, Université Paris 1 Panthéon-Sorbonne, Paris, France

Sandra Eickmeier	Deutsche Bundesbank, Frankfurt, Germany; Centre for Applied Macroeconomic Analysis (CAMA), The Australian National University, Canberra, Australia
Elisabeth Felsenstein	Institute of Statistics and Mathematical Methods in Economics, Vienna University of Technology, Vienna, Austria
Gabriele Fiorentini	School of Economics, University of Florence, Florence, Italy
Alessandro Galesi	Banco de España, Madrid, Spain
Domenico Giannone	Federal Reserve Bank of New York, New York, NY, USA; CEPR, London, UK; ECARES, Brussels, Belgium; LUISS, Roma, Italy
Alessandro Giovannelli	Department of Economics, University of Rome Tor Vergata, Rome, Italy
Eric Hillebrand	Department of Economics and Business Economics and CREATES, Aarhus University, Aarhus, Denmark
Laura E. Jackson	Department of Economics, Bentley University, Waltham, MA, USA
Lukas Koelbl	Institute of Statistics and Mathematical Methods in Economics, Vienna University of Technology, Vienna, Austria
Siem Jan Koopman	Department of Econometrics, VU University, Amsterdam, The Netherlands; CREATES
M. Ayhan Kose	World Bank, Washington, DC, USA
Johannes Tang Kristensen	Department of Business and Economics, University of Southern Denmark, Odense, Denmark; CREATES, Aarhus University, Aarhus, Denmark

Danilo Leiva-Leon Central Bank of Chile, Santiago, Chile

Michele Lenza ECARES, Brussels, Belgium; European Central Bank, Frankfurt, Germany

Massimiliano Marcellino Department of Economics, Bocconi University, Milan, Italy

Igor Masten Faculty of Economics, University of Ljubljana, Ljubljana, Slovenia; Bank of Slovenia, Ljubljana, Slovenia

Michele Modugno Board of Governors of the Federal Reserve System, Washington, DC, USA

Christopher Otrok Department of Economics, University of Missouri, Columbia, MO, USA; Federal Reserve Bank of St. Louis, St. Louis, MO, USA

Michael T. Owyang Federal Reserve Bank of St. Louis, St. Louis, MO, USA

Sait R. Ozturk Erasmus School of Economics, Erasmus University Rotterdam, Rotterdam, The Netherlands

Gabriel Perez-Quiros Bank of Spain, Madrid, Spain; CEPR, London, UK

Ivan Petrella Bank of England, Birkbeck University of London and CEPR, London, UK

Anna Petronevich Paris School of Economics, Université Paris 1 Panthéon-Sorbonne, Paris, France; Università Ca'Foscari Venezia, Venice, Italy

Riccardo Pianeti University of Bergamo, Bergamo, Italy

Pilar Poncela Department of Economic Analysis: Quantitative Economics, Universidad Autónoma de Madrid, Madrid, Spain

Tommaso Proietti Department of Economic Analysis: Quantitative Economics, Universidad Autónoma de Madrid, Madrid, Spain

Glenn D. Rudebusch Federal Reserve Bank of San Francisco, San Francisco, CA, USA

Esther Ruiz Department of Statistics, Universidad Carlos III de Madrid, Madrid, Spain

Gerhard Rünstler European Central Bank, Frankfurt, Germany

Enrique Sentana Center for Monetary and Financial Studies (CEMFI), Madrid, Spain

James H. Stock Department of Economics, Harvard University and the NEBR, Cambridge, MA, USA

Giovanni Urga Cass Business School, City University London, London, UK; University of Bergamo, Bergamo, Italy

Michel van der Wel Erasmus School of Economics, Erasmus University Rotterdam, Rotterdam, The Netherlands

Dick van Dijk Erasmus School of Economics, Erasmus University Rotterdam, Rotterdam, The Netherlands

Fabrizio Venditti Bank of Italy, Rome, Italy

Mark W. Watson Department of Economics and the Woodrow Wilson School, Princeton University and the NEBR, Princeton, NJ, USA

EDITORIAL INTRODUCTION

Dynamic factor models constitute an active and growing area of research in econometrics, in macroeconomics, and in finance. Many applications center around key policy questions that are raised by recent financial and sovereign debt crises, such as the connections between yields on government debt, credit risk, inflation, and economic growth.

We are very pleased to introduce this volume of the Advances in Econometrics series dedicated to dynamic factor models. In editing this volume, we have strived to remain true to the spirit of Advances in Econometrics to identify a current and active subject in econometrics and collect a number of contributions from both established and junior researchers to this subject. The contributions paint a picture of the state of the art of the research in the area, so that a researcher venturing into the area can get an overview of the themes being discussed from this collection.

A selection of papers in this volume reflects ongoing research into the further development of dynamic factor models with the aim to adapt them to data of different natures: Banerjee, Marcellino, and Masten consider the simultaneous treatment of integrated and stationary time series in a dynamic factor model; Koelbl, Braumann, Felsenstein, and Deistler consider time series analyses for mixed sampling frequencies. Christensen and Rudebusch consider yield-curve data, where the strong structure and natural ordering of the data allows for Nelson-Siegel polynomials as loadings. Van der Wel, Ozturk, and van Dijk consider a novel modeling approach to the implied volatility surface, where a natural ordering and strong data structure allow for a dynamic factor model with a deterministic loading matrix. Another prominent theme, also closely connected to the nature of the data, is factor structure and specification. Multi-level factor models are considered by Breitung and Eickmeier, who compare a sequential least squares algorithm, a two-step estimation approach based on canonical correlations, and existing Bayesian estimation methods. Fiorentini, Galesi, and Sentana propose a bi-factor version of their spectral EM-algorithm for dynamic factor models with loading structures. Camacho, Leiva-Leon, and Perez-Quiros consider a model for macroeconomic data with two distinct factors that disentangle real dynamics and inflation dynamics; Belvisi, Pianeti, and

Urga specify a model for a country panel of asset returns that features a global factor, a country factor, and an asset-class factor. Jackson, Kose, Otrok, and Owyang compare Bayesian and classical principal components estimators for multi-level factor models. Poncela and Ruiz compare factor estimates based on different cross-sectional dimensions, maximum likelihood estimates based on the Kalman filter and smoother, and principal components.

Time-varying parameters are a theme that has also come into focus in the dynamic factor model literature. In this volume, we present three papers where different approaches are considered for model instability. Callot and Kristensen propose a parsimonious random walk process for the loadings with probability mass at the zero-increment to capture infrequent variation. Doz and Petronevich consider a Markov-Switching specification for the mean of a factor. Delle Monache, Petrella, and Venditti investigate a score-driven dynamic model for the means of the factors, but also for variance parameters in measurement and transition equations.

Nowcasting and forecasting are still the major motivations for dynamic factor analysis. D'Agostino, Giannone, Lenza, and Modugno consider heterogeneous lead-lag structures that a forecaster usually faces at the end of the samples of many different predictors, due to their different publication lags. Giovanelli and Proietti take a forecasting point of view and consider a supervision strategy that selects principal components according to their correlation with the forecast target, with a focus on the sequential testing problem that arises. Rünstler compares Least Angle Regression and prediction weights as competing methods in selecting predictors prior to principal component estimation of the factors.

The 16th Advances in Econometrics conference, in which most of the papers collected in this volume were presented and discussed, took place in November 14–16, 2014, at CREATES (Center of Research in Econometric Analysis of Time Series), Department of Economics and Business, Aarhus University, Denmark. We gratefully acknowledge financial support from the Research Executive Agency of the European Commission under the Marie Curie CIG program, grant number 333701. We would like to thank the participants of the conference, the contributors to the volume, and the referees involved in the revision process. Special thanks are due to James Stock and Mark Watson for their spontaneous contribution on short notice. Last not least, Solveig Nygaard Sørensen provided excellent logistical and administrative support for the conference.

DYNAMIC FACTOR MODELS: A BRIEF RETROSPECTIVE

This volume pulls together an impressive collection of papers that map out the frontier of the methods and applications of dynamic factor models (DFMs) in macroeconomics. The diversity of topics in this collection shows the wide array of issues that can be elucidated using DFMs.

In this brief note, we step back to ask what accounts for the ongoing appeal of dynamic factor models as a framework for empirical macroeconomic issues. After all, DFMs are the earliest "big data" method in macroeconometrics, and perhaps the earliest big data tool in empirical economics more broadly. The toolkit for handling large data sets has grown tremendously in the past 15 years, yet DFMs remain the dominant method for handling large numbers of series in macroeconomics. Why?

We can think of six reasons.

First, and most importantly, DFMs fit the data. The starting point for DFMs is that a small number of factors can explain dynamic comovements of many series. Of course, this need not be true – but it is. Although the DFM formalization stems from Geweke (1977) and Sargent and Sims (1977), the idea that many series exhibit comovements that stem from common cyclical fundamentals is a very old one and dates at least to Burns and Mitchell (1946).[1] In their pioneering paper, Sargent and Sims (1977) showed that a single-factor model explains more than three-fourths of the variance in major monthly activity variables, and that adding a second factor explains prices. As discussed by Watson (2004), this basic fact has been remarkably stable, and is a robust finding using state space methods, principal components, or hybrid methods.

Second, the key DFM restriction of a small number of latent factors is consistent with simple equilibrium macroeconomic theories. Sargent and Sims (1977) cite Lucas (1975) for providing a theoretically motivated single-index macroeconomic model. Sargent (1989) explicitly takes a single-shock macroeconomic model and shows how, when data are measured with error, it implies a DFM that can be estimated using state space methods. Boivin and Giannoni (2006) extend Sargent's (1989) work and show that a log-linearized dynamic stochastic general equilibrium model implies a DFM with overidentifying restrictions, and then go on to estimate the resulting DFM. To be sure, a linear DFM will miss potentially important nonlinearities like the zero lower bound on interest rates, and modern macroeconomic models are in general nonlinear, however for many purposes their implied linear approximations are at least a good starting point.

Third, techniques developed in the past 15 years – the application to DFMs of principal components analysis and refinements – allow DFMs to be estimated using large data sets. As is now well-understood, the key insight that allows the extension of DFMs to high-dimensional data is that while having many series in typical regression applications is a curse, in DFMs it is a blessing because it harnesses the cross-sectional dimension to improve estimation of the unobserved factors. Geweke (1993) provided an early articulation of the intuition behind this idea (although his conference volume remarks seem to have been largely overlooked). Formal demonstrations of this point under different assumptions were provided by Connor and Korajczyk (1986), Stock and Watson (2002), Forni, Hallin, Lippi, and Reichlin (2004), and Bai and Ng (2006). Now, DFM applications with hundreds of series are commonplace, although experience has shown that in some applications the benefits of cross-sectional averaging can be achieved with fewer series.

Fourth, DFMs provide a flexible and effective framework for practical tasks of professional macroeconomists such as real-time monitoring and construction of activity indexes. Many of these applications require handling multiple series with inevitable data irregularities such as series that start, stop, are redefined, have missing values, and have mixed frequencies. State space DFMs, which build on early work by Harvey and coauthors summarized in Harvey (1989), are well-suited to handling these data irregularities. Recent developments in this area are surveyed by Bańbura, Giannone, Modugno, and Reichlin (2013). Moreover, DFMs do not need to be large to be useful for nowcasting, as was demonstrated by a small weekly DFM using non-government data that was developed by the Council of Economic Advisors to track economic activity during the shutdown of the U.S. Government in October, 2013 (Council of Economic Advisers [CEA], 2013).

Fifth, because DFMs can handle arbitrarily large data sets, forecasts based on DFMs can have rich information sets but still produce well-estimated forecasting models. In this sense, they provide a solution to the big data "large *p*" prediction problem, in which the number of predictors can exceed the number of (time series) observations. Moreover, under the assumptions of the DFM, the predictions are consistent in the sense that the difference between the forecast using the estimated and the true factor goes to zero in mean square as the number of series increases. Similar results are available using machine learning methods under alternative assumptions, for example, the LASSO achieves analogous results under sparsity assumptions. From a prediction perspective, DFMs and LASSO are both providing solutions to the many-predictor problem. We would argue that features of economic data and economic theory suggest that the properties of DFMs are more appealing than (to stick with the LASSO example) sparsity. For example, suppose three of the predictors are the 90-day, 5-year, and 30-year Treasury rates. How should they enter? In levels? As two levels and a long-short term spread? As a level, a term spread, and a curvature measure? All the DFM methods we know of are rotation invariant, so from the perspective of prediction, the practitioner does not need to choose, it suffices just to include the three interest rates. But the LASSO is not rotation invariant so it matters how these are entered, placing a difficult preliminary burden on the empirical economist – especially if there are 20 interest rates, not 3. In some applications, sparsity makes sense: in a model of neural connections, one neuron connects directly to only a few of the very many other neurons, and it is not connected to the "spread" between two other neurons, whatever that might mean. In economics, however, rotation invariance is appealing, while strict sparsity is not. More generally, the restrictions imposed by DFMs accord with the long history of empirical economic experience developed by researchers from Burns and Mitchell (1946) onwards.

Sixth, DFMs hold out the promise of solving some vexing problems of empirical macroeconomics that arise when using structural vector autoregressions (VARs). As is well known, a key identification condition for structural VAR analysis is that the space of the VAR innovations spans the structural shock(s) of interest, so that the structural shock can be recovered as a linear combination of the VAR innovations. But this condition runs into the practical problem that the space of the innovations depends on which variables are included in the VAR. One might think to ensure that the innovations span the structural shock by including very many variables in the VAR, but this introduces the problem of VAR parameter proliferation. One solution to this problem is to adopt high-dimensional structural DFM methods, which with their many variables improve the ability of the innovations to span the space of the structural shock. As Forni, Hallin, Lippi, and Reichlin (2000) show, under certain conditions DFMs provide an avenue for meeting this spanning requirement and thus resolving at least some of the issues associated with non-invertible structural VARs. The invertibility problem aside, as stressed by Bernanke, Boivin, and Eliasz (2005), structural DFMs (in their work, the closely related factor-augmented VAR) provide a simple and coherent framework for estimating structural impulse response functions for a large number of variables. As discussed in Stock and Watson (2015), the identification schemes of structural VARs carry over in a straightforward way to structural DFMs.

Looking ahead, the work in this volume points to some promising if challenging areas for further work. On the methodological side, developing tractable methods for exploiting and modeling nonlinearities in DFMs could prove important for some applications. Similarly, as is increasingly well documented, there have been important shifts in macroeconomic relations. Although there are methods for handling time variation in DFMs, an open challenge is allowing for time variation while maintaining the focus on having well-estimated factors. One area that we consider to be particularly promising for future work is that mentioned in the final point above, structural DFMs. Although there has been some work on structural DFMs (e.g., Stock and Watson (2012) use the method of external instruments), the amount of research on structural DFMs, both theory and empirical, is relatively small. We are confident that much interesting work remains to be done.

James H. Stock
Department of Economics,
Harvard University and the NBER, Cambridge, MA, USA

Mark W. Watson
Department of Economics and the
Woodrow Wilson School,
Princeton University and the NBER, Princeton, NJ, USA

NOTE

1. In fact, Burns and Mitchell (1946) adopt the perspective that the business cycle is defined by the comovements of many series: "Our definition [of business cycles] presents business cycles as a consensus among expansions in 'many' economic activities, followed by 'similarly general' recessions, contractions, and revivals. How 'general' these movements are, what types of activity share in them and what do not, how the consensus differs from one cyclical phase to another, and from one business cycle to the next, can be learned only by empirical observation" (p. 6). It is not much of a stretch to interpret this passage as saying that their research aims to

cull from multiple series a single common aggregate measure of a common latent business cycle; how different is this conceptually from estimating a latent factor using DFM methods?

REFERENCES

Bai, J., & Ng, S. (2006). Confidence intervals for diffusion index forecasts and inference for factor-augmented regressions. *Econometrica*, *74*, 1133–1150.

Bańbura, M., Giannone, D., Modugno, M., & Reichlin, L. (2013). Nowcasting and the real time data flow. In G. Elliott & A. Timmerman (Eds.), *Handbook of economic forecasting* (Vol. 2). New York, NY: Elsevier.

Bernanke, B. S., Boivin, J., & Eliasz, P. (2005). Measuring the effects of monetary policy: A Factor-Augmented Vector Autoregressive (FAVAR) approach. *Quarterly Journal of Economics*, *120*(February), 387–422.

Boivin, J., & Giannoni, M. (2006). *DSGE models in a data-rich environment*. NBER Working Paper No. 12772.

Burns, A. F., & Mitchell, W. C. (1946). *Measuring business cycles*. New York, NY: NBER.

Connor, G., & Korajczyk, R. A. (1986). Performance measurement with the arbitrage pricing theory. *Journal of Financial Economics*, *15*, 373–394.

Council of Economic Advisers. (2013). *Economic activity during the government shutdown and debt limit brinksmanship*. Council of Economic Advisers, The White House. Retrieved from http://www.whitehouse.gov/sites/default/files/docs/weekly_indicators_report_final.pdf

Forni, M., Hallin, M., Lippi, M., & Reichlin, L. (2000). The generalized factor model: Identification and estimation. *Review of Economics and Statistics*, *82*, 540–554.

Forni, M., Hallin, M., Lippi, M., & Reichlin, L. (2004). The generalized factor model: Consistency and rates. *Journal of Econometrics*, *119*, 231–255.

Geweke, J. (1977). The dynamic factor analysis of economic time series. In D. J. Aigner & A. S. Goldberger (Eds.), *Latent variables in socio-economic models*. Amsterdam: North-Holland.

Geweke, J. (1993). Comment on Quah and Sargent. In J. H. Stock & M. W. Watson (Eds.), *Business cycles, indicators, and forecasting*. (pp. 306–309). Chicago, IL: University of Chicago Press for the NBER.

Harvey, A. C. (1989). *Forecasting, structural time series models and the Kalman Filter*. Cambridge, UK: Cambridge University Press..

Lucas, R. E., Jr. (1975). An equilibrium model of the business cycle. *Journal of Political Economy*, *83*, 1113–1144.

Sargent, T. J. (1989). Two models of measurement and the investment accelerator. *Journal of Political Economy*, *97*, 251–287.

Sargent, T. J., & Sims, C. A. (1977). Business cycle modeling without pretending to have too much a-priori economic theory. In *New methods in business cycle research: Proceedings from a conference*. Minneapolis, MN: Federal Reserve Bank of Minneapolis.

Stock, J. H., & Watson, M. W. (2002). Forecasting using principal components from a large number of predictors. *Journal of the American Statistical Association*, *97*, 1167–1179.

Stock, J. H., & Watson, M. W. (2012). Disentangling the channels of the 2007–2009 recession. *Brookings Papers on Economic Activity*, Spring, 481–493.

Stock, J. H., & Watson, M. W. (2015). Factor models and structural vector autoregressions in macroeconomics. In *Handbook of macroeconomics* (Vol. 2). (forthcoming).

Watson, M. W. (2004). Comment on Giannone, Reichlin, and Sala. *NBER Macroeconomic Annual*, 216–221.

PART I
METHODOLOGY

AN OVERVIEW OF THE FACTOR-AUGMENTED ERROR-CORRECTION MODEL

Anindya Banerjee[a], Massimiliano Marcellino[b] and Igor Masten[c,d]

[a]*Department of Economics, University of Birmingham, Edgbaston, Birmingham, United Kingdom*
[b]*Department of Economics, Bocconi University, IGIER and CEPR, Milan, Italy*
[c]*Faculty of Economics, University of Ljubljana, Ljubljana, Slovenia*
[d]*Bank of Slovenia, Ljubljana, Slovenia*

ABSTRACT

The Factor-augmented Error-Correction Model (FECM) generalizes the factor-augmented VAR (FAVAR) and the Error-Correction Model (ECM), combining error-correction, cointegration and dynamic factor models. It uses a larger set of variables compared to the ECM and incorporates the long-run information lacking from the FAVAR because of the latter's specification in differences. In this paper, we review the specification and estimation of the FECM, and illustrate its use for

Dynamic Factor Models
Advances in Econometrics, Volume 35, 3–41

ISSN: 0731-9053/doi:10.1108/S0731-905320150000035001

forecasting and structural analysis by means of empirical applications based on Euro Area and US data.

Keywords: Dynamic factor models; cointegration; structural analysis; factor-augmented error correction models; FAVAR

JEL classifications: C32; E17

1. INTRODUCTION

Banerjee and Marcellino (2009) introduced the factor-augmented error correction model (FECM) as a way of bringing together two important recent strands of the econometric literature, namely, cointegration (e.g. Engle & Granger, 1987; Johansen 1995) and large dynamic factor models (e.g. Forni, Hallin, Lippi, & Reichlin, 2000; Stock & Watson, 2002a, 2002b).

Several papers have emphasized the complexity of modelling large systems of equations in which the complete cointegrating space may be difficult to identify, see, for example, Clements and Hendry (1995). At the same time, large dynamic factor models and factor augmented VARs (FAVARs, e.g. Bernanke, Boivin, & Eliasz, 2005; Stock & Watson, 2005) typically focus on variables in first differences in order to achieve stationarity. In the FECM, factors extracted from large datasets in levels, as a proxy for the non-stationary common trends, are jointly modelled with selected economic variables of interest, with which the factors can cointegrate. In this sense, the FECM nests both ECM and FAVAR models, and can be expected to produce better results, at least when the underlying conditions for consistent factor and parameter estimation are satisfied, and cointegration matters.

Banerjee, Marcellino, and Masten (2014a) assessed the forecasting performance of the FECM in comparison with the ECM and the FAVAR. Empirically, the relative ranking of the ECM, the FECM and the FAVAR depends upon the variables being modelled and the features of the processes generating the data, such as the amount and strength of cointegration, the degree of lagged dependence in the models and the forecasting horizon. However, in general, the FECM tends to perform better than both the ECM and the FAVAR.

Banerjee, Marcellino, and Masten (2014b) evaluated the use of FECM for structural analysis. Starting from a dynamic factor model for non-stationary data as in Bai (2004), they derived the moving-average representation of the

FECM and showed how the latter can be used to identify structural shocks and their propagation mechanism, using techniques similar to those adopted by the structural VAR literature.

The FECM model is related to the framework used recently to formulate testing for cointegration in panels (see, e.g. Bai, Kao, & Ng, 2009; Gengenbach, Urbain, & Westerlund, 2008). While prima facie the approaches are similar, there are several important differences. First, in panel cointegration, the dimension of the dataset (given by the set of variables amongst which cointegration is tested) remains finite and the units of the panel $i = 1, 2, \ldots, N$ provide repeated information on the cointegration vectors. By contrast in our framework the dataset in principle is infinite dimensional and driven by a finite number of common trends. Second, following from the first, the role of the factors (whether integrated or stationary) is also different as in the panel cointegrating framework the factors capture cross-section dependence while not being cointegrated with the vector of variables of interest. In our approach, this is precisely what is allowed, since the cointegration between the variables and the factors proxies for the missing cointegrating information in the whole dataset.

Another connected though different paper is Barigozzi, Lippi, and Luciani (2014). They work with a non-parametric static version of the factor model with common $I(1)$ factors only, while in our context we have a parametric representation of a fully dynamic model where the factors can be both $I(1)$ and $I(0)$, which complicates the analysis, in particular for structural applications related to permanent shocks (see Banerjee et al., 2014b). They also assume that the factors follow a VAR model, and show that their first differences admit a finite-order ECM representation, which is an interesting result. In contrast, we focus on cointegration between the factors and the observable variables. The Barigozzi et al. (2014) model is also similar to the one analysed by Bai (2004). Bai did not consider impulse responses but in his context one could easily get consistent estimates of the responses based on the factors in levels (rather than using the ECM model for the factors as in Barigozzi et al., 2014).

In this paper, we review the specification and estimation of the FECM. We then illustrate its use for forecasting and structural analysis by means of novel empirical applications based on Euro Area and US data.

For the Euro Area we use 38 quarterly macroeconomic time series from the 2013 update of the Euro Area Wide Model (AWM) dataset, over the period 1975–2012. For forecasting, we focus on two subsets of three variables each, one real and one nominal. The real set consists of real GDP, real private consumption and real exports. The nominal system, on the other hand, contains the harmonized index of consumer prices (HICP), unit labor costs and the effective nominal exchange rate of the euro. For each set, we consider forecasting over the period 2002–2013 and, to investigate the effect the Great Recession might have had on forecasting performance of competing models, we also split the forecasting sample into 2002q1–2008q3 and 2008q4–2012q4. As forecasting models, we use AR, VAR and ECM specifications, with or without factors. For the real variables, the FECM is clearly the best forecasting model, and a comparison with the FAVAR highlights the importance of including the error correction terms. For the nominal variables, the FECM also performs well if the factors are extracted from a subset of the nominal variables only, preselected as in Boivin and Ng (2006). In terms of the effects of the crisis, the performance of the FECM generally further improves.

For the United States, we use the set of monthly real and nominal macroeconomic series from the Federal Reserve Economic Database (FRED). The dataset contains over 129 macroeconomic series over the period 1960–2014. As real variables, we consider forecasting total industrial production, personal income less transfers, employment on non-agricultural payrolls, and real manufacturing trade and sales. As nominal variables we focus on the producer price index, consumer price index, consumer prices without food prices and private consumption deflator. Similar to the case of the Euro Area data, we divide our forecasting period into the Great recession (2007m7–2014m7) and before it (2000m1–2007m6). The results are again encouraging for the FECM, especially in the period of Great recession.

In both the Euro Area and US forecasting applications we compare the results from our basic FECM estimation approach that requires all the idiosyncratic errors to be $I(0)$ with an alternative method, based on variables in differences, where the idiosyncratic errors can also be $I(1)$. We find that both methods perform similarly and this finding, in addition to the outcome of formal testing procedures that generally do not reject the hypothesis of $I(0)$ idiosyncratic errors, provides support for our basic FECM estimation method.

Finally, as an illustration of the use of the FECM for structural analysis, we assess the effects of a monetary policy shock. Specifically, we replicate the FAVAR based analysis of Bernanke et al. (2005) in our FECM context, based on the extended US dataset. The shape of the impulse responses is overall similar across the models for most variables. Quantitatively, however, the responses may differ significantly due to the error-correction terms. For example, quite significant differences are observed for monetary aggregates, the yen–dollar exchange rate, consumer and commodity prices, wages and personal consumption. Omission of the error-correction terms in the FAVAR model can thus have an important impact on the empirical results.

The paper is structured as follows. Section 2 reviews the representation and estimation of the FECM model, and then specializes the results for the cases of forecasting and structural analysis. Section 3 discusses the data and the models used in the empirical applications. Section 4 presents forecasting results, while Section 5 presents the analysis of monetary policy shocks with the FECM. Section 6 concludes.

2. FACTOR-AUGMENTED ERROR-CORRECTION MODEL

In this section, we derive the FECM, relying on Banerjee et al. (2014b) to whom we refer for additional details. The starting point of our analysis is the dynamic factor model for $I(1)$ data with both $I(1)$ and $I(0)$ factors, which allows us to distinguish between common stochastic trends and stationary drivers of all variables. We start by deriving the theoretical representation of the FECM. In the empirical applications of the paper, however, the FECM is used for forecasting and structural analysis. These applications require estimable versions of the FECM, which we present in turn in two separate subsections.

2.1. Representation of the FECM

Consider the following dynamic factor model (DFM) for $I(1)$ data:

$$\begin{aligned} X_{it} &= \sum_{j=0}^{p} \lambda_{ij} F_{t-j} + \sum_{l=0}^{m} \phi_{il} c_{t-l} + \varepsilon_{it} \\ &= \lambda_i(L) F_t + \phi_i(L) c_t + \varepsilon_{it} \end{aligned} \tag{1}$$

where $i = 1, \ldots, N$, $t = 1, \ldots, T$, F_t is an r_1-dimensional vector of random walks, c_t is an r_2-dimensional vector of $I(0)$ factors, $F_t = c_t = 0$ for $t<0$, and ε_{it} is a zero-mean idiosyncratic component. $\lambda_i(L)$ and $\varphi_i(L)$ are lag polynomials of orders p and m, respectively, which are assumed to be finite.

The loadings λ_{ij} and ϕ_{ij} are either deterministic or stochastic and satisfy the following restrictions. For $\lambda_i = \lambda_i(1)$ and $\phi_i = \phi_i(1)$, we have

$E\|\lambda_i\|^4 \leqslant M < \infty$, $E\|\phi_i\|^4 \leqslant M < \infty$, and $1/N \sum_{i=0}^{N} \lambda_i \lambda_i'$, $1/N \sum_{i=0}^{N} \phi_i \phi_i'$ converge in probability to positive-definite matrices. Furthermore, we assume that $E(\lambda_{ij}\varepsilon_{is}) = E(\phi_{ij}\varepsilon_{is}) = 0$ for all i, j and s. The idiosyncratic component ε_{it} can be in principle serially and cross-correlated. Specifically, for $\varepsilon_t = [\varepsilon_{1t}, \ldots, \varepsilon_{Nt}]'$ we assume that

$$\varepsilon_t = \Gamma(L)\varepsilon_{t-1} + v_t \tag{2}$$

where v_t are orthogonal white noise errors. If the roots of $\Gamma(L)$ lie inside the unit disc for all i, the model fits the framework of Bai (2004). This assumption implies that X_{it} and F_t cointegrate. If instead ε_{it} are $I(1)$ for some i, then our model fits the framework of Bai and Ng (2004). The following derivation of the FECM representation accommodates both cases.

To derive the FECM and discuss further assumptions upon the model that ensure consistent estimation of the model's components, it is convenient to write first the DFM in static form. To this end, we follow Bai (2004) and define

$$\tilde{\lambda}_{ik} = \lambda_{ik} + \lambda_{ik+1} + \cdots + \lambda_{ip}, \quad k = 0, \ldots, p$$

Let us in addition define

$$\tilde{\Phi}_i = \left[\phi_{i0}, \ldots, \phi_{im}\right]$$

Then, we can obtain a static representation of the DFM which has the $I(1)$ factors isolated from the $I(0)$ factors:

$$X_{it} = \Lambda_i F_t + \Phi_i G_t + \varepsilon_{it} \tag{3}$$

where

$$\begin{aligned} \Lambda_i &= \tilde{\lambda}_{i0} \\ \Phi_i &= \left[\tilde{\Phi}_i, -\tilde{\lambda}_{i1}, \ldots, -\tilde{\lambda}_{ip}\right] \\ G_t &= \left[c_t', c_{t-1}', \ldots, c_{t-m}', \Delta F_t', \ldots, \Delta F_{t-p+1}'\right]' \end{aligned}$$

Introducing for convenience the notation $\Psi_i = \left[\Lambda_i', \Phi_i'\right]'$, the following assumptions are also needed for consistent estimation of both the $I(1)$

and $I(0)$ factors: $E\|\Psi_i\|^4 \leqslant M < \infty$ and $1/N \sum_{i=0}^{N} \Psi_i \Psi_i'$ converges to a $(r_1(p+1)+r_2(m+1)) \times (r_1(p+1)+r_2(m+1))$ positive-definite matrix.

Grouping across the N variables we have

$$X_t = \Lambda F_t + \Phi G_t + \varepsilon_t \tag{4}$$

where $X_t = [X_{1t}, \ldots, X_{Nt}]'$, $\Lambda = [\Lambda_1', \ldots, \Lambda_N']'$, $\Phi = [\Phi_1', \ldots, \Phi_N']'$ and $\varepsilon_t = [\varepsilon_{1t}, \ldots, \varepsilon_{Nt}]'$.

The serial correlation of the idiosyncratic component in Eq. (4) can be eliminated from the error process by pre-multiplying (3) by $I - \Gamma(L)L$. As shown in Banerjee et al. (2014b), straightforward manipulation leads to the ECM form of the DFM, which is the FECM, specified as:

$$\begin{aligned} \Delta X_t = {} & \alpha(X_{t-1} - \Lambda F_{t-1}) + \Lambda \Delta F_t + \Gamma_1(L)\Lambda \Delta F_{t-1} \\ & + \Phi G_t - \Gamma(1)\Phi G_{t-1} + \Gamma_1(L)\Phi \Delta G_{t-1} - \Gamma_1(L)\Delta X_{t-1} + v_t \end{aligned} \tag{5}$$

where $\alpha = -(I - \Gamma(1))$ and we use the factorization

$$\Gamma(L) = \Gamma(1) - \Gamma_1(L)(1 - L)$$

Equation (5) is a representation of the DFM in Eq. (1) in terms of stationary variables. From it, we can directly observe the main distinction between an FAVAR model and the FECM. The latter contains the error-correction term, $\alpha(X_{t-1} - \Lambda F_{t-1})$, while in the FAVAR model this term is omitted, leading to an omitted variables problem.

Empirically, the error-correction term can have a significant role. Banerjee et al. (2014b) report for the US data that 63 out of 77 equations for the $I(1)$ variables contain a statistically significant error-correction term. For the Euro Area dataset analysed in this paper, the score is 27 out of 32 $I(1)$ variables.

Note that it follows from Eq. (4) that

$$X_{t-1} - \Lambda F_{t-1} = \Phi G_{t-1} + \varepsilon_{t-1}$$

such that it would appear at first sight that the omitted error-correction term in the FAVAR could be approximated by including additional lags of the $I(0)$ factors. However, by substituting the previous expression into Eq. (5) and simplifying we get

$$\Delta X_t = \Lambda \Delta F_t + \Phi \Delta G_t + \Delta \varepsilon_t \tag{6}$$

which contains a non-invertible MA component. An alternative, but equivalent, way of obtaining (6) is by simply differencing (4), which is effectively achieved by differencing the $I(1)$ data. Conventional structural analysis in an FAVAR framework relies on inverting a system like (6) (see Stock & Watson, 2005 and the survey in Luetkepohl, 2014). Hence, whenever we deal with $I(1)$ data, and many macroeconomic series exhibit this feature, the standard FAVAR model produces biased results unless we use an infinite number of factors as regressors, or account explicitly for the non-invertible MA structure of the error-process.

To complete the model, we assume that the non-stationary factors follow a vector random walk process

$$F_t = F_{t-1} + \varepsilon_t^F \tag{7}$$

while the stationary factors are represented by

$$c_t = \rho c_{t-1} + \varepsilon_t^c \tag{8}$$

where ρ is a diagonal matrix with values on the diagonal in absolute term strictly less than one. ε_t^F and ε_t^c are independent of λ_{ij}, ϕ_{ij} and ε_{it} for any i, j, t. It should be noted that the error processes ε_t^F and ε_t^c need not necessarily be i.i.d. They are allowed to be serially and cross correlated and jointly follow a stable vector process:

$$\begin{bmatrix} \varepsilon_t^F \\ \varepsilon_t^c \end{bmatrix} = A(L) \begin{bmatrix} \varepsilon_{t-1}^F \\ \varepsilon_{t-1}^c \end{bmatrix} + \begin{bmatrix} u_t \\ w_t \end{bmatrix} \tag{9}$$

where u_t and w_t are zero-mean white-noise innovations to dynamic non-stationary and stationary factors, respectively. Under the stability assumption, we can express the model as

$$\begin{bmatrix} \varepsilon_t^F \\ \varepsilon_t^c \end{bmatrix} = [I - A(L)L]^{-1} \begin{bmatrix} u_t \\ w_t \end{bmatrix} \tag{10}$$

Note that, under these assumptions, we have $E\|\varepsilon_t^F\|^4 \leqslant M < \infty$, which implies that $\sum_{t=1}^T F_t F_t'$ converges at rate T^2, while $\sum_{t=1}^T G_t G_t'$ converges at the standard rate T. The cross-product matrices $\sum_{t=1}^T F_t G_t'$ and $\sum_{t=1}^T G_t' F_t$ converge at rate $T^{3/2}$. At these rates, the elements of the matrix composed of these four elements jointly converge to form a positive-definite matrix.

Using Eqs. (7), (8) in Eq. (9) implies that the latter can be rewritten as

$$\begin{bmatrix} (I-L) & 0 \\ 0 & (I-\rho L) \end{bmatrix} \begin{bmatrix} F_t \\ c_t \end{bmatrix} = A(L) \begin{bmatrix} (I-L) & 0 \\ 0 & (I-\rho L) \end{bmatrix} \begin{bmatrix} F_{t-1} \\ c_{t-1} \end{bmatrix} + \begin{bmatrix} u_t \\ w_t \end{bmatrix}$$

from which the VAR for the factors follows as

$$\begin{aligned} \begin{bmatrix} F_t \\ c_t \end{bmatrix} &= \left[\begin{bmatrix} I & 0 \\ 0 & \rho \end{bmatrix} + A(L) \right] \begin{bmatrix} F_{t-1} \\ c_{t-1} \end{bmatrix} - A(L) \begin{bmatrix} I & 0 \\ 0 & \rho \end{bmatrix} \begin{bmatrix} F_{t-2} \\ c_{t-2} \end{bmatrix} + \begin{bmatrix} u_t \\ w_t \end{bmatrix} \\ &= C(L) \begin{bmatrix} F_{t-1} \\ c_{t-1} \end{bmatrix} + \begin{bmatrix} u_t \\ w_t \end{bmatrix} \end{aligned} \tag{11}$$

where the parameter restrictions imply that $C(1)$ is a block-diagonal matrix with block sizes corresponding to the partition between F_t and c_t.

The FECM is specified in terms of static factors F and G, which calls for a corresponding VAR specification. Using the definition of G_t in Eq. (11), it is convenient to write the VAR for the static factors as

$$\begin{bmatrix} F_t \\ G_t \end{bmatrix} = M(L) \begin{bmatrix} F_{t-1} \\ G_{t-1} \end{bmatrix} + Q \begin{bmatrix} u_t \\ w_t \end{bmatrix} \tag{12}$$

where the $(r_1(p+1)+r_2(m+1)) \times (r_1+r_2)$ matrix Q accounts for dynamic singularity of G_t. This is due to the fact that the dimension of the vector process w_t is r_2, which is smaller than or equal to $r_1 p + r_2(m+1)$, the dimension of G_t. In what follows we assume that the order of the VAR in Eq. (12) is n. The matrix polynomial $M(L)$ has the following structure:

$$M(L) = \begin{bmatrix} C_{11}(L) & C_{12}(L) & 0 & \ldots & & \ldots & 0 \\ C_{21}(L) & C_{22}(L) & 0 & \ldots & & \ldots & 0 \\ 0 & I & 0 & \ldots & & \ldots & 0 \\ \vdots & & \ldots & & \ldots & & \vdots \\ 0 & \ldots & \ldots & I & 0 & \ldots & 0 \\ C_{11}(L)-I & C_{12}(L) & & \ldots & & \ldots & 0 \\ 0 & & \ldots & & I & \ldots & 0 \\ \vdots & & & & & & \vdots \\ 0 & \ldots & & \ldots & & I & 0 \end{bmatrix}$$

Note, however, that in the empirical applications we do not impose the parameter restrictions implied by the structure of $M(L)$ since the principal component estimator that we use to estimate the factors identifies only the space spanned by G_t and not c_t and ΔF_t separately.

2.2. The FECM Form for Forecasting

The specification in Eq. (5) is not a convenient forecasting model as it is heavily parameterized, which makes it very difficult or even impossible to estimate with standard techniques when N is large. Hence, we focus on forecasting a small set of variables, as in Banerjee et al. (2014b). These variables of interest, a subset of X, are denoted by X_A.

According to Eq. (5) X_{At} cointegrate with F_t, which means that we can model them with an error-correction specification. Note, however, that we need to incorporate into the model also the information in the $I(0)$ factors G_t. Given that the FECM model (5) can be re-written also as

$$\begin{aligned}\Delta X_t = &\alpha(X_{t-1} - \Lambda F_{t-1} - \Phi G_{t-1}) + \Lambda \Delta F_t + \Gamma_1(L)\Lambda\Delta F_{t-1} \\ &+ \Phi\Delta G_t + \Gamma_1(L)\Phi\Delta G_{t-1} - \Gamma_1(L)\Delta X_{t-1} + v_t\end{aligned} \tag{13}$$

this implies that G_t is best included in the cointegration space. This way the forecasting model can be written as

$$\begin{bmatrix} \Delta X_{At} \\ \Delta F_t \\ \Delta G_t \end{bmatrix} = \begin{bmatrix} \gamma_A \\ \gamma_F \\ \gamma_G \end{bmatrix} \delta' \begin{bmatrix} X_{At-1} \\ F_{t-1} \\ G_{t-1} \end{bmatrix} + B_1 \begin{bmatrix} \Delta X_{At-1} \\ \Delta F_{t-1} \\ \Delta G_{t-1} \end{bmatrix} + \cdots + B_q \begin{bmatrix} \Delta X_{At-q} \\ \Delta F_{t-q} \\ \Delta G_{t-q} \end{bmatrix} + \begin{bmatrix} \varepsilon_{At} \\ \varepsilon_t^F \\ \varepsilon_t^G \end{bmatrix} \tag{14}$$

Equation (14) is clearly an approximation of the original model in Eq. (5). Its parameterization, dictated by empirical convenience for forecasting applications, deserves a few comments. First, while in the model (5) cointegration is only between each individual variable and the factors (due to the assumed factor structure of the data), we treat the cointegration coefficients δ as unrestricted. This is because Eq. (14) is only an approximation to the original model and omits potentially many significant cross-equations correlations. For a similar reason, the loading matrices γ_A, γ_F and γ_G and short-run coefficients $B_1, \ldots, B_q$ are also left unrestricted. The lag structure of the model in such a case cannot be directly recovered from the orders of $\Gamma(L)$ and $M(L)$, in our empirical applications it is determined by suitable information criteria. Note that the extent of the potential mis-specification of Eq. (14) depends mainly on the structure of the $\Gamma(L)$ matrix in Eq. (2), which in turn depends on the extent of the cross-correlation of the idiosyncratic errors. With a diagonal $\Gamma(L)$, hence uncorrelated idiosyncratic errors, Eq. (14) is very close to Eq. (13). In such a case $\delta = [\Gamma(1) - I, \Lambda, \Phi]'$, $\gamma_F = 0$. $\gamma_G = 0$, the lag order of Eq. (14) would equal $q = \max(n, p, m)$ and the contemporaneous effects of ΔF_t and ΔG_t on ΔX_t would be lumped into the error term ε_{At} through terms Λu_t and Φw_t.

Conditional on the estimated factor space, the remaining parameters of the model can be estimated using the Johansen method (Johansen, 1995). The rank of δ can be determined, for example, either by the Johansen trace test (Johansen, 1995) or the procedure of Cheng and Phillips (2009) based on information criteria.

Estimation of the space spanned by the factors and of their number depend on the properties of the idiosyncratic components ε_{it}. Under the assumption of $I(0)$ idiosyncratic errors, the number of $I(1)$ factors r_1 can be consistently estimated using the criteria developed by Bai (2004), applied to data in levels. The overall number of static factors $r_1(p+2) + r_2(m+1)$ can be estimated using the criteria by Bai and Ng (2002), applied to the data in differences. The space spanned by the factors can be consistently estimated using principal components. F_t can be consistently estimated as the eigenvectors corresponding to the largest r_1 eigenvalues of XX' normalized such that $\tilde{F}'\tilde{F}/T^2 = I$. The stationary factors can be consistently estimated as the eigenvectors corresponding to the next q largest eigenvalues normalized such that $\tilde{G}'\tilde{G}/T = I$ (Bai, 2004).

In case some of the ε_{it} are $I(1)$, the space spanned by F_t and G_t jointly (but not separately) can be estimated consistently using the method by Bai and Ng (2004), from data in differences.

Replacing the true factors with their estimated counterparts is permitted under the assumptions discussed above and in Bai (2004) or Bai and Ng (2004), so that we do not have a generated-regressors problem. Even though the FECM can accommodate either of the assumptions about the order of integration of the idiosyncratic components, we give preference in our empirical applications to the Bai (2004) setting with $I(0)$ idiosyncratic components, but also provide results with factors obtained from the data in differences as a robustness check. There are two main reasons for our choice. First, from an economic point of view, integrated errors are unlikely as they would imply that the integrated variables can drift apart in the long run, contrary to general equilibrium arguments. This is especially so in our forecasting applications, in which we consider forecasting a small set of key observable variables. Integrated variables that drift apart are likely marginal, and as such they do not contain essential information and can be dropped from the analysis. Second, whether the idiosyncratic errors ε_{it} are stationary or not is an empirical issue. The empirical applications below use two datasets. The first one is composed of the Euro Area quarterly variables used in Fagan, Henry, and Mestre (2001), updated to cover the period 1975–2012. It contains 32 $I(1)$ series. The second uses a monthly US dataset for the period 1960–2014 with 94 $I(1)$ series. By applying the ADF unit root test to the estimated idiosyncratic components after extracting four factors from each dataset (as indicated by appropriate information criteria), the unit-root null is rejected at the 5% significance level for all series in the Euro Area dataset, and for 90 out of 94 series for the US data. Moreover, the panel unit root test (Bai & Ng, 2004) rejects the null of no panel cointegration between X_{it} and F_t for both datasets.[1] Overall, it appears that the assumption of stationary idiosyncratic errors fits well the properties of the two datasets we use.

2.3. *The FECM Form for Structural Analysis*

The identification of structural shocks in a standard VAR model relies on imposing restrictions upon the parameters of the moving-average representation of the VAR and/or the variance covariance matrix of the VAR errors. An analogous approach in the case of large-scale models entails the moving-average representation of the FAVAR. In the general case, this requires the estimation of the VAR representation of the dynamic factor model (see Lütkepohl, 2014; Stock & Watson, 2005) or, in case of large non-stationary panels with cointegration, the equations of the FECM (rather than just the approximation in Eq. (14)).

To avoid the curse of dimensionality in estimating either the FAVAR or the FECM, we need to strengthen the assumptions about the properties of the idiosyncratic components. Specifically, we assume Eq. (1) to be a strict factor model: $E(\varepsilon_{it}, \varepsilon_{js}) = 0$ for all i, j, t and s, $i \neq j$.[2] However, serial correlation of ε_{it} is still permitted in the form $\varepsilon_{it} = \gamma_i(L)\varepsilon_{it-1} + v_{it}$ with the roots of $\gamma_i(L)$ lying inside the unit disc. Under this assumption, we can write the lag polynomial $\Gamma(L)$ as

$$\Gamma(L) = \begin{bmatrix} \gamma_1(L) & \cdots & 0 \\ \vdots & \ddots & \vdots \\ 0 & \cdots & \gamma_N(L) \end{bmatrix}$$

This restriction, being stronger than Bai's assumptions, leaves all of his results directly applicable to our model, as also verified by the simulation experiments reported by Banerjee et al. (2014b). Under the strict dynamic factor assumption, the estimation of the parameters of the FECM model (5) is straightforward. Using the estimated factors and loadings, the estimates of the common components are $\tilde{\Lambda}\tilde{F}_t$, $\tilde{\Phi}\tilde{G}_t$, $\tilde{\Lambda}\Delta\tilde{F}_t$ and $\tilde{\Phi}\Delta\tilde{G}_t$, while for the cointegration relations it is $X_{t-1} - \tilde{\Lambda}\tilde{F}_{t-1}$. Finally, the estimated common components and cointegration relations can be used in Eq. (5) to estimate the remaining parameters of the FECM by OLS, equation by equation. Also in this case, replacing the true factors and their loadings with their estimated counterparts is permitted under the assumptions discussed above and in Bai (2004) so that we do not have a generated-regressors problem.

The FECM model (5) and the corresponding factor VAR representation (12) are in reduced form. The identification of structural shocks in VAR models usually rests on imposing restrictions upon the parameters of the moving-average representation of the VAR. For vector-error correction models, the derivation of the moving-average representation uses the Granger representation theorem. The generalization of the Granger representation theorem to large dynamic panels is provided by Banerjee et al. (2014b), who show that the moving-average representation of the FECM is

$$\begin{bmatrix} X_t \\ F_t \\ G_t \end{bmatrix} = \begin{bmatrix} \Lambda \\ I_{r_1} \\ 0_{r_2 \times r_1} \end{bmatrix} \omega \sum_{i=1}^{t} u_t + C_1(L) \begin{bmatrix} v_t + [\Lambda, \Phi] Q \left[u_t', w_t'\right]' \\ Q \begin{bmatrix} u_t \\ w_t \end{bmatrix} \end{bmatrix} \tag{15}$$

where $C_1(L)$ is a stable matrix polynomial and the remaining notation is as above.

Our model contains $I(1)$ and $I(0)$ factors with corresponding dynamic factors innovations. For the purpose of the identification of structural dynamic factor innovations, we assume that they are linearly related to the reduced-form innovations as

$$\tilde{\eta}_t = \begin{bmatrix} \eta_t \\ \mu_t \end{bmatrix} = H \begin{bmatrix} u_t \\ w_t \end{bmatrix} \tag{16}$$

where H is a full-rank $(r_1 + r_2) \times (r_1 + r_2)$ matrix. η_t are r_1 permanent structural dynamic factor innovations and μ_t are r_2 transitory structural dynamic factor innovations. It is assumed that $E\tilde{\eta}_t\tilde{\eta}_t' = I$ such that $H\Sigma_{u,w}H' = I$.

From the MA representation (15), we can observe that the innovations in the first group have permanent effects on X_t, while the innovations in the second group have only transitory effects, which makes the FECM a very useful model also for the discussion of long-run identifications schemes. These have been discussed in Banerjee et al. (2014b), who show how the FECM can be used to provide a large-system generalization of the structural common trends analysis of King, Plosser, Stock, and Watson (1991). In this paper, we focus on the use of contemporaneous restrictions by comparing the analysis of monetary policy shocks in the FECM to similar analysis of Bernanke et al. (2005) within the FAVAR model.

3. DATA AND EMPIRICAL APPLICATIONS

The empirical applications below illustrate the performance of the FECM in forecasting and structural analysis of monetary policy shocks. The forecasting application is based on Euro Area data, coming from the 2013 update of the Euro Area Wide Model dataset of Fagan et al. (2001). It contains 38 quarterly macroeconomic series for the period 1975–2012.[3] Thirty-two out of 38 series are $I(1)$. Data are seasonally adjusted at source. The only exception is the consumer price index, which we seasonally adjust using the X-11 procedure.

Further evidence on this matter is subsequently provided in a second forecasting application, which is based on US data, containing 132 monthly series, 94 of which are treated as $I(1)$. The same dataset is used for the analysis of the transmission of monetary policy shocks.[4] To account for the problem of measuring the monetary policy rate at the zero-lower bound we replaced the federal funds rate from 2009m7 onwards with the policy rate estimated by Wu and Xia (2014).

Bai (2004) IPC2 information criterion indicates $r_1 = 2$ for both the US and EA datasets. The choice of the total number of estimated factors for the Euro Area, r, is instead based on Bai and Ng (2004). Their PC3 criterion indicates 4 factors in total. For the US, in the choice of the total number of estimated factors r we follow Bernanke et al. (2005) and set it to 3. Including the federal funds rate as an observable factor, as in Bernanke et al. (2005), gives a total number of factors equal to 4, as for the Euro

Area application. However, as in their case, the main findings are robust to working with more factors. In addition, on both datasets we investigate whether the method of factors extraction – either from the levels or differences of the data – affects the forecasting performance of the FECM.

The datasets contain both $I(1)$ and $I(0)$ variables. The $I(0)$ variables in the panel are treated in the empirical analysis in the following way. At the stage of factor estimation all variables are used. The space spanned by F_t and G_t is estimated by the principal components of the data in levels containing both the $I(1)$ and $I(0)$ variables (Bai, 2004), whose good finite sample performance is confirmed by a simulation experiment in Banerjee et al. (2014b). The structure of the FECM equations, however, needs to be adapted for the purposes of the structural analysis.

Denote by $X_{it}^{(1)}$ the $I(1)$ variables and by $X_{it}^{(2)}$ the $I(0)$ variables. Naturally, the issue of cointegration applies only to $X_{it}^{(1)}$. As a consequence, the $I(1)$ factors load only to $X_{it}^{(1)}$ and not to $X_{it}^{(2)}$. In other words, the fact that $X_{it}^{(2)}$ are assumed to be $I(0)$ implies that the $I(1)$ factors F_t do not enter the equations for $X_{it}^{(2)}$, which is a restriction that we take into account in model estimation. Our empirical FECM is then:[5]

$$\Delta X_{it}^{(1)} = \alpha_i\left(X_{it-1}^{(1)} - \Lambda_i F_{t-1}\right) + \Lambda_i^{(1)}(L)\Delta F_t + \Phi_i^{(1)}(L)G_t + \Gamma^{(1)}(L)\Delta X_{it-1}^{(1)} + v_{it}^{(1)} \tag{17}$$

$$X_{it}^{(2)} = \Phi_i^{(2)}(L)G_t + \Gamma^{(2)}(L)\Delta X_{it-1}^{(2)} + v_{it}^{(2)} \tag{18}$$

The model for the $I(1)$ variables in Eq. (17) is the FECM, while the model for the $I(0)$ variables in Eq. (18) is an FAVAR. Note that these FAVAR equations differ from standard applications. The initial model from which we derived the FECM is the DFM for $I(1)$ data. In such a model the $I(1)$ factors by definition cannot load onto $I(0)$ variables. This restriction is explicit in Eq. (18), while the FAVAR application of Bernanke et al. (2005), for example, uses the following form of the FAVAR:

$$\Delta X_{it}^{(1)} = \Lambda_i^{(1)}(L)\Delta F_t + \Phi_i^{(1)}(L)G_t + v_{it}^{(1)} \tag{19}$$

$$X_{it}^{(2)} = \Lambda_i^{(2)}(L)\Delta F_t + \Phi_i^{(2)}(L)G_t + v_{it}^{(2)} \tag{20}$$

As discussed above, the main difference between the FECM and the FAVAR is that the latter does not contain the error-correction term.

4. FORECASTING MACROECONOMIC VARIABLES

We start with the presentation of forecasting results for selected Euro Area and US variables. For each dataset, we consider two systems of variables, one real and one nominal. The real set for the Euro Area consists of the real GDP, real private consumption and real exports. The nominal set for the Euro Area (Nominal XR) contains the HICP, unit labor costs (ULC) and the effective nominal exchange rate of the euro. The Euro Area dataset contains only 32 $I(1)$ variables on quarterly frequency from 1975 to 2012. The US panel is considerably wider and longer, which brings factor-based forecasting models closer to the large data setting they are designed for. The US dataset contains 129 monthly macroeconomic series over the period 1960–2014. As in Stock and Watson (2002b), we consider forecasting total industrial production (IP), personal income less transfers (PI), employment on non-agricultural payrolls (Empl), and real manufacturing trade and sales (ManTr) as real variables. As nominal variables we consider the producer price index (PPI), consumer price index (CPI), consumer prices without food prices (CPI no food) and private consumption deflator (PCE).

For each dataset we investigate the effects the Great Recession might have had on relative forecasting performance of competing models. To this end we split the Euro Area forecasting sample into 2002q1–2008q3 and 2008q4–2012q4. The corresponding split for the US data is 2000m1–2007m6 and 2007m7–2014m7. The tables that report statistics about relative forecasting precision of competing forecasting models across whole evaluation periods, 2002q1–2012q4 for the Euro Area and 2000m1–2014m7 for the United States, are for brevity deferred to Appendix A.

Forecasting is performed using the following set of competing models. First, we use three models that are all based on the observable variables only: an autoregressive (AR) model, a vector autoregression (VAR) and an error-correction model (ECM). In order to assess the forecasting role of the additional information, the second set of models augments the first set with factors extracted from the larger set of available variables: FAR, FAVAR and FECM specifications are factor-augmented AR, VAR and ECM models, respectively.

For the FECM model, we use two approaches to factor extraction. As argued above, our primary choice is estimation with PCA from the data in levels. As a robustness check, commented in the next subsection, we use the factors estimated from the data in differences, using the method of Bai and Ng (2004). Such an FECM model is denoted $FECM_{BN}$.

The numbers of $I(1)$ and $I(0)$ factors, both set at 2, are kept fixed over the forecasting period, but their estimates are updated recursively. Each forecasting recursion also includes model selection. The lag lengths are determined by the BIC information criterion.[6] As for the cointegration test for determining the cointegration ranks of the ECM and the FECM, we have considered two approaches: the Johansen (1995) trace test and the Cheng and Phillips (2009) semi-parametric test based on the BIC. The two methods gave very similar results (details available upon request), but, due to its lower computational burden and also its ease of implementation in practice, we gave preference to the method of Cheng and Phillips.[7]

The levels of all variables are treated as $I(1)$ with a deterministic trend, which means that the dynamic forecasts of the differences of (the logarithm of) the variables h steps ahead produced by each of the competing models are cumulated in order to obtain the forecasts of the level h steps ahead. We consider four different forecast horizons, $h = 1, 2, 4, 8$. In contrast to our use of iterated h-step-ahead forecasts (dynamic forecasts), Stock and Watson (1998, 2002a) adopt direct h-step-ahead forecasts, while Marcellino, Stock, and Watson (2006) find that iterated forecasts are often better, except in the presence of substantial misspecification.[8] In our FECM framework, such forecasts are easier to construct than their h-step-ahead equivalents, and the method of direct h-step-ahead forecasts and our iterative h-step-ahead forecasts produce similar benchmark results on a common estimation and evaluation sample.

The results of the forecast comparisons are presented in Tables 1–4, where we list the MSEs of the competing models relative to the MSE of the AR at different horizons for each variable under analysis, with asterisks indicating when the MSE differences are statistically significant according to the Clark and West (2007) test. The tables also report information on the cointegration rank selection and the number of lags in each model.

4.1. *Forecasting Results for the Euro Area*

Our basic results are presented in Table 1 for real variables and in Table 2 for nominal variables. For real variables and the period before the crisis,

Table 1. Forecasting Real Variables for the Euro Area before and in the Great Recession.

h	Log of	RMSE of AR	MSE Relative to MSE of AR model					
			FAR	VAR	FAVAR	ECM	FECM	$FECM_{BN}$
					Before crisis – 2002:1–2008:3			
	GDP	0.004	1.18	0.88*	0.95***	0.84*	0.94	0.76*
1.00	Consumption	0.003	1.09	1.06	1.55	0.87*	1.13	1.24
	Exports	0.004	1.38	1.07	1.44	0.98	1.29	1.24
	GDP	0.007	1.04	0.91*	0.96	0.82*	0.93***	0.67*
2.00	Consumption	0.006	1.12	1.17	1.47	0.90**	0.99	1.07
	Exports	0.007	1.42	1.03	1.44	0.88*	1.10	1.41
	GDP	0.011	1.02	0.98	1.02	0.90***	0.96	0.61*
4.00	Consumption	0.009	1.10	1.26	1.48	1.07	1.08	0.99
	Exports	0.018	1.27	1.06	1.29	0.82**	0.79*	1.18
	GDP	0.018	1.00	0.99	1.02	0.95	1.26	0.52**
8.00	Consumption	0.017	1.07	1.17	1.31	1.05	1.52	0.42*
	Exports	0.046	1.16	1.05	1.18	0.82*	0.28*	0.64*
					Crisis – 2008:4–2012:4			
	GDP	0.009	1.47	0.97	0.71***	0.95**	0.71*	0.61***
1.00	Consumption	0.005	1.83	1.55	1.35	1.20	0.54*	0.84***
	Exports	0.010	0.86	1.06	0.93	1.06	0.81*	0.72***
	GDP	0.019	1.19	0.92	0.84***	0.86*	0.71*	0.76**
2.00	Consumption	0.009	2.01	1.80	1.68	1.33	0.68*	1.08
	Exports	0.020	1.03	1.17	1.07	1.19	0.84**	0.76**
	GDP	0.036	1.06	0.92***	0.93	0.90*	0.76*	0.88***
4.00	Consumption	0.018	1.87	1.75	1.76	1.47	0.88*	1.35
	Exports	0.037	1.18	1.27	1.19	1.38	0.81*	0.67*
	GDP	0.058	1.01	0.93***	0.98	0.95*	0.81*	0.99
8.00	Consumption	0.034	1.67	1.62	1.66	1.46	1.02	1.50
	Exports	0.058	1.31	1.42	1.31	1.82	0.78*	0.60*

Notes: The FECM and the FAVAR contain four factors extracted from data in levels. $FECM_{BN}$ uses factors extracted from differences. Cheng and Phillips (2008) cointegration test and lag selection based on BIC. Data: 1975:1–2012:4, forecasting: 2002:1–2012:4.
*,**,*** Indicate the significance at 10%, 5%, and 1%, respectively, of the Clark and West (2007) test of equal predictive accuracy relative to AR model.

the ECM results show it to be the best model in 10 out of 12 cases. In the remaining cases, the best model is the benchmark AR model. The FECM is more precise than the AR in half of the cases. the FAVAR performs worse, outperforming the AR model only once. The FAR never beats the AR.

Table 2. Forecasting Nominal Variables for the Euro Area before and in the Great Recession, Factors Extracted from Nominal Sub-panel.

h	Variable	RMSE of AR	MSE Relative to MSE of AR model					
			FAR	VAR	FAVAR	ECM	FECM	$FECM_{BN}$
			Before crisis – 2002:1–2008:3					
	HICP	0.003	2.49	3.37	2.59	2.60	1.79	2.11
1.00	ULC	0.004	1.30	1.66	1.42	1.02	1.40	1.19
	Nominal XR	0.019	0.97***	0.98	1.05	1.07	1.25	1.01
	HICP	0.005	1.45	2.93	1.53	1.82	1.04	1.36
2.00	ULC	0.006	1.09	1.68	1.18	1.04	1.11	0.84*
	Nominal XR	0.032	0.95**	0.93***	1.05	0.95	1.19	0.97***
	HICP	0.008	1.08	2.46	1.11	1.44	0.82**	0.83*
4.00	ULC	0.012	0.86*	1.97	1.03	1.23	1.02	0.71*
	Nominal XR	0.057	0.90*	0.89*	0.94**	0.89**	1.06	0.87*
	HICP	0.010	2.25	5.20	1.65	2.37	1.52	1.07
8.00	ULC	0.020	0.84*	2.35	0.82*	1.09	1.56	0.51*
	Nominal XR	0.096	0.87*	0.83*	0.84*	0.83*	0.86***	0.81**
			Crisis – 2008:4–2012:4					
	HICP	0.005	1.68	1.96	1.67	1.96	1.49	1.41
1.00	ULC	0.008	1.01	0.78***	1.12	0.78***	0.97	0.97
	Nominal XR	0.025	1.03	0.99	0.92***	0.99	0.95	0.94***
	HICP	0.009	0.80***	1.28	0.84***	1.28	0.86*	0.82*
2.00	ULC	0.014	0.71***	0.47**	0.75***	0.47**	0.63***	0.63***
	Nominal XR	0.040	1.02	1.13	1.00	1.13	1.05	1.01
	HICP	0.018	0.76**	1.10	0.84**	1.10	0.72*	0.69*
4.00	ULC	0.028	0.38***	0.26**	0.40***	0.26**	0.37***	0.38***
	Nominal XR	0.055	1.17	1.22	1.13	1.22	1.03	1.02
	HICP	0.033	0.84*	1.06	0.91*	1.01	0.66*	0.66*
8.00	ULC	0.056	0.19*	0.25*	0.19*	0.33*	0.18*	0.19*
	Nominal XR	0.059	1.67	1.62	1.70	1.60	1.45	1.42

Notes: The FECM and the FAVAR contain four factors extracted from data in levels. $FECM_{BN}$ uses factors extracted from differences. Cheng and Phillips (2008) cointegration test and lag selection based on BIC. Data: 1975:1–2012:4, forecasting: 2002:1–2012:4.
*,**,*** Indicate the significance at 10%, 5% and 1%, respectively, of the Clark and West (2007) test of equal predictive accuracy relative to AR model. Variables pre-selected as in Boivin and Ng (2006) using 0.75 threshold for the correlation coefficient.

In the crisis period, results are fundamentally different. It can be observed from the tables that the RMSE of the benchmark AR model generally increased in the crisis both for real and nominal variables. The relative performance of the FECM, however, generally improved also for both sets of variables. For real variables, the FECM is the best performing model in 11 out of 12 cases. The gains in forecasting precision relative to the basic AR model slightly decrease across forecast horizons and are of the order of magnitude of about 20%. The maximum gain is about 45% for private consumption at one-quarter horizon.[9] Larger gains in forecasting precision of the FECM at shorter forecast horizons are in line with Clements and Hendry (1998), who show that forecast biases due to omission of the error-corrections terms are higher at shorter horizons.

Other models perform considerably worse. The VAR, the ECM and the FAVAR outperform the AR model only for GDP, but are never better than the FECM. The FAR model turns out to be worse than the AR model in all cases. The ECM, on the other hand, is never the best, and beats the AR model only for GDP. Similar observations apply to the FAVAR and the VAR models. This implies that in the crisis period, the importance of the information contained in factors increased for real variables. However, given that the FAVAR is not significantly better than the VAR model, it is important that the information embedded in factors enters via cointegration relations.

An important observation concerns the results of cointegration testing. As we can observe from Table A1 that reports also the results of cointegration tests, the Cheng and Phillips test fails to find cointegration between the three variables under evaluation. By adding factors to the system to get the FECM, the test consistently signals cointegration. Such a results is in line with the analysis of Banerjee and Marcellino (2009), who point that adding factors to the ECM proxies for the potentially missing cointegration relations. In combination with the superior information set this results in a better forecasting performance. By comparing the results of the FAVAR and the FECM we see that the FECM is consistently more precise in forecasting. This difference can be attributed to the error-correction term that the FAVAR model omits.

For forecasting nominal variables, we consider two modifications to the factor extraction procedure with the aim of improving the forecast precision of factor-based models.[10] First, we consider extracting factors from a subpanel of nominal variables only, which results in using 19 variables for factor extraction. Second, we use variable preselection based on correlation with target variables as in Boivin and Ng (2006), which shrinks the dataset by an additional 5 variables. The correlation threshold was set to 0.75.

Similar conclusions about the role of information extracted from large datasets and cointegration as for real variables can be obtained also by examining the results for nominal variables in Table 2. For the period before the crisis, the AR model is the best performing on average: 5 out of 12 cases. Among the competing models only the VAR turns out to perform similarly, being the best in three out of four cases for the nominal exchange rate. The FAR follows, being the best twice and outperforming the AR model in half of the cases. The remaining models outperform the AR in only three cases or less.

In the crisis period, the AR model remains the best performing in four cases. Similar to the period before the crisis, the FAR model outperforms the AR model in half of the cases and is best overall in one. The performance of the FAVAR model that uses the information from large panels in a system of variables improves considerably relative to the AR, outperforming it in 7 out of 12 cases. The ECM exploiting the error-correction mechanism also improves, as it outperforms the AR model in five instances (two before the crisis). The FECM that incorporates both the information from large datasets and cointegration exhibits the largest improvement. It outperforms the AR in 8 out of 12 cases (2 before the crisis) and is best in 3 cases (only once before the crisis). Some of the gains in forecasting precision relative to the AR are significant: above 30% and above 80% for the HICP and for unit labor costs, respectively, at the two-year horizon.

4.2. Forecasting Results for the United States

The results for the real US system are analogous to the those of the Euro Area and demonstrate the merits of incorporating information embedded in estimated factors through cointegration relations. As can be seen from Table 3, before the crisis the FECM never produces the best forecast and it performs better than the benchmark AR in only one instance. The best performing model turns out to be the VAR (10 out of 24 cases), followed by the AR (8 out of 24). The only factor-based model that performs well to some extent is the FAR, which is the best on five occasions.

In the crisis period the relative performance of the factor-based models improves considerably but, similar to the Euro Area case, this feature is most pronounced for the FECM. The FECM turns out to be the best performing model in 14 out of 24 cases. The improvements in forecasting precision relative to the AR exceed 20% in 8 cases and can exceed 30% in some cases. The FAVAR is the second best model producing the lowest MSE in four cases. The FAR is best three times, while the ECM produces the lowest MSE only once.

Table 3. Forecasting Real Variables for the United States before and in the Great Recession.

h	Log of	RMSE of AR	MSE Relative to MSE of AR model					
			FAR	VAR	FAVAR	ECM	FECM	$FECM_{BN}$
			Before crisis – 2000:1–2007:6					
	PI	0.006	0.94***	0.92***	0.94***	0.94***	0.97**	1.00
1.00	ManTr	0.007	0.97*	0.96*	1.00	1.09	1.10	1.06
	IP	0.005	0.87*	0.87*	0.91*	1.00***	1.13	1.11
	Empl	0.001	1.31	1.17	1.46	1.60	2.22	1.27
	PI	0.010	0.88*	0.86*	0.85*	0.95**	1.02	0.96***
3.00	ManTr	0.010	0.95*	1.08	0.98*	1.72	1.40	1.23
	IP	0.010	0.85*	0.95**	0.97*	1.58	1.35	1.40
	Empl	0.002	1.48	1.34	1.63	2.54	2.84	1.56
	PI	0.014	0.94**	0.83*	0.85**	1.10	1.08	0.96
6.00	ManTr	0.014	0.98**	1.06	0.98**	2.58	1.36	1.56
	IP	0.017	0.89*	0.94**	0.97*	2.34	1.23	1.68
	Empl	0.005	1.70	1.46	1.65	3.47	2.34	1.94
	PI	0.022	1.06	0.82**	0.94	1.45	1.03	1.05
12.00	ManTr	0.022	1.24	1.02	1.15	3.71	1.45	2.12
	IP	0.030	1.12	0.96***	1.14	3.20	1.07	2.13
	Empl	0.012	1.70	1.36	1.53	3.76	1.37	1.97
	PI	0.029	1.11	0.86***	1.03	1.73	1.13	1.12
18.00	ManTr	0.027	1.31	1.00	1.24	4.15	2.23	2.23
	IP	0.041	1.26	0.96*	1.25	3.57	1.36	2.20
	Empl	0.020	1.58	1.26	1.47	3.78	1.24	1.86
	PI	0.035	1.10	0.89***	1.06	1.96	1.18	1.10
24.00	ManTr	0.032	1.23	0.95*	1.20	4.32	2.75	2.11
	IP	0.050	1.27	0.96*	1.26	3.77	1.64	2.14
	Empl	0.027	1.40	1.17	1.35	3.63	1.25	1.71
			Crisis – 2007:7 2014:7					
	PI	0.010	0.90*	0.92*	0.89*	0.92*	0.90*	0.87*
1.00	ManTr	0.009	0.88*	0.94*	0.99**	1.08	0.88*	0.95*
	IP	0.007	0.82*	0.93*	0.82*	0.93*	0.94*	1.02
	Empl	0.001	1.13	1.08	1.44	0.98**	1.95	1.10
	PI	0.016	0.77*	0.76*	0.74*	0.75*	0.77*	0.70*
3.00	ManTr	0.017	0.91*	0.94*	1.01	1.18	0.89*	1.12
	IP	0.015	0.96*	1.02	0.97*	1.09	1.08	1.26
	Empl	0.003	1.18	1.14	1.41	1.01	2.16	1.32
	PI	0.024	0.73**	0.67**	0.71**	0.67***	0.65*	0.68**
6.00	ManTr	0.031	1.00	0.97***	1.04	1.28	0.88**	1.13
	IP	0.031	0.99*	1.05	1.01	1.17	1.10	1.27
	Empl	0.008	1.30	1.24	1.40	1.11	1.90	1.51

Table 3. *(Continued)*

h	Log of	RMSE of AR	MSE Relative to MSE of AR model					
			FAR	VAR	FAVAR	ECM	FECM	$FECM_{BN}$
	PI	0.042	0.88*	0.76***	0.82*	0.78***	0.68**	0.82***
12.00	ManTr	0.059	1.02	0.98	0.96	1.33	0.81***	1.09
	IP	0.063	0.90*	1.02	0.90*	1.20	0.88	1.20
	Empl	0.020	1.33	1.22	1.24	1.13	1.29	1.52
	PI	0.057	0.95**	0.84***	0.90**	0.90	0.71**	0.95***
18.00	ManTr	0.080	1.02	0.99	0.95**	1.43	0.90***	1.10
	IP	0.085	0.89*	1.00	0.87*	1.28	0.70	1.24
	Empl	0.032	1.28	1.16	1.18	1.11	0.88***	1.52
	PI	0.070	0.97***	0.90***	0.93***	0.99	0.77**	1.04
24.00	ManTr	0.096	1.00	0.98	0.96*	1.51	1.20	1.11
	IP	0.100	0.91*	1.00	0.89*	1.39	0.69***	1.33
	Empl	0.043	1.22	1.11	1.15	1.09	0.68**	1.57

Notes: The FECM and the FAVAR contain four factors extracted from data in levels. $FECM_{BN}$ uses factors extracted from differences. Cheng and Phillips (2008) cointegration test and lag selection based on BIC. Data: 1960:1–2014:7.
Variables: IP – Industrial production, PI – Personal income less transfers, Empl -Employees on non-aggr. payrolls, ManTr – Real manufacturing trade and sales.
*,**,*** Indicate the significance 10%, 5% and 1%, respectively, of the Clark and West (2007) test of equal predictive accuracy relative to the AR model.

The results for nominal variables, presented in Table 4, are even more in favor of the FECM model. Across both subperiods it is the best performing model in 18 out of 24 cases with improvements in forecasting precision relative to the benchmark AR model on average exceeding 10% before the crisis and 20% in the crisis. Neither the FAVAR nor the FAR ever produce the best forecast. In fact, they never improve over the AR. The only other models that tends to perform reasonably well is the ECM, which is best three times before the crisis and six times in the crisis period.

4.3. Robustness Check to I(1) Idiosyncratic Errors

As shown above, for both our datasets we cannot reject the hypothesis that the idiosyncratic components of the data are $I(0)$. Nevertheless, we also assess the forecasting performance of the FECM with the factors estimated from the data in differences, and cumulated to obtain the estimate of the space spanned by $I(1)$ and $I(0)$ factors. If our primary assumption of $I(0)$ idiosyncratic components was violated, then the estimated factor space from the data in levels as in Bai (2004) would be inconsistent. Estimating

Table 4. Forecasting Nominal Variables for the United States before and in the Great Recession.

h	Log of	RMSE of AR	MSE Relative to MSE of AR model					
			FAR	VAR	FAVAR	ECM	FECM	$FECM_{BN}$
			Before crisis – 2000:1–2007:6					
	PPI	0.006	1.08	1.07	1.18	0.93*	0.96*	1.00***
1.00	CPI all	0.003	1.09	1.03	1.13	1.01	1.03	1.00
	CPI no food	0.003	1.12	1.03	1.12	0.98*	1.00	0.99**
	PCE defl	0.002	1.18	1.02	1.20	1.21	1.07	1.17
	PPI	0.006	1.12	1.05	1.13	0.92*	0.92*	0.96**
3.00	CPI all	0.003	1.38	1.07	1.36	1.10	0.97***	0.98***
	CPI no food	0.004	1.33	1.05	1.30	1.01	0.94*	0.93*
	PCE defl	0.002	1.37	1.11	1.34	1.36	1.10	1.19
	PPI	0.006	1.13	1.03	1.14	0.89*	0.87*	0.92*
6.00	CPI all	0.003	1.24	1.02	1.23	1.00	0.89*	0.92*
	CPI no food	0.003	1.23	1.04	1.22	0.96*	0.89*	0.90*
	PCE defl	0.002	1.26	1.03	1.25	1.14	0.93**	0.98*
	PPI	0.007	1.16	1.06	1.19	0.75*	0.76*	0.81*
12.00	CPI all	0.003	1.23	1.05	1.23	0.83*	0.74*	0.76*
	CPI no food	0.004	1.25	1.07	1.23	0.81*	0.76*	0.77*
	PCE defl	0.002	1.35	1.06	1.33	1.01	0.80*	0.87**
	PPI	0.007	1.08	1.02	1.08	0.84*	0.83*	0.87*
18.00	CPI all	0.003	1.17	1.03	1.15	1.06	0.87*	0.94*
	CPI no food	0.003	1.16	1.02	1.14	1.00	0.87*	0.92*
	PCE defl	0.002	1.22	1.02	1.21	1.14	0.82*	0.92*
	PPI	0.006	1.02	1.00	1.04	0.89*	0.89*	0.90*
24.00	CPI all	0.003	1.02	1.00	1.01	1.07	0.88*	0.93*
	CPI no food	0.003	1.04	1.01	1.02	1.04	0.91*	0.94*
	PCE defl	0.002	1.13	1.00	1.11	1.17	0.83*	0.90**
			Crisis – 2007:7–2014:7					
	PPI	0.009	1.06	1.03	1.12	0.78*	0.84*	0.81*
1.00	CPI all	0.003	1.21	1.02	1.14	0.96***	0.93***	0.86***
	CPI no food	0.003	1.19	1.06	1.20	0.88**	0.92*	0.80*
	PCE defl	0.002	1.04	0.96	1.05	0.92**	0.91**	0.88*
	PPI	0.009	1.19	1.06	1.19	0.71*	0.76*	0.77*
3.00	CPI all	0.003	1.36	1.06	1.30	0.97**	0.79*	0.89***
	CPI no food	0.004	1.36	1.05	1.29	0.93**	0.78*	0.84**
	PCE defl	0.003	1.16	1.01	1.13	0.85**	0.70**	0.73**
	PPI	0.010	1.27	1.06	1.27	0.67*	0.74*	0.71*
6.00	CPI all	0.003	1.32	1.05	1.29	0.85*	0.73*	0.76**
	CPI no food	0.004	1.36	1.05	1.28	0.82*	0.70*	0.73*
	PCE defl	0.003	1.34	1.02	1.30	0.83***	0.68**	0.66**

Table 4. *(Continued)*

h	Log of	RMSE of AR	MSE Relative to MSE of AR model					
			FAR	VAR	FAVAR	ECM	FECM	$FECM_{BN}$
	PPI	0.010	1.41	1.10	1.43	0.66*	0.74*	0.70*
12.00	CPI all	0.003	1.48	1.07	1.42	0.94*	0.78*	0.77*
	CPI no food	0.004	1.55	1.08	1.43	0.94*	0.79*	0.78*
	PCE defl	0.003	1.54	1.04	1.47	0.89**	0.73*	0.69*
	PPI	0.009	1.15	1.03	1.17	0.71*	0.79*	0.73*
18.00	CPI all	0.003	1.26	1.04	1.23	0.91*	0.76*	0.74*
	CPI no food	0.004	1.31	1.06	1.27	0.93*	0.77*	0.77*
	PCE defl	0.003	1.33	1.02	1.27	0.95***	0.81*	0.70*
	PPI	0.009	1.28	1.05	1.30	0.70***	0.78**	0.73**
24.00	CPI all	0.003	1.37	1.05	1.33	0.90***	0.78*	0.76*
	CPI no food	0.004	1.41	1.06	1.34	0.90***	0.76**	0.77**
	PCE defl	0.003	1.42	1.03	1.37	0.98	0.84**	0.74***

Notes: The FECM and the FAVAR contain four factors extracted from data in levels. $FECM_{BN}$ uses factors extracted from differences. Cheng and Phillips (2008) cointegration test and lag selection based on BIC. Data: 1960:1–2014:7.
Variables: Inflations of producer price index (PPI), consumer price index of all items (CPI all), consumer price index less food (CPI no food) and personal consumption deflator (PCE defl).
*,**,*** Indicate the significance 10%, 5% and 1%, respectively, of the Clark and West (2007) test of equal predictive accuracy relative to the AR model.

the factors from differences would in such a case provide consistent estimates of the factors space, and should consequently also improve the forecasting performance.

In Tables 1–4, the relevant results are in the last columns, labeled $FECM_{BN}$. In general, the results are fairly robust to the factor estimation method, justifying the initial assumption of $I(0)$ idiosyncratic errors. For the Euro Area, there is no systematic indication that the $FECM_{BN}$ model would either outperform the FECM model or be inferior to it. Moreover, the relative performance with respect to other competing models is also virtually unchanged. Observations for the US dataset are similar. In the majority of cases of forecasting real variables, the relative MSEs of the $FECM_{BN}$ are close to those of the FECM model, but on average they are higher, which implies that the FECM using factors extracted from levels of data performs consistently better than the one with factors estimated from differences. This confirms that, from the point of view of forecasting precision, extraction of factors from levels of the data appears a valid approach. Such a conclusion can be derived also from the last column of Table 4.

5. TRANSMISSION OF MONETARY POLICY SHOCKS IN THE FECM

The first analysis of monetary policy shocks in large panels, based on an FAVAR model, was developed by Bernanke et al. (2005). The essence of their approach is in the division of variables into two blocks: slow-moving variables that do not respond contemporaneously to monetary policy shocks and fast-moving variables that do. In addition, Bernanke et al. (2005) treat the policy instrument variable, the federal funds rate, as one of the observed factors. They consider two estimation methods, namely Bayesian estimation and principal components analysis. In the latter approach, most frequently used in the literature and in practice, they estimate K factors from the whole panel and from the subset of slow-moving variables only (slow factors). They then rotate the factors estimated from the whole panel around the federal funds rate by means of a regression of these factors on the slow-factors and the federal funds rate. As a result of this rotation of the factors, the analysis proceeds with $K+1$ factors, namely the K rotated estimated factors and the federal funds rate imposed as an observable factor. We follow the basic choice of Bernanke et al. (2005) and set $K = 3$. The main findings are robust to working with more factors.

Identification of monetary policy shocks is obtained in the VAR model of rotated factors assuming a recursive ordering with the federal funds rate ordered last:

$$E\left(\tilde{\eta}_t \tilde{\eta}_t'\right) = H\Sigma_{u,w}H' = I \tag{21}$$

where H^{-1} is lower triangular. The impulse responses of the observed variables of the panel are then estimated by multiplying the impulse responses of the factors by the loadings obtained from OLS regressions of the variables on the rotated factors.

The identification scheme for the analysis of monetary policy shocks can be easily adapted to the FECM, which enables us to study the role of error-correction mechanism in propagation of monetary policy shocks. We need to introduce one modification that makes the results obtained with the FECM directly comparable to those of the FAVAR. The difference is at the stage of factor estimation. Namely, in order to capture cointegration as in Bai (2004), we estimate the factors from the data in levels, while Bernanke et al. (2005) estimate the factors from data transformed (if necessary) to $I(0)$.[11] This gives us the estimates of the space spanned by r_1 $I(1)$

factors and $r - r_1$ stationary factors. As in Bernanke et al. (2005), the federal funds rate is treated as one observable factor and the estimated factors are rotated accordingly. Because their method entails identifying the monetary policy shocks from a stationary factor VAR, the first r_1 non-stationary factors are differenced. Identification of monetary policy shocks is then obtained from a VAR of stationary factors.

The basic results are presented in Fig. 1. It contains the impulse responses obtained from the conventional FAVAR model and the FECM model. The impulse responses can be in principle for any variables in the panel but for simplicity we present them for the a similar set of variables as in Bernanke et al. (2005). The only exception is the consumer confidence indicator, which is missing from our dataset. Instead, we included the responses of real manufacturing trade and sales. In line with the lag structure chosen by Bernanke et al. (2005) we only include contemporaneous values of factors in Eqs. (17)–(20) and 13 lags in the factors VAR. The number of endogenous lags is set to 6.[12] They differ in the presence of the error-correction term for the variables that are treated as $I(1)$ in levels. Some variables are assumed to be $I(0)$. These are the interest rates, the capacity utilization rate, unemployment rate, employment, housing starts, new orders and consumer expectations. For these variables the FAVAR and the FECM also differ. Consistent with Eq. (18), the FECM for $I(0)$ variables excludes the $I(1)$ factors.

What we observe is coherence in terms of the basic shape of the impulse responses between the models. If we focus our discussion on the $I(1)$ variables, it can be observed, however, that quantitatively the responses may differ significantly due to the error-correction terms. The responses of the industrial production, production of durable goods, real manufacturing and trade sales and employment are very similar. Quite significant differences are observed for money and the yen–dollar exchange rate. A stronger response in the FECM is observed also for the commodity prices index. Pronounced differences are observed for the CPI and wages. For the former the FAVAR responses exhibit an evident price puzzle, while this is not the case for the FECM. It is true, however, that responses of prices are overall insignificant, which is also the finding of Bernanke et al. (2005). A puzzling response in the FAVAR is obtained also for wages. They slightly increase along the adjustment path while in the FECM they exhibit a significant negative response in line with economic theory. Some significant differences between the models can be observed also for real variables like personal consumption and production of non-durable goods, which in the FECM show a faster degree of mean reversion.

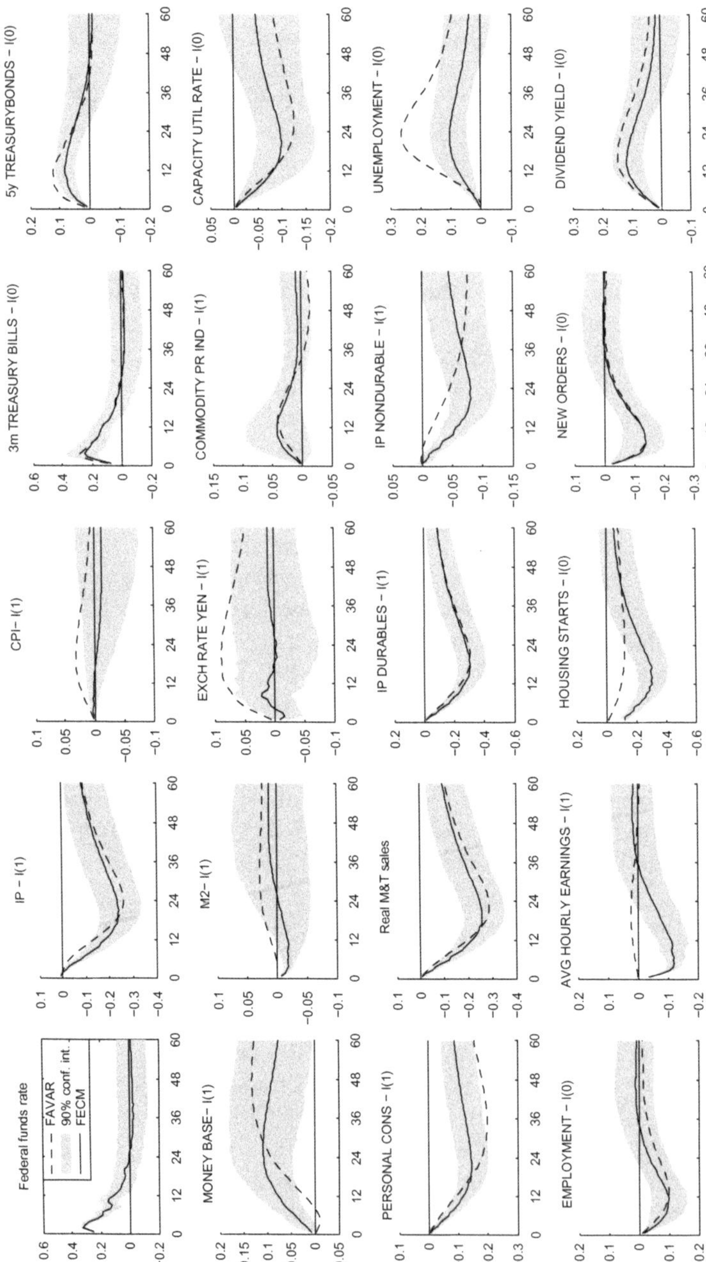

Fig. 1. Impulse Responses to Monetary Policy Shock – FAVAR versus FECM with Factors Extracted from Levels.

Omission of the error-correction terms in the FAVAR model can thus have an important impact on the empirical results. It is worth mentioning that these differences are observed conditional upon a shock that accounts for only a limited share of variance. Banerjee et al. (2014b) present an analysis of real stochastic trends, where the differences between FAVAR and FECM responses become even more pronounced, and the shock is a considerably more important source of stochastic variation in the panel.

The impulse responses of $I(0)$ variables are very similar across models, with only unemployment rate and housing starts as exceptions. This means that imposing the restriction that the differences of $I(1)$ factors do not load to $I(0)$ variables has only a limited quantitative impact, which is consistent with the FECM specification of the model. In the FECM the restriction is evident. In the FAVAR, which makes no distinction in the structure of the loadings of factors to $I(1)$ and $I(0)$ variables, such a restriction cannot be directly determined.

6. CONCLUSIONS

The FECM offers two important advantages for empirical modelling. First, the factors proxy for missing cointegration information in a standard small-scale ECM. Second, the error correction mechanism can also be inserted in the context of a large dataset. From a theoretical point of view, since the FECM nests both the FAVAR and the ECM, it can be expected to provide better empirical results, unless either the error correction terms or the factors are barely significant, or their associated coefficients are imprecisely estimated due to small sample size, or the underlying assumptions that guarantee consistent factor and parameter estimation are not satisfied.

In our forecasting application, the FECM is clearly the best forecasting model for Euro Area and US real variables, particularly in the period of the Great recession in which the relative performance of the FECM generally improves for all variables under analysis. For the US nominal variables the performance of the FECM is even better. In comparison with the FAVAR, these findings highlight the importance of including the error correction terms. For Euro Area nominal variables, the FECM also performs well if the factors are extracted from a subset of variables pre-selected as in Boivin and Ng (2006). We have also seen that the forecasting performance for the US and the Euro Area is not substantially affected if the factors are estimated from the variables in levels or in differences, with a better empirical performance in general of the former method, which suggests that the hypothesis of $I(0)$ idiosyncratic errors is not stringent.

In terms of structural analysis, we have investigated the transmission of monetary shocks, comparing the responses of several variables with those from the FAVAR based analysis of Bernanke et al. (2005). The shape of the impulse responses may differ significantly due to the error-correction terms. For example, relevant differences are observed for consumer prices, wages, monetary aggregates, the yen–dollar exchange rate, production of non-durable goods and private consumption. Omission of the error-correction terms in the FAVAR model can thus have an important impact on the empirical results.

Overall, our empirical results provide further compelling evidence that the FECM provides an important extension of classical ECM and FAVAR models both for forecasting and structural modelling. This finding, combined with the ease of estimation and use of the FECM model, suggests that it could be quite useful for empirical analyses.

NOTES

1. Results available upon request.
2. Stock and Watson (2005) show on the US dataset that the strict factor model assumption is generally rejected but is of limited quantitative importance.
3. The data and the corresponding list of variables can be downloaded from the Euro Area business cycle network webpage (www.eabcn.org/area-wide-model).
4. The structure of the Euro Area data is not rich enough to implement this structural analysis. In particular, it does not contain a sufficient number of fast-moving variables (those that react contemporaneously to the monetary policy shock).
5. Note that only levels of G_t enter (17) while also differences are present in Eq. (5). Given that G_t is $I(0)$ we can see $\Gamma^{(1)}(L)$ as coming from a repramaterization of the polynomial $\Gamma_1(L)\Phi$ in Eq. (5).
6. We have also checked and confirmed the robustness of the results when using the Hannan–Quinn (HQ) criterion (details are available upon request).
7. The simulation results provided by Cheng and Phillips (2009) show that using the BIC tends to lead to underestimation of the rank when the true rank is not very low, while it performs best when the true cointegration rank is very low (0 or 1). Given that BIC model selection is generally preferred for model selection for forecasting, we chose to use it for testing for cointegration rank as well. However, our results (available upon request) are robust to the use of HQ too.
8. Our use of iterated h-step-ahead forecasts implies that the FAR is essentially an FAVAR containing only one variable of interest and factors.

9. Across the whole evaluation period 2000m1–2014m7 we can observe for real variables (see Table A1) a very good performance of the FECM. It results to be the best performing model in 11 out of 12 cases. In all of the cases the gains over the AR model are statistically significant according to the Clark and West (2007) test. Only for private consumption at two-year horizon the forecasting precision of the FECM is lower than that of the AR model.

10. As can be seen from Table A2 over the period 2000m1–2014m7 the FECM results to be the best performing model in only one instance, HICP at two-year horizon, but the gains with respect to the AR model are small. This is in line with similar findings for the case of the United States in Banerjee et al. (2014b). The AR model is regularly beaten by the VAR, the FAVAR, the ECM and the FECM for unit labor costs, with the ECM being the best performing model overall. The gains in forecasting precision are significant, reaching also levels above 50% with the ECM model above the one-year horizon. Similar to the case of the HICP, also for the nominal exchange rate the AR model results to be consistently the best.

11. Both approaches deliver similar estimates of monetary policy shocks. However, as the factors are estimated on datasets of different order of integration, they are not numerically identical.

12. The basic shapes of impulse responses are robust to the specification of endogenous lags and lags of factors. Result available upon request.

ACKNOWLEDGEMENTS

We would like to thank the Editors, two anonymous Referees and participants at the 16th Advances in Econometrics Conference, held at CREATES, Aarhus University, for helpful comments on a previous draft.

REFERENCES

Bai, J. (2004). Estimating cross-section common stochastic trends in nonstationary panel data. *Journal of Econometrics*, *122*, 137–183.

Bai, J., & Ng, S. (2002). Determining the number of factors in approximate factor models. *Econometrica*, *70*, 191–221.

Bai, J., & Ng, S. (2004). A PANIC attack on unit roots and cointegration. *Econometrica*, *72*, 1127–1177.

Bai, J., Kao, C., & Ng, S. (2009). Panel cointegration with global stochastic trends. *Journal of Econometrics*, *149*, 82–99.

Banerjee, A., & Marcellino, M. (2009). Factor-augmented error correction models. In Castle, J. L. Shephard, N. N. (Eds.), *The methodology and practice of econometrics – A Festschrift for David Hendry* (pp. 227–254). Oxford: Oxford University Press.

Banerjee, A., Marcellino, M., & Masten, I. (2014a). Forecasting with factor-augmented error correction models. *International Journal of Forecasting*, *30*, 589–612.

Banerjee, A., Marcellino, M., & Masten, I. (2014b). Structural FECM: Cointegration in large-scale structural FAVAR models, CEPR Discussion Paper No. 9858.

Barigozzi, M., Lippi, M., & Luciani, M. (2014). Dynamic factor models, cointegration and error correction mechanisms. Working Papers ECARES 2014–14.

Bernanke, B. S., Boivin, J., & Eliasz, P. (2005). Measuring the effects of monetary policy: A factor-augmented vector autoregressive (FAVAR) approach. *Quarterly Journal of Economics*, *120*, 387–422.

Boivin, J., & Ng, S. (2006). Are more data always better for factor analysis? *Journal of Econometrics*, *132*, 169–194.

Cheng, X., & Phillips, P. C. B. (2009). Semiparametric cointegrating rank selection. *Econometrics Journal*, *12*, 83–104 s1.

Clements, M. P., & Hendry, D. F. (1995). Forecasting in cointegration systems. *Journal of Applied Econometrics*, *10*, 127–146.

Clements, M. P., & Hendry, D. F. (1998). *Forecasting economic time series*. Cambridge, MA: Cambridge University Press.

Clark, T. E., & West, K. D. (2007). Approximately normal tests for equal predictive accuracy in nested models. *Journal of Econometrics*, *138*, 291–311.

Engle, R. F., & Granger, C. W. J. (1987). Co-integration and error correction: Representation, estimation, and testing. *Econometrica*, *55*, 251–276.

Fagan, G., Henry, J., & Mestre, R. (2001). An area-wide model for the Euro area, ECB Working Paper No. 42, European Central Bank.

Forni, M., Hallin, M., Lippi, M., & Reichlin, L. (2000). The generalized dynamic-factor model: Identification and estimation. *Review of Economics and Statistics*, *82*, 540–554.

Gengenbach, C., Urbain, J-P., & Westerlund, J. (2008). Panel error correction testing with global stochastic trends, METEOR Research Memorandum 051.

Johansen, S. (1995). *Likelihood-based inference in cointegrated vector autoregressive models*. Oxford: Oxford University Press.

King, R. G., Plosser, C. I., Stock, J. H., & Watson, M. W. (1991). Stochastic trends and economic fluctuations. *American Economic Review*, *81*, 819–840.

Lütkepohl, H. (2014). Structural vector autoregressive analysis in a data rich environment: A survey, DIW Discussion Paper No. 1351.

Marcellino, M., Stock, J. H., & Watson, M. W. (2006). A comparison of direct and iterated AR methods for forecasting macroeconomic series h-steps ahead. *Journal of Econometrics*, *135*, 499–526.

Stock, J. H., & Watson, M. W. (1998). Testing for common trends. *Journal of American Statistical Association*, *83*, 1097–1107.

Stock, J. H., & Watson, M. W. (2002a). Forecasting using principal components from a large number of predictors. *Journal of American Statistical Association*, *97*, 1167–1179.

Stock, J. H., & Watson, M. W. (2002b). Macroeconomic forecasting using diffusion indexes. *Journal of Business and Economics Statistics*, *20*, 147–162.

Stock, J. H., & Watson, M. W. (2005). Implications of dynamic factor models for VAR analysis, NBER Working Paper No. 11467.

Wu, J. C., & Xia, F. D. (2014). Measuring the macroeconomic impact of monetary policy at the zero lower bound, Chicago Booth Research Paper No. 13-77.

APPENDIX A. ADDITIONAL FORECASTING RESULTS

Table A1. Forecasting Real Variables for the Euro Area over 2002–2012.

h	Variable	RMSE of AR	MSE relative to MSE of AR model					
			FAR	VAR	FAVAR	ECM	FECM	$FECM_{BN}$
	GDP	0.006	1.39	0.94***	0.75***	0.92*	0.74*	0.64**
1.00	Consumption	0.004	1.51	1.36	1.42	1.05	0.78*	1.02
	Exports	0.007	0.95	1.06	1.02	1.04	0.90**	0.82**
	GDP	0.013	1.17	0.92***	0.85***	0.85*	0.72*	0.75**
2.00	Consumption	0.007	1.71	1.59	1.60	1.18	0.76*	1.08
	Exports	0.014	1.10	1.14	1.14	1.13	0.88*	0.91**
	GDP	0.024	1.05	0.93	0.94	0.90*	0.78**	0.86***
4.00	Consumption	0.013	1.66	1.62	1.68	1.36	0.91*	1.22
	Exports	0.028	1.20	1.21	1.21	1.23	0.77*	0.80*
	GDP	0.040	1.01	0.94	0.99	0.95**	0.84***	0.95
8.00	Consumption	0.025	1.50	1.49	1.56	1.34	1.04	1.19
	Exports	0.052	1.23	1.24	1.22	1.33	0.51*	0.66*
		AR	0.81	2.51	1.24			
	Lags	FAR	0.27	0.27	1.00			
		VAR	FAVAR	ECM	FECM	$FECM_{BN}$		
		1.00	1.00	1.99	1.00	1.00		

Cointegration rank								
ECM			FECM			$FECM_{BN}$		
mean	min	max	mean	min	max	mean	min	max
0.00	0.00	0.00	1.36	1.00	2.00	1.43	1.00	2.00

Notes: The FECM and the FAVAR contain four factors extracted from data in levels. $FECM_{BN}$ uses factors extracted from differences. Cheng and Phillips (2008) cointegration test and lag selection based on BIC. Data: 1975:1–2012:4, forecasting: 2002:1–2012:4.
*,** and*** indicate the significance at 10%, 5% and 1%, respectively of the Clark and West (2007) test of equal predictive accuracy relative to AR model.

Table A2. Forecasting Nominal Variables for the Euro Area over 2002–2012.

h	Variable	RMSE of AR	MSE relative to MSE of AR model					
			FAR	VAR	FAVAR	ECM	FECM	$FECM_{BN}$
	HICP	0.004	4.05	2.46	3.08	2.15	2.08	2.14
1.00	ULC	0.006	1.61	1.03	1.57	0.85**	1.06	0.96*
	Nominal XR	0.021	1.05	0.99	1.11	1.03	1.15	1.09
	HICP	0.007	2.53	1.78	1.73	1.45	1.41	1.54
2.00	ULC	0.010	1.70	0.76*	1.12	0.61**	0.89***	0.73**
	Nominal XR	0.036	1.00	1.03	1.15	1.04	1.29	1.14
	HICP	0.012	2.07	1.44	1.46	1.19	1.04	1.56
4.00	ULC	0.020	1.76	0.63**	0.77***	0.46***	0.70***	0.59***
	Nominal XR	0.056	0.99	1.02	1.10	1.02	1.50	1.17
	HICP	0.022	2.15	1.57	1.67	1.18	0.98***	2.29
8.00	ULC	0.039	1.96	0.60**	0.58**	0.46**	0.51**	0.62**
	Nominal XR	0.084	1.00	0.99	1.02	0.98	1.53	0.98
		AR	4.93	4.00	1.00			
	Lags	FAR	1.00	0.29	0.32			
		VAR	FAVAR	ECM	FECM	$FECM_{BN}$		
		1.43	1.00	1.98	1.00	1.00		
Cointegration rank								
ECM			FECM			$FECM_{BN}$		
mean	min	max	mean	min	max	mean	min	max
0.00	0.00	0.00	1.51	1.00	2.00	1.06	0.25	2.00

Notes: The FECM and the FAVAR contain four factors extracted from data in levels. $FECM_{BN}$ uses factors extracted from differences. Cheng and Phillips (2008) cointegration test and lag selection based on BIC. Data: 1975:1–2012:4, forecasting: 2002:1–2012:4.
*,**,*** Indicate the significance at 10%, 5% and 1%, respectively, of the Clark and West (2007) test of equal predictive accuracy relative to AR model.

Table A3. Forecasting Nominal Variables for the Euro Area over 2002–2012, Factors Extracted from Nominal Subpanel with Boivin and Ng (2006) Pre-selected Variables.

h	Variable	RMSE of AR	MSE relative to MSE of AR model					
			FAR	VAR	FAVAR	ECM	FECM	$FECM_{BN}$
	HICP	0.004	1.97	2.46	1.91	2.15	1.78	1.77
1.00	ULC	0.006	1.07	1.03	1.27	0.85**	1.16	1.17
	Nominal XR	0.021	0.99	0.99	0.98***	1.03	0.99	0.99
	HICP	0.007	0.99***	1.78	0.98***	1.45	1.00	0.98**
2.00	ULC	0.010	0.78***	0.76*	0.86***	0.61**	0.76***	0.77***
	Nominal XR	0.036	0.95**	1.03	0.99	1.04	1.02	1.00
	HICP	0.012	0.81**	1.44	0.88**	1.19	0.78*	0.76*
4.00	ULC	0.020	0.48***	0.63**	0.52***	0.46***	0.50***	0.51***
	Nominal XR	0.056	0.97	1.02	0.99	1.02	0.95***	0.95***
8.00	HICP	0.022	0.98**	1.57	1.04	1.18	0.82***	0.81***
	ULC	0.039	0.32**	0.60**	0.30**	0.46**	0.29**	0.30**
	Nominal XR	0.084	1.00	0.99	1.02	0.98	0.98	0.98
		AR	4.93	4.00	1.00			
	Lags	FAR	1.00	1.00	1.00			
		VAR	FAVAR	ECM	FECM	$FECM_{BN}$		
		1.43	1.00	1.98	1.00	1.00		
Cointegration rank								
ECM			FECM			$FECM_{BN}$		
mean	min	max	mean	min	max	mean	min	max
0.00	0.00	0.00	0.34	0.00	1.00	0.32	0.00	1.00

Notes: The FECM and the FAVAR contain four factors extracted from data in levels. $FECM_{BN}$ uses factors extracted from differences. Cheng and Phillips (2008) cointegration test and lag selection based on BIC. Data: 1975:1–2012:4, forecasting: 2002:1–2012:4.
*,**,*** Indicate the significance at 10%, 5% and 1%, respectively, of the Clark and West (2007) test of equal predictive accuracy relative to AR model. Variables pre-selected as in Boivin and Ng (2006) using 0.75 threshold for the correlation coefficient.

Table A4. Forecasting US Real Variables, Evaluation Period 2000m1–2014m7.

h	Variable	RMSE of AR	MSE relative to MSE of AR model					
			FAR	VAR	FAVAR	ECM	FECM	$FECM_{BN}$
	PI	0.008	0.91*	0.92*	0.91*	0.92*	0.93*	0.90*
1.00	ManTr	0.008	0.91*	0.95*	0.99*	1.08	0.95*	1.01
	IP	0.006	0.84*	0.91*	0.85*	0.95*	1.04	1.06
	Empl	0.001	1.21	1.12	1.45	1.25	2.14	1.17
	PI	0.013	0.80*	0.78*	0.77*	0.80*	0.84*	0.76*
3.00	ManTr	0.014	0.92*	0.97*	1.00	1.32	1.01	1.16
	IP	0.013	0.92*	1.00	0.97*	1.24	1.22	1.32
	Empl	0.003	1.28	1.21	1.48	1.53	2.51	1.38
	PI	0.019	0.79**	0.71**	0.75*	0.78***	0.78*	0.75**
6.00	ManTr	0.024	1.00	0.99***	1.03	1.51	0.96*	1.20
	IP	0.025	0.97*	1.02	1.00	1.44	1.16	1.35
	Empl	0.007	1.42	1.30	1.47	1.80	2.12	1.60
	PI	0.033	0.92**	0.77**	0.84*	0.93	0.77**	0.86***
12.00	ManTr	0.044	1.05	0.98	0.99	1.63	0.87**	1.20
	IP	0.049	0.95*	1.00	0.95*	1.60	0.91***	1.36
	Empl	0.016	1.44	1.26	1.33	1.88	1.35	1.61
	PI	0.045	0.99	0.85**	0.93***	1.08	0.80***	0.98
18.00	ManTr	0.059	1.05	0.99	0.99***	1.73	1.02	1.21
	IP	0.066	0.96**	0.99	0.95**	1.74	0.81***	1.40
	Empl	0.026	1.37	1.19	1.27	1.87	0.98***	1.57

	PI	0.055	1.00	0.90**	0.96	1.19	0.84***	1.04
24.00	ManTr	0.071	1.03	0.98***	0.99***	1.81	1.33	1.19
	IP	0.079	0.99*	0.99*	0.97***	1.88	0.84***	1.46
	Empl	0.036	1.27	1.13	1.21	1.84	0.82***	1.57
		AR	0.51	3.00	3.00	3.00		
	Lags	FAR	2.00	2.00	2.00	2.00		
		VAR	FAVAR	ECM	FECM	$FECM_{BN}$		
		2.00	1.90	2.00	1.00	1.00		
Cointegration rank								
ECM			FECM			$FECM_{BN}$		
mean	min	max	mean	min	max	mean	min	max
3.13	2.00	4.00	3.01	3.00	4.00	4.00	4.00	4.00

Notes: The FECM and the FAVAR contain four factors extracted from data in levels. $FECM_{BN}$ uses factors extracted from differences. Cheng and Phillips (2008) cointegration test and lag selection based on BIC. Data: 1960:1–2014:7.

Variables: IP — Industrial production, PI — Personal income less transfers, Empl — Employees on non-aggr. payrolls, ManTr — Real manufacturing trade and sales.

*,**,*** Indicate the significance 10%, 5% and 1%, respectively, of the Clark and West (2007) test of equal predictive accuracy relative to the AR model.

Table A5. Forecasting US Nominal Variables, Evaluation Period 2000m1–2014m7.

h	Variable	RMSE of AR	MSE relative to MSE of AR model					
			FAR	VAR	FAVAR	ECM	FECM	$FECM_{BN}$
	PPI	0.008	1.07	1.04	1.15	0.84*	0.90*	0.88*
1.00	CPI all	0.003	1.14	1.02	1.14	0.98*	1.04	0.94*
	CPI no food	0.003	1.15	1.05	1.15	0.93*	1.01	0.90*
	PCE defl	0.002	1.10	0.99	1.12	1.05	1.02	1.01
	PPI	0.008	1.16	1.05	1.17	0.78*	0.81*	0.83*
3.00	CPI all	0.003	1.37	1.06	1.33	1.03	0.92*	0.94**
	CPI no food	0.004	1.34	1.05	1.29	0.97*	0.89*	0.89*
	PCE defl	0.003	1.23	1.04	1.20	1.03	0.88**	0.90**
	PPI	0.008	1.22	1.05	1.23	0.74*	0.78*	0.78*
6.00	CPI all	0.003	1.29	1.04	1.27	0.92*	0.81*	0.84*
	CPI no food	0.004	1.30	1.04	1.25	0.88*	0.79*	0.81*
	PCE defl	0.003	1.32	1.02	1.28	0.93**	0.76*	0.78*
	PPI	0.009	1.32	1.09	1.35	0.69*	0.74*	0.75*
12.00	CPI all	0.003	1.35	1.06	1.32	0.88*	0.77*	0.78*
	CPI no food	0.004	1.39	1.08	1.32	0.87*	0.78*	0.78*
	PCE defl	0.003	1.47	1.05	1.41	0.94*	0.75*	0.77*
	PPI	0.008	1.13	1.03	1.14	0.75*	0.80*	0.78*
18.00	CPI all	0.003	1.21	1.04	1.19	0.98*	0.83*	0.84*
	CPI no food	0.004	1.23	1.04	1.20	0.96*	0.83*	0.85*
	PCE defl	0.003	1.28	1.02	1.24	1.02	0.82*	0.79*

	PPI	0.008	1.19	1.03	1.22	0.76**	0.83*	0.78**
24.00	CPI all	0.003	1.21	1.03	1.18	0.98*	0.85*	0.84*
	CPI no food	0.004	1.24	1.03	1.19	0.97**	0.84*	0.86*
	PCE defl	0.003	1.31	1.02	1.27	1.05	0.85***	0.81***
		AR	5.85	5.77	6.00	5.00		
	Lags	FAR	2.00	2.00	2.00	2.00		
		VAR	FAVAR	ECM	FECM	$FECM_{BN}$		
		5.00	2.00	0.00	1.00	1.00		
Cointegration rank								
ECM			FECM			$FECM_{BN}$		
mean	min	max	mean	min	max	mean	min	max
4.00	4.00	4.00	4.00	4.00	4.50	4.00	4.00	4.33

Notes: The FECM and the FAVAR contain four factors extracted from data in levels. $FECM_{BN}$ uses factors extracted from differences. Cheng and Phillips (2008) cointegration test and lag selection based on BIC. Data: 1960:1–2014:7.

Variables: Inflations of producer price index (PPI), consumer price index of all items (CPI all), consumer price index less food (CPI no food) and personal consumption deflator (PCE defl).

*,**,*** Indicate the significance 10%, 5% and 1%, respectively, of the Clark and West (2007) test of equal predictive accuracy relative to the AR model.

ESTIMATION OF VAR SYSTEMS FROM MIXED-FREQUENCY DATA: THE STOCK AND THE FLOW CASE

Lukas Koelbl[a], Alexander Braumann[a], Elisabeth Felsenstein[a] and Manfred Deistler[a,b]

[a]*Institute of Statistics and Mathematical Methods in Economics, Vienna University of Technology, Vienna, Austria*
[b]*Institute for Advanced Studies, Vienna, Austria*

ABSTRACT

This paper is concerned with estimation of the parameters of a high-frequency VAR model using mixed-frequency data, both for the stock and for the flow case. Extended Yule–Walker estimators and (Gaussian) maximum likelihood type estimators based on the EM algorithm are considered. Properties of these estimators are derived, partly analytically and by simulations. Finally, the loss of information due to mixed-frequency data when compared to the high-frequency situation as well as the gain of information when using mixed-frequency data relative to low-frequency data is discussed.

Keywords: Dynamic models; EM estimation method; extended Yule-Walker equations; Mixed frequency data

JEL classifications: C18; C38

Dynamic Factor Models
Advances in Econometrics, Volume 35, 43–73

ISSN: 0731-9053/doi:10.1108/S0731-905320150000035002

1. INTRODUCTION

In this paper, we consider the problem of estimating the parameters of an n-dimensional high-frequency VAR model

$$y_t = \begin{pmatrix} y_t^f \\ y_t^s \end{pmatrix} = A_1 y_{t-1} + \cdots + A_p y_{t-p} + \nu_t, \quad t \in \mathbb{Z} \tag{1}$$

using mixed-frequency data. We actually observe mixed-frequency data of the form

$$\begin{pmatrix} y_t^f \\ w_t \end{pmatrix} \tag{2}$$

where

$$w_t = \sum_{i=1}^{N} c_i y_{t-i+1}^s \tag{3}$$

where $c_i \in \mathbb{R}$, $1 < N \in \mathbb{N}$ and at least one $c_i \neq 0$. Here, the n_f-dimensional, say, fast component y_t^f is observed at the highest (sampling) frequency $t \in \mathbb{Z}$ and the n_s-dimensional slow component w_t is observed only for $t \in N\mathbb{Z}$, that is, for every N-th time point. In this paper, we assume that $n_f \geqslant 1$. Equation (3) represents the general case. For the case of flow data we have that $c_i = 1$, for $i = 1, \ldots, N$, whereas for the case of stock data we have that $c_1 = 1$ and $c_i = 0$, for $i = 2, \ldots, N$.

Throughout we assume the following for the high-frequency VAR model: The system parameters $A_i \in \mathbb{R}^{n \times n}$ satisfy the stability assumption

$$\det(a(z)) \neq 0 \;\; |z| \leqslant 1 \tag{4}$$

where $a(z) = I - A_1 z - \cdots - A_p z^p$ and the polynomial order p is given or specified. Here, z is used for the complex variable as well as for the backward shift on the integers $\mathbb{Z}$. We assume that (ν_t) is white noise and we only consider the stable steady-state solution $y_t = a(z)^{-1}\nu_t$. The rank q of the innovation covariance matrix $\Sigma_\nu = \mathbb{E}\left(\nu_t \nu_t^T\right)$ is given or specified, where $q \leqslant n$ holds. When the innovation matrix Σ_ν is nonsingular, the system is called regular, otherwise it is called singular. Singular autoregressive systems are important as models for latent variables and the corresponding static factors in generalized dynamic factor models (GDFMs) (see Deistler, Anderson, Filler, Zinner, & Chen, 2010; Forni, Hallin, Lippi, & Reichlin, 2000; Forni, Hallin, Lippi, & Zaffaroni, 2015; Stock & Watson, 2002). They are also important for DSGE models for the case where the number of shocks is strictly smaller than the number of outputs (see Komunjer & Ng, 2011).

The parameter space for the high-frequency models considered is given by

$$\Theta = \left\{ (A_1, \ldots, A_p) \middle| \det(a(z)) \neq 0, |z| \leqslant 1 \right\} \times \left\{ \Sigma_\nu \middle| \Sigma_\nu = \Sigma_\nu^T, \Sigma_\nu \geqslant 0, \mathrm{rk}(\Sigma_\nu) = q \right\}$$

where $\mathrm{rk}(A)$ denotes the rank of the matrix A. Since Σ_ν is of rank $q \leqslant n$, we can write $\Sigma_\nu = bb^T$, where b is an $(n \times q)$ matrix. Accordingly, $\nu_t = b\varepsilon_t$, where $\mathbb{E}(\varepsilon_t \varepsilon_t^T) = I_q$. For given Σ_ν, b is unique up to postmultiplication by an orthogonal matrix. For a particular unique choice of b, see Filler (2010).

Model (1) can be written in companion form as

$$\underbrace{\begin{pmatrix} y_t \\ y_{t-1} \\ \vdots \\ y_{t-p+1} \end{pmatrix}}_{x_{t+1}} = \underbrace{\begin{pmatrix} A_1 & \cdots & A_{p-1} & A_p \\ I_n & & & \\ & \ddots & & \\ & & I_n & 0 \end{pmatrix}}_{\mathcal{A}} \underbrace{\begin{pmatrix} y_{t-1} \\ y_{t-2} \\ \vdots \\ y_{t-p} \end{pmatrix}}_{x_t} + \underbrace{\begin{pmatrix} b \\ 0 \\ \vdots \\ 0 \end{pmatrix}}_{\mathcal{B}} \varepsilon_t \tag{5}$$

$$y_t = \underbrace{(A_1 \cdots A_p)}_{\mathcal{C}} x_t + b\varepsilon_t \tag{6}$$

The solutions of Eq. (1) and of Eqs. (5), (6) are of the form

$$y_t = a(z)^{-1}\nu_t = a(z)^{-1}b\varepsilon_t = \left(\mathcal{C}(I - \mathcal{A}z)^{-1}\mathcal{B}z + b\right)\varepsilon_t \tag{7}$$

where $k(z) = a(z)^{-1} = \sum_{j=0}^{\infty} k_j z^j$ is the transfer function from (ν_t) to (y_t) and $k(z)b = \mathcal{C}(I - \mathcal{A}z)^{-1}\mathcal{B}z + b$ is the transfer function from (ε_t) to (y_t).

Due to the mixed-frequency structure of the observed data, the population second moments

$$\gamma^{ff}(h) = \mathbb{E}\left(y^f_{t+h}(y^f_t)^T\right), \quad h \in \mathbb{Z}$$
$$\gamma^{wf}(h) = \mathbb{E}\left(w_{t+h}(y^f_t)^T\right), \quad h \in \mathbb{Z}$$
$$\gamma^{ww}(h) = \mathbb{E}\left(w_{t+h}(w_t)^T\right), \quad h \in N\mathbb{Z} \tag{8}$$

can be directly estimated. For estimation of the high-frequency parameters, identifiability is a core issue. In our context, identifiability means that the parameters of the high-frequency system can be uniquely obtained from the population second moments given in Eq. (8). As has been discussed in Anderson et al. (2012, 2015), identifiability can only be guaranteed generically. This means that we can guarantee identifiability on a set containing an open and dense subset of the parameter space. We call this g-identifiability. For the remaining part of the paper, unless the contrary is stated explicitly, we assume that the true system is in the identifiable set.

The paper is organized as follows: In Section 2, we introduce extended Yule–Walker estimators, first for the case of stock variables (see Chen & Zadrozny, 1998) and then for the general case (3), as well as a (Gaussian) maximum likelihood type estimator based on the EM algorithm. Note that these estimators do not necessarily lead to a stable system, nor do they necessarily give a positive (semi)-definite innovations covariance matrix of rank q. For these reasons, in Section 3, algorithms are discussed for transforming these estimators to a stable and positive (semi)-definite form, respectively. In Section 4, the asymptotic properties of the extended Yule–Walker estimators are derived. In Section 5, a simulation study is presented, in which we compare the extended Yule–Walker estimators with the maximum likelihood type estimator. Furthermore, the information loss due to mixed-frequency data compared to high-frequency data on the one hand and the information gain from using mixed-frequency data compared to low-frequency data on the other hand are discussed.

2. MIXED-FREQUENCY ESTIMATORS

2.1 Extended Yule–Walker Estimators: The Stock Case

In Chen and Zadrozny (1998), extended Yule–Walker (XYW) equations have been proposed for estimation of the high-frequency parameters from mixed-frequency stock data. On a population level, these XYW equations are of the form

$$\underbrace{\mathbb{E}\left[y_t\left(\left(y_{t-1}^f\right)^T,\ldots,\left(y_{t-np}^f\right)^T\right)\right]}_{Z_1} = \underbrace{(A_1,\ldots,A_p)}_{A}\underbrace{\mathbb{E}\left[\begin{pmatrix} y_{t-1} \\ \vdots \\ y_{t-p}\end{pmatrix}\left(\left(y_{t-1}^f\right)^T,\ldots,\left(y_{t-np}^f\right)^T\right)\right]}_{Z_0} \tag{9}$$

where $Z_0 \in \mathbb{R}^{np\times n_f np}$ and $Z_1 \in \mathbb{R}^{n\times n_f np}$. In Anderson et al. (2012, 2015), it has been shown that Z_0 has full row rank np on a generic subset of the parameter space. Thus, in this case, $(A_1,\ldots,A_p)$ are uniquely determined from $(A_1,\ldots,A_p) = Z_1 Z_0^\dagger$, where $Z_0^\dagger = Z_0^T(Z_0 Z_0^T)^{-1}$ is the Moore–Penrose pseudo-inverse of Z_0.

XYW estimators are obtained by replacing the population second moments by their sample counterparts:

$$\hat{\gamma}^{ff}(h) = \frac{1}{T}\sum_{t=1}^{T-h} y_{t+h}^f\left(y_t^f\right)^T, \quad h \geqslant 0 \tag{10}$$

$$\hat{\gamma}^{ff}(h) = \hat{\gamma}^{ff}(-h)^T \tag{11}$$

$$\hat{\gamma}^{wf}(h) = \frac{1}{T/N}\sum_{t=t_1}^{t_2} w_{Nt}\left(y_{Nt-h}^f\right)^T \tag{12}$$

where the estimator of $\gamma^{wf}(h)$ has only (approximately) $1/N$-th of the summands compared to the estimator of $\gamma^{ff}(h)$ due to the missing observations and

$$t_1 = \begin{cases} 1 & N > h \\ \left\lfloor \frac{h}{N} \right\rfloor + 1 & N \leqslant h \end{cases}, \quad t_2 = \begin{cases} \left\lfloor \frac{T}{N} \right\rfloor & h \geqslant 0 \\ \left\lfloor \frac{T+h}{N} \right\rfloor & h < 0 \end{cases}$$

For the case of stock variables considered here, we have $w_{Nt} = y^s_{Nt}$.

Let $\hat{Z}_0$ and $\hat{Z}_1$ denote the corresponding estimators of Z_0 and Z_1, respectively. If Z_0 has full row rank and if we assume that its estimator is consistent, $\hat{Z}_0$ will be also of full row rank, from a certain T_0 onwards. The XYW equations with Z_0 and Z_1 replaced by $\hat{Z}_0$ and $\hat{Z}_1$ are overdetermined in general and can be solved in different ways. We may define the XYW estimator as follows:

$$\hat{A}_{\text{XYW}} = \hat{Z}_1 \hat{Z}_0^{\dagger} \tag{13}$$

where $\hat{Z}_0^{\dagger} = \hat{Z}_0^T \left(\hat{Z}_0 \hat{Z}_0^T \right)^{-1}$ is the Moore–Penrose pseudo inverse of $\hat{Z}_0$. Alternatively, the generalized method of moments (GMM) estimator (see Hansen, 1982), which is defined as

$$\hat{A}_{\text{GMM}} = \underset{A \in \mathbb{R}^{n \times np}}{\arg\min} \operatorname{vec}\left(\hat{Z}_1 - A\hat{Z}_0\right)^T Q_T \operatorname{vec}\left(\hat{Z}_1 - A\hat{Z}_0\right) \tag{14}$$

can be used. Here, (Q_T) is a sequence of random weighting matrices which converges almost surely to a constant, symmetric, positive-definite matrix Q_0. Let vec(.) and $\otimes$ denote columnwise vectorization and the Kronecker product, respectively. In addition, we assumed that $\hat{Z}_0$ has full row rank and that Q_T is nonsingular. The solution of Eq. (14) is

$$\operatorname{vec}\left(\hat{A}_{\text{GMM}}\right) = \underbrace{\left(\left(\hat{Z}_0 \otimes I_n\right) Q_T \left(\hat{Z}_0^T \otimes I_n\right) \right)^{-1} \left(\hat{Z}_0 \otimes I_n\right) Q_T}_{\hat{G}^{\dagger}_{Q_T}} \operatorname{vec}\left(\hat{Z}_1\right)$$

Note that if we set $Q_T = I_{n^2 p n_f}$, we obtain the XYW estimator.

It may be useful to exploit the information contained in covariances corresponding to higher-order lags in order to improve the quality of the estimator: For $k \geqslant 0$, we obtain

$$\begin{aligned} &\underbrace{\mathbb{E}\left(y_t \left(\left(y^f_{t-1}\right)^T, \ldots, \left(y^f_{t-np-k}\right)^T \right) \right)}_{Z_{1,k}} \\ &\quad = \left(A_1, \ldots, A_p\right) \underbrace{\mathbb{E}\left(\begin{pmatrix} y_{t-1} \\ \vdots \\ y_{t-p} \end{pmatrix} \left(\left(y^f_{t-1}\right)^T, \ldots, \left(y^f_{t-np-k}\right)^T \right) \right)}_{Z_{0,k}} \end{aligned} \tag{15}$$

Note that the XYW estimators do not use the information contained in the autocovariances of the slow process, which can be directly observed.

A symmetric estimator $\hat{\Sigma}_\nu$ for Σ_ν is obtained from

$$\operatorname{vec}\left(\hat{\Sigma}_\nu\right) = \left((\mathcal{G}\otimes\mathcal{G})\left(I_{(np)^2} - \left(\hat{\mathcal{A}}\otimes\hat{\mathcal{A}}\right)\right)^{-1}\left(\mathcal{G}^T\otimes\mathcal{G}^T\right)\right)^{-1}\operatorname{vec}(\hat{\gamma}(0)) \quad (16)$$

where $\mathcal{G} = (I_n, 0, \ldots, 0)$ and $\hat{\mathcal{A}}$ is the companion form corresponding to Eq. (5) with the true parameters replaced by their estimators. Here, we assumed that $\left((\mathcal{G}\otimes\mathcal{G})\left(I_{(np)^2} - (\mathcal{A}\otimes\mathcal{A})\right)^{-1}\left(\mathcal{G}^T\otimes\mathcal{G}^T\right)\right)$ is nonsingular, an assumption which is generically fulfilled (see Anderson et al., 2015).

2.2 Extended Yule–Walker Estimators: The General Case

Let

$$z_t = \sum_{i=1}^{N} c_i y_{t-i+1} = \begin{pmatrix} \sum_{i=1}^{N} c_i y^f_{t-i+1} \\ w_t \end{pmatrix} = \begin{pmatrix} z^f_t \\ z^s_t \end{pmatrix} \quad (17)$$

and $\gamma_z(h) = \mathbb{E}\left(z_t z^T_{t-h}\right)$. For convenience, we assume that $c_N \neq 0$. In this case, we obtain another form of the XYW equations as follows:

$$\underbrace{\mathbb{E}\left[z_t\left(\left(y^f_{t-N}\right)^T, \ldots, \left(y^f_{t-np-N+1}\right)^T\right)\right]}_{Z^g_1}$$

$$= (A_1, \ldots, A_p)\underbrace{\mathbb{E}\left[\begin{pmatrix} z_{t-1} \\ \vdots \\ z_{t-p} \end{pmatrix}\left(\left(y^f_{t-N}\right)^T, \ldots, \left(y^f_{t-np-N+1}\right)^T\right)\right]}_{Z^g_0} \quad (18)$$

In Koelbl (2015), it is shown that the matrix Z_0^{g} has full row rank on a generic subset of the parameter space and thus the XYW estimator in the general case is defined as

$$\hat{A}_{\mathrm{XYW}}^{\mathrm{g}} = \hat{Z}_1^{\mathrm{g}} \left(\hat{Z}_0^{\mathrm{g}} \right)^{\dagger} \tag{19}$$

where $\hat{Z}_0^{\mathrm{g}}$ and $\hat{Z}_1^{\mathrm{g}}$ are the estimators for Z_0^{g} and Z_1^{g}, respectively.

Let $m = \max(p, N)$, $\mathcal{H}_m = \left(c_1 I_n, c_2 I_n, \ldots, c_N I_n, 0_{n \times n(m-N)}\right)$, $\mathcal{G}_m = \left(I_n, 0_{n \times n(m-1)}\right)$, and $\mathcal{A}_m$ be the companion form with the system parameters

$$A_i^m = \begin{cases} A_i & i = 1, \ldots, p \\ 0 & i = p+1, \ldots, m \end{cases}$$

Then it is easy to see that

$$\operatorname{vec}\left(\gamma_z(0)\right) = \left((\mathcal{H}_m \otimes \mathcal{H}_m) \left(I_{(nm)^2} - (\mathcal{A}_m \otimes \mathcal{A}_m) \right)^{-1} \left(\mathcal{G}_m^T \otimes \mathcal{G}_m^T \right) \right) \operatorname{vec}(\Sigma_\nu) \tag{20}$$

Again in Koelbl (2015), it is shown that the first term on the right-hand side of Eq. (20) is generically nonsingular. Thus, we can define the symmetric estimator for the noise parameters as

$$\operatorname{vec}\left(\hat{\Sigma}_\nu\right) = \left((\mathcal{H}_m \otimes \mathcal{H}_m) \left(I_{(nm)^2} - \left(\hat{\mathcal{A}}_m \otimes \hat{\mathcal{A}}_m \right) \right)^{-1} \left(\mathcal{G}_m^T \otimes \mathcal{G}_m^T \right) \right)^{-1} \operatorname{vec}\left(\hat{\gamma}_z(0)\right)$$

2.3 *Maximum Likelihood Estimation and the EM Algorithm*

First, again, we consider the case of stock variables. Throughout this section we assume, for the sake of simplicity, that $q = n$, that is, the AR process is regular. For a given N, we define the blocked observed process $\tilde{y}_t = \left(y_t^T, \left(y_{t-1}^f \right)^T, \ldots, \left(y_{t-N+1}^f \right)^T \right)^T$, $t \in N\mathbb{Z}$ (see Anderson et al. 2015). The

(Gaussian) log-likelihood (ignoring the constant) of observations times −2 is given by (see Hannan & Deistler, 2012)

$$\tilde{L}_T(\theta) = \frac{1}{T}\log\left(\det\left(\tilde{\Gamma}_T(\theta)\right)\right) + \frac{1}{T}\tilde{Y}_{\text{obs}}^T\left(\tilde{\Gamma}_T(\theta)\right)^{-1}\tilde{Y}_{\text{obs}} \tag{21}$$

where $\tilde{Y}_{\text{obs}} = \left(\tilde{y}_N^T, \tilde{y}_{2N}^T, \ldots, \tilde{y}_T^T\right)^T$ are the observed data and $\tilde{\Gamma}_T(\theta) = \mathbb{E}\left(\tilde{Y}_{\text{obs}}(\theta)\tilde{Y}_{\text{obs}}^T(\theta)\right)$. Note that in the latter expression, the random variables $\tilde{Y}_{\text{obs}}(\theta)$ are considered to be generated by an arbitrary parameter $\theta \in \Theta$. Also note that $\tilde{\Gamma}_T(\theta)$ is a square matrix with dimension $\left(n + (N-1)n_f\right)T/N$, which is in general a very large matrix in block Toeplitz form and accordingly its inversion may cause problems, even when taking advantage of its block Toeplitz structure (see Wax & Kailath, 1983).

The EM algorithm (MLE-EM) is an iterative procedure to find the minimizer of $\tilde{L}_T(\theta)$. The idea of this algorithm is to successively minimize the conditional expectation of

$$L_T(\theta) = \frac{1}{T}\log(\det(\Gamma_T(\theta))) + \frac{1}{T}Y^T(\Gamma_T(\theta))^{-1}Y \tag{22}$$

of the complete-data $Y = \left(y_1^T, \ldots, y_T^T\right)^T$ given the observed data $\tilde{Y}_{\text{obs}}$. Here, $\Gamma_T(\theta) = \mathbb{E}\left(Y(\theta)Y^T(\theta)\right)$ where now the random variables $Y(\theta)$ are generated by an arbitrary parameter value $\theta \in \Theta$. We now follow Shumway and Stoffer (1982): Observe that for the stock case the mixed-frequency data can be represented by a time variable state-space system

$$\begin{aligned} x_{t+1} &= \mathcal{A}x_t + \mathcal{B}\varepsilon_t \\ y_t^{\times} &= \mathcal{G}_t x_{t+1} \end{aligned} \tag{23}$$

where $x_t = \left(y_{t-1}^T, \ldots, y_{t-p}^T\right)^T$ and $\mathcal{G}_{Nt} = (I_n, 0, \ldots, 0)$, $\mathcal{G}_{Nt-k} = \left(I_{n_f}, 0, \ldots, 0\right)$ for $k = 1, \ldots, N-1$. The vector $y_t^{\times}$ contains only observable components.

Throughout this section, we assume that the innovations ν_t in Eq. (1) are Gaussian white noise and that $x_1 \sim \mathcal{N}_{np}(0, V_1)$ is independent from $\nu_1, \ldots, \nu_T$. The complete-data log-likelihood, when modified with respect to the initial values (and using the same symbol as in Eq. (22)), can be factorized such that

$$L_T(\tau) = \log(\det(V_1)) + x_1^T V_1^{-1} x_1 \\ + T\log(\det(\Sigma_\nu)) + \sum_{t=1}^{T} (\mathcal{G}x_{t+1} - Ax_t)^T \Sigma_\nu^{-1} (\mathcal{G}x_{t+1} - Ax_t)$$

where $\mathcal{G} = (I_n, 0, \ldots, 0)$ and $\tau = \left(\text{vec}(A)^T, \text{vech}(\Sigma_\nu)^T, \text{vech}(V_1)^T\right)^T$.

The algorithm starts with an initial value $\tau^{(0)}$, for example, where $\left(\hat{A}_1, \ldots, \hat{A}_p\right)$ and $\hat{\Sigma}_\nu$ are XYW estimates and where V_1 is the corresponding solution of the Lyapunov equations, that is, $\Gamma_p = \mathcal{A}\Gamma_p\mathcal{A}^T + \mathcal{B}\mathcal{B}^T$ where $\Gamma_p = \mathbb{E}\left(x_t x_t^T\right)$. Let $\tau^{(k)}$ be the estimate at the k-th iteration, iteration $k+1$ is as follows:

E-step. The expected complete-data log-likelihood conditional on $\tilde{Y}_{\text{obs}}$, say $Q\left(\tau|\tau^{(k)}\right) = \mathbb{E}_{\tau^{(k)}}\left(L_T(\tau)|\tilde{Y}_{\text{obs}}\right)$, where the notation $\mathbb{E}_{\tau^{(k)}}$ indicates that the conditional expectation is taken corresponding to the parameter $\tau^{(k)}$, is given by

$$Q\left(\tau|\tau^{(k)}\right) = \log(\det(V_1)) + \text{tr}\left(V_1^{-1}\left(x_{1|T}x_{1|T}^T + P_{1|T}\right)\right) + \text{T}\log(\det(\Sigma_\nu)) \\ + \text{tr}\left(\Sigma_\nu^{-1}\left(\mathcal{G}S_{11}\mathcal{G}^T + AS_{00}A^T - \mathcal{G}S_{10}A^T - AS_{10}^T\mathcal{G}^T\right)\right)$$

with

$$\begin{aligned} S_{00} &= \sum_{t=1}^{T-1}\left(x_{t|T}x_{t|T}^T + P_{t|T}\right) \\ S_{11} &= \sum_{t=2}^{T}\left(x_{t|T}x_{t|T}^T + P_{t|T}\right) \\ S_{10} &= \sum_{t=2}^{T}\left(x_{t|T}x_{t-1|T}^T + P_{t,t-1|T}\right) \end{aligned}$$

where

$$\begin{aligned} x_{t|T} &= \mathbb{E}_{\tau^{(k)}}\left(x_t|\tilde{Y}_{\text{obs}}\right) \\ P_{t|T} &= \mathbb{E}_{\tau^{(k)}}\left(x_t - x_{t|T}\right)\left(x_t - x_{t|T}\right)^T \\ P_{t,t-1|T} &= \mathbb{E}_{\tau^{(k)}}\left(x_t - x_{t|T}\right)\left(x_{t-1} - x_{t-1|T}\right)^T \end{aligned}$$

They can be calculated by means of the Kalman filter or other smoothing algorithms (see Shumway & Stoffer, 1982).

M-step. Determine $\tau^{(k+1)}$ by minimizing the expected conditional likelihood, that is, $\tau^{(k+1)} = \arg\min_\tau Q(\tau|\tau^{(k)})$. The parameter updates are given by

$$\begin{aligned} A^{(k)} &= \mathcal{G}S_{10}S_{00}^{-1} \\ \Sigma^{(k)} &= T^{-1}\left(\mathcal{G}S_{11}\mathcal{G}^T - \mathcal{G}S_{10}S_{00}^{-1}S_{10}^T\mathcal{G}^T\right) \\ V_1^{(k)} &= P_{1|T} + x_{1|T}x_{1|T}^T \end{aligned}$$

The algorithm stops if the relative decrease of the log-likelihoods of observations is below a prespecified threshold. Here, we do not discuss convergence properties of the EM algorithm.

In the general case, the state-space system has to be written as (see also Mariano & Murasawa, 2010)

$$\begin{aligned} x_{t+1}^m &= \mathcal{A}_m x_t^m + \mathcal{B}_m \varepsilon_t \\ y_t^\times &= \mathcal{G}_t^m x_{t+1}^m \end{aligned} \tag{24}$$

where $x_t^m = \left(y_{t-1}^T, \ldots, y_{t-m}^T\right)^T$, $\mathcal{B}_m = \left(b^T, 0_{n\times n(m-1)}\right)^T$, $\mathcal{G}_{Nt-k}^m = \left(I_{n_f}, 0, \ldots, 0\right)$ for $k = 1, \ldots, N-1$ and

$$\mathcal{G}_{Nt}^m = \begin{pmatrix} I_{n_f} & 0 & \cdots & 0 & 0 & 0_{n_f\times n(m-N)} \\ 0_{n_s\times n_f} & c_1 I_{n_s} & \cdots & 0 & c_N I_{n_s} & 0_{n_s\times n(m-N)} \end{pmatrix}$$

The algorithm described above for the stock case can be analogously applied to the general case with the state-space model (5), (6) modified in a straightforward way.

3. PROJECTING THE MF ESTIMATORS ON THE PARAMETER SPACE

It is well known that in the high-frequency case, the Yule–Walker estimator always leads to a stable AR polynomial, provided that $\hat{\Gamma}_p > 0$ holds, where $\hat{\Gamma}_p$ is the high-frequency estimator of $\Gamma_p > 0$ (see Deistler & Anderson 2010). Furthermore, in the high-frequency case, the estimated covariance matrix of the noise is positive definite. In general, the XYW/GMM estimators do not fulfill these desirable properties and the same is true for the estimators obtained from the EM algorithm. Indeed in many simulations such a situation occurs. Consequently, in a second step, one has to check whether the estimated parameters, say $\hat{\theta}$, lie in the parameter space Θ. If $\hat{\theta}$ is not contained in this space, the question of finding a $\hat{\theta}_P \in \Theta$, which is sufficiently close to $\hat{\theta}$, arises. In this paper, we separate this problem in two sub-problems: The first problem is to find a stable polynomial matrix close to an unstable estimator of $a(z)$. The second problem is to find a positive (semi)-definite covariance matrix of rank q, which is close to an indefinite (symmetric) estimator of Σ_ν.

3.1 Stabilization of the Estimated System Parameters

In this section we commence from an unstable estimate for the system parameters, say $\hat{A}_{\text{un}} \in \mathbb{R}^{n \times np}$, corresponding to $\hat{a}_{\text{un}}(z)$, such that there exists a $z_0 \in \mathbb{C}$, $|z_0| \leqslant 1$, and $\det(\hat{a}_{\text{un}}(z_0)) = 0$. As $S = \{(A_1, \ldots, A_p) | \det(a(z)) \neq 0, |z| \leqslant 1\}$ is an open set, there exists no best approximation of such an $\hat{A}_{\text{un}}$, for instance in Frobenius norm, by an element of S. In addition, in general, S is non-convex. We consider the problem of finding

$$\inf_{A \in S} \|A - \hat{A}_{\text{un}}\|^2 \tag{25}$$

There exists a substantial literature dealing with finding the "nearest" stable polynomial both for the univariate (see Combettes & Trussell, 1992; Moses & Liu, 1991; Orbandexivry, Nesterov, & van Dooren, 2013; Stoica & Moses, 1992) and the multivariate case (see Balogh & Pintelon, 2008; D'haene, Pintelon, & Vandersteen, 2006). An interesting way to solve the univariate stabilization problem is proposed in Orbandexivry et al. (2013) using the so-called Dikin Ellipsoid. We will repeat the most important steps of this procedure and generalize it to the multivariate case, which can be easily done. We point out that all these methods need a stable initial value.

Problem (25) can be reformulated as in Orbandexivry et al. (2013) to

$$\inf_{A, P} \frac{1}{2} \|A - \hat{A}_{\text{un}}\|_F^2 \tag{26}$$

where minimization with respect to P runs over $P = P^T > 0$, $P - \mathcal{A}P\mathcal{A}^T > 0$, with $\mathcal{A}$ the companion form of A. For a fixed $P = P^T > 0$, we can define the set

$$S_P = \{A \in \mathbb{R}^{n \times np} : P - \mathcal{A}P\mathcal{A}^T > 0, \mathcal{A} \text{ is the companion form of } A\} \subset S$$

and the function $b_P(A) = -\log(\det(P - \mathcal{A}P\mathcal{A}^T))$, which is a barrier function. It follows from theorem 5 in Orbandexivry et al. (2013) that for $A \in S$, $P = P^T > 0$ such that $P - \mathcal{A}P\mathcal{A}^T > 0$ and any $0 \leqslant \alpha < 1$, the so-called Dikin Ellipsoid $\mathcal{E}(P, A; \alpha) = A + \{H \in \mathbb{R}^{n \times np} : \langle b_P''(A)H, H\rangle \leqslant \alpha\}$ is a subset of S_P where $\langle A, B\rangle = \mathrm{tr}(AB^T)$ and $\langle b_P''(A)H, H\rangle$ is the second derivative of $b_P(A)$ in a given direction H. Now, for given A and α, the question arises which P should be chosen such that $\mathcal{E}(P, A; \alpha)$ is maximized. In Orbandexivry et al. (2013) the authors argue that a good choice, say P^*, is given by solving

$$Q^{-1} - \mathcal{A}^T Q^{-1}\mathcal{A} = npI_{np} \tag{27}$$

$$P^* - \mathcal{A}P^*\mathcal{A}^T = Q \tag{28}$$

We are now in a position to formulate a new, restricted optimization problem for a given $0 \leqslant \alpha < 1$, $A \in S$ and a corresponding P^*:

$$\min_H \tfrac{1}{2}\|A + H - \hat{A}_{\mathrm{un}}\|_F^2 \tag{29}$$

where H is such that $\langle b_{P*}''(A)H, H\rangle \leqslant \alpha$. Note that we now have a convex optimization problem. It can be shown that $\langle b_{P*}''(A)H, H\rangle \leqslant \alpha$ can be rewritten as $\mathrm{vec}(H)^T B\mathrm{vec}(H) \leqslant \alpha$, where

$$\begin{aligned}\frac{1}{2}B = {} & (P^* \otimes \mathcal{G}Q^{-1}\mathcal{G}^T) + (P^*\mathcal{A}^T Q^{-1}\mathcal{A}P^* \otimes \mathcal{G}Q^{-1}\mathcal{G}^T) \\ & + (P^*\mathcal{A}^T Q^{-1}\mathcal{G}^T \otimes \mathcal{G}Q^{-1}\mathcal{A}P^*)K_{n,np}\end{aligned} \tag{30}$$

and $K_{n,np}$ is a commutation matrix, see Magnus and Neudecker (1979). It is easy to conclude that the matrix B is symmetric positive definite and thus can be factorized as $B = UDU^T$, where D is a diagonal matrix with positive entries d_i, $i = 1, \ldots, n^2p$, and U is an orthonormal matrix. The solution of Eq. (29) can be derived (see Orbandexivry et al., 2013, p. 1199) by finding the root of the function

$$\psi(\lambda) = \sum_{i=1}^{n^2p} \frac{d_i\left(e_i^T U^T \operatorname{vec}(\hat{A}_{\text{un}} - A)\right)^2}{(1 + \lambda d_i)^2} - \alpha = 0 \tag{31}$$

with respect to $\lambda \in (0, \infty)$, where e_i is the i-th unit vector, and then substituting into

$$\operatorname{vec}(H) = \left(I_{n^2p} + \lambda B\right)^{-1} \operatorname{vec}\left(\hat{A}_{\text{un}} - A\right) \tag{32}$$

It is worth mentioning that there exists a unique λ and therefore a unique H.

A stable initial estimator may be obtained, for example, by reflecting the unstable roots of $\hat{a}_{\text{un}}(z)$ on the unit circle (see Lippi & Reichlin, 1994). The whole stabilization procedure has to be iterated.

3.2 Positive (Semi)-Definiteness of the Noise Covariance Matrix

Under assumptions that guarantee consistency of the sample second moments and the system parameters A, Eq. (16) gives a consistent estimator for Σ_ν. This estimate is symmetric but may not be positive (semi)-definite and of rank q. Consider

$$\inf_{\Sigma_{\text{ps}} \in D} \|\Sigma_{\text{ps}} - \hat{\Sigma}_\nu\|_F^2 \tag{33}$$

where $D = \left\{\Sigma_\nu \in \mathbb{R}^{n \times n} | \, \Sigma_\nu = \Sigma_\nu^T, \Sigma_\nu \geqslant 0, \operatorname{rk}(\Sigma_\nu) = q\right\}$. The matrix $\hat{\Sigma}_\nu$ can be represented as $\hat{\Sigma}_\nu = Q\Lambda Q^T$ where Λ is the diagonal matrix containing the eigenvalues λ_i in descending order and Q is an orthonormal matrix containing the corresponding eigenvectors. For simplicity, we assume that the q-th and the $(q+1)$-th eigenvalue are distinct.

To obtain an arbitrarily close solution of problem (33), we define $\hat{\Sigma}_{\text{ps}} = Q\Lambda_+ Q^T$, where Λ_+ is a diagonal matrix with entries

$$\lambda_i^+ = \begin{cases} \max(\lambda_i, \varepsilon) & i = 1, \ldots, q \\ 0 & i = q+1, \ldots, n \end{cases}$$

for sufficiently small $\varepsilon > 0$. Note that by the so-called Wielandt–Hoffman Theorem (see Hoffman & Wielandt, 1953) $\sum_{i=1}^{n} (\lambda_i^A - \lambda_i^B)^2 \leqslant \|A - B\|_F^2$ holds for symmetric matrices $A, B \in \mathbb{R}^{n \times n}$, where λ_i^A and λ_i^B are the corresponding eigenvalues in a descending order, respectively. Thus $\hat{\Sigma}_{\text{ps}}$ gives an arbitrarily close solution of (33).

4. ASYMPTOTIC PROPERTIES OF THE XYW/GMM ESTIMATORS

In this section, we derive the asymptotic properties of the XYW estimator as well as of the generalized method of moments estimators. Whereas under suitable assumptions the Yule–Walker estimator has the same asymptotic covariance as the maximum likelihood estimator in the high-frequency case and thus is asymptotically efficient, this is not the case for the XYW/GMM estimators. The asymptotic distribution of XYW/GMM estimators is derived along the idea of first deriving the asymptotic distribution of the sample second moments of the observations, that is, deriving Bartlett's formula for the mixed-frequency case, and then, in a second step, linearizing the function attaching the parameters to the second-order moments of the observations. Throughout this section, we additionally assume that ν_t in Eq. (1) is independent identically distributed, $(\nu_t) \sim \text{IID}_n(0, \Sigma_\nu)$, and that $\eta = \mathbb{E}(\nu_t \nu_t^T \otimes \nu_t \nu_t^T)$ exists. For notational simplicity, we write (y_t) as $y_t = \sum_{j=-\infty}^{\infty} k_j \nu_{t-j}$, where $k_j = 0$ for $j < 0$. Note that (z_t) can be analogously represented as $z_t = \sum_{j=-\infty}^{\infty} \tilde{k}_j \nu_{t-j}$, where $\tilde{k}_j = \sum_{i=1}^{N} c_i k_{j-i+1}$. For convenience, we restrict ourselves mainly to the case of stock variables, that is, $w_{Nt} = y_{Nt}^s$. The general case will be discussed at the end of the section. In the following, we will use the partition $k_j = \begin{pmatrix} k_j^f \\ k_j^s \end{pmatrix}$, where k_j^f denotes the first n_f and k_j^s the last n_s rows of k_j, respectively.

Let $\kappa = \eta - \text{vec}(\Sigma_\nu)\text{vec}(\Sigma_\nu)^T - (\Sigma_\nu \otimes \Sigma_\nu) - K_{n,n}(\Sigma_\nu \otimes \Sigma_\nu)$, $\gamma^{z^f f}(h) = \mathbb{E}\left(z_t^f \left(y_t^f\right)^T\right) = \gamma^{ff}(h)$ and $\hat{\gamma}^{z^f f}(h)$ be the corresponding estimator. We denote convergence in distribution by $\xrightarrow{d}$ and convergence in probability by $\xrightarrow{p}$.

Theorem 1. Under the assumptions stated above in this section, we obtain

$$\sqrt{T}\left(\begin{pmatrix}\operatorname{vec}\left(\hat{\gamma}^{z^f f}(0)\right)\\ \operatorname{vec}\left(\hat{\gamma}^{wf}(0)\right)\\ \vdots\\ \operatorname{vec}\left(\hat{\gamma}^{z^f f}(s)\right)\\ \operatorname{vec}\left(\hat{\gamma}^{wf}(s)\right)\end{pmatrix}-\begin{pmatrix}\operatorname{vec}\left(\gamma^{z^f f}(0)\right)\\ \operatorname{vec}\left(\gamma^{wf}(0)\right)\\ \vdots\\ \operatorname{vec}\left(\gamma^{z^f f}(s)\right)\\ \operatorname{vec}\left(\gamma^{wf}(s)\right)\end{pmatrix}\right)\xrightarrow{d}\mathcal{N}_h(0,\Sigma_\gamma)$$

where $h=(s+1)nn_f$ and $s\in\mathbb{N}$. Σ_γ is obtained as described in Lemmas 1–3 in the appendix.

The proof of Theorem 1 is also given in the appendix.

Remark 1. The last theorem can be extended to any lag including negative ones. Indeed, we will use lags $(1-p,\ldots,np)$ for the XYW and GMM estimator. Also note that the assumption that the innovations are i.i.d. can be relaxed, see, for example, Hall and Heyde (1980) and Francq and Zakoian (2009).

Remark 2. Note that we do not distinguish between singular and non-singular normal distributions. For a detailed discussion about singular normal distributions, see Khatri (1961), Rao (1972), and Anderson (1994).

Having obtained the asymptotic distribution of the covariance estimators, we have to linearize the mapping attaching the system parameters to the second moments of the observations. The next theorem derives the asymptotic distribution of the XYW/GMM estimators and is related to Gingras (1985). Let Θ_{XYW} be the generic set of the system and noise parameters where Z_0 has full row rank np.

Theorem 2. Let (y_t) be the output of system (1) with inputs $(\nu_t)\sim\text{IID}_n(0,\Sigma_\nu)$, $\theta\in\Theta_{\text{XYW}}$ and assume that $\eta=\mathbb{E}\left(\nu_t\nu_t^T\otimes\nu_t\nu_t^T\right)$ exists. Then the GMM estimator

$$\begin{aligned}\operatorname{vec}\left(\hat{A}_{\text{GMM}}\right)&=\left(\left(\hat{Z}_0\otimes I_n\right)Q_T\left(\hat{Z}_0^T\otimes I_n\right)\right)^{-1}\left(\hat{Z}_0\otimes I_n\right)Q_T\operatorname{vec}\left(\hat{Z}_1\right)\\&=\hat{G}_{Q_T}^{\dagger}\operatorname{vec}\left(\hat{Z}_1\right)\end{aligned}$$

is asymptotically normal with zero mean and a covariance matrix given by

$$\Sigma_{\text{GMM}}=\left(G_{Q_0}^{\dagger}JP\right)\Sigma_\gamma\left(G_{Q_0}^{\dagger}JP\right)^T \tag{34}$$

that is,

$$\sqrt{T}\left(\text{vec}\left(\hat{A}_{\text{GMM}}\right) - \text{vec}(A)\right) \xrightarrow{d} \mathcal{N}_{n^2p}(0, \Sigma_{\text{GMM}}) \tag{35}$$

Here, $Q_T \xrightarrow{p} Q_0$ where Q_0 is constant, symmetric, and positive definite and Σ_γ is the asymptotic covariance of the mixed-frequency covariances, described in Theorem 1, for the lags $(-p+1, \ldots, np)$. Furthermore,

$$G_{Q_0}^{\dagger} = \left((Z_0 \otimes I_n) Q_0 \left(Z_0^T \otimes I_n\right)\right)^{-1} (Z_0 \otimes I_n) Q_0$$

$$J = \begin{pmatrix} D & 0_{n\times n} & & 0_{n\times n} \\ 0_{n\times n} & D & \ddots & \vdots \\ \vdots & \ddots & \ddots & 0_{n\times n} \\ 0_{n\times n} & \cdots & 0_{n\times n} & D \end{pmatrix} \in \mathbb{R}^{n^2pn_f \times n(n+1)n_fp}$$

$$D = \left(-A_p \quad 0_{n\times(n_f-1)n} \quad -A_{p-1} \quad 0_{n\times(n_f-1)n} \quad \ldots \quad -A_1 0_{n\times(n_f-1)n} I_n 0_{n\times(n_f-1)n} \right)$$

and the permutation matrix P is given as $P = \left(I_{(n+1)p} \otimes P_2\right)$, where

$$P_2 = \left(I_{n_f} \otimes \begin{pmatrix} I_{n_f} \\ 0_{n_s \times n_f} \end{pmatrix}, \qquad I_{n_f} \otimes \begin{pmatrix} 0_{n_f \times n_s} \\ I_{n_s} \end{pmatrix} \right)$$

Proof. We commence with the observation that

$$\begin{aligned}
&\sqrt{T}\left(\text{vec}\left(\hat{A}_{\text{GMM}}\right) - \text{vec}(A)\right) \\
&\quad = \sqrt{T}\left(\hat{G}_{Q_T}^{\dagger} \text{vec}\left(\hat{Z}_1\right) - \hat{G}_{Q_T}^{\dagger}\left(\hat{Z}_0^T \otimes I_n\right)\text{vec}(A)\right) \\
&\quad = \sqrt{T}\hat{G}_{Q_T}^{\dagger} \text{vec}\left(\hat{Z}_1 - A\hat{Z}_0 - Z_1 + AZ_0\right) \\
&\quad = \sqrt{T}\hat{G}_{Q_T}^{\dagger} \underbrace{\left(I_{pnn_f} \otimes \left(I_n \quad -A\right)\right)}_{J_1} \text{vec}\begin{pmatrix} \hat{Z}_1 - Z_1 \\ \hat{Z}_0 - Z_0 \end{pmatrix} \\
&\quad = \sqrt{T}\hat{G}_{Q_T}^{\dagger} \underbrace{J_1 J_2}_{J} P\left(\begin{pmatrix} \text{vec}\left(\hat{\gamma}^{zff}(i)\right) \\ \text{vec}\left(\hat{\gamma}^{wf}(i)\right) \end{pmatrix} - \begin{pmatrix} \text{vec}\left(\gamma^{zff}(i)\right) \\ \text{vec}\left(\gamma^{wf}(i)\right) \end{pmatrix}\right)
\end{aligned}$$

for $i = -p+1, \ldots, np$, where J_2 is a reordering matrix. Since under our assumptions, the sample autocovariances are consistent estimators and $Q_T \xrightarrow{p} Q_0$, the same is true for $\hat{G}^{\dagger}_{Q_T} JP$, that is, $\hat{G}^{\dagger}_{Q_T} JP \xrightarrow{p} G^{\dagger}_{Q_0} JP$. Theorem 1 and Slutsky's Lemma then directly lead to the result of the theorem. ■

Remark 3. It is well known that under our assumptions the asymptotic covariance for the high-frequency Yule–Walker estimator is of the form $\left(\Gamma_p^{-1} \otimes \Sigma_\nu\right)$ (see Hannan, 1970; Lütkepohl, 2005) and thus, in this case, the fourth moment of the innovations does not influence the asymptotic covariance of the parameter estimates. In the case discussed here, the fourth moment of the innovations does not vanish under linearization in general.

Having obtained the expression for the asymptotic covariance, we can determine the asymptotically optimal weighting matrix for the GMM estimator.

Theorem 3. Under the assumptions of Theorem 2, the optimal asymptotic weighting matrix for the GMM estimator is

$$Q_0^* = \left(JP\Sigma_\gamma P^T J^T\right)^{-1} \tag{36}$$

and the corresponding asymptotic covariance is given by

$$\Sigma^*_{\mathrm{GMM}} = \left((Z_0 \otimes I_n) Q_0^* \left(Z_0^T \otimes I_n\right)\right)^{-1} \tag{37}$$

For the XYW estimator, where $Q_0 = I_{n^2 p n_f}$, the asymptotic covariance is given by

$$\Sigma_{\mathrm{XYW}} = \left(\left(Z_0^{\dagger}\right)^T \otimes I_n\right) JP\Sigma_\gamma P^T J^T \left(Z_0^{\dagger} \otimes I_n\right) \tag{38}$$

Proof. The proof of the theorem directly follows from theorem 3.2 in Hansen (1982). ■

Remark 4. Using the blocked process for the AR(1) case as described in Anderson et al. (2015) it is easy to derive the asymptotic covariance matrix of the maximum likelihood estimator for the mixed-frequency stock case (see Koelbl, 2015). In contrast to the high-frequency case, where under our assumptions the Yule–Walker estimator is asymptotically equivalent to the maximum likelihood estimator, the mixed-frequency

XYW estimator is, in general, not equivalent to the mixed-frequency maximum likelihood estimator (see Example 2).

Remark 5. In order to estimate the asymptotic covariance matrix of the XYW/GMM estimators as well as the asymptotically optimal weighting matrix Q_0^*, we have to estimate the fourth moment of (ν_t), unless we assume that (ν_t) has a Gaussian distribution, where the fourth moment does not occur. In Koelbl (2015), it is shown that the fourth moment can be reconstructed on a generic subset of Θ from

$$\mathrm{vec}(\kappa) = \left(\mathcal{G}_2 \left(I_{(np)^4} - \mathcal{A} \otimes \mathcal{A} \otimes \mathcal{A} \otimes \mathcal{A} \right)^{-1} \mathcal{G}_2^T \right)^{-1} \mathrm{vec}(\psi)$$

where $\mathcal{G}_2 = \mathcal{G} \otimes \mathcal{G} \otimes \mathcal{G} \otimes \mathcal{G}$ and $\psi = \mathbb{E}\left(y_t y_t^T \otimes y_t y_t^T\right) - \mathrm{vec}(\gamma(0))\mathrm{vec}(\gamma(0))^T - (\gamma(0) \otimes \gamma(0)) - K_{n,n}(\gamma(0) \otimes \gamma(0))$.

Remark 6. Until now, the asymptotic results obtained in this section are only valid for the stock case. Nevertheless, an adaption to the general case is straightforward: Using an obvious notation and following the same steps as in the proof of Theorem 2, we obtain

$$\sqrt{T}\left(\mathrm{vec}\left(\hat{A}^{\mathrm{g}}_{\mathrm{GMM}}\right) - \mathrm{vec}(A)\right)$$
$$= \sqrt{T}\left(\hat{G}^{\mathrm{g}}_{Q_T}\right)^{\dagger} JP\left(\begin{pmatrix} \mathrm{vec}\left(\hat{\gamma}^{z^f f}(i)\right) \\ \mathrm{vec}\left(\hat{\gamma}^{wf}(i)\right) \end{pmatrix} - \begin{pmatrix} \mathrm{vec}\left(\gamma^{z^f f}(i)\right) \\ \mathrm{vec}\left(\gamma^{wf}(i)\right) \end{pmatrix} \right)$$

for $i = N - p, \ldots, N + np - 1$. The results concerning the asymptotic behavior of the covariance estimators, that is, Theorem 1 and Lemmas 1–3, are still valid.

5. SIMULATIONS

In our context the following issues arise: First, the comparison of the XYW estimators and the MLE-EM estimators. Second, the information loss caused by mixed-frequency data in relation to high-frequency data. Third, the information gain obtained by using mixed-frequency data when compared to low-frequency data. It should be emphasized here, that we only present preliminary results and in order to get a more complete picture further work is needed.

It is intuitively clear that the quality of the mixed-frequency estimators depends on N and on n_f, for example, because the number of summands in Eq. (12) depends on N, see Example 3. Moreover, the quality of the mixed-frequency estimators depends on the underlying parameters of the high-frequency model. In particular, if we are close to the (mixed-frequency) nonidentifiable subset in the parameter space, a large information loss due to mixed-frequency data in comparison to high-frequency data can be expected. One way to measure information loss in our context would be to compare the (asymptotic) covariances of the MLEs from mixed-frequency data (MF-MLE) with the (asymptotic) covariances of the MLEs from high-frequency data (HF-MLE). Another way would be to compare the one-step-ahead prediction error covariances. This is of particular importance for comparisons to the low-frequency case, where in general identifiability cannot be achieved. In order to demonstrate the effects of being close to the nonidentifiable subset, we consider a simple example:

Example 1. Assume that $p = 1$, $n_f = n_s = 1$, $N = 2$, $\Sigma_\nu = I_2$ and the case of stock variables. The system parameters $A_1 = \begin{pmatrix} a_{ff} & a_{fs} \\ a_{sf} & a_{ss} \end{pmatrix}$ are not identifiable if and only if they satisfy the equations $a_{fs} = 0$, $a_{sf} = 0, a_{ss} \neq 0$ (see Anderson et al., 2015). For the models

$$y_t = \begin{pmatrix} 0.9 & 0 \\ a_{sf} & 0.8 \end{pmatrix} y_{t-1} + \nu_t, \quad \nu_t \sim \mathcal{N}_2(0, I_2) \tag{39}$$

and $a_{sf} \in \{0, 0.01, 0.1, 0.25\}$, we obtain the following sections of the likelihoods as shown in Fig. 1 where we only vary over a_{ss}.

Table 1 reports the mean squared errors (MSE) of the XYW estimator corresponding to (13) and the MLE-EM, initialized by the XYW estimator, for the model class described above for four different values of a_{sf}. It shows that also close to the identifiability boundary problems for the estimators arise. The last two columns show the relative number of hits of the estimators for a_{ss} for the intervals $[-0.9, -0.7]$ and $[0.7, 0.9]$, respectively. Note that, in particular, for the nonidentifiable case $a_{sf} = 0$, the MLE-EM gives estimates close to the class of equivalent system parameters

$$\left\{ \begin{pmatrix} 0.9 & 0 \\ 0 & 0.8 \end{pmatrix}, \begin{pmatrix} 0.9 & 0 \\ 0 & -0.8 \end{pmatrix} \right\}$$

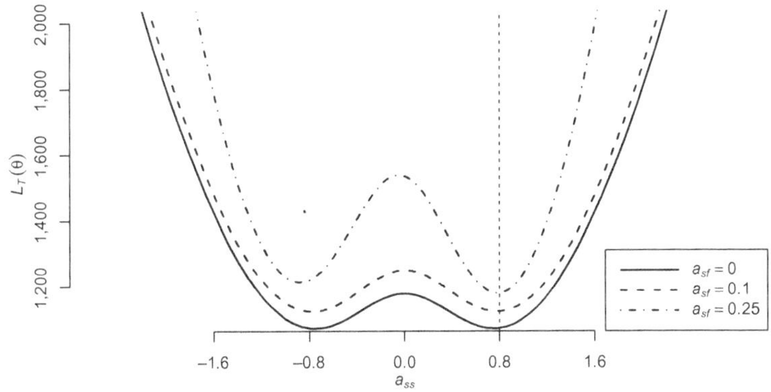

Fig. 1. Sections of the Likelihood Functions $L_T(\theta)$ for Three Different Values of a_{sf}.

Table 1. Comparison of XYW and MLE-EM Estimators in Terms of Mean-Squared Errors and Hits.

a_{sf}	Estimators	MSE $\hat{a}_{ff}$	MSE $\hat{a}_{sf}$	MSE $\hat{a}_{fs}$	MSE $\hat{a}_{ss}$	$[-0.9, -0.7]$	$[0.7, 0.9]$
0	XYW	0.131	0.086	25.335	1.698	0.04	0.05
	MLE-EM	0.004	0.007	0.001	1.160	0.43	0.51
0.01	XYW	0.124	0.085	11.135	1.589	0.05	0.06
	MLE-EM	0.001	0.006	0.001	1.100	0.42	0.55
0.1	XYW	0.002	0.016	0.016	0.061	0.00	0.26
	MLE-EM	0.001	0.001	0.001	0.001	0.00	1.00
0.25	XYW	0.001	0.030	0.003	0.028	0.00	0.39
	MLE-EM	0.001	0.003	0.001	0.004	0.00	1.00

Furthermore, in this case the matrix Z_0 is singular and in particular the solution set of Eq. (9) is given as

$$\begin{pmatrix} 0.9 & d_1 \\ 0 & d_2 \end{pmatrix}$$

where d_1 and d_2 are arbitrary. The MSEs of the XYW estimator for a_{fs} are relatively large compared to the MLE-EM in this case.

Also, on an intuitive level, the memory of the data generating process is assumedly important for the information loss discussed above. This is demonstrated in Example 2:

Example 2. Consider the following two models:
Model 1:

$$y_t = \begin{pmatrix} 0.9556 & 0.8611 \\ -0.6914 & 0.2174 \end{pmatrix} y_{t-1} + \nu_t, \quad \nu_t \sim \mathcal{N}_2(0, I_2) \tag{40}$$
$$z_{0,1} = 0.7303 \pm 0.8437i$$

Model 2:

$$y_t = \begin{pmatrix} -1.2141 & 1.1514 \\ -0.9419 & 0.8101 \end{pmatrix} y_{t-1} + \nu_t, \quad \nu_t \sim \mathcal{N}_2(0, I_2) \tag{41}$$
$$z_{0,1} = -2 \pm 2.4294i$$

where $z_{0,1}$ denotes the roots of the determinant of the autoregressive polynomial. The correlations of the two processes are depicted in Fig. 2 where the black bars are the unobserved autocorrelations.

In comparing these two models, Table 2 shows the MSE

$$\text{MSE}\left(\hat{\theta}\right) = \frac{1}{m} \sum_{j=1}^{m} \sum_{i=1}^{7} \left(\theta_i - \hat{\theta}_i^j\right)^2$$

for the parameters $\theta = \left(\text{vec}(A)^T, \text{vech}(\Sigma_\nu)^T\right)^T$ for $T = 500$, $m = 10^3$ simulation runs, $N = 2$ and the case of stock variables. Here, HF-YW is the standard Yule–Walker estimator from high-frequency data, HF-XYW is the

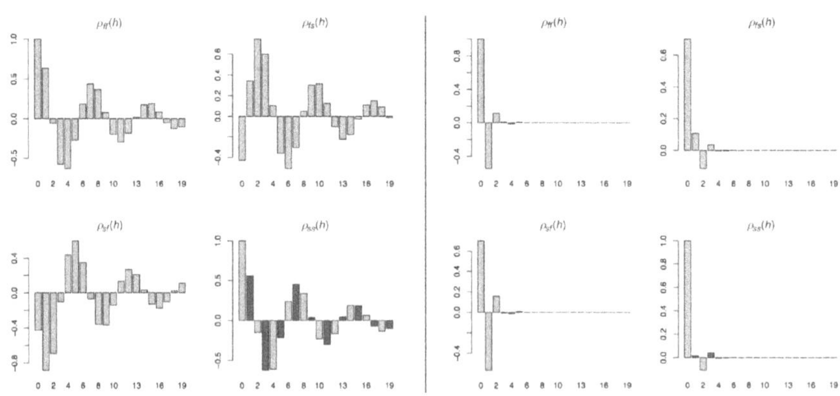

Fig. 2. Autocorrelations of Model 1 (left) and Model 2 (right).

Table 2. Absolute and Relative Mean-Squared Errors of the System and Noise Parameters.

	Estimators	Model 1		Model 2	
		Absolute	Relative	Absolute	Relative
HF	YW	0.007	1	0.009	1
	XYW	0.012	1.77	0.272	30.07
MF	MLE-EM	0.010	1.53	0.977	107.93
	XYW	0.076	11.70	2.977	329.69

Table 3. Absolute and Relative Root Mean-Squared One-Step-Ahead Forecasting Errors.

	Estimators	Model 1		Model 2	
		Absolute	Relative	Absolute	Relative
LF	YW	3.607	1	2.859	1
MF	MLE-EM	2.371	0.66	2.857	0.99
	XYW	2.388	0.66	34.149	11.94
HF	YW	1.995	0.55	1.998	0.70

estimator obtained from inserting the sample second moments obtained from high-frequency data into the XYW estimator corresponding to the pseudo-inverse, see Eq. (13). MF-MLE-EM is the estimator described in Section 2.3, initialized by the MF-XYW estimator, and MF-XYW is the XYW estimator described in Eq. (13). In addition, as a measure for the information loss, the MSE relatively to the MSE of the HF-YW estimators are presented. In particular, the relative MSE of the MF-MLE-EMs show the information loss due to mixed-frequency data. However, convergence problems due to the existence of local minima of the likelihood may arise in calculating the MF-MLE-EM. This can be mitigated by using several starting values.

In Table 3, an analogous comparison based on one-step-ahead prediction errors is given. This table relates the mixed-frequency prediction errors to the prediction errors obtained by using Yule–Walker equations in the high-frequency case as well as the prediction errors obtained by using Yule–Walker equations in the low-frequency case. In the high-frequency case, the one-step-ahead forecast of y_t, $t \in 2\mathbb{Z}$ is based on y_{t-1}, in the low-frequency case the forecast of y_t, $t \in 2\mathbb{Z}$, is based on y_{t-2} and finally in the mixed-frequency case the forecast of y_t, $t \in 2\mathbb{Z}$ is based on y^f_{t-1}, y_{t-2}.

In Table 4, the absolute and relative Frobenius norms of the asymptotic covariance matrices of the estimators of the system parameters are

Table 4. Absolute and Relative Norms of the Asymptotic Covariance Matrix of the System Parameters.

	Estimators	Model 1		Model 2	
		Absolute	Relative	Absolute	Relative
HF	YW	0.421	1	1.416	1
	XYW	0.516	1.23	13.690	9.67
	XYW $k = 1$	0.558	1.33	18.830	13.29
	GMM $k = 1$	0.514	1.22	12.674	8.95
MF	MLE	0.623	1.48	2.632	1.86
	XYW	1.504	3.57	112.907	79.72
	XYW $k = 1$	0.942	2.24	163.694	115.57
	GMM $k = 1$	0.878	2.09	85.775	60.56

presented. Note that, for instance, HF-XYW $k = 1$ corresponds to the high-frequency XYW estimator based on Eq. (15), that is, the XYW equations are extended with a further lag. We observe that the MF-XYW estimators have larger asymptotic covariances than the MF-MLE. This is the case even if the optimal weights are chosen for the MF-GMM $k = 1$ estimator.

In addition, it is clear that for the same high-frequency model increasing n_f will give better results for the mixed-frequency estimators and that the quality of the parameter estimators will decrease with increasing N. We demonstrate these effects in Example 3.

Example 3. In order to demonstrate the relations between the MF-XYW and the MF-MLE-EM estimator as well as the effects of increasing n_f and N for the stock case, we consider the following Model 3 for $T = 500$ and $m = 10^3$ simulation runs:

$$y_t = \begin{pmatrix} 0.9154 & 0.1002 & 0.2250 & -0.3594 \\ 2.7553 & 1.5950 & 3.3705 & -5.4438 \\ 0.4516 & -0.1998 & 0.8294 & -0.7917 \\ 0.7375 & 0.1185 & 0.7489 & -0.6667 \end{pmatrix} y_{t-1} + \nu_t,$$

$$\nu_t \sim \mathcal{N}_4\left(0, bb^T\right) \tag{42}$$

$$b = \begin{pmatrix} 1.1140 & 0 & 0 & 0 \\ -0.3807 & 0.6514 & 0 & 0 \\ 0.3448 & -0.3742 & 0.3103 & 0 \\ -0.1749 & -0.1389 & -0.2241 & 1.317 \end{pmatrix}$$

$$z_0 = -0.8783, \quad z_1 = -0.7983, \quad z_{2,3} = -0.1557 \pm 0.7614i$$

Table 5. Absolute and Relative Mean-Squared Errors for the Respective Parameter Estimators for Model 3.

Estimators	$N = 2, n_f = 2$		$N = 2, n_f = 3$		$N = 3, n_f = 3$		$N = 12, n_f = 3$	
	Absolute	Relative	Absolute	Relative	Absolute	Relative	Absolute	Relative
MLE-EM	0.230	1	0.014	1	0.015	1	0.984	1
XYW	0.774	3.37	0.381	28.04	1.592	104.74	5.740	5.83

Table 6. Absolute and Relative Mean-Squared Errors of the System and Noise Parameters.

	Estimators	Model 1		Model 2	
		Absolute	Relative	Absolute	Relative
HF	YW	0.007	1	0.009	1
	XYW	0.020	2.86	0.545	60.56
MF	MLE-EM	0.043	6.14	0.123	13.67
	XYW	0.070	10.70	1.472	161.77

Here, the MF-MLE-EM clearly outperforms the MF-XYW estimator for all the four cases shown in Table 5. For both estimators, the quality decreases with increasing N and increases with increasing n_f.

Example 4. In this example, we consider again Model 1 and Model 2 as introduced in Example 2 for $T = 500$ and $m = 10^3$ simulation runs but now for the case of flow variables, that is, $w_{2t} = y^s_{2t} + y^s_{2t-1}$. Here, HF-XYW is the estimator obtained from inserting the sample second moments obtained from high-frequency data, that is, $y^f_t \in \mathbb{Z}$ and $w_t \in \mathbb{Z}$, into the XYW estimator corresponding to the pseudo-inverse, see Eq. (19). As can be seen in Table 6, the estimators for the flow case do not necessarily lead to better estimates compared to the stock case.

6. OUTLOOK AND CONCLUSIONS

In this paper, we discussed and analyzed estimators for the parameters of a high-frequency VAR model from mixed-frequency data where the low-frequency data are obtained from general linear aggregation schemes including, in particular, stock and flow data. We considered estimators obtained from the XYW equations with different weighting matrices as well as Gaussian maximum likelihood type estimators based on the EM algorithm. The problem of getting estimators resulting in stable systems and positive (semi)-definite covariances of prescribed rank q has been treated. Furthermore, we derived the asymptotic distribution of the XYW/GMM estimators. Finally, we presented a simulation study comparing XYW and Gaussian maximum likelihood estimators and discussed the information loss due to mixed-frequency data compared to high-frequency data and the information gain if we use mixed-frequency data rather than low-frequency data.

In particular, the dependence of the results obtained from the point in parameter space for high-frequency AR systems chosen needs further investigation.

ACKNOWLEDGMENTS

Support by the FWF (Austrian Science Fund under contract P24198/N18) is gratefully acknowledged. We thank Prof. Tommaso Proietti, Universita di Roma "Tor Vergata", Italy, and Prof. Brian D. O. Anderson, Australian National University, Australia, for helpful comments.

REFERENCES

Anderson, B. D. O., Deistler, M., Felsenstein, E., Funovits, B., Koelbl, L., & Zamani, M. (2015). Multivariate AR systems and mixed frequency data: g-Identifiability and estimation. *Econometric Theory*, doi:10.1017/S0266466615000043:1–34. Retrieved from http://dx.doi.org/10.1017/S0266466615000043

Anderson, B. D. O., Deistler, M., Felsenstein, E., Funovits, B., Zadrozny, P. A., Eichler, M., ..., Zamani, M. (2012). Identifiability of regular and singular multivariate autoregressive models. *Proceedings of the 51th IEEE conference on decision and control (CDC)*, pp. 184–189.

Anderson, T. (1994). *The statistical analysis of time series*. New York, NY: Wiley.

Balogh, L., & Pintelon, R. (2008). Stable approximation of unstable transer function models. *IEEE Transactions on Instrumentations and Measurement*, *57*(12), 2720–2726.

Chen, B., & Zadrozny, P. A. (1998). An extended Yule-Walker method for estimating a vector autoregressive model with mixed-frequency data. *Advances in Econometrics*, *13*, 47–73.

Combettes, P. L., & Trussell, H. J. (1992). Best stable and invertible approximations for ARMA systems. *IEEE Transactions on Signal Processing*, *40*:3066–3069.

Deistler, M., Anderson, B. D. O., Filler, A., Zinner, C., & Chen, W. (2010). Generalized linear dynamic factor models – An approach via singular autoregressions. *European Journal of Control*, *16*(3), 211–224.

D'haene, T., Pintelon, R., & Vandersteen, G. (2006). An iterative method to stabilize a transfer function in the s- and z-domains. *IEEE Transactions on Instrumentations and Measurement*, *55*, 1192–1196.

Filler, A. (2010). *Generalized dynamic factor models Structure theory and estimation for single frequency and mixed frequency data*. PhD thesis, Vienna University of Technology.

Forni, M., Hallin, M., Lippi, M., & Reichlin, L. (2000). The generalized dynamic factor model: Identification and estimation. *Review of Economics and Statistics*, *82*(4), 540–554.

Forni, M., Hallin, M., Lippi, M., & Zaffaroni, P. (2015). Dynamic factor models with infinite-dimensional factor spaces: One-sided representations. *Journal of Econometrics*, *185*, 359–371.

Francq, C., & Zakoian, J.-M. (2009). *Bartlett's formula for a general class of non linear processes*.

Gingras, D. F. (1985). Asymptotic properties of high-order Yule-Walker estimates of the AR parameters of an ARMA time series. *IEEE Transactions on Acoustics Speech and Signal Processing*, *33*(4), 1095–1101.

Hall, P., & Heyde, C. (1980). *Martingal limit theory and its application*, New York, NY: Academic Press.

Hannan, E. J. (1970). *Multiple time series*, New York, NY: Wiley.

Hannan, E. J., & Deistler, M. (2012). *The statistical theory of linear systems*, Philadelphia, PA: SIAM Classics in Applied Mathematics.

Hansen, L. P. (1982). Large sample properties of generalized method of moments estimators. *Econometrica*, *50*(4), 1029–1054.

Hoffman, A., & Wielandt, H. (1953). The variation of the spectrum of a normal matrix. *Duke Mathematical Journal*, *20*, 37–39.

Khatri, C. G. (1961). Some results for the singular normal multivariate regression models. *The Indian Journal of Statistics*, *30*, 267–280.

Koelbl, L. (2015). *VAR systems: g-Identifiability and asymptotic properties of parameter estimates for the mixed-frequency case*. PhD thesis, Vienna University of Technology.

Komunjer, I., & Ng, S. (2011). Dynamic identification of dynamic stochastic general equilibrium models. *Econometrica*, *79*(6), 1995–2032.

Lippi, M., & Reichlin, L. (1994). VAR analysis, nonfundamental representations, Blaschke matrices. *Journal of Econometrics*, *63*, 307–325.

Lütkepohl, H. (2005). *New introduction to multiple time series analysis*, Berlin: Springer.

Magnus, J. R., & Neudecker, H. (1979). The commutation matrix: Some properties and applications. *Annals of Statistics*, *7*(2), 381–394.

Mariano, R. S., & Murasawa, Y. (2010). A coincident index, common factors, and monthly real GDP. *Oxford Bulletin of Economics and Statistics*, *72*(1), 27–46.

Moses, L. R., & Liu, D. (1991). Determining the closest stable polynomial to an unstable one. *IEEE Transactions on Signal Processing*, *39*(4), 901–906.

Niebuhr, T., & Kreiss, J.-P. (2013). Asymptotics for autocovariances and integrated periodograms for linear processes observed at lower frequencies. *International Statistical Review*, *82*(1), 123–140.

Orbandexivry, F.-X., Nesterov, Y., & van Dooren, P. (2013). Nearest stable system using successive convex approximations. *Automatica*, *49*, 1195–1203.

Rao, C. R. (1972). *Linear statistical inference and its applications*, New York, NY: Wiley.

Shumway, R., & Stoffer, D. (1982). An approach to time series smoothing and forecasting using the EM algorithm. *Journal of Time Series Analysis*, *3*(4), 253–264.

Stock, J. H., & Watson, M. W. (2002). Forecasting using principal components from a large number of predictors. *Journal of the American Statistical Associations*, *97*(460), 1167–1179.

Stoica, P., & Moses, L. R. (1992). On the unit circle problem: The Schur-Cohn procedure revisited. *Signal Processing*, *26*, 95–118.

Su, N., & Lund, R. (2011). Multivariate versions of Bartlett's formula. *Journal of Multivariate Analysis*, *105*, 18–31.

Wax, M., & Kailath, T. (1983). Efficient inversion of Toeplitz-Block Toeplitz matrix. *IEEE Transactions on Acoustics Speech and Signal Processing*, *31*, 1218–1221.

APPENDIX

The next lemma, see Su and Lund (2011), gives a multivariate version of Bartlett's formula.

Lemma 1. Under the assumptions of Section 4, we obtain

$$\lim_{T \to \infty} T\mathrm{Cov}\left(\mathrm{vec}\left(\hat{\gamma}^{z^f f}(p)\right), \mathrm{vec}\left(\hat{\gamma}^{z^f f}(q)\right)\right) = S_{p,q} + R_{p,q}$$

for $p, q \in \mathbb{Z}$, where

$$R_{p,q} = \sum_{k=-\infty}^{\infty} \left(\gamma^{ff}(k+q-p) \otimes \gamma^{z^f z^f}(k)\right) + K_{n_f,n_f}\left(\gamma^{z^f f}(k+q) \otimes \gamma^{f z^f}(k-p)\right)$$
$$S_{p,q} = \sum_{k=-\infty}^{\infty} \sum_{r=-\infty}^{\infty} \left(k^f_{k-p} \otimes \tilde{k}^{z^f}_k\right) \kappa \left(k^f_{r+k-q} \otimes \tilde{k}^{z^f}_{r+k}\right)^T$$

and K_{n_f,n_f}, $K_{n,n}$ are commutation matrices.

Note that in the Gaussian case $\kappa = 0$ and thus $S_{p,q}$ is zero. Using the idea of the proof of Su and Lund (2011) and taking into account that $\hat{\gamma}^{wf}(q)$ has only approximately T/N summands, we obtain:

Lemma 2. Under the assumptions of Section 4, we obtain

$$\lim_{T \to \infty} T\mathrm{Cov}\left(\mathrm{vec}\left(\hat{\gamma}^{z^f f}(p)\right), \mathrm{vec}(\hat{\gamma}^{wf}(q))\right) = \overline{S}_{p,q} + \overline{R}_{p,q}$$

for $p, q \in \mathbb{Z}$, where

$$\overline{R}_{p,q} = \sum_{k=-\infty}^{\infty} \left(\gamma^{ff}(k+q-p) \otimes \gamma^{z^f w}(k)\right) + K_{n_f,n_f}\left(\gamma^{z^f f}(k+q) \otimes \gamma^{fw}(k-p)\right)$$

$$\overline{S}_{p,q} = \sum_{k=-\infty}^{\infty} \sum_{r=-\infty}^{\infty} \left(k^f_{k-p} \otimes \tilde{k}^{f}_k\right) \kappa \left(k^f_{r+k-q} \otimes \tilde{k}^{s}_{r+k}\right)^T$$

Replacing $\hat{\gamma}^{wf}(q)$ in Lemma 2 by the "high-frequency autocovariance estimator", that is, by $(1/T)\sum_{t=1}^{T} w_t \left(y_{t-q}^f\right)^T$, the result is still valid. The following result generalizes the result given in Niebuhr and Kreiss (2013) to the multivariate case.

Lemma 3. Under the assumptions of Section 4, we obtain

$$\lim_{T\to\infty} T\mathrm{Cov}\left(\mathrm{vec}\left(\hat{\gamma}^{wf}(p)\right), \mathrm{vec}\left(\hat{\gamma}^{wf}(q)\right)\right) = N\left(\tilde{S}_{p,q} + \tilde{R}_{p,q}\right)$$

for $p, q \in \mathbb{Z}$, where

$$\tilde{R}_{p,q} = \sum_{k=-\infty}^{\infty} \left(\gamma^{ff}(Nk+q-p)\otimes\gamma^{ww}(kN)\right) + K_{n_f,n_s}\left(\gamma^{wf}(Nk+q)\otimes\gamma^{fw}(Nk-p)\right)$$

$$\tilde{S}_{p,q} = \sum_{k=-\infty}^{\infty}\sum_{r=-\infty}^{\infty} \left(k_{k-p}^f \otimes \tilde{k}_k^s\right)\kappa\left(k_{Nr+k-q}^f \otimes \tilde{k}_{Nr+k}^s\right)^T$$

Note that for $N = 1$ we still obtain Bartlett's formula for the high-frequency case. The three lemmas above are needed for the following proof of Theorem 1.

Proof of Theorem 1. W.l.o.g. let us assume that T is a multiple of N. We will prove this theorem for $\hat{\hat{\gamma}}^{zf}(i) = (1/T)\sum_{t=1}^{T} z_t^f \left(y_{t-i}^f\right)^T$ and $\hat{\hat{\gamma}}^{wf}(i) = (N/T)\sum_{t=1}^{T/N} w_{Nt}\left(y_{Nt-i}^f\right)^T$ instead of $\hat{\gamma}^{zf}(i)$ and $\hat{\gamma}^{wf}(i)$, respectively, since it can be shown that this change does not influence the asymptotic properties, see Hannan (1970). In order to apply theorem 14 in Hannan (1970, p. 228), we define a particular blocked process $u_t = \left(y_{Nt}^T \; y_{Nt-1}^T \; \cdots \; y_{Nt-(N-1)}^T\right)^T$ which satisfies the assumptions of this theorem. Now applying theorem 14 in Hannan (1970) leads to

$$\sqrt{\frac{T}{N}}\left(\mathrm{vec}\left(\hat{\gamma}_u(i)\right) - \mathrm{vec}\left(\gamma_u(i)\right)\right)_{i=0,\ldots,s} \xrightarrow{d} N_{nN(s+1)}(0, \Sigma_u)$$

where $\gamma_u(i)$ is the population autocovariance of the process (u_t) for lag i and $\hat{\gamma}_u(i)$ is its sample counterpart with T/N summands. In a last step we have to find a transformation matrix, say H, which transforms $\hat{\gamma}_u$ to the desired autocovariances. To obtain this transformation we define $H_1 = \left((1/N)I_{n_f^2} \; (1/N)I_{n_f^2} \; \cdots \; (1/N)I_{n_f^2} \right)$ and observe that for $j = 0, \ldots, s$ $\operatorname{vec}\left(\hat{\hat{\gamma}}^{wf}(j)\right) = (N/T)\sum_{t=1}^{T/N} \operatorname{vec}\left(w_{Nt}\left(y_{Nt-j}^{f}\right)^T\right) = S_j^{wf}\left(\operatorname{vec}(\hat{\gamma}_u(i))\right)_{i=0,\ldots,s}$ and

$$\operatorname{vec}\left(\hat{\hat{\gamma}}^{zf}(j)\right) = H_1 \begin{pmatrix} \frac{N}{T}\sum_{t=1}^{T/N} \operatorname{vec}\left(z_{Nt}^{f}\left(y_{Nt-j}^{f}\right)^T\right) \\ \frac{N}{T}\sum_{t=1}^{T/N} \operatorname{vec}\left(z_{Nt-1}^{f}\left(y_{Nt-1-j}^{f}\right)^T\right) \\ \vdots \\ \frac{N}{T}\sum_{t=1}^{T/N} \operatorname{vec}\left(z_{Nt-(N-1)}^{f}\left(y_{Nt-(N-1)-j}^{f}\right)^T\right) \end{pmatrix}$$
$$= H_1 S_j^{zf}\left(\operatorname{vec}(\hat{\gamma}_u(i))\right)_{i=0,\ldots,s}$$

where S_j^{zf} and S_j^{wf} are selector matrices for lag j. Finally, we can construct our particular transformation matrix $H = \left(\left(H_1 S_0^{zf}\right)^T, \left(S_0^{wf}\right)^T, \ldots, \left(H_1 S_s^{zf}\right)^T, \left(S_s^{wf}\right)^T \right)^T$ and obtain the desired result

$$\sqrt{\frac{T}{N}}\left(\begin{pmatrix} \operatorname{vec}\left(\hat{\gamma}^{zf}(i)\right) \\ \operatorname{vec}\left(\hat{\gamma}^{wf}(i)\right) \end{pmatrix} - \begin{pmatrix} \operatorname{vec}\left(\gamma^{zf}(i)\right) \\ \operatorname{vec}\left(\gamma^{wf}(i)\right) \end{pmatrix} \right)_{i=0,\ldots,s}$$
$$= \sqrt{\frac{T}{N}} H\left(\operatorname{vec}(\hat{\gamma}_u(i)) - \operatorname{vec}(\gamma_u(i))\right)_{i=0,\ldots,s} \;\xrightarrow{d}\; \mathcal{N}_h(0, \Sigma_\gamma)$$

The asymptotic covariance $\Sigma_\gamma = H\Sigma_u H^T$ can be derived using Lemmas 1–3. ■

MODELING YIELDS AT THE ZERO LOWER BOUND: ARE SHADOW RATES THE SOLUTION?

Jens H. E. Christensen and Glenn D. Rudebusch

Federal Reserve Bank of San Francisco, San Francisco, CA, USA

ABSTRACT

Recent U.S. Treasury yields have been constrained to some extent by the zero lower bound (ZLB) on nominal interest rates. Therefore, we compare the performance of a standard affine Gaussian dynamic term structure model (DTSM), which ignores the ZLB, to a shadow-rate DTSM, which respects the ZLB. Near the ZLB, we find notable declines in the forecast accuracy of the standard model, while the shadow-rate model forecasts well. However, 10-year yield term premiums are broadly similar across the two models. Finally, in applying the shadow-rate model, we find no gain from estimating a slightly positive lower bound on U.S. yields.

Keywords: Term structure modeling; zero lower bound; monetary policy

JEL classifications: G12; E43; E52; E58

Dynamic Factor Models
Advances in Econometrics, Volume 35, 75–125

ISSN: 0731-9053/doi:10.1108/S0731-905320150000035003

1. INTRODUCTION

With recent historic lows reached by nominal yields on government debt in several countries, understanding how to model the yield curve when some interest rates are near their zero lower bound (ZLB) is an issue that commands attention both for bond portfolio pricing and risk management and for macroeconomic and monetary policy analysis. Unfortunately, the workhorse representation in finance for bond pricing – the affine Gaussian dynamic term structure model – ignores the ZLB and places positive probabilities on negative interest rates. In essence, this model disregards the existence of a readily available zero-yield currency that an investor always has the option of holding and that dominates any security with a negative yield. This theoretical flaw in the standard model casts doubt on its usefulness for answering a key empirical question of this paper: how to best extract reliable market-based measures of expectations for future monetary policy when nominal interest rates are near the ZLB. Of course, as recent events have shown, at times, the ZLB can be a somewhat soft floor, and the non-negligible costs of transacting in and holding large amounts of currency have allowed government bond yields to push a bit below zero in several countries, notably Denmark and Switzerland. In our analysis below, we do not rigidly enforce a lower constraint of exactly zero on yields, but as a convenient abbreviation, we will refer to an episode of near-zero short rates as a ZLB period. The timing of this period for the United States is evident from the nominal U.S. Treasury zero-coupon yields shown in Fig. 1. The start of the ZLB period is commonly dated to December 16, 2008, when the Federal Open Market Committee (FOMC) lowered its target policy rate – the overnight federal funds rate – to a range from 0 to 1/4 percent, and it continued past the end of our sample in October 2014.

The past term structure literature offers three established frameworks to model yields near the ZLB that guarantee positive interest rates: stochastic-volatility models with square-root processes, Gaussian quadratic models, and Gaussian shadow-rate models. However, the first two of these approaches treat the ZLB as a reflecting barrier and not as an absorbing state, which seems inconsistent with the prolonged period of very low interest rates shown in Fig. 1. In contrast, shadow-rate models are completely consistent with an absorbing ZLB state for yields. In addition to these established frameworks, there is a growing literature offering new and interesting ways of accounting for the ZLB, including Filipović, Larsson, and Trolle (2014) and Monfort, Pegoraro, Renne, and Roussellet (2014).

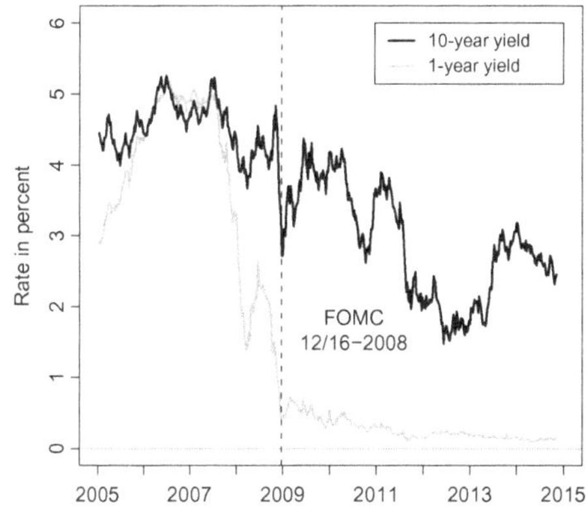

Fig. 1. Treasury Yields Since 2005. *Notes*: One- and 10-year weekly U.S. Treasury zero-coupon bond yields from January 7, 2005, to October 31, 2014.

While we consider all of these modeling approaches to be worthy of further investigation, Gaussian shadow-rate models are of particular interest because away from the ZLB they reduce exactly to standard Gaussian affine models. Therefore, the voluminous literature on affine Gaussian models remains completely applicable and relevant when given a modest shadow-rate tweak to handle the ZLB as we demonstrate. There have also been a few studies comparing the performance of these frameworks. For example, Kim and Singleton (2012) and Andreasen and Meldrum (2014) provide empirical results favoring shadow-rate representations over quadratic models. However, one important issue still requiring supporting evidence is the relative performance of the standard Gaussian affine dynamic term structure model (DTSM) versus an equivalent shadow-rate model. The standard affine DTSM is extremely well entrenched in the literature. It is both very popular and well understood. Despite its theoretical flaw noted above, could it be good enough for empirical purposes? To shed light on this issue, we compare the performance of a standard Gaussian DTSM of U.S. Treasury yields and its exact equivalent shadow-rate version. This comparison provides a clean read on the relative merits of standard and shadow-rate models during an episode of near-zero nominal yields.

For our comparative empirical analysis, we employ affine and shadow-rate versions of the arbitrage-free Nelson–Siegel (AFNS) model class that are estimated on the same data sample. The AFNS modeling structure provides an ideal framework for our analysis because of its excellent empirical properties and tractable and robust estimation.[1] For the Gaussian affine model, we use the structure identified by Christensen and Rudebusch (2012) (henceforth CR), which is referred to throughout as the CR model. Since CR only detail the model's favorable properties through 2010, our analysis provides an update through October 2014, which includes a much longer ZLB period. As for the shadow-rate model, we use the shadow-rate AFNS model class introduced in Christensen and Rudebusch (2015). This is a latent-factor model in which the state variables have standard Gaussian dynamics, but the short rate is given an interpretation of a shadow rate in the spirit of Black (1995) to respect the ZLB for bond pricing. Christensen and Rudebusch (2015) apply this structure to a sample of near-ZLB Japanese government bond yields; however, they limit their analysis to the full-sample estimated parameters and state variables. Instead, we exploit the empirical tractability of that shadow-rate AFNS model, denoted here as the B-CR model, to study real-time forecast performance.[2]

Therefore, in this paper, we combine and extend the analysis of two recent papers. We compare the results from the CR model to those obtained from the B-CR model using the same sample of U.S. Treasury yield data. We can compare model performance across normal and ZLB periods and study real-time forecast performance, short-rate projections, term premium decompositions, and the properties of its estimated parameters. We find that the B-CR model provides slightly better fit as measured by in-sample metrics such as the RMSEs of fitted yields and the quasi likelihood values. Still, it is evident that a standard three-factor Gaussian DTSM like the CR model has enough flexibility to fit the cross-section of yields fairly well at each point in time even when the short end of the yield curve is flattened by the ZLB. However, it is not the case that the Gaussian model can account for all aspects of the term structure at the ZLB. Indeed, we show that the CR model clearly fails along two dimensions. First, despite fitting the yield curve, the model cannot capture the dynamics of yields at the ZLB. One stark indication of this is the high probability the model assigns to negative future short rates – obviously a poor prediction. Second, it misses the compression of yield volatility that occurs at the ZLB as expected future short rates are pinned near zero, longer-term rates fluctuate less. The B-CR model, even without incorporating stochastic volatility, can capture this effect. In terms of forecasting future short rates, we establish that the CR model is competitive over the normal period from 1995 to 2008. Thus, this model could have been expected to continue to perform well in the most recent period, if only it had not been for the problems associated with the ZLB. However, we also show that during the most recent period the B-CR model stands out in terms of forecasting future short rates in addition to performing on par with the regular model during the normal period. Overall, a shadow-rate model shows clear empirical advantages. Still, the affine model may be a good first approximation for certain tasks. For example, we estimate 10-year term premiums that are broadly similar across the affine and shadow-rate models.

In addition, we study two empirical questions pertaining to the implementation of shadow-rate models. The first concerns the appropriate choice of the lower bound on yields. We argue that U.S. Treasury yield data point to zero as the appropriate lower bound, but throughout the paper, we consider the case with the lower bound treated as a free parameter that is determined by quasi maximum likelihood.[3] Our findings suggest that there are few if any gains in forecast performance from estimating the lower bound, and those gains come at the cost of fairly large estimated values of the lower bound. Since the estimated path of the shadow rate is sensitive to this choice (as we demonstrate), this is not an innocent parameter and should be chosen with care.[4] The second question we address is the closeness of the estimated parameters between a standard DTSM and its equivalent shadow-rate representation. If any differences between the parameter sets are small economically and statistically, this would provide a quick and efficient shortcut to avoid having to estimate shadow-rate models. Instead, one could simply rely on the estimated parameters from the matching standard model.[5] Unfortunately, we find that the differences in the estimated parameters can be sizable and economically important. Thus, while we cannot endorse the approach of relying on estimated parameters from a standard model as a way of implementing the corresponding shadow-rate model, our results still suggest that its optimal parameters do provide a reasonable guess of where to start the parameter optimization in the estimation of the shadow-rate model.

The rest of the paper is structured as follows. Section 2 describes Gaussian models in general as well as the CR model that we consider, while Section 3 details our shadow-rate model. Section 4 contains our empirical findings and discusses the implications for assessing policy expectations and term premiums in the current low-yield environment. Section 5 concludes. Three appendices contain additional technical details.

2. A STANDARD GAUSSIAN TERM STRUCTURE MODEL

In this section, we provide an overview of the affine Gaussian term structure model, which ignores the ZLB, and describe the CR model.

2.1. The General Model

Let $P_t(\tau)$ be the price of a zero-coupon bond at time t that pays \$ 1, at maturity $t+\tau$. Under standard assumptions, this price is given by

$$P_t(\tau) = E_t^P\left[\frac{M_{t+\tau}}{M_t}\right]$$

where the stochastic discount factor, M_t, denotes the value at time t_0 of a claim at a future date t, and the superscript P refers to the actual, or real-world, probability measure underlying the dynamics of M_t. (As we will discuss in the next section, there is no restriction in this standard setting to constrain $P_t(\tau)$ from rising above its par value; i.e., the ZLB is ignored.)

We follow the usual reduced-form empirical finance approach that models bond prices with unobservable (or latent) factors, here denoted as X_t, and the assumption of no residual arbitrage opportunities. We assume that X_t follows an affine Gaussian process with constant volatility, with dynamics in continuous time given by the solution to the following stochastic differential equation (SDE):

$$\mathrm{d}X_t = K^P\left(\theta^P - X_t\right)\mathrm{d}t + \Sigma \mathrm{d}W_t^P$$

where K^P is an $n\times n$ mean-reversion matrix, θ^P is an $n\times 1$ vector of mean levels, Σ is an $n\times n$ volatility matrix, and W_t^P is an n-dimensional Brownian motion. The dynamics of the stochastic discount function are given by

$$\mathrm{d}M_t = r_t M_t \mathrm{d}t + \Gamma_t' M_t \mathrm{d}W_t^P$$

and the instantaneous risk-free rate, r_t, is assumed affine in the state variables

$$r_t = \delta_0 + \delta_1' X_t$$

where $\delta_0 \in R$ and $\delta_1 \in R^n$. The risk premiums, Γ_t, are also affine as in Duffee (2002):

$$\Gamma_t = \gamma_0 + \gamma_1 X_t$$

where $\gamma_0 \in R^n$ and $\gamma_1 \in R^{n\times n}$.

Duffie and Kan (1996) show that these assumptions imply that zero-coupon yields are also affine in X_t:

$$y_t(\tau) = -\frac{1}{\tau}A(\tau) - \frac{1}{\tau}B(\tau)'X_t$$

where $A(\tau)$ and $B(\tau)$ are given as solutions to the following system of ordinary differential equations:

$$\frac{dB(\tau)}{d\tau} = -\delta_1 - \left(K^P + \Sigma\gamma_1\right)'B(\tau), \quad B(0) = 0,$$

$$\frac{dA(\tau)}{d\tau} = -\delta_0 + B(\tau)'\left(K^P\theta^P - \Sigma\gamma_0\right) + \frac{1}{2}\sum_{j=1}^{n}\left[\Sigma' B(\tau)B(\tau)'\Sigma\right]_{j,j}, \quad A(0) = 0$$

Thus, the $A(\tau)$ and $B(\tau)$ functions are calculated *as if* the dynamics of the state variables had a constant drift term equal to $K^P\theta^P - \Sigma\gamma_0$ instead of the actual $K^P\theta^P$ and a mean-reversion matrix equal to $K^P + \Sigma\gamma_1$ as opposed to the actual K^P. The probability measure with these alternative dynamics is frequently referred to as the risk-neutral, or Q, probability measure since the expected return on any asset under this measure is equal to the risk-free rate r_t that a risk-neutral investor would demand. The difference is determined by the risk premium Γ_t and reflects investors' aversion to the risks embodied in X_t.

Finally, we define the term premium as

$$\mathrm{TP}_t(\tau) = y_t(\tau) - \frac{1}{\tau}\int_t^{t+\tau} E_t^P[r_s]\,ds$$

That is, the term premium is the difference in expected return between a buy and hold strategy for a τ-year. Treasury bond and an instantaneous rollover strategy at the risk-free rate r_t.[6]

2.2. The CR Model

A wide variety of Gaussian term structure models have been estimated. Here, we describe the empirical representation identified by CR that uses high-frequency observations on U.S. Treasury yields from a sample that includes the recent ZLB period. It improves the econometric identification of the latent factors, which facilitates model estimation.[7] The CR model is an AFNS representation as introduced in Christensen, Diebold, and Rudebusch (2011) with three latent state variables, $X_t = (L_t, S_t, C_t)$. These are described by the following system of SDEs under the risk-neutral Q-measure:[8]

$$\begin{pmatrix} dL_t \\ dS_t \\ dC_t \end{pmatrix} = \begin{pmatrix} 0 & 0 & 0 \\ 0 & \lambda & -\lambda \\ 0 & 0 & \lambda \end{pmatrix} \left[\begin{pmatrix} \theta_1^Q \\ \theta_2^Q \\ \theta_3^Q \end{pmatrix} - \begin{pmatrix} L_t \\ S_t \\ C_t \end{pmatrix} \right] dt + \Sigma \begin{pmatrix} dW_t^{L,Q} \\ dW_t^{S,Q} \\ dW_t^{C,Q} \end{pmatrix}, \quad \lambda > 0 \quad (1)$$

where Σ is the constant covariance (or volatility) matrix.

In addition, the instantaneous risk-free rate is defined by

$$r_t = L_t + S_t \quad (2)$$

This specification implies that zero-coupon bond yields are given by

$$y_t(\tau) = L_t + \left(\frac{1 - e^{-\lambda\tau}}{\lambda\tau} \right) S_t + \left(\frac{1 - e^{-\lambda\tau}}{\lambda\tau} - e^{-\lambda\tau} \right) C_t - \frac{A(\tau)}{\tau}$$

where the factor loadings in the yield function match the level, slope, and curvature loadings introduced in Nelson and Siegel (1987). The final yield-adjustment term, $A(\tau)/\tau$, captures convexity effects due to Jensen's inequality.

The model is completed with a risk premium specification that connects the factor dynamics to the dynamics under the real-world P-measure as explained in Section 2.1. The maximally flexible specification of the AFNS model has P-dynamics given by[9]

$$\begin{pmatrix} dL_t \\ dS_t \\ dC_t \end{pmatrix} = \begin{pmatrix} \kappa_{11}^P & \kappa_{12}^P & \kappa_{13}^P \\ \kappa_{21}^P & \kappa_{22}^P & \kappa_{23}^P \\ \kappa_{31}^P & \kappa_{32}^P & \kappa_{33}^P \end{pmatrix} \left[\begin{pmatrix} \theta_1^P \\ \theta_2^P \\ \theta_3^P \end{pmatrix} - \begin{pmatrix} L_t \\ S_t \\ C_t \end{pmatrix} \right] dt + \begin{pmatrix} \sigma_{11} & 0 & 0 \\ \sigma_{21} & \sigma_{22} & 0 \\ \sigma_{31} & \sigma_{32} & \sigma_{33} \end{pmatrix} \begin{pmatrix} dW_t^{L,P} \\ dW_t^{S,P} \\ dW_t^{C,P} \end{pmatrix} \quad (3)$$

Using both in- and out-of-sample performance measures, CR went through a careful empirical analysis to justify various zero-value restrictions on the

K^P matrix. Imposing these restrictions results in the following dynamic system for the P-dynamics:

$$\begin{pmatrix} \mathrm{d}L_t \\ \mathrm{d}S_t \\ \mathrm{d}C_t \end{pmatrix} = \begin{pmatrix} 10^{-7} & 0 & 0 \\ \kappa_{21}^P & \kappa_{22}^P & \kappa_{23}^P \\ 0 & 0 & \kappa_{33}^P \end{pmatrix} \left(\begin{pmatrix} 0 \\ \theta_2^P \\ \theta_3^P \end{pmatrix} - \begin{pmatrix} L_t \\ S_t \\ C_t \end{pmatrix} \right) \mathrm{d}t + \Sigma \begin{pmatrix} \mathrm{d}W_t^{L,P} \\ \mathrm{d}W_t^{S,P} \\ \mathrm{d}W_t^{C,P} \end{pmatrix} \quad (4)$$

where the covariance matrix Σ is assumed diagonal and constant. Note that in this specification, the Nelson–Siegel level factor is restricted to be an independent unit-root process under both probability measures.[10] As discussed in CR, this restriction helps improve forecast performance independent of the specification of the remaining elements of K^P. Because interest rates are highly persistent, empirical autoregressive models, including DTSMs, suffer from substantial small-sample estimation bias. Specifically, model estimates will generally be biased toward a dynamic system that displays much less persistence than the true process (so estimates of the real-world mean-reversion matrix, K^P, are upward biased). Furthermore, if the degree of interest rate persistence is underestimated, future short rates would be expected to revert to their mean too quickly causing their expected longer-term averages to be too stable. Therefore, the bias in the estimated dynamics distorts the decomposition of yields and contaminates estimates of long-maturity term premiums. As described in detail in Bauer, Rudebusch, and Wu (2012), bias-corrected K^P estimates are typically very close to a unit-root process, so we view the imposition of the unit-root restriction as a simple shortcut to overcome small-sample estimation bias.

We re-estimated this CR model over a larger sample of weekly nominal U.S. Treasury zero-coupon yields from January 4, 1985, until October 31, 2014, for eight maturities: three months, six months, one year, two years, three years, five years, seven years, and 10 years.[11] The model parameter estimates are reported in Table 1. As in CR, we test the significance of the four parameter restrictions imposed on K^P in the CR model relative to the unrestricted AFNS model.[12] The four parameter restrictions are not rejected by the data at conventional levels of significance similar to what CR report; thus, the CR model appears flexible enough to capture the relevant information in the data compared with an unrestricted model.

Table 1. Parameter Estimates for the CR Model.

K^P	$K^P_{\cdot,1}$	$K^P_{\cdot,2}$	$K^P_{\cdot,3}$	θ^P		Σ
$K^P_{1,\cdot}$	10^{-7}	0	0	0	σ_{11}	0.0066
						(0.0001)
$K^P_{2,\cdot}$	0.3390	0.4157	−0.4548	0.0218	σ_{22}	0.0100
	(0.1230)	(0.1154)	(0.0843)	(0.0244)		(0.0002)
$K^P_{3,\cdot}$	0	0	0.6189	−0.0247	σ_{33}	0.0271
			(0.1571)	(0.0074)		(0.0004)

Notes: The estimated parameters of the K^P matrix, θ^P vector, and diagonal Σ matrix are shown for the CR model. The estimated value of λ is 0.4482 (0.0022). The numbers in parentheses are the estimated parameter standard deviations. The maximum log likelihood value is 70,754.54.

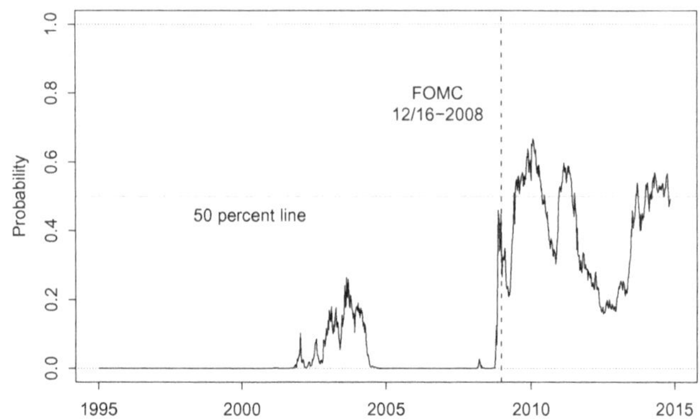

Fig. 2. Probability of Negative Short Rates. *Notes:* Illustration of the conditional probability of negative short rates three months ahead from the CR model.

2.3. Negative Short-Rate Projections in Standard Models

Before turning to the description of the shadow-rate model, it is useful to reinforce the basic motivation for our analysis by examining short-rate forecasts from the estimated CR model. With regard to short-rate forecasts, any standard affine Gaussian DTSM may place positive probabilities on future negative interest rates. Accordingly, Fig. 2 shows the probability that the short rate three months out will be negative obtained from rolling real-time weekly re-estimations of the CR model. Prior to 2008 the probabilities of future negative interest rates are negligible except for a brief period in 2003 and 2004 when the Fed's policy rate temporarily stood at 1 percent. However, near the ZLB – since late 2008 – the model is typically predicting substantial likelihoods of impossible realizations. Worse still, whenever these probabilities are above 50 percent (indicated with a solid gray horizontal line), the model's conditional expected short rate is negative, which has frequently been the case since 2009.

3. A SHADOW-RATE MODEL

In this section, we describe an option-based approach to the shadow-rate model and estimate a shadow-rate analog to the CR model with U.S. data.

3.1. The Option-Based Approach to the Shadow-Rate Model

The concept of a shadow interest rate as a modeling tool to account for the ZLB can be attributed to Black (1995). He noted that the observed nominal short rate will be nonnegative because currency is a readily available asset to investors that carries a nominal interest rate of zero. Therefore, the existence of currency sets a ZLB on yields. To account for this ZLB, Black postulated a shadow short rate, s_t, that is unconstrained by the ZLB. The usual observed instantaneous risk-free rate, r_t, which is used for discounting cash flows when valuing securities, is then given by the greater of the shadow rate or zero:

$$r_t = \max\{0, s_t\} \tag{5}$$

Accordingly, as s_t falls below zero, the observed r_t simply remains at the zero bound.

While Black (1995) described circumstances under which the zero bound on nominal yields might be relevant, he did not provide specifics for implementation. The small set of empirical research on shadow-rate models has relied on numerical methods for pricing.[13] To overcome the computational burden of numerical-based estimation that limits the use of shadow-rate models, Krippner (2013) suggested an alternative option-based approach that makes shadow-rate models almost as easy to estimate as the standard model.[14] To illustrate this approach, consider two bond-pricing situations: one without currency as an alternative asset, and the other that has a currency in circulation with a constant nominal value and no transaction costs. In the world without currency, the price of a shadow-rate zero-coupon bond, $P_t(\tau)$, may trade above par; that is, its risk-neutral expected instantaneous return equals the risk-free shadow short rate, which may be negative. In contrast, in the world with currency, the price at time t for a zero-coupon bond that pays 1 when it matures in τ years is given by $\underline{P}_t(\tau)$. This price will never rise above par, so nonnegative yields will never be observed.

Now consider the relationship between the two bond prices at time t for the shortest (say, overnight) maturity available, δ. In the presence of currency, investors can either buy the zero-coupon bond at price $P_t(\delta)$ and receive one unit of currency the following day or just hold the currency. As a consequence, this bond price, which would equal the shadow bond price, must be capped at 1:

$$\begin{aligned}\underline{P}_t(\delta) &= \min\{1, P_t(\delta)\}\\ &= P_t(\delta) - \max\{P_t(\delta) - 1, 0\}\end{aligned}$$

That is, the availability of currency implies that the overnight claim has a value equal to the zero-coupon shadow bond price minus the value of a call option on the zero-coupon shadow bond with a strike price of 1. More generally, we can express the price of a bond in the presence of currency as the price of a shadow bond minus the call option on values of the bond above par:

$$\underline{P}_t(\tau) = P_t(\tau) - C_t^A(\tau, \tau; 1)$$

where $C_t^A(\tau, \tau; 1)$ is the value of an American call option at time t with maturity in τ years and strike price 1 written on the shadow bond maturing in τ years. In essence, in a world with currency, the bond investor has had to sell off the possible gain from the bond rising above par at any time prior to maturity.

Unfortunately, analytically valuing this American option is complicated by the difficulty in determining the early exercise premium. However, Krippner (2013) argues that there is an analytically close approximation based on tractable European options. Specifically, Krippner (2013) shows that the ZLB instantaneous forward rate, $\underline{f}_t(\tau)$, is

$$\underline{f}_t(\tau) = f_t(\tau) + z_t(\tau)$$

where $f_t(\tau)$ is the instantaneous forward rate on the shadow bond, which may go negative, while $z_t(\tau)$ is an add-on term given by

$$z_t(\tau) = \lim_{\delta \to 0} \left[\frac{\partial}{\partial \delta} \frac{C_t^E(\tau, \tau + \delta; 1)}{P_t(\tau + \delta)} \right] \tag{6}$$

where $C_t^E(\tau, \tau + \delta; 1)$ is the value of a European call option at time t with maturity $t + \tau$ and strike price 1 written on the shadow discount bond maturing at $t + \tau + \delta$. Thus, the observed yield-to-maturity is

$$\begin{aligned}
\underline{y}_t(\tau) &= \frac{1}{\tau} \int_t^{t+\tau} \underline{f}_t(s) \mathrm{d}s \\
&= \frac{1}{\tau} \int_t^{t+\tau} f_t(s) \mathrm{d}s + \frac{1}{\tau} \int_t^{t+\tau} \lim_{\delta \to 0} \left[\frac{\partial}{\partial \delta} \frac{C_t^E(s, s + \delta; 1)}{P_t(s + \delta)} \right] \mathrm{d}s \\
&= y_t(\tau) + \frac{1}{\tau} \int_t^{t+\tau} \lim_{\delta \to 0} \left[\frac{\partial}{\partial \delta} \frac{C_t^E(s, s + \delta; 1)}{P_t(s + \delta)} \right] \mathrm{d}s
\end{aligned}$$

Hence, bond yields constrained at the ZLB can be viewed as the sum of the yield on the unconstrained shadow bond, denoted $y_t(\tau)$, which is modeled using standard tools, and an add-on correction term derived from the price formula for the option written on the shadow bond that provides an upward push to deliver the higher nonnegative yields actually observed.

As highlighted by Christensen and Rudebusch (2015), the Krippner (2013) framework should be viewed as not fully internally consistent and simply an approximation to an arbitrage-free model.[15] Of course, away from the ZLB, with a negligible call option, the model will match the standard arbitrage-free term structure representation. In addition, the size of the approximation error near the ZLB has been determined via simulation for Japanese yield data in Christensen and Rudebusch (2015) to be quite modest, and we provide similar evidence in Appendix A for our sample of U.S. Treasury yields.

3.2. The B-CR Model

In theory, the option-based shadow-rate result is quite general and applies to any assumptions made about the dynamics of the shadow-rate process. However, as implementation requires the calculation of the limit in Eq. (6), the option-based shadow-rate models are limited practically to the

Gaussian model class. The AFNS class is well suited for this extension.[16] In the shadow-rate AFNS model, the shadow risk-free rate is defined as the sum of level and slope as in Eq. (2) in the original AFNS model class, while the affine short rate is replaced by the nonnegativity constraint:

$$s_t = L_t + S_t, \quad r_t = \max\{0, s_t\}$$

All other elements of the model remain the same. Namely, the dynamics of the state variables used for pricing under the Q-measure remain as described in Eq. (1), so the yield on the shadow discount bond maintains the popular Nelson and Siegel (1987) factor loading structure

$$y_t(\tau) = L_t + \left(\frac{1 - e^{-\lambda\tau}}{\lambda\tau}\right) S_t + \left(\frac{1 - e^{-\lambda\tau}}{\lambda\tau} - e^{-\lambda\tau}\right) C_t - \frac{A(\tau)}{\tau}$$

where $A(\tau)/\tau$ is the same maturity-dependent yield-adjustment term.

The corresponding instantaneous shadow forward rate is given by

$$f_t(\tau) = -\frac{\partial}{\partial T} \ln P_t(\tau) = L_t + e^{-\lambda\tau} S_t + \lambda\tau e^{-\lambda\tau} C_t + A^f(\tau)$$

where the yield-adjustment term in the instantaneous forward rate function is given by

$$\begin{aligned}
A^f(\tau) &= -\frac{\partial A(\tau)}{\partial \tau} \\
&= -\frac{1}{2}\sigma_{11}^2 \tau^2 - \frac{1}{2}(\sigma_{21}^2 + \sigma_{22}^2)\left(\frac{1 - e^{-\lambda\tau}}{\lambda}\right)^2 \\
&\quad - \frac{1}{2}(\sigma_{31}^2 + \sigma_{32}^2 + \sigma_{33}^2)\left[\frac{1}{\lambda^2} - \frac{2}{\lambda^2} e^{-\lambda\tau} - \frac{2}{\lambda}\tau e^{-\lambda\tau} + \frac{1}{\lambda^2} e^{-2\lambda\tau} + \frac{2}{\lambda}\tau e^{-2\lambda\tau} + \tau^2 e^{-2\lambda\tau}\right] \\
&\quad - \sigma_{11}\sigma_{21}\tau\frac{1 - e^{-\lambda\tau}}{\lambda} - \sigma_{11}\sigma_{31}\left[\frac{1}{\lambda}\tau - \frac{1}{\lambda}\tau e^{-\lambda\tau} - \tau^2 e^{-\lambda\tau}\right] \\
&\quad - (\sigma_{21}\sigma_{31} + \sigma_{22}\sigma_{32})\left[\frac{1}{\lambda^2} - \frac{2}{\lambda^2} e^{-\lambda\tau} - \frac{1}{\lambda}\tau e^{-\lambda\tau} + \frac{1}{\lambda^2} e^{-2\lambda\tau} + \frac{1}{\lambda}\tau e^{-2\lambda\tau}\right]
\end{aligned}$$

Krippner (2013) provides a formula for the ZLB instantaneous forward rate, $\underline{f}_t(\tau)$, that applies to any Gaussian model

$$\underline{f}_t(\tau) = f_t(\tau)\Phi\left(\frac{f_t(\tau)}{\omega(\tau)}\right) + \omega(\tau)\frac{1}{\sqrt{2\pi}}\exp\left(-\frac{1}{2}\left[\frac{f_t(\tau)}{\omega(\tau)}\right]^2\right)$$

where $\Phi(\cdot)$ is the cumulative probability function for the standard normal distribution, $f_t(\tau)$ is the shadow forward rate, and $\omega(\tau)$ is related to the conditional variance, $v(\tau, \tau+\delta)$, appearing in the shadow bond option price formula as follows:

$$\omega(\tau)^2 = \frac{1}{2}\lim_{\delta\to 0}\frac{\partial^2 v(\tau, \tau+\delta)}{\partial\delta^2}$$

Within the shadow-rate AFNS model, $\omega(\tau)$ takes the following form:

$$\begin{aligned}\omega(\tau)^2 = {} & \sigma_{11}^2\tau + \left(\sigma_{21}^2 + \sigma_{22}^2\right)\frac{1-e^{-2\lambda\tau}}{2\lambda} \\ & + \left(\sigma_{31}^2 + \sigma_{32}^2 + \sigma_{33}^2\right)\left[\frac{1-e^{-2\lambda\tau}}{4\lambda} - \frac{1}{2}\tau e^{-2\lambda\tau} - \frac{1}{2}\lambda\tau^2 e^{-2\lambda\tau}\right] \\ & + 2\sigma_{11}\sigma_{21}\frac{1-e^{-\lambda\tau}}{\lambda} + 2\sigma_{11}\sigma_{31}\left[-\tau e^{-\lambda\tau} + \frac{1-e^{-\lambda\tau}}{\lambda}\right] \\ & + (\sigma_{21}\sigma_{31} + \sigma_{22}\sigma_{32})\left[-\tau e^{-2\lambda\tau} + \frac{1-e^{-2\lambda\tau}}{2\lambda}\right]\end{aligned}$$

Therefore, the zero-coupon bond yields that observe the ZLB, denoted $\underline{y}_t(\tau)$, are easily calculated as

$$\underline{y}_t(\tau) = \frac{1}{\tau}\int_t^{t+\tau}\left[f_t(s)\Phi\left(\frac{f_t(s)}{\omega(s)}\right) + \omega(s)\frac{1}{\sqrt{2\pi}}\exp\left(-\frac{1}{2}\left[\frac{f_t(s)}{\omega(s)}\right]^2\right)\right]ds$$

Table 2. Parameter Estimates for the B-CR Model.

K^P	$K^P_{\cdot,1}$	$K^P_{\cdot,2}$	$K^P_{\cdot,3}$	θ^P		Σ
$K^P_{1,\cdot}$	10^{-7}	0	0	0	σ_{11}	0.0069
						(0.0001)
$K^P_{2,\cdot}$	0.1953	0.3138	−0.4271	0.0014	σ_{22}	0.0112
	(0.1474)	(0.1337)	(0.0904)	(0.0364)		(0.0002)
$K^P_{3,\cdot}$	0	0	0.4915	−0.0252	σ_{33}	0.0257
			(0.1200)	(0.0087)		(0.0004)

Notes: The estimated parameters of the K^P matrix, θ^P vector, and diagonal Σ matrix are shown for the B-CR model. The estimated value of λ is 0.4700 (0.0026). The numbers in parentheses are the estimated parameter standard deviations. The maximum log likelihood value is 71,408.90.

As in the affine AFNS model, the shadow-rate AFNS model is completed by specifying the price of risk using the essentially affine risk premium specification introduced by Duffee (2002), so the real-world dynamics of the state variables can be expressed as in Eq. (3). Again, in an unrestricted case, both K^P and θ^P are allowed to vary freely relative to their counterparts under the Q-measure. However, we focus on the case with the same K^P and θ^P restrictions as in the CR model, that is, the P-dynamics are given by Eq. (4), on the assumption that outside of the ZLB period, the shadow-rate model would properly collapse to the standard CR form. We label this shadow-rate model as the B-CR model as already discussed in Section 1.

We estimate the B-CR model from January 4, 1985, until October 31, 2014, for eight maturities: three months, six months, one year, two years, three years, five years, seven years, and 10 years.[17] The estimated B-CR model parameters are reported in Table 2. Later on, we make a more comprehensive analysis of the estimated parameters in both models. For now, as with the CR model, we test the significance of the four parameter restrictions imposed on K^P in the B-CR model relative to the unrestricted B-AFNS model.[18] Fig. 3 shows that, for most sample cutoff points since 1995, the four parameter restrictions are not rejected by the data at conventional levels of significance. Also shown with a solid gray line are the quasi likelihood ratio tests of the six restrictions in the most parsimonious B-AFNS model with independent factors relative to the unrestricted B-AFNS model, which are clearly rejected. Thus, similar to the CR model, the B-CR model appears flexible enough to capture the relevant information in the data compared with an unrestricted model.

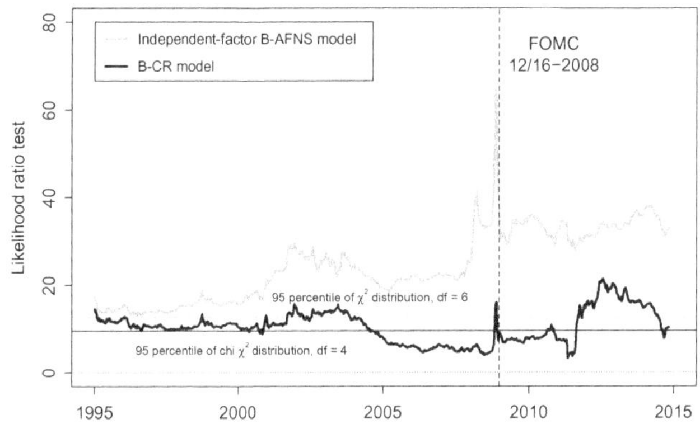

Fig. 3. Quasi Likelihood Ratio Tests of Parameter Restrictions in B-AFNS Models. *Notes:* Illustration of the value of quasi likelihood ratio tests of the restrictions imposed in the independent-factor B-AFNS model, and in the B-AFNS model underlying the B-CR model, relative to the B-AFNS model with unrestricted K^P-matrix and diagonal Σ-matrix. The analysis covers weekly re-estimations from January 6, 1995, to October 31, 2014, a total of 1,035 observations, while the full data set used in the analysis covers the period from January 4, 1985, to October 31, 2014.

3.3. Measuring the Effect of the ZLB

To provide evidence that we should anticipate to see at least some difference across the regular and shadow-rate models, we turn our focus to the value of the option to hold currency, which we define as the difference between the yields that observe the ZLB and the comparable lower shadow discount bond yields that do not. Fig. 4 shows these yield spreads at the 5- and 10-year maturity based on real-time rolling weekly re-estimations of the B-CR model starting in 1995 through October 31, 2014. Beyond a very few temporary small spikes, the option had economically insignificant value prior to the failure of Lehman Brothers in the fall of 2008.[19] However, despite the zero short rate since 2008, it is not really until after August 2011 that the option obtains significant sustained value. At its peak in the fall of 2012, the yield spread was 80 and 60 basis points at the five- and 10-year maturity, respectively. Option values at those levels suggest that it should matter for model performance whether a model accounts for the ZLB of nominal yields. Section 4 is dedicated to analyzing this question, but first we discuss the choice of lower bound in the shadow-rate model.

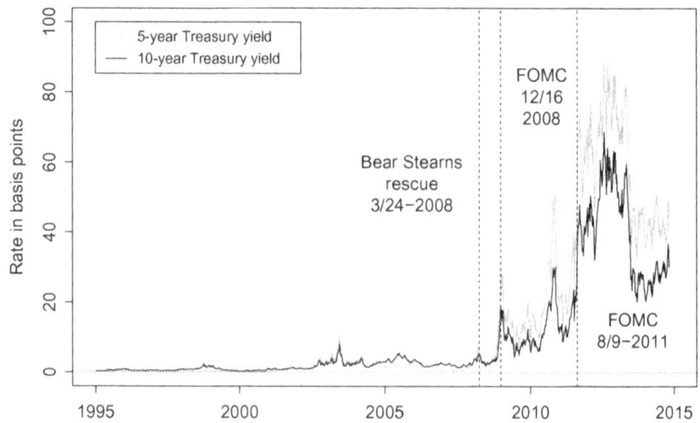

Fig. 4. Value of Option to Hold Currency. *Notes:* We show time-series plots of the value of the option to hold currency embedded in the Treasury yield curve as estimated in real time by the B-CR model. The data cover the period from January 6, 1995, to October 31, 2014.

3.4. Nonzero Lower Bound for the Short Rate

In this section, we consider a generalization of the B-CR model that allows for the lower bound of the short rate to differ from zero, that is,

$$r_t = \max\{r_{\min}, s_t\}$$

Christensen and Rudebusch (2015) provide the formula for the forward rate that respects the $r_{\min}$ lower bound:[20]

$$\underline{f}_t(\tau) = r_{\min} + (f_t(\tau) - r_{\min})\Phi\left(\frac{f_t(\tau) - r_{\min}}{\omega(\tau)}\right) + \omega(\tau)\frac{1}{\sqrt{2\pi}}\exp\left(-\frac{1}{2}\left[\frac{f_t(\tau) - r_{\min}}{\omega(\tau)}\right]^2\right)$$

where the shadow forward rate, $f_t(\tau)$, and $\omega(\tau)$ remain as before.

A few papers have used a nonzero lower bound for the short rate. In the case of U.S. Treasury yields, Wu and Xia (2014) simply fix the lower bound at 25 basis points. A similar approach applied to Japanese, U.K., and U.S. yields is followed by Ichiue and Ueno (2013).[21] As an alternative, Kim and Priebsch (2013) leave $r_{\min}$ as a free parameter to be determined in the model estimation. Using U.S. Treasury yields they report an estimated value of 14 basis points.

In theory, the lower bound should be zero because that is the nominal return on holding currency, which is a readily available alternative to holding bonds. Furthermore, U.S. Treasury yield data also supports a choice of zero for the lower bound. Specifically, in the daily H.15 database through October 31, 2014 (of which we use a weekly subsample), the zero boundary is *never* violated. The one-month yield is 0 on 51 dates, the three-month yield is 0 on 8 dates, while the six-month yield never goes below 2 basis points. In addition, since late 2008, the spread between the six- and three-month yields is always nonnegative with a single exception, October 11, 2013, when it was negative 1 basis point. Thus, with three- and six-month yields less than 10 basis points and the yield curve steep for much of the time spent near the ZLB, the choice of zero for the lower bound of the short rate appears to be a reasonable assumption that is supported by both the data and theoretical considerations.[22] Still, it is econometrically feasible to leave $r_{\min}$ as a free parameter to be determined by the data as in Kim and Priebsch (2013). Thus, ultimately, it is an empirical question what is the appropriate choice for the lower bound of the short rate in shadow-rate models. To make a comprehensive and in-depth assessment of the economical and statistical importance of this parameter, we use rolling weekly re-estimations of the B-CR model with and without restricting $r_{\min}$ to zero.

Fig. 5 illustrates the estimated value of $r_{\min}$ from the rolling re-estimations since January 6, 1995. At the start of this sample, $r_{\min}$ was estimated to be almost 285 basis points, but since then, it has been trending lower. As of October 31, 2014, the full-sample estimate of the lower bound was 11 basis points. Even that relatively low value of $r_{\min}$ censors much of the variation in the short end of the yield curve. There are 215 weekly observations of the three-month yield below that level in our sample, while the corresponding number for the six-month and one-year yields is 124 and 8, respectively. Given that we have 307 weekly observations from the ZLB period (defined as the period since December 19, 2008), more than 70 percent of the time spent in the ZLB period the model would ignore variation in the three-month yields, while it ignores variation in the six-month yield more than 40 percent of the time.

In Fig. 5(b), we show quasi likelihood ratio tests of restricting $r_{\min}$ to zero in the B-CR model relative to leaving it unrestricted. We note that the zero restriction has been systematically rejected since 1995. Most problematically, the rejection is strongest since 2010 when a lower bound of zero appears to be most appropriate according to the level of short-term yields in the data. Thus, the need to fully understand the effects of varying the lower bound in shadow-rate models is evident.

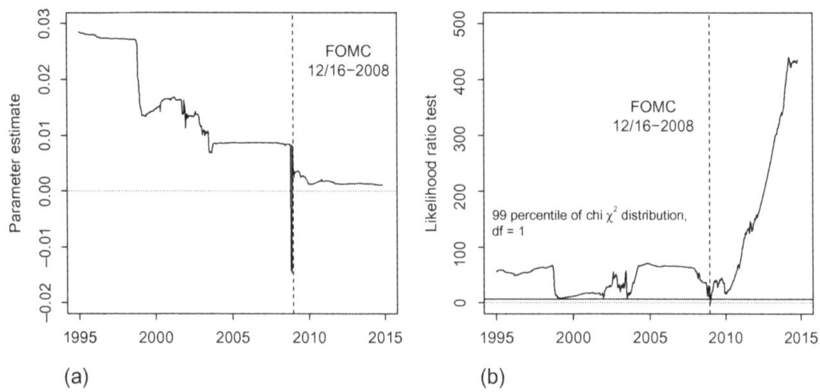

Fig. 5. Estimates of $r_{\min}$ and Quasi Likelihood Ratio Tests of its Zero Restriction. *Notes:* Panel (a) illustrates the estimated value of $r_{\min}$ in the B-CR model from weekly re-estimations from January 6, 1995, to October 31, 2014, a total of 1,035 observations. Panel (b) shows quasi likelihood ratio tests of restricting $r_{\min}$ to zero in the B-CR model, also based on weekly re-estimations from January 6, 1995, to October 31, 2014. The full data set used in the analysis covers the period from January 4, 1985, to October 31, 2014.

To demonstrate that the choice of $r_{\min}$ may not be innocuous, consider the estimated shadow-rate path. Fig. 6 shows this path for the B-CR model with and without restricting $r_{\min}$ to zero. For comparison, the estimated short-rate path from the CR model is also shown. All three models are indistinguishable along this dimension before 2008, so the figure only shows the estimated paths since then. Note that the estimated shadow-rate path is sensitive to the choice of $r_{\min}$, as was also highlighted by Bauer and Rudebusch (2014). For this reason, our results below will include B-CR model specifications with $r_{\min}$ restricted to zero and freely estimated.

4. COMPARING AFFINE AND SHADOW-RATE MODELS

In this section, we compare the empirical affine and shadow-rate models across a variety of dimensions, including parameter stability, in-sample fit, volatility dynamics, and out-of-sample forecast performance.

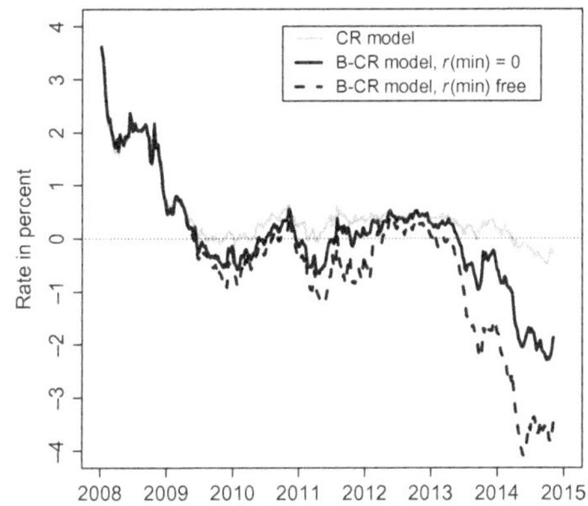

Fig. 6. Estimated Short- and Shadow-Rate Paths. *Notes:* Illustration of the estimated shadow-rate paths from the B-CR model with and without restricting $r_{\min}$ to zero with a comparison to the estimated short-rate path from the CR model. All three paths are based on full sample estimations with data covering the period from January 4, 1985, to October 31, 2014.

4.1. Analysis of Parameter Estimates

To begin, we analyze the parameter stability and similarity of the empirical affine and shadow-rate models, where the latter is estimated with and without restricting $r_{\min}$ to zero throughout. We also assess the reasonableness of using estimated parameters from the affine model in combination with the shadow-rate model as a way of avoiding the burden of making a full shadow-rate model estimation.

Fig. 7 shows the estimated parameters in the mean-reversion matrix K^P. First, we note that all three models give very similar parameter estimates before December 2008. This is not surprising since the shadow-rate models collapse to the affine model away from the lower bound, and it does not matter much whether the lower bound is fixed at zero or left as a free parameter. Second, since late 2008, we do see some larger deviations with a tendency for the shadow-rate models to produce higher persistence of the slope and curvature factors as indicated by their lower estimates of κ_{22}^P and κ_{33}^P. However, judged by the estimated parameter standard deviations reported in Tables 1 and 2, these differences in the individual parameters do not appear to be statistically significant.

In Fig. 8, we compare the estimated volatility parameters. While the estimated volatility parameters for the level factor are fairly similar across all

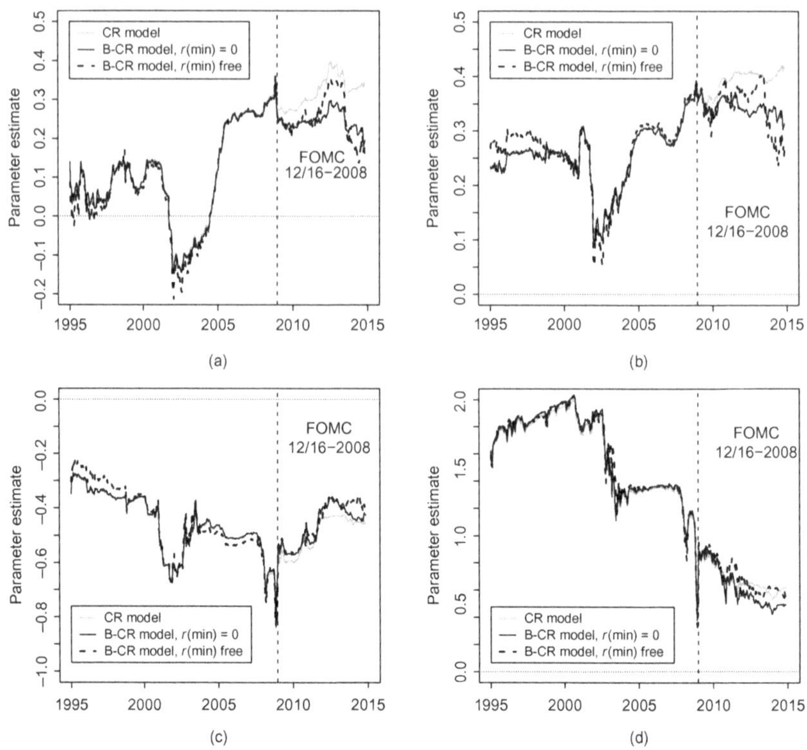

Fig. 7. Estimates of Mean-Reversion Parameters. *Notes:* Illustration of the estimated parameters in the mean-reversion K^P matrix in the CR and B-CR models, where the latter is estimated with and without restricting $r_{\min}$ to zero. The analysis covers weekly re-estimations from January 6, 1995, to October 31, 2014, a total of 1,035 observations, while the full data set used covers the period from January 4, 1985, to October 31, 2014. Estimates of (a) κ^P_{21}; (b) κ^P_{22}; (c) κ^P_{23}; and (d) κ^P_{33}.

three models throughout the entire period as can be seen in Fig. 8(a), there are larger differences in the estimated volatility of the slope and curvature factors during the most recent period as illustrated in Fig. 8(b) and (c). In the shadow-rate models, the slope factor is allowed to be more volatile in the post-crisis period compared to the standard model as it is not required to fully match the low volatility of short-term yields near the ZLB whenever the shadow rate is in negative territory.

Fig. 9 shows the estimated mean parameters since 1995. These parameters represent another area where the models "learn" something about

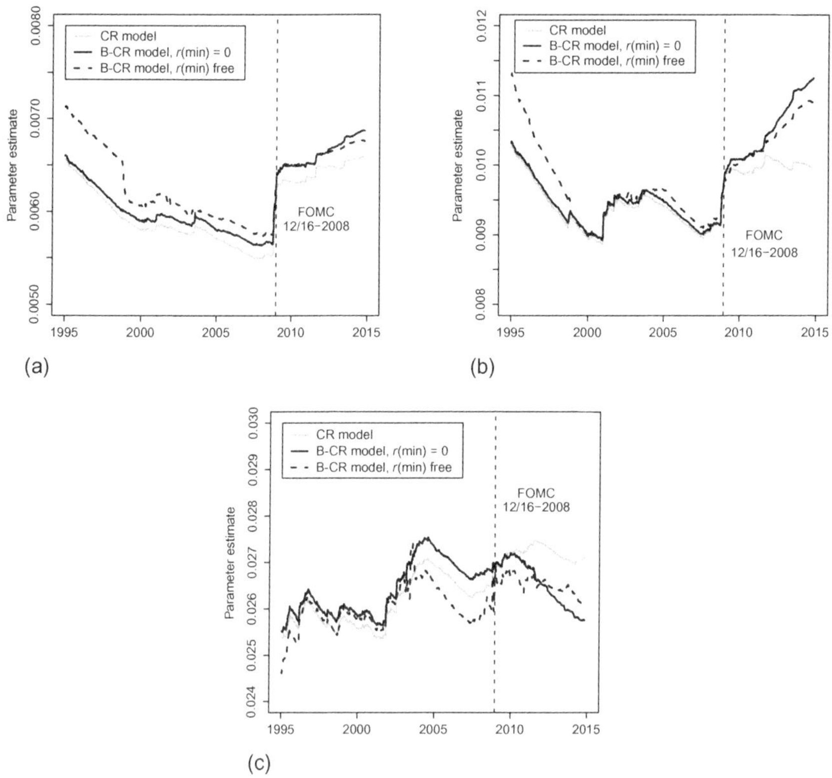

Fig. 8. Estimates of Volatility Parameters. *Notes:* Illustration of the estimated parameters in the volatility Σ matrix in the CR and B-CR models, where the latter is estimated with and without restricting $r_{\min}$ to zero. The analysis covers weekly re-estimations from January 6, 1995, to October 31, 2014, a total of 1,035 observations, while the full data set used covers the period from January 4, 1985, to October 31, 2014. Estimates of (a) σ_{11}; (b) σ_{22}; and (c) σ_{33}.

the true parameter values through the updating during the ZLB period. The low yield levels in this period translate into gradually declining estimates of the mean parameters, θ_2^P and θ_3^P, in particular the estimate of θ_3^P has declined notably since the crisis. Since the curvature factor in its role as the stochastic mean of the slope factor represents expectations for future monetary policy, a potential explanation for the decline in its estimated mean would be the anchoring of monetary policy expectations in the

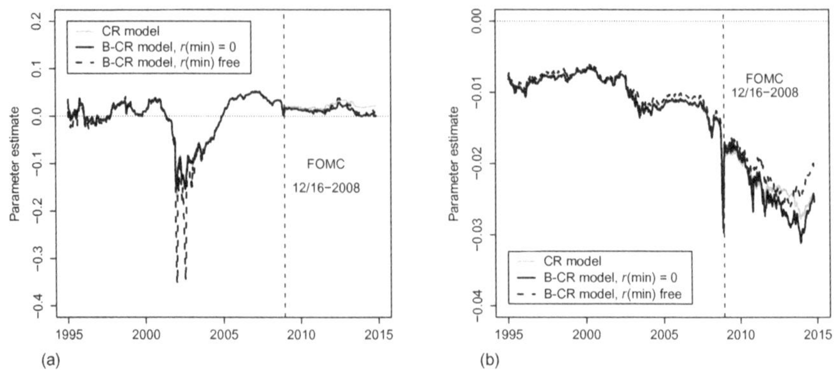

Fig. 9. Estimates of Mean Parameters. *Notes:* Illustration of the estimated parameters in the mean θ^P vector in the CR and B-CR models, where the latter is estimated with and without restricting $r_{\min}$ to zero. The analysis covers weekly re-estimations from January 6, 1995, to October 31, 2014, a total of 1,035 observations, while the full data set used covers the period from January 4, 1985, to October 31, 2014. Estimates of (a) θ_2^P and (b) θ_3^P.

medium term at a low level, perhaps reflecting various forms of policy forward guidance employed by the FOMC since late 2008.

Finally, in Fig. 10, we compare the various estimates of the λ parameter that determines the rate of decay in the yield factor loading of the slope factor and the peak maturity in the yield factor loading of the curvature factor. Here, we see a more long-term trend toward lower values. This suggests that investors' speculation about future monetary policy as represented through the variation of the curvature factor has tended to take place at longer maturities more recently than compared to two decades ago. What role the greater transparency of the FOMC's monetary policy decisions plays for this trend is an interesting question that we leave for future research.

To summarize our findings so far, overall, the differences between the standard and the shadow-rate models for individual parameters look relatively small and are in most cases not statistically significant. Still, it could be the case that the minor differences combined could add up to material differences not only statistically, but also economically. We end the section by analyzing this important question further.

The way we proceed is to take the estimated parameters from the B-CR model with $r_{\min}$ restricted to zero as of December 28, 2007, and those from the CR model as of December 28, 2007, and October 31, 2014. We then

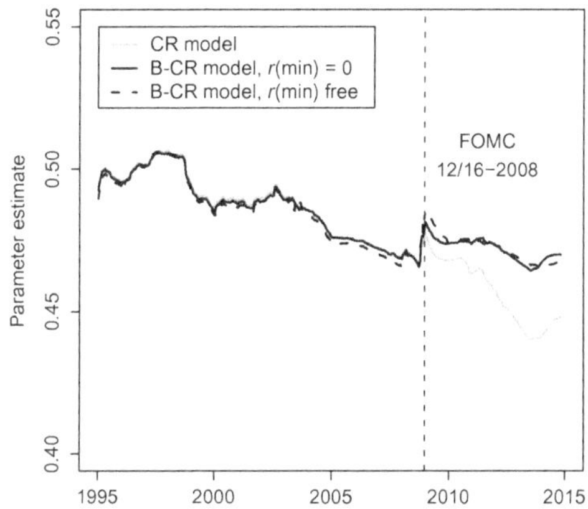

Fig. 10. Estimates of the λ Parameter. *Notes:* Illustration of the estimated λ parameter in the CR and B-CR models, where the latter is estimated with and without restricting $r_{\min}$ to zero. The analysis covers weekly re-estimations from January 6, 1995, to October 31, 2014, a total of 1,035 observations, while the full data set used covers the period from January 4, 1985, to October 31, 2014.

combine these three parameter vectors with the B-CR model with $r_{\min}$ restricted to zero to obtain the corresponding filtered state variables as of October 31, 2014. Finally, we use each pair of parameters and filtered state variables to calculate the projection of the short rate as of October 31, 2014. These projected paths are shown in Fig. 11. The benchmark in the comparison is obtained by using the B-CR model's own estimated parameters as of October 31, 2014, to filter the state variables on that date and combine them to generate the associated short-rate projection as of October 31, 2014, shown with a solid black line in Fig. 11.

It is immediately noted that the differences in the short-rate projections are huge. The parameters estimated with the CR and B-CR models as of December 28, 2007, both imply a rather quick rate of mean reversion *and* to a high-level approaching 4 percent in the long run, while the parameters estimated with the CR model as of October 31, 2014, imply an even quicker rate of mean reversion at first, but they have the short-rate leveling off near 3 percent in the long run due to lower estimated values of the mean parameters, θ_2^P and θ_3^P. In contrast, the B-CR model's own projection implies a later liftoff, a more gradual normalization of monetary policy, and to a lower long-run level of about 2 percent, due to the higher persistence (lower κ_{22}^P and κ_{33}^P estimates) and lower means of the slope and curvature factors relative to the two alternative parameter vectors.

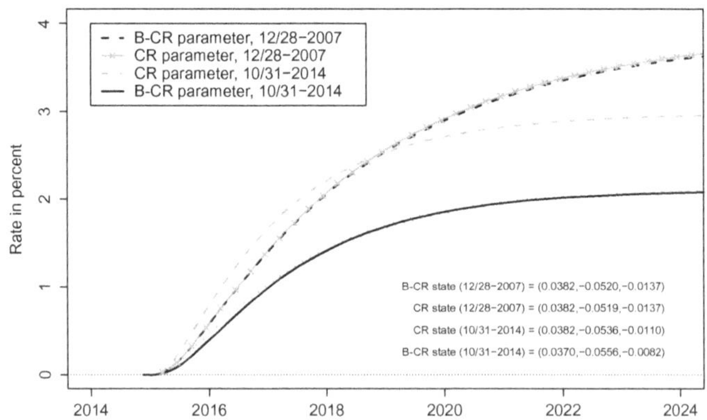

Fig. 11. Sensitivity of Short-Rate Projections to Model Parameters. *Notes:* Illustration of short-rate projections as of October 31, 2014, implied by the B-CR model with $r_{\min}$ restricted to zero using different model parameters and state variables as explained in the main text.

To shed more light on the source of the disagreement about the short-rate paths, in the lower right corner of Fig. 11, we report the four different filtered state variable vectors used as conditioning variables in the calculations of the short-rate projections. We note that they are slightly different from one another. However, to demonstrate that this is not the cause for the differences in the short-rate projections, we condition on all four state variable vectors using the B-CR model with its estimated parameters as of October 31, 2014. In addition to the solid black line, this produces the dotted gray lines in the figure, which are practically indistinguishable from the solid black line. Thus, the variation in the short-rate projections in Fig. 11 is entirely driven by differences in the parameter vectors.

Furthermore, we note that the differences are material, not just economically, but also statistically. The log likelihood values obtained from evaluating the extended Kalman filter of the B-CR model at each of the four parameter sets are 71,200.19, 71,181.75, 71,260.51, and 71,408.90, respectively, where the latter is the value of the likelihood function of the B-CR model evaluated at its own optimal parameters as of October 31, 2014. Thus, the deviations in each parameter do combine into huge likelihood differences as well.

Based on these findings we cannot recommend using pre-crisis parameter estimates to assess recent policy expectations as done in Bauer and Rudebusch (2014), and even the use of contemporaneous parameter estimates from the affine model in combination with the shadow-rate model does not seem warranted as a way to alleviate the burden of estimation. To facilitate the estimation of the shadow-rate model, we feel that, at most, what can be gained from estimating the matching affine model is to use its optimal parameters as a starting point for the parameter optimization in the estimation of the shadow-rate model.

Finally, we note that these results demonstrate the importance of undertaking rolling real-time model estimations like the ones performed in this paper when evaluating model performance near the ZLB. However, it remains an open question to what extent the state variables will maintain their recent high persistence or revert back toward the lower pre-crisis levels once the normalization of policy rates begins. Thus, we caution that there is a risk that rolling estimations might underperform during the early stages of the subsequent policy tightening cycle.

4.2. *In-Sample Fit and Yield Volatility*

The summary statistics of the fit to yield levels of the affine and shadow-rate models are reported in Table 3. They indicate a very similar fit in the normal period up until the end of 2008. However, since then, we see a notable advantage to the shadow-rate models that is also reflected in the likelihood values. Still, we conclude from this in-sample analysis that it is not in the model fit that the shadow-rate model really distinguishes itself from its regular cousin, and this conclusion is not sensitive to the choice of lower bound.

However, a serious limitation of standard Gaussian models is the assumption of constant yield volatility, which is particularly unrealistic when periods of normal volatility are combined with periods in which yields are greatly constrained in their movements near the ZLB. A shadow-rate model approach can mitigate this failing significantly. In the CR model, where zero-coupon yields are affine functions of the state variables, model-implied conditional predicted yield volatilities are given by the square root of

$$V_t^P\left[y_T^N(\tau)\right] = \frac{1}{\tau^2}B(\tau)' V_t^P[X_T]B(\tau)$$

where $T-t$ is the prediction period, τ is the yield maturity, $B(\tau)$ contains the yield factor loadings, and $V_t^P[X_T]$ is the conditional covariance matrix of the state variables.[23] In the B-CR model, on the other hand,

Table 3. Summary Statistics of the Fitted Errors.

RMSE	Maturity in Months								All Yields
	3	6	12	24	36	60	84	120	
Full sample									
CR	30.82	15.02	0.00	2.49	0.00	3.05	2.71	10.74	12.81
B-CR, $r_{\min} = 0$	29.89	14.27	0.90	2.26	0.27	2.67	2.28	9.97	12.32
B-CR, $r_{\min}$ free	29.78	14.23	0.88	2.23	0.35	2.65	2.36	9.86	12.27
Normal period (Jan. 6, 1995–Dec. 12, 2008)									
CR	32.76	15.66	0.00	2.51	0.00	3.01	2.50	10.53	13.47
B-CR, $r_{\min} = 0$	32.69	15.48	0.74	2.39	0.09	2.80	2.09	10.48	13.40
B-CR, $r_{\min}$ free	32.69	15.49	0.62	2.40	0.10	2.81	2.16	10.50	13.41
ZLB period (Dec. 19, 2008–Oct. 31, 2014)									
CR	21.13	12.05	0.00	2.42	0.00	3.17	3.41	11.56	10.07
B-CR, $r_{\min} = 0$	13.40	7.55	1.37	1.64	0.57	2.06	2.93	7.60	6.96
B-CR, $r_{\min}$ free	12.11	7.04	1.54	1.30	0.75	1.91	3.02	6.60	6.27

Notes: Shown are the root-mean-squared fitted errors (RMSEs) for the CR and B-CR models, where the latter is estimated with and without restricting $r_{\min}$ to zero. All numbers are measured in basis points. The data covers the period from January 6, 1985, to October 31, 2014.

zero-coupon yields are nonlinear functions of the state variables and conditional predicted yield volatilities have to be generated by standard Monte Carlo simulation. Fig. 12 shows the implied three-month conditional yield volatility of the three-month and two-year yields from the CR and B-CR models.

To evaluate the fit of these predicted three-month-ahead conditional yield standard deviations, they are compared to a standard measure of realized volatility based on the same data used in the model estimation, but at daily frequency. The realized standard deviation of the daily changes in the interest rates are generated for the 91-day period ahead on a rolling basis. The realized variance measure is used by Andersen and Benzoni (2010), Collin-Dufresne, Goldstein, and Jones (2009), as well as Jacobs and Karoui (2009) in their assessments of stochastic volatility models. For each observation date t the number of trading days N during the subsequent 91-day time window is determined and the realized standard deviation is calculated as

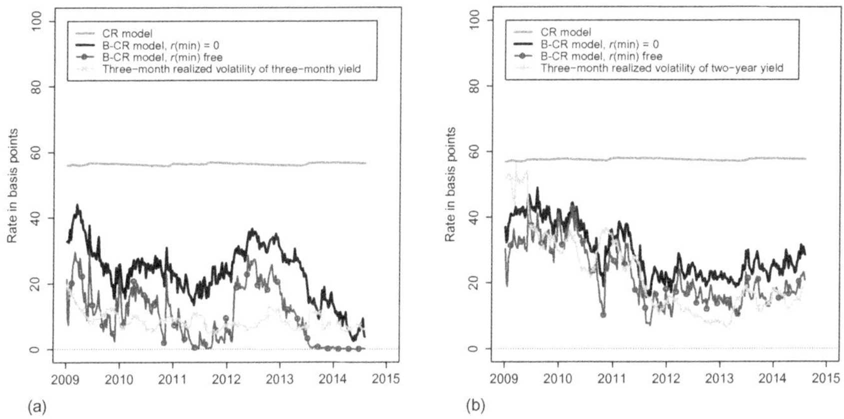

Fig. 12. Three-Month Conditional Yield Volatilities Since 2009. *Notes:* Panel (a) illustrates the three-month conditional volatility of the three-month yield implied by the estimated CR and B-CR models, where the latter is estimated with $r_{\min}$ both restricted to zero and left free. Also shown is the subsequent three-month realized volatility of the three-month yield based on daily data. Panel (b) illustrates the corresponding results for the three-month conditional volatility of the two-year yield.

$$RV_{t,\tau}^{\text{STD}} = \sqrt{\sum_{n=1}^{N} \Delta y_{t+n}^2(\tau)}$$

where $\Delta y_{t+n}(\tau)$ is the change in yield $y(\tau)$ from trading day $t+(n-1)$ to trading day $t+n$.[24]

While the conditional yield volatility from the CR model changes little (merely reflecting the updating of estimated parameters), the conditional yield volatility from the B-CR models fairly closely matches the realized volatility series.[25] As for leaving $r_{\min}$ as a free parameter, we note that it does lead to a slightly closer fit to the realized yield volatilities of medium-term yields, but it comes at the tradeoff of periodically producing effectively zero volatility of short-term yields as shown in Fig. 12(a).

4.3. Forecast Performance

In this section, we first compare the ability of standard and shadow-rate models to forecast future short rates, before we proceed to evaluate their ability to forecast the entire cross section of yields.

4.3.1. Short-Rate Forecasts

Extracting the term premiums embedded in the Treasury yield curve is ultimately an exercise in generating accurate policy rate expectations. Thus, to study bond investors' expectations in real time, we use the rolling re-estimations of the CR model and its shadow-rate equivalents on expanding samples – adding one week of observations each time, a total of 1,035 estimations. As a result, the end dates of the expanding samples range from January 6, 1995, to October 31, 2014. For each end date during that period, we project the short-rate six months, one year, and two years ahead.[26] Importantly, the estimates of these objects rely essentially only on information that was available in real time. Besides examining the full sample, we also distinguish between forecast performance in the normal period prior to the policy rate reaching its effective lower bound (the 13 years from 1995 through 2008) and the ZLB period.

For robustness, we include results from another established U.S. Treasury term structure model introduced in Kim and Wright (2005), henceforth KW), which is a standard latent three-factor Gaussian term structure model of the kind described in Section 2.1.[27] Summary statistics for the forecast errors relative to the subsequent realizations of the target overnight federal funds rate set by the FOMC are reported in Table 4, which also contains the forecast errors obtained using a random walk assumption. We note the strong forecast performance of the KW model relative to the CR model during the normal period, while it is equally obvious that the KW model underperforms grossly during the ZLB period since December 19, 2008. As expected, the CR and B-CR models exhibit fairly similar performance during the normal period, while the B-CR model stands out in the most recent ZLB period. Importantly, we note that, for forecasting future short rates near the ZLB, forecast accuracy is not improved by allowing for a nonzero lower bound in the B-CR model despite the reported in-sample statistical advantage of doing so.

Fig. 13 compares the models' one-year-ahead forecasts to the subsequent target rate realizations. The KW model's systematic overprediction of future target rates since late 2008 stands out. For the CR model, the deterioration in forecast performance is not really detectable until after the August 2011 FOMC meeting when explicit forward guidance was first introduced. Since the CR model mitigates finite-sample bias in the estimates of the mean-reversion matrix K^P by imposing a unit-root property on the Nelson–Siegel level factor, it suggests that the recent deterioration for the CR model must be caused by other more fundamental factors. Importantly, though, the shadow-rate models appear much less affected by any such issues.

Table 4. Summary Statistics for Target Federal Funds Rate Forecast Errors.

	Six-Month Forecast		One-Year Forecast		Two-Year Forecast	
	Mean	RMSE	Mean	RMSE	Mean	RMSE
Full forecast period						
Random walk	14.94	80.01	30.16	142.74	60.58	232.35
KW model	7.44	62.61	51.42	124.46	125.38	222.64
CR model	−0.51	63.27	13.99	123.55	55.01	224.25
B-CR model, $r_{min} = 0$	3.51	59.79	21.80	117.95	64.79	216.53
B-CR model, r_{min} free	10.16	57.58	29.22	116.86	72.80	213.76
Normal forecast period						
Random walk	20.71	94.19	40.73	165.87	77.47	262.75
KW model	−3.40	69.68	35.62	132.67	102.34	226.36
CR model	−0.33	72.08	16.41	141.20	57.30	250.94
B-CR model, $r_{min} = 0$	2.57	69.97	24.00	136.56	69.30	243.18
B-CR model, r_{min} free	9.45	67.07	31.45	134.89	75.99	239.22
ZLB forecast period						
Random walk	0.00	0.00	0.00	0.00	0.00	0.00
KW model	35.53	38.70	96.53	97.28	207.99	208.73
CR model	−0.97	30.17	7.08	43.93	46.82	69.37
B-CR model, $r_{min} = 0$	5.95	12.35	15.51	19.91	48.62	54.32
B-CR model, r_{min} free	11.98	15.79	22.87	26.47	61.37	65.82

Notes: Summary statistics of the forecast errors–mean and root-mean-squared errors (RMSEs)–of the target overnight federal funds rate six months, one year, and two years ahead. The forecasts are weekly. The top panel covers the full forecast period that starts on January 6, 1995, and runs until May 2, 2014, for the six-month forecasts (1,009 forecasts), until November 1, 2013, for the one-year forecasts (983 forecasts), and until November 2, 2012, for the two-year forecasts (931 forecasts). The middle panel coves the normal forecast period from January 6, 1995, to December 12, 2008, 728 forecasts The lower panel covers the zero lower bound forecast period that starts on December 19, 2008, and runs until May 2, 2014, for the six-month forecasts (281 forecasts), until November 1, 2013, for the one-year forecasts (255 forecasts), and until November 2, 2012, for the two-year forecasts (203 forecasts). All measurements are expressed in basis points.

4.3.2. Yield Forecasts

Now, we extend the analysis above and evaluate the models' yield forecast performance more broadly. Note that, due to the nonlinear yield function in the shadow-rate models, their yield forecasts are generated using Monte Carlo simulations. Also, we note that yield forecasts from the KW model are not available and therefore not included in the analysis.

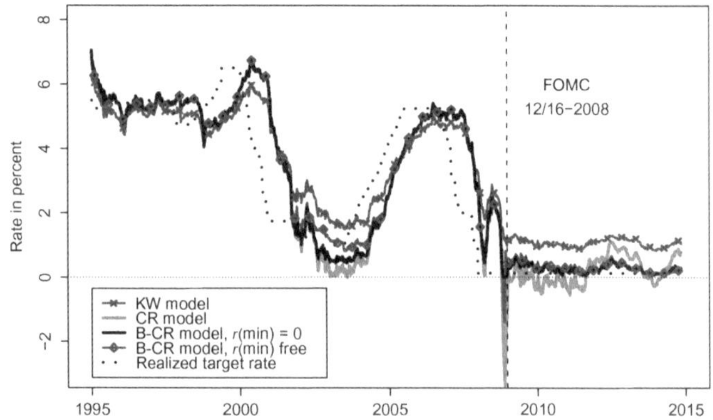

Fig. 13. Forecasts of the Target Overnight Federal Funds Rate. *Notes:* Forecasts of the target overnight federal funds rate one year ahead from the CR and B-CR models, where the latter is estimated with $r_{\min}$ both restricted to zero and left free. Also shown are the corresponding forecasts from the KW model. Subsequent realizations of the target overnight federal funds rate are included, so at date t, the figure shows forecasts as of time t and the realization from t plus one year. The forecast data are weekly observations from January 6, 1995, to October 31, 2014.

Table 5 reports the summary statistics of errors for real-time forecasts of three-month, two-year, five-year, and 10-year Treasury yields during the normal period from January 6, 1995, to December 12, 2008. First, we note that there is barely any difference in the shadow-rate models between restricting $r_{\min}$ to zero or leaving it free during the normal period. Second, in this period, the CR model is slightly worse at forecasting short- and medium-term yields than the B-CR model, but has a slight advantage at forecasting long-term yields. Finally, we note that the CR and B-CR models are competitive at forecasting yields of all maturities up to one year ahead relative to the random walk.

Table 6 reports the summary statistics of yield forecast errors during the recent ZLB period for the same four yield maturities considered in Table 5. First, because all three models have mean-reverting factor dynamics for the slope and curvature factors, they systematically underestimate how long yields would remain low in the aftermath of the financial crisis. This aspect of the data clearly benefits the random walk assumption. Second, the B-CR model dominates at forecasting short-term yields consistent with its ability to forecast future federal funds target rates reported in Table 4. On the other hand, the CR model continues to exhibit a strong performance at forecasting long-term yields. Finally, it is again the case that leaving $r_{\min}$ as a free parameter when

Table 5. Summary Statistics for Forecast Errors of U.S. Treasury Yields in the Normal Period.

	Six-Month Forecast		One-Year Forecast		Two-Year Forecast	
	Mean	RMSE	Mean	RMSE	Mean	RMSE
Three-month yield						
Random walk	−20.19	90.14	−39.28	157.11	−75.03	248.91
CR model	−29.27	89.51	−46.30	154.19	−86.86	254.06
B-CR model, $r_{\min} = 0$	−32.53	88.39	−53.73	151.32	−97.88	248.24
B-CR model, $r_{\min}$ free	−37.60	87.83	−59.48	150.69	−103.06	245.26
Two-year yield						
Random walk	−20.14	86.68	−36.74	132.02	−73.05	207.26
CR model	−19.69	86.87	−39.00	130.61	−82.26	203.61
B-CR model, $r_{\min} = 0$	−21.32	86.54	−42.01	130.03	−86.32	202.61
B-CR model, $r_{\min}$ free	−21.78	86.61	−42.41	130.04	−86.47	202.08
Five-year yield						
Random walk	−17.02	74.15	−29.19	98.38	−58.79	137.49
CR model	−24.38	73.83	−40.66	98.68	−76.62	140.18
B-CR model, $r_{\min} = 0$	−25.09	74.09	−41.93	99.08	−78.38	140.78
B-CR model, $r_{\min}$ free	−24.80	74.10	−41.48	98.98	−77.87	140.79
Ten-year yield						
Random walk	−12.76	59.13	−20.87	72.80	−42.42	84.95
CR model	−7.21	56.33	−18.34	69.85	−44.06	85.19
B-CR model, $r_{\min} = 0$	−7.81	56.52	−19.27	70.32	−45.40	86.24
B-CR model, $r_{\min}$ free	−7.68	56.52	−19.17	70.33	−45.52	86.34

Notes: Summary statistics of the forecast errors–mean and root-mean-square errors (RMSEs) – of the three-month, two-year, five-year, and 10-year U.S. Treasury yields six months, one year, and two years ahead. The forecasts are weekly during the normal period from January 6, 1995, to December 12, 2008, a total of 728 forecasts for all three forecast horizons. All measurements are expressed in basis points.

yields are near the ZLB implies notably poorer yield forecast performance at longer forecast horizons, except for the longest yield maturity.

4.4. Decomposing 10-Year Yields

One important use for affine DTSMs has been to separate longer-term yields into a short-rate expectations component and a term premium.

Table 6. Summary Statistics for Forecast Errors of U.S. Treasury Yields in the ZLB Period.

	Six-Month Forecast		One-Year Forecast		Two-Year Forecast	
	Mean	RMSE	Mean	RMSE	Mean	RMSE
Three-month yield						
Random walk	−1.34	6.45	−2.15	7.17	−4.63	7.36
CR model	−6.87	26.29	−19.02	42.16	−64.52	79.07
B-CR model, $r_{\min} = 0$	−16.98	19.68	−33.42	35.80	−74.04	78.55
B-CR model, $r_{\min}$ free	−21.04	23.78	−38.69	41.38	−85.16	89.05
Two-year yield						
Random walk	−3.45	23.04	−9.84	26.13	−22.60	41.34
CR model	−18.08	31.03	−47.88	54.65	−113.68	119.87
B-CR model, $r_{\min} = 0$	−22.12	35.08	−49.83	60.62	−110.02	123.05
B-CR model, $r_{\min}$ free	−25.01	37.33	−56.35	66.12	−122.44	133.21
Five-year yield						
Random walk	−2.98	55.31	−11.73	68.57	−38.34	103.11
CR model	−28.11	60.93	−62.11	90.52	−134.98	165.32
B-CR model, $r_{\min} = 0$	−27.41	63.31	−60.34	94.92	−132.49	169.35
B-CR model, $r_{\min}$ free	−31.41	65.58	−67.18	99.19	−142.71	175.32
Ten-year yield						
Random walk	−8.36	67.36	−19.44	88.00	−55.95	126.57
CR model	−21.59	74.46	−48.09	103.45	−115.63	167.99
B-CR model, $r_{\min} = 0$	−25.63	74.87	−52.74	106.28	−120.74	171.23
B-CR model, $r_{\min}$ free	−25.97	73.41	−55.67	105.40	−125.41	171.62

Notes: Summary statistics of the forecast errors–mean and root-mean-square errors (RMSEs)–of the three-month, two-year, five-year, and 10-year U.S. Treasury yields six months, one year, and two years ahead. The forecasts are weekly during the zero lower bound period that starts on December 19, 2008, and runs until May 2, 2014, for the six-month forecasts (281 forecasts), until November 1, 2013, for the one-year forecasts (255 forecasts), and until November 2, 2012, for the two-year forecasts (203 forecasts). All measurements are expressed in basis points.

Here, we document the different decompositions of the 10-year Treasury yield implied by the CR and B-CR models. To do so, we calculate, for each end date during our rolling re-estimation period, the average expected path for the overnight rate, $(1/\tau)\int_t^{t+\tau} E_t^P[r_s]\mathrm{d}s$, as well as the associated term premium – assuming the two components sum to the fitted bond yield, $\hat{y}_t(\tau)$.[28]

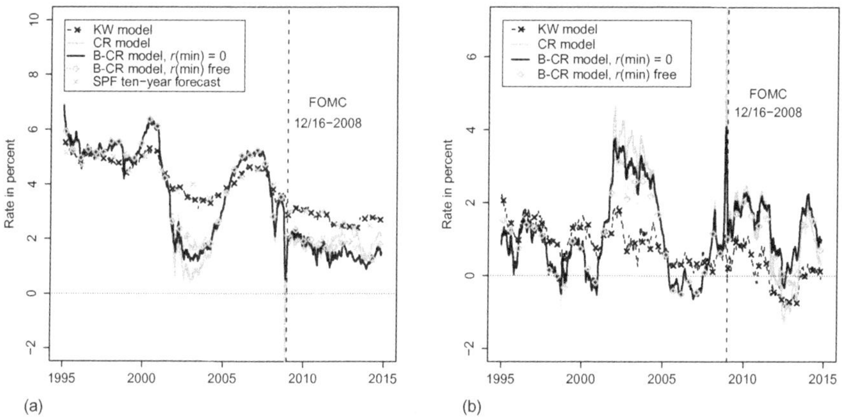

Fig. 14. Ten-Year Expected Short Rate and Term Premium. *Notes:* Panel (a) provides real-time estimates of the average policy rate expected over the next 10 years from the CR and B-CR models, where the latter is estimated with $r_{\min}$ both restricted to zero and left free. Also shown are the corresponding estimates form the KW model and the annual forecasts of the average three-month Treasury bill rate over the next 10 years from the SPF. Panel (b) shows the corresponding real-time estimated 10-year term premium series. The data cover the period from January 6, 1995, to October 31, 2014.

Fig. 14 shows the real-time decomposition of the 10-year Treasury yield into a policy expectations component and a term premium component according to the CR and B-CR models, again with a comparison to the corresponding estimates from the KW model. Studying the time-series patterns in greater detail, we first note the similar decompositions from the CR and B-CR model until December 2008 – with the notable exception of the 2002–2004 period when yields were low the last time. Second, we see some smaller discrepancies across these two model decompositions in the period between December 2008 and August 2011. Finally, we point out the sustained difference in the extracted policy expectations and term premiums in the period from August 2011 through 2012 when yields of all maturities reached historical low levels.

Also shown in Fig. 14(a) are long-term forecasts of average short rates from the Survey of Professional Forecasters (SPF) – specifically, the median of respondents' expectations for the average three-month Treasury bill rate over the next 10 years.[29] First, we note that the short-rate expectations from the survey are less variable and higher on average than those produced by the CR and B-CR models. Second, it is clear that the KW model's short-rate expectations track the survey expectations quite closely. This is not surprising since survey data, admittedly from a different source (the Blue Chip Financial Forecasts), are used as an input in its empirical implementation.

Finally, we note that Fig. 14 suggests that, at least through late 2011, the ZLB did not greatly affect the term premium decomposition of the CR model. To provide a concrete example of this, we repeat the analysis in CR of the Treasury yield response to eight key announcements by the Fed regarding its first large-scale asset purchase (LSAP) program. Table 7 shows the CR and B-CR model decompositions of the 10-year U.S. Treasury yield on these eight dates and the total changes.[30] The yield decompositions on these dates are quite similar for both of these models, though the B-CR model ascribes a bit more of the changes in yields to a signaling channel effect adjusting short-rate expectations. Hence, the conclusions of CR about the effects of the Fed's first LSAP program on U.S. Treasury yields are robust to the use of a shadow-rate model.

4.5. *Assessing Recent Shifts in Near-Term Monetary Policy Expectations*

In this section, we attempt to assess the extent to which the models are able to capture recent shifts in near-term monetary policy expectations.

To do so, we compare the variation in the models' one- and two-year short-rate forecasts since 2007 to the rates on one- and two-year federal funds futures contracts as shown in Fig. 15.[31] We note that the existence of time-varying risk premiums even in very short-term federal funds futures contracts is well documented (see Piazzesi & Swanson, 2008). However, the risk premiums in such short-term contracts are small relative to the sizeable variation over time observed in Fig. 15. As a consequence, we interpret the bulk of the variation from 2007 to 2009 as reflecting declines in short-rate expectations.

Furthermore, since August 2011, most evidence – including the low yield volatility shown in Fig. 12 – suggests that risk premiums have been significantly depressed, likely to a point that a zero-risk-premium assumption for the futures contracts discussed here is a satisfactory approximation. Combined these observations suggest that it is defensible for most of the shown eight-year period to map the models' short-rate projections to the rates on the federal funds futures contracts *without* adjusting for their risk premiums.

Table 7. Decomposition of Responses of 10-Year U.S. Treasury Yield.

Announcement Date	Model	Decomposition from Models			10-Year Treasury Yield
		Avg. target rate next 10 years	10-year term premium	Residual	
Nov. 25, 2008	CR	−20	0	−2	−21
	B-CR	−10	−10	0	
Dec. 1, 2008	CR	−10	−10	−2	−22
	B-CR	−21	2	−3	
Dec. 16, 2008	CR	−7	−7	−3	−17
	B-CR	−17	3	−3	
Jan. 28, 2009	CR	6	1	5	12
	B-CR	9	−2	5	
Mar. 18, 2009	CR	−14	−23	−15	−52
	B-CR	−17	−20	−14	
Aug. 12, 2009	CR	−1	1	6	6
	B-CR	−4	4	6	
Sep. 23, 2009	CR	−5	2	1	−2
	B-CR	−3	1	1	
Nov. 4, 2009	CR	−1	5	3	7
	B-CR	−1	5	3	
Total net change	CR	−53	−29	−7	−89
	B-CR	−65	−17	−7	

Notes: The decomposition of responses of the 10-year U.S. Treasury yield on eight LSAP announcement dates into changes in (i) the average expected target rate over the next 10 years, (ii) the 10-year term premium, and (iii) the unexplained residual based on the CR and B-CR models, where the latter is estimated with $r_{\min}$ restricted to zero. All changes are measured in basis points.

At the one- and two-year forecast horizons, the correlations between the short-rate forecasts from the models and the federal funds futures rates are all quite high. The KW model has the highest correlations, 97.7 and 95.1 percent, at the one- and two-year horizon, respectively, followed by the B-CR model with $r_{\min}$ restricted to zero, which has correlations of 97.0 and 90.5 percent at the one- and two-year horizon, respectively.[32] The CR model has the lowest correlations, 88.3 and 68.7 percent. If, instead, a distance metric is used, the performance across models is more varied. Table 8 reports the mean deviations and the root-mean-square deviations (RMSDs) from all four models relative to the rates of the federal funds futures contracts. For the CR model, the distance to the futures rates measured by RMSDs is 79.61 and 124.77 basis points at the one- and two-year horizon, respectively. The KW model also shows a

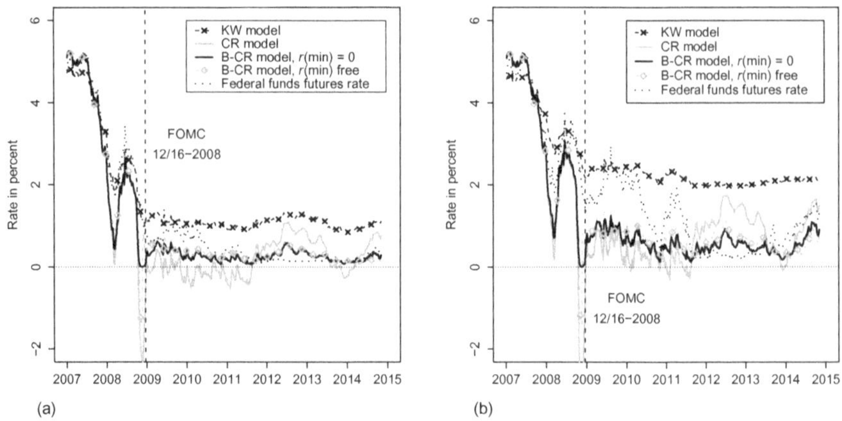

Fig. 15. Comparison of Short-Rate Projections. *Notes:* Panel (a) illustrates the one-year short-rate projections from the CR and B-CR models, where the latter is estimated with $r_{\min}$ both restricted to zero and left free. Also shown are the corresponding estimates form the KW model and the rates on one-year federal funds futures. Panel (b) shows the corresponding results for a two-year projection period with a comparison to the rates on two-year federal funds futures. The data are weekly covering the period from January 5, 2007, to October 31, 2014.

poor match with RMSDs of 66.38 and 110.12 basis points at the one- and two-year horizons, while the two shadow-rate models provide a much closer fit to the futures rates during the period under analysis. Thus, both measured by correlations and by a distance metric, the B-CR model's short-rate projections appear to be better aligned with the information reflected in rates on federal funds futures than the projections generated by the standard CR model, and this conclusion is not sensitive to the choice of lower bound.

5. CONCLUSION

In this paper, we study the performance of a standard Gaussian DTSM of U.S. Treasury yields and its equivalent shadow-rate version. This provides us with a clean read on the merits of casting a standard model as a shadow-rate model to respect the ZLB of nominal yields.

Table 8. Summary Statistics of Differences relative to Federal Funds Futures Rates.

Model	One-Year Contract		Two-Year Contract	
	Mean	RMSD	Mean	RMSD
KW model	49.54	66.38	86.84	110.12
CR model	−28.73	79.61	−46.99	124.77
B-CR, $r_{min} = 0$	−19.26	40.92	−47.11	76.17
B-CR, r_{min} free	−14.56	41.68	−39.83	75.37

Notes: The mean deviations and the root-mean-square deviations (RMSDs) between the short-rate expectations from four term structure models, on one side, and federal funds futures rates, on the other, are reported for two contract horizons. In each case, the summary statistics are calculated for the periods from January 5, 2007, to October 31, 2014. All numbers are measured in basis points.

We find that the standard model performed well until the end of 2008 but underperformed since then. In the current near-ZLB yield environment, we find that the shadow-rate model provides superior in-sample fit, matches the compression in yield volatility unlike the standard model, and delivers better real-time short-rate forecasts. Thus, while one could expect the regular model to get back on track as soon as short- and medium-term yields rise from their current low levels, our findings suggest that, in the meantime, shadow-rate models offer a tractable way of mitigating the problems related to the ZLB constraint on nominal yields. For a practical application of the shadow-rate model that takes advantage of its accurate forecasts near the ZLB, see Christensen, Lopez, and Rudebusch (2015).

However, allowing for a nonzero lower bound for the short rate determined by quasi maximum likelihood provides at best only modest gains in model performance at the cost of unrealistically large estimates of the lower bound before the financial crisis. Thus, we consider this added complexity unnecessary and strongly recommend setting the lower bound at zero for U.S. Treasury yields. Of course, as the recent experience of negative sovereign yields in Europe demonstrates, the lower bound is not in general always equal to zero. How to determine this possibly time-varying constraint on nominal yields remains an important topic for further research. In addition, differences between yield curve dynamics in normal and ZLB periods could reflect deeper nonlinearities in the factor structure, or maybe even a regime switch in the factor dynamics as argued in Christensen (2015), that are beyond the static affine dynamic structure assumed in this paper. This also remains an open question for future research.

NOTES

1. Diebold and Rudebusch (2013) and Krippner (2015) provide comprehensive discussions on the AFNS model with and without the ZLB.

2. Following Kim and Singleton (2012), the prefix "B-" refers to a shadow-rate model in the spirit of Black (1995).

3. Kim and Priebsch (2013) estimate the lower bound in their shadow-rate model, but do not make a comprehensive assessment of the implications of doing so.

4. Bauer and Rudebusch (2014) show that estimated shadow-rate paths are very sensitive to the choice of lower bound, which is consistent with our results.

5. Bauer and Rudebusch (2014) is a study that uses this approach.

6. Note that a Jensen's inequality term has been left out for the rollover strategy in this definition.

7. Difficulties in estimating Gaussian term structure models are discussed in Christensen et al. (2011), who propose using a Nelson–Siegel structure to avoid them. See Joslin, Singleton, and Zhu (2011), Hamilton and Wu (2012), and Andreasen and Christensen (2015) for alternative approaches to facilitate estimation of Gaussian DTSMs.

8. Two details regarding this specification are discussed in Christensen et al. (2011). First, with a unit root in the level factor under the Q-measure, the model is not arbitrage free with an unbounded horizon; therefore, as is often done in theoretical discussions, we impose an arbitrary maximum horizon. Second, we identify this class of models by normalizing the $\boldsymbol{\theta}^Q$ means under the Q-measure to zero without loss of generality.

9. As noted in Christensen et al. (2011), the unconstrained AFNS model has a sign restriction and three parameters less than the standard canonical three-factor Gaussian DTSM.

10. Due to the unit-root property of the first factor, we can arbitrarily fix its mean at $\theta_1^P = 0$.

11. The yield data include three- and six-month Treasury bill yields from the H.15 series from the Federal Reserve Board as well as off-the-run Treasury zero-coupon yields for the remaining maturities from the Gürkaynak, Sack, and Wright (2007) database, which is available at http://www.federalreserve.gov/pubs/feds/2006/200628/200628abs.html.

12. That is, a test of the joint hypothesis $\kappa_{12}^P = \kappa_{13}^P = \kappa_{31}^P = \kappa_{32}^P = 0$ using a standard likelihood ratio test. Note also that this test is done *before* imposing the unit-root property.

13. For example, Kim and Singleton (2012) and Bomfim (2003) use finite-difference methods to calculate bond prices, while Ichiue and Ueno (2007) employ interest rate lattices.

14. Wu and Xia (2014) derive a discrete-time version of the Krippner framework and implement a three-factor specification using U.S. Treasury data. In related research, Priebsch (2013) derives a second-order approximation to the Black (1995) shadow-rate model and estimates a three-factor version thereof, but it requires the calculation of a double integral in contrast to the single integral needed to fit the yield curve in the Krippner framework. Krippner (2015) provides a definitive treatment.

15. In particular, there is no explicit partial differential equation (PDE) that bond prices must satisfy, including boundary conditions, for the absence of arbitrage as in Kim and Singleton (2012).

16. For details of the derivations, see Christensen and Rudebusch (2015).

17. Due to the nonlinear measurement equation for the yields in the shadow-rate AFNS model, estimation is based on the standard extended Kalman filter as described in Christensen and Rudebusch (2015) and referred to as quasi maximum likelihood. We also estimated unrestricted and independent factor shadow-rate AFNS models and obtained similar results to those reported below.

18. That is, a test of the joint hypothesis $\kappa_{12}^{P} = \kappa_{13}^{P} = \kappa_{31}^{P} = \kappa_{32}^{P} = 0$ using a quasi likelihood ratio test. As with the CR model, this test is performed *before* imposing the unit-root property.

19. Consistent with our series for the 2003-period, Bomfim (2003) in his calibration of a two-factor shadow-rate model to U.S. interest rate swap data reports a probability of hitting the zero-boundary within the next two years equaling 3.6 percent as of January 2003. Thus, it appears that bond investors did not perceive the risk of reaching the ZLB during the 2003–2004 period of low interest rates to be material.

20. Note that for $r_{\min} \to -\infty$ it holds that $\underline{f}_t(\tau) \to f_t(\tau)$ for any t and all $\tau > 0$.

21. For Japan, Ichiue and Ueno (2013) impose a lower bound of nine basis points from January 2009 to December 2012 and reduce it to five basis points thereafter. For the United States, they use a lower bound of 14 basis points starting in November 2009. Finally, for the United Kingdom, they assume the standard ZLB for the short rate.

22. Christensen and Rudebusch (2015) report support for zero as a lower bound in Japanese government bond yields.

23. The conditional covariance matrix is calculated using the analytical solutions provided in Fisher and Gilles (1996).

24. Note that other measures of realized volatility have been used in the literature, such as the realized mean absolute deviation measure as well as fitted GARCH estimates. Collin-Dufresne et al. (2009) also use option-implied volatility as a measure of realized volatility.

25. In their analysis of Japanese government bond yields, Kim and Singleton (2012) also report a close match to yield volatilities for their Gaussian shadow-rate model.

26. Appendix B contains the formulas used to calculate short-rate projections.

27. The KW model is estimated using one-, two-, four-, seven-, and 10-year off-the-run Treasury zero-coupon yields from the Gürkaynak et al. (2007) database, as well as three- and six-month Treasury bill yields. To facilitate empirical implementation, model estimation includes monthly data on the six- and 12-month-ahead forecasts of the three-month T-bill yield from Blue Chip Financial Forecasts and semiannual data on the average expected three-month T-bill yield six to 11 years hence from the same source. For updated data provided by the staff of the Federal Reserve Board, see http://www.federalreserve.gov/econresdata/researchdata/feds200533.html.

28. The details of these calculations for both the CR and B-CR model are provided in appendices B and C.

29. The SPF is normally performed quarterly and only include questions about shorter-term expectations, but once a year respondents are asked about their long-term expectations. It is the median of the responses to this question that is shown in the figure. The data are available at http://www.philadelphiafed.org/research-and-data/real-time-center/survey-of-professional-forecasters/.

30. Due to the computational burden of estimating the B-CR model on this daily yield sample, we only perform this exercise for the B-CR model with r_{min} restricted to zero.

31. The futures data are from Bloomberg. The one-year futures rate is the weighted average of the rates on the 12- and 13-month federal funds futures contracts, while the two-year futures rate is the rate on the 24-month federal funds futures contract through 2010, and the weighted average of the rates on the 24- and 25-month contracts since then. The absence of data on the 24-month contracts prior to 2007 determines the start date for the analysis.

32. The B-CR model without restrictions on r_{min} has one- and two-year correlations of 96.4 and 89.0 percent, respectively.

33. Of course, away from the ZLB, with a negligible call option, the model will match the standard arbitrage-free term structure representation.

34. We calculate the conditional covariance matrix using the analytical solutions provided in Fisher and Gilles (1996).

ACKNOWLEDGMENTS

We thank two anonymous referees, Martin Møller Andreasen as well as conference participants at the FRBSF Workshop on "Term Structure Modeling at the Zero Lower Bound," the 20th International Conference on Computing in Economics and Finance, the First Annual Conference of the International Association for Applied Econometrics, and the Banque de France Workshop on "Term Structure Modeling and the Zero Lower Bound" – especially Don Kim and Jean-Paul Renne – for helpful comments. The views in this paper are solely the responsibility of the authors and should not be interpreted as reflecting the views of the Federal Reserve Bank of San Francisco or the Board of Governors of the Federal Reserve System. We thank Lauren Ford and Simon Riddell for excellent research assistance.

REFERENCES

Andersen, T. G., & Benzoni, L. (2010). Do bonds span volatility risk in the U.S. treasury market? A specification test for affine term structure models. *Journal of Finance*, *65*(2), 603–653.

Andreasen, M. M., & Christensen, B. J. (2015). The SR approach: A new estimation procedure for non-linear and non-gaussian dynamic term structure models. *Journal of Econometrics*, *184*(2), 420–451.
Andreasen, M. M., & Meldrum, A. (2014). *Dynamic term structure models: The best way to enforce the zero lower bound.* CREATES Research Paper No. 2014-47, Aarhus University, Department of Economics and Business.
Bauer, M. D., & Rudebusch, G. D. (2014). *Monetary policy expectations at the zero lower bound.* Federal Reserve Bank of San Francisco, Working Paper No. 2013-18.
Bauer, M. D., Rudebusch, G. D., & Wu, J. C. (2012). Correcting estimation bias in dynamic term structure models. *Journal of Business Economics and Statistics*, *30*(3), 454–467.
Black, F. (1995). Interest rates as options. *Journal of Finance*, *50*(7), 1371–1376.
Bomfim, A. N. (2003). Interest rates as options: 'Assessing the markets' view of the liquidity trap, Working Paper No. 2003-45. *Finance and Economics Discussion Series*, Federal Reserve Board, Washington, DC.
Christensen, J. H. E. (2015). *A regime-switching model of the yield curve at the zero bound.* Working Paper No. 2013-34, Federal Reserve Bank of San Francisco.
Christensen, J. H. E., Diebold, F. X., & Rudebusch, G. D. (2011). The affine arbitrage-free class of Nelson-Siegel term structure models. *Journal of Econometrics*, *164*(1), 4–20.
Christensen, J. H. E., Lopez, J. A., & Rudebusch, G D. (2015). A probability-based stress test of Federal Reserve assets and income. *Journal of Monetary Economics, first published online*, *8*, doi: 10.1016/j.jmoneco.2015.03.007.
Christensen, J. H. E., & Rudebusch, G. D. (2012). The response of interest rates to U.S. and U.K. quantitative easing. *Economic Journal*, *122*, 385–414.
Christensen, J. H. E., & Rudebusch, G. D. (2015). Estimating shadow-rate term structure models with near-zero yields. *Journal of Financial Econometrics*, *13*(2), 226–259.
Collin-Dufresne, P., Goldstein, R S., & Jones, C S. (2009). Can interest rate volatility be extracted from the cross-section of bond yields? *Journal of Financial Economics*, *94*(1), 47–66.
Diebold, F. X., & Rudebusch, G. D. (2013). *Yield curve modeling and forecasting: The dynamic Nelson-Siegel approach*, Princeton, NJ: Princeton University Press.
Duffee, G. R. (2002). Term premia and interest rate forecasts in affine models. *Journal of Finance*, *57*(1), 405–443.
Duffie, D., & Kan, R. (1996). A yield-factor model of interest rates. *Mathematical Finance*, *6*(4), 379–406.
Filipović, D., Larsson, M., & Trolle, A. (2014). Linear-rational term structure models. *Manuscript Swiss Finance Institute.*
Fisher, M., & Gilles, C. (1996). Term premia in exponential-affine models of the term structure. *Manuscript Board of Governors of the Federal Reserve System*
Gürkaynak, R. S., Sack, B., & Wright, J. H. (2007). The U.S. treasury yield curve: 1961 to the present. *Journal of Monetary Economics*, *54*(8), 2291–2304.
Hamilton, J. D., & Wu, J. C. (2012). Identification and estimation of Gaussian affine term structure models. *Journal of Econometrics*, *168*(2), 315–331.
Ichiue, H., & Ueno, Y. (2007). *Equilibrium interest rates and the yield curve in a low interest rate environment*, Working Paper No. 2007-E-18, Bank of Japan.
Ichiue, H., & Ueno, Y. (2013). *Estimating term premia at the zero bound: An analysis of Japanese, US, and UK Yields*, Working Paper No. 2013-E-8, Bank of Japan.

Jacobs, K., & Karoui, L. (2009). Conditional volatility in affine term structure models: Evidence from treasury and swap markets. *Journal of Financial Economics*, *91*(3), 288–318.

Joslin, S., Singleton, K J., & Zhu, H. (2011). A new perspective on Gaussian dynamic term structure models. *Review of Financial Studies*, *24*(3), 926–970.

Kim, D. H., & Priebsch, M. (2013). *Estimation of multi-factor shadow-rate term structure models*, Washington, DC: Federal Reserve Board.

Kim, D. H., & Singleton, K. J. (2012). Term structure models and the zero bound: An empirical investigation of Japanese yields. *Journal of Econometrics*, *170*(1), 32–49.

Kim, D. H., & Wright, J H. (2005). *An arbitrage-free three-factor term structure model and the recent behavior of long-term yields and distant-horizon forward rates*. Working Finance and Economics Discussion Series 2055-33, Board of Governors of the Federal Reserve System.

Krippner, L. (2013). *A tractable framework for zero lower bound Gaussian term structure models*. Discussion Paper No. 2013-02, Reserve Bank of New Zealand.

Krippner, L. (2015). *Term structure modeling at the zero lower bound: A practitioner's guide*, New York, NY: Palgrave-Macmillan.

Monfort, A., Pegoraro, F., Renne, J-P., & Roussellet, G. (2014). Staying at zero with affine processes: A new dynamic term structure model. *Manuscript Banque de France*.

Nelson, C. R., & Siegel, A. F. (1987). Parsimonious modeling of yield curves. *Journal of Business*, *60*(4), 473–489.

Piazzesi, M., & Swanson, E. T. (2008). Futures Prices as Risk-Adjusted Forecasts of Monetary Policy. *Journal of Monetary Economics*, *55*(4), 677–691.

Priebsch, M. (2013). *Computing arbitrage-free yields in multi-factor gaussian shadow-rate term structure models*. Finance and Economics Discussion Series Working Paper No. 2013-63, Board of Governors of the Federal Reserve System.

Wu, C ., & Xia, F D. (2014). Measuring the macroeconomic impact of monetary policy at the zero lower bound. *Manuscript, University of California at San Diego*.

APPENDIX A: HOW GOOD IS THE OPTION-BASED APPROXIMATION?

As noted in Section 3.1, Krippner (2013) does not provide a formal derivation of arbitrage-free pricing relationships for the option-based approach. Therefore, in this appendix, we analyze how closely the option-based bond pricing from the estimated B-CR model matches an arbitrage-free bond pricing that is obtained from the same model using Black's (1995) approach based on Monte Carlo simulations. The simulation-based shadow yield curve is obtained from 50,000 10-year long factor paths generated using the estimated Q-dynamics of the state variables in the B-CR model, which, ignoring the nonnegativity equation (5), are used to construct 50,000 paths for the shadow short rate. These are converted into a corresponding number of shadow discount bond paths and averaged for each maturity before the resulting shadow discount bond prices are converted into yields. The simulation-based yield curve is obtained from the same underlying 50,000 Monte Carlo factor paths, but at each point in time in the simulation, the resulting short rate is constrained by the nonnegativity equation (5) as in Black (1995). The shadow-rate curve from the B-CR model can also be calculated analytically via the usual affine pricing relationships, which ignore the ZLB. Thus, any difference between these two curves is simply numerical error that reflects the finite number of simulations.

To document that the close match between the option-based and the simulation-based yield curves is not limited to any specific date where the ZLB of nominal yields is likely to have mattered, we undertake this simulation exercise for the last observation date in each year since 2006.[33] Table A1 reports the resulting shadow yield curve differences and yield curve differences for various maturities on these nine dates. Note that the errors for the shadow yield curves solely reflect simulation error as the model-implied shadow yield curve is identical to the analytical arbitrage-free curve that would prevail without currency in circulation. These simulation errors in Table A1 are typically very small in absolute value, and they increase only slowly with maturity. Their average absolute value – shown in the bottom row – is less than one basis point even at a 10-year maturity. This implies that using simulations with a large number of draws (N = 50,000) arguably delivers enough accuracy for the type of inference we want to make here.

Given this calibration of the size of the numerical errors involved in the simulation, we can now assess the more interesting size of the approximation

Table A1. Approximation Errors in Yields for Shadow-Rate Model.

Dates		Maturity in Months				
		12	36	60	84	120
12/29/06	Shadow yields	0.30	−0.73	−1.22	−1.08	−1.10
	Yields	0.33	−0.71	−1.13	−0.87	−0.52
12/28/07	Shadow yields	0.17	0.02	0.65	0.86	0.79
	Yields	0.22	0.16	0.85	1.18	1.28
12/26/08	Shadow yields	−0.01	0.51	0.56	0.45	0.28
	Yields	0.09	0.76	1.51	1.93	2.26
12/31/09	Shadow yields	0.10	0.63	1.18	1.28	1.08
	Yields	0.04	0.74	1.46	1.69	1.68
12/31/10	Shadow yields	−0.21	−0.11	−0.10	−0.16	0.00
	Yields	−0.10	0.45	0.86	1.03	1.39
12/30/11	Shadow yields	0.19	0.88	1.11	1.51	1.92
	Yields	−0.02	0.68	1.80	3.11	4.56
12/28/12	Shadow yields	0.11	−0.35	−0.48	−0.26	−0.09
	Yields	0.13	0.35	1.28	2.48	3.79
12/27/13	Shadow yields	−0.17	−0.17	0.25	0.25	0.14
	Yields	0.06	0.70	1.60	2.00	2.20
10/31/14	Shadow yields	0.30	0.22	−0.10	−0.25	−0.13
	Yields	0.21	0.66	0.84	1.13	1.92
Average	Shadow yields	0.16	0.46	0.76	0.80	0.75
abs. diff	Yields	0.13	0.55	1.27	1.76	2.21

Notes: At each date, the table reports differences between the analytical shadow yield curve obtained from the option-based estimates of the B-CR model and the shadow yield curve obtained from 50,000 simulations of the estimated factor dynamics under the Q-measure in that model. The table also reports for each date the corresponding differences between the fitted yield curve obtained from the B-CR model and the yield curve obtained via simulation of the estimated B-CR model with imposition of the ZLB. The bottom two rows give averages of the absolute differences across the 9 dates. All numbers are measured in basis points.

error in the option-based approach to valuing yields in the presence of the ZLB. In Table A1, the errors of the fitted B-CR model yield curves relative to the simulated results are only slightly larger than those reported for the shadow yield curve. In particular, for maturities up to five years, the errors tend to be less than 1 basis point, so the option-based approximation error adds very little if anything to the numerical simulation error. At the 10-year maturity, the approximation errors are understandably larger, but even the largest errors at the 10-year maturity do not exceed 4 basis points in absolute value and the average absolute value is less than 2 basis points. Overall, the option-based approximation errors in our three-factor setting appear relatively small. Indeed, they are smaller than the fitted errors to be reported in Table 3. That is, for the B-CR model analyzed here, the gain from using a numerical estimation approach instead of the option-based approximation would in all likelihood be negligible.

APPENDIX B: FORMULA FOR POLICY EXPECTATIONS IN AFNS AND B-AFNS MODELS

In this appendix, we detail how conditional expectations for future policy rates are calculated within AFNS and B-AFNS models.

In affine models, in general, the conditional expected value of the state variables is calculated as

$$E_t^P[X_{t+\tau}] = \left(I - \exp\left(-K^P\tau\right)\right)\theta^P + \exp\left(-K^P\tau\right)X_t$$

In AFNS models, the instantaneous short rate is defined as

$$r_t = L_t + S_t$$

Thus, the conditional expectation of the short rate is

$$E_t^P[r_{t+\tau}] = E_t^P[L_{t+\tau} + S_{t+\tau}] = \begin{pmatrix} 1 & 1 & 0 \end{pmatrix} E_t^P[X_{t+\tau}]$$

In B-AFNS models, the instantaneous shadow rate is defined as

$$s_t = L_t + S_t$$

In turn, the conditional expectation of the shadow-rate process is

$$E_t^P[s_{t+\tau}] = E_t^P[L_{t+\tau} + S_{t+\tau}] = \begin{pmatrix} 1 & 1 & 0 \end{pmatrix} E_t^P[X_{t+\tau}]$$

Now, the conditional covariance matrix of the state variables is given by[34]

$$V_t^P[X_{t+\tau}] = \int_0^\tau \mathrm{e}^{-K^P s}\Sigma\Sigma' \mathrm{e}^{-(K^P)'s}\mathrm{d}s$$

Hence, the conditional covariance of the shadow-rate process is

$$V_t^P[s_{t+\tau}] = \begin{pmatrix} 1 & 1 & 0 \end{pmatrix} V_t^P[X_{t+\tau}] \begin{pmatrix} 1 \\ 1 \\ 0 \end{pmatrix}$$

Finally, following equation (65) in Kim and Singleton (2012), the conditional expectation of the short rate in the B-AFNS models,

$$r_t = \max(r_{\min}, s_t)$$

is given by

$$\begin{aligned} E^P[r_{t+\tau}] &= \int_{-\infty}^{\infty} r_{t+\tau} f(r_{t+\tau}|X_t)\mathrm{d}r_{t+\tau} = r_{\min} + \int_{r_{\min}}^{\infty} (s_{t+\tau} - r_{\min}) f(s_{t+\tau}|X_t)\mathrm{d}s_{t+\tau} \\ &= r_{\min} + \left(E_t^P[s_{t+\tau}] - r_{\min}\right) N\left(\frac{\left(E_t^P[s_{t+\tau}] - r_{\min}\right)}{\sqrt{V_t^P[s_{t+\tau}]}}\right) \\ &\quad + \frac{1}{\sqrt{2\pi}}\sqrt{V_t^P[s_{t+\tau}]} \;\; \exp\left(-\frac{1}{2}\frac{\left(E_t^P[s_{t+\tau}] - r_{\min}\right)^2}{V_t^P[s_{t+\tau}]}\right) \end{aligned}$$

APPENDIX C: ANALYTICAL FORMULAS FOR AVERAGES OF POLICY EXPECTATIONS AND FOR TERM PREMIUMS IN THE CR MODEL

In this appendix, we derive the analytical formulas for averages of policy expectations and for term premiums within the CR model.

For a start, the term premium is defined as

$$TP_t(\tau) = y_t(\tau) - \frac{1}{\tau}\int_t^{t+\tau} E_t^P[r_s]\mathrm{d}s$$

In the CR model, as in any AFNS model, the instantaneous short rate is defined as

$$r_t = L_t + S_t$$

while the specification of the P-dynamics is given by

$$\begin{pmatrix} dL_t \\ dS_t \\ dC_t \end{pmatrix} = \begin{pmatrix} 10^{-7} & 0 & 0 \\ \kappa_{21}^P & \kappa_{22}^P & \kappa_{23}^P \\ 0 & 0 & \kappa_{33}^P \end{pmatrix} \left[\begin{pmatrix} 0 \\ \theta_2^P \\ \theta_3^P \end{pmatrix} - \begin{pmatrix} L_t \\ S_t \\ C_t \end{pmatrix} \right] dt + \begin{pmatrix} \sigma_{11} & 0 & 0 \\ 0 & \sigma_{22} & 0 \\ 0 & 0 & \sigma_{33} \end{pmatrix} \begin{pmatrix} dW_t^{L,P} \\ dW_t^{S,P} \\ dW_t^{C,P} \end{pmatrix}$$

Thus, the mean-reversion matrix is given by

$$K^P = \begin{pmatrix} 10^{-7} & 0 & 0 \\ \kappa_{21}^P & \kappa_{22}^P & \kappa_{23}^P \\ 0 & 0 & \kappa_{33}^P \end{pmatrix}$$

Its matrix exponential can be calculated analytically:

$$\exp(-K^P\tau) = \begin{pmatrix} 1 & 0 & 0 \\ -\kappa_{21}^P \dfrac{1-e^{-\kappa_{22}^P\tau}}{\kappa_{22}^P} & e^{-\kappa_{22}^P\tau} & -\kappa_{23}^P \dfrac{e^{-\kappa_{33}^P\tau}-e^{-\kappa_{22}^P\tau}}{\kappa_{22}^P-\kappa_{33}^P} \\ 0 & 0 & e^{-\kappa_{33}^P\tau} \end{pmatrix}$$

Now, the conditional mean of the state variables is

$$\begin{aligned} E_t^P[X_{t+\tau}] &= \theta^P + \begin{pmatrix} 1 & 0 & 0 \\ -\kappa_{21}^P \dfrac{1-e^{-\kappa_{22}^P\tau}}{\kappa_{22}^P} & e^{-\kappa_{22}^P\tau} & -\kappa_{23}^P \dfrac{e^{-\kappa_{33}^P\tau}-e^{-\kappa_{22}^P\tau}}{\kappa_{22}^P-\kappa_{33}^P} \\ 0 & 0 & e^{-\kappa_{33}^P\tau} \end{pmatrix} \begin{pmatrix} L_t \\ S_t-\theta_2^P \\ C_t-\theta_3^P \end{pmatrix} \\ &= \begin{pmatrix} L_t \\ \theta_2^P - \kappa_{21}^P \dfrac{1-e^{-\kappa_{22}^P\tau}}{\kappa_{22}^P} L_t + e^{-\kappa_{22}^P\tau}(S_t-\theta_2^P) - \kappa_{23}^P \dfrac{e^{-\kappa_{33}^P\tau}-e^{-\kappa_{22}^P\tau}}{\kappa_{22}^P-\kappa_{33}^P}(C_t-\theta_3^P) \\ \theta_3^P + e^{-\kappa_{33}^P\tau}(C_t-\theta_3^P) \end{pmatrix} \end{aligned}$$

In order to get back to the term premium formula, we note that the conditional expectation of the instantaneous short-rate process is given by

$$\begin{aligned}
E_t^P[r_s] &= E_t^P[L_s + S_s] \\
&= \left(1 - \kappa_{21}^P \frac{1 - e^{-\kappa_{22}^P(s-t)}}{\kappa_{22}^P}\right) L_t + \theta_2^P + e^{-\kappa_{22}^P(s-t)}\left(S_t - \theta_2^P\right) \\
&\quad - \kappa_{23}^P \frac{e^{-\kappa_{33}^P(s-t)} - e^{-\kappa_{22}^P(s-t)}}{\kappa_{22}^P - \kappa_{33}^P}\left(C_t - \theta_3^P\right)
\end{aligned}$$

Next, we integrate from t to $t+\tau$:

$$\begin{aligned}
\int_t^{t+\tau} E_t^P[r_s]\mathrm{d}s &= \int_t^{t+\tau} \left(\left[1 - \kappa_{21}^P \frac{1 - e^{-\kappa_{22}^P(s-t)}}{\kappa_{22}^P}\right] L_t + \theta_2^P + e^{-\kappa_{22}^P(s-t)}\left(S_t - \theta_2^P\right) \right. \\
&\quad \left. - \kappa_{23}^P \frac{e^{-\kappa_{33}^P(s-t)} - e^{-\kappa_{22}^P(s-t)}}{\kappa_{22}^P - \kappa_{33}^P}\left(C_t - \theta_3^P\right) \right) \mathrm{d}s \\
&= \theta_2^P \tau + \left(1 - \frac{\kappa_{21}^P}{\kappa_{22}^P}\right) \tau L_t + \frac{\kappa_{21}^P}{\kappa_{22}^P} L_t \int_t^{t+\tau} e^{-\kappa_{22}^P(s-t)}\mathrm{d}s \\
&\quad + \left(S_t - \theta_2^P\right) \int_t^{t+\tau} e^{-\kappa_{22}^P(s-t)}\mathrm{d}s \\
&\quad - \frac{\kappa_{23}^P}{\kappa_{22}^P - \kappa_{33}^P}\left(C_t - \theta_3^P\right) \int_t^{t+\tau} \left(e^{-\kappa_{33}^P(s-t)} - e^{-\kappa_{22}^P(s-t)}\right)\mathrm{d}s \\
&= \theta_2^P \tau + \left(1 - \frac{\kappa_{21}^P}{\kappa_{22}^P}\right) \tau L_t - \frac{\kappa_{21}^P}{\kappa_{22}^P} L_t \left[\frac{1}{\kappa_{22}^P} e^{-\kappa_{22}^P(s-t)}\right]_t^{t+\tau} \\
&\quad + \left(S_t - \theta_2^P\right) \left[\frac{-1}{\kappa_{22}^P} e^{-\kappa_{22}^P(s-t)}\right]_t^{t+\tau} \\
&\quad - \frac{\kappa_{23}^P}{\kappa_{22}^P - \kappa_{33}^P}\left(C_t - \theta_3^P\right) \left[\frac{-1}{\kappa_{33}^P} e^{-\kappa_{33}^P(s-t)} + \frac{1}{\kappa_{22}^P} e^{-\kappa_{22}^P(s-t)}\right]_t^{t+\tau} \\
&= \theta_2^P \tau + \left(1 - \frac{\kappa_{21}^P}{\kappa_{22}^P}\right) \tau L_t + \frac{\kappa_{21}^P}{\kappa_{22}^P} \frac{1 - e^{-\kappa_{22}^P \tau}}{\kappa_{22}^P} L_t + \frac{1}{\kappa_{22}^P}\left(S_t - \theta_2^P\right)\left(1 - e^{-\kappa_{22}^P \tau}\right) \\
&\quad - \frac{\kappa_{23}^P}{\kappa_{22}^P - \kappa_{33}^P}\left(C_t - \theta_3^P\right) \left(\frac{1}{\kappa_{33}^P}\left[1 - e^{-\kappa_{33}^P \tau}\right] - \frac{1}{\kappa_{22}^P}\left[1 - e^{-\kappa_{22}^P \tau}\right]\right)
\end{aligned}$$

The relevant term to go into the term premium formula is

$$\frac{1}{\tau}\int_t^{t+\tau} E_t^P[r_s]\mathrm{d}s = \theta_2^P + \left(1 - \frac{\kappa_{21}^P}{\kappa_{22}^P}\right) L_t + \frac{\kappa_{21}^P}{\kappa_{22}^P} \frac{1 - \mathrm{e}^{-\kappa_{22}^P \tau}}{\kappa_{22}^P \tau} L_t + \frac{1 - \mathrm{e}^{-\kappa_{22}^P \tau}}{\kappa_{22}^P \tau}(S_t - \theta_2^P)$$
$$- \frac{\kappa_{23}^P}{\kappa_{22}^P - \kappa_{33}^P}\left(\frac{1 - \mathrm{e}^{-\kappa_{33}^P \tau}}{\kappa_{33}^P \tau} - \frac{1 - \mathrm{e}^{-\kappa_{22}^P \tau}}{\kappa_{22}^P \tau}\right)(C_t - \theta_3^P)$$

The final expression for the term premium is then given by

$$\begin{aligned}
TP_t(\tau) &= y_t(\tau) - \frac{1}{\tau}\int_t^{t+\tau} E_t^P[r_s]\mathrm{d}s \\
&= L_t + \frac{1-\mathrm{e}^{-\lambda\tau}}{\lambda\tau} S_t + \left(\frac{1-\mathrm{e}^{-\lambda\tau}}{\lambda\tau} - \mathrm{e}^{-\lambda\tau}\right) C_t - \frac{A(\tau)}{\tau} \\
&\quad - \theta_2^P - \left(1 - \frac{\kappa_{21}^P}{\kappa_{22}^P}\right) L_t - \frac{\kappa_{21}^P}{\kappa_{22}^P}\frac{1-\mathrm{e}^{-\kappa_{22}^P\tau}}{\kappa_{22}^P\tau} L_t - \frac{1-\mathrm{e}^{-\kappa_{22}^P\tau}}{\kappa_{22}^P\tau}(S_t - \theta_2^P) \\
&\quad + \frac{\kappa_{23}^P}{\kappa_{22}^P - \kappa_{33}^P}\left(\frac{1-\mathrm{e}^{-\kappa_{33}^P\tau}}{\kappa_{33}^P\tau} - \frac{1-\mathrm{e}^{-\kappa_{22}^P\tau}}{\kappa_{22}^P\tau}\right)(C_t - \theta_3^P) \\
&= \frac{\kappa_{21}^P}{\kappa_{22}^P}\left(1 - \frac{1-\mathrm{e}^{-\kappa_{22}^P\tau}}{\kappa_{22}^P\tau}\right) L_t + \left(\frac{1-\mathrm{e}^{-\lambda\tau}}{\lambda\tau} - \frac{1-\mathrm{e}^{-\kappa_{22}^P\tau}}{\kappa_{22}^P\tau}\right) S_t \\
&\quad + \left(\frac{1-\mathrm{e}^{-\lambda\tau}}{\lambda\tau} - \mathrm{e}^{-\lambda\tau} + \frac{\kappa_{23}^P}{\kappa_{22}^P - \kappa_{33}^P}\left[\frac{1-\mathrm{e}^{-\kappa_{33}^P\tau}}{\kappa_{33}^P\tau} - \frac{1-\mathrm{e}^{-\kappa_{22}^P\tau}}{\kappa_{22}^P\tau}\right]\right) C_t \\
&\quad - \left(1 - \frac{1-\mathrm{e}^{-\kappa_{22}^P\tau}}{\kappa_{22}^P\tau}\right)\theta_2^P - \frac{\kappa_{23}^P}{\kappa_{22}^P - \kappa_{33}^P}\left(\frac{1-\mathrm{e}^{-\kappa_{33}^P\tau}}{\kappa_{33}^P\tau} - \frac{1-\mathrm{e}^{-\kappa_{22}^P\tau}}{\kappa_{22}^P\tau}\right)\theta_3^P - \frac{A(\tau)}{\tau}
\end{aligned}$$

In the B-CR model, $\frac{1}{\tau}\int_t^{t+\tau} E_t^P[r_s]\mathrm{d}s$ is not available in analytical form, instead it has to be approximated by numerically integrating the formula for $E_t^P[r_s]$ provided in Appendix B.

DYNAMIC FACTOR MODELS FOR THE VOLATILITY SURFACE☆

Michel van der Wel, Sait R. Ozturk and Dick van Dijk

Erasmus University Rotterdam, Erasmus School of Economics, Rotterdam, The Netherlands

ABSTRACT

The implied volatility surface is the collection of volatilities implied by option contracts for different strike prices and time-to-maturity. We study factor models to capture the dynamics of this three-dimensional implied volatility surface. Three model types are considered to examine desirable features for representing the surface and its dynamics: a general dynamic factor model, restricted factor models designed to capture the key features of the surface along the moneyness and maturity dimensions, and in-between spline-based methods. Key findings are that: (i) the restricted and spline-based models are both rejected against the

☆Van der Wel is from Erasmus University Rotterdam, CREATES, Tinbergen Institute and ERIM. Ozturk is from Erasmus University Rotterdam and the Tinbergen Institute. Van Dijk is from Erasmus University Rotterdam, Tinbergen Institute and ERIM.

Dynamic Factor Models
Advances in Econometrics, Volume 35, 127–174

ISSN: 0731-9053/doi:10.1108/S0731-905320150000035004

general dynamic factor model, (ii) the factors driving the surface are highly persistent, and (iii) for the restricted models option Δ is preferred over the more often used strike relative to spot price as measure for moneyness.

Keywords: Dynamic factor models; implied volatility surface; Kalman filter; maximum likelihood

JEL classifications: C32; C58; G13

1. INTRODUCTION

The value and pay-off of an option depend on the price of the underlying asset relative to the strike price (also called the moneyness) and the remaining time-to-maturity (or simply maturity). The maturity and strike are given in the option contract and for a given underlying asset typically a range of options with different maturities and strikes can be traded in financial markets. Because the prices of these different option contracts with the same underlying are difficult to interpret and compare, option prices often are converted into implied volatility. The implied volatility is obtained by backing out the volatility in such a way that the observed market price of the option contract matches the price implied by a certain pricing model, usually the Black and Scholes (1973) model or a binomial tree (introduced by Cox, Ross, & Rubinstein, 1979). The collection of implied volatilities across both the maturity and moneyness dimensions is referred to as the implied volatility surface.[1] An extant literature, pioneered by Rubinstein (1994), shows that at any given point of time there are typical and common patterns for the implied volatility across the strike price (or moneyness more generally) and maturity dimensions. The pattern for given maturity across different strikes is often referred to as the volatility smile (because of its U-shape). The pattern for given moneyness across different maturities is referred to as the volatility term structure.

Since the pay-offs of option contracts for different strikes and maturities ultimately depend on the same underlying, a strong comovement in the different option prices is expected. As the implied volatility is a transformation of the prices, this feature carries over to the implied volatility surface. It is a natural idea to represent the comovement of different parts of the volatility surface in terms of common factors. However, there is no clear guidance in the literature on what type of factor model to use for this purpose. On one side of the spectrum, Fengler, Härdle, and Mammen (2007) suggest a flexible semiparametric factor model. On the other side of the spectrum, Christoffersen, Fournier, and Jacobs (2003) suggest a restricted factor model that decomposes the volatility surface into three factors representing the implied volatility level, smile, and term structure.

The contribution of this paper is to compare different factor model specifications for the implied volatility surface and examine desirable and undesirable features of such models. We use three different setups. First, we consider a general dynamic factor model (DFM) where only identification restrictions are imposed. Second, we examine restricted model specifications, where the factors are forced to represent comovement in the implied volatilities along the moneyness and maturity dimensions. Such a setup is commonly used in the economics and finance literature, see, for example, Dumas, Fleming, and Whaley (1998), Christoffersen et al. (2013), and Christoffersen, Goyenko, Jacobs, and Karoui (2012). Third, we propose spline-based models that offer a flexible approach to capture the shape of the implied volatility surface. For this third setup, we follow the smooth dynamic factor modeling approach of Jungbacker, Koopman, and van der Wel (2014). As in the second setup, the factor loadings are structured in such a way that the corresponding factors represent smile and term structure effects, but the restrictions imposed on the loadings are much less strict. The spline-based models can be seen as a likelihood-based alternative of the semiparametric implementation using basis functions, as in Fengler et al. (2007) and Park, Mammen, Härdle, and Borak (2009). All the different DFM specifications are estimated with maximum likelihood adopting the framework of Jungbacker and Koopman (2015). Since both the restricted economic models and spline-based models are nested in the general DFM, we use likelihood-ratio tests (LR-tests) to compare the models, besides comparing models based on information criteria.

We examine the merits of the different DFM setups in an empirical setting, using daily implied volatilities for European options on the S&P500 index from 1999 through 2013. This is one of the most actively traded derivative securities, with contracts being available for a wide range of strike prices and maturities. We construct the implied volatility surface using six different groups of moneyness (measured by the Δ of the option[2]) and four different groups of time-to-maturity. On each day, we find a contract nearest to the midpoint of each of the 24 moneyness–maturity pairs. We consider the balanced panel of the 24 daily selected contracts to capture the implied volatility surface.

Our results provide three key implications. First, the economic and spline-based models are both rejected against the general DFM; although the spline-based models perform much better than the restricted economic models. In all three model specifications, we find that the level of the volatility surface is the most important factor. The second and the third factors differ, however, and the preference for the general DFM suggests that the remaining comovement in the volatility surface does not correspond with the (economically plausible) smile and term structure effects as imposed by the restricted models. Second, for all estimated models the factors driving the surface are highly persistent. Third, to capture moneyness in the restricted economic models, the option Δ performs better than the strike price relative to the spot price. This is an important implication, as many of the existing models use the strike relative to the spot price rather than Δ. An explanation for this finding is that, unlike Δ, the strike relative to the spot price does not take into account that the likelihood of the option being in-the-money at expiration depends on the (current) volatility of the stock and remaining time-to-maturity, as Bollen and Whaley (2004) point out.

We consider four extensions to our baseline model setup. First, an alternative surface construction strategy is used based on moneyness measured by strike relative to spot price. Even when the data are constructed based on this measure, the models favor using Δ for moneyness. Second, while our main analysis is based on three-factor models we also examine higher-dimensional restricted models. We consider all models of Dumas et al. (1998), which have up to six factors. Only two of the six-factor models provide a likelihood that is better to the general three-factor DFM. Third, we consider alternatives for the factor dynamics and report similar results when random walks are taken for the factors. Fourth, the importance of the crisis period is examined by taking a log-transformation of the data and considering a sub-sample that omits the crisis period. While some factors are less persistent, the factor structure is also strong outside of the crisis period and when using logs.

We contribute to two strands of literature. First, our approach contributes to the dynamic factor modeling literature by studying a surface. Many applications of factor models are "two-dimensional" in nature and can be categorized as to whether the data can be logically structured in a particular way. Popular data sets as provided by Stock and Watson (2002) and related studies are a collection of macroeconomic variables where no immediate logical or natural ordering of the variables exists. However, in other cases such as Treasury yields, as in Jungbacker et al. (2014), a natural ordering exists as the three-month bond logically comes before the six-month bond. Intermediate data sets are somewhat organized, such as housing prices as studied in Mönch and Ng (2011). We offer an approach to deal with organized three-dimensional data, by stacking the surface to get back to the case of two-dimensional data but benefiting from the resulting block-structure in the factor models.

Second, we contribute to the literature on the modeling of volatility surface. To the best of our knowledge, we are the first to model the surface using a likelihood-based general dynamic factor approach. Alternative statistical implementations of factor models for the surface include Skiadopoulos, Hodges, and Clewlow (1999), Cont and da Fonseca (2002), and Fengler, Härdle, and Villa (2003). The restricted setting based on Dumas et al. (1998) has been extended by Gonçalves and Guidolin (2006) to include factor dynamics. The analysis is however done in a two-step framework, where the factors are first obtained by OLS and then modeled using vector-autoregressions. We provide an efficient approach to estimate the factors and their dynamics in one step. The closest paper to the spline-based setup is Bedendo and Hodges (2009), who decompose the volatility smile using cubic polynomials and treat the knot values of these polynomials as factors. This can be seen as a restricted version of our approach and as the polynomials are based on moneyness it is fairly close to the restricted economic setting. Moreover, we provide a unified framework for the surface, whereas the surface extension in their work is a separate treatment of the smile for different maturity groups.

The rest of the paper is organized as follows. Section 2 describes the data used in the analysis, summary statistics, and a preliminary analysis of the data based on principal component analysis. Section 3 details the three different setups of the DFM. Section 4 provides the main estimation results. Section 5 discusses the outcomes of various robustness checks and extensions of the main modeling approach. Section 6 concludes.

2. VOLATILITY SURFACE DATA

This section describes how we construct the implied volatility surface and provides a preliminary investigation of its characteristics. Knowledge of the key empirical features of the volatility surface provides useful input for the specification of the factor models as discussed in the next section. Section 2.1 discusses the data construction. Section 2.2 provides summary statistics and some insights into the possible usefulness of factor models based on principal component analysis.

2.1. Constructing the Volatility Surface

We use a daily data set of European options on the S&P500 index traded on the Chicago Board Options Exchange (CBOE). This is one of the most actively traded derivative securities. On an average day, over a thousand option contracts are quoted. These vary along several dimensions: option type (call or put), expiry/maturity date, and strike price. The data set, retrieved from OptionMetrics, consists of end-of-day values for all available option bid and ask quotes, as well as the corresponding time-to-maturity and strike price values. For each option contract, OptionMetrics calculates the implied volatility and other relevant characteristics, such as Δ and the strike price relative to spot price. Our sample period spans almost 15 years, from January 4, 1999, through August 30, 2013. Days on which the market is closed due to holidays or other reasons (such as the week following 9/11) are excluded, resulting in 3,688 daily observations.

The data are filtered to remove options that are inactive or may contain data errors. Our filtering procedures follow Barone-Adesi, Engle, and Mancini's (2008). Specifically, we delete options (i) with a maturity longer than 360 days or shorter than 10 days, (ii) with an implied volatility above 70%,[3] (iii) with a price below \$0.05, or (iv) with missing values for either the implied volatility or Δ. Moreover, we only consider out-of-the-money put and call options as these are more actively traded than in-the-money options. Because of the put-call parity considering out-of-the-money options is identical to studying in-the-money options or both types. Every in-the-money call (put) option can be matched to an out-of-the-money put (call) option, where the Δ of the call option is always one plus the Δ of the put option. For example, an in-the-money call option with a Δ of 0.75 should have the same implied volatility as an out-of-the-money put with a Δ of -0.25.

We create daily implied volatility surfaces spanning the maturity and moneyness dimensions. We divide the data into four maturity groups, separated by maturities of 45, 90 and 180 days, and six moneyness groups of at-the-money, out-of-the-money, and deep out-of-the-money options for both call and put options. Following Bollen and Whaley (2004), we define moneyness in terms of Δ because this also considers the volatility of the underlying asset, unlike the ratio of the strike price to the spot index price. We consider put options with $-0.125<\Delta<0$ as deep out-of-the-money (which we abbreviate with DOTM Put), with $-0.375<\Delta<-0.125$ as out-of-the-money (OTM Put), and with $-0.5<\Delta<-0.375$ as at-the-money (ATM Put). Similarly, we label call options with $0.375<\Delta<0.5$ as at-the-money (ATM Call), with $0.125 < \Delta < 0.375$ as out-of-the-money (OTM Call) and with $0 < \Delta < 0.125$ as deep out-of-the-money (DOTM Call). The combination of the four maturity and six moneyness groups provides 24 different groups, each containing a subset of all option contracts quoted on a day. For each maturity–moneyness group, we select the contract closest to the midpoint of both dimensions.[4] On an average day there are 17 contracts in each group. Only if there is no option data within a group, we consider the closest contract across all groups. The approach to consider different groups in the large cross-section of data follows the literature, see, for example, Bollen and Whaley (2004) and Barone-Adesi et al. (2008). We consider a fairly large number of 24 groups, which is chosen to strike a balance between obtaining a balanced panel of similar contracts over the entire sample period and representing overall movements in the large cross-section of options.[5]

2.2. *Summary Statistics and Preliminary Analysis*

Fig. 1 highlights some of the stylized facts about the implied volatility surface by displaying it for two specific days. Fig. 1(a) shows the surface for June 7, 1999. The implied volatility slopes downward along the moneyness dimension for each of the four maturity groups. It is highest for deep out-of-the-money puts having low strike prices relative to the spot price and lowest for deep out-of-the-money calls with high strike prices relative to the spot price. This is a usual pattern and is commonly referred to as the volatility smile.[6] The same pattern is indeed also found for September 19, 2008, a day during the height of the financial crisis, as shown in Fig. 1(b). All implied volatilities are substantially higher compared to June 7, 1999, but the same smile is observed for all maturity groups. Interestingly, the implied volatility slopes (slightly) upward along the maturity dimension for June 7, 1999, while it slopes downward for September 19, 2008. A common explanation for this difference in the term structure is the mean-reversion of volatility: at relatively high levels of volatility the term structure slopes downward, while at low levels of volatility it slopes upward.

Table 1 provides summary statistics of the implied volatility surface. The table shows the time series mean and standard deviation of five variables for each of the 24 moneyness–maturity groups. For each group, statistics are reported for the mid-quote (in US dollars), the implied volatility, the option Δ, the maturity (in days), and the strike relative to spot price (K/S). The mid-quote increases with time-to-maturity and generally displays an

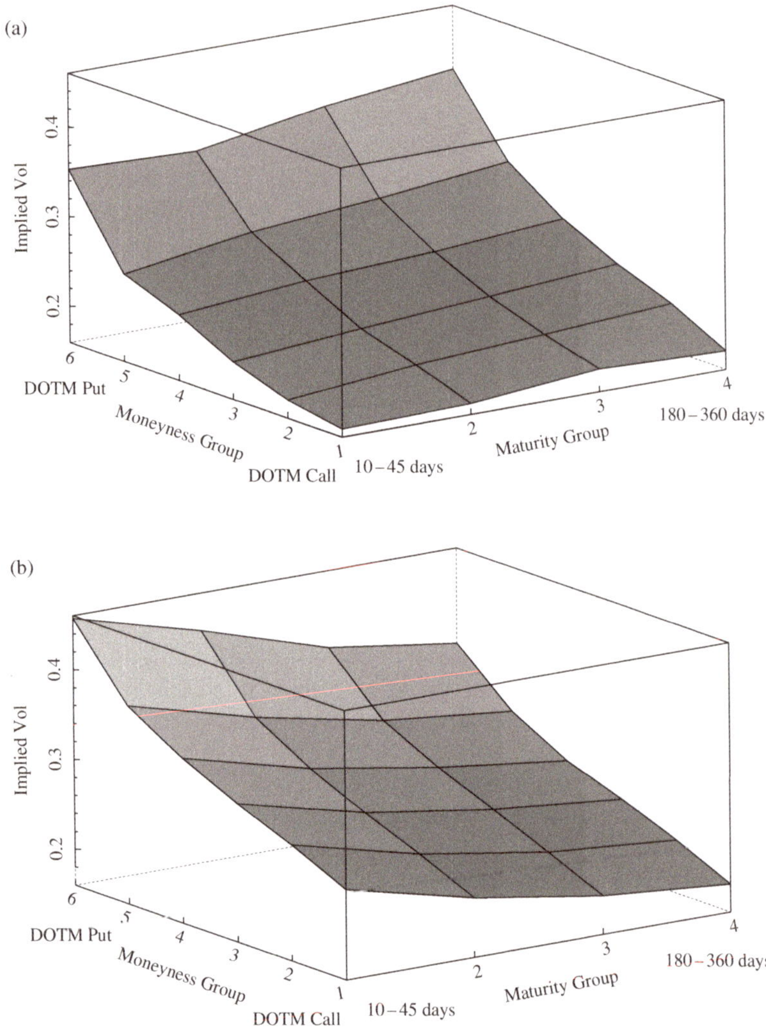

Fig. 1. Volatility Surface on Two Days. *Notes:* This figure shows the volatility surface on two days: June 7, 1999 and September 19, 2008. We show the implied volatility across four maturity groups and six moneyness groups. The maturity groups are 10–45 (group 1 in the figure), 45–90 (group 2), 90–180 (group 3), and 180–360 (group 4) days. The moneyness groups are $-0.125 < \Delta < 0$ (deep out-of-the-money put options, group 6 in the figure), $-0.375 < \Delta < -0.125$ (out-of-the-money puts, group 5), $-0.5 < \Delta < -0.375$ (at-the-money puts, group 4), $0.375 < \Delta < 0.5$ (at-the-money call options, group 3), $0.125 < \Delta < 0.375$ (out-of-the-money calls, group 2), and $0 < \Delta < 0.125$ (deep out-of-the-money calls, group 1). On each day, we select the option that is closest to the middle of each maturity–moneyness group. (a) June 7, 1999 and (b) September 19, 2008.

Table 1. Summary Statistics.

		Summary Statistics							
		10–45 days		45–90 days		90–180 days		180–360 days	
		Mean	SD	Mean	SD	Mean	SD	Mean	SD
DOTM Put	Mid-Quote	2.63	2	4.47	2.15	6.48	2.23	9.99	3.42
	Impl Vol	0.288	0.11	0.3	0.101	0.306	0.0925	0.308	0.0827
	Δ	−0.0627	0.0075	−0.0631	0.00862	−0.0627	0.00855	−0.0643	0.0119
	Maturity	27.9	10.7	67.8	9.45	133	23.3	269	26.2
	K/S	0.895	0.0422	0.834	0.0489	0.775	0.0578	0.705	0.0684
OTM Put	Mid-Quote	10.9	4.72	18.6	6.18	27.6	8.22	42.1	11.9
	Impl Vol	0.228	0.0957	0.234	0.086	0.239	0.0769	0.243	0.0674
	Δ	−0.25	0.0147	−0.249	0.0154	−0.25	0.0155	−0.25	0.0178
	Maturity	27.4	8.54	67.4	8.66	133	23.8	270	25.1
	K/S	0.964	0.0159	0.943	0.0199	0.923	0.0252	0.899	0.0326
ATM Put	Mid-Quote	21.2	8.58	35.7	11.5	52.8	15.4	80.3	21.3
	Impl Vol	0.204	0.0868	0.206	0.0764	0.209	0.0678	0.211	0.0587
	Δ	−0.437	0.0175	−0.437	0.0185	−0.437	0.0205	−0.438	0.0162
	Maturity	27.5	8.42	67.3	8.72	133	23	270	25.1
	K/S	0.994	0.00378	0.993	0.00643	0.993	0.0112	1	0.0195
ATM Call	Mid-Quote	18.6	6.75	29.9	8.43	42.1	10.6	59.8	13.2
	Impl Vol	0.191	0.0812	0.191	0.0698	0.193	0.0612	0.195	0.0528
	Δ	0.437	0.0182	0.438	0.0201	0.438	0.0226	0.438	0.0188
	Maturity	27.5	8.46	67.2	9.08	133	22.9	270	25.2
	K/S	1.01	0.00629	1.02	0.00984	1.03	0.0156	1.05	0.0247

Table 1. *(Continued)*

Summary Statistics									
		10–45 days		45–90 days		90–180 days		180–360 days	
		Mean	SD	Mean	SD	Mean	SD	Mean	SD
OTM Call	Mid-Quote	7.91	2.88	12.6	3.68	17.8	4.8	25.4	6.32
	Impl Vol	0.177	0.075	0.174	0.063	0.175	0.0551	0.176	0.0473
	Δ	0.25	0.0191	0.25	0.0219	0.25	0.0233	0.251	0.0215
	Maturity	27.3	8.2	67.5	8.47	134	22.8	270	24.8
	K/S	1.03	0.0166	1.06	0.0235	1.08	0.0314	1.13	0.0429
DOTM Call	Mid-Quote	1.32	0.495	2.08	0.707	2.91	1.08	4.43	1.78
	Impl Vol	0.165	0.068	0.158	0.0561	0.157	0.0485	0.156	0.0412
	Δ	0.0616	0.00906	0.0617	0.0108	0.0617	0.013	0.0656	0.0168
	Maturity	27.3	8.2	67.3	8.74	133	23.1	267	28.9
	K/S	1.07	0.0336	1.11	0.0451	1.17	0.0591	1.25	0.078

Notes: This table shows summary statistics for the option data. The table provides the mean and standard deviation (Sd) over time for the mid-quote (in dollars), implied volatility (Impl Vol), option Δ, maturity (in days), and strike relative to stock price (K/S). We show these numbers across four maturity groups and six moneyness groups. The maturity groups are 10–45, 45–90, 90–180, and 180–360 days. The moneyness groups are $-0.125<\Delta<0$ (deep out-of-the-money put options, DOTM Put), $-0.375<\Delta<-0.125$ (out-of-the-money puts, OTM Put), $-0.5<\Delta<-0.375$ (at-the-money puts, ATM Put), $0.375<\Delta<0.5$ (at-the-money call options, ATM Call), $0.125<\Delta<0.375$ (out-of-the-money calls, OTM Call), and $0<\Delta<0.125$ (deep out-of-the-money calls, DOTM Call). On each day we find the option that is closest to the middle of each maturity–moneyness group. The numbers represent averages over time for the selected contracts in each group.

inverse U-shape across the moneyness groups. The first feature reflects the options' time value, that is, options with longer maturity are traded at a higher price than short maturity options with comparable moneyness. The second feature is due to our definition of the moneyness groups and reflects the fact that options that are closer to being in-the-money are priced higher. The numbers for implied volatility confirm the observation based on Fig. 1 that implied volatility slopes downward with moneyness. The average implied volatility for the DOTM Put group is almost the double of the average implied volatility for the DOTM Call group. The volatility term structure is fairly flat on average. The average option Δ is close to the midpoint of the relevant ranges defining the different moneyness groups and the same is the case for the average maturity. It is interesting to note that, whereas Δ is fairly constant across maturity groups, this is not the case for the strike price relative to spot price, illustrating that moneyness is measured differently by both variables.

Figs. 2 and 3 provide time series plots of the volatility surface. Fig. 2(a) plots the average implied volatility across the 24 groups together with the CBOE Market Volatility Index (VIX, in short). Particularly the high volatility of the financial crisis during the second half of 2008 stands out, but also the increased uncertainty following 9/11, the second Gulf War in 2002–2003, the European sovereign debt crisis in 2010, and the debt-ceiling crisis of 2011. Fig. 2(b) of the figure plots the time series of implied volatility for all 24 moneyness–maturity groups. We observe a strong comovement across the groups. During times of high volatility, the dispersion across the groups is larger than the dispersion during times of low volatility. This is borne out more clearly in Fig. 3, plotting the slope of the volatility smile (Fig. 3(a)) and the term structure (Fig. 3(b)). The slope of the volatility smile in each maturity group is defined as the implied volatility of the deep out-of-the-money put options minus the implied volatility of the deep out-of-the-money calls. The slope of the volatility term structure is defined as the implied volatility of the longest maturity group minus the implied volatility of the shortest maturity group. Fig. 3(a) shows that for all days in the sample the slope of the smile is positive for each of the maturity groups. Its magnitude varies considerably though, between 0.05 and 0.30, approximately. The smile is larger when the level of volatility is higher and is strongest during and following the financial crisis. The time series of the slope of the term structure in Fig. 3(b) indicates the term structure can be both negative and positive. Consistent with the mean-reversion interpretation, during times of high volatility the term structure slopes downward, while it tends to slope upward during times of low volatility.

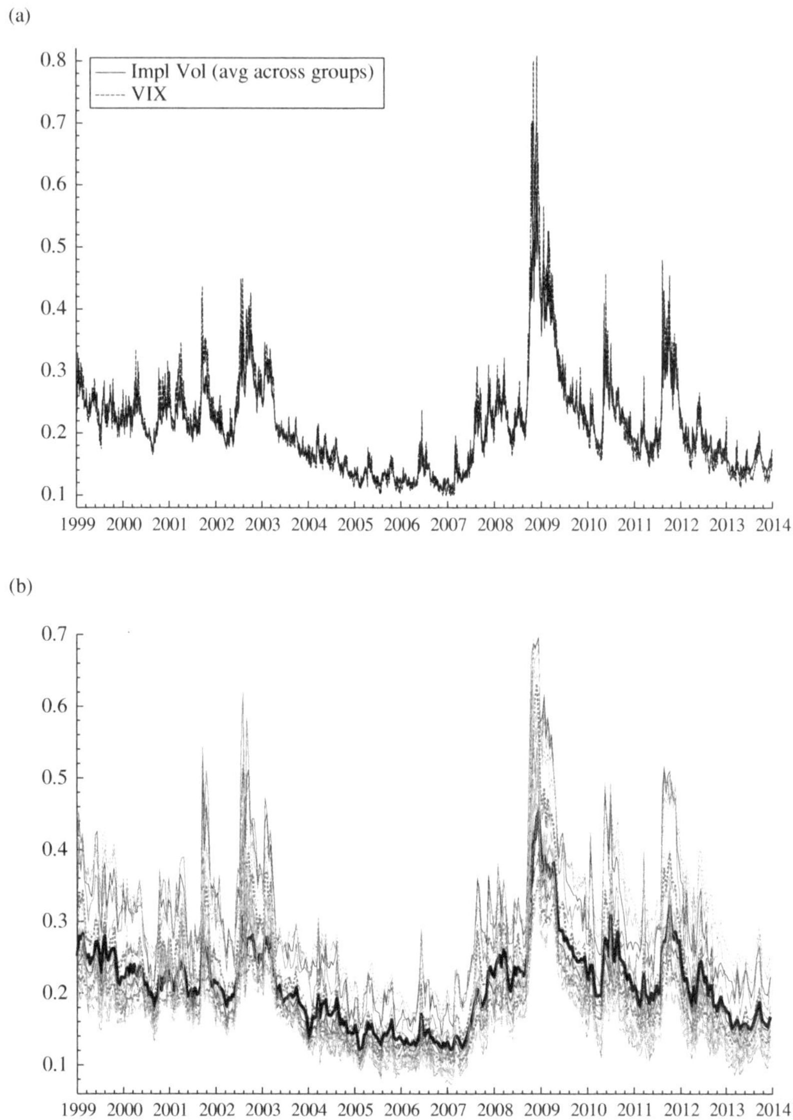

Fig. 2. Volatility Time Series. *Notes:* This figure shows the time series of implied volatility. Panel (a) plots the average implied volatility across the maturity–moneyness groups, together with the VIX. Panel (b) plots the time series for each of the 24 maturity–moneyness groups. To compress space, Panel (b) shows snapshots of every 10 observations.

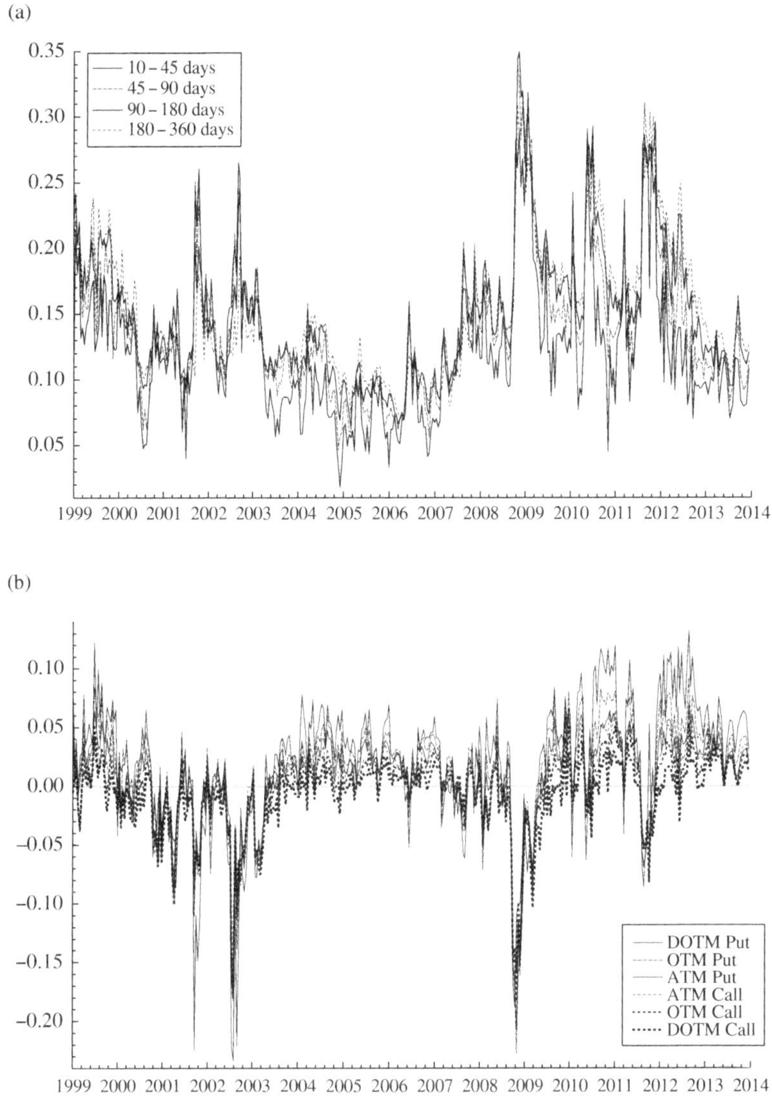

Fig. 3. Slope of the Volatility Smile and Term Structure. *Notes:* This figure shows the time series of the implied volatility smile and term structure. Panel (a) plots the slope of the implied volatility smile. For each of the four maturity groups, we define the slope in that maturity group as the implied volatility of the deep out-of-the-money put minus the implied volatility of the deep out-of-the-money call. This is the implied volatility of the option with the lowest strike relative to stock price minus the implied volatility of the highest strike relative to the stock price. Panel (b) plots the slope of the volatility term structure. For each of the six moneyness groups, we define the slope as the implied volatility of the longest maturity minus the shortest maturity. To compress space, the figure shows snapshots of every 10 observations.

Similar to the pattern of the average implied volatility, both the smile and term structure slopes show strong persistence.

To investigate the degree of comovement and persistence in the implied volatilities further, we analyze cross-correlations between the moneyness–maturity groups and (partial) autocorrelations within each group. Table A1 shows cross-correlations between the different moneyness groups for the shortest (10–45 days) and longest (180–360 days) maturity categories. Without exception, the cross-correlations are very high: Within a maturity category they all exceed 0.9, but even across maturity categories they do not fall below 0.8. Also, cross-correlations decline as the difference in moneyness between the groups becomes larger. Table A2 reports (partial) autocorrelations at various lags for a selection of implied volatilities, for the slope of the volatility smile at different maturities and for the slope of the term structure at different moneyness levels. We observe strong autocorrelations in each case, particularly for contracts with longer maturities. At the 20th lag, the autocorrelation of the implied volatility stays above 0.8 for all contracts, while those of the slopes of the volatility smile and the term structure remain above 0.6 in all cases. For implied volatilities, the partial autocorrelations with the first lag are around 0.98 (they are by definition equal to the autocorrelations with the first lag) and they drop dramatically to the 0.1–0.25 range for the second lag. The two slope variables follow a similar shape in their partial autocorrelations.

The strong cross-correlations of the implied volatility across the moneyness–maturity groups suggest that it might be useful to employ common factors to describe the features of the implied volatility surface. To examine this further, we run a principal component analysis. Table A3 provides the percentage of variation explained by the first 10 of the 24 principal components and the (partial) autocorrelation structure of the first three principal components. The bulk of the variation in the implied volatilities is captured by the first three principal components. Combined, they explain nearly 99% of the variation in the panel of 24 series. Fig. A1 shows the time series of the first three principal components. The first principal component corresponds to the level of the implied volatility panel. The second and third component cannot immediately be linked to the volatility smile and term structure though. The principal component analysis confirms the presence of a strong factor structure in the implied volatility surface and motivates our setup of dynamic factor modeling.

3. MODELS FOR THE VOLATILITY SURFACE

We now turn to the DFMs for the implied volatility surface. We propose three different setups. Section 3.1 describes a general DFM, which nests all subsequent specifications. Section 3.2 discusses a restricted model recently proposed by Christoffersen et al. (2013). This setup explicitly takes the structure of the implied volatility surface data into account. Specifically, in addition to a factor representing the level of the surface, the model is forced to contain factors representing the smile and the (slope of the) term structure. This is achieved by setting the factor loadings equal to the moneyness and maturity of the corresponding option contract. Finally, Section 3.3 introduces related but more flexible specifications using splines to set the factor loadings along the moneyness and maturity dimensions.

3.1. General DFM

The observation vector in our factor models consists of the vectorized volatility surface. Stacking the vector of implied volatilities for the T different maturities for each of the M moneyness groups results in the $(TM \times 1)$ vector of observations

$$y_t = \begin{pmatrix} \mathrm{IV}_{\tau_1,m_1,t} \\ \vdots \\ \mathrm{IV}_{\tau_T,m_1,t} \\ \mathrm{IV}_{\tau_1,m_2,t} \\ \vdots \\ \vdots \\ \mathrm{IV}_{\tau_T,m_M,t} \end{pmatrix}$$

where $\mathrm{IV}_{\tau_i,m_j,t}$ denotes the implied volatility on day t for an option with time-to-maturity τ_i, with $i = 1, 2, \ldots, T$, and moneyness m_j, with $j = 1, 2, \ldots, M$. In our application, we have $T = 4$ maturities and $M = 6$ moneyness categories. It is useful to note that because of the seasonality in option expiry dates and the variation in the option Δ (as shown in Table 1), it is not possible to use option contracts with exactly the same moneyness and maturity on each day and across the moneyness and maturity groups. For notational convenience, we suppress here the time-subscripts on the maturity τ and moneyness m as well as the moneyness subscripts for maturity and the maturity subscripts for moneyness.

We start with a general DFM, given by

$$\begin{aligned} y_t &= \Lambda f_t + \varepsilon_t, & \varepsilon_t &\sim N(0, \Sigma_\varepsilon) \\ f_{t+1} &= \mu + \Phi(f_t - \mu) + \eta_{t+1}, & \eta_{t+1} &\sim N(0, \Sigma_\eta) \end{aligned} \tag{1}$$

where y_t is the observation vector as defined above, f_t are the r latent dynamic factors that are loaded onto the implied volatilities using the $(TM \times r)$ factor loading matrix Λ and ε_t is the $(TM \times 1)$ vector of measurement errors with covariance matrix Σ_ε. Following standard practice and in line with the results of our (partial) autocorrelation analysis, we adopt a vector autoregressive (VAR) model of order one for the factors, with intercept μ, VAR-coefficient matrix Φ, and factor innovations η_{t+1}, which are assumed to be normally distributed with covariance matrix Σ_η.

The model is estimated using likelihood-based methods along the lines of Jungbacker and Koopman (2014). Motivated by the principal component analysis in Section 2.2, we focus on models with $r = 3$ factors. To enable estimation, we impose identification restrictions. One possibility is to restrict the top-square part of the loading matrix Λ and set it equal to the identity matrix, following Geweke and Zhou (1996). We implement a slight variation of this identification motivated by the empirical analysis of the volatility surface. We impose restrictions on four rows of Λ, corresponding to the shortest and longest maturities in the third and fourth moneyness groups (ATM Put and ATM Call), by setting

$$\begin{pmatrix} \lambda_{2T+1,1} & \lambda_{2T+1,2} & \lambda_{2T+1,3} \\ \lambda_{3T,1} & \lambda_{3T,2} & \lambda_{3T,3} \\ \lambda_{3T+1,1} & \lambda_{3T+1,2} & \lambda_{3T+1,3} \\ \lambda_{4T,1} & \lambda_{4T,2} & \lambda_{4T,3} \end{pmatrix} = \begin{pmatrix} 1 & -1 & 1 \\ 1 & 1 & 1 \\ 1 & -1 & -1 \\ 1 & 1 & -1 \end{pmatrix}$$

The restrictions force the first latent factor to capture the level of implied volatility across the four selected groups, the second factor to capture the term structure for the two moneyness groups and the third factor to capture the smile for the two maturity groups. Furthermore, we impose diagonality on the covariance matrix Σ_ε of the measurement errors, implying that all comovement of the implied volatilities is attributed to the latent factors. The drop of the partial autocorrelations after the first lag of the principal components as reported in Table A3 motivates our use of a VAR with one lag. The VAR parameters in μ and Φ are left unrestricted and the parameters in Σ_η are estimated using an LDL-decomposition. We initialize the latent factors from a standard normal distribution.[7]

3.2. Restricted Economic DFMs

Given the structure of the implied volatility surface, in the context of DFMs it is quite natural to think of the latent factors as components representing the moneyness and maturity dimensions of the surface (or, in other words, the volatility smile and (the slope of) the term structure). Following this intuitively plausible idea, we adopt the setting of Christoffersen et al. (2013), who extract the implied volatility level, moneyness slope, and maturity slope by regressing the implied volatility cross-section at a given time t on the moneyness and the maturity:

$$\mathrm{IV}_{\tau_i,m_j,t} = l_t + \tau_i c_t + m_j s_t + \varepsilon_{i,j,t} \tag{2}$$

with l_t the implied volatility level, c_t the implied volatility maturity curve (along the maturity dimension), and s_t the implied volatility smile (along the moneyness dimension). This approach can easily be cast in the dynamic factor framework. To do so, we collect the level, curve, and smile into the latent state vector to obtain $f_t = (l_t, c_t, s_t)'$. This results in a special case of the DFM in Eq. (1) with restricted factor loading matrix

$$\Lambda = \begin{pmatrix} 1 & \tau_1 & m_1 \\ \vdots & \vdots & \vdots \\ 1 & \tau_T & m_1 \\ 1 & \tau_1 & m_2 \\ \vdots & \vdots & \vdots \\ \vdots & \vdots & \vdots \\ 1 & \tau_T & m_M \end{pmatrix}$$

This loading matrix is deterministic and contains no parameters that need to be estimated.

Also here we have suppressed variation over time and across moneyness and maturity groups of the maturity τ_i and moneyness m_j. Because this variation may be useful for capturing the shape of the volatility surface, we run two variants. First, we consider specification with time-varying loading matrix Λ_t containing the actual time-to-maturity and moneyness of each contract in the 24 moneyness–maturity groups on day t. Second, we consider a specification with a constant loading matrix Λ, where we take the time-series average of the maturities and moneyness for each of the 24 groups.

We also consider two different definitions of moneyness m_j. First, following Christoffersen et al. (2013), we define moneyness as the ratio of the option's strike price K and the spot price S. Second, following the constructing of our volatility surface data, we take moneyness equal to the option Δ. As discussed in Section 2.1, the latter takes more properties of the data into account for defining the relative likelihood of the option ending up in-the-money. We consider both variables for both the constant and time-varying loading case.[8] In all four of these restricted cases, we de-mean the second and third column to let the first factor capture movements in the level of the volatility surface (for the time-varying cases, the columns are de-meaned on a daily basis).

3.3. *Spline-Based DFMs*

While the idea that the latent factors in a DFM should capture the key features of the volatility surface along the maturity and moneyness dimensions is intuitively plausible, the approach of Christoffersen et al. (2013) discussed above is quite restrictive in terms of the specification of the factor loadings. The third and final setup is a hybrid approach, aiming to combine the flexibility of the general DFM of Section 3.1 with the economically plausible factor interpretation of the restricted model in Section 3.2. We propose to specify the factor loadings using splines in order to capture the shape of the volatility surface along both the maturity and moneyness dimensions in a flexible manner. The splines we use follow Poirier (1976) and have been used before in a dynamic factor modelling framework by Jungbacker et al. (2014).

In a first variant of this approach, we impose relatively little structure on the loadings, in the sense that we do not impose restrictions on the loadings along the maturity dimension across the different moneyness groups, and vice versa. This is achieved by considering four separate splines across the moneyness dimension (one spline for each maturity group) and six separate splines across the maturity dimension (one for each moneyness group). Written in a manner similar to the restricted models from the previous section, we have

$$\text{IV}_{\tau_i,m_j,t} = l_t + f^j(\tau_i)c_t + g^i(m_j)s_t + \varepsilon_{i,j,t}$$

with $f^j(\tau_i)$ the spline for maturity i in moneyness group j, and $g^i(m_j)$ the spline for moneyness group j with maturity i. The maturity splines $f^j(\cdot)$

capture the shape of the volatility term structure for the different moneyness groups, while the moneyness splines $g^i(\cdot)$ capture the shape of the volatility smile for the different maturity groups. Also this setup can be written in the DFM framework, by using the factor loading matrix

$$\Lambda = \begin{pmatrix} 1 & f^1(\tau_1) & g^1(m_1) \\ \vdots & \vdots & \vdots \\ 1 & f^1(\tau_T) & g^T(m_1) \\ 1 & f^2(\tau_1) & g^1(m_2) \\ \vdots & \vdots & \vdots \\ \vdots & \vdots & \vdots \\ 1 & f^M(\tau_T) & g^T(m_M) \end{pmatrix}$$

For each of the splines, a number of knots has to be selected. Given that $M = 6$ the moneyness splines consist of at most six knots. The maturity splines consist of at most four knots, because $T = 4$. The parameters to be estimated in the loading matrix are the knot values for each of the splines. For example, suppose x knots are chosen for the j-th maturity spline $f^j(\cdot)$. Then the spline interpolation for all elements of this spline gives

$$f^j(\tau_i) = w_{ij}'\lambda_j$$

where w^{ij} is an $(x \times 1)$ vector containing the spline weights and λ_j is an $(x \times 1)$ vector with the coefficients associated to each of the knots. The spline weights are determined by the knot positions and the restrictions associated with the cubic spline being a third-order polynomial and being twice continuously differentiable at the knots (Poirier, 1973 and 1976, provides the exact functional form). The moneyness splines are similarly defined. For comparability with the restricted models of the previous section, we restrict the average knot values for each spline to be equal to zero. Besides the comparability, an advantage of this setting is that it imposes enough restrictions for the model to be identified. Note that in the case of six and four knots, the model is identical to the general DFM of Section 3.1, apart from the identification restriction.

In the second variant of this approach, we restrict the moneyness splines $g^i(\cdot)$ to be the same across all maturity groups $i = 1, \ldots, T$ and the maturity splines $f^j(\cdot)$ to be the same across all moneyness groups $j = 1, \ldots, M$. This leads to the restricted specification for the observed implied volatility given by

$$\mathrm{IV}_{\tau_i,m_j,t} = l_t + f(\tau_i)c_t + g(m_j)s_t + \varepsilon_{i,j,t}$$

with $f(\tau_i)$ the spline value for maturity i and $g^i(m_j)$ the spline value for moneyness group j. In this case, the factor loading matrix is given by

$$\Lambda = \begin{pmatrix} 1 & f(\tau_1) & g(m_1) \\ \vdots & \vdots & \vdots \\ 1 & f(\tau_T) & g(m_1) \\ 1 & f(\tau_1) & g(m_2) \\ \vdots & \vdots & \vdots \\ \vdots & \vdots & \vdots \\ 1 & f(\tau_T) & g(m_M) \end{pmatrix}$$

Also for this variant, a number of knots have to be selected for each spline and the coefficients to be estimated are called as the knot values. Similarly, we impose the restriction that the average of the knot values is equal to zero.

A special case of this second variant is the specification with $T = 4$ knots for the maturity spline and $M = 6$ knots for the moneyness spline. In this case, the knot values are simply the levels of each of the elements of the spline and the model is a block-version of the general DFM with

$$\Lambda = \begin{pmatrix} 1 & \lambda_{1,1} & \lambda_{2,1} \\ \vdots & \vdots & \vdots \\ 1 & \lambda_{1,T} & \lambda_{2,1} \\ 1 & \lambda_{1,1} & \lambda_{2,2} \\ \vdots & \vdots & \vdots \\ \vdots & \vdots & \vdots \\ 1 & \lambda_{1,T} & \lambda_{2,M} \end{pmatrix}$$

with $T + M$ parameters to be estimated in the loading matrix, which reduces to $T + M - 2$ parameters after imposing the restriction that the averages of the second and third columns must be equal to zero.

4. MAIN RESULTS

Table 2 provides key statistics concerning the fit of the three different DFM setups, estimated using the full sample of daily S&P500 implied volatility surfaces over the period January 4, 1999–August 30, 2013. The

Table 2. Comparing Dynamic Factor Models.

Comparison of Factor Models						
Model and Moneyness		Const/TV	Loglik	LR-test	#Pars	AIC
General DFM (only ident restr)			302,132.8		102	−604,061.6
Restr DFM	K/S	Const	257,089.0	90,087.6	42	−514,094.0
		TV	261,552.8	81,182.4	42	−522,999.2
	Δ	Const	263,989.1	76,287.4	42	−542,237.8
		TV	271,160.9	61,943.8	42	−542,260.8
Spline DFM	Block DFM		291,583.0	21,099.6	50	−583,066.0
	Mat and Mon Splines		287,181.4	29,902.8	47	−574,268.8
	Separate Splines		293,206.4	17,852.8	66	−586,280.8

Notes: This table provides statistics concerning the fit of the three dynamic factor models, which are estimated using the full sample of daily S&P500 implied volatility surfaces over the period January 4, 1999–August 30, 2013. The first model is the general dynamic factor model (General DFM), where only identification restrictions have been imposed. The second set of models are the restricted factor models based on the economic literature (Restr DFM). We estimate four variants, depending on whether the loading matrix is constant (Const) or time varying (TV) and whether strike relative to stock price (K/S) or Δ is used for moneyness. The third set of models are spline-based (Spline DFM). We consider the block dynamic factor model with knots on all places (Block DFM), a model with one spline for the maturity and one spline for the moneyness dimensions (Mat and Mon Splines) and a model where there are separate splines for all maturity and moneyness groups (Separate Splines). For each model, we provide the log-likelihood value (Loglik), the number of parameters (#Pars) and the Akaike Information Criterion (AIC). For the restricted economic and spline models, we also provide a likelihood-ratio test (LR-test) relative to the general dynamic factor model.

baseline model is the general DFM of Section 3.1. With three factors, the model contains 102 parameters in total. The estimated log-likelihood is 302,133, which serves as a benchmark for comparison with the other models.

For the restricted DFM discussed in Section 3.2, we consider four variants, where moneyness is measured either by the strike price relative to the spot price or Δ and where the loading matrix is either constant or time varying. For all four variants, we find much lower log-likelihood values, ranging between 257,089 and 271,160. This makes the general DFM specification preferred by both an LR-test and the Akaike Information Criterion.[9] These measures take into account that the restricted models have far less parameters, namely, only 42 compared to 102 in the general DFM. Among the four restricted models, the time-varying loading matrix is preferred over the constant loading matrix for both moneyness measures. For both the time varying and constant loading case, Δ is preferred as measure of moneyness. This is striking, as many of the existing models use the strike relative to the spot price rather than Δ to measure moneyness.

As discussed in Section 3.3, we consider three variants of the spline-based DFMs. First, the most flexible case contains a separate moneyness spline for each maturity group and a separate maturity spline for each moneyness group. Second, a more restrictive case has a single spline for moneyness and a single spline for maturity. And third, the block DFM constitutes a special case obtained when the number of knots is equal to the number of elements in the case of a single spline for moneyness and a single spline for maturity. All three variants perform a lot better than the restricted models. For example, the log-likelihood of the block DFM is more than 20,000 points higher than the best restricted model even though the loading matrix is constant. For each of the splines, we select three knots for the maturity spline (out of maximum four) and four for the moneyness spline (out of maximum six). Due to the restriction that the average of the knot values is zero, two and three parameters have to be estimated for the splines to model the maturity and moneyness dimensions, respectively. The maturity and moneyness spline variant performs worse than the block DFM, even though the difference in the number of parameters is very small. The separate splines variant performs best. Out of all restricted and spline models, it comes closest to the general DFM. Nevertheless, it is still rejected by the LR-test and also not preferred by the Akaike Information Criterion.[10]

In the remainder of this section, we show some further estimation output for a selection of the models. We focus on one variant for each of the three model types. Besides the general DFM, we also consider the restricted model where Δ is used as the measure of moneyness. For the spline-based models, we consider the most flexible case with separate splines for maturity (moneyness) across moneyness (maturity) groups. Our motivation for this selection is that these models provided the highest log-likelihood within each model class. For the restricted model, we consider the constant factor loading case, for comparability with the other models.

Fig. 4 presents the estimated latent factors for the three selected models. The first factor is very similar across the models and captures the overall level of the volatility surface. This is expected, because in all cases, the second and third columns of the loading matrix are de-meaned. The second

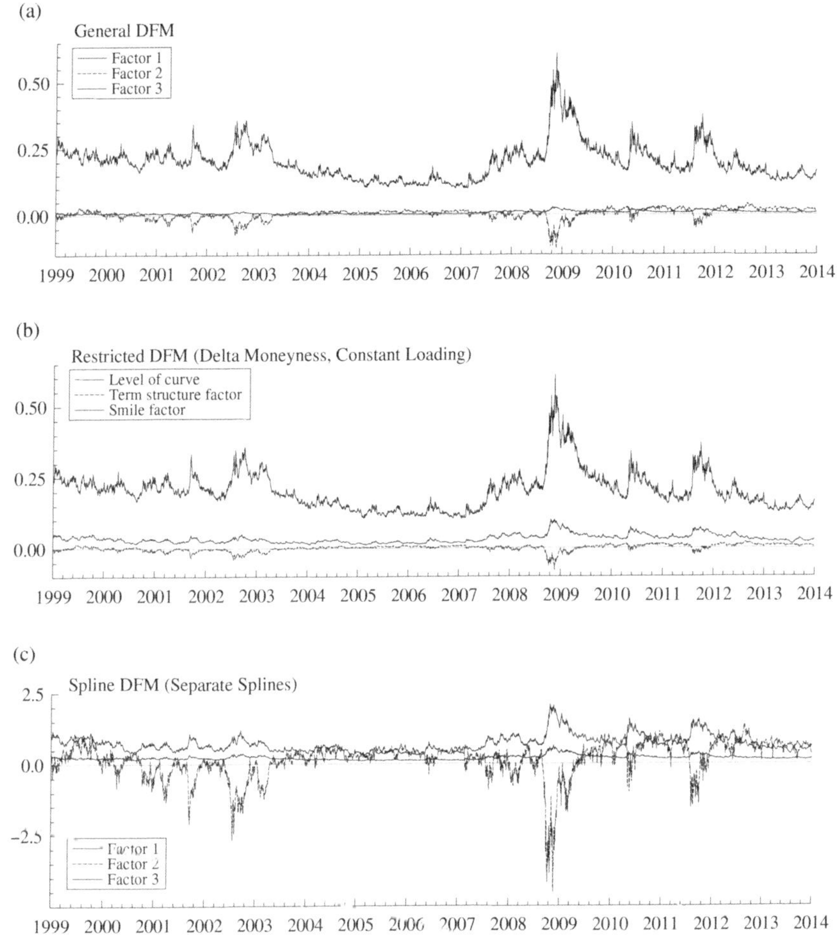

Fig. 4. Estimated Factors. *Notes:* This figure shows the three dynamic factors for three of the factor models we consider. We show the smoothed factors for the general dynamic factor model (Panel a), the restricted dynamic factor model with a constant loading matrix where moneyness is measured with Δ (Panel b), and the spline dynamic factor model where there is a separate spline for each moneyness and maturity group (Panel c). The solid line represents the first factor, the dotted line the second factor and the dashed line the third factor.

factor also seems fairly similar across the three model types, though the scaling varies. The variation of the second factor is particularly strong for the spline-based model. Due to the identification restriction for the splines, the factor loading matrix has smaller entries in the second (and third) column and hence the factors have larger variation. The third factor differs even more across the models, particularly, for the general DFM compared to the two restricted models.

Due to the different model structure and loading matrix, the factors may seem different, but this might just be due to a rotation. To examine this possibility, we rotate the general DFM to each of the two other models.[11] Fig. A2(a) confirms that the first and second factors of the general and restricted factor models are similar. However, the third factor remains different, even after rotation. From Fig. A2(b) it is clear that the spline model is much closer to the general DFM, something which was already suggested by the log-likelihood values.

Table 3 documents the factor dynamics. All factors are very persistent, with diagonal elements very close to one. The off-diagonals are mostly zero. There is a strong correlation between the factor innovations for all models. For the general DFM, this correlation is −0.95 between the first and second factor, 0.72 between the first and third factor, and −0.70 between the second and third factor.[12] The correlations are similar for the other models. The different scaling of the second and third factor for the spline-based model compared to the other models is also clear from the estimated covariance matrix. While the variance of the first factor is roughly in the same order of size as it is in the other models, the variance of the second and third factors is more than a 100 times higher.

Fig. 5 provides further explanation for the different findings from the three models. Based on the factor loading matrix, for all the models an implied term structure is constructed for all six moneyness groups and an implied smile is constructed for all four maturity groups. The restricted models allow for very little variation in the type of term structure and smile. The unrestricted and spline-based factor models offer a lot more variation. This is a likely reason for the relatively poor performance of the restricted economic DFMs. Also the difference in scaling between the spline-based model compared to the other two models is visible, explaining the larger variance for the second and third factors in the spline-based case.

Finally, Fig. 6 provides some plots for the fit of the general DFM. Out of the 24 groups in our data, we show a selection of six. The top two plots are fairly in the center of the surface, while the bottom four plots are on the corners. Overall the model fits the data fairly well for the points in the

Table 3. Factor Dynamics.

Panel a: Factor Dynamics General DFM

		Φ			$\Sigma_\eta(\times 10^{-4})$		
	μ	$f_{1,t-1}$	$f_{2,t-1}$	$f_{3,t-1}$	$f_{1,t}$	$f_{2,t}$	$f_{3,t}$
$f_{1,t}$	0.196	0.995	0.028	−0.005	1.059	−0.453	0.051
$f_{2,t}$	0.003	−0.002	0.971	0.056	−0.453	0.212	−0.022
$f_{3,t}$	0.008	0.001	0.001	0.967	0.051	−0.022	0.005

Panel b: Factor Dynamics Restricted DFM

		Φ			$\Sigma_\eta(\times 10^{-4})$		
	μ	l_{t-1}	c_{t-1}	s_{t-1}	l_t	c_t	s_t
l_t	0.200	0.995	0.054	0.006	0.939	−0.220	0.129
c_t	0.001	−0.001	0.969	0.003	−0.220	0.057	−0.027
s_t	0.032	0.001	0.002	0.990	0.129	−0.027	0.029

Panel c: Factor Dynamics Spline DFM

		Φ			$\Sigma_\eta(\times 10^{-2})$		
	μ	$f_{1,t-1}$	$f_{2,t-1}$	$f_{3,t-1}$	$f_{1,t}$	$f_{2,t}$	$f_{3,t}$
$f_{1,t}$	0.207	0.995	0.001	0.000	0.012	−0.169	0.035
$f_{2,t}$	0.135	−0.094	0.967	0.025	−0.169	2.548	−0.442
$f_{3,t}$	0.694	0.017	0.001	0.988	0.035	−0.442	0.143

Notes: This table provides the estimates of the factor dynamics for three of the factor models we consider. We show the coefficients for the general dynamic factor model (Panel a), the restricted dynamic factor model with a constant loading matrix where moneyness is measured with Δ (Panel b), and the spline dynamic factor model where there is a separate spline for each moneyness and maturity group (Panel c). For all three models, we plot the intercept μ, VAR coefficient matrix Φ, and the innovation variance Σ_η.

center of the surface. Fitting the corners appears to be more problematic, since the residuals do not look like merely white noise. In Fig. A3, we provide similar figures for the restricted and spline-based factor models. The same conclusion is drawn for these models, though the problems in fitting the corners of the surface are much more serious for the restricted economic model. Fig. A4 provides a detailed comparison of the fit for the different three models. The four panels on the left represent the fitting error, measured by root mean-squared error, for each of the four maturity groups. Each panel contains the fitting error for each of the moneyness groups in the maturity group. The panels on the right display the root

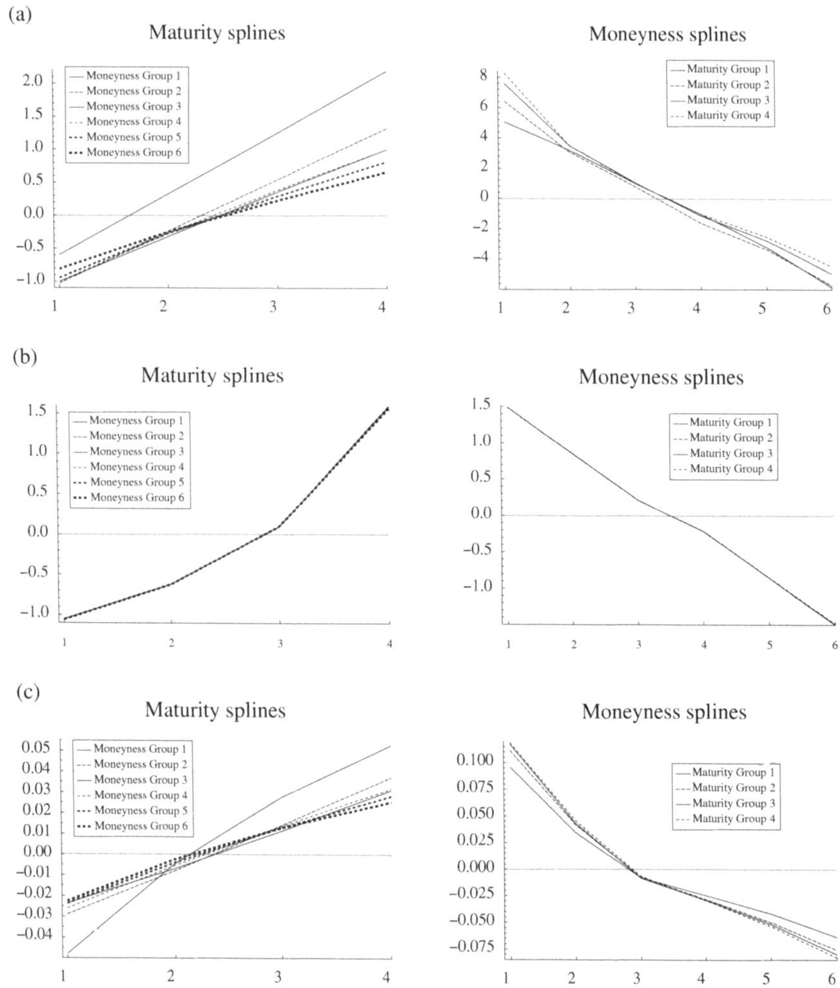

Fig. 5. Shape of Implied Volatility Term Structure and Smile. *Notes:* This figure shows the volatility term structure and volatility smile that is implied by three of the factor models we consider. We show the term structure and smile for the general dynamic factor model (Panel a), the restricted dynamic factor model with a constant loading matrix where moneyness is measured with Δ (Panel b), and the spline dynamic factor model where there is a separate spline for each moneyness and maturity group (Panel c). In all panels, the left figure depicts the volatility term structure and the right picture the volatility smile. The term structure and smile are implied by the factor loading matrix for each model.

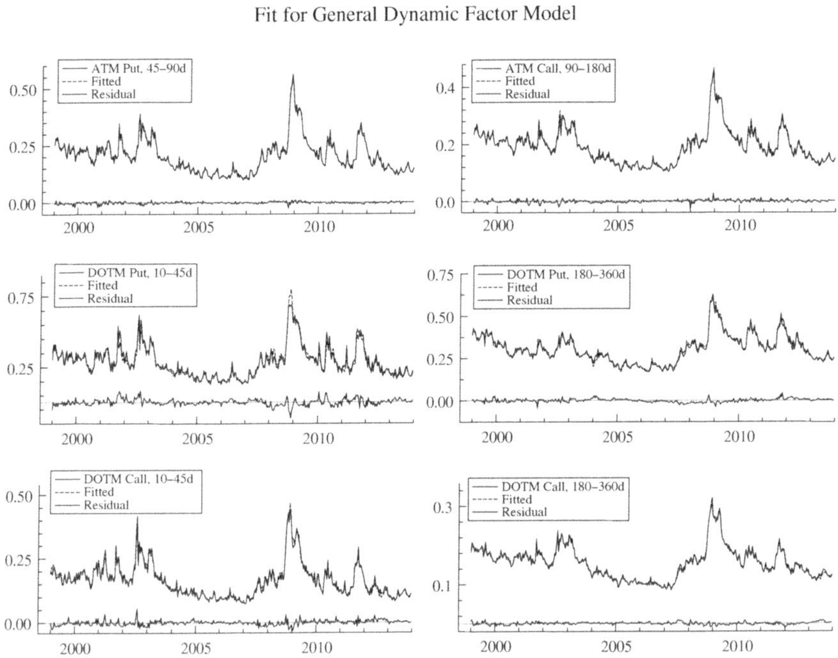

Fig. 6. Fit. *Notes:* This figure documents the fit of the volatility surface based on the general dynamic factor model. We show the smoothed fitted values, which are obtained by pre-multiplying the smoothed factors with the estimated factor loading matrix. In the figure, the solid line is the actual implied volatility, the dotted line the fitted, and the dashed line the residual. We show six different maturity–moneyness combinations: the top plots depict two points in the middle of the surface and the bottom plots four outer points on the surface. To compress space, the figure shows snapshots of every 10 observations.

mean-squared percentage error in the same manner. The figure confirms that the fitting errors are largest for the restricted economic model, and smallest for the general DFM. The spline-based factor model is much closer to the general DFM than the restricted economic model is. The restricted economic model has the largest root mean-squared error particularly for the deep out-of-the-money put options and the short maturity options.

To study the economic significance, we convert the fitted implied volatilities back to fitted prices. We do this using the Black-Scholes model for European options, adjusted for the dividend rate on the S&P500 index.[13]

Using this model each fitted implied volatility provides a fitted price. This fitted price can then be compared to the actual price of the option, for which we use the mid-point of the bid and ask. Fig. 7 is similar to Fig. A4 and provides the fitted pricing error for the general DFM, the spline-based model, and the restricted economic model. Over all 24 option groups, the root mean-squared error in terms of prices is $1.22 for the general DFM, $1.41 for the spline-based model, and $2.28 for the restricted economic model. The difference of more than a dollar suggests the difference between the general and restricted economic model is also significant in an economic sense.[14] Fig. 7 further shows that the percentage pricing error is largest for the deep out-of-the-money options (both call and put). This is the case for all models. An explanation is that these deep out-of-the-money options are least liquid, with relatively wider bid-ask spreads and more erratic trading patterns than the other groups of options. It might actually be a sign of strength of the restricted model to be able to single out these contracts as being the least appropriately priced ones relative to the remaining universe of outstanding option contracts. However, all three models show higher fitting errors for the deep out-of-the-money categories, and the mispricing of the restricted economic model is largest also for the other categories, particularly for the shortest maturity options.

5. ROBUSTNESS AND EXTENSIONS

Here we consider a number of robustness checks and extensions of our main modeling approach. First, Section 5.1 considers an alternative construction of the volatility surface based on moneyness measured by strike relative to spot price. Second, Section 5.2 examines some of the higher-dimensional restricted models that have been proposed in the literature. Third, Section 5.3 considers random walks for the factor dynamics. Finally, to examine to what extent the financial crisis dominates our results, Section 5.4 shows results when the analysis is done using log-transformed data or a sub-sample omitting the crisis.

5.1. Alternative Surface Construction

One of our main findings discussed in Section 4 is that Δ is a better measure of moneyness for building models of the volatility surface compared to

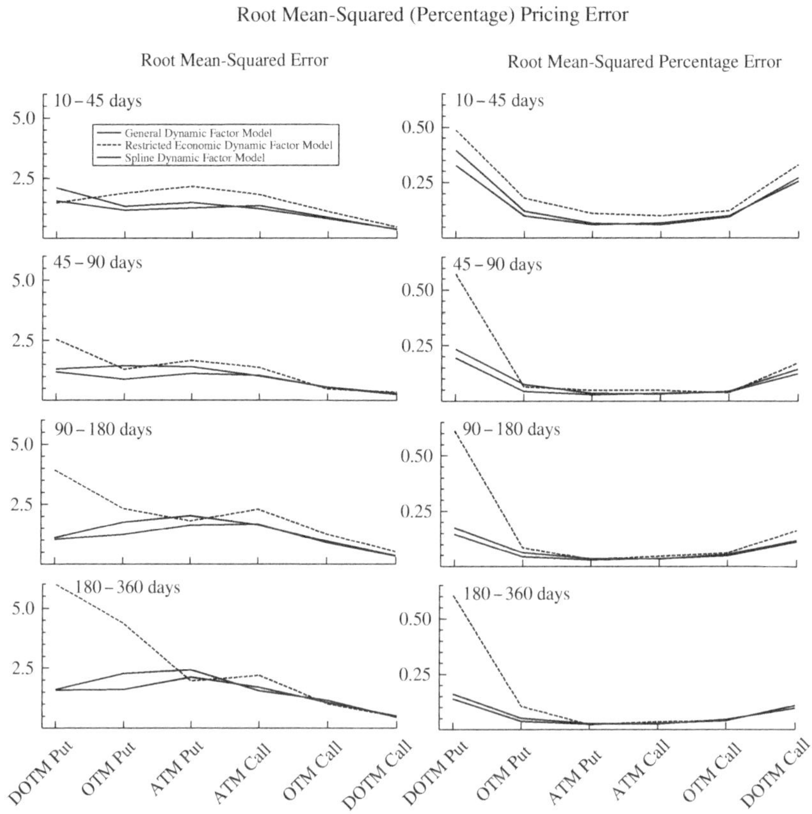

Fig. 7. Root Mean-Squared (Percentage) Error Based on Prices. *Notes:* This figure shows the fitted pricing error that is implied by three of the factor models we consider. We translate the fitted implied volatility of each model at each time into an option price and compare this to the actual option price. The left figures provide the root mean-squared pricing error and the right figures the root mean-squared percentage pricing error (as a fraction, so 0.10 is 10%). In each column, the four different plots provide the different maturity categories. Each plot provides the root mean-squared (percentage) pricing error for each of the moneyness categories within the maturity category. We show the fitted pricing error for the general dynamic factor model (solid line), the restricted dynamic factor model with a constant loading matrix where moneyness is measured with Δ (dotted line), and the spline dynamic factor model where there is a separate spline for each moneyness and maturity group (dashed line).

the strike relative to the spot price. One possible explanation for this finding is that in fact it is spurious, since the construction of the surface is based on Δ. To examine this possibility and the sensitivity of our results to the surface construction, we re-estimate all DFM specifications with the surface constructed using the strike relative to the spot price. Similar to before, we consider six moneyness groups, but they are now defined based on K/S rather than Δ. The ranges of the strike-to-spot ratio that define the six groups are: $0<K/S<0.9$ (deep out-of-the-money put options), $0.9<K/S<0.95$ (out-of-the-money puts), $0.95<K/S<1$ (at-the-money puts), $1<K/S<1.05$ (at-the-money call options), $1.05<\Delta<1.1$ (out-of-the-money calls), and $K/S>1.1$ (deep out-of-the-money calls). We keep the definitions for the maturity groups the same, as given in Section 2.1. On each day, we again find the option that is closest to the middle of each maturity–moneyness group, except for the deep out-of-the-money groups, where for moneyness we consider options that are closest to 0.85 and 1.15 for put and call options, respectively.[15] Table A4 provides summary statistics for the obtained surface.

Table 4 reports the estimation results based on this alternative volatility surface data. Focusing first on the restricted models, the table shows that even when the data are constructed based on K/S, using Δ in the model for the volatility surface still provides a better fit. An explanation of the robust finding that Δ should be preferred over the strike relative to the spot price is that the latter measure does not take into account that the likelihood of the option being in-the-money at expiration depends on the (current) volatility and remaining time-to-maturity, as Bollen and Whaley (2004) point out. Broadening our view to all models in the table, the ranking of the models remains unchanged relative to the baseline results in Table 2. Also for this alternative construction of the surface, all restricted and spline-based models are rejected. The spline models still perform better than the restricted models, though the difference has become smaller and a model based on Δ with time-varying loadings outperforms two of the three spline models.

5.2. Higher-Dimensional Models

So far we have focused on three-factor models. Dumas et al. (1998) consider models of up to six factors. In this section, we consider these higher-order models to examine how they compare to the general DFM. We base our analysis on the model setup from Section 3.2. We extend Eq. (2) with additional factors to match the higher-dimensional models. Specifically, to

Table 4. Comparing Dynamic Factor Models – Surface Based on K/S.

Comparison of Factor Models – Surface based on K/S						
Model and Moneyness		Const/TV	Loglik	LR-test	#Pars	AIC
General DFM (only ident restr)			288,112.8		102	−576,021.6
Restr DFM	K/S	Const	252,946.3	70,333.0	42	−505,808.6
		TV	258,878.7	58,468.2	42	−517,673.4
	Δ	Const	255,382.1	65,461.4	42	−510,680.2
		TV	262,205.1	51,815.4	42	−524,326.2
Spline DFM	Block DFM		260,100.6	56,024.4	50	−520,101.2
	Mat and Mon Splines		259,325.6	57,574.4	47	−518,557.2
	Separate Splines		279,941.7	16,342.2	66	−559,751.4

Notes: This table provides statistics concerning the fit of the three dynamic factor models, which are estimated using the full sample of daily S&P500 implied volatility surfaces over the period January 4, 1999–August 30, 2013. The surface is constructed based on the strike price relative to the spot price. The first model is the general dynamic factor model (General DFM), where only identification restrictions have been imposed. The second set of models are the restricted factor models based on the economic literature (Restr DFM). We estimate four variants, depending on whether the loading matrix is constant (Const) or time varying (TV) and whether strike relative to stock price (K/S) or Δ is used for moneyness. The third set of models are spline-based (Spline DFM). We consider the block dynamic factor model with knots on all places (Block DFM), a model with one spline for the maturity and one spline for the moneyness dimensions (Mat and Mon Splines) and a model where there are separate splines for all maturity and moneyness groups (Separate Splines). For each model, we provide the log-likelihood value (Loglik), the number of parameters (#Pars) and the Akaike Information Criterion (AIC). For the restricted economic and spline models, we also provide a likelihood-ratio test (LR-test) relative to the general dynamic factor model.

obtain a specification with four factors, we add the square of moneyness as additional factor. For the five-factor case, we further add an interaction between moneyness and maturity and, finally, to arrive at the model with six factors we add the square of maturity. Similar to before when considering the restricted models, we study four variants depending on the type of the moneyness measure and whether the loading matrix is constant or time varying.

Table 5 provides the output from this analysis. Obviously, the log-likelihood increases with the number of factors. The increase is not that large though, leading to the conclusion that a substantial number of factors is needed to fit the volatility surface well using these restricted models. In fact, out of all the higher-order models, only the six-factor time-varying

Table 5. Comparing Higher-Order Restricted Dynamic Factor Models.

Comparison of Restricted Factor Models					
#Factors	Moneyness	Const/TV	Loglik	#Pars	AIC
4	K/S	Const	260,723.0	54	−521,338.0
		TV	266,423.5	54	−532,739.0
	Δ	Const	278,090.2	54	−556,072.4
		TV	287,811.3	54	−575,514.6
5	K/S	Const	284,037.4	69	−567,936.8
		TV	292,112.6	69	−584,087.2
	Δ	Const	284,469.8	69	−568,801.6
		TV	296,794.7	69	−593,451.4
6	K/S	Const	294,499.2	87	−588,824.4
		TV	305,685.8	87	−611,197.6
	Δ	Const	296,505.8	87	−592,837.6
		TV	311,932.4	87	−623,690.8

Notes: This table provides statistics concerning the fit of the higher-dimensional restricted dynamic factor models, which are estimated using the full sample of daily S&P500 implied volatility surfaces over the period January 4, 1999–August 30, 2013. The model with four factors explains the surface with a constant, moneyness, the square of moneyness, and maturity. The model with five factors adds an interaction between moneyness and maturity, and the six-factor model adds the square of maturity as additional factor. In all cases, we estimate four variants, depending on whether the loading matrix is constant (Const) or time varying (TV) and whether strike relative to stock price (K/S) or Δ is used for moneyness. For each model, we provide the log-likelihood value (Loglik), the number of parameters (#Pars), and the Akaike Information Criterion (AIC).

loading models perform better than a three-factor general DFM with constant loadings. For all dimensions, the time-varying loading matrix case outperforms the constant loading case. Also the conclusion that Δ is preferred over strike relative to stock price holds up irrespective of the number of factors.

5.3. *Alternative Factor Dynamics*

The results in Table 3 document a very strong persistence in the factor dynamics. We examine this aspect in more detail by considering two alternatives for the factor dynamics. As the persistence of the first factor is strongest in all model specifications, the first alternative considers a random walk for this factor while retaining an unrestricted first-order autoregressive specification for the second and third factors. The resulting VAR-coefficient model has a block-structure with 1 on the top-diagonal and a square block on the bottom-part with the coefficients for the second and third factor. As also the second and third factors seem persistent, we also run an alternative where we consider random walks for all three factors. In this case the VAR-coefficient matrix is simply an identity matrix. In these alternatives the intercept in the dynamics are estimated as μ, rather than $(I - \Phi)\mu$ as in Eq. (1).[16]

The alternatives are considered for all three main models that are reported in Table 3. In all cases, the likelihood decreases for the alternative factor dynamics. LR-tests reject the imposed restrictions on the factor dynamics. Table A5 reports the factor dynamics for the regular case and the two alternatives mentioned above. In both of the alternatives, the intercept is estimated at zero. When the level is modeled as a random walk, the lower-block with the VAR-coefficients for the second and third factor remains unchanged. Also the covariance matrix is very similar. When all factors are modeled as random walks, the covariance matrix is again very similar. These findings are similar for the restricted and spline-based models. Overall, the likelihood-ratio tests reject the imposed restrictions, but the similarity in findings when random walks are used for the factors hints that they are at least close to being nonstationary.

5.4. *Alternative Sample Period and Log-Transformation*

Fig. 2 highlights the extreme volatility during the financial crisis in 2008–2009. We examine to what extent this special period determines our key findings. First, as alternative data we consider the log of implied volatility. Second, we consider a sub-sample from 1999 through 2007 and omit the crisis period altogether. As a first examination, Table A3 documents the percentage of explained variation for these two alternative data sets. In both cases, the first three factors combined explain close to 99% of the variation. The importance of the first factor decreases somewhat in the sub-sample analysis, but overall the factor structure is also very strong outside of the crisis period.

We also re-run the main models with these two alternative data sets (results unreported). The relative importance of the models does not change and also here Δ is preferred over strike relative to spot price as measure of moneyness. The persistence of the factors remains high, but decreases somewhat for the second and third factors. In the case of the logarithmic transformation the second diagonal element of Φ is equal to 0.97 and the third one equal to 0.95, while for the sub-sample analysis these are 0.95 and 0.85, respectively.

6. CONCLUSION

This paper considers factor models for the implied volatility surface. Three main model types are considered. First, a general DFM where only identification restrictions are imposed. Second, restricted factor models that are inspired by linear models commonly used in the economic and financial literature. Third, spline-based models that offer a smooth alternative in-between the restrictive second model class and the heavily parameterized general DFM. The second and third model setups explicitly take into account that the data modeled is a three-dimensional surface.

We report three key findings. First, the economic and spline-based models are both rejected as restricted versions of the general DFM. The spline-based models are able to outperform the economic models, but even the best model is rejected by a likelihood-ratio test and not preferred based on information criteria. Second, the factors driving the surface are highly persistent. The VAR coefficient matrix has diagonal elements close to one. Third, for the restricted models Δ is preferred as a measure for moneyness over the strike relative to the spot price. Even if the surface is constructed based on the strike relative to the spot, the option Δ provides a better fit.

There are four main directions for further research. First, the spline-based models can be improved further by considering model selection in more detail. For example, in the context of term structure modeling, Jungbacker et al. (2014) offer testing procedures to select optimal "smooth loadings." A similar method can be used in the current setting. Second, due to the strong persistence it may be of interest to study nonstationary (factor) models for the volatility surface. Third, we study the factor structure using a balanced panel of 24 selected contracts each day based on pre-defined moneyness and maturity groups. An alternative is to consider the entire cross-section of individual option contracts.

A fourth and final direction for further research is to study the forecasting performance of the various models. There are a number of issues to pay attention to when using the methodology of this paper to forecast actual option implied volatilities, and thus prices. First, our current model is for 24 points on a surface, while forecasting would be most relevant at the individual option level. Second, for the restricted economic models the loading matrix depends on moneyness, which would then also need to be forecast, while for the DFM the surface would somehow need to be interpolated to price contracts not immediately in the model. Recent literature provides supportive evidence that various existing models have forecasting power for the volatility surface, see, for example, Konstantinidi et al. (2008), Chalamandaris and Tsekrekos (2010), and Bernales and Guidolin (2014). These studies consider a subset of our models and do not include the spline-based models that combine the flexibility of the general DFM with the economically plausible factor interpretation of the restricted models. A somewhat related issue is whether the option market and shape of the volatility surface can be used to forecast the level of the underlying. There are supportive results for such predictive relationships. For example, Bollerslev, Tauchen, and Zhou (2009) document that the variance risk premium (i.e., the difference between the level of implied and realized variation) has predictive ability for S&P500 returns, especially at intermediate forecast horizons (of approximately one quarter). Xing, Zhang, and Zhao (2010) find that the volatility skew can predict future equity returns. The latent factors in the models we study capture both the level and smile/skew. The benefit of the models in this paper for forecasting option prices itself and predictive power for the underlying are our main suggestions for further research.

NOTES

1. The model used to obtain the volatility surface is not crucial. Important is that a (nonlinear) unique transformation of the prices is taken, to be able to compare options using a common unit of measurement. Another interpretation of the volatility surface is that outstanding options on the same underlying with the same maturity can be used to extract an implied risk-neutral distribution, which in turn implies a particular shape of the volatility surface. An often cited reason for the existence of the volatility smile is "crashophobia," see Rubinstein (1994). We are agnostic about the causes of the volatility surface and solely focus on the econometric modeling of the surface using factor modeling techniques.

2. The Δ is defined as the sensitivity of the option price with respect to movements in the price of the underlying. The Δ is close to the implied probability that an option will end up in-the-money and is thus often used as measure for moneyness.

3. During the financial crisis, there are a few days that volatility exceeds this level. We run a robustness analysis where we remove this second criteria and all results are similar.

4. Closeness is defined by the summed squared distance for both Δ and maturity, where we put ten times more weight on Δ because the smaller values compared to maturity.

5. An alternative would be to consider the unbalanced set of all option contracts. While this poses no serious challenges for our modeling approach, we leave this for further research.

6. In stricter terms, a smile requires the implied volatility to slope upward again for high strike prices relative to the spot price. The observed pattern in Fig. 1 is better described as a smirk due to the lack of such symmetry. In spite of this, we continue referring to the pattern as a smile due to the widespread usage of this term.

7. We also implemented variants of a diffuse initialization with large variance and an initialization from the implied steady-state of the VAR. Results are qualitatively similar, though optimization for the steady-state initialization was less stable due to the strong persistence in the data.

8. To keep Δ comparable for put and call options we consider $1+\Delta$ for puts. This follows from the put–call parity, which states that the Δ of a call option minus the Δ of a put option with the same strike and maturity should be one. The necessity of this simple transformation is apparent from Table 1, where it can be seen that average Δ jumps from -0.437 to $+0.437$ when going from at-the-money put options to at-the-money call options.

9. Besides the Akaike Information Criterion we also consider the Schwartz Information Criterion which penalizes the general DFM more heavily due to its larger number of parameters. In all cases, identical conclusions are drawn based on this alternative information criterion.

10. As expected, the model fit improves as the number of knots increases. For example, when considering three knots for the maturity spline and five for the moneyness spline, the likelihood for the model with one maturity and one moneyness spline increases from 287,181.4 to 291,485.0. The location of knots also matters. A variant with the same number of knots (three and four) but on different location provides a likelihood of 288,064.8. We leave selection of the number of knots and the location of knots as a topic for further research.

11. This is done by regressing the columns of one loading matrix on the columns of the other loading matrix. The resulting square rotation matrix can be used to construct a best-fitting rotated loading matrix. By pre-multiplying the factors with the inverse of this rotation matrix the rotated factors are found.

12. These correlations are based on the covariance matrix reported in Table 3.

13. Other ingredients are the S&P500 level and the term structure. For the term structure, we use LIBOR rates with maturity 30, 60, 90, 180, and 360 days, which are linearly interpolated to get the interest rate for each option contract. The LIBOR rate is obtained from Bloomberg, the dividend yield from OptionMetrics and the index from CRSP. The maturity date in OptionMetrics is set to Saturday. We subtract two days from the time-to-maturity calculated with the OptionMetrics date to get closer to the actual maturity day of the option contract.

14. The difference between the restricted economic model and the other two displayed models is largest for the deep out-of-the-money and out-of-the-money put options for 90–180 days and 180–360 days. Ignoring these four groups gives a root mean-squared error in terms of prices of \$1.19 for the general DFM, \$1.34 for the spline-based model, and \$1.56 for the restricted economic model. The difference between the general DFM and the restricted economic model is smaller in economic terms, 37 cents, but still more than twice as large as the difference between the general DFM and the spline-based model of 15 cents.

15. When using the middle of the deep out-of-the-money groups, the options are very deep out-of-the-money and fairly illiquid. This is less the case for Δ, as this is a nonlinear measure. We also estimate all models by selecting the deep out-of-the-money groups in the middle and all results are qualitatively similar (such as the relative fit of the models), but the overall fit decreases.

16. An alternative would be to apply formal bias correction procedures to deal with the persistence of the series. A similar issue often arises in the analysis of the term structure of interest rates, see Bauer, Rudebusch, and Wu (2012) for a discussion and possible solutions.

ACKNOWLEDGMENTS

We would like to thank the editors, two referees and participants at the 16th Annual Advances in Econometrics conference for their comments. Michel van der Wel is grateful to Netherlands Organisation for Scientific Research (NWO) for a Veni grant; and acknowledges support from CREATES, funded by the Danish National Research Foundation.

REFERENCES

Barone-Adesi, G., Engle, R. F., & Mancini, L. (2008). A GARCH option pricing model with filtered historical simulation. *Review of Financial Studies*, *21*, 1223–1258.

Bauer, M., Rudebusch, G., & Wu, J. (2012) Correcting estimation bias in dynamic term structure models. *Jornal of Business and Economics Statistics*, *30*, 454–467.

Bedendo, M., & Hodges, S. D. (2009). The dynamics of the volatility skew: A Kalman filter approach. *Journal of Banking and Finance*, *33*, 1156–1165.

Bernales, A., & Guidolin, M. (2014). Can we forecast the implied volatility surface dynamics of equity options? Predictability and economic value tests. *Journal of Banking and Finance*, *46*, 326–342.

Black, F., & Scholes, M. (1973). The pricing of options and corporate liabilities. *Journal of Political Economy*, *81*, 637–654.

Bollen, N. P., & Whaley, R. E. (2004). Does net buying pressure affect the shape of implied volatility functions? *Journal of Finance*, *59*, 711–753.

Bollerslev, T., Tauchen, G., & Zhou, H. (2009). Expected stock returns and variance risk premia. *Review of Financial Studies*, *22*, 4463–4492.

Chalamandaris, G., & Tsekrekos, A. E. (2010). Predictable dynamics in implied volatility surfaces from OTC currency options. *Journal of Banking and Finance*, *34*, 1175–1188.

Christoffersen, P. F., Fournier, M., & Jacobs, K. (2013). The factor structure in equity options. Working Paper.

Christoffersen, P. F., Goyenko, R., Jacobs, K., & Karoui, M. (2012). Illiquidity premia in the equity options market. Working Paper.

Cont, R., & da Fonseca, J. (2002). Dynamics of implied volatility surfaces. *Quantitative Finance*, *2*, 45–60.

Cox, J. C., Ross, S. A., & Rubinstein, M. (1979). Option pricing: A simplified approach. *Journal of Financial Economics*, *7*, 229–263.

Dumas, B., Fleming, J., & Whaley, R. E. (1998). Implied volatility functions: Empirical tests. *Journal of Finance*, *53*, 2059–2106.

Fengler, M. R., Härdle, W. K., & Mammen, E. (2007). A semiparametric factor model for implied volatility surface dynamics. *Journal of Financial Econometrics*, *5*, 189–218.

Fengler, M. R., Härdle, W. K., & Villa, C. (2003). The dynamics of implied volatilities: A common principal components approach. *Review of Derivatives Research*, *6*, 179–202.

Geweke, J., & Zhou, G. (1996). Measuring the pricing error of the arbitrage pricing theory. *Review of Financial Studies*, *9*, 557–587.

Gonçalves, S., & Guidolin, M. (2006). Predictable dynamics in the S&P500 index options implied volatility surface. *Journal of Business*, *79*, 1591–1635.

Jungbacker, B., & Koopman, S. J. (2015). Likelihood-based dynamic factor analysis for measurement and forecasting. *Econometrics Journal*, *18*, C1–C21.

Jungbacker, B., Koopman, S. J., & van der Wel, M. (2014). Smooth dynamic factor analysis with application to the U.S. term structure of interest rates. *Journal of Applied Econometrics*, *29*, 65–90.

Konstantinidi, E., Skiadopoulos, G., & Tzagkaraki, E. (2008). Can the evolution of implied volatility be forecasted? Evidence from European and US implied volatility indices. *Journal of Banking and Finance*, *32*, 2401–2411.

Mönch, E., & Ng, S. (2011). A hierarchical factor analysis of U.S. housing market dynamics. *Econometrics Journal*, *14*, 1–24.

Park, B., Mammen, E., Härdle, W., & Borak, S. (2009). Time series modelling with semiparametric factor dynamics. *Journal of American Statistical Association*, *104*, 284–298.

Poirier, D. J. (1973). Piecewise regression using cubic spline. *Journal of American Statistical Association*, *68*, 515–524.

Poirier, D. J. (1976). *The econometrics of structural change: With special emphasis on spline functions*. Amsterdam: North-Holland.

Rubinstein, M. (1994). Implied binomial trees. *Journal of Finance*, *49*, 771–818.

Skiadopoulos, G., Hodges, S., & Clewlow, L. (1999). The Dynamics of the S&P500 Implied Volatility Surface. *Review of Derivatives Research*, *3*, 263–282.

Stock, J. H., Watson, M. W. (2002). Macroeconomic forecasting using diffusion indexes. *Jornal of Business and Economics Statistics*, *20*, 147–162.

Xing, Y., Zhang, X., Zhao, R. (2010). What does the individual option volatility smirk tell us about future equity returns? *Journal of Financial and Quantitative Analysis*, *45*, 641–662.

APPENDIX: TABLES & FIGURES

Table A1. Cross-Correlations of Implied Volatilities.

Maturity	Moneyness	10–45						180–360					
		DOTM	OTM	ATM	ATM	OTM	DOTM	DOTM	OTM	ATM	ATM	OTM	DOTM
		Put	Put	Put	Call	Call	Call	Put	Put	Put	Call	Call	Call
10–45	DOTM Put	0.012	0.981	0.969	0.955	0.947	0.926	0.862	0.904	0.915	0.903	0.908	0.899
	OTM Put		0.009	0.996	0.982	0.977	0.958	0.852	0.906	0.922	0.908	0.915	0.909
	ATM Put			0.008	0.983	0.982	0.968	0.843	0.903	0.922	0.906	0.916	0.914
	ATM Call				0.007	0.995	0.981	0.830	0.891	0.910	0.910	0.921	0.920
	OTM Call					0.006	0.992	0.816	0.882	0.906	0.905	0.920	0.925
	DOTM Call						0.005	0.789	0.861	0.890	0.890	0.911	0.926
180–360	DOTM Put							0.007	0.981	0.964	0.955	0.935	0.896
	OTM Put								0.005	0.993	0.985	0.975	0.947
	ATM Put									0.003	0.989	0.986	0.968
	ATM Call										0.003	0.995	0.977
	OTM Call											0.002	0.989
	DOTM Call												0.002

Notes: This table reports the cross correlations and variances of implied volatilities belonging to contracts with the shortest (10–45 days) and longest (180–360) maturities at each moneyness category. The diagonal gives the variances and the upper triangle gives the correlations.

Table A2. Persistence of the Volatility Surface.

Panel a: Implied Volatilities

Moneyness	Maturity	ACF					PACF				
		1	2	3	5	20	1	2	3	5	20
DOTM Put	10–45	0.975	0.955	0.941	0.919	0.803	0.975	0.112	0.106	0.045	−0.014
	180–360	0.992	0.986	0.980	0.971	0.902	0.992	0.145	0.056	0.045	−0.036
OTM Put	10–45	0.975	0.960	0.948	0.930	0.808	0.975	0.167	0.111	0.027	−0.011
	180–360	0.991	0.985	0.980	0.970	0.900	0.991	0.142	0.073	0.050	−0.022
ATM Put	10–45	0.974	0.958	0.947	0.929	0.812	0.974	0.183	0.131	0.048	−0.033
	180–360	0.991	0.986	0.981	0.971	0.905	0.991	0.180	0.054	0.036	0.002
ATM Call	10–45	0.967	0.952	0.942	0.923	0.811	0.967	0.250	0.165	0.070	−0.035
	180–360	0.991	0.986	0.981	0.972	0.915	0.991	0.234	0.082	0.044	0.037
OTM Call	10–45	0.972	0.957	0.944	0.925	0.818	0.972	0.219	0.101	0.085	−0.024
	180–360	0.991	0.986	0.981	0.972	0.918	0.991	0.190	0.069	0.052	−0.001
DOTM Call	10–45	0.971	0.955	0.942	0.922	0.825	0.971	0.205	0.096	0.093	−0.016
	180–360	0.991	0.986	0.981	0.972	0.917	0.991	0.213	0.088	0.033	−0.014

Panel b: Slope of Volatility Smile

Maturity	ACF					PACF				
	1	2	3	5	20	1	2	3	5	20
10–45	0.930	0.882	0.846	0.799	0.655	0.930	0.130	0.085	0.098	−0.012
45–90	0.971	0.953	0.937	0.912	0.774	0.971	0.189	0.065	0.036	−0.018
90–180	0.978	0.968	0.959	0.943	0.829	0.978	0.255	0.090	0.054	−0.008
180–360	0.984	0.976	0.970	0.958	0.874	0.984	0.252	0.098	0.058	−0.011

Panel c: Slope of Volatility Term Structure

Moneyness	ACF					PACF				
	1	2	3	5	20	1	2	3	5	20
DOTM Put	0.934	0.889	0.861	0.821	0.648	0.934	0.137	0.131	0.065	−0.007
OTM Put	0.933	0.899	0.875	0.843	0.640	0.933	0.213	0.128	0.035	−0.015
ATM Put	0.933	0.895	0.873	0.839	0.631	0.933	0.193	0.157	0.046	−0.043
ATM Call	0.923	0.888	0.871	0.835	0.635	0.923	0.242	0.185	0.069	−0.054
OTM Call	0.932	0.897	0.869	0.831	0.632	0.932	0.219	0.091	0.088	−0.022
DOTM Call	0.925	0.886	0.853	0.811	0.631	0.925	0.209	0.085	0.092	−0.014

Notes: This table provides (partial) autocorrelations for the volatility surface. Panel (a) presents the persistence of each moneyness category for the shortest (10–45 days) and longest (180–360) maturities. Panels (b) and (c) show the persistence for the slope of the volatility smile at different maturities and the slope of the term structure at different moneyness levels, respectively. The slope of the volatility smile in each of the four maturity groups is defined as the implied volatility of the deep out-of-the-money put minus the implied volatility of the deep out-of-the-money call. The slope of the volatility term structure for each of the six moneyness groups is defined as the implied volatility of the longest maturity minus the shortest maturity. (Partial) autocorrelations are given for lags 1, 2, 3, 5, and 20.

Table A3. Principal Component Analysis.

Panel a: Explained Variation

	Regular Data		Log-Transformation		1999–2007	
	%	Cum. %	%	Cum. %	%	Cum. %
PC 1	95.19	95.19	96.00	96.00	94.72	94.72
PC 2	2.69	97.88	2.08	98.08	2.80	97.52
PC 3	1.09	98.97	1.03	99.11	1.29	98.81
PC 4	0.29	99.26	0.19	99.30	0.30	99.11
PC 5	0.22	99.49	0.17	99.48	0.24	99.35
PC 6	0.15	99.63	0.14	99.61	0.15	99.50
PC 7	0.09	99.72	0.13	99.74	0.14	99.63
PC 8	0.07	99.79	0.05	99.80	0.10	99.73
PC 9	0.04	99.84	0.03	99.83	0.05	99.78
PC 10	0.03	99.87	0.03	99.86	0.04	99.82

Panel b: Persistence of Principal Components (Regular Data)

	ACF					PACF				
	1	2	3	5	20	1	2	3	5	20
PC1	0.988	0.978	0.970	0.957	0.868	0.988	0.108	0.080	0.046	−0.018
PC2	0.958	0.934	0.917	0.891	0.747	0.958	0.189	0.132	0.046	0.002
PC3	0.879	0.826	0.787	0.729	0.508	0.879	0.236	0.118	0.085	−0.040

Notes: This table shows results of a principal component analysis on the panel of 24 implied volatilities that represent the volatility surface at each point in time. Panel a reports the percentage of variation explained by each individual principal component series (first column) and the cumulative percentage (second column). Panel b reports (partial) autocorrelations of the first three principal components for lags 1, 2, 3, 5, and 20. In addition, Panel a reports the explained variation for two alternative data sets, where either the log of implied volatility is considered or only the 1999–2007 sub-sample.

Table A4. Summary Statistics – Surface Construction Based on K/S.

		Summary Statistics							
		10–45 days		45–90 days		90–180 days		180–360 days	
		Mean	SD	Mean	SD	Mean	SD	Mean	SD
DOTM Put	Mid-Quote	1.82	3.6	6.44	5.85	14.7	9.38	31.1	15
	Impl Vol	0.344	0.0798	0.29	0.0739	0.271	0.0675	0.258	0.0599
	Δ	−0.0321	0.0323	−0.079	0.0437	−0.127	0.0488	−0.18	0.0476
	Maturity	27.5	8.67	67.4	8.55	133	23.1	270	24.7
	K/S	0.85	0.00513	0.85	0.00682	0.85	0.0108	0.85	0.0112
OTM Put	Mid-Quote	5.57	5.8	15.3	9.16	28.5	13.1	50.4	18.8
	Impl Vol	0.265	0.0824	0.245	0.0753	0.238	0.0674	0.234	0.0594
	Δ	−0.111	0.0607	−0.192	0.0576	−0.246	0.0515	−0.287	0.0461
	Maturity	27.4	8.34	67.3	8.66	133	23.8	270	26.2
	K/S	0.925	0.00359	0.925	0.00494	0.925	0.00689	0.925	0.00732
ATM Put	Mid-Quote	14.2	8.57	28.4	11.4	44.9	15.1	69.1	21
	Impl Vol	0.219	0.085	0.215	0.0756	0.216	0.0674	0.218	0.0591
	Δ	−0.293	0.0623	−0.354	0.0426	−0.377	0.0406	−0.384	0.0431
	Maturity	27.4	8.25	67.3	8.6	133	23	269	25.7
	K/S	0.975	0.00279	0.975	0.00446	0.975	0.00729	0.975	0.00724
ATM Call	Mid-Quote	11.2	8.45	25.8	12.3	44.1	17.2	73	24
	Impl Vol	0.182	0.0838	0.188	0.0734	0.194	0.0648	0.201	0.0565
	Δ	0.283	0.0889	0.386	0.0576	0.441	0.0503	0.487	0.0493
	Maturity	27.3	8.21	67.4	8.49	134	22.8	269	25.2
	K/S	1.03	0.00274	1.02	0.00441	1.03	0.0063	1.03	0.00713

Table A4. *(Continued)*

		Summary Statistics							
		10–45 days		45–90 days		90–180 days		180–360 days	
		Mean	SD	Mean	SD	Mean	SD	Mean	SD
OTM Call	Mid-Quote	2.18	3.96	8.55	8.26	21.2	14	46	21.3
	Impl Vol	0.177	0.0707	0.169	0.07	0.178	0.0639	0.188	0.056
	Δ	0.0629	0.0661	0.157	0.0902	0.258	0.0864	0.36	0.068
	Maturity	27.3	8.21	67.4	8.5	134	22.9	269	25
	K/S	1.07	0.00435	1.07	0.00511	1.07	0.00664	1.07	0.00767
DOTM Call	Mid-Quote	0.41	1.32	1.66	3.24	6.15	7.17	20.6	15.1
	Impl Vol	0.22	0.0675	0.169	0.0549	0.163	0.0574	0.172	0.0542
	Δ	0.0129	0.0218	0.0366	0.0441	0.0914	0.0705	0.195	0.0852
	Maturity	29.2	10.7	67.2	10.4	134	23	270	25.3
	K/S	1.14	0.0191	1.15	0.0143	1.15	0.0126	1.15	0.0105

Notes: This table shows summary statistics for the option data when the surface is constructed on the strike price relative to the stock price. The table provides the mean and standard deviation (SD) over time for the mid-quote (in dollars), implied volatility (Impl Vol), option Δ, maturity (in days), and strike relative to stock price (K/S). We show these numbers across four maturity groups and six moneyness groups. The maturity groups are 10–45, 45–90, 90–180, and 180–360 days. The moneyness groups are $0<K/S<0.9$ (deep out-of-the-money put options, DOTM Put), $0.9<K/S<0.95$ (out-of-the-money puts, OTM Put), $0.95<K/S<1$ (at-the-money puts, ATM Put), $1<K/S<1.05$ (at-the-money call options, ATM Call), $1.05<\Delta<1.1$ (out-of-the-money calls, OTM Call) and $K/S>1.1$ (deep out-of-the-money calls, DOTM Call). On each day, we find the option that is closest to the middle of each maturity–moneyness group, except for the DOTM groups, where for moneyness, we consider options that are closest to 0.85 and 1.15 for put and call options, respectively. The numbers in the table represent averages over time for the selected contracts in each group.

Table A5. Alternative Factor Dynamics Specifications.

	μ	Φ			$\Sigma_\eta(\times 10^{-4})$		
		$f_{1,t-1}$	$f_{2,t-1}$	$f_{3,t-1}$	$f_{1,t}$	$f_{2,t}$	$f_{3,t}$
Panel a: Free factor dynamics general DFM							
$f_{1,t}$	0.196	0.995	0.028	−0.005	1.059	−0.453	0.051
$f_{2,t}$	0.003	−0.002	0.971	0.056	−0.453	0.212	−0.022
$f_{3,t}$	0.008	0.001	0.001	0.967	0.051	−0.022	0.005
Panel b: Restricted factor dynamics DFM – Level random walk							
$f_{1,t}$	0.000	1	0	0	1.064	−0.454	0.051
$f_{2,t}$	0.000	0	0.989	−0.002	−0.454	0.211	−0.022
$f_{3,t}$	0.000	0	−0.002	0.985	0.051	−0.022	0.005
Panel c: Restricted factor dynamics DFM – All random walks							
$f_{1,t}$	0.000	1	0	0	1.064	−0.454	0.051
$f_{2,t}$	0.000	0	1	0	−0.454	0.212	−0.022
$f_{3,t}$	0.000	0	0	1	0.051	−0.022	0.005

Notes: This table provides estimates of the factor dynamics for general dynamic factor model under three alternative specifications. We show the coefficients for the general dynamic factor model where the factor dynamics are estimated as usual (Panel a), when the level factor is restricted to be a unit root (Panel b), and when all factors are unit roots (Panel c). For all three models, we plot the intercept μ, VAR coefficient matrix Φ, and the innovation variance Σ_η.

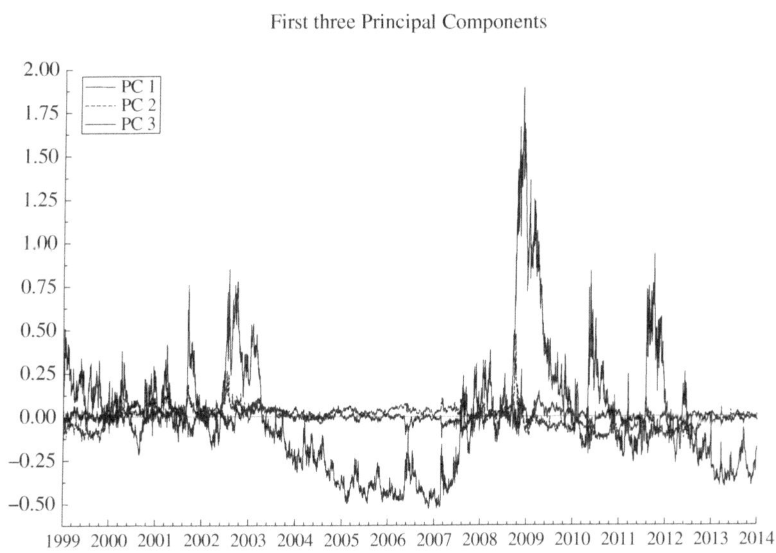

Fig. A1. Principal Components of Volatility Surface. *Notes:* This figure shows the first three principal components of the volatility surface. We run a principal component analysis on the panel of 24 implied volatilities that represent the volatility surface at each point in time. The solid line represents the first principal component, the dotted line the second principal component and the dashed line the third principal component.

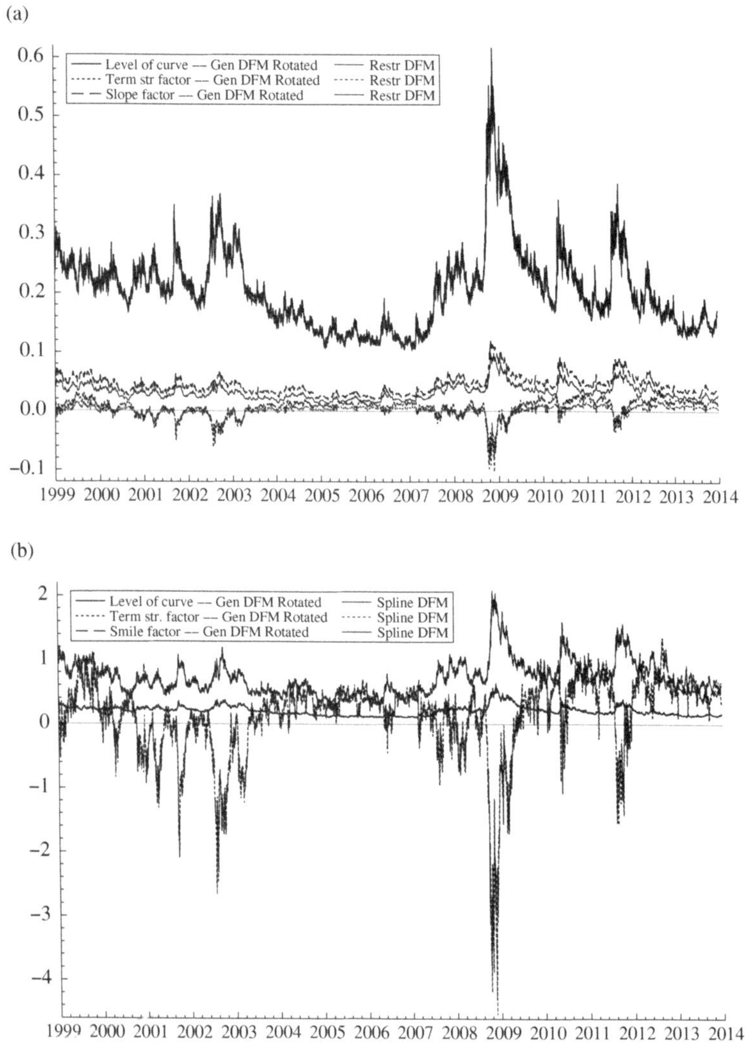

Fig. A2. Estimated Factors – Rotated to Restricted Factors. *Notes:* This figure shows rotated dynamics factors for the factor models we consider. By regressing one factor loading matrix on another factor loading matrix we bring the loading matrices and the models closer together and obtain rotated factors. In this figure, we show the resulting rotated factors of the general dynamic factor model, which are brought close to the restricted dynamic factor model with a constant loading matrix where moneyness is measured with Δ (Panel a), and close to the spline dynamic factor model where there is a separate spline for each moneyness and maturity group (Panel b). The solid line represents the first factor, the dotted line the second factor and the dashed line the third factor. The thick lines denote the rotated factors, while the thin lines denote the factors of the restricted and spline models in Panels a and b, respectively.

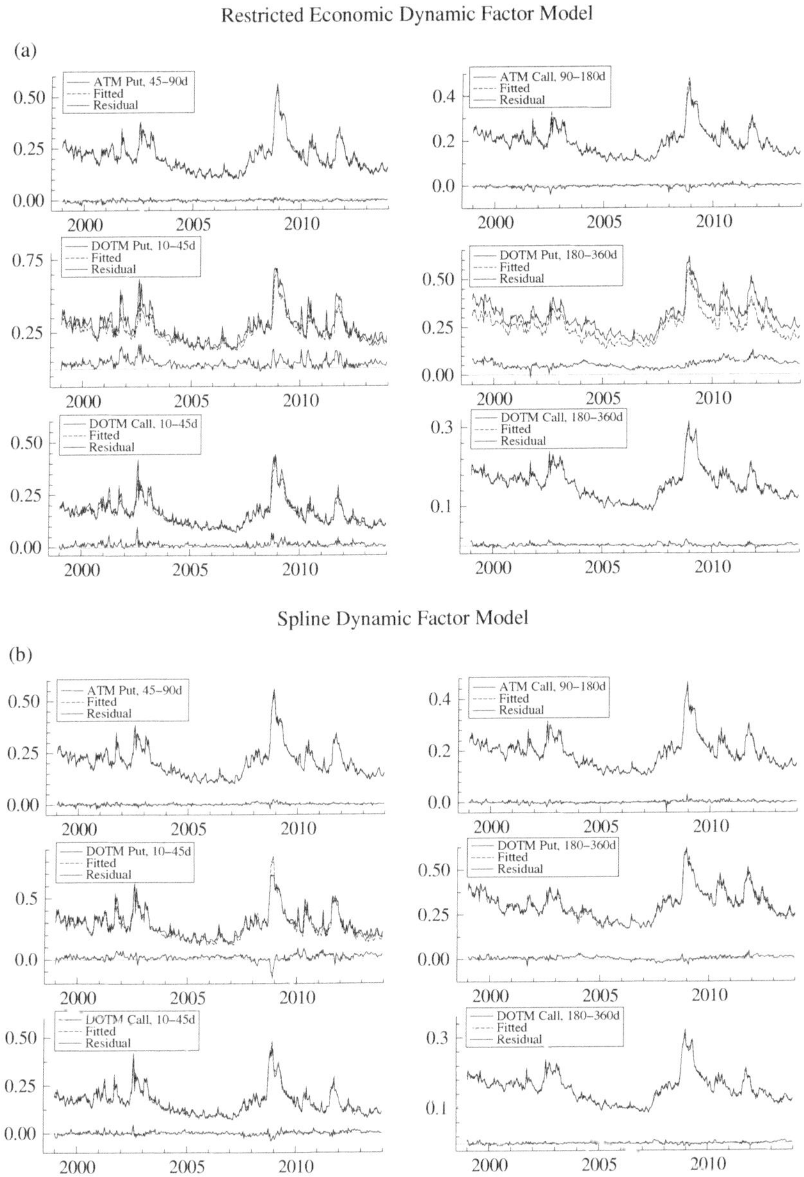

Fig. A3. Fit. *Notes:* This figure documents the fit of the volatility surface based on two dynamic factor models. The models we consider are the restricted dynamic factor model with a constant loading matrix where moneyness is measured with Δ (Panel a), and the spline dynamic factor model where there is a separate spline for each moneyness and maturity group (Panel b). We show the smoothed fitted values, which are obtained by pre-multiplying the smoothed factors with the factor loading matrix. In the figure the solid line is the actual implied volatility, the dotted line the fitted and the dashed line the residual. We show six different maturity–moneyness combinations: the top plots depict two points in the middle of the surface and the bottom plots four outer points on the surface. To compress space, the figure shows snapshots of every 10 observations.

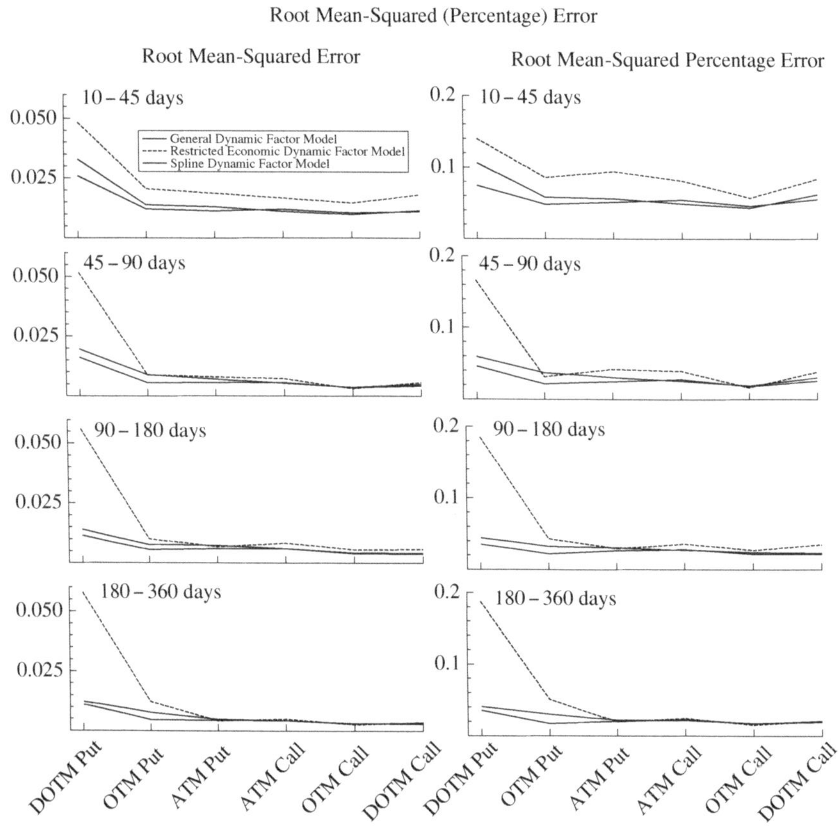

Fig. A4. Root Mean-Squared (Percentage) Error. *Notes:* This figure shows the fitting errors of three of the factor models we consider. The left figures provide the root mean-squared pricing error and the right figures the root mean-squared percentage pricing error (as a fraction, so 0.10 is 10%). In each column, the four different plots provide the different maturity categories. Each plot provides the root mean-squared (percentage) error for each of the moneyness categories within the maturity category. We show the fitting errors for the general dynamic factor model (solid line), the restricted dynamic factor model with a constant loading matrix where moneyness is measured with Δ (dotted line), and the spline dynamic factor model where there is a separate spline for each moneyness and maturity group (dashed line).

PART II
FACTOR STRUCTURE AND SPECIFICATION

ANALYZING INTERNATIONAL BUSINESS AND FINANCIAL CYCLES USING MULTI-LEVEL FACTOR MODELS: A COMPARISON OF ALTERNATIVE APPROACHES

Jörg Breitung[a,b] and Sandra Eickmeier[b,c]

[a]*Institute of Econometrics and Statistics, University of Cologne, Cologne, Germany*

[b]*Deutsche Bundesbank, Frankfurt, Germany*

[c]*Centre for Applied Macroeconomic Analysis (CAMA), The Australian National University, Canberra, Australia*

ABSTRACT

This paper compares alternative estimation procedures for multi-level factor models which imply blocks of zero restrictions on the associated matrix of factor loadings. We suggest a sequential least squares algorithm for minimizing the total sum of squared residuals and a two-step approach based on canonical correlations that are much simpler and faster than Bayesian approaches previously employed in the literature. An additional advantage is that our approaches can be used to estimate

Dynamic Factor Models
Advances in Econometrics, Volume 35, 177–214

ISSN: 0731-9053/doi:10.1108/S0731-905320150000035005

more complex multi-level factor structures where the number of levels is greater than two. Monte Carlo simulations suggest that the estimators perform well in typical sample sizes encountered in the factor analysis of macroeconomic data sets. We apply the methodologies to study international comovements of business and financial cycles.

Keywords: Factor models; canonical correlations; international business cycles; financial cycles; business cycle asymmetries

JEL classifications: C38; C55

1. INTRODUCTION

In recent years (dynamic) factor models have become increasingly popular for macroeconomic analysis and forecasting in a data-rich environment.[1] A serious limitation of the standard approximate factor model is that it assumes the common factors to affect all variables of the system. As argued by Boivin and Ng (2006), the efficiency of the Principal Component (PC) estimator may deteriorate substantially if groups of variables are included that do not provide any information about the factors, that is, the corresponding factor loadings of some subgroups of variables are equal to zero. Similarly, if factors are ignored that affect a subset of variables only, the respective idiosyncratic components may be highly correlated, resulting in poor (PC) estimates of those factors which load on all variables.

There are various examples for models with factors loading on subgroups of variables only. In an international context, for example, factors may represent regional characteristics, and it may be of independent interest to analyze these 'regional factors' in addition to 'global factors' linking all variables in the model. Alternatively, a block structure may represent economic, cultural or other characteristics. A natural way to deal with such block structures is to extract 'regional factors' from various subgroups of data (data associated with specific 'regions') separately. However, if there exist at the same time global factors that affect all regions in the sample, a separate analysis of the regions will mix up regional and global factors which hampers identification of the factors and involves a severe loss of efficiency.

A characterizing feature of such model structures is that the loading matrix of the common factors is subject to blocks of zero restrictions and the technical challenge is to take into account such restrictions when estimating the common factors. Estimating the state-space representation of the model employing Bayesian methods is most popular (see, e.g. Kose, Otrok, & Whiteman, 2003; Moench, Ng, & Potter, 2013; Kaufmann & Schumacher, 2012; Francis, Owyang, & Savascin, 2012).[2] Other recent papers adapt frequency domain PCs (Hallin & Liska, 2011), a two-step quasi maximum likelihood (ML) estimator (Banbura, Giannone, & Reichlin, 2010; Cicconi, 2012), two-stage PC approaches (e.g. Beck et al., 2009; Beck, Hubrich, & Marcellino, 2011; Aastveit, Bjoernland, & Thorsrud, 2011) or a sequential PC approach (Wang, 2010).[3]

In this paper, we make several contributions. First, we provide a comprehensive comparison of existing estimation approaches for multi-level factor models and propose two very simple alternative estimation techniques based on sequential least squares (LS) and canonical correlations. The sequential LS algorithm is equivalent to the (quasi) ML estimator assuming Gaussian i.i.d. errors and treating the common factors as unknown parameters. It is closely related to Wang (2010)'s sequential PC approach and the quasi ML approach of Banbura et al. (2010). The estimator based on a canonical correlation analysis (CCA) avoids any iterations and can be computed in two steps. In particular, we employ this computationally convenient and consistent estimator for initializing the LS algorithm in order to ensure that the procedure starts in the neighbourhood of the global minimum.

These estimation techniques provide (point) estimates in less than 0.02 seconds (in typical macroeconomic settings) compared to a Bayesian estimator that requires several hours. Moreover, our Monte Carlo simulations suggest that, in some circumstances, the sequential LS and the CCA estimators tend to outperform alternative estimation methods such as the two-step PC estimator and the quasi ML estimator based on the EM algorithm.

An additional advantage with the LS approach compared to the two-step PC approach is that it requires less stringent assumptions. The two-step PC estimator involves estimating, in the first step, the global factors as the first PCs of the full dataset. In a second step, the global factors are purged of all variables and the regional factors are extracted applying region-specific PC analyses to the residuals. In Section 2.4.1, we argue that for the consistency of this estimator we need to assume that the number of regions tends to infinity, whereas in empirical practice the number of groups is typically small (often less than 10). In such cases, the largest eigenvalues may correspond to dominating regional factors so that identification of the global factors by the largest eigenvalues breaks down.

We also extend the sequential LS estimation approach to a three-level factor model (with, e.g. global factors, regional factors and factors specific to types of variables) with overlapping blocks of factors. Such factor structures are challenging as they cannot be estimated one level after another (which is the rationale for Wang's, 2010, sequential PC approach).

A final contribution are two applications in which we study international comovements of business and financial cycles. The first application uses the two-level factor model, while the second application uses the three-level factor model assuming an overlapping factor structure.

In the first application we (basically) replicate the study by Hirata, Kose, and Otrok (forthcoming) and apply several estimation methodologies for two-level factor models to an annual real activity dataset of more than 100 countries between 1960 and 2010. We estimate global and regional factors which turn out to be similar across methods. We confirm Hirata et al. (forthcoming)'s main finding that regional (business cycle) factors have become more important and global factors less important over time.

In the second application, we use a large quarterly macro-financial dataset for 24 countries between 1995 and 2011. We estimate a global factor, regional factors, as well as factors specific to types of variables (i.e. macro and financial factors). We find that financial variables strongly comove internationally, to a similar extent as macroeconomic variables. Macroeconomic and financial dynamics share common factors, but financial factors independent from macro factors also matter for financial variables. Finally, the temporal evolution of the estimated financial factors looks plausible.

The remainder of this paper is organized as follows. We first present the two-level factor model in Section 2.1. In Sections 2.2 and 3, we then suggest a sequential LS estimator for the two-level factor model and, as an extension, the three-level factor model. In Section 2.3, we propose a CCA estimator. We show in Section 2.4.1 that the two-stage PC approach that has been used in the literature works well only under specific conditions. In Sections 2.4.2 and 2.4.3, we compare the sequential LS approach with the sequential PC and the quasi ML approaches. In Section 4, we investigate the relative performance of alternative estimators by means of Monte Carlo simulations. For ease of exposition, we assume in the methodological sections that we work with a large international dataset. We label factors associated with all variables as 'global factors' and factors associated with specific groups 'regional factors' and/or 'variable type-specific factors'. However, the models are, of course, more general and can be applied to other empirical setups with variables being associated with other groups as well. In Section 5, we present our applications, and we conclude in Section 6.

2. THE TWO-LEVEL FACTOR MODEL

2.1. The Model

Consider the following two-level factor model

$$y_{r,it} = \gamma'_{r,i} G_t + \lambda'_{r,i} F_{r,t} + u_{r,it} \tag{1}$$

where $r = 1, \ldots, R$ indicates the region, the index $i = 1, \ldots, n_r$ denotes the ith variable of region r and $t = 1, \ldots, T$ stands for the time period. The vector $G_t = \left(g_{1,t}, \ldots, g_{m_0,t}\right)'$ comprises m_0 global factors and the $m_r \times 1$ vector $F_{r,t}$ collects the m_r regional factors in region r. The idiosyncratic component is denoted by $u_{r,it}$, where the usual assumptions of an approximate factor model (e.g. Bai, 2003) apply. In vector notation, the factor model for region r is written as

$$y_{r,\cdot t} = \Gamma_r G_t + \Lambda_r F_{r,t} + u_{r,\cdot t} \tag{2}$$

$$= \begin{pmatrix} \Gamma_r & \Lambda_r \end{pmatrix} \begin{pmatrix} G_t \\ F_{r,t} \end{pmatrix} + u_{r,\cdot t} \tag{3}$$

where $y_{r,\cdot t} = \left(y_{r,1t}, \ldots, y_{r,n_r t}\right)'$ and Γ_r, Λ_r and $u_{r,\cdot t}$ are defined conformably. The entire system representing all R regions results as

$$\begin{pmatrix} y_{1,\cdot t} \\ \vdots \\ y_{R,\cdot t} \end{pmatrix} = \begin{pmatrix} \Gamma_1 & \Lambda_1 & 0 & \cdots & 0 \\ \Gamma_2 & 0 & \Lambda_2 & \cdots & 0 \\ \vdots & & & \ddots & \vdots \\ \Gamma_R & 0 & 0 & \cdots & \Lambda_R \end{pmatrix} \begin{pmatrix} G_t \\ F_{1,t} \\ F_{2,t} \\ \vdots \\ F_{R,t} \end{pmatrix} + \begin{pmatrix} u_{1,\cdot t} \\ u_{2,\cdot t} \\ \vdots \\ u_{R,\cdot t} \end{pmatrix} \tag{4}$$

$$y_t = \Lambda^* F_t^* + u_t \tag{5}$$

where $F_t^* = (G'_t, F'_{1,t}, \ldots, F'_{R,t})'$. Define the $T \times N$ matrices $Y = (y_1, \ldots, y_T)'$ and $U = (u_1, \ldots, u_T)'$. The $T \times r$ matrix of factors is given by $F = (F_1, \ldots, F_T)'$. With this notation we write the full system as

$$Y = F\Lambda^{*\prime} + U \tag{6}$$

Assume that the idiosyncratic components are identically and independent normally distributed[4] (i.i.d.) across i, t, and r with $\mathbb{E}(u_{r,it}^2) = \sigma^2$ for all r, i, t. Treating the factors and factor loadings as unknown parameters yields the log-likelihood function

$$\mathcal{L}\left(F^*, \Lambda^*, \sigma^2\right) = \text{const} - \frac{NT}{2}\log\left(\sigma^2\right) - \frac{1}{2\sigma^2}\text{tr}\left[(Y - F^*\Lambda^{*\prime})(Y - F^*\Lambda^{*\prime})'\right] \quad (7)$$

where $N = \sum_{r=1}^{R} n_r$. Since the matrix F is unrestricted, we can concentrate out these parameters with $\hat{F} = Y\Lambda^*(\Lambda^{*\prime}\Lambda^*)^{-1}$ yielding the concentrated likelihood function

$$\mathcal{L}_c\left(\Lambda^*, \sigma^2\right) = \text{const} - \frac{NT}{2}\log\left(\sigma^2\right) - \frac{1}{2\sigma^2}\text{tr}\left[Y\Lambda^*(\Lambda^{*\prime}\Lambda^*)^{-1}\Lambda^{*\prime}Y'\right]$$

Obviously, the likelihood function is invariant to any restriction preserving transformation of the loading matrix given by $\Lambda^* Q$ with

$$Q = \begin{pmatrix} Q_{00} & 0 & 0 & \cdots & 0 \\ Q_{10} & Q_{11} & 0 & \cdots & 0 \\ \vdots & & \ddots & & 0 \\ Q_{R0} & 0 & 0 & \cdots & Q_{RR} \end{pmatrix}$$

In order to identify the factors, we need to choose some nonsingular matrix Q. In what follows, we adapt the normalization common in the PC analysis, that is,

(i) $Q_{00} = \left(T^{-1}\sum_{t=1}^{T} G_t G_t'\right)^{-1/2}$ and $Q_{rr} = \left(T^{-1}\sum_{t=1}^{T} F_{r,t} F_{r,t}'\right)^{-1/2}$ for $r = 1, \ldots, R$ yielding orthonormal global and regional factors within each of the $R+1$ blocks.
(ii) The matrices $N^{-1}\Gamma_r'\Gamma_r$ and $N^{-1}\Lambda_r'\Lambda_r$ are diagonal which corresponds to the respective assumption of the PC estimator, see, for example, Breitung and Choi (2013).
(iii) The matrices Q_{k0}, $k = 1, \ldots, R$ are chosen such that the R blocks of regional factors are uncorrelated with the block of global factors.

Note that we do not need to assume that the regional factors from different regions are uncorrelated. This assumption is often imposed for a

Bayesian analysis of the multi-level factor model (e.g. Kose, Otrok, et al., 2003) and it implies an over-identified model structure.

2.2. The Sequential Least-Squares Estimator

The maximization of the likelihood function (7) is equivalent to minimizing the sum of squared residuals (RSS)

$$S(F^*, \Lambda^*) = \sum_{t=1}^{T} \left(y_t - \Lambda^* F_t^*\right)' \left(y_t - \Lambda^* F_t^*\right) \tag{8}$$

$$= \sum_{r=1}^{R} \sum_{i=1}^{n_r} \sum_{t=1}^{T} \left(y_{r,it} - \gamma_{r,i}' G_t - \lambda_{r,i}' F_{r,t}\right)^2 \tag{9}$$

Assume that we have available suitable initial estimators of the global and regional factors, denoted by $\hat{G}^{(0)} = \left(\hat{G}_1^{(0)}, \ldots, \hat{G}_T^{(0)}\right)'$ and $\hat{F}_r^{(0)} = \left(\hat{F}_{r,1}^{(0)}, \ldots, \hat{F}_{r,T}^{(0)}\right)'$. The associated loading coefficients are estimated from $\sum_{r=1}^{R} n_r$ time series regressions of the form

$$y_{r,it} = \gamma_{r,i}' \hat{G}_t^{(0)} + \lambda_{r,i}' \hat{F}_{r,t}^{(0)} + \tilde{u}_{r,it} \tag{10}$$

Denote the resulting estimates as $\hat{\gamma}_{r,i}^{(0)}$, $\hat{\lambda}_{r,i}^{(0)}$ and the respective matrices as $\hat{\Gamma}_r^{(0)} = \left(\hat{\gamma}_{r,1}^{(0)}, \ldots, \hat{\gamma}_{r,n_r}^{(0)}\right)'$ and $\hat{\Lambda}_r^{(0)} = \left(\hat{\lambda}_{r,1}^{(0)}, \ldots, \hat{\lambda}_{r,n_r}^{(0)}\right)'$. The loading matrix for the full system is constructed as

$$\hat{\Lambda}_{(0)}^* = \begin{pmatrix} \hat{\Gamma}_1^{(0)} & \hat{\Lambda}_1^{(0)} & 0 & \cdots & 0 \\ \hat{\Gamma}_2^{(0)} & 0 & \hat{\Lambda}_2^{(0)} & \cdots & 0 \\ \vdots & & & \ddots & \vdots \\ \hat{\Gamma}_R^{(0)} & 0 & 0 & \cdots & \hat{\Lambda}_R^{(0)} \end{pmatrix}$$

An updated estimator for the vector of factors is obtained from the least-squares regression of y_t on $\hat{\Lambda}_{(0)}^*$ yielding

$$\hat{F}^{*}_{t,(1)} = \begin{pmatrix} \hat{G}^{(1)}_{t} \\ \hat{F}^{(1)}_{1,t} \\ \vdots \\ \hat{F}^{(1)}_{R,t} \end{pmatrix} = \left(\hat{\Lambda}^{*\prime}_{(0)}\hat{\Lambda}^{*}_{(0)}\right)^{-1}\hat{\Lambda}^{*\prime}_{(0)}y_t \tag{11}$$

where in each step the factors are normalized to have a unit variance by multiplying the vector of factors with the matrix $\left(T^{-1}\sum_{t=1}^{T}\hat{F}^{*}_{t,(1)}\hat{F}^{*\prime}_{t,(1)}\right)^{-1/2}$.

Next, the updated factors can be used to obtain the associated loading coefficients from the least-squares regression (10), yielding the updated estimator $\hat{\Lambda}^{*}_{(1)}$ which in turn yields the updated factors $\hat{F}^{*}_{t,(2)}$. It is easy to see that

$$\mathcal{S}\left(\hat{F}^{*}_{(0)}, \hat{\Lambda}^{*}_{(0)}\right) \geq \mathcal{S}\left(\hat{F}^{*}_{(1)}, \hat{\Lambda}^{*}_{(0)}\right) \geq \mathcal{S}\left(\hat{F}^{*}_{(1)}, \hat{\Lambda}^{*}_{(1)}\right) \geq \cdots$$

since in each step the previous estimators are contained in the parameter space of the subsequent least-squares estimators. Hence, the next estimation step cannot yield a larger RSS. Any fixed point is characterized by the condition

$$\hat{\Lambda}^{*\prime}Y'Y\left(I_N - \hat{\Lambda}^{*}\left(\hat{\Lambda}^{*\prime}\hat{\Lambda}^{*}\right)^{-1}\hat{\Lambda}^{*\prime}\right) = 0 \tag{12}$$

which results from the fact that the sum of squared residuals does no longer decrease whenever the estimated factors and factor loadings are orthogonal to the residuals of the previous step.[5] Since the objective function is bounded from below by zero, there exists a set of fixed points associated with the space spanned by $\hat{\Lambda}^{*}Q$, where Q is a restriction preserving transformation matrix defined in Section 2.1. If the parameter space is identified (e.g. by using the restrictions given in Section 2.1), then the fixed point is unique and corresponds to a minimum.

To ensure that the iterative algorithm converges quickly to the global minimum, we initialize the algorithm with suitable starting values for the factors. In our Monte Carlo experiments and in the empirical applications, we employ the CCA estimator, which is considered in Section 2.3.

So far we have assumed that the idiosyncratic variances are identical for all variables and regions. Although the resulting LS estimator is consistent in the case of heteroskedastic errors (since the LS estimators are robust against heteroskedastic errors), the asymptotic efficiency may be improved by using a generalized least-squares (GLS) approach (cf. Breitung and Tenhofen (2011)).

It is important to notice that the proposed algorithm does not impose a particular normalization. Therefore, the vector of common components $\xi_t = \Lambda^* F_t^*$ is identified and consistently estimated, whereas the factors and loading matrices are estimated consistently up to some arbitrary rotation. In order to impose the normalization proposed in Section 2.1, we first regress the final estimators of the regional factors $\hat{F}_{r,t}$ $(r = 1, \ldots, R)$ on $\hat{G}_t$. The residuals from these regressions yield the orthogonalized regional factors. In order to adopt the same normalization as in the PC analysis, the normalized global factors can be obtained as the r_g PCs of the estimated common components resulting from the nonzero eigenvalues and the associated eigenvectors of the matrix

$$\hat{\Gamma}\left(\frac{1}{T}\sum_{t=1}^{T} \hat{G}_t \hat{G}_t'\right)\hat{\Gamma}'$$

The PC normalization of the regional factors can be imposed in a similar manner by using the covariance matrix of the respective common components.

Confidence intervals of the factors (or factor loadings) can be obtained from a simple bootstrap procedure. The artificial data are generated according to the estimated analog of model (4)

$$y_t^* = \hat{\Lambda}^* \hat{F}_t^* + u_t^*$$

where the errors are drawn from the empirical distribution of the idiosyncratic residuals. In order to account for the serial correlation of the errors, a block bootstrap scheme may be employed.

2.3. *The CCA Estimator*

We start with estimating the $m = m_0 + m_r$ global and regional factors in each region separately by a PC analysis yielding the vector of factors $\hat{F}_{r,t}^{+}$ which is a consistent estimator for the factor space of the $m \times 1$ vector of factors $(G_t', F_{r,t}')'$. Since the PCs of two different regions share a common component (the global factor), we apply a CCA to determine the linear combination $\hat{\mathcal{G}}_{r,t} = \tau_r' \hat{F}^+_{r,t}$ that is most correlated with the linear combination $\hat{\mathcal{G}}_{s,t} = \tau_s' \hat{F}^+_{s,t}$ of some other region s. This problem is equivalent to solving the generalized eigenvalue problem

$$\left| \mu \sum_{t=1}^{T} \hat{F}^+_{r,t} \hat{F}^{+\prime}_{r,t} - \sum_{t=1}^{T} \hat{F}^+_{r,t} \hat{F}^{+\prime}_{s,t} \left(\sum_{t=1}^{T} \hat{F}^+_{s,t} \hat{F}^{+\prime}_{s,t} \right)^{-1} \sum_{t=1}^{T} \hat{F}^+_{s,t} \hat{F}^{+\prime}_{r,t} \right| = 0$$

The eigenvectors associated with the m_0 largest eigenvalues provide the weights of the linear combination $\hat{\mathcal{G}}_{r,t} = \tau_r' \hat{F}^+_{r,t}$ which serves as an estimator of the global factors G_t. As in the appendix of Breitung and Pigorsch (2013) it can be shown that as $N \to \infty$ and $T \to \infty$, the linear combination $\hat{\mathcal{G}}_{r,t}$ (or $\hat{\mathcal{G}}_{s,t}$) converges in probability to HG_t, where H is some regular $m_0 \times m_0$ matrix. Hence, $\hat{\mathcal{G}}_{r,t}$ yields a consistent estimator of the space spanned by G_t.

Obviously, there are $R(R-1)/2$ possible pairs $(\hat{F}^+_{r,t}, \hat{F}^+_{s,t})$ that can be employed for a CCA. We suggest to choose the linear combination with the largest canonical correlation (resp. eigenvalue) as the preferred estimate of G_t. In the next step, the estimated global factors are purged of all variables and from R region-specific PC analyses the regional factors are extracted.

Formal tests or information criteria for the number of factors are, to the best of our knowledge, not yet available. Most of the empirical literature that applies multi-level factor models just fixes the number of regional and global factors to be one. We believe that it would be desirable to either extend criteria designed to determine the number of factors in one-level factor models (such as the one recently suggested by Ahn & Hornstein, 2013) to the multi-level factor model. Alternatively, one may apply the criteria of Breitung and Pigorsch (2013), developed in order to estimate the number of dynamic factors from static factor estimates based on CCA. The idea is to estimate (global and regional) factors from regional datasets and to apply CCA to the sets of estimated factors and count the number of canonical correlations that are close to unity. This should provide us with a guess of the number of global factors. Our first (admittedly tentative) experiences are however not very promising. The rather poor performance of such selection criteria may be due to the fact that the variables assigned to regional factors are typically smaller than required for a reliable choice of the respective factor dimension. Furthermore, the choice of *combinations* of m_0 and m_r seems to be much more challenging than the choice of just a single number of factors, which is already a considerable challenge in empirical practice. In our applications below, we therefore experiment with different numbers of factors. For a similar approach in the context of a one-level factor model, see, for example, Boivin and Giannoni (2008), Boivin, Giannoni, and Mojon (2008) and Buch, Eickmeier, and Prieto (2014).

2.4. Relation to Existing (Non-Bayesian) Approaches

2.4.1. Two-Step PC Estimators

Since the set of regional factors $\{F_{1,t}, \ldots, F_{R,t}\}$ is assumed to be uncorrelated with the vector of global factors G_t, the regional factors may be treated as idiosyncratic components yielding the reduced factor model

$$y_{r,it} = \gamma'_{r,i} G_t + e_{r,it}$$

where $e_{r,it} = \lambda'_{r,i} F_{r,t} + u_{r,it}$. Accordingly, the global factors may be estimated by the first m_0 PCs of the matrix $T^{-1} Y'Y$, where $Y = (Y_1, \ldots, Y_R)$ and $Y_r = (y_{r,it})$ is the $T \times n_r$ data matrix of region r. In a second step, the regional factors may be estimated again with PCs from the covariance matrix of the resulting idiosyncratic components associated with a specific region. We refer to this estimator as the 'top-down PC estimator' as this estimator starts from a PC analysis of the entire system. In empirical studies, this top-down PC estimator is employed by Beck et al. (2011), Beck et al. (2009), Aastveit et al. (2011) and Thorsrud (2013). A problem with this estimator is that the regional factors give rise to a strong correlation among the regional clusters of idiosyncratic components. Let $\tau_{r,ij} = \max_t \mathbb{E}(|e_{r,it} e_{r,jt}|)$. Since the errors possess a factor structure it follows that $\sum_{i=1}^{n_r} \sum_{j=1}^{n_r} \tau_{r,ij} = O(n_r^2)$ and, therefore,

$$\sum_{r=1}^{R} \sum_{i=1}^{n_r} \sum_{j=1}^{n_r} \tau_{r,ij} = O\left(\sum_{r=1}^{R} n_r^2\right)$$

As shown by Bai (2003) consistent estimation of the factors requires that

$$\frac{1}{\left(\sum_{r=1}^{R} n_r\right)} \sum_{r=1}^{R} \sum_{i=1}^{n_r} \sum_{j=1}^{n_r} \tau_{r,ij} \leq M < \infty$$

and, thus, $\sum_{r=1}^{R} n_r^2 / \sum_{r=1}^{R} n_r$ needs to be bounded. Obviously, this condition is fulfilled if n_r is fixed and $R \to \infty$. In empirical practice, however, n_r is large relative to R so that an asymptotic framework assuming $n_r/R \to 0$ is inappropriate in typical empirical applications.

An identical estimator would be obtained by an alternative two-stage PC estimator. Let $\hat{F}_{r,t}^{+}$ denote the vector of the first $m_0 + m_r$ PCs of the region-specific covariance matrix $T^{-1} Y_r' Y_r$. The global factor is estimated by a second PC analysis of the covariance matrix of the estimated factors $T^{-1} \sum_{i=1}^{T} \tilde{F}_t' \tilde{F}_t$ where $\tilde{F}_t = \left(\hat{F}_{1,t}^{+}, \ldots, \hat{F}_{R,t}^{+}\right)'$. This estimator may be refereed to as the 'bottom-up PC estimator'. A problem with the last PC step is that the number of regions is often too small in practice, violating the conditions established by Bai (2003) and Bai and Ng (2002) for consistent estimation of the (global) factors.

For illustration, consider the model with a single global factor G_t. The equivalence to the bottom-up and top-down PC estimators results from the fact that for the eigenvalue problem we have

$$\begin{aligned} \max_{\upsilon'\upsilon=1} \frac{\upsilon' Y' Y \upsilon}{\upsilon' \upsilon} &= \max_{\upsilon'\upsilon=1} \frac{\sum_{r=1}^{R} \upsilon_r' Y_r' Y_r \upsilon_r}{\sum_{r=1}^{R} \upsilon_r' \upsilon_r} \\ &= \max_{a_1,\ldots,a_R} \sum_{r=1}^{R} a_r^2 \max_{\upsilon_r'\upsilon_r=1} \frac{\upsilon_r' Y_r' Y_r \upsilon_r}{\upsilon_r' \upsilon_r} \quad | \text{ subject to } \sum_{r=1}^{R} a_r^2 = 1 \end{aligned} \tag{13}$$

where $\upsilon = (\upsilon_1', \ldots, \upsilon_R')'$ and $a_i^2 = \upsilon_i' \upsilon_i / \sum_{r=1}^{R} \upsilon_r' \upsilon_r$. Since $\max_{\upsilon_r'\upsilon_r=1} \upsilon_r' Y_r Y_r' \upsilon_r / \upsilon_r' \upsilon_r = \hat{F}_{1r}' \hat{F}_{1r}$ is identical to the largest region specific eigenvalue, where $\hat{F}_{1r}$ is the $T \times 1$ vector of the first PC of region r, it follows that

$$\max_{\upsilon'\upsilon=1} \frac{\upsilon' Y' Y \upsilon}{\upsilon' \upsilon} = \max_{a'a=1} a' \tilde{F}' \tilde{F} a \tag{14}$$

where $a = (a_1, \ldots, a_R)'$ and $\tilde{F} = (\hat{F}_{11}, \ldots, \hat{F}_{1R})$ is a $T \times R$ matrix of the first PCs of all regions. Accordingly, the first PC of the full sample results as the (maximum variance) linear combination of the R region-specific PCs. Thus, the bottom-up and the top-down PC estimators are equivalent.

2.4.2. The Sequential PC Approach

Wang (2010) proposes a sequential PC estimator for maximizing the log-likelihood estimator which is also based on the minimization of the RSS of the two-level factor model.

$$\mathcal{S}(F^*, \Lambda^*) = \sum_{r=1}^{R} \sum_{i=1}^{n_r} \sum_{t=1}^{T} \left(y_t - \Lambda^* F_t^*\right)' \left(y_t - \Lambda^* F_t^*\right)$$

with respect to $F^* = \left(F_1^*, \ldots, F_T^*\right)'$ and Λ^*.

Assume that we have a suitable initial estimator of the global factors, denoted by $\hat{G}^{(0)} = (\hat{G}_1^{(0)}, \ldots, \hat{G}_T^{(0)})'$. Conditional on these initial estimates, it is straightforward to obtain initial estimators of the regional factors in region r. All variables are purged of the global factor by running regressions of the variables on the estimated global factor. Then regional factors are estimated as the first m_r PCs of the sample covariance matrix

$$\hat{\Sigma}_r^{(0)} = \frac{1}{T} \sum_{t=1}^{T} Y_r' M_{\hat{G}^{(0)}} Y_r \tag{15}$$

where $Y_r = \left(y_{r,it}\right)$ is the $T \times n_r$ matrix of observations from region r and $M_{\hat{G}^0} = I_T - \hat{G}^{(0)} (\hat{G}^{(0)\prime} \hat{G}^{(0)})^{-1} \hat{G}^{(0)\prime}$. Let $\hat{F}_r^{(0)}$ be the $T \times m_r$ matrix of the resulting PCs. To eliminate the regional factors from the sample, the following R regressions are performed:

$$Y_r = \hat{F}_r^{(0)} B_r + W_r \quad \text{for } r = 1, 2, \ldots, R \tag{16}$$

Note that at this stage the assumption is imposed that the regional factors are orthogonal to the global factor (which enters the residual of this regression). Let $\hat{W} = \left(\hat{W}_1, \ldots, \hat{W}_R\right)$ denote the $T \times N$ matrix of residuals from the R regressions (16). The updated estimates of the global factors $\hat{G}_t^{(1)}$ are obtained as the first m_0 PCs obtained from the sample covariance matrix

$$\hat{\Omega}^{(1)} = \frac{1}{T} \hat{W}' \hat{W}$$

With the updated estimate of the global factors, the matrix $\hat{\Sigma}_r^{(0)}$ can be computed as in Eq. (15) but using $M_{\hat{G}^{(1)}}$ instead of $M_{\hat{G}^{(0)}}$ in order to obtain the updated estimate $\hat{F}_r^{(1)}$. These steps are repeated until convergence.

Wang (2010) initializes the algorithm either with global factors obtained with the top-down PC approach considered in Section 2.4.1 or, alternatively, with a confirmatory factor analysis given the set of admissible rotations of the regional PCs.

Since both sequential LS and PC approaches minimize the sum of squared errors, the fixed point is identical, and the approaches should yield the same estimates. The main advantage of the LS estimator is that it can be straightforwardly generalized to more than two factor levels with overlapping factor structures (as we will show in Section 3), whereas the sequential PC estimator is confined to hierarchical factor models. Second, the LS estimator is computationally less demanding and tends to be faster. Third, in models with heteroskedastic or autocorrelated errors, the sequential LS technique can be used to compute the implied ML estimator that is equivalent to minimizing the weighted sum of squared residuals (cf. Breitung & Tenhofen, 2011), which is equivalent to the (pseudo) ML estimator. It is unclear how this could be achieved with the sequential PC approach.

2.4.3. *The Quasi ML Approach*

A related estimation procedure based on quasi ML is employed in Banbura et al. (2010). The conceptual difference to the sequential LS approach is that their approach treats the factors as normally distributed random variables yielding a log-likelihood function which includes – besides the RSS (9) – an additional expression that is due to the distribution of the vector of factors. To maximize the likelihood function, an EM algorithm is adapted. It was originally proposed for the standard factor model without block structures. This approach gives rise to a shrinkage estimator of the vector of factors given by

$$\hat{F}_t^* = \left(\hat{\Lambda}^{*\prime}\hat{\Lambda}^* + \overline{\sigma}^2 I_N\right)^{-1}\hat{\Lambda}^{*\prime} y_t \tag{17}$$

where $\overline{\sigma}^2 = (NT)^{-1}S(\hat{F}^*, \hat{\Lambda}^*)$.

As $N \to \infty$ we have $\hat{\Lambda}^{*\prime}\hat{\Lambda}^*/N + \left(\overline{\sigma}^2/N\right)I_N \to \lim_{N\to\infty} \Lambda^{*\prime}\Lambda^*/N$ and, therefore, the estimators (11) and (17) are asymptotically equivalent for large N.

The computing time for the QML estimator is considerably larger than the least-squares estimator due to the different method to estimate the common factors and – as demonstrated in Section 4 – it tends to perform slightly worse. Furthermore, no extension to more complicated factor structures is available yet.

3. THE THREE-LEVEL FACTOR MODEL

The factor model can be extended to include further (overlapping) levels. Assume an international macro-financial panel, where the variables are clustered according to some additional criteria. For example, the variables may be grouped into output-related variables (e.g. production indices, employment), price variables (e.g. consumer prices, producer prices, wages) and financial variables (e.g. interest rates, stock returns). Accordingly, an additional index $k = 1, \ldots, K$ is introduced and the factor model is written as

$$y_{rk,it} = \gamma'_{rk,i} G_t + \lambda'_{rk,i} F_{r,t} + \theta'_{rk,i} H_{k,t} + u_{rk,it} \tag{18}$$

where $H_{k,t}$ is an $m_k \times 1$ vector of additional factors. The system can be cast (period-wise) as

$$\begin{pmatrix} y_{11,\cdot t} \\ \vdots \\ y_{R1,\cdot t} \\ y_{12,\cdot t} \\ \vdots \\ y_{R2,\cdot t} \\ \vdots \\ y_{1K,\cdot t} \\ \vdots \\ y_{RK,\cdot t} \end{pmatrix} = \begin{pmatrix} \Gamma_{11} & \Lambda_{11} & 0 & \cdots & 0 & \Theta_{11} & 0 & \cdots & 0 \\ \Gamma_{21} & 0 & \Lambda_{21} & \cdots & 0 & \Theta_{21} & 0 & \cdots & 0 \\ \vdots & & & \ddots & & \vdots & \vdots & & \vdots \\ \Gamma_{R1} & 0 & 0 & \cdots & \Lambda_{R1} & \Theta_{R1} & 0 & \cdots & 0 \\ \Gamma_{12} & \Lambda_{12} & 0 & \cdots & 0 & 0 & \Theta_{12} & \cdots & 0 \\ \Gamma_{22} & 0 & \Lambda_{22} & \cdots & 0 & 0 & \Theta_{22} & \cdots & 0 \\ \vdots & & & \ddots & & \vdots & \vdots & & \vdots \\ \Gamma_{R2} & 0 & 0 & \cdots & \Lambda_{R2} & 0 & \Theta_{R2} & \cdots & 0 \\ \vdots & & & \ddots & & \vdots & \vdots & & \vdots \\ \Gamma_{1K} & \Lambda_{1K} & 0 & \cdots & 0 & 0 & 0 & \cdots & \Theta_{1K} \\ \Gamma_{2K} & 0 & \Lambda_{2K} & \cdots & 0 & 0 & 0 & \cdots & \Theta_{2K} \\ \vdots & & & \ddots & & \vdots & \vdots & & \vdots \\ \Gamma_{RK} & 0 & 0 & \cdots & \Lambda_{RK} & 0 & 0 & \cdots & \Theta_{RK} \end{pmatrix} \begin{pmatrix} G_t \\ F_{1,t} \\ \vdots \\ F_{R,t} \\ H_{1,t} \\ \vdots \\ H_{K,t} \end{pmatrix} + \begin{pmatrix} u_{11,\cdot t} \\ \vdots \\ u_{R1,\cdot t} \\ \vdots \\ u_{1K,\cdot t} \\ \vdots \\ u_{RK,\cdot t} \end{pmatrix}$$

$$y_t = \Lambda^{**} F_t^{**} + u_t \tag{19}$$

To identify the parameters, we assume that $\mathbb{E}(H_{k,t}H'_{k,t}) = I_{m_k}$ as well as $\mathbb{E}(H_{k,t}G'_t) = 0$ and $\mathbb{E}(H_{k,t}F'_{r,t}) = 0$. The least-squares principle can be applied to estimate the factors and factor loadings, where the iteration adopts a sequential estimation of the factors G_t, $F_{1,t}, \ldots, F_{R,t}$, and $H_{1,t}, \ldots, H_{K,t}$. In what follows, we focus on the sequential LS procedure which is convenient to implement. Consistent starting values can be obtained from a CCA of the relevant subfactors (see below). Let $\hat{G}_t^{(0)}$, $\hat{F}_{1,t}^{(0)}, \ldots, \hat{F}_{R,t}^{(0)}$, and $\hat{H}_{1,t}^{(0)}, \ldots, \hat{H}_{K,t}^{(0)}$ denote the initial estimators. The elements of the loading matrices can be estimated by running regressions of $y_{rk,it}$ on the initial factor estimates $\hat{G}_t^{(0)}$, $\hat{F}_{1,t}^{(0)}, \ldots, \hat{F}_{R,t}^{(0)}$, and $\hat{H}_{1,t}^{(0)}, \ldots, \hat{H}_{K,t}^{(0)}$. The resulting least-squares estimators for the loading coefficients are organized as in the matrix Λ^{**}, yielding the estimator $\hat{\Lambda}^{**}$. An update of the factor estimates is obtained by running a regression of y_t on $\hat{\Lambda}^{**}$ yielding the updated vector of factors, $\hat{G}_t^{(1)\prime}, \hat{F}_{1,t}^{(1)\prime}, \ldots, \hat{F}_{R,t}^{(1)\prime}$, and $\hat{H}_{1,t}^{(1)}, \ldots, \hat{H}_{K,t}^{(1)}$. With this updated estimates of the factors, we are able to obtain improved estimates of the loading coefficients by running again regressions of $y_{rk,it}$ on the estimated factors. This sequential LS estimation procedure continues until convergence.

The last step involves orthogonalizing the two vectors of factors $\left(\hat{F}'_{1,t}, \ldots, \hat{F}'_{R,t}\right)$ and $\left(\hat{H}'_{1,t}, \ldots, \hat{H}'_{K,t}\right)$. Although this orthogonalization step is not necessary for identification of the factors, it enables us to perform a variance decomposition of individual variables with respect to the factors. Orthogonalizing the factors can be achieved by regressing $\left(\hat{F}'_{1,t}, \ldots, \hat{F}'_{R,t}\right)$ on $\left(\hat{H}'_{1,t}, \ldots, \hat{H}'_{K,t}\right)$ (or vice versa) and taking the residuals as new estimates of $\left(F'_{1,t}, \ldots, F'_{R,t}\right)$ $\left(\text{or of } \left(H'_{1,t}, \ldots, H'_{K,t}\right)\right)$. We note that the results may depend on whether we regress $\left(\hat{F}'_{1,t}, \ldots, \hat{F}'_{R,t}\right)$ on $\left(\hat{H}'_{1,t}, \ldots, \hat{H}'_{K,t}\right)$ or $\left(\hat{H}'_{1,t}, \ldots, \hat{H}'_{K,t}\right)$ on $\left(\hat{F}'_{1,t}, \ldots, \hat{F}'_{R,t}\right)$.

The initialization for the three-level factor model works as follows.

We first estimate the global factor as the first m_0 PCs and the global factors are eliminated from the variables by running least-square regressions of the variables on the estimated global factors.[6] In the next step, the CCA is employed to extract the common component among the $m_r + m_k$ estimated factors from *region* r, *group* k and the estimated vectors from the same *region* r but different *group* k'. This common component is the estimated regional factor. Similarly, the estimated factor $H_{k,t}$ is obtained from a CCA of the factor of region r, group k and a different region r' but the same group. These initial estimates are used to start the sequential LS procedure.

The overall estimation procedure outlined for the three-level factor model with an overlapping factor structure can be generalized straightforwardly to allow for further levels of factors (provided that the number of units in each group is sufficiently large). Furthermore, the levels may be specified as a hierarchical structure (e.g. Moench et al. (2013)), that is, the second level of factors (e.g. regions) is divided into a third level of factors (e.g. countries) such that each third-level group is uniquely assigned to one second-level group. For such hierarchical structures, the CCA can be adapted to yield a consistent initial estimator for a sequential estimation procedure that switches between estimating the factors and (restricted) loadings.

4. MONTE CARLO SIMULATIONS

4.1. Two-Level Factor Model

In this section, we first examine the small sample properties of the LS estimation procedure for the two-level factor model (Section 2.2). We compare them to those of the simple CCA approach (which provides us with starting values for the sequential LS approach), the two-step PC estimation procedure considered in Section 2.4.1[7] and the quasi ML approach.[8] An advantage of all these approaches is that they do not take long. By contrast, the Bayesian method requires many hours for a single estimation. Therefore, we are not able to include Bayesian methods in our Monte Carlo study, but we will compare the sequential LS estimation and the other procedures to the Bayesian approach in the first empirical application in Section 5.1.

The Monte Carlo set-up is as follows. The data are generated according to model (1) with factors following autoregressive processes with

$$G_t = 0.5G_{t-1} + \varepsilon_t, \quad \text{with } \varepsilon_t \overset{\text{iid}}{\sim} \mathcal{N}(0, 1),$$
$$F_{r,t} = 0.5F_{r,t-1} + \epsilon_t, \quad \text{with } \epsilon_t \overset{\text{iid}}{\sim} \mathcal{N}\left(0, s_f^2\right)$$

and an autoregressive idiosyncratic component:

$$u_{r,it} = 0.1u_{r,it-1} + w_{r,it}, \quad \text{with } w_{r,it} \overset{\text{iid}}{\sim} \mathcal{N}(0,1).^{9} \tag{20}$$

The two-level factor model results as

$$y_{r,it} = \gamma'_{r,i}G_t + \lambda'_{r,i}F_{r,t} + u_{r,it} \tag{21}$$

We vary the standard deviation of the regional factors $s_f \in \{0.5, 1, 2\}$ to study the effect of the importance of the regional factors relative to the global factor(s). Following Boivin and Ng (2006), factor loadings are generated as

$$\gamma_{r,i} \overset{\text{iid}}{\sim} \mathcal{N}(1,1)$$
$$\lambda_{r,i} \overset{\text{iid}}{\sim} \mathcal{N}(1,1)$$

We finally multiply the idiosyncratic components by a confirmable common factor such that the unconditional variances of the idiosyncratic component $E(u^2_{r,it})$ and the common component $E[(y_{r,it} - u_{r,it})^2]$ are identical. Accordingly, the commonality (i.e. the fraction of explained variance) is 0.5. We note that all results improve as the idiosyncratic component gets less important relative to the common component. However, the relative performances of the different methodologies remain unchanged.

We consider $R \in \{2,4\}$ regions with $n_r \in \{20, 50, 80\}$ variables in each region, and one global and one regional factor in each region. The time dimension is $T \in \{50, 200\}$. For each of the experiments, we determine the R^2 (or trace R^2) of a regression of the actual on the estimated factors based on 1,000 replications of the model.

From the Monte Carlo experiments presented in Tables 1 and 2, it turns out that the performance of the two-step PC estimator crucially depends on the relative importance of the global and regional factors. Only if the variance of the global factor is large relative to the variance of the regional factors ($s_f = 0.5$), the two-step PC estimator yields reliable estimates for the global factors, whereas the global factors are not well estimated if the regional factors dominate ($s_f = 2$).

The CCA estimator for the global factor is less sensitive to the relative importance of the global and regional factors and performs reasonably well

Table 1. Monte Carlo Simulation Results: R^2 (or Trace R^2) of a Regression of Actual on Estimates Factors (Based on 1,000 Replications) – Two-Level Factor Model.

n_r	T	R	s_f	Two-Step PC		CCA		Sequential LS		Quasi ML	
				G	F	G	F	G	F	G	F
20	50	2	0.5	0.92	0.67	0.91	0.64	0.95	0.69	0.95	0.66
20	50	2	1	0.68	0.73	0.85	0.82	0.92	0.85	0.92	0.83
20	50	2	2	0.18	0.69	0.64	0.85	0.79	0.89	0.72	0.83
20	50	4	0.5	0.96	0.73	0.97	0.71	0.98	0.72	0.98	0.71
20	50	4	1	0.86	0.85	0.95	0.86	0.96	0.86	0.96	0.85
20	50	4	2	0.33	0.84	0.84	0.89	0.89	0.90	0.89	0.89
50	50	2	0.5	0.95	0.82	0.96	0.82	0.98	0.84	0.98	0.83
50	50	2	1	0.74	0.82	0.94	0.91	0.97	0.92	0.97	0.90
50	50	2	2	0.18	0.77	0.85	0.93	0.92	0.94	0.87	0.86
50	50	4	0.5	0.98	0.86	0.99	0.85	0.99	0.85	0.99	0.85
50	50	4	1	0.88	0.91	0.98	0.92	0.98	0.93	0.98	0.91
50	50	4	2	0.34	0.89	0.94	0.94	0.96	0.94	0.95	0.90
80	50	2	0.5	0.95	0.87	0.98	0.88	0.99	0.89	0.99	0.87
80	50	2	1	0.74	0.83	0.96	0.93	0.98	0.94	0.98	0.90
80	50	2	2	0.19	0.78	0.90	0.94	0.95	0.95	0.92	0.85
80	50	4	0.5	0.98	0.90	0.99	0.90	0.99	0.90	0.99	0.89
80	50	4	1	0.89	0.92	0.99	0.94	0.99	0.94	0.99	0.92
80	50	4	2	0.35	0.90	0.96	0.95	0.97	0.95	0.97	0.88
20	200	2	0.5	0.93	0.73	0.92	0.70	0.96	0.74	0.96	0.74
20	200	2	1	0.72	0.79	0.87	0.86	0.93	0.88	0.93	0.88
20	200	2	2	0.17	0.74	0.72	0.90	0.84	0.92	0.81	0.90
20	200	4	0.5	0.96	0.77	0.97	0.76	0.98	0.76	0.98	0.77
20	200	4	1	0.88	0.88	0.95	0.88	0.96	0.89	0.97	0.88
20	200	4	2	0.35	0.87	0.88	0.92	0.91	0.92	0.91	0.92

Table 1. *(Continued)*

n_r	T	R	s_f	Two-Step PC		CCA		Sequential LS		Quasi ML	
				G	F	G	F	G	F	G	F
50	200	2	0.5	0.95	0.86	0.97	0.86	0.98	0.88	0.98	0.88
50	200	2	1	0.76	0.86	0.95	0.94	0.97	0.94	0.97	0.93
50	200	2	2	0.17	0.81	0.87	0.96	0.93	0.96	0.92	0.91
50	200	4	0.5	0.98	0.89	0.99	0.89	0.99	0.89	0.99	0.89
50	200	4	1	0.90	0.94	0.98	0.95	0.99	0.95	0.99	0.94
50	200	4	2	0.39	0.92	0.95	0.96	0.97	0.96	0.96	0.94
80	200	2	0.5	0.95	0.90	0.98	0.91	0.99	0.92	0.99	0.91
80	200	2	1	0.77	0.88	0.97	0.96	0.98	0.96	0.98	0.93
80	200	2	2	0.17	0.83	0.92	0.97	0.96	0.97	0.95	0.90
80	200	4	0.5	0.98	0.92	0.99	0.92	0.99	0.93	0.99	0.92
80	200	4	1	0.90	0.95	0.99	0.96	0.99	0.96	0.99	0.95
80	200	4	2	0.40	0.93	0.97	0.97	0.98	0.98	0.98	0.92

Note: For details on the simulation design, see the text. G: global factor, F: regional factor.

Table 2. Monte Carlo Simulation Results: R^2 (or Trace R^2) of a Regression of Actual on Estimates Factors (Based on 1,000 Replications) – Three-Level Factor Model (LS Method).

n_r	T	s_f	s_h	G	F	H	n_r	T	s_f	s_h	G	F	H
20	50	0.5	0.5	0.91	0.51	0.51	20	200	0.5	0.5	0.94	0.66	0.67
20	50	0.5	1	0.86	0.35	0.74	20	200	0.5	1	0.90	0.53	0.84
20	50	0.5	2	0.61	0.13	0.71	20	200	0.5	2	0.76	0.18	0.85
20	50	1	0.5	0.86	0.74	0.34	20	200	1	0.5	0.90	0.84	0.52
20	50	1	1	0.82	0.71	0.70	20	200	1	1	0.87	0.81	0.81
20	50	1	2	0.56	0.42	0.71	20	200	1	2	0.74	0.63	0.87
20	50	2	0.5	0.62	0.72	0.13	20	200	2	0.5	0.76	0.84	0.17
20	50	2	1	0.54	0.70	0.42	20	200	2	1	0.74	0.86	0.62
20	50	2	2	0.41	0.66	0.66	20	200	2	2	0.64	0.82	0.82
20	50	0.5	0.5	0.96	0.60	0.68	20	200	0.5	0.5	0.97	0.71	0.81
20	50	0.5	1	0.91	0.40	0.80	20	200	0.5	1	0.95	0.57	0.91
20	50	0.5	2	0.60	0.17	0.68	20	200	0.5	2	0.78	0.16	0.85
20	50	1	0.5	0.94	0.81	0.53	20	200	1	0.5	0.95	0.86	0.73
20	50	1	1	0.91	0.77	0.83	20	200	1	1	0.94	0.83	0.89
20	50	1	2	0.55	0.46	0.69	20	200	1	2	0.80	0.64	0.89
20	50	2	0.5	0.84	0.85	0.19	20	200	2	0.5	0.89	0.90	0.37
20	50	2	1	0.79	0.85	0.62	20	200	2	1	0.88	0.90	0.81
20	50	2	2	0.54	0.78	0.72	20	200	2	2	0.80	0.86	0.90
50	50	0.5	0.5	0.97	0.77	0.77	50	200	0.5	0.5	0.98	0.84	0.85
50	50	0.5	1	0.95	0.63	0.88	50	200	0.5	1	0.96	0.78	0.93
50	50	0.5	2	0.85	0.26	0.88	50	200	0.5	2	0.91	0.53	0.95
50	50	1	0.5	0.95	0.89	0.63	50	200	1	0.5	0.96	0.93	0.79
50	50	1	1	0.94	0.88	0.87	50	200	1	1	0.95	0.92	0.91
50	50	1	2	0.84	0.74	0.90	50	200	1	2	0.91	0.85	0.95
50	50	2	0.5	0.84	0.87	0.24	50	200	2	0.5	0.91	0.95	0.50
50	50	2	1	0.83	0.90	0.73	50	200	2	1	0.90	0.95	0.85
50	50	2	2	0.71	0.86	0.86	50	200	2	2	0.86	0.93	0.93
50	50	0.5	0.5	0.98	0.80	0.86	50	200	0.5	0.5	0.99	0.86	0.91

Table 2. *(Continued)*

n_r	T	s_f	s_h	G	F	H	n_r	T	s_f	s_h	G	F	H
50	50	0.5	1	0.97	0.67	0.91	50	200	0.5	1	0.98	0.81	0.96
50	50	0.5	2	0.85	0.29	0.85	50	200	0.5	2	0.95	0.55	0.97
50	50	1	0.5	0.98	0.91	0.77	50	200	1	0.5	0.98	0.94	0.88
50	50	1	1	0.97	0.89	0.92	50	200	1	1	0.98	0.92	0.95
50	50	1	2	0.87	0.77	0.90	50	200	1	2	0.95	0.86	0.97
50	50	2	0.5	0.93	0.92	0.36	50	200	2	0.5	0.96	0.96	0.71
50	50	2	1	0.92	0.93	0.84	50	200	2	1	0.95	0.96	0.92
50	50	2	2	0.79	0.90	0.88	50	200	2	2	0.93	0.94	0.96
80	50	0.5	0.5	0.98	0.84	0.85	80	200	0.5	0.5	0.98	0.90	0.90
80	50	0.5	1	0.97	0.76	0.92	80	200	0.5	1	0.98	0.85	0.95
80	50	0.5	2	0.90	0.37	0.91	80	200	0.5	2	0.95	0.69	0.97
80	50	1	0.5	0.97	0.92	0.76	80	200	1	0.5	0.98	0.96	0.85
80	50	1	1	0.96	0.91	0.91	80	200	1	1	0.97	0.94	0.94
80	50	1	2	0.91	0.84	0.94	80	200	1	2	0.94	0.90	0.97
80	50	2	0.5	0.90	0.91	0.37	80	200	2	0.5	0.94	0.97	0.68
80	50	2	1	0.91	0.93	0.83	80	200	2	1	0.94	0.97	0.90
80	50	2	2	0.84	0.91	0.91	80	200	2	2	0.91	0.96	0.96
80	50	0.5	0.5	0.99	0.87	0.90	80	200	0.5	0.5	0.99	0.91	0.94
80	50	0.5	1	0.98	0.79	0.94	80	200	0.5	1	0.99	0.87	0.97
80	50	0.5	2	0.92	0.41	0.90	80	200	0.5	2	0.97	0.72	0.98
80	50	1	0.5	0.99	0.93	0.85	80	200	1	0.5	0.99	0.96	0.92
80	50	1	1	0.98	0.92	0.94	80	200	1	1	0.98	0.95	0.97
80	50	1	2	0.94	0.86	0.94	80	200	1	2	0.97	0.91	0.98
80	50	2	0.5	0.95	0.94	0.49	80	200	2	0.5	0.97	0.97	0.82
80	50	2	1	0.95	0.95	0.90	80	200	2	1	0.97	0.97	0.94
80	50	2	2	0.90	0.93	0.93	80	200	2	2	0.95	0.96	0.97

Note: For details on the simulation design, see the text. G: global factor, F: regional factor, H: variable-specific factor.

for all values of s_f. This is due to the fact that if the regional factors are more important than the global factor, the largest eigenvalue may correspond to a regional factor instead of the global factor and the two-step PC estimator may confound global with regional factors. In contrast, our two-step estimator identifies the global factors by CCA of the (standardized) factors, which does not depend on the relative importance of the factors.

While the simple CCA approach performs already well, iterations tend to lead to small improvements on average. The sequential LS estimator (which uses CCA-based starting values) produces even more realiable estimates of the global factor in sample sizes typically encountered in macroeconomic datasets. The quasi ML approach, finally, also delivers reliable global factor estimates. The average correlation between true and estimated global factors is never smaller than 0.79 and in general larger than 0.9.

In small samples, the regional factors are less precisely estimated by all methods when they are less important than the global factors. Those estimates tend to improve substantially as the sample size increases. However, as the standard deviation of the regional factors relative to the global factor increases and as the sample size grows, the regional factors are less precisely estimated with the quasi ML approach.

For $n_r = 50$, $T = 200$ and $R = 2$, we also compare the estimated density functions of the R^2 (resp. trace R^2) of the global and regional factors as well as the computing time across methods (on average over the simulations). The estimated densities of the sample R^2 from 1,000 replications presented in Fig. 1 show that not only the correspondence between the factors obtained with the two-step PC approach tends to be smaller than the one obtained with the other methods, but also the variance is larger. The sequential procedures yield better factor estimates, but in few cases (for $s_f = 2$), the quasi ML approach delivers rather inaccurate solutions.

We have also looked at the average (trace) R^2 (means and distributions) of the sequential PC approach suggested by Wang (2010) and of the sequential LS approach, where we employ the two-step PC approach to generate the starting values for the factor estimates. As expected, we obtain virtually the same results as for the sequential LS approach with CCA-based starting values and, hence, do not show them here.

Among the two methods with no iteration, the two-step PC approach is slightly faster than the CCA approach. It takes, on average over all iterations, between 0.006 and 0.007 seconds (depending on s_f) compared to 0.008–0.009 seconds with the CCA. Notwithstanding, the sequential LS approach with CCA starting values tends to be faster (between 0.016 and

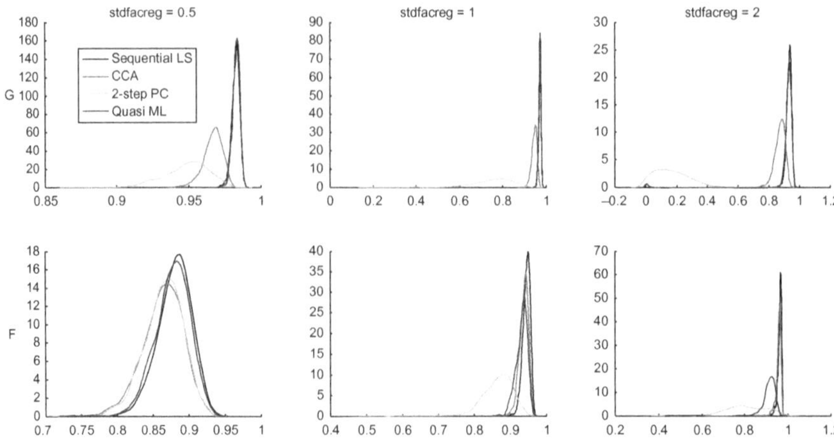

Fig. 1. Estimated Densities of R^2 (or Trace R^2) from a Regression of True on Estimated Factors (or Trace R^2) (for $T=200$, $n_g=50$, $R=2$) (1,000 Replications). *Notes*: First row: global factor, second row: regional factor. Estimated densities are obtained by smoothing with a Gaussian kernel.

0.021 seconds) than the sequential LS approach starting with the two-step PC approach (between 0.019 and 0.027 seconds). This suggests that, although the starting values do not seem to matter for the precision of the factor estimates, using improved (CCA-based) starting values leads to faster convergence of the algorithm. The sequential PC approach takes longer than both sequential LS approaches, especially as the regional factors become more important (between 0.04 and 0.12 seconds). The quasi ML approach is slower than the other methods. It takes between 0.30 and 1.08 seconds.

4.2. *Three-Level Factor Model*

We next carry out simulations for the three-level factor model using the sequential LS approach. The data are generated according to model (18). Global and regional factors are generated as before. Third-level factors (e.g. factors specific to certain types of variables) and idiosyncratic components are generated as

$$H_{k,t} = 0.5\, H_{k,t-1} + \vartheta_t, \quad \text{with } \vartheta_t \overset{\text{iid}}{\sim} \mathcal{N}\left(0, s_h^2\right)$$

$$u_{rk,it} = 0.1\, u_{rk,it-1} + w_{rk,it}, \quad \text{with } w_{rk,it} \overset{\text{iid}}{\sim} \mathcal{N}(0,1)$$

and the factor loadings $\gamma_{rk,i}$, $\lambda_{rk,i}$ and $\theta_{rk,i}$ are independently drawn from an $\mathcal{N}(1,1)$ distribution. We consider $s_f \in \{0.5, 1, 2\}$ and $s_h \in \{0.5, 1, 2\}$ to study the importance of the regional factors and the factors specific to certain types of variables relative to the global factors, respectively. We further assume that each variable is driven by $m_0 = 1$ global factor, $m_r = 1$ regional factor and $m_k = 1$ variable type-specific factor. We consider $R=2$ regions with $n_r \in \{20, 50, 80\}$ variables in each region and $N/2$ variables in each of the $K=2$ groups. The time dimensions are $T \in \{50, 200\}$. Again, we multiply the idiosyncratic components by a common scalar such that the unconditional variances of idiosyncratic and common components are similar.

Overall, our simulation results suggest that in reasonably large samples, the LS approach yields very precise estimators of the factors. In small samples, global factors are also quite precisely estimated, whereas the precision of regional and variable type-specific factor estimates depends on the importance of those factors.

5. APPLICATIONS

In this section, we provide two applications of our methodology to study international business and financial comovements. The first application serves to compare the methods for estimating a two-level factor model presented in Section 2 with the Bayesian approach. The second application makes use of the three-level factor model with an overlapping factor structure as outlined in Section 3.

5.1. Comovement of International Business Cycle

There is an interest reaching far back in describing and understanding the international synchronization of business cycles. Examples for key questions that have been addressed in the literature are: Does increased trade and financial integration lead to more or less synchronization of business cycles (something which is theoretically unclear)?[9] Has there been a decoupling of emerging economies from advanced economies in recent years, for instance, due to regional or bilateral integration agreements or similar policies within regions and, hence, emergence of regional cycles?[10]

We basically replicate the analysis conducted by Hirata et al. (forthcoming) using the sequential LS and CCA methodologies, in comparison to their Bayesian approach (and to the two-step PC and the quasi ML approaches).

From their dataset of annual consumption, investment and GDP growth for 106 countries,[11] we estimate global and regional factors for the entire period 1960–2010 and separately for 1960–1984 and 1985–2010. We initially follow Hirata et al. (forthcoming) and estimate one global factor and one factor for each of seven regions (North America, Europe, Oceania, Latin America and the Carribean, Asia, Sub-Saharan Africa, Middle East and North Africa). Hirata et al. (forthcoming) also estimate country factors. We use a simplified model with no country factors given the small number of series available for each country. Nevertheless, the assumptions on the idiosyncratic components in our model are fairly flexible to account for weak correlation across variables (also within a country).

To apply the LS approach, we do not need to make assumptions on the processes for the factors and idiosyncratic components nor do we need to choose priors for the parameters. When adopting the Bayesian approach, we specify our model as in Hirata et al. (forthcoming) and refer to their study for details. The regional factors are normalized to be positively correlated with GDP growth in a large country in each region (here, the United States, Germany, Australia, Brazil, Japan, South Africa and Morocco), and the global factor is normalized to be positively correlated with US GDP growth.

Fig. 2 shows the global and regional factors estimated over the entire period 1960–2010 obtained using the different methodologies. Overall, the sets of factor estimates are similar. The LS approach suggests a somewhat less severe global recession at the end of the sample than the other approaches. All approaches attribute some of the Great Recession to the global factor. At the same time, for all regions but Africa and the Middle East another important part is attributed to the regional factors. There are also some minor differences between the levels of the African factors estimated using the Bayesian and the other methodologies over parts of the sample period.

Table 3 reveals that the regional factors estimated based on the LS approach are notably correlated across regions. The highest correlations of above 0.4 in absolute terms are found for the pairs North America with Europe and with Oceania and Africa with Latin America. Most correlations are positive, some are negative, but rather small.[12]

Table 4 shows the variance decomposition of GDP growth estimated based on the sequential LS method on average over all countries in each region for the entire sample period and the two subsamples. We find that the regional factors have become more important over time in almost all regions, and in the second subsample, they are more important than the

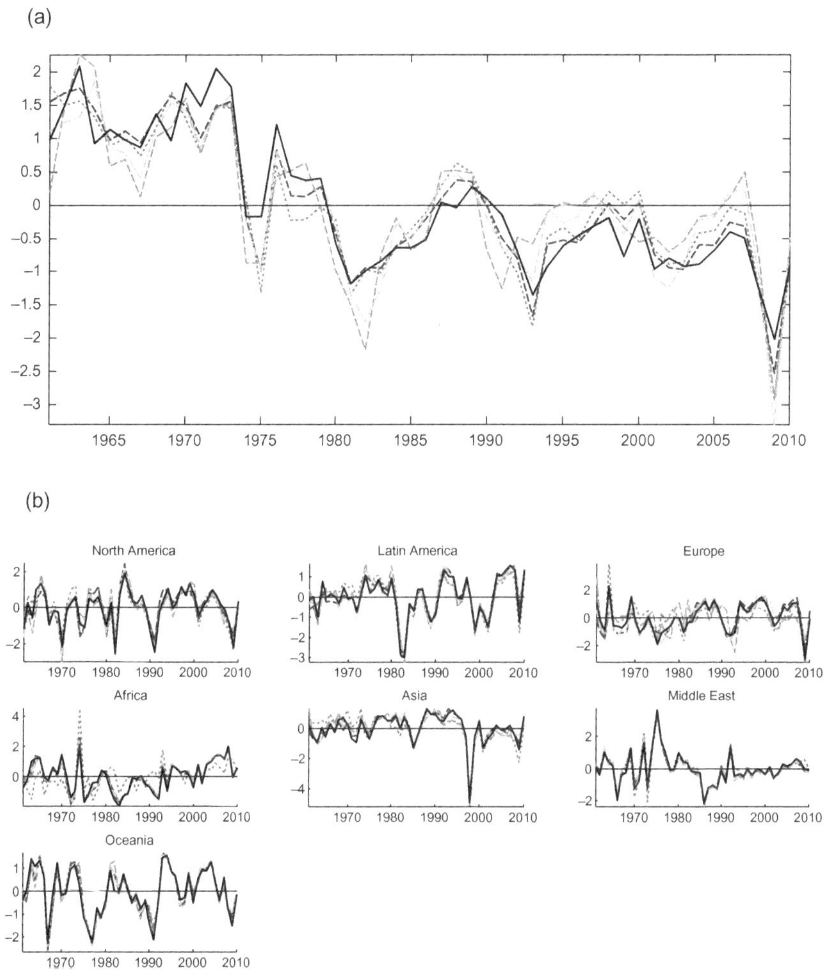

Fig. 2. Estimates for Global and Regional Factors of International Business Cycles (Dark Solid: Sequential LS, Dark Dashed: CCA, Dotted: Bayesian (Posterior Mean), Light Dashed: Quasi ML, Dotted Dashed: Two-Step PC) (Model with 1 Global Factor and 1 Regional Factor for Each Country) (Application 1). (a) Global Factor G and (b) Regional Factors F.

Table 3. Correlation between Regional Factors 1961–2010 (Sequential LS Methodology) (Application 1).

	(1)	(2)	(3)	(4)	(5)	(6)	(7)
North America (1)	1						
Latin America (2)	0.16	1					
Europe (3)	0.47	0.12	1				
Africa (4)	0.02	0.44	0.22	1			
Asia (5)	−0.01	0.14	0.02	−0.12	1		
Middle East (6)	−0.13	0.15	−0.28	−0.09	−0.12	1	
Oceania (7)	0.46	−0.02	0.33	0.25	0.00	−0.05	1

Table 4. Variance Shares of GDP Growth Explained by Global and Regional Factors in Percent (1 Global Factor and 1 Regional Factor, Sequential LS Methodology) (Application 1).

	1960–2010			1960–1984			1985–2010		
	G	*F*	*G*+*F*	*G*	*F*	*G*+*F*	*G*	*F*	*G*+*F*
World	15	16	31	11	19	30	9	28	37
North America	28	58	86	35	47	82	19	73	92
Latin America	12	18	31	13	25	38	7	23	30
Europe	35	19	54	18	31	49	9	52	61
Africa	7	8	15	7	9	16	12	13	24
Asia	18	23	41	13	14	27	7	38	45
Midle East	9	16	25	4	17	21	9	22	31
Oceania	8	44	52	18	43	60	13	54	67

Note: *G*: global factor, *F*: regional factor.

global factor. Moreover, the importance of the global factor has declined over time in most regions except for the Middle East and Africa. In the latter two regions, the shares accounted for by the global factor have broadly doubled (from low shares though).

The shares explained by the common global and regional factors tend to be larger than those estimated by Hirata et al. (forthcoming). A reason might be that Hirata et al. (forthcoming) also estimate country factors while comovements among variables within a country in our approach are only implicitly accounted for by cross-correlated idiosyncratic components.

As a robustness check, we also estimated the model using the sequential LS method by allowing for two global and two regional factors. The overall commonality rises by 15 percentage points compared to the model with one global factor and one regional factor in both subsamples. A comparison between the two subsamples confirms the main result from the model with one global factor and one regional factor, that is, that higher business cycle synchronization is due to a greater variance share explained by regional factors.[13] Hence, overall we confirm Hirata et al. (forthcoming)'s main result that the increased business cycle synchronization we have observed in the last decades is due to 'regionalization' rather than 'globalization'.

5.2. International Financial Linkages

In the second application, we broadly extend the previous analysis to financial cycles at the global level. We address the following main questions. (i) How strongly do financial variables in different countries comove? (ii) Are macroeconomic and financial dynamics at the global level driven by the same common factor(s)? Or are there (global) financial factors independent of macroeconomic factors? (iii) Is there something like a 'financial cycle', that is, do different groups of financial variables share a common factor, or are there factors specific to individual groups of financial variables? (iv) Are financial factors associated with financial developments in advanced or rather emerging economies or both?

It is far from clear what answers we should expect. While the global financial crisis affected financial markets and economic growth worldwide, other financial crises (such as the Asian crisis in 1997 or the Argentinian crisis in 1999–2002) only mainly affected the neighbouring emerging countries. Financial variables do not only move together during financial busts, but also in boom periods. For example, prior to the latest crisis, many countries experienced simultaneously housing and credit booms. The strong international comovement among financial variables can be explained with financial globalization having led to capital flows, an equalization of asset prices through arbitrage and confidence effects, and cross-border lending and global banks. Moreover, monetary policy has become increasingly similar, at least in advanced countries.[14]

We broadly use the dataset built by Eickmeier, Gambacorta, and Hofmann (2014). It comprises overall 348 quarterly series from 11 advanced and 13 emerging market economies over 1995–2011. Two hundred and seven series are financial and 141 are macroeconomic. The macroeconomic block includes, for each country (if available for a sufficiently long time span) price series (consumer prices, producer prices, GDP deflator) and output series (GDP, consumption, investment). The financial block contains stock and house prices, domestic and cross-border credit, interest rates (money market rates, long-term government bond yields), monetary aggregates M0 and M2 as well as implied stock market volatility. All series enter in year-on-year growth rates, except for interest rates and implied stock market volatility which enter in levels. Also, each series is demeaned, and its variance is normalized to one.[15]

We now apply our three-level factor model to the dataset. We estimate a 'global factor' G_t, which is common to all variables in our dataset. Moreover, we estimate regional factors F_t, that is, a factor specific to all variables in advanced countries (advanced economies' factor) and one specific to all variables in emerging economies (emerging economies' factor).[16] We consider only two regions because we have fewer countries in our sample than in the previous application.[17] Finally, we estimate variable type-specific factors H_t. It is unclear a priori how to divide the variables. Hence, we consider several (variable-wise) splits of the data leading to different models:[18]

- real activity series; price series; financial series (all other variables) (model 1)
- real activity series; price series; financial price series (comprising house and stock prices and implied volatility); financial quantities (comprising money and credit aggregates) (model 2)
- real activity series; price series; interest rates; stock prices; house prices; credit; monetary aggregates; implied stock market volatility.[19] (model 3)

The orthogonalization of variable type-specific and regional factors is achieved by regressing the regional factors on the variable type-specific factors.

One advantage of a finer level of disaggregation is that factors are more easily interpretable. Fig. 3, hence, shows the financial factor estimates from model 3 (which are estimates conditional on the global and regional factors). The temporal evolution looks broadly plausible. The financial boom in the mid-2000s is characterized by below average interest rate and implied volatility factors and an above average stock price factor early in the boom, as well as above average credit, money and – to a less clearer extent – house price factors later in the boom. This is consistent with various explanations for the boom and subsequent crisis, including loose monetary policy (in the United States and worldwide) (Taylor, 2009; Hofmann & Bogdanova, 2012), the 'global saving glut' (Bernanke, 2005) (which may have led to lower bond yields), strong credit growth due to deregulation on financial markets (Eickmeier et al., 2014) and major changes in the housing sector. It is interesting that the housing boom is

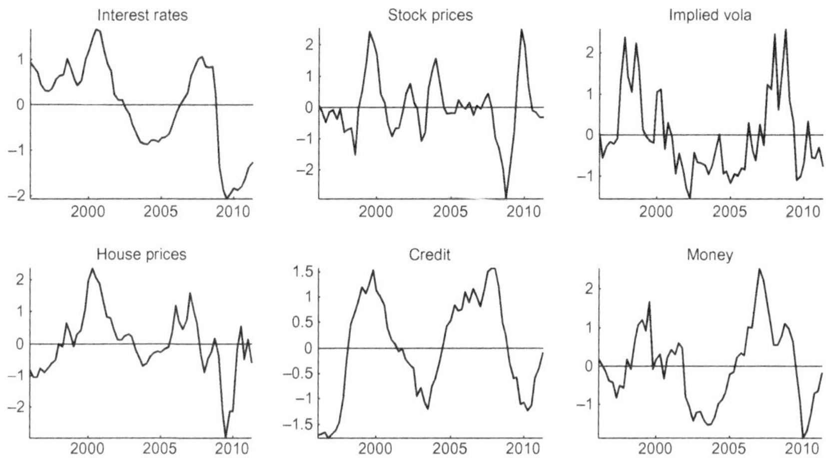

Fig. 3. Variable-Specific Factor Estimates from Model 3 (Sequential LS Methodology, Application 2). *Note*: The factors are normalized as described in the main text.

indeed reflected somewhat in the global housing factor, even though the increase in house prices was not shared by some major emerging and advanced countries (e.g. Thailand, Malaysia, Germany, Japan and Korea) (Andre, 2010; Ferrero, 2012). During the global financial crisis, the implied stock market volatility factor shows the greatest peak, and stock and house price factors display the deepest troughs. At the end of the sample period, we observe that the interest rate factor is still far below average, suggesting a very loose monetary policy stance. The evolution of all factors indicate sharp reversals towards improvements in financial markets, but only conditions in global stock markets seem to have fully recovered after the global financial crisis (at least temporarily).

We are now ready to answer the questions raised at the beginning of this section.

(i) Financial variables worldwide strongly comove, with variance shares explained by common factors of more than 40 percent on average over all financial variables (Table 5). The degree of synchronization among financial variables worldwide is similar to the degree of synchronization among macroeconomic variables. There is, however, a lot of heterogeneity across variables. The commonality is particularly high for fast-moving financial variables such as stock prices and interest rates and considerably lower for

Table 5. Variance Shares Explained by the Common Factors on Average Over All and Over Groups of Variables in Percent for 1995–2011 (Sequential LS Methodology) (Application 2).

	G	*F*	*H*	*u*
Model 1				
All variables	14	13	17	56
Advanced countries	10	12	20	58
Emerging market economies	18	14	13	54
All macro	11	10	26	53
All financial	17	15	10	58
All interest rates	33	17	9	41
All stock prices	11	11	38	40
All house prices	9	12	4	75
All credit	17	16	11	56
All money	11	19	4	66
All implied volatility	12	13	6	69
Model 2				
All variables	15	12	18	56
Advanced countries	12	10	20	58
Emerging market economies	18	13	15	54
All macro	11	9	25	55
All financial	18	13	13	56
All interest rates	38	18	5	39
All stock prices	11	18	26	45
All house prices	5	11	1	83
All credit	15	14	14	57
All money	12	11	18	59
All implied volatility	13	10	12	64
Model 3				
All variables	17	4	20	59
Advanced countries	15	5	22	58
Emerging market economies	19	3	18	60
All macro	19	4	15	62
All financial	16	4	24	57
All interest rates	7	2	40	50
All stock prices	32	1	34	33
All house prices	25	2	31	41
All credit	16	8	13	63
All money	17	4	15	64
All implied volatility	11	5	11	73

Note: *G*: global factor, *F*: regional factor, *H*: variable-specific factor, *u*: idiosyncratic component.

monetary and credit aggregates as well as house prices. The finding for house prices is not surprising given that houses are not tradable and that regulation and financing in housing markets differ across countries. Interestingly, the commonality is relatively low for stock price volatility. One possible explanation is that the high observed degree of worldwide comovement of financial stress or general uncertainty, which should be reflected in the volatility series, is already captured by other common (global or regional) factors.

(ii) Macroeconomic and financial dynamics are driven by the same (global and regional) factors, which explain together more than 20 percent and roughly 30 percent of the variation in macro and financial variables, respectively. This is in line with Claessens, Kose, and Terrones (2012) who illustrate strong linkages between different phases of macro and financial cycles. We find, however, that financial factors independent from macro factors also matter for financial variables, explaining between 10 and 24 percent, depending on the model. Global factors tend to be more important for financial variables than regional factors.

(iii) The overall commonality in the data (all data, but also only financial data) (i.e. the data fit) remains remarkably similar if financial variables are explained by factors specific to individual types of financial variables rather than by one single common financial factor. This is remarkable, given that we would have expected more disaggregated factors to be more highly correlated with individual series and, hence, the explained part to increase with a higher level of disaggregation. (The disaggregated financial factors in model 3 explain indeed a larger share of fluctuations in interest rates and asset prices compared to model 2, but the overall commonality does not increase, because the shares explained by the regional factors are lower in model 3 compared to model 2.) That this is not the case might suggest that it is sufficient to split the data into real activity, prices and financial variables (model 1) or real activity, prices, financial quantities and prices (model 2) (or, put differently, it might suggest existence of a 'financial cycle' or a 'financial quantity cycle' and a 'financial price cycle') and that a finer split may not be necessary. This is useful information for modellers who study the international synchronization of financial variables.

(iv) We also find that the financial factors load highly on variables from many advanced and emerging countries simultaneously with no clear regional pattern (results are not shown, but available upon request). This underlines the global nature of financial market developments.

Our main results are broadly robust once we let the sample end before the global financial crisis and once we alter the last estimation step and orthogonalize regional and variable type-specific factors by regressing variable type-specific on regional factors rather than regressing regional on variable type-specific factors, as before.

6. CONCLUDING REMARKS

In this paper, we have compared alternative estimation procedures for multi-level factor models which impose blocks of zero restrictions on the associated matrix of factor loadings. For the two-level factor model, we have suggested an estimator based on CCA and a simple sequential LS algorithm that minimizes the total sum of squared residuals. The latter estimator is related to Wang (2010)'s sequential PC estimator and to Banbura et al. (2010)'s quasi ML approach, and it is much simpler and faster than Bayesian approaches previously employed in the literature. The sequential LS and CCA estimation approaches can be applied to block structures of two or higher levels of factors (with either overlapping or hierarchical factor structures). Monte Carlo simulations suggest that the estimators perform well (in terms of precision of factor estimates and computing time) in typical sample sizes encountered in the factor analysis of macroeconomic data sets.

We have applied the methodologies to study international comovements of business and financial cycles. We first basically replicate the study by Hirata et al. (forthcoming) and also find that regional cycles have become more important and global cycles less important over time. Our factor estimates (based on sequential LS or CCA) and their (Bayesian) factor estimates are similar. We then move on to analyze the comovement of financial variables at the global level. We find that the estimated financial factors plausibly evolve over time. The international synchronization of financial variables is comparable to the comovement of macro variables. Both types of variables share common factors, but independent financial factors also seem to play an important role.

NOTES

1. For forecasting applications see, for example, Stock and Watson (2002a), Stock and Watson (2002b), Eickmeier and Ziegler (2008). For structural macro applications see, for example, Bernanke, Boivin and Eliasz (2005), Eickmeier (2007), Eickmeier and Hofmann (2013), Beck, Hubrich and Marcellino (2009), Kose, Otrok, et al. (2003).

2. The latter two papers assume that the groups of variables are unknown and are determined endogenously in the model. By contrast, the two former papers as well as the present paper determine a priori which variables are associated with which group.

3. Other studies estimate small-dimensional multi-level factor models (e.g. Gregory & Head, 1999). In our paper we focus, however, on large-dimensional models and, therefore, do not differences between those papers and ours further.

4. The assumption that the errors are i.i.d. is a simplifying assumption that is used to obtain a simple (quasi) likelihood function. The estimator remains consistent if the errors are heteroskedastic and autocorrelated, cf. Wang (2010).

5. Note that this condition is equivalent to the first-order condition for the PC estimator. In our case, however, the loading matrix is subject to zero restrictions.

6. Alternatively, a CCA between (i) the variables in region r and group k and (ii) the variables in group r' and k' with $r \neq r'$ and $k \neq k'$ may be employed to extract the common factors. In our experience the two-step top-down estimator used in our simulation performs similarly and has the advantage that the starting values are invariant with respect to a reorganization of the levels (that is interchanging regions and groups).

7. We have shown in Section 2.4.1 equivalence of the top-down and the bottom-up PC approaches and therefore, only show Monte Carlo results for one of them.

8. We are grateful to Domenico Giannone for providing us with his Matlab codes.

9. See, for example, Kose, Otrok, et al. (2003), Kose, Otrok, and Prasad (2012), Kose, Prasad, and Terrones (2003, 2007).

10. See, for example, Hirata et al. (forthcoming) and Kose et al. (2012).

11. We are grateful to Ayhan Kose for kindly sharing his dataset with us.

12. We have also verified correlations between regional factors estimated with the Bayesian approach. Those are correlated to a similar extent (which is not surprising given that factor estimates are similar), although uncorrelated factors are assumed in the underlying model. The explanation is that the Bayesian approach involves overidentifying assumptions (namely, that the regional factors are uncorrelated across regions), which are generally not satisfied by the *estimated factors*.

13. One global factor looks almost identical to the one estimated before. The other one seems to match oil price movements fairly well. It has its largest trough around the first oil price shock in 1973–1974 and another deep trough around the second oil price shock in 1979–1980 (there are no major troughs around the Gulf war and the war with Iraq 1991 and 2003, respectively). Factor plots and variance shares are available upon request.

14. There has been a general change in the strategy towards inflation targeting. Central banks now tend to react to output growth and inflation which comove

internationally. And recently, monetary policy was coordinated explicitly or implicitly to fight the crisis.

15. The dataset used originally by Eickmeier et al. (2014) also comprises lots of – less standard – US financial series as well as overnight rates and lending rates for different countries, which are not included here. Overnight and lending rates are not included in order not to give interest rates in our dataset a too large weight. Asset prices are included here, but not in the baseline model of Eickmeier et al. (2014). For more details on the dataset and transformations, we refer to their analysis.

16. Those factors are normalized to be positively correlated with US GDP (global and advanced economies' factors) and GDP of Hong Kong (emerging economies' factor).

17. Our application is an extension of Eickmeier et al. (2014) who extract factors common to all financial variables and identify them as a global monetary policy factor, a global credit supply factor and a global credit demand factor, but do not consider regional factors.

18. The variables can certainly be split also in other ways. We leave systematic assessment of the best split to future research.

19. The factors were normalized to be positively correlated with US GDP (macro factor and real activity factor), the US GDP deflator (price factor), US stock prices (stock price factor), US house prices (house price factor), US domestic credit (credit factor), US M2 (money factor) and Chinese GDP (emerging factor), US money market rate (interest rate factor), US implied stock market volatility (implied stock market volatility factor).

ACKNOWLEDGEMENTS

The opinions expressed in this paper are those of the authors and do not necessarily reflect the views of the Deutsche Bundesbank. This paper has been presented at seminars at the universities Duisburg-Essen, Padova and Tübingen and the Bundesbank, a joint Norges Bank-Bundesbank modelling workshop (Oslo), the 4th International Carlo Giannini conference (Pavia) and a factor modelling workshop (Frankfurt). We thank Knut Are Aastveit, Giovanni Caggiano, Carolina Castagnetti, Efrem Castelnuovo, Christian Schumacher and two referees for helpful comments and suggestions.

REFERENCES

Aastveit, K., Bjoernland, H., & Thorsrud, L. (2011). *The world is not enough! Small open economies and regional dependence*, Norges Bank Working Paper No. 2011/16.

Ahn, S., & Hornstein, A. (2013). Eigenvalue ratio test for the number of factors. *Econometrica*, *81*, 1203–1227.

Andre, C. (2010). *A bird's eye view of OECD housing markets*, OECD Economics Department Working Paper No. 746.

Bai, J. (2003). Inferential theory for factor models of large dimensions. *Econometrica*, *71*, 135–171.

Bai, J., & Ng, S. (2002). Determining the number of factors in approximate factor models. *Econometrica*, *70*(1), 191–221.

Banbura, M., Giannone, D., & Reichlin, L. (2010). Nowcasting. In M. P. Clements & D. F. Hendry (Eds.), *Oxford handbook of economic forecasting*. Oxford: Oxford University Press.

Beck, G., Hubrich, K., & Marcellino, M. (2009). Regional inflation dynamics within and across euro area countries and a comparison with the US. *Economic Policy*, *57*, 141–184.

Beck, G., Hubrich, K., & Marcellino, M. (2011). *On the importance of regional and sectoral shocks for price-setting*, ECB Working Paper No. 1134.

Bernanke, B. (2005). *The global saving glut and the U.S. current account deficit*, Speech at the Sandridge Lecture, Virginia Association of Economists, Richmond, VA.

Bernanke, B., Boivin, J., & Eliasz, P. (2005). Measuring the effects of monetary policy: A factor-augmented vector autoregressive (FAVAR) approach. *The Quarterly Journal of Economics*, *120*(1), 387.

Boivin, J., & Giannoni, M. (2008). Global forces and monetary effectiveness. In J. Gal & M. J. Gertler (Eds.), *International dimensions of monetary policy* (pp. 429–478). Chicago, IL: University of Chicago Press.

Boivin, J., Giannoni, M., & Mojon, B. (2008). How has the euro changed the monetary transmission? *NBER Macroeconomic Annual*, *23*, 77–125.

Boivin, J., & Ng, S. (2006). Are more data always better for factor analysis? *Journal of Econometrics*, *132*(1), 169–194.

Breitung, J., & Choi, I. (2013). Factor models. In N. Hashimzade & M. A. Thornton (Eds.), *Empirical macroeconomics*. Cheltenham: Edward Elgar.

Breitung, J., & Pigorsch, U. (2013). A canonical correlation approach for selecting the number of dynamic factors. *Oxford Bulletin of Economics and Statistics*, *75*, 23–36.

Breitung, J., & Tenhofen, J. (2011). GLS estimation of dynamic factor models. *Journal of the American Statistical Association*, *106*, 1150–1166.

Buch, C., Eickmeier, S., & Prieto, E. (2014). Macroeconomic factors and microlevel bank behavior. *Journal of Money, Credit and Banking*, *46*(4), 715–751.

Cicconi, C. (2012). *Essays on macroeconometric short-term forecasting*, Dissertation, Université Libre de Bruxelles, Chapter 3.

Claessens, S., Kose, M., & Terrones, M. (2012). How do business and financial cycles interact? *Journal of International Economics*, *87*, 178–190.

Eickmeier, S. (2007). Business cycle transmission from the US to Germany – A structural factor approach. *European Economic Review*, *51*(3), 521–551.

Eickmeier, S., Gambacorta, L., & Hofmann, B. (2014). Understanding global liquidity. *European Economic Review*, *68*, 1–18.

Eickmeier, S., & Hofmann, B. (2013). Monetary policy, housing booms and financial (im)balances. *Macroeconomic Dynamics*, *17*(4), 830–860.

Eickmeier, S., & Ziegler, C. (2008). How successful are dynamic factor models at forecastin output and ination? A meta-analytic approach. *Journal of Forecasting*, *27*, 237–265.

Ferrero, A. (2012). *House price booms, current account deficits, and low interest rates*. Federal Reserve Bank of New York, Staff Reports 541.

Francis, N., Owyang, M., & Savascin. (2012). *An endogenously clustered factor approach to international business cycles*, Federal Reserve Bank of St. Louis Working Paper No. 2012-014A.

Gregory, A., & Head, A. (1999). Common and country-specific fluctuations in productivity, investment, and the current account. *Journal of Monetary Economics*, *44*(3), 423–452.

Hallin, M., & Liska, R. (2011). Dynamic factors in the presence of blocks. *Journal of Econometrics*, *163*(1), 29–41.

Hirata, H., Kose, M., & Otrok, C. (forthcoming). Regionalization vs. globalization. In *Global interdependence, decoupling and recoupling*. Cambridge, MA: MIT Press.

Hofmann, B., & Bogdanova, B. (2012). Taylor rules and monetary policy: A global "Great deviation"? *BIS Quarterly Review*, *June*, 37–49.

Kaufmann, S., & Schumacher, C. (2012). *Finding relevant variables in sparse Bayesian factor models: Economic applications and simulation results*. Bundes bank Discussion Paper No. 29/2012.

Kose, A., Otrok, C., & Prasad, E. (2012). *Global business cycles: Convergence or decoupling? International Economic Review*, *53*(2), 511–538.

Kose, M., Otrok, C., & Whiteman, C. (2003). International business cycles: World, region, and country-specific factors. *The American Economic Review*, *93*(4), 1216–1239.

Kose, M., Prasad, E., & Terrones, M. (2003). How does globalization affect the synchronization of business cycles? *The American Economic Review*, *93*(2), 57–62.

Kose, M., Prasad, E., & Terrones, M. (2007). *How does financial globalization affect risk sharing? Patterns and channels*, IMF Working Paper No. WP/07/238.

Moench, E., Ng, S., & Potter, S. (2013). Dynamic hierarchical factor models. *Review of Economics and Statistics*, *95*(5), 1811–1817.

Stock, J., & Watson, M. (2002a). Forecasting using principal components from a large number of predictors. *Journal of the American Statistical Association*, *97*(460), 1167–1179.

Stock, J., & Watson, M. (2002b). Macroeconomic forecasting using diffusion indexes. *Journal of Business & Economic Statistics*, *20*(2), 147.

Taylor, J. B. (2009). Housing and monetary policy, *Housing, housing finance, and monetary policy. Proceedings of the Federal Reserve Bank of Kansas City Symposium in Jackson Hole*, Wyoming, pp. 463–476.

Thorsrud, L. (2013). *Global and regional business cycles. Shocks and propagations*, Norges Bank Working Paper No. 2013/08.

Wang, P. (2010). *Large dimensional factor models with a multi-level factor structure*. Mimeo, Hong Kong University of Science and Technology.

FAST ML ESTIMATION OF DYNAMIC BIFACTOR MODELS: AN APPLICATION TO EUROPEAN INFLATION

Gabriele Fiorentini[a], Alessandro Galesi[b] and Enrique Sentana[b]

[a]*Università di Firenze and RCEA, Firenze, Italy*
[b]*Center for Monetary and Financial Studies (CEMFI), Madrid, Spain*

ABSTRACT

We generalise the spectral EM algorithm for dynamic factor models in Fiorentini, Galesi, and Sentana (2014) to bifactor models with pervasive global factors complemented by regional ones. We exploit the sparsity of the loading matrices so that researchers can estimate those models by maximum likelihood with many series from multiple regions. We also derive convenient expressions for the spectral scores and information matrix, which allows us to switch to the scoring algorithm near the optimum. We explore the ability of a model with a global factor and

Dynamic Factor Models
Advances in Econometrics, Volume 35, 215–282

ISSN: 0731-9053/doi:10.1108/S0731-905320150000035006

three regional ones to capture inflation dynamics across 25 European countries over 1999–2014.

Keywords: Euro area; inflation convergence; spectral maximum likelihood; Wiener–Kolmogorov filter

JEL: C32; C38; E37; F45

1. INTRODUCTION

The dynamic factor models introduced by Geweke (1977) and Sargent and Sims (1977) constitute a flexible tool for capturing the cross-sectional and dynamic correlations between multiple series in a parsimonious way. Although single factor versions of those models prevail because their ease of interpretation and the fact that they provide a reasonable first approximation to many data sets, there is often the need to add more common factors to adequately capture the off-diagonal elements of the autocovariance matrices. When the cross-sectional dimension, N, is commensurate with the time series dimension, T, one possibility is to rely on the approximate factor structures originally introduced by Chamberlain and Rothschild (1983) in the static case, which allow for some mild contemporaneous and dynamic correlation between idiosyncratic terms. This has led many authors to rely on static, cross-sectional principal component methods (see, e.g. Bai & Ng, 2008, and the references therein), which are consistent under certain assumptions. There are two closely related issues, though. First, the cross-sectional asymptotic boundedness conditions on the eigenvalues of the autocovariance matrices of the idiosyncratic terms underlying those approximate factor models are largely meaningless in empirical situations in which N is small relative to T. And second, although the factors could be regarded as a set of parameters in any given realisation, efficiency considerations indicate that a signal extraction approach which exploits the serial correlation of common and specific factors would be more appropriate for such data sets.

In those situations in which it is natural to group the N series into R homogeneous blocks, an attractive solution is bifactor models with two types of factors:

1. Pervasive common factors that affect all N series.
2. Block factors that only affect a subset of the series, such as the ones belonging to the same country or region.

In principle, Gaussian pseudo maximum likelikelihood estimators (PMLEs) of the parameters can be obtained from the usual time domain version of the log-likelihood function computed as a by-product of the Kalman filter prediction equations or from Whittle's (1962) frequency domain asymptotic approximation. Further, once the parameters have been estimated, the Kalman smoother or its Wiener–Kolmogorov counterpart provide optimally filtered estimates of the latent factors. These estimation and filtering issues are well understood (see, e.g. Harvey, 1989), and the same can be said of their numerical implementation (see Jungbacker & Koopman, 2015). In practice, though, researchers may be reluctant to use ML because of the heavy computational burden involved, which is disproportionately larger as the number of series considered increases.

In the context of standard dynamic factor models, Shumway and Stoffer (1982), Watson and Engle (1983) and Quah and Sargent (1993) applied the EM algorithm of Dempster, Laird, and Rubin (1977) to the time domain versions of these models, thereby avoiding the computation of the likelihood function and its score. This iterative algorithm has been very popular in various areas of applied econometrics (see, e.g. Hamilton, 1990 in a different time series context). Its popularity can be attributed mainly to the efficiency of the procedure, as measured by its speed, and also to the generality of the approach, and its convergence properties (see Ruud, 1991). However, the time domain version of the EM algorithm has only been derived for dynamic factor models in which all the latent variables follow pure AR processes (see Doz, Giannone, & Reichlin, 2012, for a recent example), and works best when the effects of the common factors on the observed variables are contemporaneous, which substantially limits the class of models to which it can be successfully applied.

In a recent companion paper (Fiorentini, Galesi, & Sentana, 2014), we introduced a frequency domain version of the EM algorithm for general dynamic factor models with latent ARMA processes. We showed there that our algorithm reduces the computational burden so much that researchers can estimate such models by maximum likelihood with a large number of series even without good initial values. Instead, the emphasis of the current paper is to consider the application of the spectral EM algorithm to dynamic versions of bifactor models. In that regard, our approach differs from both the Bayesian procedures considered by Kose, Otrok, and Whiteman (2003) among many others, and the sequential procedures put forward by Breitung and Eickmeier (2014) and others.

We illustrate our algorithm with an empirical application in which we study the dynamics of European inflation rates since the creation of the

European Monetary Union (EMU). Specifically, we consider a dynamic bifactor model with a single global factor and three regional factors representing core, new entrant and outside EMU countries.

The rest of this paper is organised as follows. In Section 2, we review the properties of dynamic factor models and their filters, as well as maximum likelihood estimation in the frequency domain. Then, we derive our estimation algorithm and present a numerical evaluation of its finite sample behaviour in Section 3. This is followed by the empirical application in Section 4 and our conclusions in Section 5. Auxiliary results are gathered in appendices.

2. THEORETICAL BACKGROUND

2.1. Dynamic Bifactor Models

Let $\mathbf{y}_t$ denote a finite-dimensional vector of N observed series, which can be grouped into R different categories or blocks as follows:

$$\mathbf{y}'_t = \left(\mathbf{y}'_{1t} \cdots \mathbf{y}'_{rt} \cdots \mathbf{y}'_{Rt}\right)$$

where $\mathbf{y}_{1t}$ is of dimension N_1, $\mathbf{y}_{rt}$ of dimension N_r and $\mathbf{y}_{Rt}$ is of dimension N_R, with $N_1 + \cdots + N_r + \cdots + N_R = N$. Henceforth, we shall refer to each category as a 'region' even though they could represent alternative groupings.

To keep the notation to a minimum, we focus on models with a single global factor and a single factor per region, which suffice to illustrate our procedures. Specifically, we assume that $\mathbf{y}_t$ can be defined in the time domain by the system of dynamic stochastic difference equations

$$\left.\begin{array}{c} \mathbf{y}_{rt} = \boldsymbol{\mu}_r + \mathbf{c}_{rg}(L)x_{gt} + \mathbf{c}_{rr}(L)x_{rt} + \mathbf{u}_{rt}, \quad r = 1, \ldots, R \\ \alpha_{x_g}(L)x_{gt} = \beta_{x_g}(L)f_{gt}, \\ \alpha_{x_r}(L)x_{rt} = \beta_{x_r}(L)f_{rt}, \quad r = 1, \ldots, R \\ \alpha_{u_i}(L)u_{i,t} = \beta_{u_i}(L)v_{i,t}, \quad i = 1, \ldots, N, \\ \left(f_{gt}, f_{1t}, \ldots, f_{Rt}, v_{1t}, \ldots, v_{Nt}\right)|I_{t-1}; \boldsymbol{\mu}, \boldsymbol{\theta} \sim N\left[0, \operatorname{diag}\left(1, 1, \ldots, 1, \psi_1, \ldots, \psi_N\right)\right] \end{array}\right\} \quad (1)$$

where x_{gt} is the global factor, x_{rt} $(r = 1, \ldots, R)$ the rth regional factor, $\mathbf{u}_t = \left(\mathbf{u}'_{1t}, \ldots, \mathbf{u}'_{rt}, \ldots, \mathbf{u}'_{Rt}\right)'$ the N-specific factors,

$$\mathbf{c}_{rg}(L) = \sum_{k=-m_g}^{n_g} \mathbf{c}_{rgk} L^k \tag{2}$$

$$\mathbf{c}_{rr}(L) = \sum_{l=-m_r}^{n_r} \mathbf{c}_{rrl} L^k \tag{3}$$

for $(r = 1, \ldots, R)$ are $N_R \times 1$ vectors of possibly two-sided polynomials in the lag operator $c_{ig}(L)$ and $c_{ir}(L)$, $\alpha_{x_g}(L)$, $\alpha_{x_r}(L)$ and $\alpha_{u_i}(L)$ are one-sided polynomials of orders p_{x_g}, p_{x_r} and p_{u_i}, respectively, while $\beta_{x_g}(L)$, $\beta_{x_r}(L)$ and $\beta_{u_i}(L)$ are one-sided polynomials of orders q_{x_g}, q_{x_r} and q_{u_i}, coprime with $\alpha_{x_g}(L)$, $\alpha_{x_r}(L)$ and $\alpha_{u_i}(L)$, respectively, I_{t-1} is an information set that contains the values of $\mathbf{y}_t$ and $\mathbf{f}_t = (f_{gt}, f_{1t}, \ldots, f_{Rt})'$ up to, and including time $t-1$, $\boldsymbol{\mu}$ is the mean vector and $\boldsymbol{\theta}$ refers to all the remaining model parameters.

A specific example for a series y_{it} in region r would be

$$\left.\begin{aligned} y_{it} &= \mu_i + c_{i0g}x_{gt} + c_{i1g}x_{gt-1} + c_{i0r}x_{rt} + c_{i1r}x_{rt-1} + u_{it} \\ x_{gt} &= \alpha_{1x_g}x_{gt-1} + f_{gt} \\ x_{rt} &= \alpha_{1x_r}x_{rt-1} + \alpha_{2x_r}x_{rt-2} + f_{rt} \\ u_{it} &= \alpha_{1u_i}u_{it-1} + v_{it} \end{aligned}\right\} \tag{4}$$

Note that the dynamic nature of the model is the result of three different characteristics:

1. The serial correlation of the global and regional factors $\mathbf{x}_t' = (x_{gt}, x_{1t}, \ldots, x_{Rt})$
2. The serial correlation of the idiosyncratic factors $\mathbf{u}_t$
3. The heterogeneous dynamic impact of the global and regional factors on each of the observed variables through the country-specific distributed lag polynomials $c_{ig}(L)$ and $c_{ir}(L)$.

To some extent, characteristics 1 and 3 overlap, as one could always write any dynamic factor model in terms of white noise common factors with dynamic loadings. In this regard, the inclusion of AR polynomials in the dynamics of global and regional factors can be regarded as a parsimonious way of modelling a common infinite distributed lag in those loadings.

The main difference with respect to the standard dynamic factor models considered in Fiorentini, Galesi, and Sentana (2014) is the presence of regional factors, which allow for richer covariance relationships between series that belong to the same region (see, e.g. Stock & Watson, 2009).[1] As we shall see below, though, the covariance between series in different regions depends exclusively on the pervasive common factor.

Model (1) differs from the dynamic hierarchical factor model considered by Moench, Ng, and Potter (2013) in an important aspect. In their model, the common factor affects the observed series only through its effect on the regional factor. As a result, the autocovariance matrices of each block have a single factor structure and the dynamic impact of the common factor in the observed variables must involve longer distributed lags than the dynamic impact of the regional factor. As usual, the increase in parsimony involves a reduction in flexibility.

2.2. Spectral Density Matrix

Under the assumption that $\mathbf{y}_t$ is a covariance stationary process, possibly after suitable transformations as in Section 4, the spectral density matrix of the observed variables will be proportional to

$$\mathbf{G}_{\mathbf{yy}}(\lambda)=\begin{bmatrix} \mathbf{G}_{\mathbf{y}_1\mathbf{y}_1}(\lambda) & \dots & \mathbf{G}_{\mathbf{y}_1\mathbf{y}_r}(\lambda) & \dots & \mathbf{G}_{\mathbf{y}_1\mathbf{y}_R}(\lambda) \\ \vdots & \ddots & \vdots & \ddots & \vdots \\ \mathbf{G}_{\mathbf{y}_r\mathbf{y}_1}(\lambda) & \dots & \mathbf{G}_{\mathbf{y}_r\mathbf{y}_r}(\lambda) & \dots & \mathbf{G}_{\mathbf{y}_r\mathbf{y}_R}(\lambda) \\ \vdots & \ddots & \vdots & \ddots & \vdots \\ \mathbf{G}_{\mathbf{y}_R\mathbf{y}_1}(\lambda) & \dots & \mathbf{G}_{\mathbf{y}_R\mathbf{y}_r}(\lambda) & \dots & \mathbf{G}_{\mathbf{y}_R\mathbf{y}_R}(\lambda) \end{bmatrix} = \mathbf{C}\left(e^{-i\lambda}\right)\mathbf{G}_{xx}(\lambda)\mathbf{C}'\left(e^{i\lambda}\right)+\mathbf{G}_{\mathbf{uu}}(\lambda) \tag{5}$$

where

$$\mathbf{C}(z)=\begin{bmatrix} \mathbf{c}_{1g}(z) & \mathbf{c}_{11}(z) & \dots & \mathbf{0} & \dots & \mathbf{0} \\ \vdots & \vdots & \ddots & \vdots & \ddots & \vdots \\ \mathbf{c}_{rg}(z) & \mathbf{0} & \dots & \mathbf{c}_{rr}(z) & \dots & \mathbf{0} \\ \vdots & \vdots & \ddots & \vdots & \ddots & \vdots \\ \mathbf{c}_{Rg}(z) & \mathbf{0} & \dots & \mathbf{0} & \dots & \mathbf{c}_{RR}(z) \end{bmatrix} = \begin{bmatrix} \mathbf{c}_g(z) & \mathbf{C}_r(z) \end{bmatrix} \tag{6}$$

$$\mathbf{G_{xx}}(\lambda) = \text{diag}\left[G_{x_g x_g}(\lambda), G_{x_1 x_1}(\lambda), \ldots, G_{xrx_r}(\lambda) \ldots, G_{x_R x_R}(\lambda)\right],$$
$$G_{x_g x_g}(\lambda) = \frac{\beta_{x_g}(\mathrm{e}^{-\mathrm{i}\lambda})\beta_{x_g}(\mathrm{e}^{\mathrm{i}\lambda})}{\alpha_{x_g}(\mathrm{e}^{-\mathrm{i}\lambda})\alpha_{x_g}(\mathrm{e}^{\mathrm{i}\lambda})}, \quad G_{x_r x_r}(\lambda) = \frac{\beta_{x_r}(\mathrm{e}^{-\mathrm{i}\lambda})\beta_{x_r}(\mathrm{e}^{\mathrm{i}\lambda})}{\alpha_{xr}(\mathrm{e}^{-\mathrm{i}\lambda})\alpha_{x_r}(\mathrm{e}^{\mathrm{i}\lambda})}$$

and

$$\mathbf{G_{uu}}(\lambda) = \text{diag}\left[G_{u_1 u_1}(\lambda), \ldots, G_{u_N u_N}(\lambda)\right],$$
$$G_{u_i u_i}(\lambda) = \psi_i \frac{\beta_{u_i}(\mathrm{e}^{-\mathrm{i}\lambda})\beta_{u_i}(\mathrm{e}^{\mathrm{i}\lambda})}{\alpha_{u_i}(\mathrm{e}^{-\mathrm{i}\lambda})\alpha_{u_i}(\mathrm{e}^{\mathrm{i}\lambda})}$$

Thus, the matrix $\mathbf{G_{yy}}(\lambda)$ inherits the restricted $(R+1)$-factor structure of the unconditional covariance matrix of a static bifactor model with a common global factor and an additional factor per region. As a result, the cross-covariances between two series within one region will depend on the influence of both the global and regional factors on each of the series since

$$\mathbf{G}_{\mathbf{y}_r\mathbf{y}_r}(\lambda) = \mathbf{c}_{rg}(\mathrm{e}^{-\mathrm{i}\lambda})G_{x_g x_g}(\lambda)\mathbf{c}'_{rg}(\mathrm{e}^{\mathrm{i}\lambda}) + \mathbf{c}_{rr}(\mathrm{e}^{-\mathrm{i}\lambda})G_{x_r x_r}(\lambda)\mathbf{c}_{rr}(\mathrm{e}^{\mathrm{i}\lambda}) + \mathbf{G}_{\mathbf{u}_r\mathbf{u}_r}(\lambda)$$

In this regard, the assumption that the regional factors are orthogonal at all leads and lags to the global factor can be regarded as a convenient identification condition because we could easily transform a model with dynamic correlation between them by orthogonalising x_{rt} with respect to x_{gt} on a frequency-by-frequency basis.

In contrast, the cross-covariances between two series that belong to different regions will only depend on their dynamic sensitivities to the common factor because

$$\mathbf{G}_{\mathbf{y}_r\mathbf{y}_k}(\lambda) = \mathbf{c}_{rg}(\mathrm{e}^{-\mathrm{i}\lambda})G_{x_g x_g}(\lambda)\mathbf{c}'_{r'g}(\mathrm{e}^{\mathrm{i}\lambda}), \quad r \neq r'$$

For the model presented in Eq. (4),

$$G_{x_g x_g}(\lambda) = \frac{1}{\alpha_{x_g}(\mathrm{e}^{-\mathrm{i}\lambda})\alpha_{x_g}(\mathrm{e}^{\mathrm{i}\lambda})} = \frac{1}{1+\alpha_{1x_g}^2 - 2\alpha_{1x_g}\cos\lambda}$$
$$G_{x_r x_r}(\lambda) = \frac{1}{\alpha_{x_r}(\mathrm{e}^{-\mathrm{i}\lambda})\alpha_{x_r}(\mathrm{e}^{\mathrm{i}\lambda})} = \frac{1}{1+\alpha_{1x_r}^2 + \alpha_{2x_r}^2 - 2\alpha_{1x_r}(1-\alpha_{2x_r})\cos\lambda - 2\alpha_{2x_r}\cos 2\lambda}$$

where we have exploited the fact that the variances of f_{gt} and f_{rt} can be normalised to 1 for identification purposes.[2]

Similarly,

$$G_{u_i u_i}(\lambda) = \frac{\psi_i}{\alpha_{u_i}(\mathrm{e}^{-\mathrm{i}\lambda})\alpha_{u_i}(\mathrm{e}^{\mathrm{i}\lambda})} = \frac{\psi_i}{1+\alpha_{u_i}^2 - 2\alpha_{u_i}\cos\lambda}$$

Finally,

$$c_{ig}\left(\mathrm{e}^{-\mathrm{i}\lambda}\right) = c_{ig0} + c_{ig1}\mathrm{e}^{-\mathrm{i}\lambda}$$
$$c_{ir}\left(\mathrm{e}^{-\mathrm{i}\lambda}\right) = c_{ir0} + c_{ir1}\mathrm{e}^{-\mathrm{i}\lambda}$$

The fact that the idiosyncratic impact of the common factors on each of the observed variables is in principle dynamic implies that the spectral density matrix of $\mathbf{y}_t$ will generally be complex but Hermitian, even though the spectral densities of x_{gt}, x_{rt} and u_{it} are all real because they correspond to univariate processes.

2.3. *Identification*

The identification by means of homogeneous restrictions of linear dynamic models with latent variables such as Eq. (1) was discussed by Geweke (1977) and Geweke and Singleton (1981), and more recently by Scherrer and Deistler (1998) and Heaton and Solo (2004). These authors extend well-known results from static factor models and simultaneous equation systems to the spectral density matrix (5) on a frequency-by-frequency basis. Thus, two models will be observationally equivalent if and only if they generate exactly the same spectral density matrix for the observed variables at all frequencies. As in the traditional case, there are two different identification issues:

1. the non-parametric identification of global, regional and specific components,
2. the parametric identification of dynamic loadings and factor dynamics within the common components.

The answer to the first question is easy when $\mathbf{G}_{\mathbf{uu}}(\lambda)$ is a diagonal, full rank matrix.[3] Specifically, we can show that for the bifactor model (1), non-parametric identification of global, regional and idiosyncratic terms is guaranteed when $R \geq 3$ and $N_r \geq 3$ provided that at least three series in each region load on its regional factor and at least three series from three different regions load on the global factor. The intuition is as follows. We know that $N>3$ is the so-called Ledermann bound for single factor models (see, e.g. Scherrer & Deistler, 1998). If we select a single series with non-zero loadings on the global factor from each of the regions, the resulting vector will follow a single factor structure with orthogonal 'idiosyncratic' components that will be the sum of the relevant regional factors and the true idiosyncratic components for each series. Since it is not possible to transfer variance from the global to the idiosyncratic components (or vice versa) in those circumstances, and any model with more than one global factor will lead to some singular idiosyncratic variance, we can uniquely decompose $\mathbf{G}_{\mathbf{y}_r\mathbf{y}_r}(\lambda)$ into the rank one matrix $\mathbf{c}_{rg}(\mathrm{e}^{-\mathrm{i}\lambda})G_{x_gx_g}(\lambda)\mathbf{c}'_{rg}(\mathrm{e}^{\mathrm{i}\lambda})$ and the full rank matrix $\mathbf{c}_{rr}(\mathrm{e}^{-\mathrm{i}\lambda})G_{x_rx_r}(\lambda)\mathbf{c}_{rr}(\mathrm{e}^{\mathrm{i}\lambda})+\mathbf{G}_{\mathbf{u}_r\mathbf{u}_r}(\lambda)$ in this way. To separate this second component into its two constituents on a region-by-region basis, we can use the same arguments but this time applied to series within each region.

The separate identification of $\mathbf{c}_{rg}(\mathrm{e}^{-\mathrm{i}\lambda})$, $\mathbf{c}_{rr}(\mathrm{e}^{-\mathrm{i}\lambda})$, $G_{x_gx_g}(\lambda)$ and $G_{x_rx_r}(\lambda)$ is trickier, as we could always write any dynamic factor model (up to time shifts) in terms of white noise common factors. But it can be guaranteed (up to scaling and sign changes) if in addition the dynamic loading polynomials $c_{ir}(.)$ are one-sided of finite order and coprime, so that they do not share a common root within block r, and the dynamic loading polynomials $c_{ig}(.)$ are also one-sided of finite order and coprime, so they do not share a common root across all N countries (see theorem 3 in Heaton and Solo (2004) for a more formal argument along these lines).

To avoid dealing with non-sensical situations, henceforth we maintain the assumption that the model that has to be estimated is identified. This will indeed be the case in model (4), which forms the basis for our empirical application in Section 4.

2.4. *Wiener–Kolmogorov Filter*

By working in the frequency domain, we can easily obtain smoothed estimators of the latent variables. Specifically, let

$$\mathbf{y}_t-\boldsymbol{\mu}=\int_{-\pi}^{\pi}\mathrm{e}^{\mathrm{i}\lambda t}\,\mathrm{d}\mathbf{Z}^{\mathbf{y}}(\lambda),$$

$$V[\mathrm{d}\mathbf{Z}^{\mathbf{y}}(\lambda)]=\mathbf{G}_{\mathbf{yy}}(\lambda)\mathrm{d}\lambda$$

denote the spectral decomposition of the observed vector process.

Assuming that $\mathbf{G_{yy}}(\lambda)$ is not singular at any frequency, the Wiener–Kolmogorov two-sided filter for the $(R+1)$ 'common' factors $\mathbf{x}_t$ at each frequency is given by

$$d\mathbf{Z}^{\mathbf{x}^K}(\lambda) = \mathbf{G_{xx}}(\lambda)\mathbf{C}'\left(e^{i\lambda}\right)\mathbf{G}_{\mathbf{yy}}^{-1}(\lambda)d\mathbf{Z}^{\mathbf{y}}(\lambda) \tag{7}$$

where

$$\mathbf{G_{xx}}(\lambda)\mathbf{C}'\left(e^{i\lambda}\right)\mathbf{G}_{\mathbf{yy}}^{-1}(\lambda)$$

is known as the transfer function of the common factors' smoother. As a result, the spectral density of the smoothed values of the common factors, $\mathbf{x}_{t|\infty}^K$, is

$$\mathbf{G}_{\mathbf{x}^K\mathbf{x}^K}(\lambda) = \mathbf{G_{xx}}(\lambda)\mathbf{C}'\left(e^{i\lambda}\right)\mathbf{G}_{\mathbf{yy}}^{-1}(\lambda)\mathbf{C}\left(e^{-i\lambda}\right)\mathbf{G_{xx}}(\lambda)$$

thanks to the Hermitian nature of $\mathbf{G_{yy}}(\lambda)$, while the spectral density of the final estimation errors $\mathbf{x}_t - \mathbf{x}_{t|\infty}^K$ will be given by

$$\mathbf{G_{xx}}(\lambda) - \mathbf{G_{xx}}(\lambda)\mathbf{C}'\left(e^{i\lambda}\right)\mathbf{G}_{\mathbf{yy}}^{-1}(\lambda)\mathbf{C}\left(e^{-i\lambda}\right)\mathbf{G_{xx}}(\lambda) = \mathbf{\Omega}(\lambda)$$

Similarly, the Wiener–Kolmogorov smoother for the N-specific factors will be

$$\begin{aligned} d\mathbf{Z}^{\mathbf{u}^K}(\lambda) &= \mathbf{G_{uu}}(\lambda)\mathbf{G}_{\mathbf{yy}}^{-1}(\lambda)d\mathbf{Z}^{\mathbf{y}}(\lambda) \\ &= \left[\mathbf{I}_N - \mathbf{C}\left(e^{-i\lambda}\right)\mathbf{G_{xx}}(\lambda)\mathbf{C}'\left(e^{i\lambda}\right)\mathbf{G}_{\mathbf{yy}}^{-1}(\lambda)\right]d\mathbf{Z}^{\mathbf{y}}(\lambda) \\ &= d\mathbf{Z}^{\mathbf{y}}(\lambda) - \mathbf{C}\left(e^{-i\lambda}\right)d\mathbf{Z}^{\mathbf{x}^K}(\lambda) \end{aligned}$$

Hence, the spectral density matrix of the smoothed values of the specific factors will be given by

$$\mathbf{G}_{\mathbf{u}^K\mathbf{u}^K}(\lambda) = \mathbf{G_{uu}}(\lambda)\mathbf{G}_{\mathbf{yy}}^{-1}(\lambda)\mathbf{G_{uu}}(\lambda)$$

while the spectral density of their final estimation errors $\mathbf{u}_t - \mathbf{u}_{t|\infty}^K$ is

$$\mathbf{G_{uu}}(\lambda) - \mathbf{G}_{\mathbf{u}^K\mathbf{u}^K}(\lambda) = \mathbf{G_{uu}}(\lambda) - \mathbf{G_{uu}}(\lambda)\mathbf{G}_{\mathbf{yy}}^{-1}(\lambda)\mathbf{G_{uu}}(\lambda) = \mathbf{C}\left(e^{-i\lambda}\right)\mathbf{\Omega}(\lambda)\mathbf{C}'\left(e^{i\lambda}\right) = \mathbf{\Xi}(\lambda)$$

Finally, the co-spectrum between $\mathbf{x}^K_{t|\infty}$ and $\mathbf{u}^K_{t|\infty}$ will be

$$\mathbf{G}_{\mathbf{x}^K\mathbf{u}^K}(\lambda) = \mathbf{G}_{\mathbf{xx}}(\lambda)\mathbf{C}'\left(e^{i\lambda}\right)\mathbf{G}_{\mathbf{yy}}^{-1}(\lambda)\mathbf{G}_{\mathbf{uu}}(\lambda)$$

Computations can be considerably speeded up by exploiting the Woodbury formula under the assumption that neither $\mathbf{G}_{xx}(\lambda)$ nor $\mathbf{G}_{\mathbf{uu}}(\lambda)$ are singular at any frequency (see Sentana, 2000, for a generalisation):

$$\begin{aligned}
|\mathbf{G}_{\mathbf{yy}}(\lambda)| &= |\mathbf{G}_{\mathbf{uu}}(\lambda)|\cdot|\mathbf{G}_{\mathbf{xx}}(\lambda)|\cdot|\mathbf{\Omega}^{-1}(\lambda)| \\
\mathbf{G}_{\mathbf{yy}}^{-1}(\lambda) &= \mathbf{G}_{\mathbf{uu}}^{-1}(\lambda) - \mathbf{G}_{\mathbf{uu}}^{-1}(\lambda)\mathbf{C}\left(e^{-i\lambda}\right)\mathbf{\Omega}(\lambda)\mathbf{C}'\left(e^{i\lambda}\right)\mathbf{G}_{\mathbf{uu}}^{-1}(\lambda), \\
\mathbf{\Omega}(\lambda) &= \left[\mathbf{G}_{\mathbf{xx}}^{-1}(\lambda) + \mathbf{C}'\left(e^{i\lambda}\right)\mathbf{G}_{\mathbf{uu}}^{-1}(\lambda)\mathbf{C}\left(e^{-i\lambda}\right)\right]^{-1}
\end{aligned}$$

The advantage of this expression is that $\mathbf{G}_{\mathbf{uu}}(\lambda)$ is a diagonal matrix and $\mathbf{\Omega}(\lambda)$ of dimension $(R+1)$, much smaller than N, which greatly simplifies the computations.

On this basis, the transfer function of the Wiener–Kolmogorov common factor smoother becomes

$$\mathbf{G}_{\mathbf{xx}}(\lambda)\mathbf{C}'\left(e^{i\lambda}\right)\mathbf{G}_{\mathbf{yy}}^{-1}(\lambda) = \mathbf{\Omega}(\lambda)\mathbf{C}'\left(e^{i\lambda}\right)\mathbf{G}_{\mathbf{uu}}^{-1}(\lambda)$$

so

$$\begin{aligned}
\mathbf{G}_{\mathbf{x}^K\mathbf{x}^K}(\lambda) &= \mathbf{\Omega}(\lambda)\mathbf{C}'\left(e^{i\lambda}\right)\mathbf{G}_{\mathbf{uu}}^{-1}(\lambda)\mathbf{C}\left(e^{-i\lambda}\right)\mathbf{G}_{\mathbf{xx}}(\lambda) \\
&= \mathbf{G}_{\mathbf{xx}}(\lambda)\mathbf{C}'\left(e^{i\lambda}\right)\mathbf{G}_{\mathbf{uu}}^{-1}(\lambda)\mathbf{C}\left(e^{-i\lambda}\right)\mathbf{\Omega}(\lambda) \\
&= \mathbf{G}_{\mathbf{xx}}(\lambda)\left\{\mathbf{G}_{\mathbf{xx}}(\lambda) + \left[\mathbf{C}'\left(e^{i\lambda}\right)\mathbf{G}_{\mathbf{uu}}^{-1}(\lambda)\mathbf{C}\left(e^{-i\lambda}\right)\right]^{-1}\right\}^{-1}\mathbf{G}_{\mathbf{xx}}(\lambda) \\
&= \mathbf{G}_{\mathbf{xx}}(\lambda) - \mathbf{\Omega}(\lambda)
\end{aligned} \tag{8}$$

where we have used the fact that

$$\mathbf{\Omega}(\lambda)\mathbf{C}'\left(e^{i\lambda}\right)\mathbf{G}_{\mathbf{uu}}^{-1}(\lambda)\mathbf{C}\left(e^{-i\lambda}\right) = \mathbf{I}_{R+1} - \mathbf{\Omega}(\lambda)\mathbf{G}_{\mathbf{xx}}^{-1}(\lambda) \tag{9}$$

which can be easily proved by premultiplying both sides by $\mathbf{\Omega}^{-1}(\lambda)$.

Similarly, the transfer function of the Wiener–Kolmogorov-specific factors smoother will be

$$\mathbf{G}_{\mathbf{uu}}(\lambda)\mathbf{G}_{\mathbf{yy}}^{-1}(\lambda) = \mathbf{I}_N - \mathbf{C}\left(e^{-i\lambda}\right)\boldsymbol{\Omega}(\lambda)\mathbf{C}'\left(e^{i\lambda}\right)\mathbf{G}_{\mathbf{uu}}^{-1}(\lambda)$$

so

$$\mathbf{G}_{\mathbf{u}^K\mathbf{u}^K}(\lambda) = \mathbf{G}_{\mathbf{uu}}(\lambda) - \mathbf{C}\left(e^{-i\lambda}\right)\boldsymbol{\Omega}(\lambda)\mathbf{C}'\left(e^{i\lambda}\right) \tag{10}$$

Finally,

$$\mathbf{G}_{\mathbf{x}^K\mathbf{u}^K}(\lambda) = \boldsymbol{\Omega}(\lambda)\mathbf{C}'\left(e^{i\lambda}\right) \tag{11}$$

In addition, we can exploit the special structure of the matrix $\mathbf{C}(z)$ in Eq. (6) to further speed up the calculations. Specifically, tedious algebraic manipulations show that the $(R+1)\times(R+1)$ Hermitian matrix $\boldsymbol{\Omega}^{-1}(\lambda) = \mathbf{G}_{\mathbf{xx}}^{-1}(\lambda) + \mathbf{C}'\left(e^{i\lambda}\right)\mathbf{G}_{\mathbf{uu}}^{-1}(\lambda)\mathbf{C}\left(e^{-i\lambda}\right)$ can be easily computed as

$$\begin{bmatrix} \omega^{gg}(\lambda) & \omega^{g1}(\lambda) & \cdots & \omega^{gr}(\lambda) & \cdots & \omega^{gR}(\lambda) \\ \omega^{1g}(\lambda) & \omega^{11}(\lambda) & \cdots & 0 & \cdots & 0 \\ \vdots & \vdots & \ddots & \vdots & \ddots & \vdots \\ \omega^{rg}(\lambda) & 0 & \cdots & \omega^{rr}(\lambda) & \cdots & 0 \\ \vdots & \vdots & \ddots & \vdots & \ddots & \vdots \\ \omega^{Rg}(\lambda) & 0 & \cdots & 0 & \cdots & \omega^{RR}(\lambda) \end{bmatrix} \tag{12}$$

with

$$\omega^{gg}(\lambda) = G_{x_g x_g}^{-1}(\lambda) + \mathbf{c}'_{rg}\left(e^{i\lambda}\right)\mathbf{G}_{\mathbf{uu}}^{-1}(\lambda)\mathbf{c}_{rg}\left(e^{-i\lambda}\right),$$
$$\omega^{rr}(\lambda) = G_{x_r x_r}^{-1}(\lambda) + \mathbf{c}'_{rr}\left(e^{i\lambda}\right)\mathbf{G}_{\mathbf{u}_r\mathbf{u}_r}^{-1}(\lambda)\mathbf{c}_{rr}\left(e^{-i\lambda}\right)$$

and

$$\omega^{rg}(\lambda) = \mathbf{c}'_{rr}\left(e^{i\lambda}\right)\mathbf{G}_{\mathbf{u}_r\mathbf{u}_r}^{-1}(\lambda)\mathbf{c}_{rg}\left(e^{-i\lambda}\right) = \omega^{gr*}(\lambda)$$

where * denotes the complex conjugate transpose.

Interestingly, we can write Eq. (12) as

$$\mathbf{A}(\lambda) + \mathbf{B}(\lambda)\mathbf{D}^*(\lambda)$$

where

$$\mathbf{A}(\lambda) = \mathrm{diag}\left[\omega^{gg}(\lambda), \omega^{11}(\lambda), \ldots, \omega^{rr}(\lambda), \ldots, \omega^{RR}(\lambda)\right]$$

$$\mathbf{B}(\lambda) = \begin{bmatrix} 1 & 0 \\ 0 & \omega^{1g}(\lambda) \\ \vdots & \vdots \\ 0 & \omega^{rg}(\lambda) \\ \vdots & \vdots \\ 0 & \omega^{Rg}(\lambda) \end{bmatrix}$$

and

$$\mathbf{D}^{*}(\lambda) = \begin{bmatrix} 0 & \omega^{g1}(\lambda) & \cdots & \omega^{gr}(\lambda) & \cdots & \omega^{gR}(\lambda) \\ 1 & 0 & \cdots & 0 & \cdots & 0 \end{bmatrix}$$

are two rank 2 matrices.

The advantage of this formulation is that the Woodbury formula for complex matrices implies that

$$\mathbf{\Omega}(\lambda) = [\mathbf{A}(\lambda) + \mathbf{B}(\lambda)\mathbf{D}^{*}(\lambda)]^{-1} = \mathbf{A}^{-1}(\lambda) - \mathbf{A}^{-1}(\lambda)\mathbf{B}(\lambda)\mathbf{F}^{-1}(\lambda)\mathbf{D}^{*}(\lambda)\mathbf{A}^{-1}(\lambda)$$

where

$$\mathbf{F}(\lambda) = \mathbf{I}_2 + \mathbf{D}^{*}(\lambda)\mathbf{A}^{-1}(\lambda)\mathbf{B}(\lambda) = \begin{bmatrix} 1 & \omega_{+g}(\lambda) \\ \dfrac{1}{\omega^{gg}(\lambda)} & 1 \end{bmatrix}$$

with

$$\omega_{+g}(\lambda) = \sum_{r=1}^{R} \frac{\|\omega^{rg}(\lambda)\|^2}{\omega^{rr}(\lambda)}$$

where we have exploited the fact that $\omega^{rg}(\lambda)$ and $\omega^{gr}(\lambda)$ are complex conjugates so that the matrix $\mathbf{F}(\lambda)$ is actually real.

If we put all the pieces together we will end up with

$$\boldsymbol{\Omega}(\lambda) = \begin{bmatrix} \omega_{gg}(\lambda) & \omega_{g1}(\lambda) & \cdots & \omega_{gr}(\lambda) & \cdots & \omega_{gR}(\lambda) \\ \omega_{1g}(\lambda) & \omega_{11}(\lambda) & \cdots & \omega_{1r}(\lambda) & \cdots & \omega_{1R}(\lambda) \\ \vdots & \vdots & \ddots & \vdots & \ddots & \vdots \\ \omega_{rg}(\lambda) & \omega_{r1}(\lambda) & \cdots & \omega_{rr}(\lambda) & \cdots & \omega_{rR}(\lambda) \\ \vdots & \vdots & \ddots & \vdots & \ddots & \vdots \\ \omega_{Rg}(\lambda) & \omega_{R1}(\lambda) & \cdots & \omega_{Rr}(\lambda) & \cdots & \omega_{RR}(\lambda) \end{bmatrix} = \begin{bmatrix} \omega_{gg}(\lambda) & \boldsymbol{\omega}^{*}_{rg}(\lambda) \\ \boldsymbol{\omega}_{rg}(\lambda) & \boldsymbol{\Omega}_{rr}(\lambda) \end{bmatrix} \tag{13}$$

where

$$\omega_{gg}(\lambda) = \frac{1}{\omega^{gg}(\lambda)} + \frac{1}{\omega^{gg}(\lambda)} \frac{\omega_{+g}(\lambda)}{\omega^{gg}(\lambda) - \omega_{+g}(\lambda)} = \frac{1}{\omega^{gg}(\lambda) - \omega_{+g}(\lambda)}$$

$$\omega_{rr}(\lambda) = \frac{1}{\omega^{rr}(\lambda)} \left(1 + \frac{\|\omega^{rg}(\lambda)\|^2}{\omega^{rr}(\lambda)} \omega_{gg}(\lambda) \right)$$

$$\omega_{rg}(\lambda) = -\frac{\omega^{rg}(\lambda)}{\omega^{rr}(\lambda)} \omega_{gg}(\lambda) = \omega^{*}_{rg}(\lambda)$$

and

$$\omega_{rk}(\lambda) = \frac{\omega^{rg}(\lambda)\omega^{gk}(\lambda)}{\omega^{rr}(\lambda)\omega^{kk}(\lambda)} \omega_{gg}(\lambda) = \omega^{*}_{kr}(\lambda)$$

It is of some interest to compare these expressions to the corresponding expressions in the case of a model with a single global factor but no regional factors and a model with regional factors but no global factor.

In the first case, we would have

$$\omega(\lambda) = \frac{1}{\omega^{gg}(\lambda)}$$

while in the second case

$$\omega_{rr}(\lambda) = \frac{1}{\omega^{rr}(\lambda)}$$

As expected, the existence of regional factors makes more difficult the estimation of the common factor and vice versa.

The Woodbury formula also implies that

$$|\mathbf{\Omega}(\lambda)| = |\mathbf{A}(\lambda)||\mathbf{F}(\lambda)|$$

with

$$|\mathbf{F}(\lambda)| = 1 - \frac{\omega_{+g}(\lambda)}{\omega^{gg}(\lambda)}$$

The bifactor structure can also be used to speed up the filtering procedure. Specifically,

$$\mathbf{\Omega}(\lambda)\mathbf{C}'(e^{i\lambda}) = \begin{bmatrix} \omega_{gg}(\lambda) & \boldsymbol{\omega}^*_{rg}(\lambda) \\ \boldsymbol{\omega}_{rg}(\lambda) & \mathbf{\Omega}_{rr}(\lambda) \end{bmatrix} \begin{bmatrix} \mathbf{c}'_{rg}(e^{i\lambda}) \\ \mathbf{C}'_r(e^{i\lambda}) \end{bmatrix} = \begin{bmatrix} \omega_{gg}(\lambda)\mathbf{c}'_{rg}(e^{i\lambda}) + \boldsymbol{\omega}^*_{rg}(\lambda)\mathbf{C}'_r(e^{i\lambda}) \\ \boldsymbol{\omega}_{rg}(\lambda)\mathbf{c}'_{rg}(e^{i\lambda}) + \mathbf{\Omega}_{rr}(\lambda)\mathbf{C}'_r(e^{i\lambda}) \end{bmatrix}$$

and

$$\begin{aligned} \mathbf{C}(e^{-i\lambda})\mathbf{\Omega}(\lambda)\mathbf{C}'(e^{i\lambda}) &= \mathbf{c}_{rg}(e^{i\lambda})\omega_{gg}(\lambda)\mathbf{c}'_{rg}(e^{i\lambda}) + \mathbf{C}_r(e^{-i\lambda})\mathbf{\Omega}_{rr}(\lambda)\mathbf{C}'_r(e^{i\lambda}) \\ &\quad + \mathbf{c}_{rg}(e^{-i\lambda})\boldsymbol{\omega}^*_{rg}(\lambda)\mathbf{C}'_r(e^{i\lambda}) + \mathbf{C}_r(e^{-i\lambda})\boldsymbol{\omega}_{rg}(\lambda)\mathbf{c}'_{rg}(e^{i\lambda}) \end{aligned}$$

which can be computed rather quickly by exploiting the block diagonal nature of $\mathbf{C}_r(z)$ in Eq. (6).

2.5. The Minimal Sufficient Statistics for $\{\mathbf{x}_t\}$

Define $\mathbf{x}^G_{t|\infty}$ as the spectral GLS estimator of $\mathbf{x}_t$ through the transformation

$$d\mathbf{Z}^{\mathbf{x}^G}(\lambda) = \left[\mathbf{C}'(e^{i\lambda})\mathbf{G}^{-1}_{\mathbf{uu}}(\lambda)\mathbf{C}(e^{-i\lambda})\right]^{-1}\mathbf{C}'(e^{i\lambda})\mathbf{G}^{-1}_{\mathbf{uu}}(\lambda)d\mathbf{Z}^{\mathbf{y}}(\lambda)$$

Similarly, define $\mathbf{u}^G_{t|\infty}$ through

$$d\mathbf{Z}^{\mathbf{u}^G}(\lambda) = \left\{\mathbf{I}_N - \left[\mathbf{C}'(e^{i\lambda})\mathbf{G}^{-1}_{\mathbf{uu}}(\lambda)\mathbf{C}(e^{-i\lambda})\right]^{-1}\mathbf{C}'(e^{i\lambda})\mathbf{G}^{-1}_{\mathbf{uu}}(\lambda)\right\}d\mathbf{Z}^{\mathbf{y}}(\lambda)$$

It is then easy to see that the joint spectral density of $\mathbf{x}^G_{t|\infty}$ and $\mathbf{u}^G_{t|\infty}$ will be block-diagonal, with the (1,1) block being

$$\mathbf{G}_{\mathbf{xx}}(\lambda) + \left[\mathbf{C}'\left(e^{i\lambda}\right)\mathbf{G}^{-1}_{\mathbf{uu}}(\lambda)\mathbf{C}\left(e^{-i\lambda}\right)\right]^{-1}$$

and the (2,2) block

$$\mathbf{G}_{\mathbf{yy}}(\lambda) - \mathbf{C}\left(e^{-i\lambda}\right)\left[\mathbf{C}'\left(e^{i\lambda}\right)\mathbf{G}^{-1}_{\mathbf{uu}}(\lambda)\mathbf{C}\left(e^{-i\lambda}\right)\right]^{-1}\mathbf{C}'\left(e^{i\lambda}\right)$$

whose rank is $N-(R+1)$.

This block-diagonality allows us to factorise the spectral log-likelihood function of $\mathbf{y}_t$ as the sum of the log-likelihood function of $\mathbf{x}^G_{t|\infty}$, which is of dimension $(R+1)$, and the log-likelihood function of $\mathbf{u}^G_{t|\infty}$. Importantly, the parameters characterising $\mathbf{G}_{\mathbf{xx}}(\lambda)$ only enter through the first component. In contrast, the remaining parameters affect both components. Moreover, we can easily show that

1. $\mathbf{x}^G_{t|\infty} = \mathbf{x}_t + \boldsymbol{\zeta}^G_{t|\infty}$, with $\mathbf{x}_t$ and $\boldsymbol{\zeta}^G_{t|\infty}$ orthogonal at all leads and lags.
2. The smoothed estimator of $\mathbf{x}_t$ obtained by applying the Wiener–Kolmogorov filter to $\mathbf{x}^G_{t|\infty}$ coincides with $\mathbf{x}^K_{t|\infty}$.

This confirms that $\mathbf{x}^G_{t|\infty}$ constitute minimal sufficient statistics for $\mathbf{x}_t$, thereby generalising earlier results by Jungbacker and Koopman (2015), who considered models in which $\mathbf{C}(e^{-i\lambda}) = \mathbf{C}$ for all λ, and Fiorentini, Sentana, and Shephard (2004), who looked at the related class of factor models with time-varying volatility (see also Gouriéroux, Monfort, & Renault, 1991). In addition, the degree of unobservability of $\mathbf{x}_t$ depends exclusively on the 'size' of $\left[\mathbf{C}'(e^{i\lambda})\mathbf{G}^{-1}_{\mathbf{uu}}(\lambda)\mathbf{C}(e^{-i\lambda})\right]^{-1}$ relative to $\mathbf{G}_{\mathbf{xx}}(\lambda)$ (see Sentana, 2004, for a closely related discussion).

2.6. Maximum Likelihood Estimation in the Frequency Domain

Let

$$\mathbf{I}_{\mathbf{yy}}(\lambda) = \frac{1}{2\pi T}\sum_{t=1}^{T}\sum_{s=1}^{T}\left(\mathbf{y}_t - \boldsymbol{\mu}\right)\left(\mathbf{y}_s - \boldsymbol{\mu}\right)' e^{-i(t-s)\lambda} \quad (14)$$

denote the periodogram matrix and $\lambda_j = 2\pi j/T$ $(j = 0, \ldots, T-1)$ the usual Fourier frequencies. If we assume that $\mathbf{G}_{\mathbf{yy}}(\lambda)$ is not singular at any of those frequencies, the so-called Whittle (discrete) spectral approximation to the log-likelihood function is[4]

$$N\varkappa - \frac{1}{2}\sum_{j=0}^{T-1} \ln|\mathbf{G}_{\mathbf{yy}}(\lambda_j)| - \frac{1}{2}\sum_{j=0}^{T-1} \mathrm{tr}\left\{\mathbf{G}_{\mathbf{yy}}^{-1}(\lambda_j)\left[2\pi\mathbf{I}_{\mathbf{yy}}(\lambda_j)\right]\right\} \tag{15}$$

with $\varkappa = -(T/2)\ln(2\pi)$ (see, e.g. Hannan, 1973; Dunsmuir & Hannan, 1976).

Expression (14), though, is far from ideal from a computational point of view, and for that reason we make use of the fast Fourier transform (FFT). Specifically, given the $T \times N$ original real data matrix $\mathbf{Y} = (\mathbf{y}_1, \ldots, \mathbf{y}_t, \ldots, \mathbf{y}_T)'$, the FFT creates the centered and orthogonalised $T \times N$ complex data matrix $\mathbf{Z}^{\mathbf{y}} = (\mathbf{z}_0^{\mathbf{y}}, \ldots, \mathbf{z}_j^{\mathbf{y}}, \ldots, \mathbf{z}_{T-1}^{\mathbf{y}})'$ by effectively premultiplying $\mathbf{Y} - \ell_T\boldsymbol{\mu}'$ by the $T \times T$ Fourier matrix $\mathbf{W}$. On this basis, we can easily compute $\mathbf{I}_{\mathbf{yy}}(\lambda_j)$ as $2\pi\mathbf{z}_j^{\mathbf{y}}\mathbf{z}_j^{\mathbf{y}*}$, where $\mathbf{z}_j^{\mathbf{y}*}$ is the complex conjugate transpose of $\mathbf{z}_j^{\mathbf{y}}$. Hence, the spectral approximation to the log-likelihood function (15) becomes

$$N\varkappa - \frac{1}{2}\sum_{j=0}^{T-1} \ln|\mathbf{G}_{\mathbf{yy}}(\lambda_j)| - \frac{2\pi}{2}\sum_{j=0}^{T-1} \mathbf{z}_j^{\mathbf{y}*}\mathbf{G}_{\mathbf{yy}}^{-1}(\lambda_j)\mathbf{z}_j^{\mathbf{y}}$$

which can be regarded as the log-likelihood function of T independent but heteroskedastic complex Gaussian observations.

But since $\mathbf{z}_j^{\mathbf{y}}$ does not depend on $\boldsymbol{\mu}$ for $j = 1, \ldots, T-1$ because ℓ_T is proportional to the first column of the orthogonal Fourier matrix and $\mathbf{z}_0^{\mathbf{y}} = (\bar{\mathbf{y}}_T - \boldsymbol{\mu})$, where $\bar{\mathbf{y}}_T$ is the sample mean of $\mathbf{y}_t$, it immediately follows that the ML of $\boldsymbol{\mu}$ will be $\bar{\mathbf{y}}_T$, so in what follows we focus on demeaned variables. As for the remaining parameters, the score function will be given by:

$$\mathbf{d}(\theta) = \frac{1}{2}\sum_{j=0}^{T-1} \mathbf{d}(\lambda_j;\theta)$$

$$\begin{aligned}\mathbf{d}(\lambda_j;\theta) &= \frac{1}{2}\frac{\partial \mathrm{vec}'\left[\mathbf{G}_{\mathbf{yy}}(\lambda_j)\right]}{\partial\theta}\left[\mathbf{G}_{\mathbf{yy}}^{-1}(\lambda_j)\otimes\mathbf{G}_{\mathbf{yy}}^{\prime -1}(\lambda_j)\right]\mathrm{vec}\left[2\pi\mathbf{z}_j^{\mathbf{y}c}\mathbf{z}_j^{\mathbf{y}\prime} - \mathbf{G}_{\mathbf{yy}}'(\lambda_j)\right] \\ &= \frac{1}{2}\frac{\partial \mathrm{vec}'\left[\mathbf{G}_{\mathbf{yy}}(\lambda_j)\right]}{\partial\boldsymbol{\theta}}\mathbf{M}(\lambda_j)\mathbf{m}(\lambda_j)\end{aligned} \tag{16}$$

where $\mathbf{z}_j^{\mathbf{y}c} = \mathbf{z}_j^{\mathbf{y}*\prime}$ is the complex conjugate of $\mathbf{z}_j^{\mathbf{y}}$,

$$\mathbf{m}(\lambda_j) = \text{vec}\left[2\pi \mathbf{z}_j^{\mathbf{y}c}\mathbf{z}_j^{\mathbf{y}\prime} - \mathbf{G}'_{\mathbf{yy}}(\lambda_j)\right] \tag{17}$$

and

$$\mathbf{M}(\lambda_j) = \mathbf{G}_{\mathbf{yy}}^{-1}(\lambda_j) \otimes \mathbf{G}_{\mathbf{yy}}^{\prime -1}(\lambda_j) \tag{18}$$

The information matrix is block diagonal between μ and the elements of θ, with the (1,1)-element being $\mathbf{G}_{\mathbf{yy}}(0)$ and the (2,2)-block being

$$\mathbf{Q}(\theta) = \frac{1}{4\pi}\int_{-\pi}^{\pi} \mathbf{Q}(\lambda;\theta)\mathrm{d}\lambda = \frac{1}{4\pi}\int_{-\pi}^{\pi} \frac{\partial \text{vec}'\left[\mathbf{G}_{\mathbf{yy}}(\lambda)\right]}{\partial \boldsymbol{\theta}} \mathbf{M}(\lambda)\left\{\frac{\partial \text{vec}'\left[\mathbf{G}_{\mathbf{yy}}(\lambda)\right]}{\partial \boldsymbol{\theta}}\right\}^{*} \mathrm{d}\lambda \tag{19}$$

a consistent estimator of which will be provided by either by the outer product of the score or by

$$\boldsymbol{\Phi}(\boldsymbol{\theta}) = \frac{1}{2}\sum_{j=0}^{T-1} \frac{\partial \text{vec}'\left[\mathbf{G}_{\mathbf{yy}}(\lambda_j)\right]}{\partial \boldsymbol{\theta}} \mathbf{M}(\lambda_j)\left\{\frac{\partial \text{vec}'\left[\mathbf{G}_{\mathbf{yy}}(\lambda_j)\right]}{\partial \boldsymbol{\theta}}\right\}^{*}$$

Formal results showing the strong consistency and asymptotic normality of the resulting ML estimators of dynamic latent variable models under suitable regularity conditions were provided by Dunsmuir (1979), who generalised earlier results for VARMA models by Dunsmuir and Hannan (1976). These authors also show the asymptotic equivalence between time and frequency domain ML estimators.[5]

Appendix A provides detailed expressions for the Jacobian of $\text{vec}\left[\mathbf{G}_{\mathbf{yy}}(\lambda)\right]$ and the spectral score of dynamic bifactor models, while Appendix B includes numerically reliable and efficient formulae for their information matrix. Those expressions make extensive use of the complex version of the Woodbury formula described in Section 2.4. We can also exploit the same formula to compute the quadratic form $\mathbf{z}_j^{\mathbf{y}*}\mathbf{G}_{\mathbf{yy}}^{-1}(\lambda_j)\mathbf{z}_j^{\mathbf{y}}$ as

$$\begin{aligned}&\mathbf{z}_j^{\mathbf{y}*}\mathbf{G}_{\mathbf{uu}}^{-1}(\lambda_j)\mathbf{z}_j^{\mathbf{y}} - \mathbf{z}_j^{\mathbf{y}*}\mathbf{G}_{\mathbf{uu}}^{-1}(\lambda)\mathbf{C}(\mathrm{e}^{-\mathrm{i}\lambda})\boldsymbol{\Omega}(\lambda_j)\mathbf{C}'(\mathrm{e}^{\mathrm{i}\lambda})\mathbf{G}_{\mathbf{uu}}^{-1}(\lambda)\mathbf{z}_j^{\mathbf{y}}\\ &= \mathbf{z}_j^{\mathbf{y}*}\mathbf{G}_{\mathbf{uu}}^{-1}(\lambda_j)\mathbf{z}_j^{\mathbf{y}} - \mathbf{z}_j^{\mathbf{x}^K*}(\theta)\boldsymbol{\Omega}^{-1}(\lambda_j)\mathbf{z}_j^{\mathbf{x}^K}(\theta)\end{aligned}$$

where

$$\mathbf{z}_j^{\mathbf{x}^K}(\theta) = E\left[\mathbf{z}_j^{\mathbf{x}}|\mathbf{Z}^{\mathbf{y}},\theta\right] = \mathbf{G}_{\mathbf{xx}}(\lambda_j)\mathbf{C}'(\mathrm{e}^{\mathrm{i}\lambda_j})\mathbf{G}_{\mathbf{yy}}^{-1}(\lambda_j)\mathbf{z}_j^{\mathbf{y}} = \boldsymbol{\Omega}(\lambda_j)\mathbf{C}'(\mathrm{e}^{\mathrm{i}\lambda_j})\mathbf{G}_{\mathbf{uu}}^{-1}(\lambda_j)\mathbf{z}_j^{\mathbf{y}} \tag{20}$$

denotes the filtered value of $\mathbf{z}_j^{\mathbf{x}}$ given the observed series and the current parameter values from Eq. (7).

Nevertheless, when N is large the number of parameters is huge, and the direct maximisation of the log-likelihood function becomes excruciatingly slow, especially without good initial values. For that reason, in the next section, we describe a much faster alternative to obtain the maximum likelihood estimators of all the model parameters.

3. SPECTRAL EM ALGORITHM

The EM algorithm of Dempster, Laird, and Rubin (1977) adapted to static factor models by Rubin and Thayer (1982) was successfully employed to handle a very large dataset of stock returns by Lehmann and Modest (1988). Shumway and Stoffer (1982), Watson and Engle (1983) and Quah and Sargent (1993) also applied the algorithm in the time domain to dynamic factor models and some generalisations, while Demos and Sentana (1998) adapted it to conditionally heteroskedastic factor models in which the common factors followed GARCH-type processes.

We saw before that the spectral density matrix of a dynamic single factor model has the structure of the unconditional covariance matrix of a static factor model, but with different common and idiosyncratic variances for each frequency. This idea led us to propose a spectral version of the EM algorithm for dynamic factor models with only pervasive factors in a companion paper (see Fiorentini, Galesi, & Sentana, 2014). In order to apply the same idea to bifactor models, we need to do some additional algebra.

3.1. Complete Log-Likelihood Function

Consider a situation in which the $(R+1)$ common factors $\mathbf{x}_t$ were also observed. The joint spectral density of $\mathbf{y}_t$ and $\mathbf{x}_t$, which is given by

$$\begin{bmatrix} \mathbf{G}_{\mathbf{yy}}(\lambda) & \mathbf{G}_{\mathbf{yx}}(\lambda) \\ \mathbf{G}^*_{\mathbf{yx}}(\lambda) & \mathbf{G}_{\mathbf{xx}}(\lambda) \end{bmatrix} = \begin{bmatrix} \mathbf{C}(e^{-i\lambda})\mathbf{G}_{\mathbf{xx}}(\lambda)\mathbf{C}'(e^{i\lambda}) + \mathbf{G}_{\mathbf{uu}}(\lambda) & \mathbf{C}(e^{-i\lambda})\mathbf{G}_{\mathbf{xx}}(\lambda) \\ \mathbf{G}_{\mathbf{xx}}(\lambda)\mathbf{C}'(e^{i\lambda}) & \mathbf{G}_{\mathbf{xx}}(\lambda) \end{bmatrix}$$

could be diagonalised as

$$\begin{bmatrix} \mathbf{I}_N & \mathbf{C}(e^{-i\lambda}) \\ \mathbf{0} & \mathbf{I}_{R+1} \end{bmatrix} \begin{bmatrix} \mathbf{G}_{\mathbf{uu}}(\lambda) & \mathbf{0} \\ \mathbf{0} & \mathbf{G}_{\mathbf{xx}}(\lambda) \end{bmatrix} \begin{bmatrix} \mathbf{I}_N & \mathbf{0} \\ \mathbf{C}'(e^{i\lambda}) & \mathbf{I}_{R+1} \end{bmatrix}$$

with

$$\left| \begin{bmatrix} \mathbf{I}_N & \mathbf{0} \\ \mathbf{C}'(e^{i\lambda}) & \mathbf{I}_{R+1} \end{bmatrix} \right| = 1$$

and

$$\begin{bmatrix} \mathbf{I}_N & \mathbf{0} \\ \mathbf{C}'(e^{i\lambda}) & \mathbf{I}_{R+1} \end{bmatrix}^{-1} = \begin{bmatrix} \mathbf{I}_N & \mathbf{0} \\ -\mathbf{C}'(e^{i\lambda}) & \mathbf{I}_{R+1} \end{bmatrix}$$

Let us define as $[\mathbf{Z}^{\mathbf{y}}|\mathbf{Z}^{\mathbf{x}}]$ as the Fourier transform of the $T \times (N+1+R)$ matrix

$$[\mathbf{y}_1, \ldots, \mathbf{y}_N, \mathbf{x}_g, \mathbf{x}_1, \ldots, \mathbf{x}_R] = [\mathbf{Y}|\mathbf{X}]$$

so that the joint periodogram of $\mathbf{y}_t$ and $\mathbf{x}_t$ at frequency λ_j could be quickly computed as

$$2\pi \begin{pmatrix} \mathbf{z}_j^{\mathbf{y}} \\ \mathbf{z}_j^{\mathbf{x}} \end{pmatrix} \begin{pmatrix} \mathbf{z}_j^{\mathbf{y}*} & \mathbf{z}_j^{\mathbf{x}*} \end{pmatrix}$$

where we have implicitly assumed that either the elements of $\mathbf{y}$ have zero mean, or else that they have been previously demeaned by subtracting their sample averages.

In this notation, the spectral approximation to the joint log-likelihood function would become

$$l(\mathbf{y},\mathbf{x}) = (N+R+1)\varkappa - \frac{1}{2}\sum_{j=0}^{T-1}\ln\left|\begin{bmatrix}\mathbf{G_{yy}}(\lambda) & \mathbf{G_{yx}}(\lambda_j)\\ \mathbf{G^*_{yx}}(\lambda_j) & \mathbf{G_{xx}}(\lambda_j)\end{bmatrix}\right|$$

$$-\frac{2\pi}{2}\sum_{j=0}^{T-1}\begin{pmatrix}\mathbf{z}_j^{\mathbf{y}*} & \mathbf{z}_j^{x*}\end{pmatrix}\begin{bmatrix}\mathbf{I}_N & \mathbf{0}\\ -\mathbf{C}'(e^{i\lambda_j}) & 1\end{bmatrix}\begin{bmatrix}\mathbf{G}^{-1}_{\mathbf{uu}}(\lambda_j) & \mathbf{0}\\ \mathbf{0} & \mathbf{G}^{-1}_{\mathbf{xx}}(\lambda_j)\end{bmatrix}\begin{bmatrix}\mathbf{I}_N & \mathbf{C}(e^{-i\lambda_j})\\ \mathbf{0} & 1\end{bmatrix}\begin{pmatrix}\mathbf{z}_j^{\mathbf{y}}\\ \mathbf{z}_j^{x}\end{pmatrix}$$

$$= N\varkappa - \frac{1}{2}\sum_{j=0}^{T-1}\ln|\mathbf{G_{uu}}(\lambda_j)| - \frac{2\pi}{2}\sum_{j=0}^{T-1}\mathbf{z}_j^{\mathbf{u}*}\mathbf{G}^{-1}_{\mathbf{uu}}(\lambda_j)\mathbf{z}_j^{\mathbf{u}}$$

$$+(R+1)\varkappa - \frac{1}{2}\sum_{j=0}^{T-1}\ln|\mathbf{G_{xx}}(\lambda_j)| - \frac{2\pi}{2}\sum_{j=0}^{T-1}\mathbf{z}_j^{\mathbf{x}*}\mathbf{G}^{-1}_{\mathbf{xx}}(\lambda_j)\mathbf{z}_j^{\mathbf{x}}$$

$$= \sum_{i=1}^{N}\left[\varkappa - \frac{1}{2}\sum_{j=0}^{T-1}\ln G_{u_iu_i}(\lambda_j) - \frac{2\pi}{2}\sum_{j=0}^{T-1}G^{-1}_{u_iu_i}(\lambda_j)z_j^{u_i}z_j^{u_i*}\right] \quad (21)$$

$$+\varkappa - \frac{1}{2}\sum_{j=0}^{T-1}\ln G_{x_gx_g}(\lambda_j) - \frac{2\pi}{2}\sum_{j=0}^{T-1}G^{-1}_{x_gx_g}(\lambda_j)z_j^{x_g}z_j^{x_g*} \quad (22)$$

$$+\sum_{r=1}^{R}\left[\varkappa - \frac{1}{2}\sum_{j=0}^{T-1}\ln G_{x_rx_r}(\lambda_j) - \frac{2\pi}{2}\sum_{j=0}^{T-1}G^{-1}_{x_rx_r}(\lambda_j)z_j^{u_r}z_j^{u_r*}\right] \quad (23)$$

$$= \sum_{i=1}^{N} l(\mathbf{y}_i|\mathbf{X}) + l(\mathbf{x}_g) + \sum_{j=1}^{R} l(\mathbf{x}_j) - l(\mathbf{Y}|\mathbf{X}) + l(\mathbf{X})$$

where[6] if country i belongs to region r we have that

$$z_j^{u_i} = z_j^{y_i} - c_{ig}(e^{-i\lambda_j})z_j^{x_g} - c_{ir}(e^{-i\lambda_j})z_j^{x_r} = z_j^{y_i} - \sum_{k=-m_g}^{n_g} c_{ikg}e^{-ik\lambda}z_j^{x_g} - \sum_{l=-m_r}^{n_r} c_{ilr}e^{-il\lambda}z_j^{x_r} \quad (24)$$

so that

$$\begin{aligned}
z_j^{u_i} z_j^{u_i *} &= z_j^{y_i} z_j^{y_i *} - c_{ig}(e^{-i\lambda_j}) z_j^{x_g} z_j^{y_i *} - c_{ir}(e^{-i\lambda_j}) z_j^{x_r} z_j^{y_i *} - c_{ig}(e^{i\lambda_j}) z_j^{y_i} z_j^{x_g *} - c_{ir}(e^{i\lambda_j}) z_j^{y_i} z_j^{x_r *} \\
&\quad + c_{ig}(e^{-i\lambda_j}) c_{ig}(e^{i\lambda_j}) z_j^{x_g} z_j^{x_g *} + c_{ir}(e^{-i\lambda_j}) c_{ir}(e^{i\lambda_j}) z_j^{x_r} z_j^{x_r *} \\
&\quad + c_{ig}(e^{-i\lambda_j}) c_{ir}(e^{i\lambda_j}) z_j^{x_g} z_j^{x_r *} + c_{ir}(e^{-i\lambda_j}) c_{ig}(e^{i\lambda_j}) z_j^{x_r} z_j^{x_g *} \\
&= I_{y_i y_i}(\lambda_j) - c_{ig}(e^{-i\lambda_j}) I_{x_g y_i}(\lambda_j) - c_{ir}(e^{-i\lambda_j}) I_{x_r y_i}(\lambda_j) - c_{ig}(e^{i\lambda_j}) I_{y_i x_g}(\lambda_j) \\
&\quad - c_{ir}(e^{i\lambda_j}) I_{y_i x_r}(\lambda_j) + c_{ig}(e^{-i\lambda_j}) c_{ig}(e^{i\lambda_j}) I_{x_g x_g}(\lambda_j) + c_{ir}(e^{-i\lambda_j}) c_{ir}(e^{i\lambda_j}) I_{x_r x_r}(\lambda_j) \\
&\quad + c_{ig}(e^{-i\lambda_j}) c_{ir}(e^{i\lambda_j}) I_{x_g x_r}(\lambda_j) + c_{ir}(e^{-i\lambda_j}) c_{ig}(e^{i\lambda_j}) I_{x_r x_g}(\lambda_j) = I_{u_i u_i}(\lambda_j)
\end{aligned}$$

In this way, we have decomposed the joint log-likelihood function of $\mathbf{y}_1, \ldots, \mathbf{y}_N$ and $\mathbf{x}$ as the sum of the marginal log-likelihood of $\mathbf{x}$, $l(\mathbf{X})$, and the log-likelihood function of $\mathbf{y}_1, \ldots, \mathbf{y}_N$ given $\mathbf{x}$, $l(\mathbf{Y}|\mathbf{X})$. In turn, each of those components can be decomposed as the sum of univariate log-likelihoods. Specifically, $l(\mathbf{Y}|\mathbf{X})$ can be computed as in Eq. (21) by exploiting the diagonality of $\mathbf{G}_{\mathbf{uu}}(\lambda_j)$, while $l(\mathbf{X})$ coincides with the sum of Eqs. (22) and (23) by the diagonality of $\mathbf{G}_{\mathbf{xx}}(\lambda_j)$.

Importantly, all the above expressions can be computed using real arithmetic only since

$$\begin{aligned}
c_{ig}(e^{-i\lambda_j}) I_{x_g y_i}(\lambda_j) + c_{ig}(e^{i\lambda_j}) I_{y_i x_g}(\lambda_j) &= 2\Re\left[c_{ig}(e^{-i\lambda_j}) I_{x_g y_i}(\lambda_j)\right] \\
c_{ir}(e^{-i\lambda_j}) I_{x_r y_i}(\lambda_j) + c_{ir}(e^{i\lambda_j}) I_{y_i x_r}(\lambda_j) &= 2\Re\left[c_{ir}(e^{-i\lambda_j}) I_{x_r y_i}(\lambda_j)\right] \\
c_{ig}(e^{-i\lambda_j}) c_{ir}(e^{i\lambda_j}) I_{x_g x_r}(\lambda_j) \qquad & \\
+ c_{ir}(e^{-i\lambda_j}) c_{ig}(e^{i\lambda_j}) I_{x_r x_g}(\lambda_j) &= 2\Re\left[c_{ig}(e^{-i\lambda_j}) c_{ir}(e^{i\lambda_j}) I_{x_g x_r}(\lambda_j)\right] \\
c_{ig}(e^{-i\lambda_j}) c_{ig}(e^{i\lambda_j}) I_{x_g x_g}(\lambda_j) &= \|c_{ig}(e^{-i\lambda_j})\|^2 I_{x_g x_g}(\lambda_j)
\end{aligned}$$

and

$$c_{ir}(e^{-i\lambda_j}) c_{ir}(e^{i\lambda_j}) I_{x_r x_r}(\lambda_j) = \|c_{ir}(e^{-i\lambda_j})\|^2 I_{x_r x_r}(\lambda_j)$$

Let us classify the parameters into three blocks:

1. the parameters that characterise the spectral density of $\mathbf{x}_t : \theta_x = \left(\theta'_{x_g}, \theta'_{x_1}, \ldots, \theta'_{x_R}\right)'$
2. the parameters that characterise the spectral density of u_{it} $(i = 1, \ldots, N) : \psi = (\psi_1, \ldots, \psi_N)'$ and $\theta_{\mathbf{u}} = \left(\theta'_{u_i}, \ldots, \theta'_{u_N}\right)'$

3. the parameters that characterise the dynamic idiosyncratic impact of the global and regional factor on each observed variable: $\mathbf{c}_{ig} = \left(c_{i,-m_g,g}, \ldots, c_{i,0,g}, \ldots, c_{i,n_g,g}\right)'$ and $\mathbf{c}_{ir} = \left(c_{i,-m_r,r}, \ldots, c_{i,0,r}, \ldots, c_{i,n_r,r}\right)'$.

Importantly, θ_{x_g} only appear in Eq. (22), θ_{x_r} in Eq. (23), while $\boldsymbol{\theta}_{\mathbf{u}_i}$, $\mathbf{c}_{ig}$ and $\mathbf{c}_{ir}$ appear in Eq. (21). This sequential cut on the joint spectral density confirms that z^{x_g} and z^{x_r}, and therefore x_{gt} and x_{rt}, would be weakly exogenous for ψ_i, $\theta_{\mathbf{u}_i}$, $\mathbf{c}_{ig}$ and $\mathbf{c}_{ir}$ (see Engle, Hendry, & Richard, 1983). Moreover, the fact that f_{gt} and f_{rt} are uncorrelated at all leads and lags with v_{it} implies that x_{gt} and x_{rt} would be strongly exogenous too.

We can also exploit the aforementioned log-likelihood decomposition to obtain the score of the complete log-likelihood function. In this way, we can write

$$\frac{\partial l(\mathbf{Y},\mathbf{x})}{\partial \boldsymbol{\theta}_{x_g}} = \frac{\partial l\left(\mathbf{x}_g\right)}{\partial \boldsymbol{\theta}_{x_g}} = \frac{1}{2}\sum_{j=0}^{T-1} \frac{\partial G_{x_g x_g}\left(\lambda_j\right)}{\partial \boldsymbol{\theta}_{x_g}} G_{x_g x_g}^{-2}\left(\lambda_j\right)\left[2\pi z_j^{x_g} z_j^{x_g *} - G_{x_g x_g}\left(\lambda_j\right)\right] \tag{25a}$$

$$\frac{\partial l(\mathbf{Y},\mathbf{x})}{\partial \boldsymbol{\theta}_{x_r}} = \frac{\partial l(\mathbf{x}_r)}{\partial \boldsymbol{\theta}_{x_r}} = \frac{1}{2}\sum_{j=0}^{T-1} \frac{\partial G_{x_r x_r}\left(\lambda_j\right)}{\partial \boldsymbol{\theta}_{x_r}} G_{x_r x_r}^{-2}\left(\lambda_j\right)\left[2\pi z_j^{x_r} z_j^{x_r *} - G_{x_r x_r}\left(\lambda_j\right)\right] \tag{25b}$$

$$\frac{\partial l(\mathbf{Y},\mathbf{x})}{\partial \boldsymbol{\theta}_{u_i}} = \frac{\partial l\left(\mathbf{y}_i|\mathbf{X}\right)}{\partial \boldsymbol{\theta}_{u_i}} = \frac{1}{2}\sum_{j=0}^{T-1} \frac{\partial G_{u_i u_i}\left(\lambda_j\right)}{\partial \boldsymbol{\theta}_{u_i}} G_{u_i u_i}^{-2}\left(\lambda_j\right)\left[2\pi z_j^{u_i} z_j^{u_i *} - G_{u_i u_i}\left(\lambda_j\right)\right] \tag{25c}$$

$$\begin{aligned}\frac{\partial l(\mathbf{Y},\mathbf{x})}{\partial c_{ikg}} &= \frac{\partial l\left(\mathbf{y}_i|\mathbf{X}\right)}{\partial c_{ikg}} = \frac{2\pi}{2}\sum_{j=0}^{T-1} G_{u_i u_i}^{-1}\left(\lambda_j\right)\left[z_j^{u_i} \mathrm{e}^{ik\lambda_j} z_j^{x_g *} + \mathrm{e}^{-ik\lambda_j} z_j^{x_g} z_j^{u_i *}\right] \\ &= \frac{2\pi}{2}\sum_{j=0}^{T-1} G_{u_i u_i}^{-1}\left(\lambda_j\right)\left[\begin{array}{l}\left(z_j^{y_i} - \sum\limits_{k=-m_g}^{n_g} c_{ikg}\mathrm{e}^{-ik\lambda} z_j^{x_g} - \sum\limits_{l=-m_r}^{n_r} c_{ilr}\mathrm{e}^{-il\lambda} z_j^{x_r}\right)\mathrm{e}^{ik\lambda_j} z_j^{x_g *} \\ +\mathrm{e}^{-ik\lambda_j} z_j^{x_g}\left(z_j^{y_i *} - \sum\limits_{k=-m_g}^{n_g} c_{ikg}\mathrm{e}^{ik\lambda} z_j^{x_g *} - \sum\limits_{l=-m_r}^{n_r} c_{ilr}\mathrm{e}^{il\lambda} z_j^{x_r *}\right)\end{array}\right]\end{aligned} \tag{25d}$$

$$\frac{\partial l(\mathbf{Y},\mathbf{x})}{\partial c_{ilr}}=\frac{\partial l\left(\mathbf{y}_i|\mathbf{X}\right)}{\partial c_{ikr}}=\frac{2\pi}{2}\sum_{j=0}^{T-1}G_{u_iu_i}^{-1}\left(\lambda_j\right)\left[z_j^{u_i}\mathrm{e}^{\mathrm{i}l\lambda_j}z_j^{x_r*}+\mathrm{e}^{-\mathrm{i}l\lambda_j}z_j^{xr}z_j^{u_i*}\right]$$

$$=\frac{2\pi}{2}\sum_{j=0}^{T-1}G_{u_iu_i}^{-1}\left(\lambda_j\right)\left[\begin{array}{c}\left(z_j^{y_i}-\sum_{k=-m_g}^{n_g}c_{ikg}\mathrm{e}^{-\mathrm{i}k\lambda}z_j^{x_g}-\sum_{l=-m_r}^{n_r}c_{ilr}\mathrm{e}^{-\mathrm{i}l\lambda}z_j^{x_r}\right)\mathrm{e}^{\mathrm{i}l\lambda_j}z_j^{x_r*}\\+\mathrm{e}^{-\mathrm{i}l\lambda_j}z_j^{x_r}\left(z_j^{y_i*}-\sum_{k=-m_g}^{n_g}c_{ikg}\mathrm{e}^{\mathrm{i}k\lambda}z_j^{x_g*}-\sum_{l=-m_r}^{n_r}c_{ilr}\mathrm{e}^{\mathrm{i}l\lambda}z_j^{x_r*}\right)\end{array}\right] \tag{25e}$$

where we have used the fact that

$$\frac{\partial z_j^{u_i}}{\partial c_{ikg}}=-\mathrm{e}^{-\mathrm{i}k\lambda}z_j^{x_g}$$

$$\frac{\partial z_j^{u_i}}{\partial c_{ilr}}=-\mathrm{e}^{-\mathrm{i}l\lambda}z_j^{x_r}$$

in view of Eq. (24).

Expression (25a) confirms that the MLE of $\boldsymbol{\theta}_{x_g}$ would be obtained from a univariate time series model for x_{gt}, and the same applies to $\boldsymbol{\theta}_{x_r}$. However, since $G_{x_gx_g}\left(\lambda_j\right)$ also depends on $\boldsymbol{\theta}_{x_g}$, there are no closed-form solutions for models with MA components. Although it would be straightforward to adapt the indirect inference procedures we have developed in our companion paper (see Fiorentini, Galesi, & Sentana, 2014) to deal with general ARMA processes without resorting to the numerical maximisation of Eq. (22), in what follows we only consider pure autoregressions. Obviously, the same comments apply to $\boldsymbol{\theta}_{x_r}$.

In this regard, if we consider the AR(2) example for x_r in Eq. (4), the derivatives of $G_{x_rx_r}(\lambda)$ with respect to α_{1x_r} and α_{2x_g} would be

$$\frac{\partial G_{x_rx_r}(\lambda)}{\partial\alpha_{1x_r}}=\frac{2\left(\cos\lambda-\alpha_{1x_r}-\alpha_{2x_r}\cos\lambda\right)}{\left(1+\alpha_{1x_r}^2+\alpha_{2x_r}^2-2\alpha_{1x_r}\left(1-\alpha_{2x_r}\right)\cos\lambda-2\alpha_{2x_r}\cos2\lambda\right)^2}$$

$$\frac{\partial G_{x_rx_r}(\lambda)}{\partial\alpha_{2x_r}}=\frac{2\left(\cos2\lambda-\alpha_{1x_r}\cos\lambda-\alpha_{2x_r}\right)}{\left(1+\alpha_{1x_r}^2+\alpha_{2x_r}^2-2\alpha_{1x_r}\left(1-\alpha_{2x_r}\right)\cos\lambda-2\alpha_{2x_r}\cos2\lambda\right)^2}$$

Hence, the log-likelihood scores would become

$$\begin{aligned}
\frac{\partial l(\mathbf{x}_r)}{\partial \alpha_{1x_r}} &= \frac{1}{2}\sum_{j=0}^{T-1} \frac{2\left(\cos\lambda_j - \alpha_{1x_r} - \alpha_{2x_r}\cos\lambda_j\right)}{\left(1+\alpha_{1x_r}^2+\alpha_{2x_r}^2-2\alpha_{1x_r}\left(1-\alpha_{2x_r}\right)\cos\lambda_j - 2\alpha_{2x_r}\cos2\lambda_j\right)^2} \\
&\quad \times \left(1+\alpha_{1x_r}^2+\alpha_{2x_r}^2-2\alpha_{1x_r}\left(1-\alpha_{2x_r}\right)\cos\lambda_j - 2\alpha_{2x_r}\cos2\lambda_j\right)^2 \\
&\quad \times \left[2\pi z_j^{x_r} z_j^{x_r*} - \frac{1}{\left(1+\alpha_{1x_r}^2+\alpha_{2x_r}^2-2\alpha_{1x_r}\left(1-\alpha_{2x_r}\right)\cos\lambda_j - 2\alpha_{2x_r}\cos2\lambda_j\right)}\right] \\
&= 2\pi\sum_{j=0}^{T-1}\left(\cos\lambda_j - \alpha_{1x_r} - \alpha_{2x_r}\cos\lambda_j\right) z_j^{x_r} z_j^{x_r*}
\end{aligned}$$

and

$$\begin{aligned}
\frac{\partial l(\mathbf{x}_r)}{\partial \alpha_{2x_r}} &= \frac{1}{2}\sum_{j=0}^{T-1} \frac{2\left(\cos2\lambda_j - \alpha_{1x_r}\cos\lambda_j - \alpha_{2x_r}\right)}{\left(1+\alpha_{1x_r}^2+\alpha_{2x_r}^2-2\alpha_{1x_r}\left(1-\alpha_{2x_r}\right)\cos\lambda_j - 2\alpha_{2x_r}\cos2\lambda_j\right)^2} \\
&\quad \times \left(1+\alpha_{1x_r}^2+\alpha_{2x_r}^2-2\alpha_{1x_r}\left(1-\alpha_{2x_r}\right)\cos\lambda_j - 2\alpha_{2x_r}\cos2\lambda_j\right)^2 \\
&\quad \times \left[2\pi z_j^{x_r} z_j^{x_r*} - \frac{1}{\left(1+\alpha_{1x_r}^2+\alpha_{2x_r}^2-2\alpha_{1x_r}\left(1-\alpha_{2x_r}\right)\cos\lambda_j - 2\alpha_{2x_r}\cos2\lambda_j\right)}\right] \\
&= 2\pi\sum_{j=0}^{T-1}2\left(\cos2\lambda_j - \alpha_{1x_r}\cos\lambda_j - \alpha_{2x_r}\right) z_j^{x_r} z_j^{x_r*}
\end{aligned}$$

where we have exploited the Yule–Walker equations to show that

$$\begin{aligned}
&\sum_{j=0}^{T-1} \frac{\left(\cos\lambda - \alpha_{1x_r} - \alpha_{2x_r}\cos\lambda\right)}{\left(1+\alpha_{1x_r}^2+\alpha_{2x_r}^2-2\alpha_{1x_r}\left(1-\alpha_{2x_r}\right)\cos\lambda - 2\alpha_{2x_r}\cos2\lambda\right)} \\
&\quad = \gamma_{x_r x_r}(1) - \alpha_{1x_r}\gamma_{x_r x_r}(0) - \alpha_{2x_r}\gamma_{x_r x_r}(1) = 0, \\
&\sum_{j=0}^{T-1} \frac{\left(\cos2\lambda - \alpha_{1x_r}\cos\lambda - \alpha_{2x_r}\right)}{\left(1+\alpha_{1x_r}^2+\alpha_{2x_r}^2-2\alpha_{1x_r}\left(1-\alpha_{2x_r}\right)\cos\lambda - 2\alpha_{2x_r}\cos2\lambda\right)} \\
&\quad = \gamma_{x_r x_r}(2) - \alpha_{1x_r}\gamma_{x_r x_r}(1) - \alpha_{2x_r}\gamma_{x_r x_r}(0) = 0
\end{aligned}$$

As a result, when we set both scores to 0 we would be left with the system of equations

$$\sum_{j=0}^{T-1}\left[z_j^{x_r}z_j^{x_r*}\otimes\begin{pmatrix}1 & \cos\lambda_j\\ \cos\lambda_j & 1\end{pmatrix}\right]\begin{pmatrix}\hat{\alpha}_{1x_r}\\ \hat{\alpha}_{2x_r}\end{pmatrix}=\sum_{j=0}^{T-1}\left[z_j^{x_r}z_j^{x_r*}\otimes\begin{pmatrix}\cos\lambda_j\\ \cos 2\lambda_j\end{pmatrix}\right]$$

But since

$$I_{x_r x_r}(\lambda_j)=\hat{\gamma}_{x_r x_r}(0)+2\sum_{k=1}^{T-1}\hat{\gamma}_{x_r x_r}(k)\cos(k\lambda_j)$$

we would have that

$$\sum_{j=0}^{T-1}2\pi I_{x_r x_r}(\lambda_j)=T\hat{\gamma}_{x_r x_r}(0)$$

$$\sum_{j=0}^{T-1}\cos\lambda_j\left[2\pi I_{x_r x_r}(\lambda_j)\right]=T\left[\hat{\gamma}_{x_r x_r}(1)+\hat{\gamma}_{x_r x_r}(T-1)\right]$$

and

$$\sum_{j=0}^{T-1}\cos 2\lambda_j\left[2\pi I_{x_r x_r}(\lambda_j)\right]=T\left[\hat{\gamma}_{x_r x_r}(2)+\hat{\gamma}_{x_r x_r}(T-2)\right]$$

which are the sample (circulant) autocovariances of x_{rt} of orders 0, 1 and 2, respectively. Therefore, the spectral estimators for $\hat{\alpha}_{1x_r}$ and $\hat{\alpha}_{2x_r}$ are (almost) identical to the ones we would obtain in the time domain, which will be given by the solution to the system of equations

$$\begin{pmatrix}\hat{\gamma}_{x_r x_r}(0) & \hat{\gamma}_{x_r x_r}(1)\\ \hat{\gamma}_{x_r x_r}(1) & \hat{\gamma}_{x_r x_r}(0)\end{pmatrix}\begin{pmatrix}\hat{\alpha}_{1x_r}\\ \hat{\alpha}_{2x_r}\end{pmatrix}=\begin{pmatrix}\hat{\gamma}_{x_r x_r}(1)\\ \hat{\gamma}_{x_r x_r}(2)\end{pmatrix}$$

because both $\hat{\gamma}_{x_r x_r}(T-1)=T^{-1}x_{rT}x_{r1}$ and $\hat{\gamma}_{x_r x_r}(T-2)=T^{-1}(x_{rT}x_{r2}+x_{rT-1}x_{r1})$ are $o_p(1)$.

Similar expressions would apply to the dynamic parameters that appear in θ_{u_i} for a given value of $\mathbf{c}_{ig}$ and $\mathbf{c}_{ir}$ in view of Eq. (25c), since in this case it would be possible to estimate the variances of the innovations ψ_i in closed form.

Specifically, for an AR(1) example in Eq. (4), the partial derivatives of $G_{u_iu_i}(\lambda)$ with respect to ψ_i and α_{1u_i} would be

$$\frac{\partial G_{u_iu_i}(\lambda)}{\partial \psi_i} = \frac{1}{1+\alpha_{1u_i}^2 - 2\alpha_{1u_i}\cos\lambda}$$

$$\frac{\partial G_{u_iu_i}(\lambda)}{\partial \alpha_{1u_i}} = \frac{2(\cos\lambda - \alpha_{1u_i})\psi_i}{\left(1+\alpha_{1u_i}^2 - 2\alpha_{1u_i}\cos\lambda\right)^2}$$

Hence, the corresponding log-likelihood scores would be

$$\begin{aligned}\frac{\partial l(\mathbf{y}_i|\mathbf{X})}{\partial \psi_i} &= \frac{1}{2}\sum_{j=0}^{T-1}\frac{\left(1+\alpha_{1u_i}^2 - 2\alpha_{1u_i1}\cos\lambda_j\right)^2}{\left(1+\alpha_{1u_i}^2 - 2\alpha_{1u_i}\cos\lambda_j\right)\psi_i^2}\left[2\pi z_j^{u_i}z_j^{u_i*} - \frac{\psi_i}{1+\alpha_{u_i1}^2 - 2\alpha_{u_i1}\cos\lambda_j}\right]\\ &= \frac{1}{2\psi_i^2}\sum_{j=0}^{T-1}\left[\left(1+\alpha_{1u_i}^2 - 2\alpha_{1u_i}\cos\lambda_j\right)2\pi z_j^{u_i}z_j^{u_i*} - \psi_i\right]\end{aligned}$$

$$\begin{aligned}\frac{\partial l(\mathbf{y}_i|\mathbf{X})}{\partial \alpha_{1u_i}} &= \frac{1}{2}\sum_{j=0}^{T-1}\frac{2(\cos\lambda_j - \alpha_{1u_i})\psi_i\left(1+\alpha_{u_i1}^2 - 2\alpha_{u_i1}\cos\lambda_j\right)^2}{\left(1+\alpha_{1u_i}^2 - 2\alpha_{1u_i}\cos\lambda_j\right)^2\psi_i^2}\\ &\times\left[2\pi z_j^{u_i}z_j^{u_i*} - \frac{\psi_i}{\left(1+\alpha_{1u_i}^2 - 2\alpha_{1u_i}\cos\lambda_j\right)}\right] = \frac{2\pi}{\psi_i}\sum_{j=0}^{T-1}(\cos\lambda_j - \alpha_{1u_i})z_j^{u_i}z_j^{u_i*}\end{aligned}$$

As a result, the spectral ML estimators of ψ_i and α_{u_i1} for fixed values of $\mathbf{c}_{ig}$ and $\mathbf{c}_{ir}$ would satisfy

$$\tilde{\psi}_i = \frac{2\pi}{T}\sum_{j=0}^{T-1}\left(1+\tilde{\alpha}_{1u_i1}^2 - 2\tilde{\alpha}_{1u_i}\cos\lambda_j\right)z_j^{u_i}z_j^{u_i*}$$

$$\tilde{\alpha}_{1u_i} = \frac{\sum_{j=0}^{T-1}\cos\lambda_j z_j^{u_i}z_j^{u_i*}}{\sum_{j=0}^{T-1}z_j^{u_i}z_j^{u_i*}}$$

Intuitively, these parameter estimates are, respectively, the sample analogues to the variance of v_{it}, which is the residual variance in the regression of u_{it} on u_{it-1}, and the slope coefficient in the same regression.

Finally, Eqs. (25d) and (25e) would allow us to obtain the ML estimators of $\mathbf{c}_{ig}$ and $\mathbf{c}_{ir}$ for given values of $\boldsymbol{\theta}_{u_i}$. In particular, if we write together the derivatives for c_{ikg} $\left(k=-m_g,\ldots,0,\ldots,n_g\right)$ and c_{ikr} $(k=-m_r,\ldots,0,\ldots,n_r)$ we end up with the 'weighted' normal equations:

$$\sum_{j=0}^{T-1}\left[G_{u_iu_i}^{-1}(\lambda_j)\begin{pmatrix}
e^{im_g\lambda_j}z_j^{x_g}z_j^{x_g*}e^{-im_g\lambda_j}+e^{im_g\lambda_j}z_j^{x_g}z_j^{x_g*}e^{-im_g\lambda_j} & \cdots & e^{-in_g\lambda_j}z_j^{x_g}z_j^{x_g*}e^{-im_g\lambda_j}+e^{im_g\lambda_j}z_j^{x_g}z_j^{x_g*}e^{in_g\lambda_j} & e^{im_g\lambda_j}z_j^{x_g}z_j^{x_r*}e^{-im_r\lambda_j}+e^{im_r\lambda_j}z_j^{x_r}z_j^{x_g*}e^{-im_g\lambda_j} & \cdots & e^{im_g\lambda_j}z_j^{x_g}z_j^{x_r*}e^{in_r\lambda_j}+e^{-in_r\lambda_j}z_j^{x_r}z_j^{x_g*}e^{-im_g\lambda_j} \\
\vdots & \ddots & \vdots & \vdots & \ddots & \vdots \\
e^{im_g\lambda_j}z_j^{x_g}z_j^{x_g*}e^{in_g\lambda_j}+e^{-in_g\lambda_j}z_j^{x_g}z_j^{x_g*}e^{-im_g\lambda_j} & \cdots & e^{-in_g\lambda_j}z_j^{x_g}z_j^{x_g*}e^{in_g\lambda_j}+e^{-in_g\lambda_j}z_j^{x_g}z_j^{x_g*}e^{in_g\lambda_j} & e^{-in_g\lambda_j}z_j^{x_g}z_j^{x_r*}e^{-im_r\lambda_j}+e^{im_r\lambda_j}z_j^{x_r}z_j^{x_g*}e^{in_g\lambda_j} & \cdots & e^{-in_g\lambda_j}z_j^{x_g}z_j^{x_r*}e^{in_r\lambda_j}+e^{-in_r\lambda_j}z_j^{x_r}z_j^{x_g*}e^{in_g\lambda_j} \\
e^{im_g\lambda_j}z_j^{x_g}z_j^{x_r*}e^{-im_r\lambda_j}+e^{im_r\lambda_j}z_j^{x_r}z_j^{x_g*}e^{-im_g\lambda_j} & \cdots & e^{-in_g\lambda_j}z_j^{x_g}z_j^{x_r*}e^{-im_r\lambda_j}+e^{im_r\lambda_j}z_j^{x_r}z_j^{x_g*}e^{in_g\lambda_j} & e^{im_r\lambda_j}z_j^{x_r}z_j^{x_r*}e^{-im_r\lambda_j}+e^{im_r\lambda_j}z_j^{x_r}z_j^{x_r*}e^{-im_r\lambda_j} & \cdots & e^{im_r\lambda_j}z_j^{x_r}z_j^{x_r*}e^{in_r\lambda_j}+e^{-in_r\lambda_j}z_j^{x_r}z_j^{x_r*}e^{-im_r\lambda_j} \\
\vdots & \ddots & \vdots & \vdots & \ddots & \vdots \\
e^{im_g\lambda_j}z_j^{x_g}z_j^{x_r*}e^{in_r\lambda_j}+e^{-in_r\lambda_j}z_j^{x_r}z_j^{x_g*}e^{-im_g\lambda_j} & \cdots & e^{-in_g\lambda_j}z_j^{x_g}z_j^{x_r*}e^{in_r\lambda_j}+e^{-in_r\lambda_j}z_j^{x_r}z_j^{x_g*}e^{in_g\lambda_j} & e^{-in_r\lambda_j}z_j^{x_r}z_j^{x_r*}e^{-im_r\lambda_j}+e^{im_r\lambda_j}z_j^{x_r}z_j^{x_r*}e^{in_r\lambda_j} & \cdots & e^{-in_r\lambda_j}z_j^{x_r}z_j^{x_r*}e^{in_r\lambda_j}+e^{-in_r\lambda_j}z_j^{x_r}z_j^{x_r*}e^{in_r\lambda_j}
\end{pmatrix}\right]\begin{pmatrix}\tilde{c}_{i,-m_g,g}\\ \vdots \\ \tilde{c}_{i,n_g g}\\ \tilde{c}_{i,-m_r,r}\\ \vdots \\ \tilde{c}_{i,n_r,r}\end{pmatrix}$$

$$=\sum_{j=0}^{T-1}G_{u_iu_i}^{-1}(\lambda_j)\begin{pmatrix}
z_j^{y_i}z_j^{x_g*}e^{-im_g\lambda_j}+z_j^{y_i*}z_j^{x_g}e^{im_g\lambda_j}\\
\vdots\\
z_j^{y_i}z_j^{x_g*}e^{in_g\lambda_j}+z_j^{y_i*}z_j^{x_g}e^{-in_g\lambda_j}\\
z_j^{y_i}z_j^{x_r*}e^{-im_r\lambda_j}+z_j^{y_i*}z_j^{x_r}e^{im_r\lambda_j}\\
\vdots\\
z_j^{y_i}z_j^{x_r*}e^{in_r\lambda_j}+z_j^{y_i*}z_j^{x_r}e^{-in_r\lambda_j}
\end{pmatrix}$$

Thus, unrestricted MLE's of $\mathbf{c}_{ig}$ and $\mathbf{c}_{ir}$ could be obtained from N univariate distributed lag weighted least-squares regressions of each y_{it} on x_{gt} and the appropriate x_{rt} that take into account the residual serial correlation in u_{it}. Interestingly, given that $G_{u_i u_i}(\lambda_j)$ is real, the above system of equations would not involve complex arithmetic. In addition, the terms in ψ_i would cancel, so the WLS procedure would only depend on the dynamic elements in $\boldsymbol{\theta}_{u_i}$.

Let us derive these expressions for the model in Eq. (4). In that case, the matrix on the left hand of the normal equations becomes

$$\sum_{j=0}^{T-1} G_{u_i u_i}^{-1}(\lambda_j) \begin{pmatrix} 2z_j^{x_g} z_j^{x_g *} & \left(e^{-i\lambda_j} + e^{i\lambda_j}\right) z_j^{x_g} z_j^{x_g *} & z_j^{x_g} z_j^{x_r *} + z_j^{x_r} z_j^{x_g *} & z_j^{x_g} z_j^{x_r *} e^{i\lambda_j} + e^{-i\lambda_j} z_j^{x_r} z_j^{x_g *} \\ \left(e^{i\lambda_j} + e^{-i\lambda_j}\right) z_j^{x_g} z_j^{x_g *} & 2z_j^{x_g} z_j^{x_g *} & e^{-i\lambda_j} z_j^{x_g} z_j^{x_r *} + z_j^{x_r} z_j^{x_g *} e^{i\lambda_j} & z_j^{x_g} z_j^{x_r *} + z_j^{x_r} z_j^{x_g *} \\ \left(z_j^{x_g} z_j^{x_r *} + z_j^{x_r} z_j^{x_g *}\right) & e^{-i\lambda_j} z_j^{x_g} z_j^{x_r *} + z_j^{x_r} z_j^{x_g *} e^{i\lambda_j} & 2z_j^{x_r} z_j^{x_r *} & z_j^{x_r} z_j^{x_r *} e^{i\lambda_j} + e^{-i\lambda_j} z_j^{x_r} z_j^{x_r *} \\ z_j^{x_g} z_j^{x_r *} e^{i\lambda_j} + e^{-i\lambda_j} z_j^{x_r} z_j^{x_g *} & z_j^{x_g} z_j^{x_r *} + z_j^{x_r} z_j^{x_g *} & e^{-i\lambda_j} z_j^{x_r} z_j^{x_r *} + z_j^{x_r} z_j^{x_r *} e^{i\lambda_j} & 2z_j^{x_r} z_j^{x_r *} \end{pmatrix}$$

while the vector on the right-hand side will be

$$\sum_{j=0}^{T-1} G_{u_i u_i}^{-1}(\lambda_j) \begin{pmatrix} z_j^{y_i} z_j^{x_g *} + z_j^{y_i *} z_j^{x_g} \\ e^{i\lambda_j} z_j^{y_i} z_j^{x_g *} + e^{-i\lambda_j} z_j^{y_i *} z_j^{x_g} \\ z_j^{y_i} z_j^{x_r *} + z_j^{y_i *} z_j^{x_r} \\ e^{i\lambda_j} z_j^{y_i} z_j^{x_r *} + e^{-i\lambda_j} z_j^{y_i *} z_j^{x_r} \end{pmatrix}$$

In principle, we could carry out a zig-zag procedure that would estimate $\mathbf{c}_{ig}$ and $\mathbf{c}_{ir}$ for given $\boldsymbol{\theta}_{u_i}$, and then $\boldsymbol{\theta}_{u_i}$ for a given $\mathbf{c}_{ig}$ and $\mathbf{c}_{ir}$. This would correspond to the spectral analogue to the Cochrane–Orcutt (1949) procedure. Obviously, iterations would be unnecessary when $\mathbf{G}_{\mathbf{uu}}(\lambda_j)$ is in fact constant, so that the idiosyncratic terms are static. In that case, the above equations could be written in terms of the elements of the covariance and the first autocovariance matrices of y_t, x_{gt} and x_{rt}.

3.2. Expected Log-Likelihood Function

In practice, of course, we do not observe $\mathbf{x}_t$. Nevertheless, the EM algorithm can be used to obtain values for $\boldsymbol{\theta}$ as close to the optimum as desired. At each iteration, the EM algorithm maximises the expected value of $l(\mathbf{Y}|\mathbf{X})+l(\mathbf{X})$ conditional on $\mathbf{Y}$ and the current parameter estimates, $\boldsymbol{\theta}^{(n)}$. The rationale stems from the fact that $l(\mathbf{Y},\mathbf{X})$ can also be factorised as $l(\mathbf{Y})+l(\mathbf{X}|\mathbf{Y})$. Since the expected value of the latter, conditional on $\mathbf{Y}$ and $\boldsymbol{\theta}^{(n)}$, reaches a maximum at $\boldsymbol{\theta}=\boldsymbol{\theta}^{(n)}$, *any increase* in the expected value of $l(\mathbf{Y},\mathbf{X})$ must represent an increase in $l(\mathbf{Y})$. This is the generalised EM principle.

In the E step, we must compute

$$E\left[l(\mathbf{x}_g)|\mathbf{Z}^{\mathbf{y}},\boldsymbol{\theta}^{(n)}\right]=\varkappa-\frac{1}{2}\sum_{j=0}^{T-1}\ln G_{x_g x_g}(\lambda_j)-\frac{2\pi}{2}\sum_{j=0}^{T-1}G^{-1}_{x_g x_g}(\lambda_j)E\left[z_j^{x_g}z_j^{x_g *}|\mathbf{Z}^{\mathbf{y}},\boldsymbol{\theta}^{(n)}\right],$$

$$E\left[l(\mathbf{x}_r)|\mathbf{Z}^{\mathbf{y}},\boldsymbol{\theta}^{(n)}\right]=\varkappa-\frac{1}{2}\sum_{j=0}^{T-1}\ln G_{x_r x_r}(\lambda_j)-\frac{2\pi}{2}\sum_{j=0}^{T-1}G^{-1}_{x_r x_r}(\lambda_j)E\left[z_j^{x_r}z_j^{x_r *}|\mathbf{Z}^{\mathbf{y}},\boldsymbol{\theta}^{(n)}\right],$$

$$E\left[l(\mathbf{y}_i|\mathbf{X})|\mathbf{Z}^{\mathbf{y}},\boldsymbol{\theta}^{(n)}\right]=\varkappa-\frac{1}{2}\sum_{j=0}^{T-1}\ln G_{u_i u_i}(\lambda_j)-\frac{2\pi}{2}\sum_{j=0}^{T-1}G^{-1}_{u_i u_i}(\lambda_j)E\left[z_j^{u_i}z_j^{u_i *}|\mathbf{Z}^{\mathbf{y}},\boldsymbol{\theta}^{(n)}\right]$$

But

$$\begin{aligned}E\left[\mathbf{z}_j^{\mathbf{x}}\mathbf{z}_j^{\mathbf{x}*}|\mathbf{Z}^{\mathbf{y}},\boldsymbol{\theta}^{(n)}\right]&=\mathbf{z}_j^{\mathbf{x}^K}(\boldsymbol{\theta}^{(n)})\mathbf{z}_j^{\mathbf{x}^K *}(\boldsymbol{\theta}^{(n)})\\&\quad+E\left\{\left[\mathbf{z}_j^{\mathbf{x}}-\mathbf{z}_j^{\mathbf{x}^K}(\boldsymbol{\theta}^{(n)})\right]\left[\mathbf{z}_j^{\mathbf{x}*}-\mathbf{z}_j^{\mathbf{x}^K *}(\boldsymbol{\theta}^{(n)})\right]|\mathbf{z}_j^{\mathbf{y}},\boldsymbol{\theta}^{(n)}\right\}\\&=\mathbf{I}^{(n)}_{\mathbf{x}^K\mathbf{x}^K}(\lambda_j)+\boldsymbol{\Omega}^{(n)}(\lambda_j)\end{aligned}$$

where

$$\begin{aligned}\mathbf{I}_{\mathbf{x}^K\mathbf{x}^K}(\lambda)&=2\pi\mathbf{G}_{\mathbf{xx}}(\lambda)\mathbf{C}'(e^{i\lambda})\mathbf{G}^{-1}_{\mathbf{yy}}(\lambda)\mathbf{I}_{\mathbf{yy}}(\lambda)\mathbf{G}^{-1}_{\mathbf{yy}}(\lambda)\mathbf{C}(e^{-i\lambda})\mathbf{G}_{\mathbf{xx}}(\lambda)\\&=2\pi\boldsymbol{\Omega}(\lambda)\mathbf{C}'(e^{i\lambda})\mathbf{G}^{-1}_{\mathbf{uu}}(\lambda)\mathbf{I}_{\mathbf{yy}}(\lambda)\mathbf{G}^{-1}_{\mathbf{uu}}(\lambda)\mathbf{C}(e^{-i\lambda})\boldsymbol{\Omega}(\lambda)\end{aligned}\tag{26}$$

is the periodogram of the smoothed values of the $R+1$ common factors $\mathbf{x}$ and

$$E\left\{\left[\mathbf{z}_j^{\mathbf{x}}-\mathbf{z}_j^{\mathbf{x}^K}(\boldsymbol{\theta})\right]\left[\mathbf{z}_j^{\mathbf{x}*}-\mathbf{z}_j^{\mathbf{x}^K *}(\boldsymbol{\theta})\right]|\mathbf{Z}^{\mathbf{y}},\boldsymbol{\theta}\right\}=\boldsymbol{\Omega}(\lambda_j)$$

In turn, if we define

$$\mathbf{I}_{\mathbf{yx}^K}(\lambda) = \mathbf{I}_{\mathbf{yy}}(\lambda)\mathbf{G}_{\mathbf{yy}}^{-1}(\lambda)\mathbf{C}(\mathrm{e}^{-\mathrm{i}\lambda})\mathbf{G}_{\mathbf{xx}}(\lambda) = \mathbf{I}_{\mathbf{yy}}(\lambda)\mathbf{G}_{\mathbf{uu}}^{-1}(\lambda)\mathbf{C}(\mathrm{e}^{-\mathrm{i}\lambda})\mathbf{\Omega}(\lambda)$$

as the cross-periodogram between the observed series $\mathbf{y}$ and the smoothed values of the common factors $\mathbf{x}$, we will have that

$$\begin{aligned}
\mathbf{I}_{\mathbf{uu}}^{(n)}(\lambda_j) &= E\left[\mathbf{z}_j^{\mathbf{u}}\mathbf{z}_j^{\mathbf{u}*}|\mathbf{Z}^{\mathbf{y}},\boldsymbol{\theta}^{(n)}\right] = E\left\{\left[\mathbf{z}_j^{\mathbf{y}} - \mathbf{C}(\mathrm{e}^{-\mathrm{i}\lambda_j})\mathbf{z}_j^{\mathbf{x}}\right]\left[\mathbf{z}_j^{\mathbf{y}*} - \mathbf{z}_j^{\mathbf{x}*}\mathbf{C}'(\mathrm{e}^{\mathrm{i}\lambda_j})\right]|\mathbf{Z}^{\mathbf{y}},\boldsymbol{\theta}^{(n)}\right\} \\
&= \left[\mathbf{z}_j^{\mathbf{y}} - \mathbf{C}(\mathrm{e}^{-\mathrm{i}\lambda_j})\mathbf{z}_j^{\mathbf{x}^K}(\boldsymbol{\theta}^{(n)})\right]\left[\mathbf{z}_j^{\mathbf{y}*} - \mathbf{z}_j^{\mathbf{x}^K*}(\boldsymbol{\theta}^{(n)})\mathbf{C}'(\mathrm{e}^{\mathrm{i}\lambda_j})\right] \\
&\quad + \mathbf{C}(\mathrm{e}^{-\mathrm{i}\lambda_j})\mathbf{\Omega}^{(n)}(\lambda_j)\mathbf{C}'(\mathrm{e}^{\mathrm{i}\lambda_j}) \\
&= \mathbf{I}_{\mathbf{yy}}(\lambda_j) - \mathbf{I}_{\mathbf{yx}^K}^{(n)}(\lambda)\mathbf{C}'(\mathrm{e}^{\mathrm{i}\lambda_j}) - \mathbf{C}(\mathrm{e}^{-\mathrm{i}\lambda_j})\mathbf{I}_{\mathbf{x}^K\mathbf{y}}^{(n)}(\lambda) \\
&\quad + \mathbf{C}(\mathrm{e}^{-\mathrm{i}\lambda_j})\left[\mathbf{I}_{\mathbf{x}^K\mathbf{x}^K}^{(n)}(\lambda_j) + \mathbf{\Omega}^{(n)}(\lambda_j)\right]\mathbf{C}'(\mathrm{e}^{\mathrm{i}\lambda_j})
\end{aligned}$$

which resembles the expected value of $\mathbf{I}_{\mathbf{uu}}(\lambda_j)$ but the values at which the expectations are evaluated are generally different from the values at which the distributed lags are computed.

The assumed bifactor structure implies that for the ith series, the above expression reduces to

$$\begin{aligned}
I_{u_iu_i}^{(n)}(\lambda_j) &= E\left[z_j^{u_i}z_j^{u_i*}|\mathbf{Z}^{\mathbf{y}},\boldsymbol{\theta}^{(n)}\right] \\
&= I_{y_iy_i}(\lambda_j) - c_{ig}(\mathrm{e}^{-\mathrm{i}\lambda_j})I_{x_g^Ky_i}^{(n)}(\lambda_j) - c_{ir}(\mathrm{e}^{-\mathrm{i}\lambda_j})I_{x_r^Ky_i}^{(n)}(\lambda_j) - I_{y_ix_g^K}^{(n)}(\lambda_j)c_{ig}(\mathrm{e}^{\mathrm{i}\lambda_j}) \\
&\quad - I_{y_ix_r^K}^{(n)}(\lambda_j)c_{ir}(\mathrm{e}^{\mathrm{i}\lambda_j}) + \left[I_{x_g^Kx_g^K}^{(n)}(\lambda_j) + \omega_{gg}^{(n)}(\lambda_j)\right]c_{ig}(\mathrm{e}^{-\mathrm{i}\lambda_j})c_{ig}(\mathrm{e}^{\mathrm{i}\lambda_j}) \\
&\quad + \left[I_{x_r^Kx_r^K}^{(n)}(\lambda_j) + \omega_{rr}^{(n)}(\lambda_j)\right]c_{ir}(\mathrm{e}^{-\mathrm{i}\lambda_j})c_{ir}(\mathrm{e}^{\mathrm{i}\lambda_j}) \\
&\quad + \left[I_{x_g^Kx_r^K}^{(n)}(\lambda_j) + \omega_{gr}^{(n)}(\lambda_j)\right]c_{ig}(\mathrm{e}^{-\mathrm{i}\lambda_j})c_{ir}(\mathrm{e}^{\mathrm{i}\lambda_j}) \\
&\quad + \left[I_{x_r^Kx_g^K}^{(n)}(\lambda_j) + \omega_{rg}^{(n)}(\lambda_j)\right]c_{ir}(\mathrm{e}^{-\mathrm{i}\lambda_j})c_{ig}(\mathrm{e}^{\mathrm{i}\lambda_j})
\end{aligned}$$

Therefore, if we put all these expressions together, we end up with

$$E\left[l(\mathbf{x}_g)|\mathbf{Y},\boldsymbol{\theta}^{(n)}\right] = \varkappa - \frac{1}{2}\sum_{j=0}^{T-1}\ln|G_{x_gx_g}(\lambda_j)| - \frac{2\pi}{2}\sum_{j=0}^{T-1}G_{x_gx_g}^{-1}(\lambda_j)\left[I_{x_g^Kx_g^K}^{(n)}(\lambda_j) + \omega_{gg}^{(n)}(\lambda_j)\right] \tag{27}$$

$$E\left[l(\mathbf{x}_r)|\mathbf{Y},\boldsymbol{\theta}^{(n)}\right] = \varkappa - \frac{1}{2}\sum_{j=0}^{T-1}\ln|G_{x_rx_r}(\lambda_j)| - \frac{2\pi}{2}\sum_{j=0}^{T-1}G_{x_rx_r}^{-1}(\lambda_j)\left[I_{x_r^Kx_r^K}^{(n)}(\lambda_j) + \omega_{rr}^{(n)}(\lambda_j)\right] \tag{28}$$

$$E\left[l(\mathbf{y}_i|\mathbf{X})|\mathbf{Y},\boldsymbol{\theta}^{(n)}\right] = \varkappa - \frac{1}{2}\sum_{j=0}^{T-1}\ln|G_{u_iu_i}(\lambda_j)| - \frac{2\pi}{2}\sum_{j=0}^{T-1}G_{u_iu_i}^{-1}(\lambda_j)I_{u_iu_i}^{(n)}(\lambda_j) \tag{29}$$

We can then maximise $E\left[l(\mathbf{x}_g)|\mathbf{Y},\boldsymbol{\theta}^{(n)}\right]$ in Eq. (27) with respect to $\boldsymbol{\theta}_{x_g}$ to update those parameters, and the same applies to Eq. (28) and $\boldsymbol{\theta}_{x_r}$. Similarly, we can maximise $E\left[l(\mathbf{y}_i|\mathbf{X})|\mathbf{Y},\boldsymbol{\theta}^{(n)}\right]$ with respect to $\mathbf{c}_{ig}$, $\mathbf{c}_{ir}$, ψ_i and $\boldsymbol{\theta}_{u_i}$ to update those parameters.

In order to conduct those maximisations, we need the scores of the expected log-likelihood functions.

Given the similarity between Eqs. (27) and (22), it is easy to see that

$$\frac{\partial E\left[l(\mathbf{x}_g)|\mathbf{Y},\boldsymbol{\theta}^{(n)}\right]}{\partial\boldsymbol{\theta}_{x_g}} = \frac{1}{2}\sum_{j=0}^{T-1}\frac{\partial G_{x_gx_g}(\lambda_j)}{\partial\boldsymbol{\theta}_{x_g}}G_{x_gx_g}^{-2}(\lambda_j)\left\{2\pi\left[I_{x_g^Kx_g^K}^{(n)}(\lambda_j)+\omega_{gg}^{(n)}(\lambda_j)\right]-G_{x_gx_g}(\lambda_j)\right\}$$

which, not surprisingly, coincides with the the expected value of Eq. (25a) given $\mathbf{Y}$ and the current parameter estimates, $\boldsymbol{\theta}^{(n)}$. As a result, for the AR(1) process for x_g in Eq. (4) we will have

$$\frac{\partial E\left[l(\mathbf{x}_g)|\mathbf{Y},\boldsymbol{\theta}^{(n)}\right]}{\partial\alpha_{1x_g}} = 2\pi\sum_{j=0}^{T-1}\left(\cos\lambda_j - \alpha_{x1}\right)\left[I_{x_g^Kx_g^K}^{(n)}(\lambda_j) + \omega_{gg}^{(n)}(\lambda_j)\right]$$

whence

$$\hat{\alpha}_{1x_g}^{(n+1)} = \frac{\sum_{j=0}^{T-1}\cos\lambda_j\left[I_{x_g^Kx_g^K}^{(n)}(\lambda_j) + \omega_{gg}^{(n)}(\lambda_j)\right]}{\sum_{j=0}^{T-1}\left[I_{x_g^Kx_g^K}^{(n)}(\lambda_j) + \omega_{gg}^{(n)}(\lambda_j)\right]}$$

Likewise, we will have that

$$\frac{\partial E\left[l(\mathbf{x}_r)|\mathbf{Y},\boldsymbol{\theta}^{(n)}\right]}{\partial\boldsymbol{\theta}_{x_r}}=\frac{1}{2}\sum_{j=0}^{T-1}\frac{\partial G_{x_r x_r}(\lambda_j)}{\partial\boldsymbol{\theta}_{x_r}}G_{x_r x_r}^{-2}(\lambda_j)\left\{2\pi\left[I_{x_r^K x_r^K}^{(n)}(\lambda_j)+\omega_{rr}^{(n)}(\lambda_j)\right]-G_{x_r x_r}(\lambda_j)\right\}$$

Hence, in the case of the AR(2) process for x_{rt} in Eq. (4), the expected log-likelihood scores become

$$\frac{\partial E\left[l(\mathbf{x}_r)|\mathbf{Y},\boldsymbol{\theta}^{(n)}\right]}{\partial\alpha_{1x_r}}=2\pi\sum_{j=0}^{T-1}(\cos\lambda_j-\alpha_{1x_r}-\alpha_{2x_r}\cos\lambda_j)\left[I_{x_r^K x_r^K}^{(n)}(\lambda_j)+\omega_{rr}^{(n)}(\lambda_j)\right],$$

$$\frac{\partial E\left[l(\mathbf{x}_r)|\mathbf{Y},\boldsymbol{\theta}^{(n)}\right]}{\partial\alpha_{2x_r}}=2\pi\sum_{j=0}^{T-1}2(\cos 2\lambda_j-\alpha_{1x_r}\cos\lambda_j-\alpha_{2x_r})\left[I_{x_r^K x_r^K}^{(n)}(\lambda_j)+\omega_{rr}^{(n)}(\lambda_j)\right]$$

so that the updated autoregressive coefficients will be the solution to the system of equations

$$\sum_{j=0}^{T-1}\left\{\left[I_{x_r^K x_r^K}^{(n)}(\lambda_j)+\omega_{rr}^{(n)}(\lambda_j)\right]\otimes\begin{pmatrix}1 & \cos\lambda_j\\ \cos\lambda_j & 1\end{pmatrix}\right\}\begin{pmatrix}\hat{\alpha}_{1x_r}\\ \hat{\alpha}_{2x_r}\end{pmatrix}$$
$$=\sum_{j=0}^{T-1}\left\{\left[I_{x_r^K x_r^K}^{(n)}(\lambda_j)+\omega_{rr}^{(n)}(\lambda_j)\right]\otimes\begin{pmatrix}\cos\lambda_j\\ \cos 2\lambda_j\end{pmatrix}\right\}$$

Similar expressions would apply to the dynamic parameters that appear in $\boldsymbol{\theta}_{u_i}$ and ψ_i for given values of $\mathbf{c}_{ig}$ and $\mathbf{c}_{ir}$. Specifically, when the idiosyncratic terms follow AR(1) processes

$$\frac{\partial E\left[l(\mathbf{y}_i|\mathbf{X})|\mathbf{Y},\boldsymbol{\theta}^{(n)}\right]}{\partial\psi_i}=\frac{1}{2\psi_i^2}\sum_{j=0}^{T-1}\left(1+\alpha_{u_i1}^2-2\alpha_{u_i1}\cos\lambda\right)\left\{2\pi I_{u_iu_i}^{(n)}(\lambda_j)-\psi_i\right\},$$

$$\frac{E\left[l(\mathbf{y}_i|\mathbf{X})|\mathbf{Y},\boldsymbol{\theta}^{(n)}\right]}{\partial\alpha_{u_i1}}=\frac{2\pi}{\psi_i}\sum_{j=0}^{T-1}(\cos\lambda_j-\alpha_{1u_i})I_{u_iu_i}^{(n)}(\lambda_j)$$

As a result, the spectral ML estimators of ψ_i and α_{u_i1} given $\mathbf{c}_{ig}$ and $\mathbf{c}_{ir}$ will satisfy

$$\hat{\psi}_i^{(n+1)} = \frac{2\pi}{T}\sum_{j=0}^{T-1}\left[1+\left(\hat{\alpha}_{1u_i}^{(n+1)}\right)^2 - 2\hat{\alpha}_{1u_i}^{(n+1)}\cos\lambda_j\right] I_{u_iu_i}^{(n)}(\lambda_j),$$

$$\hat{\alpha}_{1u_i}^{(n+1)} = \frac{\sum_{j=0}^{T-1}\cos\lambda_j I_{u_iu_i}^{(n)}(\lambda_j)}{\sum_{j=0}^{T-1} I_{u_iu_i}^{(n)}(\lambda_j)}$$

Finally, the derivatives of Eq. (29) with respect to c_{ikg} $(k = -m_g, \ldots, 0, \ldots, n_g)$ and c_{ilr} $(l = -m_r, \ldots, 0, \ldots, n_r)$ for fixed values of $\boldsymbol{\theta}_{u_i}$ will give rise to a set of modified 'weighted' normal equations analogous to the ones in the previous section but with cross-product terms of the form $z_j^{x_g} z_j^{x_r *}$ replaced by $\left[I_{x_g^K x_r^K}^{(n)}(\lambda_j) + \omega_{gr}^{(n)}(\lambda_j)\right]$.

For the example in Eq. (4), the matrix on the left hand of the normal equations becomes

$$2\sum_{j=0}^{T-1} G_{u_iu_i}^{-1}(\lambda_j)\begin{pmatrix}
[I_{x_g^K x_g^K}^{(n)}(\lambda_j) + \omega_{gg}^{(n)}(\lambda_j)] & \cos\lambda_j[I_{x_g^K x_g^K}^{(n)}(\lambda_j) + \omega_{gg}^{(n)}(\lambda_j)] & \Re[I_{x_g^K x_r^K}^{(n)}(\lambda_j) + \omega_{gr}^{(n)}(\lambda_j)] & \cos\lambda_j\Re[I_{x_g^K x_r^K}^{(n)}(\lambda_j)] - \sin\lambda_j\Im[I_{x_g^K x_r^K}^{(n)}(\lambda_j)] \\
\cos\lambda_j[I_{x_g^K x_g^K}^{(n)}(\lambda_j) + \omega_{gg}^{(n)}(\lambda_j)] & [I_{x_g^K x_g^K}^{(n)}(\lambda_j) + \omega_{gg}^{(n)}(\lambda_j)] & \cos\lambda_j\Re[I_{x_g^K x_r^K}^{(n)}(\lambda_j)] + \sin\lambda_j\Im[I_{x_g^K x_r^K}^{(n)}(\lambda_j)] & \Re[I_{x_g^K x_r^K}^{(n)}(\lambda_j) + \omega_{gr}^{(n)}(\lambda_j)] \\
\Re[I_{x_g^K x_r^K}^{(n)}(\lambda_j) + \omega_{gr}^{(n)}(\lambda_j)] & \cos\lambda_j\Re[I_{x_g^K x_r^K}^{(n)}(\lambda_j)] + \sin\lambda_j\Im[I_{x_g^K x_r^K}^{(n)}(\lambda_j)] & [I_{x_r^K x_r^K}^{(n)}(\lambda_j) + \omega_{rr}^{(n)}(\lambda_j)] & \cos\lambda_j[I_{x_r^K x_r^K}^{(n)}(\lambda_j) + \omega_{rr}^{(n)}(\lambda_j)] \\
\cos\lambda_j\Re[I_{x_g^K x_r^K}^{(n)}(\lambda_j)] - \sin\lambda_j\Im[I_{x_g^K x_r^K}^{(n)}(\lambda_j)] & \Re[I_{x_g^K x_r^K}^{(n)}(\lambda_j) + \omega_{gr}^{(n)}(\lambda_j)] & \cos\lambda_j\left[I_{x_r^K x_r^K}^{(n)}(\lambda_j) + \omega_{rr}^{(n)}(\lambda_j)\right] & [I_{x_r^K x_r^K}^{(n)}(\lambda_j) + \omega_{rr}^{(n)}(\lambda_j)]
\end{pmatrix}$$

while the vector on the right-hand side will be

$$2\sum_{j=0}^{T-1} G_{u_iu_i}^{-1}(\lambda_j)\begin{pmatrix} \Re[I_{y_ix_g^K}^{(n)}(\lambda_j)] \\ \cos\lambda_j\Re[I_{y_ix_g^K}^{(n)}(\lambda_j)] - \sin\lambda_j\Im[I_{y_ix_g^K}^{(n)}(\lambda_j)] \\ \Re[I_{y_ix_r^K}^{(n)}(\lambda_j)] \\ \cos\lambda_j\Re[I_{y_ix_r^K}^{(n)}(\lambda_j)] - \sin\lambda_j\Im[I_{y_ix_r^K}^{(n)}(\lambda_j)] \end{pmatrix}$$

In principle, we could carry out a zig-zag procedure that would estimate $\mathbf{c}_{ig}$, $\mathbf{c}_{ir}$ and ψ_i for given $\boldsymbol{\theta}_{u_i}$ and $\boldsymbol{\theta}_{u_i}$ for given $\mathbf{c}_{ig}$, $\mathbf{c}_{ir}$ and ψ_i, although it is not clear that we really need to fully maximise the expected log-likelihood function at each EM iteration since the generalised EM principle simply requires us to increase it. Obviously, such iterations would be unnecessary when the idiosyncratic terms are static.

3.3. Alternative Marginal Scores

As is well known, the EM algorithm slows down considerably near the optimum. At that point, the best practical strategy would be to switch to a first derivative-based method. Fortunately, the EM principle can also be exploited to simplify the computation of the score. Since the Kullback inequality implies that $E[l(\mathbf{X}|\mathbf{Y};\boldsymbol{\theta})|\mathbf{Y};\boldsymbol{\theta}] = 0$, it is clear that $\partial l(\mathbf{Y};\boldsymbol{\theta})/\partial\boldsymbol{\theta}$ can be obtained as the expected value (given $\mathbf{Y}$ and $\boldsymbol{\theta}$) of the sum of the unobservable scores corresponding to $l(\mathbf{y}_1, \ldots, \mathbf{y}_N|\mathbf{X})$ and $l(\mathbf{X})$. This yields

$$\frac{\partial l(\mathbf{Y})}{\partial\boldsymbol{\theta}_{x_g}} = \frac{1}{2}\sum_{j=0}^{T-1}\frac{\partial G_{x_gx_g}(\lambda_j)}{\partial\boldsymbol{\theta}_{x_g}}G_{x_gx_g}^{-2}(\lambda_j)\left[2\pi E\left[z_j^{x_g}z_j^{x_g*}|\mathbf{Z}^{\mathbf{y}},\boldsymbol{\theta}\right] - G_{x_gx_g}(\lambda_j)\right]$$

$$\frac{\partial l(\mathbf{Y})}{\partial\boldsymbol{\theta}_{x_r}} = \frac{1}{2}\sum_{j=0}^{T-1}\frac{\partial G_{x_rx_r}(\lambda_j)}{\partial\boldsymbol{\theta}_{x_r}}G_{x_rx_r}^{-2}(\lambda_j)\left[2\pi E\left[z_j^{x_r}z_j^{x_r*}|\mathbf{Z}^{\mathbf{y}},\boldsymbol{\theta}\right] - G_{xx}(\lambda_j)\right]$$

$$\frac{\partial l(\mathbf{Y})}{\partial\boldsymbol{\theta}_{u_i}} = \frac{1}{2}\sum_{j=0}^{T-1}\frac{\partial G_{u_iu_i}(\lambda_j)}{\partial\boldsymbol{\theta}_{u_i}}G_{u_iu_i}^{-2}(\lambda_j)\left[2\pi E\left[z_j^{u_i}z_j^{u_i*}|\mathbf{Z}^{\mathbf{y}},\boldsymbol{\theta}\right] - G_{u_iu_i}(\lambda_j)\right]$$

$$\frac{\partial l(\mathbf{Y})}{\partial c_{ikg}} = \frac{2\pi}{2}\sum_{j=0}^{T-1}G_{u_iu_i}^{-1}(\lambda_j)\left[e^{ik\lambda_j}E\left[z_j^{u_i}z_j^{x_g*}|\mathbf{Z}^{\mathbf{y}},\boldsymbol{\theta}\right] + e^{-ik\lambda_j}E\left[z_j^{x_g}z_j^{u_i*}|\mathbf{Z}^{\mathbf{y}},\boldsymbol{\theta}\right]\right]$$

$$\frac{\partial l(\mathbf{Y})}{\partial c_{ilr}} = \frac{2\pi}{2}\sum_{j=0}^{T-1}G_{u_iu_i}^{-1}(\lambda_j)\left[e^{il\lambda_j}E\left[z_j^{u_i}z_j^{x_r*}|\mathbf{Z}^{\mathbf{y}},\boldsymbol{\theta}\right] + e^{-il\lambda_j}E\left[z_j^{x_r}z_j^{u_i*}|\mathbf{Z}^{\mathbf{y}},\boldsymbol{\theta}\right]\right]$$

But since the scores are now evaluated at the values of the parameters at which the expectations are computed, we will have that

$$
\begin{aligned}
E\left[\mathbf{z}_j^{\mathbf{x}}\mathbf{z}_j^{\mathbf{x}*}|\mathbf{Z}^{\mathbf{y}},\boldsymbol{\theta}\right] &= \mathbf{I}_{\mathbf{x}^K\mathbf{x}^K}(\lambda_j)+\boldsymbol{\Omega}(\lambda_j),\\
E\left[\mathbf{z}_j^{\mathbf{u}}\mathbf{z}_j^{\mathbf{u}*}|\mathbf{Z}^{\mathbf{y}},\boldsymbol{\theta}\right] &= E\left[\mathbf{z}_j^{\mathbf{u}}|\mathbf{Z}^{\mathbf{y}},\boldsymbol{\theta}\right]E\left[\mathbf{z}_j^{\mathbf{u}*}|\mathbf{Z}^{\mathbf{y}},\boldsymbol{\theta}\right]\\
&\quad +E\left[\left\{\mathbf{z}_j^{\mathbf{u}}-E\left[\mathbf{z}_j^{\mathbf{u}}|\mathbf{Z}^{\mathbf{y}},\boldsymbol{\theta}\right]\right\}\left\{\mathbf{z}_j^{\mathbf{u}*}-E\left[\mathbf{z}_j^{\mathbf{u}*}|\mathbf{Z}^{\mathbf{y}},\boldsymbol{\theta}\right]\right\}|\mathbf{Z}^{\mathbf{y}},\boldsymbol{\theta}\right]\\
&= \mathbf{I}_{\mathbf{u}^K\mathbf{u}^K}(\lambda_j)+\mathbf{C}(\mathrm{e}^{-\mathrm{i}\lambda_j})\boldsymbol{\Omega}(\lambda_j)\mathbf{C}'(\mathrm{e}^{\mathrm{i}\lambda_j}).\\
E\left[\mathbf{z}_j^{\mathbf{u}}\mathbf{z}_j^{\mathbf{x}*}|\mathbf{Z}^{\mathbf{y}},\boldsymbol{\theta}\right] &= E\left[\mathbf{z}_j^{\mathbf{u}}|\mathbf{Z}^{\mathbf{y}},\boldsymbol{\theta}\right]E\left[\mathbf{z}_j^{\mathbf{x}*}|\mathbf{Z}^{\mathbf{y}},\boldsymbol{\theta}\right]\\
&\quad +E\left[\left\{\mathbf{z}_j^{\mathbf{u}}-E\left[\mathbf{z}_j^{\mathbf{u}}|\mathbf{Z}^{\mathbf{y}},\boldsymbol{\theta}\right]\right\}\left\{\mathbf{z}_j^{\mathbf{x}*}-E\left[\mathbf{z}_j^{\mathbf{x}*}|\mathbf{Z}^{\mathbf{y}},\boldsymbol{\theta}\right]\right\}|\mathbf{Z}^{\mathbf{y}},\boldsymbol{\theta}\right]\\
&= \mathbf{I}_{\mathbf{u}^K\mathbf{x}^K}(\lambda_j)-\mathbf{C}(\mathrm{e}^{-\mathrm{i}\lambda_j})\boldsymbol{\Omega}(\lambda_j)
\end{aligned}
$$

where

$$
\begin{aligned}
\mathbf{z}_j^{\mathbf{u}^K} &= E\left[\mathbf{z}_j^{\mathbf{u}}|\mathbf{Z}^{\mathbf{y}},\boldsymbol{\theta}\right] = \mathbf{G}_{\mathbf{uu}}(\lambda_j)\mathbf{G}_{\mathbf{yy}}^{-1}(\lambda_j)\mathbf{z}_j^{\mathbf{y}} = \mathbf{z}_j^{\mathbf{y}}-\mathbf{C}(\mathrm{e}^{-\mathrm{i}\lambda})\mathbf{z}_j^{\mathbf{x}^K}\\
E\left[\left(\mathbf{z}_j^{\mathbf{u}}-\mathbf{z}_j^{\mathbf{u}^K}\right)\left(\mathbf{z}_j^{\mathbf{u}*}-\mathbf{z}_j^{\mathbf{u}^K*}\right)|\mathbf{Z}^{\mathbf{y}},\boldsymbol{\theta}\right] &= \mathbf{C}(\mathrm{e}^{-\mathrm{i}\lambda_j})\boldsymbol{\Omega}(\lambda_j)\mathbf{C}'(\mathrm{e}^{\mathrm{i}\lambda_j})\\
E\left[\left(\mathbf{z}_j^{\mathbf{u}}-\mathbf{z}_j^{\mathbf{u}_K}\right)\left(\mathbf{z}_j^{\mathbf{x}*}-\mathbf{z}_j^{\mathbf{x}^K*}\right)|\mathbf{Z}^{\mathbf{y}},\boldsymbol{\theta}\right] &= \mathbf{C}(\mathrm{e}^{-\mathrm{i}\lambda_j})\boldsymbol{\Omega}(\lambda_j)
\end{aligned}
$$

$$
\begin{aligned}
\mathbf{I}_{\mathbf{u}^K\mathbf{u}^K}(\lambda) &= 2\pi\mathbf{G}_{\mathbf{uu}}(\lambda)\mathbf{G}_{\mathbf{yy}}^{-1}(\lambda)\mathbf{I}_{\mathbf{yy}}(\lambda)\mathbf{G}_{\mathbf{yy}}^{-1}(\lambda)\mathbf{G}_{\mathbf{uu}}(\lambda)\\
&= 2\pi\left[\mathbf{I}_N-\mathbf{C}(\mathrm{e}^{-\mathrm{i}\lambda})\boldsymbol{\Omega}(\lambda)\mathbf{C}'(\mathrm{e}^{\mathrm{i}\lambda})\mathbf{G}_{\mathbf{uu}}^{-1}(\lambda)\right]\mathbf{I}_{\mathbf{yy}}(\lambda)\\
&\quad\times\left[\mathbf{I}_N-\mathbf{G}_{\mathbf{uu}}^{-1}(\lambda)\mathbf{C}(\mathrm{e}^{\mathrm{i}\lambda})\boldsymbol{\Omega}(\lambda)\mathbf{C}'(\mathrm{e}^{\mathrm{i}\lambda})\right]
\end{aligned}
\tag{30}
$$

is the periodogram of the smoothed values of the specific factors, and

$$
\begin{aligned}
\mathbf{I}_{\mathbf{x}^K\mathbf{u}^K}(\lambda) &= 2\pi\mathbf{G}_{\mathbf{xx}}(\lambda)\mathbf{C}'(\mathrm{e}^{\mathrm{i}\lambda})\mathbf{G}_{\mathbf{yy}}^{-1}(\lambda)\mathbf{I}_{\mathbf{yy}}(\lambda)\mathbf{G}_{\mathbf{yy}}^{-1}(\lambda)\mathbf{G}_{\mathbf{uu}}(\lambda)\\
&= 2\pi\boldsymbol{\Omega}(\lambda)\mathbf{C}'(\mathrm{e}^{\mathrm{i}\lambda})\mathbf{G}_{\mathbf{uu}}^{-1}(\lambda)\mathbf{I}_{\mathbf{yy}}(\lambda)\left[\mathbf{I}_N-\mathbf{G}_{\mathbf{uu}}^{-1}(\lambda)\mathbf{C}(\mathrm{e}^{-\mathrm{i}\lambda})\boldsymbol{\Omega}(\lambda)\mathbf{C}'(\mathrm{e}^{\mathrm{i}\lambda})\right]
\end{aligned}
\tag{31}
$$

is the co-periodogram between $\mathbf{x}_{t|\infty}^K$ and $\mathbf{u}_{t|\infty}^K$.

Tedious algebra shows that these scores coincide with the expressions in Appendix A. They also closely related to the scores of the expected log-likelihoods in the previous subsection, but the difference is that the expectations were taken there with respect to the conditional distribution of $\mathbf{x}$ given $\mathbf{Y}$ evaluated at $\boldsymbol{\theta}^{(n)}$, not $\boldsymbol{\theta}$.

4. INFLATION DYNAMICS ACROSS EUROPEAN COUNTRIES

4.1. Introduction

Increasing economic and financial integration implies that nowadays countries are more sensitive to shocks originating outside their frontiers. In particular, national price levels may be affected by external shocks such as fluctuations in global commodity prices, shifts in global demand, exchange rate swings, or variations in the prices of competing countries. Understanding the extent to which foreign factors determine the temporal evolution of domestic inflation is a key question for macroeconomic policy.

A recent growing literature tackles this question by employing factor analysis techniques. Ciccarelli and Mojon (2010) estimate a static single factor model for 22 OECD economies over the period 1960–2008 and document that the estimated global factor accounts for about 70 percent of the variance of CPI inflation in those countries. Mumtaz and Surico (2012) estimate a dynamic factor model with drifting coefficients and stochastic volatility for a panel of 164 inflation indicators for the G7 countries, Australia, New Zealand and Spain. These authors find that the historical decline in the level of inflation is shared by most countries in their sample, which is consistent with the idea that a global factor drives the bulk of inflation movements across economies.

At the same time, the inflation rates of closely integrated economies tend to be more correlated with each other than with other countries, which is difficult to square with a single factor model. Motivated by this, we explore the ability of the dynamic bifactor models discussed in Section 2.1 to capture inflation dynamics across European countries. The European case is of particular interest because whether EMU has played a decisive role in the observed convergence of inflation rates across its member economies remains an open question. In this regard, Estrada, Galí, and López-Salido (2013) examine the extent to which the inflation rates of the original 11 euro area countries and other OECD economies have become synchronised over the period 1999–2012, reporting strong evidence of convergence towards low inflation rates. They also show that other advanced non-euro countries experience similar levels of convergence, which suggests that EMU may not be responsible for the generalised decline in inflation.

4.2. Model Set-up and Estimation Results

We use monthly data on Harmonised Indices of Consumer Prices (HICP) for 25 European economies over the period 1998:1–2014:12.[7] In particular, we consider three groups of countries:

1. the original[8] euro area members: Austria, Belgium, Finland, France, Germany, Greece, Ireland, Italy, Luxembourg, Netherlands, Portugal and Spain;
2. the new euro area participants: Cyprus, Estonia, Latvia, Lithuania, Malta and Slovakia;
3. other non-EMU countries: Bulgaria, Denmark, Iceland, Norway, Poland, Sweden and the United Kingdom.

We focus on year-on-year growth rates of HICP indices excluding energy and unprocessed food, which are widely viewed as the relevant measure to track for inflation targeting purposes; see, for example, Galí (2002). As a result, we are left with $T = 192$ time series observations. Fig. 1, which contains the inflation rates for each country (solid line) together with the inflation rate of the European Union (dashed line), confirms the generalised downward trend in inflation.

The econometric specification that we consider is essentially identical to the example consider in Section 2.1. Specifically, we assume that the inflation rate of country i in region r follows

$$\begin{aligned} y_{it} &= \mu_i + c_{i0g}x_{gt} + c_{i1g}x_{gt-1} + c_{i0r}x_{rt} + c_{i1r}x_{rt-1} + u_{it} \\ x_{gt} &= \alpha_g x_{gt-1} + f_{gt} \\ x_{rt} &= \alpha_r x_{rt-1} + f_{rt} \\ u_{it} &= \alpha_i u_{it-1} + v_{it} \end{aligned}$$

where x_g is a global factor which affects all European countries, x_r is an orthogonal region-specific factor which affects all countries within a region, u_i is the idiosyncratic term of country i and μ_i denotes its mean inflation rate.

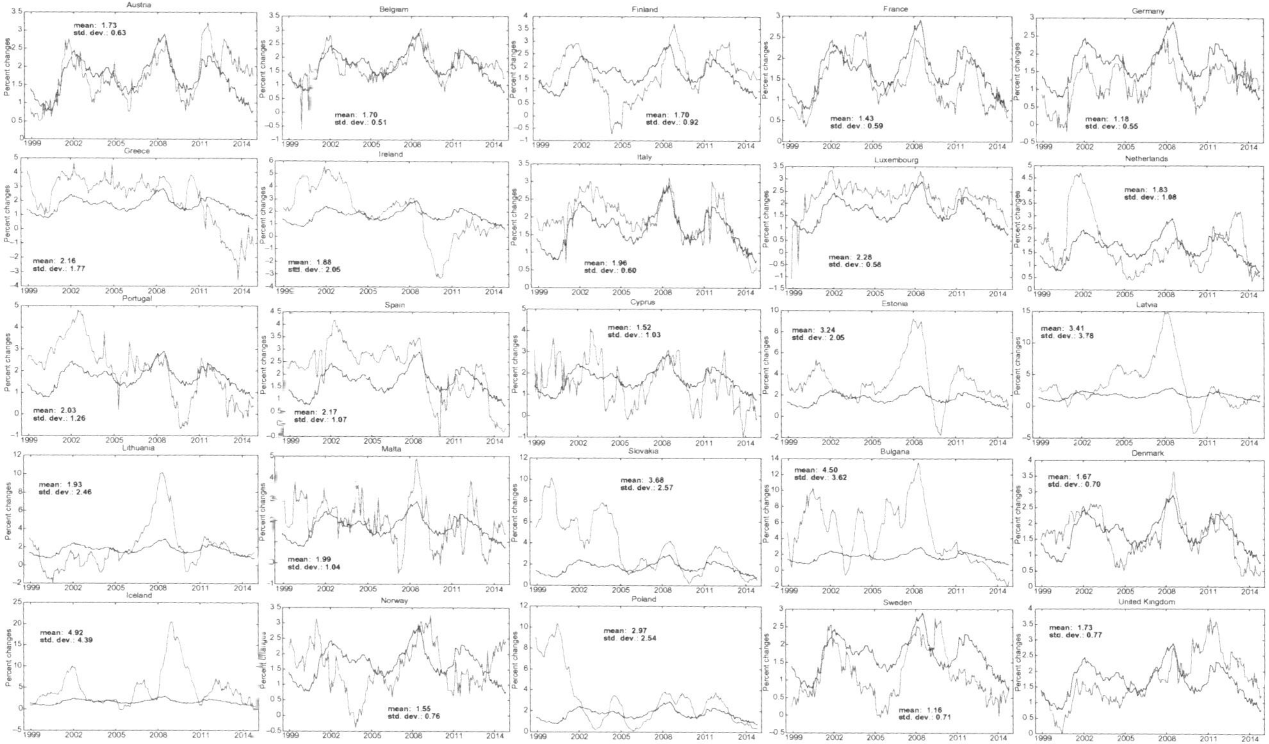

Fig. 1. European Inflation Rates. *Notes:* Inflation series are HICP excluding energy and unprocessed food. Dashed line refers to HICP Inflation of European Union (EU12 until 2004, EU15 until 2006, EU27 until 2013, then EU28). Mean and standard deviations refer to country-specific series. Mean and standard deviation for European Union inflation are 1.69 and 0.50, respectively.

In this regard, it is important to emphasise that since we effectively work with demeaned inflation rates, our dynamic bifactor model is silent about cross-country differences in average inflation rates, which are taken as given.

We also assume that the global and regional factors affect the inflation rate of a country not only through their contemporaneous values but also via their one-month lagged values with country-specific loadings. Further, we assume that all factors (global, regional and idiosyncratic) follow orthogonal AR(1) processes. Despite the apparent simplicity of our model, each series is effectively the sum of three components: an ARMA(1,1) global component, another ARMA(1,1) regional component and an idiosyncratic AR(1) term.

We estimate our dynamic bifactor model using the EM algorithm developed in previous sections. As starting values, we assume unit loadings on the contemporaneous and lagged values of both common and regional factors, unit-specific variances, autoregressive coefficients set to 0.5 for both common and idiosyncratic factors and 0.3 for regional factors. Importantly, the scoring algorithm fails to achieve convergence from these initial values, which are very far away from the optimum. To speed up the EM iterations, we employ just five Cochrane–Orcutt iterations instead of continuing until convergence. Despite the large amount of parameters involved (154), the algorithm performs remarkably well, as shown in Fig. 2. The first EM iteration yields a massive increase in the log-likelihood function, while subsequent iterations also provide noticeable gains. As expected, though, after 200 iterations the improvements become minimal. For that reason, we switched to a scoring algorithm with line searches at that stage, which converged rather smoothly to the parameter estimates reported in Tables 1 and 2, together with standard errors obtained on the basis of the analytical expressions for the information matrix in appendix B.

Table 3 contains the results of joint significance tests for the dynamic loading coefficients associated to the global (columns 1 and 2) and regional (columns 3 and 4) factors for each country. Those tests confirm that with the possible exception of Iceland, all countries in our sample are dynamically correlated. More importantly, they also show that some clusters of countries are more correlated with each other than what a single factor model would allow for, thereby confirming the need for a bifactor model. This is particularly noticeable for the Baltic countries, but it also affects Norway, Sweden and the United Kingdom among those countries which have never belonged to EMU.

From an empirical point of view, it is of substantive interest to look at the evolution and persistence of those latent factors. Unfortunately, it is

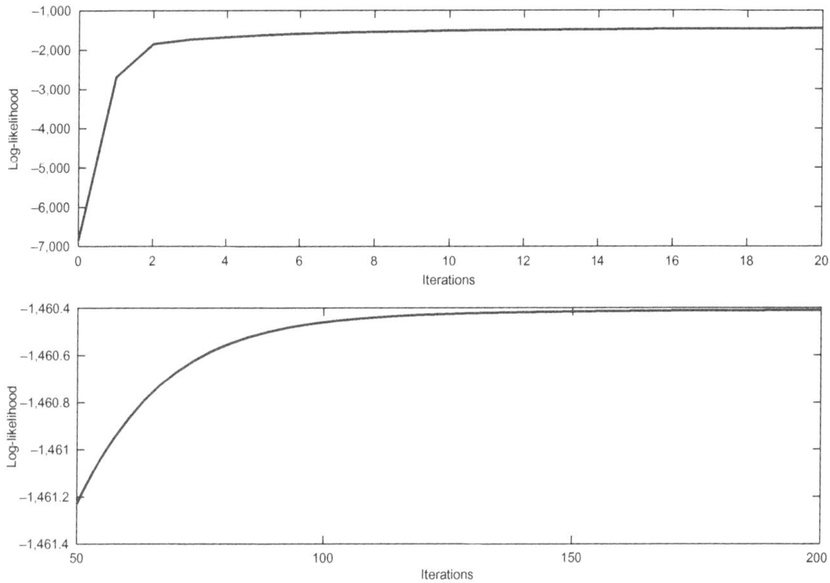

Fig. 2. EM Algorithm Log-Likelihood Evolution.

well known that the usual Wiener–Kolmogorov filter can lead to filtering distortions at both ends of the sample. For that reason, we wrote the model in a state-space form and applied the standard Kalman fixed interval smoother in the time domain with exact initial conditions derived from the stationary distribution of the 33 state variables (2 for the common factor and each of the regional factors and 1 for each of the idiosyncratic ones; see Appendix C for details).[9]

Smoothed versions of the global and regional factors are displayed in Fig. 3. In Fig. 3(a), we plot the estimated global factor jointly with the unweighted average of inflation rates across countries in our sample, and the inflation rate of the European Union countries. For ease of comparison, we re-scale both the global factor and the equally weighted inflation average to have the same mean and variance as the European Union inflation. The smoothed global factor, which with an estimated autocorrelation of 0.97 is rather persistent, tracks fairly well these two measures over the sample. The main exception is the period 1999–2002, when the global factor is significantly higher than the inflation rate of the European Union countries. Such discrepancies are explained by two facts: (i) the European Union HICP is a consumption-weighed average of country-specific price

Table 1. Dynamic Loadings Estimates.

Country	$c_{gi,0}$	Std. Err.	$c_{gi,1}$	Std. Err.	$c_{ri,0}$	Std. Err.	$c_{ri,1}$	Std. Err.
Core euro area								
Austria	−0.024	(0.017)	0.021	(0.017)	−0.058	(0.018)	0.021	(0.019)
Belgium	0.041	(0.021)	0.000	(0.021)	−0.170	(0.026)	0.000	(0.033)
Finland	−0.001	(0.016)	0.054	(0.016)	−0.043	(0.016)	0.054	(0.016)
France	0.041	(0.012)	0.011	(0.012)	0.019	(0.011)	0.011	(0.012)
Germany	−0.001	(0.018)	0.013	(0.018)	−0.006	(0.020)	0.013	(0.020)
Greece	0.357	(0.039)	−0.070	(0.039)	0.083	(0.036)	−0.070	(0.036)
Ireland	0.160	(0.023)	0.022	(0.023)	0.049	(0.022)	0.022	(0.022)
Italy	0.117	(0.017)	−0.001	(0.017)	0.047	(0.021)	−0.001	(0.021)
Luxembourg	−0.153	(0.019)	0.206	(0.020)	0.044	(0.020)	0.206	(0.020)
Netherlands	0.093	(0.019)	−0.005	(0.019)	−0.065	(0.019)	−0.005	(0.019)
Portugal	0.185	(0.026)	0.021	(0.026)	0.014	(0.026)	0.021	(0.026)
Spain	0.187	(0.023)	0.007	(0.023)	0.036	(0.023)	0.007	(0.023)
New entrants euro area								
Cyprus	0.286	(0.036)	−0.145	(0.036)	−0.063	(0.047)	−0.145	(0.047)
Estonia	0.269	(0.031)	−0.033	(0.030)	0.117	(0.049)	−0.033	(0.046)
Latvia	0.148	(0.037)	0.086	(0.037)	0.215	(0.076)	0.086	(0.087)
Lithuania	0.162	(0.034)	0.013	(0.033)	0.166	(0.059)	0.013	(0.057)
Malta	0.148	(0.036)	−0.015	(0.036)	0.019	(0.050)	−0.015	(0.050)
Slovakia	0.390	(0.035)	0.000	(0.035)	−0.022	(0.042)	0.000	(0.041)
Outside euro area								
Bulgaria	0.472	(0.060)	−0.098	(0.060)	0.036	(0.065)	−0.098	(0.064)
Denmark	0.077	(0.015)	0.028	(0.015)	0.035	(0.018)	0.028	(0.018)
Iceland	0.078	(0.065)	0.063	(0.065)	0.038	(0.074)	0.063	(0.073)
Norway	−0.006	(0.021)	−0.006	(0.021)	−0.046	(0.031)	−0.006	(0.027)
Poland	0.546	(0.043)	−0.149	(0.043)	−0.005	(0.044)	−0.149	(0.042)
Sweden	−0.019	(0.017)	0.025	(0.017)	0.007	(0.025)	0.025	(0.021)
United Kingdom	0.026	(0.016)	−0.019	(0.015)	0.038	(0.027)	−0.019	(0.021)

Table 2. Autoregressive Coefficients Estimates.

Country	α	Std. Err.	ψ	Std. Err.
Global	0.9736	(0.017)	1.000	
Core euro area	0.2810	(0.207)	1.000	
New entrants euro area	0.9828	(0.013)	1.000	
Outside euro area	−0.1392	(0.302)	1.000	
Core euro area				
Austria	0.936	(0.025)	0.049	(0.005)
Belgium	0.912	(0.033)	0.033	(0.007)
Finland	0.974	(0.016)	0.041	(0.004)
France	0.948	(0.023)	0.022	(0.002)
Germany	0.887	(0.033)	0.063	(0.006)
Greece	0.941	(0.025)	0.194	(0.022)
Ireland	0.983	(0.011)	0.079	(0.009)
Italy	0.663	(0.071)	0.051	(0.006)
Luxembourg	0.852	(0.039)	0.049	(0.006)
Netherlands	0.970	(0.017)	0.055	(0.006)
Portugal	0.898	(0.034)	0.107	(0.011)
Spain	0.899	(0.035)	0.080	(0.009)
New entrants euro area				
Cyprus	0.805	(0.046)	0.213	(0.024)
Estonia	0.956	(0.028)	0.106	(0.013)
Latvia	0.977	(0.024)	0.113	(0.027)
Lithuania	0.960	(0.026)	0.147	(0.018)
Malta	0.799	(0.045)	0.268	(0.028)
Slovakia	0.981	(0.013)	0.135	(0.016)
Outside euro area				
Bulgaria	0.968	(0.018)	0.505	(0.055)
Denmark	0.918	(0.030)	0.036	(0.004)
Iceland	0.980	(0.013)	0.705	(0.072)
Norway	0.940	(0.025)	0.066	(0.009)
Poland	0.986	(0.010)	0.171	(0.023)
Sweden	0.953	(0.022)	0.044	(0.005)
United Kingdom	0.973	(0.016)	0.032	(0.004)

indices, and (ii) there are differences between our sample of countries and the set of economies used to construct the European Union HICP.[10] Since 2002, the global factor generally trends downwards, in line with the other two measures. The other panels of Fig. 3 plot the estimated regional factors, which are scaled so that their innovations have unit variance. Interestingly, the factor for the new entrants to the euro area is even more persistent than the global factor (its autocorrelation is 0.98). In contrast, we do not observe statistically significant persistence in the evolution of the

Table 3. Significance of Dynamic Loadings.

Country	$H_0: c_{gi,0} = c_{gi,1} = 0$		$H_0: c_{ri,0} = c_{ri,1} = 0$	
	Wald test	*p*-Value	Wald test	*p*-Value
Core euro area				
Austria	3.07	(0.216)	15.44	(0.000)
Belgium	5.38	(0.068)	56.38	(0.000)
Finland	11.26	(0.004)	7.92	(0.019)
France	13.88	(0.001)	4.29	(0.117)
Germany	0.55	(0.760)	5.83	(0.054)
Greece	86.60	(0.000)	5.99	(0.050)
Ireland	47.22	(0.000)	6.40	(0.041)
Italy	61.32	(0.000)	12.23	(0.002)
Luxembourg	119.75	(0.000)	6.42	(0.041)
Netherlands	23.51	(0.000)	16.88	(0.000)
Portugal	53.15	(0.000)	0.42	(0.812)
Spain	65.92	(0.000)	5.68	(0.058)
New entrants euro area				
Cyprus	64.54	(0.000)	2.21	(0.330)
Estonia	78.72	(0.000)	25.96	(0.000)
Latvia	17.35	(0.000)	66.20	(0.000)
Lithuania	22.60	(0.000)	30.37	(0.000)
Malta	19.21	(0.000)	0.40	(0.817)
Slovakia	125.00	(0.000)	0.47	(0.790)
Outside euro area				
Bulgaria	64.18	(0.000)	0.88	(0.644)
Denmark	30.05	(0.000)	5.75	(0.057)
Iceland	2.36	(0.308)	0.68	(0.710)
Norway	0.18	(0.915)	13.52	(0.001)
Poland	164.30	(0.000)	2.51	(0.285)
Sweden	3.18	(0.204)	8.32	(0.016)
United Kingdom	3.78	(0.151)	11.84	(0.003)

other two regional factors. These results suggest that some of the new entrant economies share a regional factor which drives the medium-term trends in inflation, while other regional factors have a predominant role at higher frequencies. We revisit this question below.

Given the estimated factors and factor loadings, we can compute the contributions of global, regional and idiosyncratic factors in driving the observed changes in prices across countries. Fig. 4 plots the results for all the countries in our sample. The global factor clearly drives the downward trend in inflation for many countries, including Cyprus, Denmark, France, Italy, Poland, Slovakia and Spain, among others. We also observe a sizeable role for the regional factor for Estonia, Latvia, and Lithuania. For

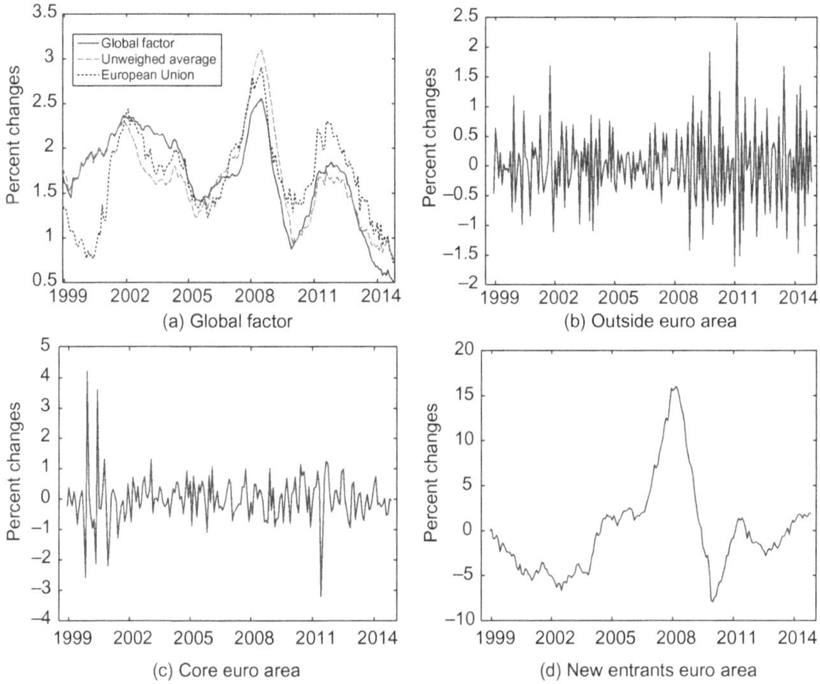

Fig. 3. Smoothed Inflation Factors. *Notes:* The series Global factor and Unweighed average are re-scaled to have same mean and variance as the European Union inflation. Regional factors are re-scaled so that their innovations have unit variance.

these Baltic economies, inflation dramatically swings over the period 2005–2011. Conversely, the regional factor only plays a marginal role for the other new entrants, which did not experience such swings over the same period. In this regard, it is worth noticing that the Baltic countries adopted the euro in the late part of the sample (Estonia in 2011, Latvia in 2014 and Lithuania in 2015), while the other three entrants joined the euro area earlier (Cyprus and Malta in 2008, Slovakia in 2009). Although the observed differences in the volatility of inflation among the group of new entrant countries may be due to their different timings in fulfilling the monetary union accession criteria, these results suggest that EMU may have had a dampening effect on inflation fluctuations for all the new entrant countries.

We complement our time domain results by decomposing the spectral density of each country inflation series into the corresponding global, regional, and idiosyncratic components. Fig. 5 shows for each frequency the

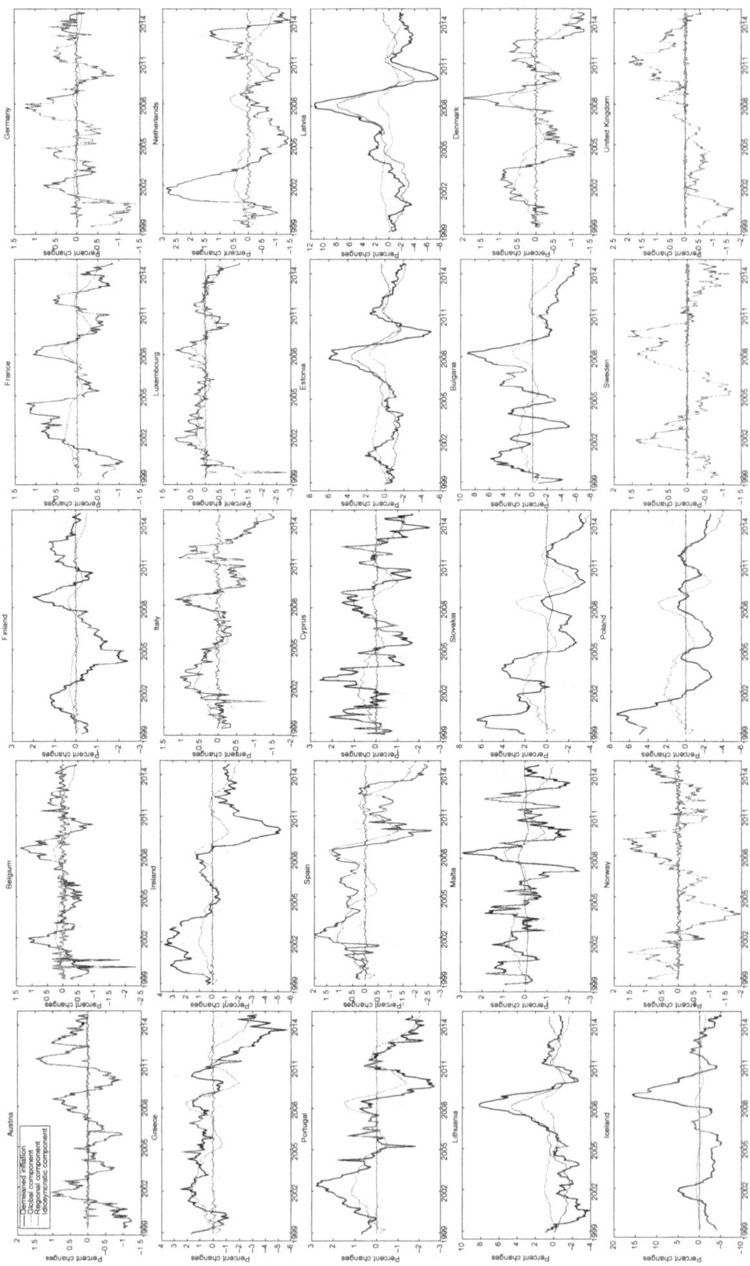

Fig. 4. Contributions of Global, Regional, and Idiosyncratic Factors to Observed HICP Inflation. *Notes:* Inflation series are HICP excluding energy and unprocessed food.

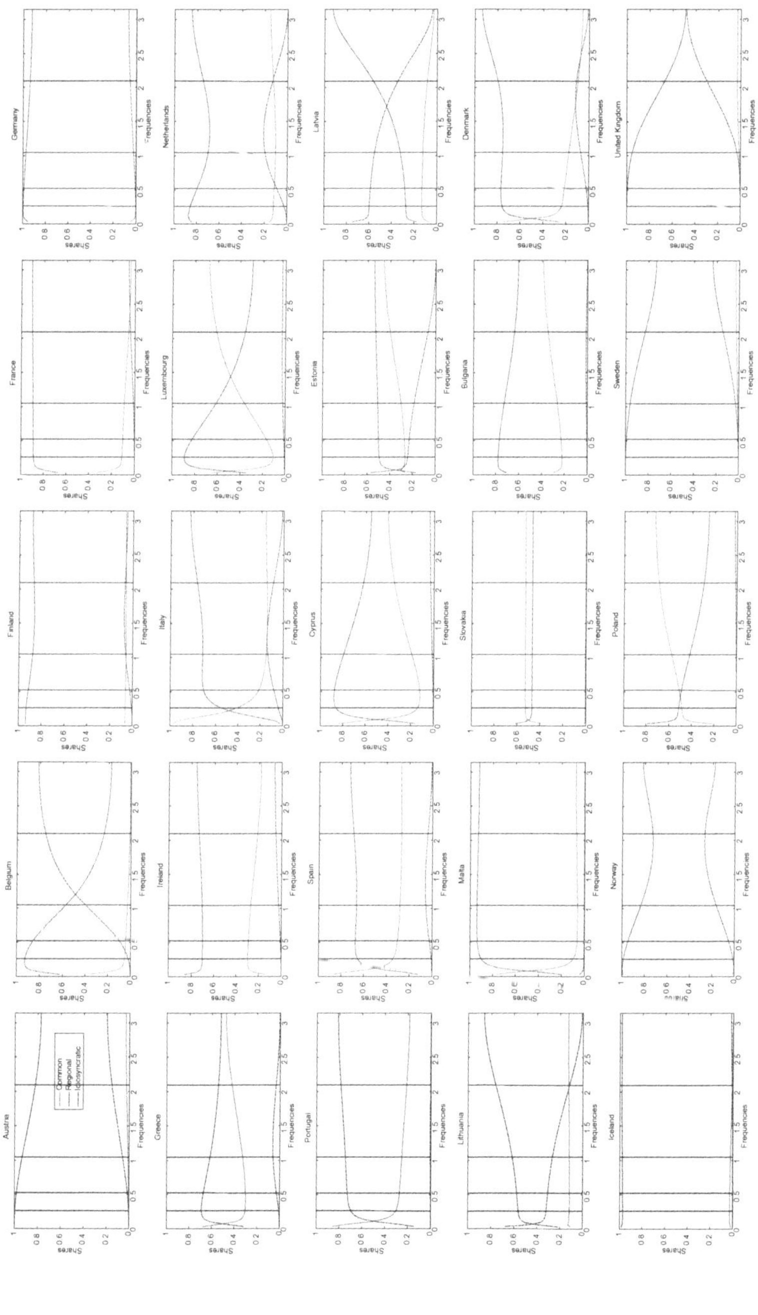

Fig. 5. Spectral Decompositions. *Notes:* The vertical lines correspond to those frequencies which reflect movements in the series at cycles of 2 and 1 years, and 6 and 3 months.

fraction of variance explained by each of those components. To aid in the interpretation of the results, we have added vertical lines at those frequencies which capture movements in the series at 2 and 1 years, and 6 and 3 months. As can be seen, the global factor explains an important fraction of variance across many economies, especially at lower frequencies. This result confirms the view that most countries experience a common downward trend in inflation. Nevertheless, we also observe that the global factor plays virtually no role in other economies such as Norway, Sweden and the United Kingdom, whose correlations are mostly driven by the third regional factor. This somewhat surprising result may be partly explained by the fact that energy and food components are by construction excluded from our analysis. The regional factor of new entrants affects particularly Estonia, Latvia, and Lithuania, which confirms our previous time domain findings. In contrast, regional factors do not seem to influence medium-term trends for most other countries.

4.3. Robustness Analysis

To assess the reliability of the results described in the previous section, we conduct three robustness exercises. First, we considered a version of the model with just a global factor and no regional factors. Fig. 6(a) shows that the new smoothed global factor tracks fairly well its counterpart in the baseline model with regional factors. Hardly surprisingly, though, the single factor model leads to a markedly worse fit: its log-likelihood function at the optimum is $-1,571.2$, while it is $-1,460.4$ for the bifactor model.

Second, we also considered an alternative model with a subdivision of the core euro area region to single out those countries which experienced the most dramatic drops in interest rates prior to their accession to EMU. This is an important distinction to explore as there has been considerable debate on whether the conduct of monetary policy by the ECB since its inception has resulted in unwanted effects on those economies; see Estrada and Saurina (2014) for a discussion of the Spanish case. By looking at the evolution of real interest differentials between 1995 and 1999, we interestingly find that the additional group is composed by Portugal, Ireland, Italy, Greece and Spain (the so-called PIIGS). Unlike what happened in the case of the single factor model, we find that a dynamic bifactor model with four regions, including two within the core euro area, does not lead to such a huge improvement in fit. In addition, the interpretation of the new regional factors is inconclusive. Fig. 6(c) and (d) plot the smoothed factors

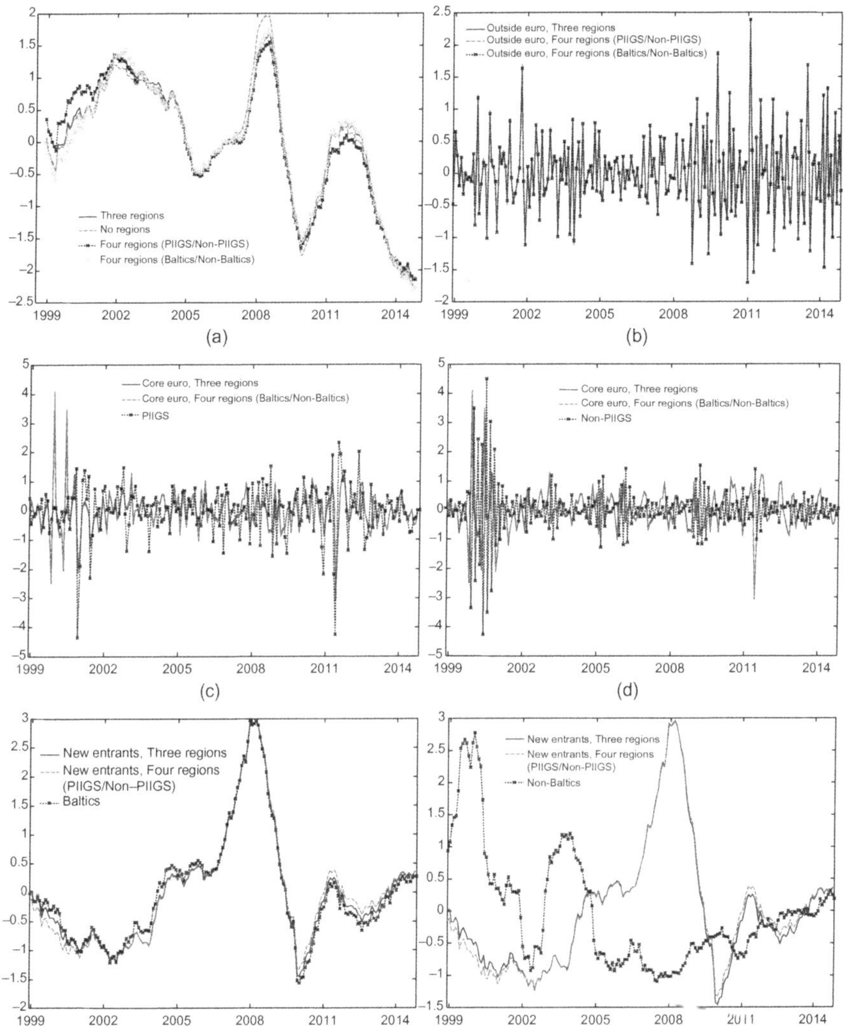

Fig. 6. Smoothed Common and Regional Inflation Factors. (a) Global Factor; (b) Outside Euro Area; (c) Core Euro Area and PIIGS; (d) Core Euro Area and Non-PIIGS; (e) New Entrants Euro Area and Baltics; and (f) New Entrants Euro Area and Non-Baltics.

for PIIGS and Non-PIIGS, jointly with the core euro area factor obtained in the baseline model with only three regions. While the correlation coefficient of the smoothed hard core euro area countries with its baseline counterpart is .28, the analogous coefficient for the PIIGS factor is −0.46.

Finally, we have also experimented with an alternative model in which we subdivided instead the new entrants euro area region into two subregions: Baltic countries (Estonia, Latvia, and Lithuania) and the rest (Cyprus, Malta, and Slovakia). This model provides a substantial better fit. This is confirmed by Fig. 6(e) and (f), which plot the smoothed factors for Baltic and Non-Baltic countries, jointly with the new entrants factor in the baseline specification. As can be seen, the Baltic countries factor tracks very well the original new entrants factor, while the Non-Baltic countries factor is markedly unsynchronised, especially over the early 2000s, which is in line with the results we discussed in the previous section.

5. CONCLUSIONS AND EXTENSIONS

We generalise the frequency domain version of the EM algorithm for dynamic factor models in Fiorentini, Galesi, and Sentana (2014) to bifactor models in which pervasive common factors are complemented by block factors. We explain how to efficiently exploit the sparsity of the loading matrix to reduce the computational burden so much that researchers can estimate such models by maximum likelihood with a large number of series from multiple regions. We find that the EM algorithm leads to substantial likelihood gains starting from arbitrary initial values. Unfortunately, it slows down considerably near the optimum. For that reason, we also derive convenient expressions for the frequency domain scores and information matrix that allow us to switch to the scoring method at that point.

In an empirical application, we explore the ability of a bifactor model to capture inflation dynamics across European countries. Specifically, we apply our procedure to year-on-year core inflation rates for 25 European countries over the period 1999:1–2014:12. We estimate a model with a common factor and three regional factors: original euro area members, new entrants and others. Overall, our results suggest that a global factor drives the medium–long-term trends of inflation across most European economies, which is consistent with the evidence in the previous literature. But we also find a persistent regional factor driving the inflation trends of the Baltic countries, which are new entrants to the euro area. In contrast, we find that the regional factors for most other countries affect mainly their short run movements.

An extension of our algorithm to models with ARMA latent variables along the lines of Fiorentini, Galesi, and Sentana (2014) would be conceptually straightforward, but its successful practical implementation would require some experimentation. It would also be interesting to compare the forecasting accuracy of the dynamic bifactor model relative to its single-factor counterpart. Our empirical results suggest that regional factors affect the short run movements of inflation for most countries, hence the inclusion of regional factors in a forecasting model might yield more accurate inflation forecasts. Another empirically relevant extension would be to modify our procedures to deal with unbalanced data sets with different time spans for different series (see Bańbura & Modugno, 2014, for an extension of the time domain version of the EM algorithm that can deal with those cases). It would also be very useful to develop a clustering algorithm that would automatically assign individual series to blocks (see Francis, Owyang, & Savaşçin 2012, as well as Bonhomme & Manresa, 2015, for some related work in a panel data context). Finally, it would be convenient to extend our algorithm to dynamic trifactor models, in which each block has a bifactor structure of its own. Such models would be particularly well suited to the analysis of international business cycles using a large set of country-specific macro-variables. All these important issues deserve further investigation.

NOTES

1. Static versions of bifactor models have a long tradition in psychometrics after their introduction by Holzinger and Swineford (1937) as an important special case of confirmatory factor analysis (see Reise, 2012, for an up to date list of references).

2. Other symmetric scaling assumptions would normalise the unconditional variance of x_{gt} and x_{rt} $(r = 1, \ldots, R)$, or some norm of the vectors of impact multipliers $\mathbf{c}_{g0} = \left(\mathbf{c}'_{1g0}, \ldots, \mathbf{c}'_{Rg0}\right)$ and $\mathbf{c}_{rr0}$ $(r = 1, \ldots, R)$ or their long run counterparts $\mathbf{c}_g(1)$ and $\mathbf{c}_{rr}(1)$. Alternatively, we could asymmetrically fix one element of $\mathbf{c}_{g0}$ and $\mathbf{c}_{rr0}$ (or $\mathbf{c}_g(1)$ and $\mathbf{c}_{rr}(1)$) $(r = 1, \ldots, R)$ to 1.

3. Scherrer and Deistler (1998) refer to this situation as the Frisch case.

4. There is also a continuous version which replaces sums by integrals (see Dusmuir & Hannan, 1976).

5. This equivalence is not surprising in view of the contiguity of the Whittle measure in the Gaussian case (see Choudhuri, Ghosal, & Roy, 2004).

6. Note that we could have expressed those log-likelihood in terms of $\mathbf{I}_{\mathbf{xx}}(\lambda_j) = \mathbf{z}_j^{\mathbf{x}}\mathbf{z}_j^{\mathbf{x}*}$, $\mathbf{I}_{\mathbf{uu}}(\lambda) = \mathbf{z}_j^{\mathbf{u}}\mathbf{z}_j^{\mathbf{u}*}$ and $\mathbf{I}_{\mathbf{ux}}(\lambda) = \mathbf{z}_j^{\mathbf{u}}\mathbf{z}_j^{\mathbf{x}*}$, but for the EM algorithm it is more convenient to work with the underlying complex random variables.

7. Since our aim is to maximise the time span of our balanced sample, we exclude several countries for which data start at later dates: Czech Republic and Slovenia (1999:12-), Hungary and Romania (2000:12-) and Croatia and Switzerland (2004:12-).

8. We include Greece among the original euro area even though its accession year was 2001.

9. The main difference between the Wiener–Kolmogorov filtered values, $\mathbf{x}_{t|\infty}^{K}$, and the Kalman filter smoothed values, $\mathbf{x}_{t|T}^{K}$, results from the implicit dependence of the former on a doubly infinite sequence of past and future observations. As shown by Fiorentini (1995) and Gómez (1999), though, they can be made numerically identical by replacing both pre- and post-sample observations by their least-squares projections onto the linear span of the sample observations.

10. Specifically, the weight of a country is its share of household final monetary consumption expenditure in the total. The European Union HICP is constructed as the weighed average of the original 12 countries until 2004, then it extends to 15 countries until 2006, 27 countries until 2013, and finally 28 countries until the end of the sample.

ACKNOWLEDGEMENTS

We are grateful to Ángel Estrada and Albert Satorra, as well as to audiences at the Advances in Econometrics Conference on Dynamic Factor Models (Aarhus 2014), the EC^2 Advances in Forecasting Conference (UPF 2014), the Italian Congress of Econometrics and Empirical Economics (Salerno 2015), the V Workshop in Time Series Econometrics (Zaragoza 2015) and the XVIII Meeting of Applied Economics (Alicante 2015) for helpful comments and suggestions. Detailed comments from two anonymous referees have also substantially improved the paper. Of course, the usual caveat applies. Financial support from MIUR through the project 'Multivariate statistical models for risk assessment' (Fiorentini) and the Spanish Ministry of Science and Innovation through grant 2014-59262 (Sentana) is gratefully acknowledged.

REFERENCES

Bai, J., & Ng, S. (2008). Large dimensional factor analysis. *Foundations and Trends in Econometrics*, *3*, 89–163.

Bańbura, M., & Modugno, M. (2014). Maximum likelihood estimation of factor models on datasets with arbitrary pattern of missing data. *Journal of Applied Econometrics*, *29*, 133–160.

Bonhomme, S., & Manresa, E. (2015). Grouped patterns of heterogeneity in panel data. *Econometrica*, *83*, 1147–1184.

Breitung, J., & Eickmeier, S. (2014). *Analyzing international business and financial cycles using multi-level factor models*. Deutsche Bundesbank Research Centre Discussion Paper No. 11/2014.

Chamberlain, G., & Rothschild, M. (1983). Arbitrage, factor structure, and mean-variance analysis on large asset markets. *Econometrica*, *51*, 1281–1304.

Choudhuri, N., Ghosal, S., & Roy, A. (2004). Contiguity of the Whittle measure for a Gaussian time series. *Biometrika*, *91*, 211–218.

Ciccarelli, M., & Mojon, B. (2010). Global inflation. *Review of Economics and Statistics*, *92*, 524–535.

Cochrane, D., & Orcutt, G. H. (1949). Application of least squares regression to relationships containing auto-correlated error terms. *Journal of the American Statistical Association*, *44*, 32–61.

Demos, A., & Sentana, E. (1998). An EM algorithm for conditionally heteroskedastic factor models. *Journal of Business and Economic Statistics*, *16*, 357–361.

Dempster, A., Laird, N., & Rubin, D. (1977). Maximum likelihood from incomplete data via the EM algorithm. *Journal of the Royal Statistical Society B*, *39*, 1–38.

Doz, C., Giannone, D., & Reichlin, L. (2012). A quasi-maximum likelihood approach for large, approximate dynamic factor models. *Review of Economics and Statistics*, *94*, 1014–1024.

Dunsmuir, W. (1979). A central limit theorem for parameter estimation in stationary vector time series and its application to models for a signal observed with noise. *Annals of Statistics*, *7*, 490–506.

Dunsmuir, W., & Hannan, E. J. (1976). Vector linear time series models. *Advances in Applied Probability*, *8*, 339–364.

Engle, R. F., Hendry, D. F., & Richard, J.-F. (1983). Exogeneity. *Econometrica*, *51*, 277–304.

Estrada, A., Gal, J., & López-Salido, D. (2013). Patterns of convergence and divergence in the euro area. *IMF Economic Review*, *61*, 601–630.

Estrada, A., & Saurina, J. (2014). *Spanish boom and bust: Some lessons for macroprudential policy*. Mimeo, Bank of Spain.

Fiorentini, G. (1995). *Conditional heteroskedasticity: Some results on estimation, inference and signal extraction, with an application to seasonal adjustment*. Unpublished Doctoral Dissertation, European University Institute.

Fiorentini, G., Galesi, A., & Sentana, E. (2014). *A spectral EM algorithm for dynamic factor models*. CEMFI Working Paper No. 1411.

Fiorentini, G., Sentana, E., & Shephard, N. (2004). Likelihood estimation of latent generalised Arch structures. *Econometrica*, *72*, 1481–1517.

Francis, N., Owyang, M., & Savaşçin, O. (2012). *An endogenously clustered factor approach to international business cycles*, Federal Reserve Bank of St. Louis Working Paper No. 2012-14A.

Gali, J. (2002). Monetary policy in the early years of EMU. In M. Buti & A. Sapir (Eds.), *Economic and monetary union and economic policy in Europe* (pp. 41–72). Cheltenham: Edward Elgar.

Geweke, J. F. (1977). The dynamic factor analysis of economic time series models. In D. Aigner & A. Goldberger (Eds.), *Latent variables in socioeconomic models* (pp. 365–383). Amsterdam: North-Holland.

Geweke, J. F., & Singleton, K. J. (1981). Maximum likelihood "confirmatory" factor analysis of economic time series. *International Economic Review*, *22*, 37–54.

Gómez, V. (1999). Three equivalent methods for filtering finite nonstationary time series. *Journal of Business and Economic Statistics*, *17*, 109–116.

Gouriéroux, C., Monfort, A., & Renault, E. (1991). *A general framework for factor models*. Mimeo, INSEE.

Hamilton, J. (1990). Analysis of time series subject to changes in regime. *Journal of Econometrics*, *45*, 39–70.

Hannan, E. J. (1973). The asymptotic theory of linear time series models. *Journal of Applied Probability*, *10*, 130–145 (Corrigendum 913).

Harvey, A. C. (1989). *Forecasting, structural models and the Kalman filter*. Cambridge: Cambridge University Press.

Heaton, C., & Solo, V. (2004). Identification of causal factor models of stationary time series. *Econometrics Journal*, *7*, 618–627.

Holzinger, K. J., & Swineford, S. (1937). The bi-factor method. *Psychometrika*, *47*, 41–54.

Jungbacker, B., & Koopman, S. J. (2015). Likelihood-based dynamic factor analysis for measurement and forecasting. *Econometrics Journal*, *17*, 1–21.

Kose, M., Otrok, C., & Whiteman, C. (2003). International business cycles: World, region and country-specific factors. *American Economic Review*, *93*, 1216–1239.

Lehmann, B. N., & Modest, D. M. (1988). The empirical foundations of the arbitrage pricing theory. *Journal of Financial Economics*, *21*, 213–254.

Magnus, J. R. (1988). *Linear structures*. New York, NY: Oxford University Press.

Magnus, J. R., & Neudecker, H. (1988). *Matrix differential calculus with applications in statistics and econometrics*. Chichester: Wiley.

Moench, E., Ng, S., & Potter, S. (2013). Dynamic hierarchical factor models. *Review of Economics and Statistics*, *95*, 1811–1817.

Mumtaz, H., & Surico, P. (2012). Evolving international inflation dynamics: World and country-specific factors. *Journal of the European Economic Association*, *10*, 716–734.

Quah, D., & Sargent, T. (1993). A dynamic index model for large cross sections. In J. H. Stock & M. W. Watson (Eds.), *Business cycles, indicators and forecasting* (pp. 285–310). Chicago, IL: University of Chicago Press.

Reise, S. P. (2012). The rediscovery of bifactor measurement models. *Multivariate Behavioral Research*, *47*, 667–696.

Rubin, D., & Thayer, D. (1982). EM algorithms for ML factor analysis. *Psychometrika*, *47*, 69–76.

Ruud, P. (1991). Extensions of estimation methods using the EM algorithm. *Journal of Econometrics*, *49*, 305–341.

Sargent, T. J., & Sims, C. A. (1977). Business cycle modeling without pretending to have too much a priori economic theory. *New Methods in Business Cycle Research*, *1*, 145–168.

Scherrer, W., & Deistler, M. (1998). A structure theory for linear dynamic errors-in-variables models. *SIAM Journal on Control and Optimization*, *36*(6), 2148–2175.

Sentana, E. (2000). The likelihood function of conditionally heteroskedastic factor models. *Annales d'Economie et de Statistique*, *58*, 1–19.

Sentana, E. (2004). Factor representing portfolios in large asset markets. *Journal of Econometrics*, *119*, 257–289.

Shumway, R., & Stoffer, D. (1982). An approach to time series smoothing and forecasting using the EM algorithm. *Journal of Time Series Analysis*, *3*, 253–264.

Stock, J. H., & Watson, M. (2009). The evolution of national and regional factors in U.S. housing construction. In T. Bollerslev, J. Russell, & M. Watson (Eds.), *Volatility and time series econometrics: Essays in honour of Robert F. Engle*. Oxford: Oxford University Press.

Watson, M. W., & Engle, R. F. (1983). Alternative algorithms for estimation of dynamic MIMIC, factor, and time varying coefficient regression models. *Journal of Econometrics*, *23*, 385–400.

Whittle, P. (1962). Gaussian estimation in stationary time series. *Bulletin of the International Statistical Institute*, *39*, 105–129.

APPENDIX A. SPECTRAL SCORES

The score function for all the parameters other than the mean is given by Eq. (16). Since

$$\begin{aligned}\mathrm{d}\mathbf{G}_{\mathbf{yy}}(\lambda) = {} & \left[\mathrm{d}\mathbf{C}\left(\mathrm{e}^{-\mathrm{i}\lambda}\right)\right]\mathbf{G}_{\mathbf{xx}}(\lambda)\mathbf{C}'\left(\mathrm{e}^{\mathrm{i}\lambda}\right) + \mathbf{C}\left(\mathrm{e}^{-\mathrm{i}\lambda}\right)[\mathrm{d}\mathbf{G}_{\mathbf{xx}}(\lambda)]\mathbf{C}'\left(\mathrm{e}^{\mathrm{i}\lambda}\right) \\ & + \mathbf{C}\left(\mathrm{e}^{-\mathrm{i}\lambda}\right)\mathbf{G}_{\mathbf{xx}}(\lambda)\left[\mathrm{d}\mathbf{C}'\left(\mathrm{e}^{\mathrm{i}\lambda}\right)\right] + \mathrm{d}\mathbf{G}_{\mathbf{uu}}(\lambda)\end{aligned}$$

(see Magnus & Neudecker, 1988), it immediately follows that

$$\begin{aligned}\mathrm{dvec}\left[\mathbf{G}_{\mathbf{yy}}(\lambda)\right] = {} & \left[\mathbf{C}\left(\mathrm{e}^{\mathrm{i}\lambda}\right)\mathbf{G}_{\mathbf{xx}}(\lambda)\otimes\mathbf{I}_N\right]\mathrm{dvec}\left[\mathbf{C}\left(\mathrm{e}^{-\mathrm{i}\lambda}\right)\right] \\ & + \left[\mathbf{I}_N\otimes\mathbf{C}\left(\mathrm{e}^{-\mathrm{i}\lambda}\right)\mathbf{G}_{\mathbf{xx}}(\lambda)\right]\mathbf{K}_{N,R+1}\mathrm{dvec}\left[\mathbf{C}\left(\mathrm{e}^{\mathrm{i}\lambda}\right)\right] \\ & + \left[\mathbf{C}\left(\mathrm{e}^{\mathrm{i}\lambda}\right)\otimes\mathbf{C}\left(\mathrm{e}^{-\mathrm{i}\lambda}\right)\right]\mathbf{E}_{R+1}\mathrm{dvecd}[\mathbf{G}_{\mathbf{xx}}(\lambda)] + \mathbf{E}_N\mathrm{dvecd}[\mathbf{G}_{\mathbf{uu}}(\lambda)] \\ = {} & \left[\mathbf{C}\left(\mathrm{e}^{\mathrm{i}\lambda}\right)\mathbf{G}_{\mathbf{xx}}(\lambda)\otimes\mathbf{I}_N\right]\mathrm{dvec}\left[\mathbf{C}\left(\mathrm{e}^{-\mathrm{i}\lambda}\right)\right] \\ & + \mathbf{K}_{NN}\left[\mathbf{C}\left(\mathrm{e}^{-\mathrm{i}\lambda}\right)\mathbf{G}_{\mathbf{xx}}(\lambda)\otimes\mathbf{I}_N\right]\mathrm{dvec}\left[\mathbf{C}\left(\mathrm{e}^{\mathrm{i}\lambda}\right)\right] \\ & + \left[\mathbf{C}\left(\mathrm{e}^{\mathrm{i}\lambda}\right)\otimes\mathbf{C}\left(\mathrm{e}^{-\mathrm{i}\lambda}\right)\right]\mathbf{E}_{R+1}\mathrm{dvecd}[\mathbf{G}_{\mathbf{xx}}(\lambda)] + \mathbf{E}_N\mathrm{dvecd}[\mathbf{G}_{\mathbf{uu}}(\lambda)]\end{aligned}$$

where

$$\begin{aligned}\mathbf{E}'_m &= \left(\mathbf{e}_{1m}\mathbf{e}'_{1m}|\ldots|\mathbf{e}_{mm}\mathbf{e}'_{mm}\right) \\ (\mathbf{e}_{1m}|\ldots|\mathbf{e}_{mm}) &= \mathbf{I}_m\end{aligned} \tag{A.1}$$

is the unique $m^2\times m$ 'diagonalisation' matrix that transforms vec($\mathbf{A}$) into vecd($\mathbf{A}$) as vecd($\mathbf{A}$) $=$ $\mathbf{E}'_m$vec($\mathbf{A}$) and $\mathbf{K}_{mn}$ is the commutation matrix of orders m and n (see Magnus, 1988). Further, we can use Eq. (6) to express dvec[$\mathbf{C}(z)$] in terms of its non-zero elements d$\mathbf{c}(z)$ by means of the following linear transformation

$$
\begin{aligned}
\begin{pmatrix}
\mathrm{d}\mathbf{c}_{1g}(z) \\ \vdots \\ \mathrm{d}\mathbf{c}_{rg}(z) \\ \vdots \\ \mathrm{d}\mathbf{c}_{Rg}(z) \\ \mathrm{d}\mathbf{c}_{11}(z) \\ \vdots \\ \mathbf{0} \\ \vdots \\ \mathbf{0} \\ \vdots \\ \mathbf{0} \\ \vdots \\ \mathrm{d}\mathbf{c}_{rr}(z) \\ \vdots \\ \mathbf{0} \\ \vdots \\ \mathbf{0} \\ \vdots \\ \mathbf{0} \\ \vdots \\ \mathrm{d}\mathbf{c}_{RR}(z)
\end{pmatrix}
&=
\begin{pmatrix}
\mathbf{I}_{N_1} & \dots & \mathbf{0} & \dots & \mathbf{0} & \mathbf{0} & \dots & \mathbf{0} & \dots & \mathbf{0} \\
\vdots & \ddots & \vdots & \ddots & \mathbf{0} & \mathbf{0} & \dots & \mathbf{0} & \dots & \mathbf{0} \\
\mathbf{0} & \dots & \mathbf{I}_{N_r} & \dots & \vdots & \vdots & \ddots & \vdots & \ddots & \vdots \\
\mathbf{0} & \ddots & \vdots & \ddots & \vdots & \mathbf{0} & \dots & \mathbf{0} & \dots & \mathbf{0} \\
\mathbf{0} & \dots & \mathbf{0} & \dots & \mathbf{I}_{N_R} & \mathbf{0} & \dots & \mathbf{0} & \dots & \mathbf{0} \\
\mathbf{0} & \dots & \mathbf{0} & \dots & \mathbf{0} & \mathbf{I}_{N_1} & \dots & \vdots & \dots & \vdots \\
\vdots & \ddots & \vdots & \ddots & \vdots & \vdots & \ddots & \vdots & \ddots & \vdots \\
\mathbf{0} & \dots & \mathbf{0} & \dots & \mathbf{0} & \mathbf{0} & \dots & \mathbf{0} & \dots & \mathbf{0} \\
\vdots & \ddots & \vdots & \ddots & \vdots & \vdots & \ddots & \vdots & \ddots & \vdots \\
\mathbf{0} & \dots & \mathbf{0} & \dots & \mathbf{0} & \mathbf{0} & \dots & \mathbf{0} & \dots & \mathbf{0} \\
\vdots & \ddots & \vdots & \ddots & \vdots & \vdots & \ddots & \vdots & \ddots & \vdots \\
\mathbf{0} & \dots & \mathbf{0} & \dots & \mathbf{0} & \mathbf{0} & \dots & \mathbf{0} & \dots & \mathbf{0} \\
\vdots & \ddots & \vdots & \ddots & \vdots & \vdots & \ddots & \vdots & \ddots & \vdots \\
\mathbf{0} & \dots & \mathbf{0} & \dots & \mathbf{0} & \mathbf{0} & \dots & \mathbf{I}_{N_r} & \dots & \mathbf{0} \\
\vdots & \ddots & \vdots & \ddots & \vdots & \vdots & \ddots & \vdots & \ddots & \vdots \\
\mathbf{0} & \dots & \mathbf{0} & \dots & \mathbf{0} & \mathbf{0} & \dots & \mathbf{0} & \dots & \mathbf{0} \\
\vdots & \ddots & \vdots & \ddots & \vdots & \vdots & \ddots & \vdots & \ddots & \vdots \\
\mathbf{0} & \dots & \mathbf{0} & \dots & \mathbf{0} & \mathbf{0} & \dots & \mathbf{0} & \dots & \mathbf{0} \\
\vdots & \ddots & \vdots & \ddots & \vdots & \vdots & \ddots & \vdots & \ddots & \vdots \\
\mathbf{0} & \dots & \mathbf{0} & \dots & \mathbf{0} & \mathbf{0} & \dots & \mathbf{0} & \dots & \mathbf{0} \\
\vdots & \ddots & \vdots & \ddots & \vdots & \vdots & \ddots & \vdots & \ddots & \vdots \\
\mathbf{0} & \dots & \mathbf{0} & \dots & \mathbf{0} & \mathbf{0} & \dots & \mathbf{0} & \dots & \mathbf{I}_{N_R}
\end{pmatrix}
\begin{pmatrix}
\mathrm{d}\mathbf{c}_{1g}(z) \\ \vdots \\ \mathrm{d}\mathbf{c}_{rg}(z) \\ \vdots \\ \mathrm{d}\mathbf{c}_{Rg}(z) \\ \mathrm{d}\mathbf{c}_{11}(z) \\ \vdots \\ \mathrm{d}\mathbf{c}_{rr}(z) \\ \vdots \\ \mathrm{d}\mathbf{c}_{RR}(z)
\end{pmatrix} \\
&= \left(\mathfrak{e}_{1g}, \dots, \mathfrak{e}_{rg}, \dots, \mathfrak{e}_{Rg}, \mathfrak{e}_{11}, \dots, \mathfrak{e}_{rr}, \dots, \mathfrak{e}_{RR}\right)\mathrm{d}\mathbf{c}(z) = \mathfrak{E}\mathrm{d}\mathbf{c}(z)
\end{aligned}
$$

where $\mathfrak{E}$ contains a block analogue to the diagonalisation matrix above. Consequently, the Jacobian of $\text{vec}\left[\mathbf{G}_{\mathbf{yy}}(\lambda)\right]$ will be

$$\frac{\partial \text{vec}\left[\mathbf{G}_{\mathbf{yy}}(\lambda)\right]}{\partial \boldsymbol{\theta}'_x} = \left[\mathbf{C}\left(e^{i\lambda}\right) \otimes \mathbf{C}\left(e^{-i\lambda}\right)\right] \mathbf{E}_{R+1} \frac{\partial \text{vecd}[\mathbf{G}_{\mathbf{xx}}(\lambda)]}{\partial \boldsymbol{\theta}'_x}$$

$$\frac{\partial \text{vec}\left[\mathbf{G}_{\mathbf{yy}}(\lambda)\right]}{\partial \boldsymbol{\psi}'} = \mathbf{E}_N \frac{\partial \text{vecd}[\mathbf{G}_{\mathbf{uu}}(\lambda)]}{\partial \boldsymbol{\psi}'}$$

$$\frac{\partial \text{vec}\left[\mathbf{G}_{\mathbf{yy}}(\lambda)\right]}{\partial \boldsymbol{\theta}'_{\mathbf{u}}} = \mathbf{E}_N \frac{\partial \text{vecd}[\mathbf{G}_{\mathbf{uu}}(\lambda)]}{\partial \boldsymbol{\theta}'_{\mathbf{u}}}$$

$$\frac{\partial \text{vec}\left[\mathbf{G}_{\mathbf{yy}}(\lambda)\right]}{\partial \mathbf{c}'_{rgk}} = \left\{ \begin{array}{c} \left[e^{-ik\lambda}\mathbf{C}\left(e^{i\lambda}\right)\mathbf{G}_{\mathbf{xx}}(\lambda) \otimes \mathbf{I}_N\right] \\ +\mathbf{K}_{NN}\left[e^{ik\lambda}\mathbf{C}\left(e^{-i\lambda}\right)\mathbf{G}_{\mathbf{xx}}(\lambda) \otimes \mathbf{I}_N\right] \end{array} \right\} \mathfrak{e}_{rg}$$

$$\frac{\partial \text{vec}\left[\mathbf{G}_{\mathbf{yy}}(\lambda)\right]}{\partial \mathbf{c}'_{rrl}} = \left\{ \begin{array}{c} \left[e^{-il\lambda}\mathbf{C}\left(e^{i\lambda}\right)\mathbf{G}_{\mathbf{xx}}(\lambda) \otimes \mathbf{I}_N\right] \\ +\mathbf{K}_{NN}\left[e^{il\lambda}\mathbf{C}\left(e^{-i\lambda}\right)\mathbf{G}_{\mathbf{xx}}(\lambda) \otimes \mathbf{I}_N\right] \end{array} \right\} \mathfrak{e}_{rr}$$

where we have used the fact that

$$\frac{\partial \text{vec}[\mathbf{C}(z)]}{\partial \mathbf{c}'_{rgk}} = \mathfrak{E} \begin{pmatrix} \mathbf{0} \\ \vdots \\ \mathbf{I}_{N_r} \\ \vdots \\ \mathbf{0} \\ \mathbf{0} \\ \vdots \\ \mathbf{0} \\ \vdots \\ \mathbf{0} \end{pmatrix} z^k = \mathfrak{e}_{rg} z^k$$

and

$$\frac{\partial \text{vec}[\mathbf{C}(z)]}{\partial \mathbf{c}'_{rrl}} = \mathfrak{E} \begin{pmatrix} \mathbf{0} \\ \mathbf{0} \\ \vdots \\ \mathbf{0} \\ \mathbf{0} \\ \vdots \\ \vdots \\ \mathbf{I}_{N_r} \\ \vdots \\ \mathbf{0} \end{pmatrix} z^l = \mathfrak{e}_{rr} z^l$$

since

$$\frac{\partial \mathbf{c}_{rg}(z)}{\partial \mathbf{c}'_{rgk}} = z^k \mathbf{I}_{N_r}$$
$$\frac{\partial \mathbf{c}_{rr}(z)}{\partial \mathbf{c}'_{rrl}} = z^l \mathbf{I}_{N_r}$$

in view of Eqs. (2) and (3).

If we combine those expressions with the fact that

$$\left[\mathbf{G}_{\mathbf{yy}}^{-1}(\lambda_j)\otimes\mathbf{G}'^{-1}_{\mathbf{yy}}(\lambda_j)\right]\mathrm{vec}\left[\mathbf{z}_j^{\mathbf{y}c}\mathbf{z}_j^{\mathbf{y}\prime}-\mathbf{G}'_{\mathbf{yy}}(\lambda_j)\right]$$
$$= \mathrm{vec}\left[2\pi\mathbf{G}'^{-1}_{\mathbf{yy}}(\lambda)\mathbf{z}_j^{\mathbf{y}c}\mathbf{z}_j^{\mathbf{y}\prime}\mathbf{G}'^{-1}_{\mathbf{yy}}(\lambda)-\mathbf{G}'^{-1}_{\mathbf{yy}}(\lambda)\right]$$

and $\mathbf{I}'_{\mathbf{yy}}(\lambda)=\mathbf{z}_j^{\mathbf{y}c}\mathbf{z}_j^{\mathbf{y}\prime}$ we obtain:

$$\begin{aligned}
2\mathbf{d}_{\boldsymbol{\theta}_\mathbf{x}}(\lambda;\boldsymbol{\theta}) &= \frac{\partial \mathrm{vecd}'[\mathbf{G}_{\mathbf{xx}}(\lambda)]}{\partial\boldsymbol{\theta}_\mathbf{x}}\mathbf{E}'_{R+1}\left[\mathbf{C}'(e^{i\lambda})\otimes\mathbf{C}'(e^{-i\lambda})\right]\\
&\quad\times\mathrm{vec}\left[2\pi\mathbf{G}'^{-1}_{\mathbf{yy}}(\lambda)\mathbf{I}'_{\mathbf{yy}}(\lambda)\mathbf{G}'^{-1}_{\mathbf{yy}}(\lambda)-\mathbf{G}'^{-1}_{\mathbf{yy}}(\lambda)\right]\\
&= \frac{\partial \mathrm{vecd}'[\mathbf{G}_{\mathbf{xx}}(\lambda)]}{\partial\boldsymbol{\theta}_\mathbf{x}}\mathrm{vecd}\left[\begin{array}{r}2\pi\mathbf{C}'(e^{-i\lambda})\mathbf{G}'^{-1}_{\mathbf{yy}}(\lambda)\mathbf{I}'_{\mathbf{yy}}(\lambda)\mathbf{G}'^{-1}_{\mathbf{yy}}(\lambda)\mathbf{C}(e^{i\lambda})\\-\mathbf{C}'(e^{-i\lambda})\mathbf{G}'^{-1}_{\mathbf{yy}}(\lambda)\mathbf{C}(e^{i\lambda})\end{array}\right]\\
2\mathbf{d}_{\psi}(\lambda;\boldsymbol{\theta}) &= \frac{\partial \mathrm{vecd}'[\mathbf{G}_{\mathbf{uu}}(\lambda)]}{\partial\psi}\mathrm{vecd}\left[2\pi\mathbf{G}'^{-1}_{\mathbf{yy}}(\lambda)\mathbf{I}'_{\mathbf{yy}}(\lambda)\mathbf{G}'^{-1}_{\mathbf{yy}}(\lambda)-\mathbf{G}'^{-1}_{\mathbf{yy}}(\lambda)\right]\\
2\mathbf{d}_{\boldsymbol{\theta}_\mathbf{u}}(\lambda;\boldsymbol{\theta}) &= \frac{\partial \mathrm{vecd}'[\mathbf{G}_{\mathbf{uu}}(\lambda)]}{\partial\boldsymbol{\theta}_\mathbf{u}}\mathrm{vecd}\left[2\pi\mathbf{G}'^{-1}_{\mathbf{yy}}(\lambda)\mathbf{I}'_{\mathbf{yy}}(\lambda)\mathbf{G}'^{-1}_{\mathbf{yy}}(\lambda)-\mathbf{G}'^{-1}_{\mathbf{yy}}(\lambda)\right]\\
2\mathbf{d}_{\mathbf{c}_{rgk}}(\lambda;\boldsymbol{\theta}) &= \mathfrak{e}'_{rg}\left\{\begin{array}{c}\left[\mathbf{G}_{\mathbf{xx}}(\lambda)\mathbf{C}'(e^{i\lambda})e^{-ik\lambda}\otimes\mathbf{I}_N\right]\\+\left[\mathbf{G}_{\mathbf{xx}}(\lambda)\mathbf{C}'(e^{-i\lambda})e^{ik\lambda}\otimes\mathbf{I}_N\right]\mathbf{K}_{NN}\end{array}\right\}\\
&\quad\times\mathrm{vec}\left[2\pi\mathbf{G}'^{-1}_{\mathbf{yy}}(\lambda)\mathbf{I}'_{\mathbf{yy}}(\lambda)\mathbf{G}'^{-1}_{\mathbf{yy}}(\lambda)-\mathbf{G}'^{-1}_{\mathbf{yy}}(\lambda)\right]\\
&= \mathfrak{e}'_{rg}\left\{\begin{array}{l}e^{-ik\lambda}\mathrm{vec}\left[\begin{array}{r}2\pi\mathbf{G}'^{-1}_{\mathbf{yy}}(\lambda)\mathbf{I}'_{\mathbf{yy}}(\lambda)\mathbf{G}'^{-1}_{\mathbf{yy}}(\lambda)\mathbf{C}(e^{i\lambda})\mathbf{G}_{\mathbf{xx}}(\lambda)\\-\mathbf{G}'^{-1}_{\mathbf{yy}}(\lambda)\mathbf{C}(e^{i\lambda})\mathbf{G}_{\mathbf{xx}}(\lambda)\end{array}\right]\\+e^{ik\lambda}\mathrm{vec}\left[\begin{array}{r}2\pi\mathbf{G}^{-1}_{\mathbf{yy}}(\lambda)\mathbf{I}_{\mathbf{yy}}(\lambda)\mathbf{G}^{-1}_{\mathbf{yy}}(\lambda)\mathbf{C}(e^{-i\lambda})\mathbf{G}_{\mathbf{xx}}(\lambda)\\-\mathbf{G}^{-1}_{\mathbf{yy}}(\lambda)\mathbf{C}(e^{-i\lambda})\mathbf{G}_{\mathbf{xx}}(\lambda)\end{array}\right]\end{array}\right\}\\
2\mathbf{d}_{\mathbf{c}_{rrl}}(\lambda;\boldsymbol{\theta}) &= \mathfrak{e}'_{rr}\left\{\begin{array}{c}\left[\mathbf{G}_{\mathbf{xx}}(\lambda)\mathbf{C}'(e^{i\lambda})e^{-il\lambda}\otimes\mathbf{I}_N\right]\\+\left[\mathbf{G}_{\mathbf{xx}}(\lambda)\mathbf{C}'(e^{-i\lambda})e^{il\lambda}\otimes\mathbf{I}_N\right]\mathbf{K}_{NN}\end{array}\right\}\\
&\quad\times\mathrm{vec}\left[2\pi\mathbf{G}'^{-1}_{\mathbf{yy}}(\lambda)\mathbf{I}'_{\mathbf{yy}}(\lambda)\mathbf{G}'^{-1}_{\mathbf{yy}}(\lambda)-\mathbf{G}'^{-1}_{\mathbf{yy}}(\lambda)\right]\\
&= \mathfrak{e}'_{rr}\left\{\begin{array}{l}e^{-il\lambda}\mathrm{vec}\left[\begin{array}{r}2\pi\mathbf{G}'^{-1}_{\mathbf{yy}}(\lambda)\mathbf{I}'_{\mathbf{yy}}(\lambda)\mathbf{G}'^{-1}_{\mathbf{yy}}(\lambda)\mathbf{C}(e^{i\lambda})\mathbf{G}_{\mathbf{xx}}(\lambda)\\-\mathbf{G}'^{-1}_{\mathbf{yy}}(\lambda)\mathbf{C}(e^{i\lambda})\mathbf{G}_{\mathbf{xx}}(\lambda)\end{array}\right]\\+e^{il\lambda}\mathrm{vec}\left[\begin{array}{r}2\pi\mathbf{G}^{-1}_{\mathbf{yy}}(\lambda)\mathbf{I}_{\mathbf{yy}}(\lambda)\mathbf{G}^{-1}_{\mathbf{yy}}(\lambda)\mathbf{C}(e^{-i\lambda})\mathbf{G}_{\mathbf{xx}}(\lambda)\\-\mathbf{G}^{-1}_{\mathbf{yy}}(\lambda)\mathbf{C}(e^{-i\lambda})\mathbf{G}_{\mathbf{xx}}(\lambda)\end{array}\right]\end{array}\right\}
\end{aligned}$$

where we have used the fact that $\mathbf{K}'_{NN} = \mathbf{K}_{NN} = \mathbf{K}^{-1}_{NN}$ (see again Magnus, 1988).

Let us now try to interpret the different components of this expression. To do so, it is convenient to further assume that $\mathbf{G_{xx}}(\lambda) > 0$ and $\mathbf{G_{uu}}(\lambda) > \mathbf{0}$.

The first thing to note is that

$$2\pi\mathbf{C}'(\mathrm{e}^{-\mathrm{i}\lambda})\mathbf{G}'^{-1}_{\mathbf{yy}}(\lambda)\mathbf{I}'_{\mathbf{yy}}(\lambda)\mathbf{G}'^{-1}_{\mathbf{yy}}(\lambda)\mathbf{C}(\mathrm{e}^{\mathrm{i}\lambda}) - \mathbf{C}'(\mathrm{e}^{-\mathrm{i}\lambda})\mathbf{G}'^{-1}_{\mathbf{yy}}(\lambda)\mathbf{C}(\mathrm{e}^{\mathrm{i}\lambda})$$
$$= \mathbf{G}^{-1}_{\mathbf{xx}}(\lambda)\left[2\pi\mathbf{I}'_{\mathbf{x}^K\mathbf{x}^K}(\lambda) - \mathbf{G}'_{\mathbf{x}^K\mathbf{x}^K}(\lambda)\right]\mathbf{G}^{-1}_{\mathbf{xx}}(\lambda)$$

Given that

$$\frac{\partial \mathrm{vecd}[\mathbf{G_{xx}}(\lambda)]}{\partial \boldsymbol{\theta}'_{x_g}} = \frac{\partial G_{x_g x_g}(\lambda)}{\partial \boldsymbol{\theta}'_{x_g}}\mathbf{e}_{1,R+1}$$

the component of the score associated to the parameters that determine $G_{x_g x_g}(\lambda)$ will be the cross-product across frequencies of the product of the derivatives of the spectral density of x_{gt} with the difference between the periodogram and spectrum of x^K_{gt} inversely weighted by the squared spectral density of x_{gt}. Thus, we can interpret this term as arising from a marginal log-likelihood function for x_{gt} that takes into account the unobservability of x_{gt}. Exactly the same comments apply to the scores of the parameters that determine $G_{x_r x_r}(\lambda)$ for $r = 1, \ldots, R$ in view of the fact that

$$\frac{\partial \mathrm{vecd}[\mathbf{G_{xx}}(\lambda)]}{\partial \boldsymbol{\theta}'_{x_r}} = \frac{\partial G_{x_r x_r}(\lambda)}{\partial \boldsymbol{\theta}'_{x_r}}\mathbf{e}_{r+1,R+1}$$

Similarly, given that

$$2\pi\mathbf{G}'^{-1}_{\mathbf{yy}}(\lambda)\mathbf{I}'_{\mathbf{yy}}(\lambda)\mathbf{G}'^{-1}_{\mathbf{yy}}(\lambda) - \mathbf{G}'^{-1}_{\mathbf{yy}}(\lambda) = \mathbf{G}'^{-1}_{\mathbf{uu}}(\lambda)\left[2\pi\mathbf{I}'_{\mathbf{u}^K\mathbf{u}^K}(\lambda) - \mathbf{G}'_{\mathbf{u}^K\mathbf{u}^K}(\lambda)\right]\mathbf{G}'^{-1}_{\mathbf{uu}}(\lambda)$$

$$\frac{\partial \mathrm{vecd}[\mathbf{G_{uu}}(\lambda)]}{\partial \psi_i} = \frac{\partial G_{u_i u_i}(\lambda)}{\partial \psi_i}\mathbf{e}_{iN}$$

and

$$\frac{\partial \mathrm{vecd}[\mathbf{G_{uu}}(\lambda)]}{\partial \boldsymbol{\theta}'_{u_i}} = \frac{\partial G_{u_i u_i}(\lambda)}{\partial \boldsymbol{\theta}'_{u_i}}\mathbf{e}_{iN}$$

the component of the score associated to the parameters that determine $G_{u_i u_i}(\lambda)$ will be the cross-product across frequencies of the product of the derivatives of the spectral density of u_{it} with the difference between the periodogram and spectrum of u_{it}^K inversely weighted by the squared spectral density of u_{it}. Once again, we can interpret this term as arising from the conditional log-likelihood function of u_{it} given $\mathbf{x}_t$ that takes into account the unobservability of u_{t_i}.

Finally, to interpret the scores of the distributed lag coefficients it is worth noting that

$$e^{-ik\lambda}\text{vec}\left[2\pi\mathbf{G}_{\mathbf{yy}}^{\prime -1}(\lambda)\mathbf{I}_{\mathbf{yy}}^{\prime}(\lambda)\mathbf{G}_{\mathbf{yy}}^{\prime -1}(\lambda)\mathbf{C}\left(e^{i\lambda}\right)\mathbf{G}_{\mathbf{xx}}(\lambda)-\mathbf{G}_{\mathbf{yy}}^{\prime -1}(\lambda)\mathbf{C}\left(e^{i\lambda}\right)\mathbf{G}_{\mathbf{xx}}(\lambda)\right]$$

and

$$e^{ik\lambda}\text{vec}\left[2\pi\mathbf{G}_{\mathbf{yy}}^{-1}(\lambda)\mathbf{I}_{\mathbf{yy}}(\lambda)\mathbf{G}_{\mathbf{yy}}^{-1}(\lambda)\mathbf{C}\left(e^{-i\lambda}\right)\mathbf{G}_{\mathbf{xx}}(\lambda)-\mathbf{G}_{\mathbf{yy}}^{-1}(\lambda)\mathbf{C}\left(e^{-i\lambda}\right)\mathbf{G}_{\mathbf{xx}}(\lambda)\right]$$

are complex conjugates because $\mathbf{G}_{\mathbf{yy}}^{-1}(\lambda)$ is Hermitian and the conjugate of a product is the product of the conjugates, so it suffices to analyse one of them. On this basis, if we write

$$\begin{aligned}&2\pi\mathbf{G}_{\mathbf{yy}}^{\prime -1}(\lambda)\mathbf{I}_{\mathbf{yy}}^{\prime}(\lambda)\mathbf{G}_{\mathbf{yy}}^{\prime -1}(\lambda)\mathbf{C}\left(e^{i\lambda}\right)\mathbf{G}_{\mathbf{xx}}(\lambda)-\mathbf{G}_{\mathbf{yy}}^{\prime -1}(\lambda)\mathbf{C}\left(e^{i\lambda}\right)\mathbf{G}_{\mathbf{xx}}(\lambda)\\&\quad=\mathbf{G}_{\mathbf{uu}}^{\prime -1}(\lambda)\left[2\pi\mathbf{I}_{\mathbf{x}^K\mathbf{u}^K}^{\prime}(\lambda)-\mathbf{G}_{\mathbf{x}^K\mathbf{u}^K}^{\prime}(\lambda)\right]\end{aligned}$$

the components of the score associated to $\mathbf{c}_{rgk}$ will be the sum across frequencies of terms of the form

$$\mathbf{G}_{\mathbf{uu}}^{\prime -1}(\lambda)\left[2\pi\mathbf{I}_{\mathbf{x}^K\mathbf{u}^K}^{\prime}(\lambda)-\mathbf{G}_{\mathbf{x}^K\mathbf{u}^K}^{\prime}(\lambda)\right]e^{-ik\lambda}$$

(and their conjugate transposes), which capture the difference between the cross-periodogram and cross-spectrum of x_{gt-r}^K and u_{it}^K inversely weighted by the spectral density of u_{it}. Exactly the same comments apply to the scores of $\mathbf{c}_{rrl}$. Therefore, we can understand those terms as arising from the normal equation in the spectral regression of y_{it} onto $x_{g,t+m_g}, \ldots, x_{g,t-n_g}$ and $x_{r,t+m_r}, \ldots, x_{rt-n_r}$ but taking into account the unobservability of the regressors.

As usual, we can exploit the Woodbury formula, as in expressions (8), (10), (11), (26), (30) and (31), to greatly speed up the computations.

APPENDIX B. SPECTRAL INFORMATION MATRIX

Given the expression for the Jacobian matrix in derived in Appendix A, we will have that

$$\frac{\partial \mathrm{vec}'\left[\mathbf{G}_{\mathbf{yy}}(\lambda)\right]}{\partial \boldsymbol{\theta}_x} = \frac{\partial \mathrm{vecd}'[\mathbf{G}_{\mathbf{xx}}(\lambda)]}{\partial \boldsymbol{\theta}_x}\mathbf{E}'_{R+1}\left[\mathbf{C}'\left(e^{i\lambda}\right)\otimes\mathbf{C}'\left(e^{-i\lambda}\right)\right]$$

$$\frac{\partial \mathrm{vec}'\left[\mathbf{G}_{\mathbf{yy}}(\lambda)\right]}{\partial \boldsymbol{\psi}} = \frac{\partial \mathrm{vecd}'[\mathbf{G}_{\mathbf{uu}}(\lambda)]}{\partial \boldsymbol{\psi}'}\mathbf{E}'_N$$

$$\frac{\partial \mathrm{vec}'\left[\mathbf{G}_{\mathbf{yy}}(\lambda)\right]}{\partial \boldsymbol{\theta}_{\mathbf{u}}} = \frac{\partial \mathrm{vecd}'[\mathbf{G}_{\mathbf{uu}}(\lambda)]}{\partial \boldsymbol{\theta}_{\mathbf{u}}}\mathbf{E}'_N$$

$$\frac{\partial \mathrm{vec}'\left[\mathbf{G}_{\mathbf{yy}}(\lambda)\right]}{\partial \mathbf{c}_{rgk}} = \mathfrak{e}'_{rg}\left\{\begin{array}{c}\left[e^{-ik\lambda}\mathbf{G}_{\mathbf{xx}}(\lambda)\mathbf{C}'\left(e^{i\lambda}\right)\otimes\mathbf{I}_N\right]\\ +\left[e^{ik\lambda}\mathbf{G}_{\mathbf{xx}}(\lambda)\mathbf{C}'\left(e^{-i\lambda}\right)\otimes\mathbf{I}_N\right]\mathbf{K}_{NN}\end{array}\right\}$$

$$\frac{\partial \mathrm{vec}'\left[\mathbf{G}_{\mathbf{yy}}(\lambda)\right]}{\partial \mathbf{c}_{rrl}} = \mathfrak{e}'_{rr}\left\{\begin{array}{c}\left[e^{-il\lambda}\mathbf{G}_{\mathbf{xx}}(\lambda)\mathbf{C}'\left(e^{i\lambda}\right)\otimes\mathbf{I}_N\right]\\ +\left[e^{il\lambda}\mathbf{G}_{\mathbf{xx}}(\lambda)\mathbf{C}'\left(e^{-i\lambda}\right)\otimes\mathbf{I}_N\right]\mathbf{K}_{NN}\end{array}\right\}$$

and

$$\left\{\frac{\partial \mathrm{vec}'\left[\mathbf{G}_{\mathbf{yy}}(\lambda)\right]}{\partial \boldsymbol{\theta}_x}\right\}^* = \left[\mathbf{C}\left(e^{-i\lambda}\right)\otimes\mathbf{C}\left(e^{i\lambda}\right)\right]\mathbf{E}_{R+1}\frac{\partial \mathrm{vecd}[\mathbf{G}_{\mathbf{xx}}(\lambda)]}{\partial \boldsymbol{\theta}'_x}$$

$$\left\{\frac{\partial \mathrm{vec}'\left[\mathbf{G}_{\mathbf{yy}}(\lambda)\right]}{\partial \boldsymbol{\psi}}\right\}^* = \mathbf{E}_N\frac{\partial \mathrm{vecd}[\mathbf{G}_{\mathbf{uu}}(\lambda)]}{\partial \boldsymbol{\psi}'}$$

$$\left\{\frac{\partial \mathrm{vec}'\left[\mathbf{G}_{\mathbf{yy}}(\lambda)\right]}{\partial \boldsymbol{\theta}_{\mathbf{u}}}\right\}^* = \mathbf{E}_N\frac{\partial \mathrm{vecd}[\mathbf{G}_{\mathbf{uu}}(\lambda)]}{\partial \boldsymbol{\theta}'_{\mathbf{u}}}$$

$$\left\{\frac{\partial \mathrm{vec}'\left[\mathbf{G}_{\mathbf{yy}}(\lambda)\right]}{\partial \mathbf{c}_{rgk}}\right\}^* = \left\{\begin{array}{c}\left[e^{ik\lambda}\mathbf{C}\left(e^{-i\lambda}\right)\mathbf{G}_{\mathbf{xx}}(\lambda)\otimes\mathbf{I}_N\right]\\ +\mathbf{K}_{NN}\left[e^{-ik\lambda}\mathbf{C}\left(e^{i\lambda}\right)\mathbf{G}_{\mathbf{xx}}(\lambda)\otimes\mathbf{I}_N\right]\end{array}\right\}\mathfrak{e}_{rg}$$

$$\left\{\frac{\partial \mathrm{vec}'\left[\mathbf{G}_{\mathbf{yy}}(\lambda)\right]}{\partial \mathbf{c}_{rrl}}\right\}^* = \left\{\begin{array}{c}\left[e^{il\lambda}\mathbf{C}\left(e^{-i\lambda}\right)\mathbf{G}_{\mathbf{xx}}(\lambda)\otimes\mathbf{I}_N\right]\\ +\mathbf{K}_{NN}\left[e^{-il\lambda}\mathbf{C}\left(e^{i\lambda}\right)\mathbf{G}_{\mathbf{xx}}(\lambda)\otimes\mathbf{I}_N\right]\end{array}\right\}\mathfrak{e}_{rr}$$

Hence, it is straightforward to see that the elements of the block of the information matrix (19) corresponding to the autoregressive parameters for the common factors will be

$$\mathbf{Q}_{\theta_x\theta_x}(\lambda;\boldsymbol{\theta}) = \frac{\partial \mathrm{vec}'\left[\mathbf{G}_{\mathbf{yy}}(\lambda)\right]}{\partial \boldsymbol{\theta}_x}\left[\mathbf{G}_{\mathbf{yy}}^{-1}(\lambda)\otimes\mathbf{G}_{\mathbf{yy}}^{\prime -1}(\lambda)\right]\left\{\frac{\partial \mathrm{vec}'\left[\mathbf{G}_{\mathbf{yy}}(\lambda)\right]}{\partial \boldsymbol{\theta}_x}\right\}^*$$

$$= \frac{\partial \mathrm{vecd}'[\mathbf{G}_{\mathbf{xx}}(\lambda)]}{\partial \boldsymbol{\theta}_x}\mathbf{E}_{R+1}'\left[\mathbf{C}'(\mathrm{e}^{\mathrm{i}\lambda})\otimes\mathbf{C}'(\mathrm{e}^{-\mathrm{i}\lambda})\right]\left[\mathbf{G}_{\mathbf{yy}}^{-1}(\lambda)\otimes\mathbf{G}_{\mathbf{yy}}^{\prime -1}(\lambda)\right]$$

$$\times\left[\mathbf{C}(\mathrm{e}^{-\mathrm{i}\lambda})\otimes\mathbf{C}(\mathrm{e}^{\mathrm{i}\lambda})\right]\mathbf{E}_{R+1}\frac{\partial \mathrm{vecd}[\mathbf{G}_{\mathbf{xx}}(\lambda)]}{\partial \boldsymbol{\theta}_x'}$$

$$= \frac{\partial \mathrm{vecd}'[\mathbf{G}_{\mathbf{xx}}(\lambda)]}{\partial \boldsymbol{\theta}_x}\left\{\left[\mathbf{C}'(\mathrm{e}^{\mathrm{i}\lambda})\mathbf{G}_{\mathbf{yy}}^{-1}(\lambda_j)\mathbf{C}(\mathrm{e}^{-\mathrm{i}\lambda})\right]\right.$$

$$\left.\odot\left[\mathbf{C}'(\mathrm{e}^{-\mathrm{i}\lambda})\mathbf{G}_{\mathbf{yy}}^{\prime -1}(\lambda_j)\mathbf{C}(\mathrm{e}^{\mathrm{i}\lambda})\right]\right\}\frac{\partial \mathrm{vecd}[\mathbf{G}_{\mathbf{xx}}(\lambda)]}{\partial \boldsymbol{\theta}_x'},$$

where $\odot$ denotes the Hadamard (or element by element) product of two matrices of equal size.

Similarly,

$$\mathbf{Q}_{\theta_u\theta_u}(\lambda;\boldsymbol{\theta}) = \frac{\partial \mathrm{vec}'\left[\mathbf{G}_{\mathbf{yy}}(\lambda)\right]}{\partial \boldsymbol{\theta}_{\mathbf{u}}}\left[\mathbf{G}_{\mathbf{yy}}^{-1}(\lambda)\otimes\mathbf{G}_{\mathbf{yy}}^{\prime -1}(\lambda)\right]\left\{\frac{\partial \mathrm{vec}'\left[\mathbf{G}_{\mathbf{yy}}(\lambda)\right]}{\partial \boldsymbol{\theta}_{\mathbf{u}}}\right\}^*$$

$$= \frac{\partial \mathrm{vecd}'[\mathbf{G}_{\mathbf{uu}}(\lambda)]}{\partial \boldsymbol{\theta}_{\mathbf{u}}}\mathbf{E}_N'\left[\mathbf{G}_{\mathbf{yy}}^{-1}(\lambda)\otimes\mathbf{G}_{\mathbf{yy}}^{\prime -1}(\lambda)\right]\mathbf{E}_N\frac{\partial \mathrm{vecd}[\mathbf{G}_{\mathbf{uu}}(\lambda)]}{\partial \boldsymbol{\theta}_{\mathbf{u}}'}$$

$$= \frac{\partial \mathrm{vecd}'[\mathbf{G}_{\mathbf{uu}}(\lambda)]}{\partial \boldsymbol{\theta}_{\mathbf{u}}}\left[\mathbf{G}_{\mathbf{yy}}^{-1}(\lambda)\odot\mathbf{G}_{\mathbf{yy}}^{\prime -1}(\lambda)\right]\frac{\partial \mathrm{vecd}[\mathbf{G}_{\mathbf{uu}}(\lambda)]}{\partial \boldsymbol{\theta}_{\mathbf{u}}'}$$

with an almost identical expression for $\mathbf{Q}_{\psi\psi}(\lambda;\boldsymbol{\theta})$.

Also,

$$\mathbf{Q}_{\boldsymbol{c}_{rgk}\boldsymbol{c}_{rrl}}(\lambda;\boldsymbol{\theta})=\frac{\partial \mathrm{vec}'\left[\mathbf{G}_{\mathbf{yy}}(\lambda)\right]}{\partial \boldsymbol{c}_{rgk}}\left[\mathbf{G}_{\mathbf{yy}}^{-1}(\lambda)\otimes\mathbf{G}_{\mathbf{yy}}'^{-1}(\lambda)\right]\left\{\frac{\partial \mathrm{vec}'\left[\mathbf{G}_{\mathbf{yy}}(\lambda)\right]}{\partial \mathbf{c}_{rrl}}\right\}^{*}$$

$$=\mathfrak{e}_{rg}'\left\{\begin{array}{c}\left[\mathrm{e}^{-\mathrm{i}k\lambda}\mathbf{G}_{\mathbf{xx}}(\lambda)\mathbf{C}'\left(\mathrm{e}^{\mathrm{i}\lambda}\right)\otimes\mathbf{I}_N\right]\\+\left[\mathrm{e}^{\mathrm{i}k\lambda}\mathbf{G}_{\mathbf{xx}}(\lambda)\mathbf{C}'\left(\mathrm{e}^{-\mathrm{i}\lambda}\right)\otimes\mathbf{I}_N\right]\mathbf{K}_{NN}\end{array}\right\}\left[\mathbf{G}_{\mathbf{yy}}^{-1}(\lambda)\otimes\mathbf{G}_{\mathbf{yy}}'^{-1}(\lambda)\right]$$

$$\times\left\{\begin{array}{c}\left[\mathrm{e}^{\mathrm{i}l\lambda}\mathbf{C}\left(\mathrm{e}^{-\mathrm{i}\lambda}\right)\mathbf{G}_{\mathbf{xx}}(\lambda)\otimes\mathbf{I}_N\right]\\+\mathbf{K}_{NN}\left[\mathrm{e}^{-\mathrm{i}l\lambda}\mathbf{C}\left(\mathrm{e}^{\mathrm{i}\lambda}\right)\mathbf{G}_{\mathbf{xx}}(\lambda)\otimes\mathbf{I}_N\right]\end{array}\right\}\mathfrak{e}_{rr}$$

$$=\mathfrak{e}_{rg}'\left\{\begin{array}{c}\mathrm{e}^{-\mathrm{i}(k+l)\lambda}\left[\mathbf{G}_{\mathbf{xx}}(\lambda)\mathbf{C}'\left(\mathrm{e}^{\mathrm{i}\lambda}\right)\otimes\mathbf{I}_N\right]\left[\mathbf{G}_{\mathbf{yy}}^{-1}(\lambda)\otimes\mathbf{G}_{\mathbf{yy}}'^{-1}(\lambda)\right]\mathbf{K}_{NN}\left[\mathbf{C}\left(\mathrm{e}^{\mathrm{i}\lambda}\right)\mathbf{G}_{\mathbf{xx}}(\lambda)\otimes\mathbf{I}_N\right]\\ \mathrm{e}^{\mathrm{i}(k+l)\lambda}\left[\mathbf{G}_{\mathbf{xx}}(\lambda)\mathbf{C}'\left(\mathrm{e}^{-\mathrm{i}\lambda}\right)\otimes\mathbf{I}_N\right]\mathbf{K}_{NN}\left[\mathbf{G}_{\mathbf{yy}}^{-1}(\lambda)\otimes\mathbf{G}_{\mathbf{yy}}'^{-1}(\lambda)\right]\left[\mathbf{C}\left(\mathrm{e}^{-\mathrm{i}\lambda}\right)\mathbf{G}_{\mathbf{xx}}(\lambda)\otimes\mathbf{I}_N\right]\\ \mathrm{e}^{-\mathrm{i}(k-l)\lambda}\left[\mathbf{G}_{\mathbf{xx}}(\lambda)\mathbf{C}'\left(\mathrm{e}^{\mathrm{i}\lambda}\right)\otimes\mathbf{I}_N\right]\left[\mathbf{G}_{\mathbf{yy}}^{-1}(\lambda)\otimes\mathbf{G}_{\mathbf{yy}}'^{-1}(\lambda)\right]\left[\mathbf{C}\left(\mathrm{e}^{-\mathrm{i}\lambda}\right)\mathbf{G}_{\mathbf{xx}}(\lambda)\otimes\mathbf{I}_N\right]\\ \mathrm{e}^{\mathrm{i}(k-l)\lambda}\left[\mathbf{G}_{\mathbf{xx}}(\lambda)\mathbf{C}'\left(\mathrm{e}^{-\mathrm{i}\lambda}\right)\otimes\mathbf{I}_N\right]\mathbf{K}_{NN}\left[\mathbf{G}_{\mathbf{yy}}^{-1}(\lambda)\otimes\mathbf{G}_{\mathbf{yy}}'^{-1}(\lambda)\right]\mathbf{K}_{NN}\left[\mathbf{C}\left(\mathrm{e}^{\mathrm{i}\lambda}\right)\mathbf{G}_{\mathbf{xx}}(\lambda)\otimes\mathbf{I}_N\right]\end{array}\right\}\mathfrak{e}_{rr}$$

$$=\mathfrak{e}_{rg}'\left\{\begin{array}{c}\mathrm{e}^{-\mathrm{i}(k+l)\lambda}\left[\mathbf{G}_{\mathbf{xx}}(\lambda)\mathbf{C}'\left(\mathrm{e}^{\mathrm{i}\lambda}\right)\otimes\mathbf{I}_N\right]\left[\mathbf{G}_{\mathbf{yy}}^{-1}(\lambda)\otimes\mathbf{G}_{\mathbf{yy}}'^{-1}(\lambda)\right]\left[\mathbf{I}_N\otimes\mathbf{C}\left(\mathrm{e}^{\mathrm{i}\lambda}\right)\mathbf{G}_{\mathbf{xx}}(\lambda)\right]\mathbf{K}_{N,R+1}\\ \mathrm{e}^{\mathrm{i}(k+l)\lambda}\left[\mathbf{G}_{\mathbf{xx}}(\lambda)\mathbf{C}'\left(\mathrm{e}^{-\mathrm{i}\lambda}\right)\otimes\mathbf{I}_N\right]\left[\mathbf{G}_{\mathbf{yy}}'^{-1}(\lambda)\otimes\mathbf{G}_{\mathbf{yy}}^{-1}(\lambda)\right]\left[\mathbf{I}_N\otimes\mathbf{C}\left(\mathrm{e}^{-\mathrm{i}\lambda}\right)\mathbf{G}_{\mathbf{xx}}(\lambda)\right]\mathbf{K}_{N,R+1}\\ \mathrm{e}^{-\mathrm{i}(k-l)\lambda}\left[\mathbf{G}_{\mathbf{xx}}(\lambda)\mathbf{C}'\left(\mathrm{e}^{\mathrm{i}\lambda}\right)\otimes\mathbf{I}_N\right]\left[\mathbf{G}_{\mathbf{yy}}^{-1}(\lambda)\otimes\mathbf{G}_{\mathbf{yy}}'^{-1}(\lambda)\right]\left[\mathbf{C}\left(\mathrm{e}^{-\mathrm{i}\lambda}\right)\mathbf{G}_{\mathbf{xx}}(\lambda)\otimes\mathbf{I}_N\right]\\ \mathrm{e}^{\mathrm{i}(k-l)\lambda}\left[\mathbf{G}_{\mathbf{xx}}(\lambda)\mathbf{C}'\left(\mathrm{e}^{-\mathrm{i}\lambda}\right)\otimes\mathbf{I}_N\right]\left[\mathbf{G}_{\mathbf{yy}}'^{-1}(\lambda)\otimes\mathbf{G}_{\mathbf{yy}}^{-1}(\lambda)\right]\left[\mathbf{C}\left(\mathrm{e}^{\mathrm{i}\lambda}\right)\mathbf{G}_{\mathbf{xx}}(\lambda)\otimes\mathbf{I}_N\right]\end{array}\right\}\mathfrak{e}_{rr}$$

$$=\mathfrak{e}_{rg}'\left\{\begin{array}{c}\mathrm{e}^{-\mathrm{i}(k+l)\lambda}\left[\mathbf{G}_{\mathbf{xx}}(\lambda)\mathbf{C}'\left(\mathrm{e}^{\mathrm{i}\lambda}\right)\mathbf{G}_{\mathbf{yy}}^{-1}(\lambda)\otimes\mathbf{G}_{\mathbf{yy}}'^{-1}(\lambda)\mathbf{C}\left(\mathrm{e}^{\mathrm{i}\lambda}\right)\mathbf{G}_{\mathbf{xx}}(\lambda)\right]\mathbf{K}_{N,R+1}\\ \mathrm{e}^{\mathrm{i}(k+l)\lambda}\left[\mathbf{G}_{\mathbf{xx}}(\lambda)\mathbf{C}'\left(\mathrm{e}^{-\mathrm{i}\lambda}\right)\mathbf{G}_{\mathbf{yy}}'^{-1}(\lambda)\otimes\mathbf{G}_{\mathbf{yy}}^{-1}(\lambda)\mathbf{C}\left(\mathrm{e}^{-\mathrm{i}\lambda}\right)\mathbf{G}_{\mathbf{xx}}(\lambda)\right]\mathbf{K}_{N,R+1}\\ \mathrm{e}^{-\mathrm{i}(k-l)\lambda}\left[\mathbf{G}_{\mathbf{xx}}(\lambda)\mathbf{C}'\left(\mathrm{e}^{\mathrm{i}\lambda}\right)\mathbf{G}_{\mathbf{yy}}^{-1}\left(\lambda_j\right)\mathbf{C}\left(\mathrm{e}^{-\mathrm{i}\lambda}\right)\mathbf{G}_{\mathbf{xx}}(\lambda)\otimes\mathbf{G}_{\mathbf{yy}}'^{-1}(\lambda)\right]\\ \mathrm{e}^{\mathrm{i}(k-l)\lambda}\left[\mathbf{G}_{\mathbf{xx}}(\lambda)\mathbf{C}'\left(\mathrm{e}^{-\mathrm{i}\lambda}\right)\mathbf{G}_{\mathbf{yy}}'^{-1}\left(\lambda_j\right)\mathbf{C}\left(\mathrm{e}^{\mathrm{i}\lambda}\right)\mathbf{G}_{\mathbf{xx}}(\lambda)\otimes\mathbf{G}_{\mathbf{yy}}^{-1}(\lambda)\right]\end{array}\right\}\mathfrak{e}_{rr},$$

where we have made use of the properties of the commutation matrix.

Further

$$
\begin{aligned}
\mathbf{Q}_{\boldsymbol{\theta}_{\mathrm{x}}\boldsymbol{\theta}_{\mathrm{u}}}(\lambda;\boldsymbol{\theta}) &= \frac{\partial \mathrm{vec}'\left[\mathbf{G}_{\mathbf{yy}}(\lambda)\right]}{\partial \boldsymbol{\theta}_{x}}\left[\mathbf{G}_{\mathbf{yy}}^{-1}(\lambda)\otimes\mathbf{G}_{\mathbf{yy}}'^{-1}(\lambda)\right]\frac{\partial \mathrm{vec}\left[\mathbf{G}_{\mathbf{yy}}(\lambda)\right]}{\partial \boldsymbol{\theta}_{\mathbf{u}}'} \\
&= \frac{\partial \mathrm{vecd}'[\mathbf{G}_{\mathbf{xx}}(\lambda)]}{\partial \boldsymbol{\theta}_{x}}\mathbf{E}_{R+1}'\left[\mathbf{C}'\left(\mathrm{e}^{\mathrm{i}\lambda}\right)\otimes\mathbf{C}'\left(\mathrm{e}^{-\mathrm{i}\lambda}\right)\right] \\
&\quad\times\left[\mathbf{G}_{\mathbf{yy}}^{-1}(\lambda)\otimes\mathbf{G}_{\mathbf{yy}}'^{-1}(\lambda)\right]\mathbf{E}_{N}\frac{\partial \mathrm{vecd}[\mathbf{G}_{\mathbf{uu}}(\lambda)]}{\partial \boldsymbol{\theta}_{\mathbf{u}}'} \\
&= \frac{\partial \mathrm{vecd}'[\mathbf{G}_{\mathbf{xx}}(\lambda)]}{\partial \boldsymbol{\theta}_{x}}\left[\mathbf{C}'\left(\mathrm{e}^{\mathrm{i}\lambda}\right)\mathbf{G}_{\mathbf{yy}}^{-1}(\lambda)\odot\mathbf{C}'\left(\mathrm{e}^{-\mathrm{i}\lambda}\right)\mathbf{G}_{\mathbf{yy}}'^{-1}(\lambda)\right]\frac{\partial \mathrm{vecd}[\mathbf{G}_{\mathbf{uu}}(\lambda)]}{\partial \boldsymbol{\theta}_{\mathbf{u}}'}
\end{aligned}
$$

$$
\begin{aligned}
\mathbf{Q}_{\boldsymbol{\theta}_{\mathrm{x}}c_{rrl}}(\lambda;\boldsymbol{\theta}) &= \frac{\partial \mathrm{vec}'\left[\mathbf{G}_{\mathbf{yy}}(\lambda)\right]}{\partial \boldsymbol{\theta}_{\mathbf{x}}}\left[\mathbf{G}_{\mathbf{yy}}^{-1}(\lambda)\otimes\mathbf{G}_{\mathbf{yy}}'^{-1}(\lambda)\right]\frac{\partial \mathrm{vec}\left[\mathbf{G}_{\mathbf{yy}}(\lambda)\right]}{\partial \mathbf{c}_{rrl}'} \\
&= \frac{\partial \mathrm{vecd}'[\mathbf{G}_{\mathbf{xx}}(\lambda)]}{\partial \boldsymbol{\theta}_{\mathbf{x}}}\mathbf{E}_{R+1}'\left[\mathbf{C}'\left(\mathrm{e}^{\mathrm{i}\lambda}\right)\otimes\mathbf{C}'\left(\mathrm{e}^{-\mathrm{i}\lambda}\right)\right] \\
&\quad\times\left[\mathbf{G}_{\mathbf{yy}}^{-1}(\lambda)\otimes\mathbf{G}_{\mathbf{yy}}'^{-1}(\lambda)\right]\left\{\begin{array}{c}\left[\mathrm{e}^{\mathrm{i}l\lambda}\mathbf{C}\left(\mathrm{e}^{-\mathrm{i}\lambda}\right)\mathbf{G}_{\mathbf{xx}}(\lambda)\otimes\mathbf{I}_{N}\right] \\ +\mathbf{K}_{NN}\left[\mathrm{e}^{-\mathrm{i}l\lambda}\mathbf{C}\left(\mathrm{e}^{\mathrm{i}\lambda}\right)\mathbf{G}_{\mathbf{xx}}(\lambda)\otimes\mathbf{I}_{N}\right]\end{array}\right\}\mathfrak{e}_{rr} \\
&= \frac{\partial \mathrm{vecd}'[\mathbf{G}_{\mathbf{xx}}(\lambda)]}{\partial \boldsymbol{\theta}_{\mathbf{x}}}\mathbf{E}_{R+1}' \\
&\quad\times\left[\begin{array}{c}\mathrm{e}^{\mathrm{i}l\lambda}\mathbf{C}'\left(\mathrm{e}^{\mathrm{i}\lambda}\right)\mathbf{G}_{\mathbf{yy}}^{-1}(\lambda)\mathbf{C}\left(\mathrm{e}^{-\mathrm{i}\lambda}\right)\mathbf{G}_{\mathbf{xx}}(\lambda)\otimes\mathbf{C}'\left(\mathrm{e}^{-\mathrm{i}\lambda}\right)\mathbf{G}_{\mathbf{yy}}'^{-1}(\lambda) \\ +\mathrm{e}^{-\mathrm{i}l\lambda}\mathbf{C}'\left(\mathrm{e}^{-\mathrm{i}\lambda}\right)\mathbf{G}_{\mathbf{yy}}'^{-1}(\lambda)\mathbf{C}\left(\mathrm{e}^{\mathrm{i}\lambda}\right)\mathbf{G}_{\mathbf{xx}}(\lambda)\otimes\mathbf{C}'\left(\mathrm{e}^{\mathrm{i}\lambda}\right)\mathbf{G}_{\mathbf{yy}}^{-1}(\lambda)\end{array}\right]\mathfrak{e}_{rr}
\end{aligned}
$$

and

$$
\begin{aligned}
\mathbf{Q}_{\boldsymbol{\theta}_{\mathrm{u}}c_{rrl}}(\lambda;\boldsymbol{\theta}) &= \frac{\partial \mathrm{vec}'\left[\mathbf{G}_{\mathbf{yy}}(\lambda)\right]}{\partial \boldsymbol{\theta}_{\mathbf{u}}}\left[\mathbf{G}_{\mathbf{yy}}^{-1}(\lambda)\otimes\mathbf{G}_{\mathbf{yy}}'^{\ 1}(\lambda)\right]\frac{\partial \mathrm{vec}\left[\mathbf{G}_{\mathbf{yy}}(\lambda)\right]}{\partial \mathbf{c}_{rrl}'} \\
&= \frac{\partial \mathrm{vecd}'[\mathbf{G}_{\mathbf{xx}}(\lambda)]}{\partial \boldsymbol{\theta}_{\mathbf{u}}}\mathbf{E}_{N}'\left[\mathbf{G}_{\mathbf{yy}}^{-1}(\lambda)\otimes\mathbf{G}_{\mathbf{yy}}'^{-1}(\lambda)\right] \\
&\quad\times\left\{\begin{array}{c}\left[\mathrm{e}^{\mathrm{i}l\lambda}\mathbf{C}\left(\mathrm{e}^{-\mathrm{i}\lambda}\right)\mathbf{G}_{\mathbf{xx}}(\lambda)\otimes\mathbf{I}_{N}\right] \\ +\mathbf{K}_{NN}\left[\mathrm{e}^{-\mathrm{i}l\lambda}\mathbf{C}\left(\mathrm{e}^{\mathrm{i}\lambda}\right)\mathbf{G}_{\mathbf{xx}}(\lambda)\otimes\mathbf{I}_{N}\right]\end{array}\right\}\mathfrak{e}_{rr} \\
&= \frac{\partial \mathrm{vecd}'[\mathbf{G}_{\mathbf{xx}}(\lambda)]}{\partial \boldsymbol{\theta}_{\mathbf{u}}}\mathbf{E}_{N}'\left[\begin{array}{c}\mathrm{e}^{\mathrm{i}l\lambda}\mathbf{G}_{\mathbf{yy}}^{-1}(\lambda)\mathbf{C}\left(\mathrm{e}^{-\mathrm{i}\lambda}\right)\mathbf{G}_{\mathbf{xx}}(\lambda)\otimes\mathbf{G}_{\mathbf{yy}}'^{-1}(\lambda) \\ +\mathrm{e}^{-\mathrm{i}l\lambda}\mathbf{G}_{\mathbf{yy}}'^{-1}(\lambda)\mathbf{C}\left(\mathrm{e}^{\mathrm{i}\lambda}\right)\mathbf{G}_{\mathbf{xx}}(\lambda)\otimes\mathbf{G}_{\mathbf{yy}}^{-1}(\lambda)\end{array}\right]\mathfrak{e}_{rr}
\end{aligned}
$$

where we have used the properties of the diagonalisation and commutation matrices, and in particular, that $\mathbf{E}'_m \mathbf{K}_{mm} = \mathbf{E}'_m$. In fact, further simplification can be achieved by exploiting (A.1). The formulae for the remaining elements are entirely analogous. In this regard, it is important to note that all the above expressions can be written as the sum of some matrix and its complex conjugate transpose, as one would expect given that the information matrix is real. If we assume that both $\mathbf{G}_{\mathbf{xx}}(\lambda)$ and $\mathbf{G}_{\mathbf{uu}}(\lambda)$ are strictly positive, we can use again the Woodbury formula to considerably simplify the previous expressions.

Given that

$$\begin{aligned} \mathbf{G}_{\mathbf{yy}}^{-1}(\lambda_j) &= \left[\mathbf{G}_{\mathbf{uu}}^{-1}(\lambda) - \mathbf{G}_{\mathbf{uu}}^{-1}(\lambda)\mathbf{C}(e^{-i\lambda})\mathbf{\Omega}(\lambda)\mathbf{C}'(e^{i\lambda})\mathbf{G}_{\mathbf{uu}}^{-1}(\lambda)\right], \\ \mathbf{G}_{\mathbf{yy}}'^{-1}(\lambda_j) &= \left[\mathbf{G}_{\mathbf{uu}}^{-1}(\lambda) - \mathbf{G}_{\mathbf{uu}}^{-1}(\lambda)\mathbf{C}(e^{i\lambda})\mathbf{\Omega}'(\lambda)\mathbf{C}'(e^{-i\lambda})\mathbf{G}_{\mathbf{uu}}^{-1}(\lambda)\right] \end{aligned}$$

we will have that

$$\begin{aligned} \mathbf{C}'(e^{i\lambda})\mathbf{G}_{\mathbf{yy}}^{-1}(\lambda) &= \mathbf{C}'(e^{i\lambda})\mathbf{G}_{\mathbf{uu}}^{-1}(\lambda) - \mathbf{C}'(e^{i\lambda})\mathbf{G}_{\mathbf{uu}}^{-1}(\lambda)\mathbf{C}(e^{-i\lambda})\mathbf{\Omega}(\lambda)\mathbf{C}'(e^{i\lambda})\mathbf{G}_{\mathbf{uu}}^{-1}(\lambda) \\ &= \mathbf{G}_{\mathbf{xx}}^{-1}(\lambda)\mathbf{\Omega}(\lambda)\mathbf{C}'(e^{i\lambda})\mathbf{G}_{\mathbf{uu}}^{-1}(\lambda) \\ \mathbf{C}'(e^{-i\lambda})\mathbf{G}_{\mathbf{yy}}'^{-1}(\lambda_j) &= \mathbf{C}'(e^{-i\lambda})\mathbf{G}_{\mathbf{uu}}^{-1}(\lambda) - \mathbf{C}'(e^{-i\lambda})\mathbf{G}_{\mathbf{uu}}^{-1}(\lambda)\mathbf{C}(e^{i\lambda})\mathbf{\Omega}'(\lambda)\mathbf{C}'(e^{-i\lambda})\mathbf{G}_{\mathbf{uu}}^{-1}(\lambda) \\ &= \mathbf{G}_{\mathbf{xx}}^{-1}(\lambda)\mathbf{\Omega}'(\lambda)\mathbf{C}'(e^{-i\lambda})\mathbf{G}_{\mathbf{uu}}^{-1}(\lambda) \end{aligned}$$

where we have used the fact that

$$\mathbf{C}'(e^{i\lambda})\mathbf{G}_{\mathbf{uu}}^{-1}(\lambda)\mathbf{C}(e^{-i\lambda})\mathbf{\Omega}(\lambda) = \mathbf{I}_{R+1} - \mathbf{G}_{\mathbf{xx}}^{-1}(\lambda)\mathbf{\Omega}(\lambda)$$

and

$$\mathbf{C}'(e^{-i\lambda})\mathbf{G}_{\mathbf{uu}}^{-1}(\lambda)\mathbf{C}(e^{i\lambda})\mathbf{\Omega}'(\lambda) = \mathbf{I}_{R+1} - \mathbf{G}_{\mathbf{xx}}^{-1}(\lambda)\mathbf{\Omega}'(\lambda)$$

As a result, and

$$\begin{aligned} \mathbf{C}'(e^{i\lambda})\mathbf{G}_{\mathbf{yy}}^{-1}(\lambda_j)\mathbf{C}(e^{-i\lambda}) &= \mathbf{G}_{\mathbf{xx}}^{-1}(\lambda)\mathbf{\Omega}(\lambda)\mathbf{C}'(e^{i\lambda})\mathbf{G}_{\mathbf{uu}}^{-1}(\lambda)\mathbf{C}(e^{-i\lambda}), \\ \mathbf{G}_{\mathbf{xx}}(\lambda)\mathbf{C}'(e^{i\lambda})\mathbf{G}_{\mathbf{yy}}^{-1}(\lambda) &= \mathbf{\Omega}(\lambda)\mathbf{C}'(e^{i\lambda})\mathbf{G}_{\mathbf{uu}}^{-1}(\lambda) \\ \mathbf{G}_{\mathbf{xx}}(\lambda)\mathbf{C}'(e^{-i\lambda})\mathbf{G}_{\mathbf{yy}}'^{-1}(\lambda_j) &= \mathbf{\Omega}'(\lambda)\mathbf{C}'(e^{-i\lambda})\mathbf{G}_{\mathbf{uu}}^{-1}(\lambda) \end{aligned}$$

and

$$\mathbf{G_{xx}}(\lambda)\mathbf{C}'\left(e^{i\lambda}\right)\mathbf{G}_{\mathbf{yy}}^{-1}\left(\lambda_j\right)\mathbf{C}\left(e^{-i\lambda}\right)\mathbf{G_{xx}}(\lambda) = \mathbf{\Omega}(\lambda)\mathbf{C}'\left(e^{i\lambda}\right)\mathbf{G}_{\mathbf{uu}}^{-1}(\lambda)\mathbf{C}\left(e^{-i\lambda}\right)\mathbf{G_{xx}}(\lambda)$$

In addition, the special structure of $\mathbf{C}(z)$ in Eq. (6) can also be successfully exploited to speed up the calculations. In particular,

$$\mathbf{C}'\left(e^{i\lambda}\right)\mathbf{G}_{\mathbf{uu}}^{-1}(\lambda)\mathbf{C}\left(e^{i\lambda}\right) = \mathbf{\Omega}^{-1}(\lambda) - \mathbf{G}_{\mathbf{xx}}^{-1}(\lambda)$$

where $\mathbf{\Omega}^{-1}(\lambda)$ has been defined in Eq. (12). Further speed gains can be achieved by noticing that

$$\mathbf{c}'_{rr}\left(e^{i\lambda}\right)\mathbf{G}_{\mathbf{uu}}^{-1}(\lambda)\mathbf{c}_{rr}\left(e^{-i\lambda}\right) = \sum_{j\in N_r} \frac{\|c_j\left(e^{i\lambda}\right)\|^2}{G_{u_j u_j}(\lambda)}$$

APPENDIX C. STATE-SPACE REPRESENTATION OF DYNAMIC BIFACTOR MODELS WITH AR(1) FACTORS

There are several ways of casting the dynamic factor model in Eq. (4) into state-space format, but the most straightforward one is to consider a state vector of dimension $2(R+1)+N$ in which the AR(1) processes for both global and regional factors are written as a bivariate VAR(1) in (x_t, x_{t-1}), and the N AR(1) processes for the specific factors are written as first order ARS in u_{it}. As a result, we can write the measurement equation without an error term as

$$\mathbf{y}_t = \mathbf{Z}\boldsymbol{\alpha}_t$$

where the state vector is

$$\begin{aligned}
\boldsymbol{\alpha}_t &= \left(\mathbf{x}'_t, \mathbf{x}'_{t-1}, \mathbf{u}'_t\right)' \\
\mathbf{x}_t &= \left(x_{gt}, x_{1t}, \ldots, x_{Rt}\right)' \\
\mathbf{u}_t &= \left(u_{1t}, \ldots, u_{it}, \ldots, u_{Nt}\right)'
\end{aligned}$$

and $\mathbf{Z}$ is the $N \times (N+2R+2)$ matrix.

$$\mathbf{Z} = [\mathbf{C}_0 | \mathbf{C}_1 | \mathbf{I}_N]$$

with $\mathbf{C}_0, \mathbf{C}_1$ being $N \times (R+1)$ sparse matrices of contemporaneous and lagged loadings.

Consequently, the transition equation is simply

$$\begin{bmatrix} \mathbf{x}_t \\ \mathbf{x}_{t-1} \\ \mathbf{u}_t \end{bmatrix} = \begin{bmatrix} \boldsymbol{\rho}_\mathbf{x} & \mathbf{0} & \mathbf{0} \\ \mathbf{I}_{R+1} & \mathbf{0} & \mathbf{0} \\ \mathbf{0} & \mathbf{0} & \boldsymbol{\rho}_\mathbf{u} \end{bmatrix} \begin{bmatrix} \mathbf{x}_{t-1} \\ \mathbf{x}_{t-2} \\ \mathbf{u}_{t-1} \end{bmatrix} + \begin{bmatrix} \mathbf{f}_t \\ \mathbf{0} \\ \mathbf{v}_t \end{bmatrix}$$

with

$$\begin{aligned} \boldsymbol{\rho}_\mathbf{x} &= \text{diag}\left(\rho_{x_g}, \rho_{x_1}, \ldots, \rho_{x_R}\right), \\ \boldsymbol{\rho}_\mathbf{u} &= \text{diag}\left(\rho_{u_1}, \ldots, \rho_{u_N}\right) \end{aligned}$$

$$\begin{aligned} \text{Cov}(\mathbf{f}_t) &= \mathbf{I}_{R+1}, \\ \text{Cov}(\mathbf{v}_t) &= \boldsymbol{\Psi} = \text{diag}\left(\psi_1, \ldots, \psi_N\right) \end{aligned}$$

Given our covariance stationarity conditions, the initial condition for the state variables will trivially be $\boldsymbol{\alpha}_{1|0} = \mathbf{0}_{(N+2R+2)x1}$, and

$$\mathbf{P}_{1|0} = \begin{bmatrix} \mathbf{Q}_{\mathbf{x}0} & \mathbf{Q}_{\mathbf{x}1} & \mathbf{0} \\ \mathbf{Q}_{\mathbf{x}1} & \mathbf{Q}_{\mathbf{x}0} & \mathbf{0} \\ \mathbf{0} & \mathbf{0} & \mathbf{Q}_{\mathbf{u}0} \end{bmatrix}$$

where $\mathbf{Q}_{\mathbf{x}0}$ and $\mathbf{Q}_{\mathbf{u}0}$ are diagonal matrices with the unconditional variance of the corresponding AR(1) processes along the main diagonal, while $\mathbf{Q}_{\mathbf{x}1}$ is also diagonal with the first autocovariance of the global and regional factors AR(1) processes on the main diagonal.

COUNTRY SHOCKS, MONETARY POLICY EXPECTATIONS AND ECB DECISIONS. A DYNAMIC NON-LINEAR APPROACH

Maximo Camacho[a], Danilo Leiva-Leon[b] and Gabriel Perez-Quiros[c,d]

[a]*University of Murcia, Murcia, Spain*
[b]*Central Bank of Chile, Santiago, Chile*
[c]*Bank of Spain, Madrid, Spain*
[d]*CEPR, London, UK*

ABSTRACT

Previous studies have shown that the effectiveness of monetary policy depends, to a large extent, on the market expectations of its future actions. This paper proposes an econometric framework to address the effect of the current state of the economy on monetary policy expectations. Specifically, we study the effect of contractionary (or expansionary) demand (or supply) shocks hitting the euro area countries on the expectations of the ECB's monetary policy in two stages. In the first stage, we construct indexes of real activity and inflation dynamics for each country,

Dynamic Factor Models
Advances in Econometrics, Volume 35, 283–316

ISSN: 0731-9053/doi:10.1108/S0731-905320150000035007

based on soft and hard indicators. In the second stage, we use those indexes to provide assessments on the type of aggregate shock hitting each country and assess its effect on monetary policy expectations at different horizons. Our results indicate that expectations are responsive to aggregate contractionary shocks, but not to expansionary shocks. Particularly, contractionary demand shocks have a negative effect on short-term monetary policy expectations, while contractionary supply shocks have negative effect on medium- and long-term expectations. Moreover, shocks to different economies do not have significantly different effects on expectations, although some differences across countries arise.

Keywords: Business cycles; inflation cycles; monetary policy

JEL classifications: E32; C22; E27

1. INTRODUCTION

Since the seminal work of Taylor (1993), many papers have tried to relate the endogenous component of monetary policy with the different shocks that affect the economy. Several versions of the well-known Taylor rule, where decisions on interest rates are related to inflation developments and output gap, have been estimated in reduced form for different countries and sample periods. However, the use of these rules in general equilibrium models and the importance that the identification of the shocks has on the effects of monetary policy have changed the focus, from simply imposing identifying restrictions on impulse response functions to monetary shocks to carefully analyze the monetary equation.

Among many others, Leeper, Sims, and Zha (1996), Leeper and Zha (2003), and Sims and Zha (2006a), or very recently, Arias, Caldara, and Rubio Ramirez (2015) estimate behavioral relations of monetary policy decisions in the context of an endogenous relation with the rest of the variables that describe the economic conditions. In particular, Leeper et al. (1996) show that most of the movements in monetary policy instruments are responses to the state of the economy, not random behavior of the monetary authorities. Leeper and Zha (2003) analyze the effect of modest policy interventions within frameworks where agents perceive that policy is composed of a regular response to the state of the economy and a random part. In a switching framework, Sims and Zha (2006b) show that most of the variation in policy variables reflects the systematic part of monetary policy in response to the changing state of the economy. Arias et al. (2015) find that, imposing sign and zero restrictions on the systematic component of monetary policy, there exist a contractionary effect of an exogenous increase in the fed funds rate.

In the case of the euro area, the attention has been focused on the transmission of monetary policy shocks, that obviously implies a proper specification of the monetary policy rule although some papers have concentrated on estimating carefully the policy rule. To quote a few, Thanassis and Elias (2011) use a threshold model to quantify the attitude of the ECB toward inflation expectations reductions; Dieppe, Kuster, and McAdam (2004) conclude that forecast-based rules are more optimal than outcome-based policies; and Stracca (2007) shows that a "speed limit" monetary policy rule, which relates decisions to output gap changes (not to the level), performs well as a guideline for policy in the euro area.

Notably, all of these contributions consider the euro area as a whole. Therefore, these approaches are precluded from capturing the implications of the monetary policy rule for each of the euro-area members. A significant exception to these aggregate approaches is Benigno and Lopez-Salido (2006), who show the heterogeneity in the dynamics of inflation in each of the euro-area members. These authors find that maintaining an aggregate HICP targeting rule is still optimal, although it could imply significant welfare losses for some countries and welfare gains for others.

Our paper relates to Benigno and Lopez-Salido (2006) because we focus on the idiosyncratic dynamics of some euro-area members: Germany, France, Italy, and Spain. However, we focus on determining which country-specific shocks (demand or supply) have more influence on the determination of the final decisions of the ECB. Following Aruoba and Diebold (2010), we identify the shocks by analyzing the interaction between activity and inflation cycles. Within this framework, we construct indexes of real activity and inflation dynamics for each country, based on soft and hard indicators. Then, we use those indexes to assess the effect of aggregate shocks on monetary policy expectations at different horizons. To identify the shocks and compute their impact on expectations, we propose a multi-state regime-switching framework that assesses the relationship between observed variables (expectations) and latent variables (shocks).

In the empirical analysis, we find that monetary policy expectations are responsive to aggregate contractionary shocks, but not to expansionary shocks. In addition, we find that negative demand shocks affect short-term expectations of interest rates, but that negative supply shocks have medium and long-run impact. Finally, we find that these results are robust across countries in the case of demand shocks, while we find more heterogeneity in the case of supply shocks. Supply shocks are more related to expectation of future rates in this case of Germany, France, and Italy than in the case of Spain.

The structure of this paper is as follows. Section 2 develops indexes of real activity and inflation dynamics for euro area countries. Section 3 assesses the state of the economy and the aggregate shocks. Section 4 studies the effect of aggregate shocks on monetary policy expectations. Section 5 concludes.

2. REAL ACTIVITY AND INFLATION CYCLES

2.1. The Model

The purpose of this section is to propose a method to construct indexes of real activity and inflation dynamics for the four main economies of the euro area (Germany, France, Italy, and Spain) from hard and soft economic indicators. To handle real activity and inflation developments in a specific country, we estimate both indexes simultaneously from a unified framework, allowing for potential interdependence between the two concepts.

Following Leiva-Leon (2015), we use a set of N economic real activity and inflation indicators, y_{it}, which are collected in the vector y_t, to extract two common factors. The first factor, $f_{a,t}$, captures the evolution of the real activity, while the second factor, $f_{b,t}$, captures the inflation dynamics. For each of the four countries, the factor structure would read as

$$y_{it} = \gamma_i^a f_{a,t} + \gamma_i^b f_{b,t} + e_{it} \tag{1}$$

where γ_i^a and γ_i^b refer to the factor loadings and $i = 1, 2\ldots, N$. The factors, $f_{a,t}$ and $f_{b,t}$, and the idiosyncratic terms, e_{it}, are assumed to evolve according to the following autoregressive dynamics:

$$f_{r,t} = \sum_{h=1}^{k} b_{rh} f_{r,t-h} + \omega_t^r \tag{2}$$

$$e_{it} = \sum_{h=1}^{m} \phi_{ih} e_{it-h} + \varepsilon_{it} \tag{3}$$

where the errors, ω_t^r, are distributed as $N(0, \sigma_r^2)$, $r = a, b$, and ε_{it} are distributed as $N(0, \sigma_i^2)$, with $i = 1, \ldots, N$.[1] Finally, all the shocks, ε_{it} and ω_t^r, are assumed to be mutually uncorrelated in cross-section and time-series dimensions. Leiva-Leon (2015) shows that this model can easily be stated in state-space form and can easily be estimated by means of a Kalman filter. In the empirical application, we set $r = m = 2$.

The state-space specification of the model is the following. The measurement equation is

$$y_t = H\beta_t \tag{4}$$

and the transition equation is

$$\beta_t = F\beta_{t-1} + \upsilon_t, \quad \upsilon_t \sim i.i.d.N(0, Q) \tag{5}$$

where $\beta_t = (f_{a,t}, f_{a,t-1}, f_{b,t}, f_{b,t-1}, e_{1,t}, e_{1,t-1}, \ldots, e_{N,t}, e_{N,t-1})'$, H contains the restrictions in Eq. (1). F contains the restrictions in Eqs. (2) and (3) and Q is the variance–covariance matrix of $(\omega_t^1, \omega_t^2, \varepsilon_{1,t} \ldots \varepsilon_{N,t})'$, enlarged with 0s to take into accounts the identities in F.

We apply the Kalman filter to extract optimal the inference on the state vector β_t. For this purpose, we compute the prediction equations as

$$\beta_{t|t-1} = F\beta_{t-1|t-1} \tag{6}$$

$$P_{t|t-1} = FP_{t-1|t-1}F' + Q \tag{7}$$

$$\eta_{t|t-1} = y_t - H\beta_{t|t-1} \tag{8}$$

$$f_{t|t-1} = HP_{t|t-1}H' + R \tag{9}$$

and the updating equations as

$$\beta_{t|t} = \beta_{t|t-1} + P_{t|t-1}H'\left[f_{t|t-1}\right]^{-1}\eta_{t|t-1} \tag{10}$$

$$P_{t|t} = \left(I - P_{t|t-1}H'\left[f_{t|t-1}\right]^{-1}H\right)P_{t|t-1} \tag{11}$$

In this paper, we consider that only two factors are required to describe the dynamics of all the economic indicators. Using only two factors goes in line with the literature of small-scale models, where the common dynamics of a small set of economic variables can successfully be described with one dynamic factor, usually related to activity, as in Aruoba and Diebold (2010), Aruoba, Diebold, and Scotti (2009), or Camacho and Martinez-Martin (2014), among many others. When the activity variables are complemented with price indicators, it makes sense to consider an additional factor, which is expected to capture the common inflation dynamics. Assuming more than two factors would create problems in the interpretation of the results and would difficult the identification assumptions.

2.2. Data

We estimate the factor model in Eqs. (1)–(3) for each of the four main economies of the euro area. For each country, we select the same four monthly indicators of real activity and the same four indicators of inflation. For the case of real activity, we use three hard indicators, Industrial Production (IP), Retail Sales (SALES) and Registered Unemployment (UNEM). In addition, we use one soft indicator, Purchase Manager Index (PMI). The selection of the variables follows Stock and Watson (1991) since it is the more parsimonious representation mimicking the way in which national accounts are constructed: one time series from the supply side, one time series from the demand side and one indicator of employment, which we substitute for unemployment since employment is not available at monthly frequency for all countries.[2] In addition, to capture expectations, we use one of the most popular expectation series available for all countries, the PMI index.

For the case of price indicators, we use two hard indicators, Consumer Price Index (CPI) and Producer Price Index (PPI), and two soft indicators, Selling Price Expectations (EXPE) and 12-months price trends (TREN). The last two indicators are based on surveys and computed by the European Commission. Again, the indicators cover the most representative series for both price developments and expectations. To avoid unit root problems, we take the first log differences to all the hard indicators and the first differences to all the soft indicators. To relate our results to the ECB's monetary policy, our sample period goes from January 1999 to April 2014. Since the specifications are linear small-scale dynamic factor models, we estimate the parameters by maximum likelihood.

2.3. Factors' Dynamics

Since both factors are estimated simultaneously from the same set of real activity and inflation indicators, we impose an identification restriction on the loading factors of all countries. The restriction relies on the hypothesis of money neutrality, which postulates that changes in the stock of money affect only nominal variables and do not affect real (inflation-adjusted) variables. Therefore, we do not allow the factor $f_{a,t}$ to be loaded by the most representative indicator of inflation, CPI. Consequently, we label $f_{a,t}$ as the real activity factor and $f_{b,t}$ as the inflation factor.

Prior to using the factors, we develop the following robustness check. First, we extract the real activity factor from a model that uses only the real activity indicators, $\tilde{f}_{a,t}$, and the inflation factor, $\tilde{f}_{b,t}$, from a model that uses only the inflation indicators. Second, we run OLS regressions to assess the explanatory power of the factors estimated from the separated models on the factors estimated from the unified model. Specifically, we regress $f_{a,t}$ on $\tilde{f}_{a,t}$, obtaining the standard goodness-of-fit measure $R^2_{a,a}$, and then we regress $f_{a,t}$ on $\tilde{f}_{b,t}$, obtaining a fitting measure of $R^2_{a,b}$. If the factor $f_{a,t}$ is properly identified, $R^2_{a,a}$ should be higher than $R^2_{a,b}$. The analogous procedure is performed for $f_{b,t}$, which would be properly identified when $R^2_{b,b} > R^2_{b,a}$. We ran this robustness check for the models of each country and we found that the factors were properly labeled in all the cases.[3]

For each country, the estimated factors of real activity and inflation are plotted in Fig. 1. Some features deserve attention. First, there are strong time variations in the comovements across these variables. Using a five-year window for all the countries, we find that the comovements vary between a maximum of +0.51 and a minimum of −0.60, with positive and negative numbers for all the countries. In addition, there are also changes in the leading and lagging behavior of the variables. For some periods, the highest cross-correlation is the contemporaneous correlation, but sometimes the maximum is captured with up to six lags of leading between real activity and inflation. Second, real activity factors decrease during the euro area recession, especially for Italy and Germany. Third, the inflation factors decrease during the last part of the sample, especially for Italy and Spain. Fourth, the lack of recovery in Spain after the 2008–2009 recession is remarkable, leading to a double dip recession.

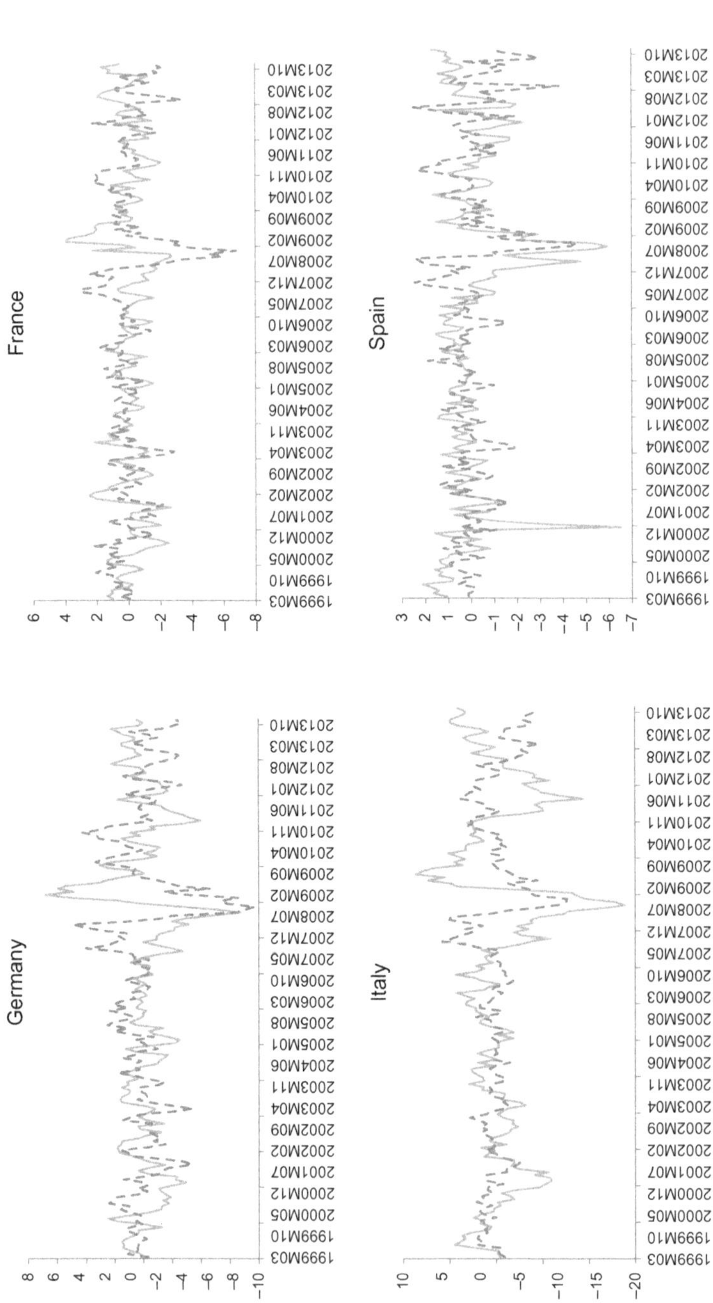

Fig. 1. Real Activity and Inflation Indexes. *Note*: The figure plots the real activity index (solid line) and the inflation index (dashed line) for each country.

3. ASSESSING THE STATE OF THE ECONOMY

3.1. The Model

To account for the interactions between high and low real activity and inflation regimes, we rely on the framework proposed by Leiva-Leon (2014). Specifically, assume that the autocorrelation in the dynamics of the factors can be approximated by regime-switching mean Markov-switching specification. Accordingly, we consider the following tractable bivariate two-state Markov-switching model:

$$\begin{bmatrix} f_{a,t} \\ f_{b,t} \end{bmatrix} = \begin{bmatrix} \mu_{a,0} + \mu_{a,1} S_{a,t} \\ \mu_{b,0} + \mu_{b,1} S_{b,t} \end{bmatrix} + \begin{bmatrix} \varepsilon_{a,t} \\ \varepsilon_{b,t} \end{bmatrix} \tag{12}$$

where

$$\begin{bmatrix} \varepsilon_{a,t} \\ \varepsilon_{b,t} \end{bmatrix} \sim i.i.d.N\left(\begin{bmatrix} 0 \\ 0 \end{bmatrix}, \begin{bmatrix} \sigma_a^2 & \sigma_{ab} \\ \sigma_{ab} & \sigma_b^2 \end{bmatrix} \right) \tag{13}$$

In this expression, the state variable $S_{k,t}$ indicates that the common factor f_{kt} is in regime 0 with a mean equal to $\mu_{k,0}$, when $S_{k,t} = 0$, and that f_{kt} is in regime 1 with a mean equal to $\mu_{k,0} + \mu_{k,1}$, when $S_{k,t} = 1$, for $k = a, b$. Moreover, we assume that $S_{a,t}$ and $S_{b,t}$ evolve as irreducible two-state Markov chains, whose transition probabilities are given by

$$\Pr\left(S_{k,t} = j | S_{k,t-1} = i\right) = p_{k,ij} \tag{14}$$

for $i, j = 0, 1$ and $k = a, b, ab$.

Within this framework, we define three Markov processes to capture the dynamics of the unobserved state. The first Markov process, $S_{a,t}$, captures the dynamics of the economic activity. The second Markov process, $S_{b,t}$, captures the dynamics of inflation. In addition, we also define a Markov process $S_{ab,t}$ that captures the dynamics of the factors in the case they would evolve with perfect synchronization. To account for the interrelation between $f_{a,t}$ and $f_{b,t}$, we allow for time-varying interdependence between $S_{a,t}$ and $S_{b,t}$. Specifically, the joint probability of the model's regimes is given by

$$\begin{aligned} \Pr\left(S_{a,t} = j, S_{b,t} = j\right) = {} & \Pr(V_t = 1)\Pr\left(S_{ab,t} = j\right) \\ & + (1 - \Pr(V_t = 1))\Pr\left(S_{a,t} = j\right)\Pr\left(S_{b,t} = j\right) \end{aligned} \tag{15}$$

where

$$V_t = \begin{cases} 0 & \text{If } S_{a,t} \text{ and } S_{b,t} \text{ are unsynchronized} \\ 1 & \text{If } S_{a,t} \text{ and } S_{b,t} \text{ are synchronized} \end{cases} \tag{16}$$

and the latent variable V_t also evolves according to an irreducible two-state Markov chain whose transition probabilities are given by

$$\Pr(V_t = j_v | V_{t-1} = i_v) = p_{ij,v} \tag{17}$$

for $ij, v = 0, 1$. Interestingly, the joint dynamics of $S_{a,t}$ and $S_{b,t}$ are a weighted average between the extreme dependent and the extreme independent cases, where the weights assigned to each of them are endogenously determined by

$$\delta_t^{ab} = \Pr(V_t = 1) \tag{18}$$

Therefore, the term δ_t^{ab} can be interpreted as the time-varying degree of synchronization between $S_{a,t}$ and $S_{b,t}$. Since the likelihood function of this model is conditional on several states, the estimation of parameters obtained with the maximum likelihood approach could become cumbersome. Therefore, we rely on a Bayesian approach to estimate this model. This approach also allows us to provide a measure of uncertainty about the parameter estimates. The Gibbs sampler used in the estimation procedure, which is detailed in the appendix, can be summarized by iterating the following four steps:

Step 1: Generate the latent variables $S_{a,t}$, $S_{b,t}$, $S_{ab,t}$ and V_t, conditional on the factors and the vector of parameters, denoted by θ.

Step 2: Generate the transition probabilities associated with each latent variable, $p_{00,a}$, $p_{11,a}$, $p_{00,b}$, $p_{11,b}$, $p_{00,ab}$, $p_{11,ab}$, $p_{00,v}$, $p_{11,v}$, conditional on $S_{a,t}$, $S_{a,t}$, $S_{ab,t}$ and V_t.

Step 3: Generate the means associated with the factors, $\mu_{a,0}$, $\mu_{a,1}$, $\mu_{b,0}$, $\mu_{b,1}$, conditional on σ_a^2, σ_b^2, σ_{ab}, $S_{a,t}$, $S_{b,t}$, $S_{ab,t}$, V_t and the factors.

Step 4: Generate the variance–covariance matrix, with elements σ_a^2, σ_b^2, σ_{ab}, conditional on $\mu_{a,0}$, $\mu_{a,1}$, $\mu_{b,0}$, $\mu_{b,1}$, $S_{a,t}$, $S_{b,t}$, $S_{ab,t}$, V_t and the factors.

Table 1 presents the estimated coefficients of the model for all countries. The table shows that real fluctuations are higher in Italy and Spain than in France and Germany, with lower growth rates in recession and higher growth rates in expansion (see values of $\mu_{a,0}$ and $\mu_{a,1}$). However, prices

Table 1. Parameter Estimates Coefficients of Eq. (12).

	Germany	France	Italy	Spain
$\mu_{a,0}$	−2.56	−1.50	−5.12	−3.37
	(0.55)	(0.39)	(0.39)	(0.26)
$\mu_{a,1}$	2.89	1.65	6.62	3.66
	(0.46)	(0.39)	(0.40)	(0.26)
$\mu_{b,0}$	−3.74	−2.93	−3.03	−2.19
	(0.54)	(0.37)	(0.31)	(0.31)
$\mu_{b,1}$	4.10	3.25	3.81	2.40
	(0.50)	(0.35)	(0.31)	(0.30)
$p_{a,11}$	0.98	0.97	0.97	0.98
	(0.02)	(0.02)	(0.01)	(0.01)
$p_{a,00}$	0.84	0.79	0.91	0.84
	(0.08)	(0.09)	(0.04)	(0.08)
$p_{b,11}$	0.98	0.97	0.98	0.97
	(0.01)	(0.02)	(0.01)	(0.01)
$p_{b,00}$	0.83	0.79	0.91	0.80
	(0.08)	(0.08)	(0.05)	(0.09)
$p_{ab,11}$	0.99	0.98	0.98	0.98
	(0.01)	(0.01)	(0.01)	(0.01)
$p_{ab,00}$	0.81	0.78	0.85	0.80
	(0.09)	(0.09)	(0.07)	(0.09)
$p_{v,11}$	0.84	0.86	0.87	0.82
	(0.10)	(0.10)	(0.09)	(0.11)
$p_{v,00}$	0.92	0.88	0.96	0.90
	(0.08)	(0.10)	(0.03)	(0.07)
σ_a^2	2.28	1.15	5.90	0.89
	(0.33)	(0.16)	(0.65)	(0.10)
σ_b^2	2.26	1.09	2.31	0.82
	(0.30)	(0.14)	(0.26)	(0.10)
σ_{ab}	−0.25	−0.05	0.91	0.07
	(0.46)	(0.13)	(0.36)	(0.07)

Notes: Parameter estimates for the coefficients of the two means, of each factor, transition probabilities of the MS models and variance covariance matrices. Factor "*a*" relates to real activity, while factor "*b*" relates to inflation developments. Standard errors are in parenthesis.

oscillate more similarly across countries. Fig. 2 plots the inference on real activity (left-hand-side graphs) and inflation (right-hand-side graphs) regimes for all the countries. The results indicate high deflationary pressures during 2008–2009 for all countries and since the early 2013 for France, Italy, and Spain. Regarding real activity, the results indicate high probability of recession around 2000–2001 and 2007–2009 in all countries, and in 2011 (mainly) in Germany and Italy.

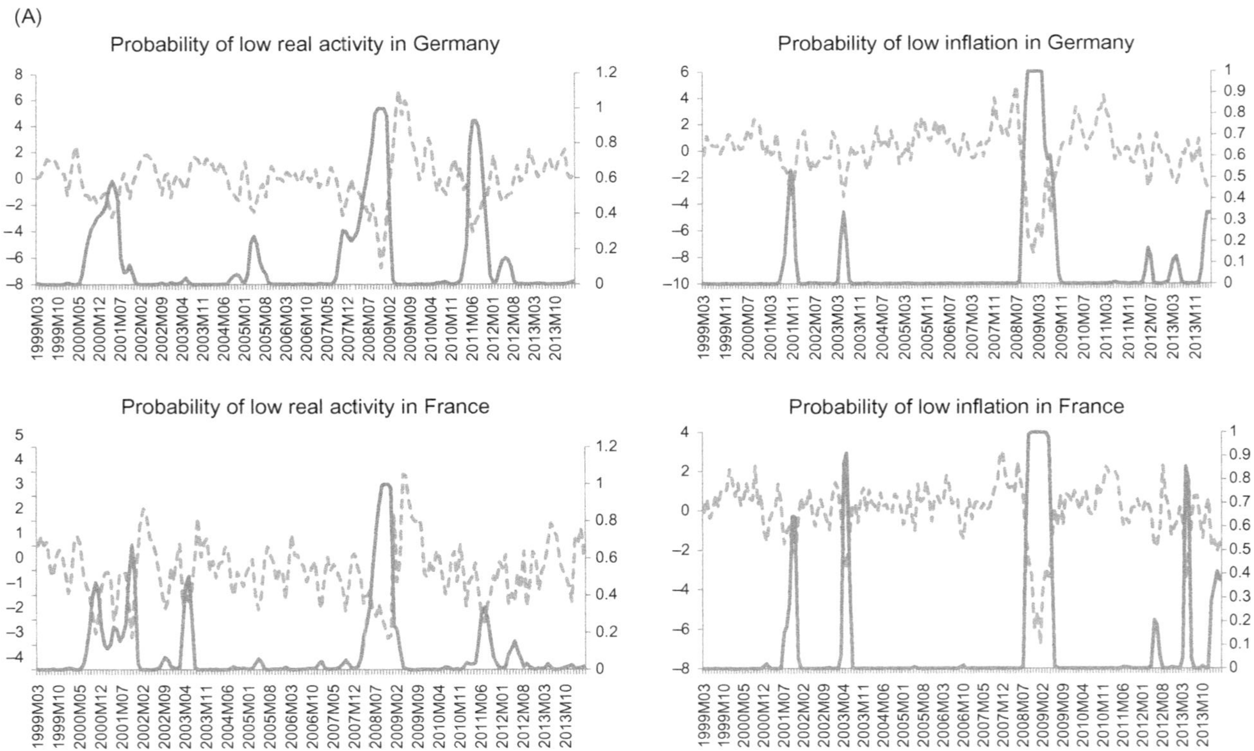

Fig. 2. Regime Inferences of Real Activity and Inflation. *Notes*: (a) Each chart plots the probability of low mean (solid line) and the corresponding index (dashed line); (b) top charts plot the probabilities of low real activity and of low inflation. Bottom charts show their synchronization.

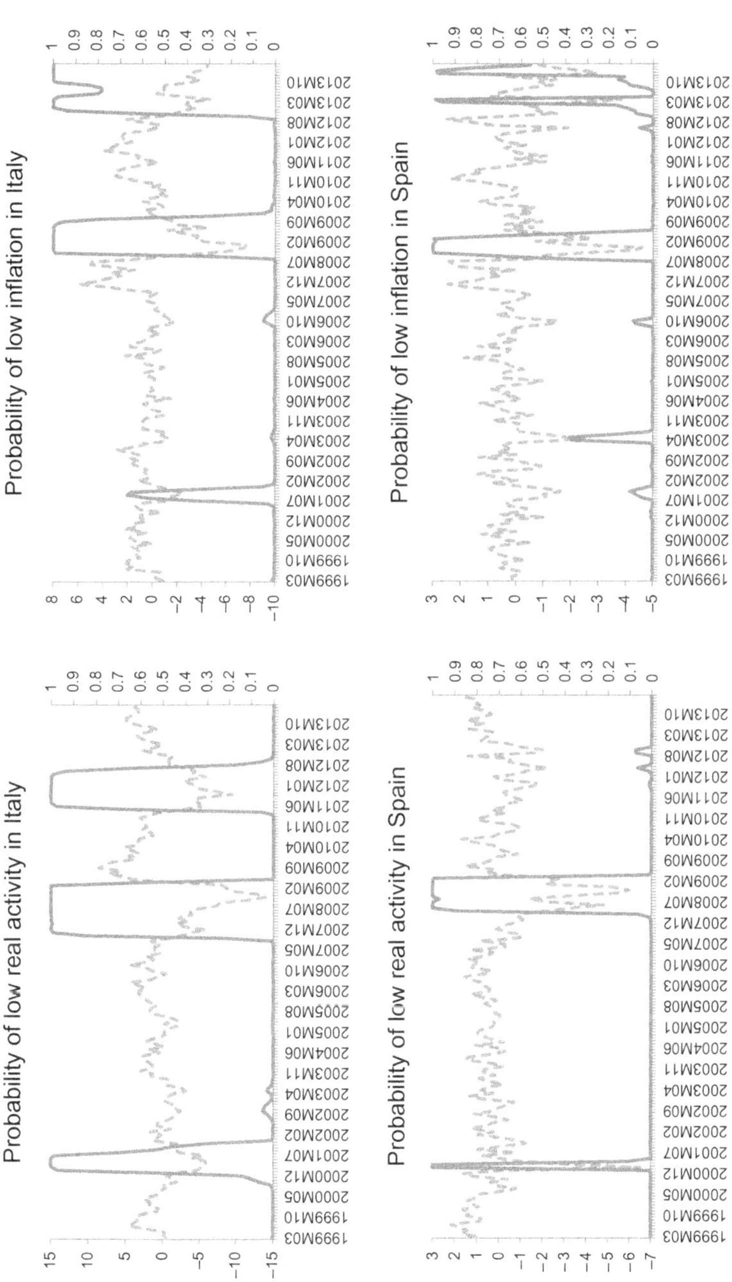

Fig. 2. (*Continued*)

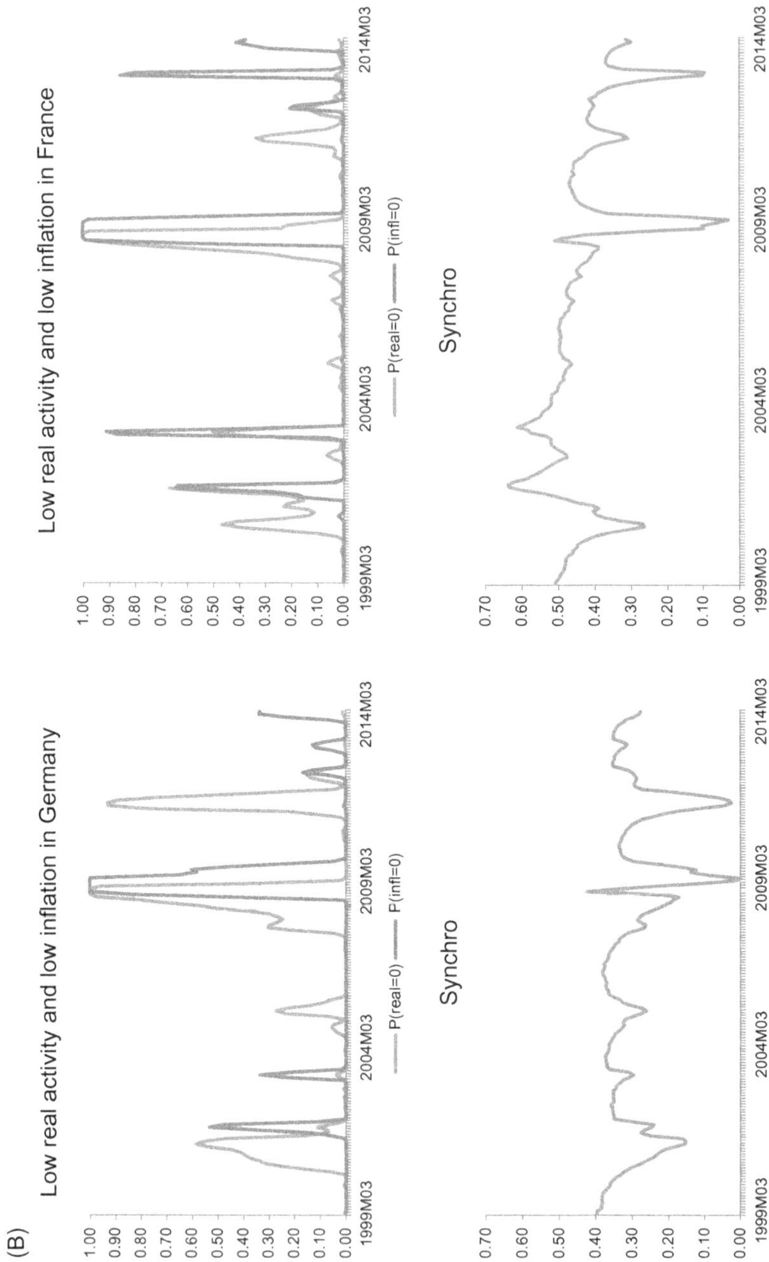

Fig. 2. (*Continued*)

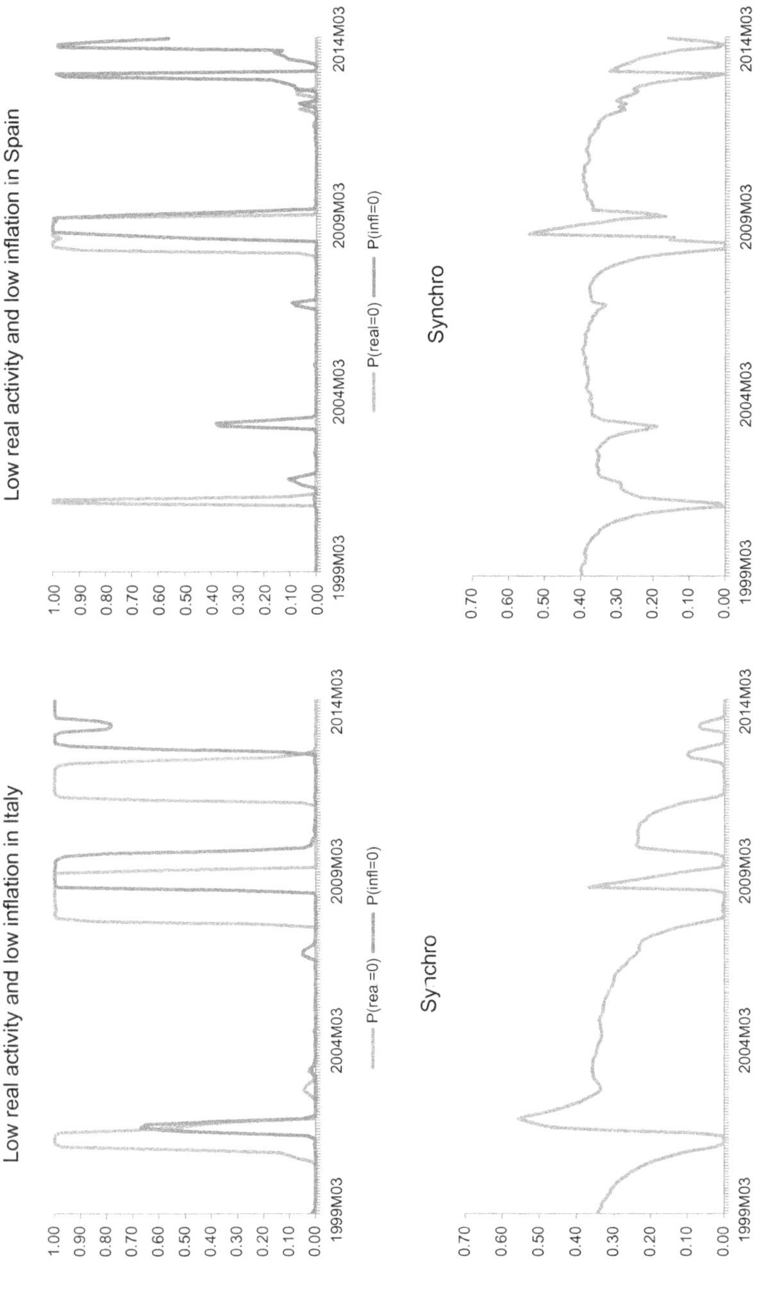

Fig. 2. (*Continued*)

Fig. 2(b) (lower row) plots the inference on the time-varying synchronization by country. As can be seen, the synchronicity changes over time, reaching up to 0.6 for France or 0.55 for Italy or 0 in other periods. France presents the highest synchronization of the real and nominal cycle with an average of δ_t^{ab} of 0.43, while Italy presents the lowest with 0.20. The changes in the synchronization over time will be the key to identify the nature of the shocks.

3.2. *Aggregate Demand and Supply Shocks*

Aruoba and Diebold (2010) showed that prices and quantities are related over the business cycle, and that the nature of this relationship contains information about the sources of shocks. While adverse demand shocks lead to periods of business cycle downturns and low inflation, adverse supply shocks lead to reductions in economic activity along with inflationary pressures. Equivalently, expansionary demand shocks increase economic activity and prices, and expansionary supply shocks lead to periods of business cycle upturns and low inflation.

Accordingly, inferences on contractionary versus expansionary and demand versus supply shocks can be computed from expression (15) as follows:

$$\Pr\left(S_{a,t}=1, S_{b,t}=1\right) \rightarrow \text{High real and high inflation} \rightarrow \text{Expansionary Demand}$$
$$\Pr\left(S_{a,t}=1, S_{b,t}=0\right) \rightarrow \text{High real and low inflation} \rightarrow \text{Expansionary Supply}$$
$$\Pr\left(S_{a,t}=0, S_{b,t}=1\right) \rightarrow \text{Low real and high inflation} \rightarrow \text{Contractionary Supply}$$
$$\Pr\left(S_{a,t}=0, S_{b,t}=0\right) \rightarrow \text{Low real and low inflation} \rightarrow \text{Contractionary Demand}$$

The results of the inferences on aggregate shocks for Germany, France, Italy, and Spain are shown in Figs. 3–6, respectively. For illustrative purposes, we include in these figures the changes in the ECB main refinancing operations, the minimum bid rate. Before analyzing the relation between the shocks and the ECB rates, it is interesting to analyze the role that the comovements have in explaining the evolution of the shocks.

Equation (15) basically states the total probability theorem applied to our specification. The probability of expansionary (or contractionary) demand (or supply) shocks is a weighted average of those shocks assuming

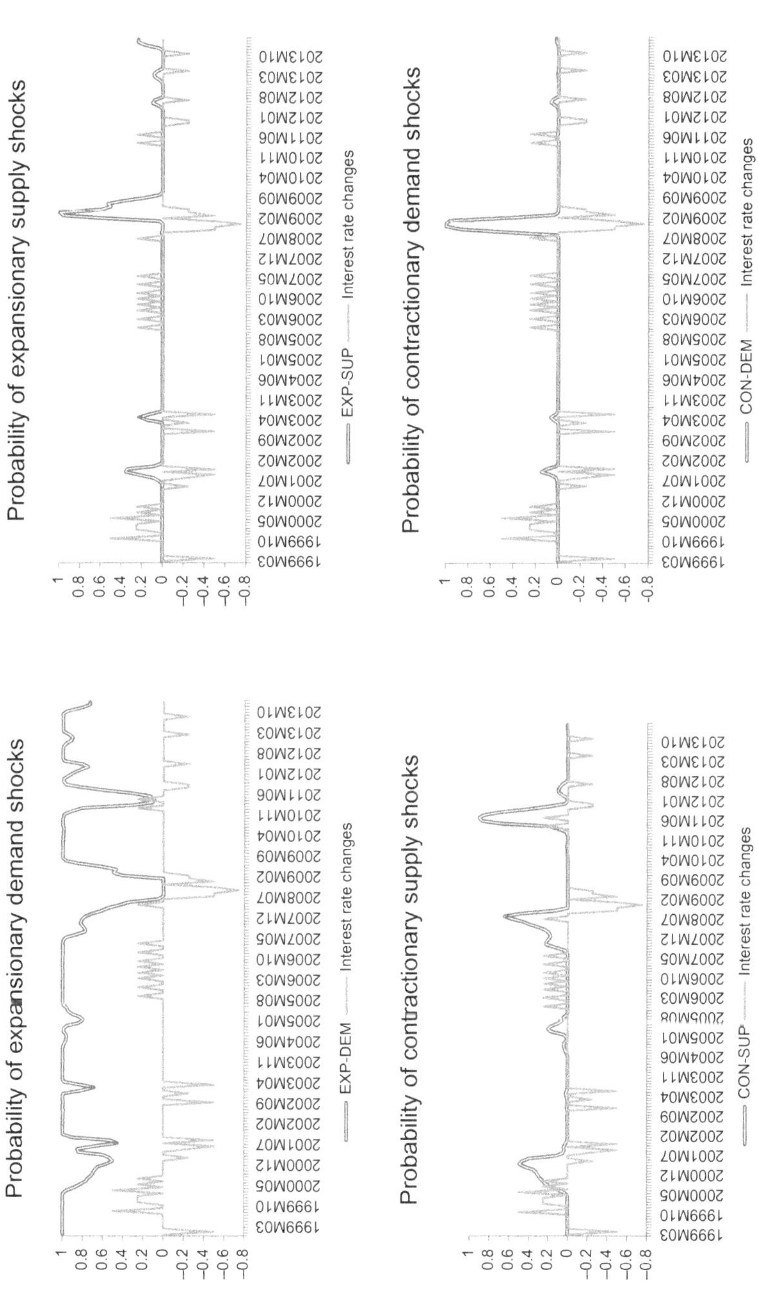

Fig. 3. Regime Inferences on Aggregate Shocks in Germany. *Note*: Each chart plots the probability of aggregate shocks (double line) and the main refinancing operations: minimum bid rate in first differenced (solid line).

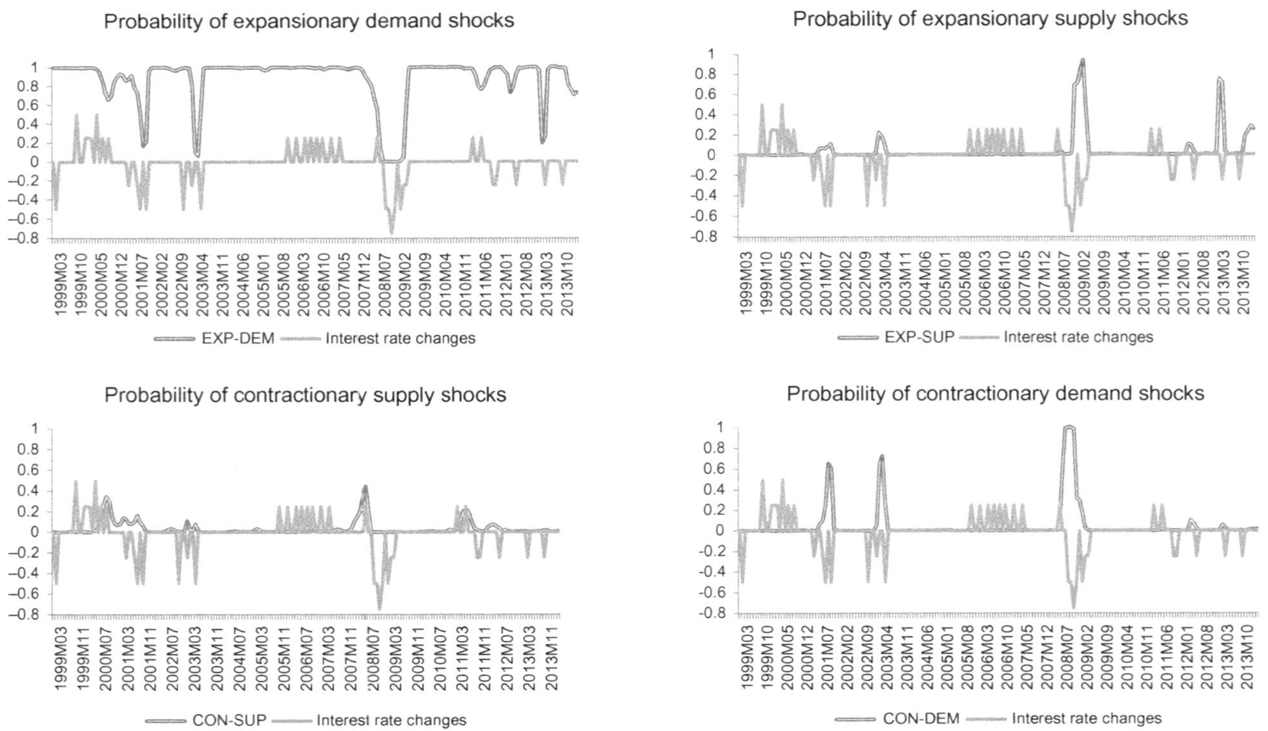

Fig. 4. Regime Inferences on Aggregate Shocks in France. *Note*: Each chart plots the probability of aggregate shocks (double line) and the main refinancing operations: minimum bid rate in first differenced (solid line).

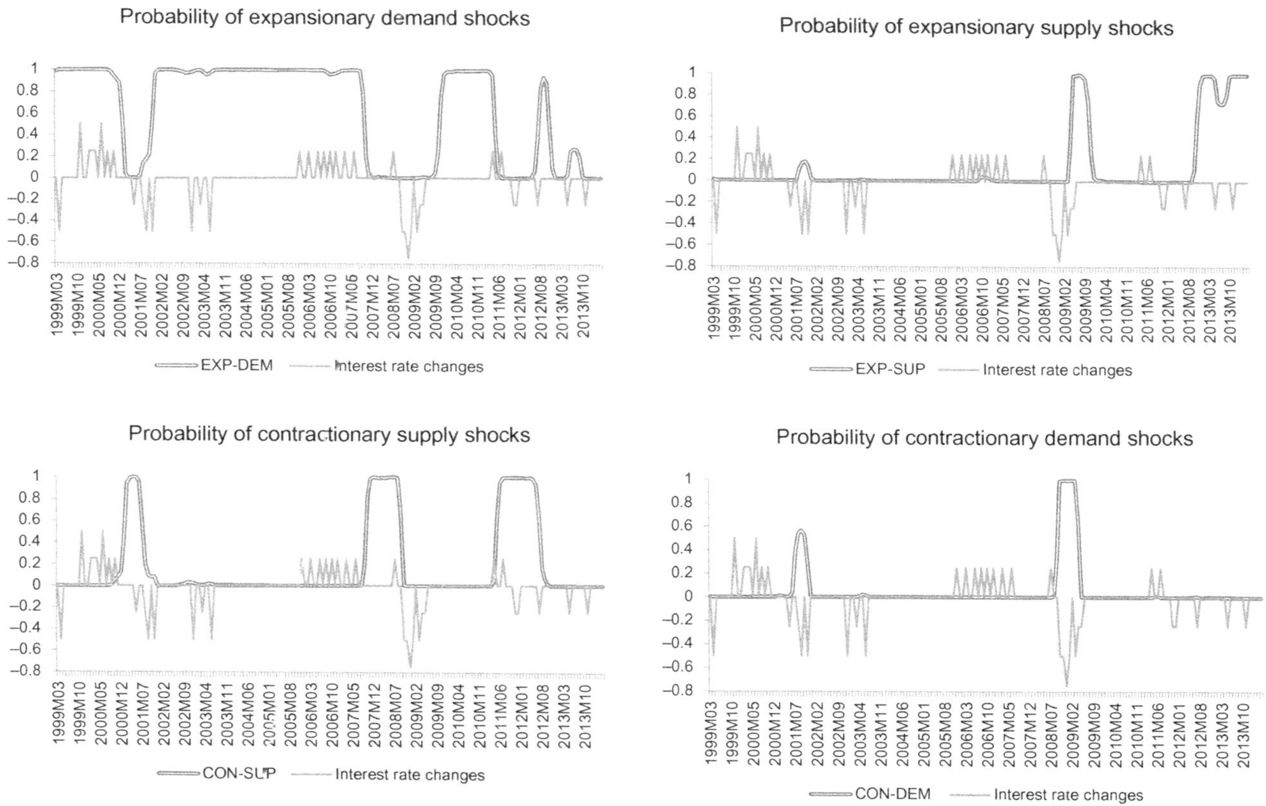

Fig. 5. Regime Inferences on Aggregate Shocks in Italy. *Note*: Each chart plots the probability of aggregate shocks (double line) and the main refinancing operations: minimum bid rate in first differenced (solid line).

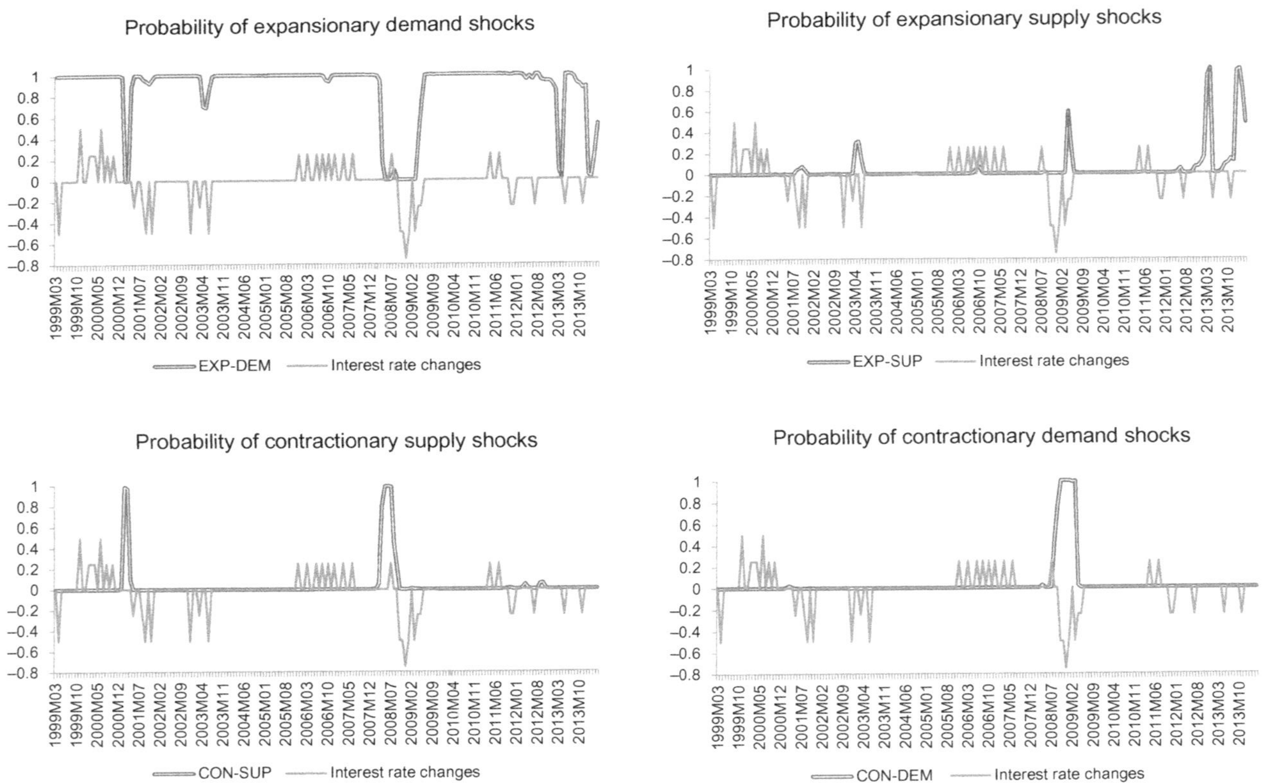

Fig. 6. Regime Inferences on Aggregate Shocks in Spain. *Note*: Each chart plots the probability of aggregate shocks (double line) and the main refinancing operations: minimum bid rate in first differenced (solid line).

perfect comovements times the probability of perfect comovement plus the probability of those shocks in the perfect independent case times the probability of independence. The question now is which kind of shocks would imply a higher level of comovements. To answer this question, we compute a weighted average of the synchronization measure according to the probability of each type of shock. In particular, we compute

$$\delta_{i,j}^{ab} = \frac{\sum_{t=1}^{T} \delta_t^{ab} \Pr(S_{a,t} = i, S_{b,t} = j)}{1/T \sum_{t=1}^{T} \Pr(S_{a,t} = i, S_{b,t} = j)} \tag{19}$$

Using this expression, we find that expansionary supply shocks present up to 40% more of comovement in Germany. In Italy, the increase in comovement is associated with higher probability of expansionary demand shocks. For all the countries, the degree of comovement is lower in the presence of contractionary shocks. In addition, for all the countries, expansionary demand shocks are predominant for most of the sample period.

It is worth emphasizing that, specially for Germany, periods of high probabilities of expansionary supply shocks and contractionary demand shocks display a negative relation to the changes in the ECB's interest rate. In France and Italy, such negative relationship is even more evident, especially for contractionary demand shocks. In Spain, inferences on contractionary demand shocks and the ECB's interest rate seem negatively related only during the 2008 recession. This provides evidence that the ECB tends to react with expansionary monetary policy (interest rate falls) during episodes of low inflation regardless of whether they appear in high growth (expansionary supply) or in low growth (contractionary demand) periods. These reactions are compatible with the mandate of the Statute of the ECB (Article 2): "The primary objective of the european Central Bank is to maintain price stability within the eurozone."

The fact that inferences on contractionary demand and expansionary supply shocks are negatively correlated with the ECB monetary policy indicates that the ECB reacts to the state of the main economies in the euro area. This is not new in the literature since this result is a standard feature of any new Keynesian model. Dees, Pesaran, Smith, and Smith (2010), just to quote a recent contribution, show impulse response functions of the interest rate associated to demand and supply shocks for different countries, including the euro area, with similar results. However, the current state of the economy leads not only to current monetary policies changes but also to a market assessment about future changes in monetary policy, in the short, medium or even long run. Markets understand the reaction function of the ECB and the relative importance of the different countries of the system and act accordingly. Markets also understand that the monetary policy transmission mechanism may take several periods of time to achieve the central bank's goals and markets understand that in the medium to long run the response of the ECB might be different than the one given in the short run and that price and real developments have different impact on interest rates.[4] Examining the way in which markets react to different types of shocks in different countries is the purpose of the next section.

4. AGGREGATE SHOCKS AND MONETARY POLICY EXPECTATIONS

In this section, we assess the effect of aggregate contractionary–expansionary demand-supply shocks in the main euro area countries, on the ECB's monetary policy expectations at different horizons. In order to assess the relationship between aggregate shocks (based on the interaction of real activity and inflation regimes) and monetary policy expectations under a unified framework, we include a measure of expectations in the set of information. In particular, we use the first differences of the j-year nominal interest rate swaps, $\hat{r}_{j,t}$, which provides information about market's expectations of the ECB's monetary policy j years ahead. We use data of swaps that span from March 2000 until October 2014, and from 1 to 17 years ahead expectations, due to data availability constraints. Interest rate swaps are the best measure of monetary policy expectations because, given that there is no transaction in period "t", they are not contaminated by liquidity premium.[5]

We assume that market agents infer the current state of the economy with the available information and then construct expectations about future monetary policy actions. Accordingly, the unified model reads as follows:

$$\begin{bmatrix} f_{a,t} \\ f_{b,t} \\ \hat{r}_{j,t} \end{bmatrix} = \begin{bmatrix} \mu_{a,0}+\mu_{a,1}S_{a,t} \\ \mu_{b,0}+\mu_{b,1}S_{b,t} \\ \mu_{j,1}S_{a,t}S_{b,t}+\mu_{j,2}S_{a,t}(1-S_{b,t})+\mu_{j,3}(1-S_{a,t})S_{b,t}+\mu_{j,4}(1-S_{a,t})(1-S_{b,t}) \end{bmatrix} + \begin{bmatrix} \varepsilon_{a,t} \\ \varepsilon_{b,t} \\ \varepsilon_{j,t} \end{bmatrix} \tag{20}$$

The main difference with respect to the model described in expression (12) is the inclusion of the equation for $\hat{r}_{j,t}$. The monetary policy shock is modeled as a function of a time-varying mean and an error term. This time-varying mean is assumed to depend on the type of shock hitting the economy, that is, expansionary demand ($S_{a,t} = 1, S_{b,t} = 1$), expansionary supply ($S_{a,t} = 1, S_{b,t} = 0$), contractionary supply ($S_{a,t} = 0, S_{b,t} = 1$), and contractionary demand ($S_{a,t} = 0, S_{b,t} = 0$). Since the Gibbs sampler, described in the appendix, generates draws of the latent variables, in every iteration the unobserved becomes "observed" and the shocks can be used as any other regressor. Therefore, this framework allows us to assess the relationship between observed continuous and unobserved discrete variables.

Our main focus regarding the unified model is assessing how responsive are expectations of the ECB's monetary policy to aggregate shocks hitting

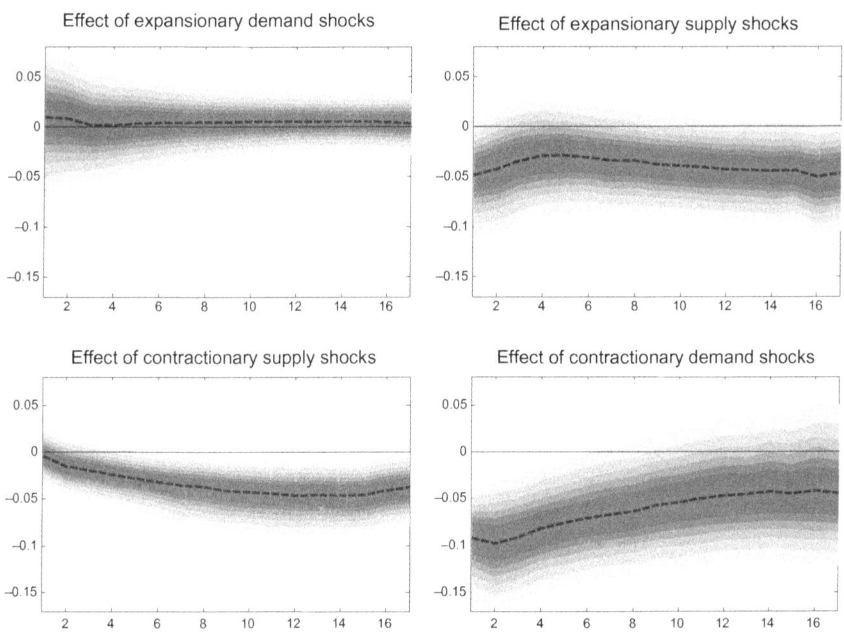

Fig. 7. Effects of Aggregate Shocks in Germany on Monetary Policy Expectations. *Notes*: In each chart, the vertical axis represents the effect of a specific aggregate shock on monetary policy expectations at different horizons (dashed line). Horizontal axis represents the horizon of expectations in years. The bands (fan chart) represent up to the 0.90 quantile of the corresponding estimate's distribution, as a measure of uncertainty.

the economies of Germany, France, Italy, and Spain. Therefore, the parameters of interest are $\mu_{j,1}$, $\mu_{j,2}$, $\mu_{j,3}$ and $\mu_{j,4}$ for $j = 1, \ldots, 17$, since they measure the responses in monetary policy expectations to aggregate shocks of each country. The estimation of model (21) follows the lines suggested to estimate model (12).[6]

The parameters estimated along with their fan charts are plotted in Figs. 7–10. The figures represent the estimated coefficients $\mu_{j,1}$, $\mu_{j,2}$, $\mu_{j,3}$ and $\mu_{j,4}$ for $j = 1, \ldots, 17$, representing the estimated values of each coefficient for every horizon of the swap rates. For example, the first estimated value $\mu_{1,1}$ represents the swap response at year 1 of an expansionary demand shock, $\mu_{1,2}$ is the swap response at year 1 of an expansionary supply shock and so on.

Overall, the figures show that the ECB's monetary policy expectations react negatively to contractionary shocks. Contractionary demand shocks

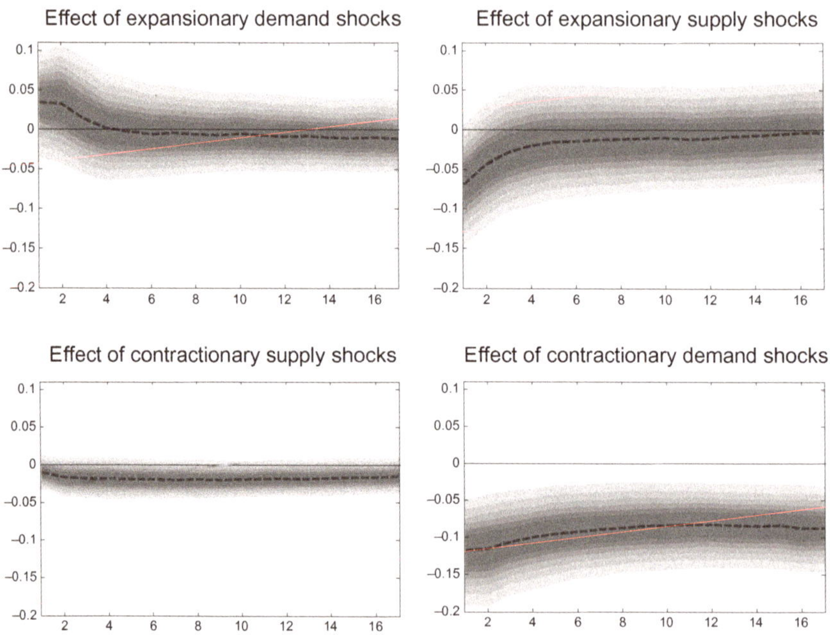

Fig. 8. Effects of Aggregate Shocks in France on Monetary Policy Expectations. *Notes*: In each chart, the vertical axis represents the effect of a specific aggregate shock on monetary policy expectations at different horizons (dashed line). Horizontal axis represents the horizon of expectations in years. The bands (fan chart) represent up to the 0.90 quantile of the corresponding estimate's distribution, as a measure of uncertainty.

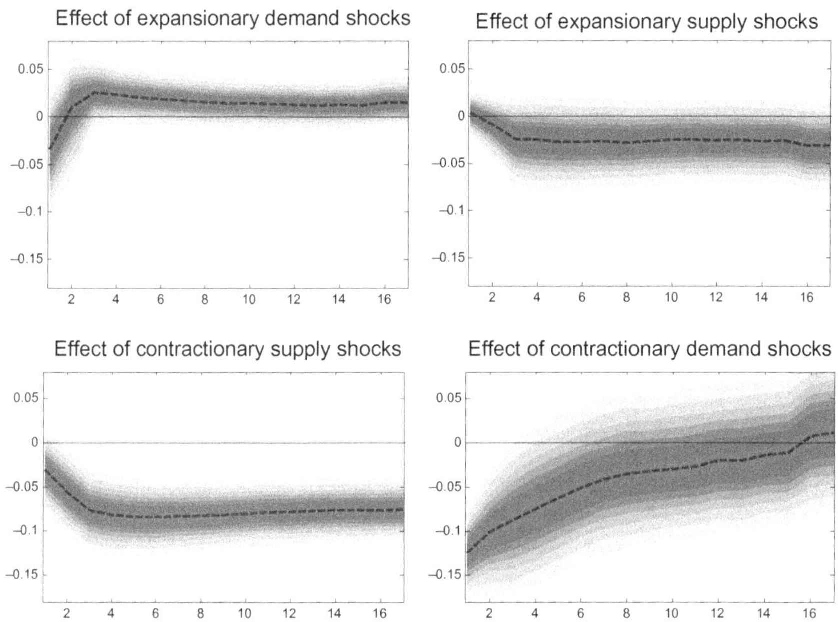

Fig. 9. Effects of Aggregate Shocks in Italy on Monetary Policy Expectations. *Notes*: In each chart, the vertical axis represents the effect of a specific aggregate shock on monetary policy expectations at different horizons (dashed line). Horizontal axis represents the horizon of expectations in years. The bands (fan chart) represent up to the 0.90 quantile of the corresponding estimate's distribution, as a measure of uncertainty.

affect monetary policy expectations at short horizons, while contractionary supply shocks affect medium to long-term expectations. In addition, the effect on monetary policy expectations of expansionary shocks is not significant.

The results for Germany are plotted in Fig. 7. The figure shows that both contractionary supply and demand shocks have a significantly negative effect on monetary policy expectations. Specifically, contractionary supply shocks affect medium- and long-term expectations, while contractionary demand shocks affect short-term expectations. The figure also shows that expansionary supply shocks and demand shocks have no effect on expectations of interest rates. For the case of France, Fig. 8 indicates that only contractionary demand shocks have a negative effect on short-, medium-, and long-term monetary policy expectations. Moreover, the

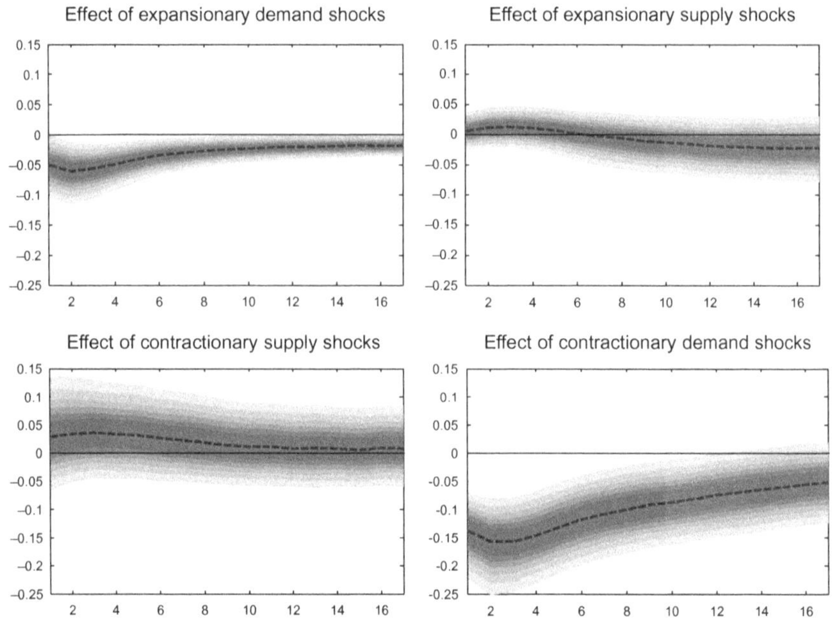

Fig. 10. Effects of Aggregate Shocks in Spain on Monetary Policy Expectations. *Notes*: In each chart the vertical axis represents the effect of a specific aggregate shock on monetary policy expectations at different horizons (dashed line). Horizontal axis represents the horizon of expectations in years. The bands (fan chart) represent up to the 0.90 quantile of the corresponding estimate's distribution, as a measure of uncertainty.

figure shows that contractionary supply shocks lead to a slightly negative effect on the expectations of the ECB's monetary policy. Fig. 9 reveals that, as in the case of Germany, contractionary supply shocks in Italy affect medium- and long-term monetary policy expectations, while contractionary demand shocks only affect short-term expectations. As in the other countries, in this case, expansionary shocks have no significant effect on the ECB's monetary policy expectations. Finally, Fig. 10 shows that only contractionary demand shocks in Spain have significant negative effect on short- and medium-term monetary policy expectations.

The differences in the significance of the long- and short-term effects could be explained as follows. When contractionary supply shocks hit the economy, the high inflation may be a bulwark against an immediate action by the ECB in decreasing the interest rate, and thus the market expects only an ECB action in the medium- to long-term. By contrast, when both real activity and inflation experience a downturn, that is, a contractionary demand, the market may expect a soon reaction by the ECB in cutting rates to stimulate the economy and to keep inflation close to the target.

In sum, we find that the market assessment to the response to a monetary policy shock depends on the nature of the shock. If markets read monetary policy correctly, they believe that the monetary policy reaction function is more aggressive to negative demand shocks than to any other type of shocks. This reaction is immediate and significant for all countries, and of similar magnitude. In the case of negative supply shocks, this effect varies across countries (it is not significant in the case of Spain) and is more related to the long run than to short run expectations. The effect of expansionary demand and supply shock is more diffuse across countries and across time delays.

5. CONCLUSIONS

This paper addresses the effect of the current state of the economy on monetary policy expectations. In particular, we study the effect of contractionary (or expansionary) demand (or supply) shocks hitting the euro area countries on the expectations of the ECB's monetary policy. The results indicate that expectations are responsive to aggregate contractionary shocks, but not to expansionary shocks. Contractionary demand shocks have a negative effect on short-term monetary policy expectations, while contractionary supply shocks have negative effect on medium- and long-term expectations. We also find that, for the case of demand shocks, these results are robust across countries. However, this is not the case for supply shocks, for which markets seem to discount more the German, French or Italian shocks than the shocks for Spain.

NOTES

1. To identify the factor model, the variances σ_a^2 and σ_b^2 are assumed to be one.
2. Unfortunately, we do not have income series for these countries.
3. To save space, these results are omitted. They are available upon request.
4. A simple panel analysis shows that the price factor is always significant when analyzing spot rates development, while activity factor is not significant. The activity factor only become significant in the medium to long run analysis. This could

explain some leading behavior of price developments over real developments. ECB reacts faster to price developments and it takes more time to react to real activity developments.

5. Some examples of using interest rate swaps to measure monetary policy expectations can be found in Gürkaynak, Sack, and Swanson (2007) or Sack (2002) among many others.

6. To avoid imposing judgement, we choose totally uninformative priors for $\mu_{j,i}$, that is, $\mu_{j,i} = 0$ for $i = 1, \ldots, 4$ and $j = 1, \ldots, 17$.

ACKNOWLEDGMENTS

We thank participants at the Advances in Econometrics Conference on Dynamic Factor Models in Aahrus and the two anonymous referees. Maximo Camacho thanks CICYT for its support through grant ECO2013-45698. The views expressed in this paper are those of the authors and do not represent the views of the Central Bank of Chile, Airef, Bank of Spain or the Eurosystem.

REFERENCES

Arias, J., Caldara, D., & Rubio Ramirez, J. (2015). *The systematic component of monetary policy in SVARs: An agnostic identification procedure*. International Finance Discussion Papers No. 1131.

Aruoba, B., & Diebold, F. (2010). Real-time macroeconomic monitoring: Real activity, inflation, and interactions. *American Economic Review*, *100*, 20–24.

Aruoba, B., Diebold, F., & Scotti, C. (2009). Real-time measurement of business conditions. *Journal of Business and Economic Statistics*, *27*, 417–427.

Benigno, P., & Lopez Salido, D. (2006). Inflation persistence and optimal monetary policy in the euro area. *Journal of Money Credit and Banking*, *38*, 587–614.

Camacho, M., & Martinez-Martin, J. (2014). Real-time forecasting US GDP from small factor models. *Empirical Economics*, *47*, 347–364.

Camacho, M., & Perez-Quiros, G. (2010). Introducing the euro-sting: Short-term indicator of euro area growth. *Journal of Applied Econometrics*, *25*, 663–694.

Dees, S. Pesaran, H., Smith, V., & Smith, R. (2010). *Supply, demand and Monetary Policy Shocks in a multi-country new keynesian model*. ECB WP Series 1230.

Dieppe, A., Kuster, K., & McAdam, P. (2004). *Optimal monetary policy rules for the euro area: An analysis using the area wide model*. ECBWP Series 360.

Gürkaynak, R., Sack, B., & Swanson, E. (2007). Market-based measures of monetary policy expectations. *Journal of Business and Economic Statistics*, *25*, 201–212.

Kim, C., & Nelson, C. (1998). Business cycle turning points, a new coincident index, and tests of duration dependence based on a dynamic factor model with regime switching. *Review of Economics and Statistics*, *80*(2), 188–201.

Leeper, E., Sims, C., & Zha, T. (1996). What does monetary policy do? *Brookings Papers on Economic Activity, 27*, 1–78.

Leeper, E., & Zha, T. (2003). Modest policy interventions. *Journal of Monetary Economics, 50*, 1673–1700.

Leiva-Leon, D. (2014). *A new approach to infer changes in the synchronization of business cycle phases*. Working Paper No. 2014-38, Bank of Canada.

Leiva-Leon, D. (2015). Real vs. nominal cycles: A multistate Markov-switching bi-factor approach. *Studies in Nonlinear Dynamics and Econometrics, 18*, 557–580.

Sack, B. (2002). *Extracting the expected path of monetary policy from futures rates*. Board of Governors of the Federal Reserve, Monetary and Financial Market Analysis Section. FEDS Working Paper No. 2002-56.

Sims, C., & Zha, T. (2006a). Does monetary policy generate recessions? *Macroeconomic Dynamics, 10*, 231–272.

Sims, C., & Zha, T. (2006b). Were there regime switches in US monetary policy? *American Economic Review, 96*, 54–81.

Stock, J., & Watson, M. (1991). A probability model of the coincident economic indicators. In K. Lahiri & G. Moore (Eds.), *Leading economic indicators: New approaches and forecasting records*. Cambridge: Cambridge University Press.

Stracca, L. (2007). A speed limit monetary policy rule for the euro area. *International Finance, 10*, 21–41.

Taylor, J. (1993). Discretion vs policy rules in practice. *Carnegie-Rochester Conference Series on Public Policy, 39*, 195–214.

Thanassis, K., & Elias, T. (2011). *Unveiling the monetary policy rule in the euro-area*. Bank of Greece WP Series, May.

APPENDIX: BAYESIAN PARAMETER ESTIMATION

The approach to estimate θ will be relied on a bivariate extended version of the multimove Gibbs-sampling procedure implemented by Kim and Nelson (1998) for Bayesian estimation of univariate Markov-switching models. In this setting, both the parameters of the model θ and the Markov-switching variables $\tilde{S}_{k,T} = \{S_{k,t}\}_1^T$ for $k = a, b$, $\tilde{S}_{ab,T} = \{S_{ab,t}\}_1^T$ and $\tilde{V}_T = \{V_t\}_1^T$ are treated as random variables given the data in $\tilde{y}_T = \{f_{a,t}, f_{b,t}\}_1^T$. The purpose of this Markov chain Monte Carlo simulation method is to approximate the joint and marginal distributions of these random variables by sampling from conditional distributions.

A.1. Priors

For the mean and variance parameters in vector θ, the Independent Normal-Wishart prior distribution is used

$$p(\mu, \Sigma^{-1}) = p(\mu)p(\Sigma^{-1}) \tag{A.1}$$

where

$$\mu \sim N\left(\underline{\mu}, \underline{V}_{\mu}\right)$$
$$\Sigma^{-1} \sim W\left(\underline{S}^{-1}, \underline{\upsilon}\right)$$

and the associated hyperparameters are given by $\underline{\mu} = \left(\alpha_0^a, \alpha_1^a - \alpha_0^a, \alpha_0^b, \alpha_1^b - \alpha_0^b\right)'$, $\underline{V}_{\mu} = I/10$, $\underline{S}^{-1} = I$, $\underline{\upsilon} = 0$. Due to the business cycle heterogeneity across euro area countries, we adjust the mean hyperparameters for each factor to the magnitude of the corresponding fluctuation. Specifically, α_0^a is the sample average among all the negative growth rates of f_t^a, while α_1^a is the sample average among all the positive growth rates of f_t^a. The same procedure is followed to obtain α_0^b and α_1^b. In this way, provide an estimation in a more "data-driven" way. It is important to mention that when the Gibbs sampler is applied to estimate the trivariate unified model in Section 4, all the hyperparameters of the coefficients associated to the aggregate shocks, $\mu_{j,1}$, $\mu_{j,2}$, $\mu_{j,3}$, $\mu_{j,4}$, are equal to zero, meaning that we follow noninformative priors to provide an estimation robust to judgement.

For the transition probabilities, Beta distributions are used as conjugate priors

$$p_{00,\iota} \sim \mathrm{Be}(u_{11,\iota}, u_{10,\iota}),\ p_{11,\iota} \sim \mathrm{Be}(u_{00,\iota}, u_{01,\iota}), \quad \text{for } \iota = a, b, ab, v \tag{A.2}$$

where the hyperparameters are given by $u_{\iota,01} = 2$, $u_{\iota,00} = 8$, $u_{\iota,10} = 1$ and $u_{\iota,11} = 9$, for $\iota = a, b, ab, v$. For each pairwise model, 6,000 iterations were performed, discarding the first 1,000.

A.2. Drawing $\tilde{S}_{a,T}$, $\tilde{S}_{b,T}$, $\tilde{S}_T$ and $\tilde{V}_T$ Given θ and $\tilde{y}_T$

Inferences on the dynamics of the state variables, $\tilde{S}_{a,T}$, $\tilde{S}_{b,T}$, $\tilde{S}_T$ and $\tilde{V}_T$, can be done following the results in Kim and Nelson (1998) by first computing draws from the conditional distributions

$$g(\tilde{S}_{k,T}|\theta, \tilde{y}_T) = g(S_{k,T}|\tilde{y}_T)\prod_{t=1}^{T} g(S_{k,t}|S_{k,t+1}, \tilde{y}_t), \quad \text{for } k = a, b \tag{A.3}$$

$$g(\tilde{S}_{ab,T}|\theta, \tilde{y}_T) = g(S_{ab,T}|\tilde{y}_T)\prod_{t=1}^{T} g(S_{ab,t}|S_{ab,t+1}, \tilde{y}_t) \tag{A.4}$$

$$g(\tilde{V}_T|\theta, \tilde{y}_T) = g(V_T|\tilde{y}_T)\prod_{t=1}^{T} g(V_t|V_{t+1}, \tilde{y}_t) \tag{A.5}$$

In order to obtain the two terms in the right-hand side of Eqs. (A.3) and (A.5), the following two steps can be employed:

Step 1: The first term can be obtained by running the filtering algorithm to compute $g(\tilde{S}_{k,t}|\tilde{y}_t)$ for $k = a, b$, $g(\tilde{S}_{ab,t}|\tilde{y}_t)$ and $g(\tilde{V}_{k,t}|\tilde{y}_t)$ for $t = 1, 2, \ldots, T$, saving them and taking the elements for which $t = T$.

Step 2: The product in the second term can be obtained for $t = T-1$, $T-2, \ldots, 1$, by following the result:

$$\begin{aligned} g(S_{ab,t}|\tilde{y}_t, S_{ab,t+1}) &= \frac{g(S_{ab,t}, S_{ab,t+1}|\tilde{y}_t)}{g(S_{ab,t+1}|\tilde{y}_t)} \\ &\propto g(S_{ab,t+1}|S_{ab,t})\, g(S_{ab,t}|\tilde{y}_t) \end{aligned} \tag{A.6}$$

where $g\left(S_{ab,t+1}|S_{ab,t}\right)$ corresponds to the transition probabilities of $S_{ab,t}$ and $g\left(S_{ab,t}|\tilde{y}_t\right)$ were saved in Step 1.

Then, it is possible to compute

$$\Pr\left[S_{ab,t}=1|S_{ab,t+1},\tilde{y}_t\right]=\frac{g\left(S_{ab,t+1}|S_{ab,t}=1\right)g(S_{ab,t}=1|\tilde{y}_t)}{\sum_{j=0}^{1}g\left(S_{ab,t+1}|S_{ab,t}=j\right)g\left(S_{ab,t}=j|\tilde{y}_t\right)} \tag{A.7}$$

and generate a random number from a $U[0,1]$. If that number is less than or equal to $\Pr\left[S_{ab,t}=1|S_{ab,t+1},\tilde{y}_t\right]$, then $S_{ab,t}=1$, otherwise $S_{ab,t}=0$. The same procedure applies for $S_{a,t}$, $S_{b,t}$ and V_t.

A.3. Drawing $p_{00,a}$, $p_{11,a}$, $p_{00,b}$, $p_{11,b}$, $p_{00,ab}$, $p_{11,ab}$, $p_{00,v}$, $p_{11,v}$ Given $\tilde{S}_{a,T}$, $\tilde{S}_{a,T}$, $\tilde{S}_{ab,T}$ and $\tilde{V}_T$

Conditional on $\tilde{S}_{k,T}$ for $k=a,b$, $\tilde{S}_{ab,T}$ and $\tilde{V}_T$, the transition probabilities are independent on the data set and the model's parameters. Hence, focusing on the case of $\tilde{S}_{ab,T}$, the likelihood function of $p_{00,ab}$, $p_{11,ab}$ is given by:

$$L\left(p_{00,ab},p_{11,ab}|\tilde{S}_{ab,T}\right)=p_{00,ab}^{n_{00,ab}}\left(1-p_{00,ab}^{n_{01,ab}}\right)p_{11,ab}^{n_{11,ab}}\left(1-p_{11,ab}^{n_{10,ab}}\right) \tag{A.8}$$

where $n_{ij,ab}$ refers to the transitions from state i to j, accounted for in $\tilde{S}_{ab,T}$.

Combining the prior distribution in Eq. (A.2) with the likelihood, the posterior distribution is given by

$$\begin{aligned}p\left(p_{00,ab},p_{11,ab}|\tilde{S}_T\right)\propto\ & p_{00,ab}^{u_{00,ab}+n_{00,ab}-1}\left(1-p_{00,ab}\right)^{u_{01,ab}+n_{01,ab}-1}\\ & p_{11,ab}^{u_{11,ab}+n_{11,ab}-1}\left(1-p_{11,ab}\right)^{u_{10,ab}+n_{10,ab}-1}\end{aligned} \tag{A.9}$$

which indicates that draws of the transition probabilities will be taken from

$$\begin{aligned}p_{00,ab}|\tilde{S}_{ab,T}&\sim Be\left(u_{00,ab}+n_{00,ab},u_{01,ab}+n_{01,ab}\right),\\ p_{11,ab}|\tilde{S}_{ab,T}&\sim Be\left(u_{11,ab}+n_{11,ab},u_{10,ab}+n_{10,ab}\right)\end{aligned} \tag{A.10}$$

The same procedure applies for the cases of $\tilde{S}_{k,T}$ for $k=a,b$ and $\tilde{V}_T$.

A.4. Drawing $\mu_{a,0}, \mu_{a,1}, \mu_{b,0}, \mu_{b,1}$ Given $\sigma_a^2, \sigma_b^2, \sigma_{ab}, \tilde{S}_{a,T}, \tilde{S}_{b,T}, \tilde{S}_{ab,T}, \tilde{V}_T$ and $\tilde{y}_T$

The model in Eq. (12) can be compactly expressed as

$$\begin{bmatrix} y_{a,t} \\ y_{b,t} \end{bmatrix} = \begin{bmatrix} 1 & S_{a,t} & 0 & 0 \\ 0 & 0 & 1 & S_{b,t} \end{bmatrix} \begin{bmatrix} \mu_{a,0} \\ \mu_{a,1} \\ \mu_{b,0} \\ \mu_{b,1} \end{bmatrix} + \begin{bmatrix} \varepsilon_{a,t} \\ \varepsilon_{b,t} \end{bmatrix}, \quad \begin{bmatrix} \varepsilon_{a,t} \\ \varepsilon_{b,t} \end{bmatrix} \sim N\left(\begin{bmatrix} 0 \\ 0 \end{bmatrix}, \begin{bmatrix} \sigma_a^2 & \sigma_{ab} \\ \sigma_{ab} & \sigma_b^2 \end{bmatrix} \right) \quad \text{(A.11)}$$

$$y_t = \overline{S}_t \mu + \xi_t, \quad \xi_t \sim N(\mathbf{0}, \Sigma)$$

stacking as:

$$y = \begin{bmatrix} y_1 \\ y_2 \\ \vdots \\ y_T \end{bmatrix}, \quad \overline{S} = \begin{bmatrix} \overline{S}_1 \\ \overline{S}_2 \\ \vdots \\ \overline{S}_T \end{bmatrix}, \quad \text{and } \xi = \begin{bmatrix} \xi_1 \\ \xi_2 \\ \vdots \\ \xi_T \end{bmatrix}$$

the model in Eq. (A.11) remains written as a normal linear regression model with an error covariance matrix of a particular form:

$$y = S\mu + \xi, \quad \xi \sim N(\mathbf{0}, I \otimes \Sigma) \quad \text{(A.12)}$$

Conditional on the covariance matrix parameters, state variables and the data, by using the corresponding likelihood function, the conditional posterior distribution $p(\mu | \tilde{S}_{a,T}, \tilde{S}_{b,T}, \tilde{S}_{ab,T}, \tilde{V}_T, \Sigma^{-1}, \tilde{y}_T)$ takes the form

$$\mu | \tilde{S}_{a,T}, \tilde{S}_{b,T}, \tilde{S}_{ab,T}, \tilde{V}_T, \Sigma^{-1}, \tilde{y}_T \sim N(\overline{\mu}, \overline{V}_\mu) \quad \text{(A.13)}$$

where

$$\overline{V}_\mu = \left(\underline{V}_\mu^{-1} + \sum_{t=1}^{T} \overline{S}_t' \Sigma^{-1} \overline{S}_t \right)^{-1}$$

$$\overline{\mu} = \overline{V}_\mu \left(\underline{V}_\mu^{-1} \underline{\mu} + \sum_{t=1}^{T} \overline{S}_t' \Sigma^{-1} y_t \right)$$

After drawing $\mu = (\mu_{a,0}, \mu_{a,1}, \mu_{b,0}, \mu_{b,1})'$ from the above multivariate distribution, if the generated value of $\mu_{a,1}$ or $\mu_{b,1}$ is less than or equal to 0, that draw is discarded, otherwise it is saved, this is in order to ensure that $\mu_{a,1} > 0$ and $\mu_{b,1} > 0$.

A.5. Drawing σ_a^2, σ_b^2, σ_{ab} Given $\mu_{a,0}$, $\mu_{a,1}$, $\mu_{b,0}$, $\mu_{b,1}$, $\tilde{S}_{a,T}$, $\tilde{S}_{b,T}$, $\tilde{S}_{ab,T}$, $\tilde{V}_T$ and $\tilde{y}_T$

Conditional on the mean parameters, state variables and the data, by using the corresponding likelihood function, the conditional posterior distribution

$$p\left(\Sigma^{-1} | \tilde{S}_{a,T}, \tilde{S}_{b,T}, \tilde{S}_{ab,T}, \tilde{V}_T, \mu, \tilde{y}_T\right)$$

takes the form

$$\Sigma^{-1} | \tilde{S}_{a,T}, \tilde{S}_{b,T}, \tilde{S}_{ab,T}, \tilde{V}_T, \mu, \tilde{y}_T \sim W\left(\overline{S}^{-1}, \overline{\upsilon}\right) \tag{A.14}$$

where

$$\overline{\upsilon} = T + \underline{\upsilon}$$

$$\overline{S} = \underline{S} + \sum_{t=1}^{T} (y_t - \overline{S}_t\mu)(y_t - \overline{S}_t\mu)'$$

after Σ^{-1} is generated the elements is Σ are recovered.

MODELLING FINANCIAL MARKETS COMOVEMENTS DURING CRISES: A DYNAMIC MULTI-FACTOR APPROACH

Martin Belvisi[a], Riccardo Pianeti[b] and Giovanni Urga[b,c]

[a]*KNG Securities, London, UK*
[b]*University of Bergamo, Bergamo, Italy*
[c]*Cass Business School, City University London, London, UK*

ABSTRACT

We propose a novel dynamic factor model to characterise comovements between returns on securities from different asset classes from different countries. We apply a global-class-country latent factor model and allow time-varying loadings. We are able to separate contagion (asset exposure driven) and excess interdependence (factor volatility driven). Using data from 1999 to 2012, we find evidence of contagion from the US stock market during the 2007–2009 financial crisis, and of excess interdependence during the European debt crisis from May 2010

Dynamic Factor Models
Advances in Econometrics, Volume 35, 317–360

ISSN: 0731-9053/doi:10.1108/S0731-905320150000035008

onwards. Neither contagion nor excess interdependence is found when the average measure of model implied comovements is used.

Keywords: Dynamic factor models; comovements; contagion; Kalman filter; autometrics

JEL classifications: C3; C5; G1

1. INTRODUCTION

The study of financial market comovements is of paramount importance for its implications in both theoretical and applied economics and finance. The practical relevance of a thorough understanding of the mechanisms governing market correlations lies in the benefits that this induces in the processes of asset allocation and risk management.

In particular, recent crisis episodes have shifted the focus of the literature on the characterization of financial market comovements during periods of financial distress. Most of the crises that have hit the financial markets in the past decades are the result of the propagation of a shock which originally broke out in a specific market. This phenomenon has been extensively explored in the literature and has led to the use of the term "contagion" to denote the situation in which a crisis originated in a specific market infects other interconnected markets. For a review of the contributions at the heart of the literature on contagion, see the papers by Karolyi (2003), Dungey, Fry, González-Hermosillo, and Martin (2005) and Billio and Caporin (2010).

A well-documented phenomenon linked to a situation of contagion is an increase in the observed correlations amongst the affected markets. The origins of this empirical evidence trace back to the contributions of King and Wadhwani (1990), Engle, Ito, and Lin (1990) and Bekaert and Hodrick (1992). Longin and Solnik (2001) and, in particular, the influential paper by Forbes and Rigobon (2002) criticize the common practice to identify periods of contagion using testing procedures based on market correlations. Forbes and Rigobon (2002) show that the presence of heteroscedasticity biases this type of testing procedure, leading to over-acceptance of the hypothesis of the presence of contagion. Bae, Karolyi, and Stulz (2003), Pesaran and Pick (2007) and Fry, Martin, and Tang (2010) propose testing procedures robust to the presence of heteroscedasticity.

In this paper, we bring together the literature on contagion with the literature on market integration in that we associate a situation of *contagion* to a prolonged episode of market distress altering the functioning of the financial system. On the contrary, a situation of *excess interdependence* is a short lasting phenomenon. Being able to distinguish between contagion and excess interdependence has a crucial information content as to how a crisis develops and spreads out. We propose a modelling framework which allows to contrast a situation of *contagion,* in the Forbes and Rigobon (2002) sense, as opposed to the case in which *excess interdependence* in financial markets is triggered by spiking market volatility. Contagion is no longer thought as correlation in excess of what is implied by an economic model (as in Bekaert, Ehrmann, Fratzscher, & Mehl, 2011; Bekaert, Harvey, & Ng, 2005), it instead corresponds to a specific market situation entailing a persistent change in financial linkages between markets. On the contrary, conditional heteroscedasticy of financial time series does not display trending behaviour (Brandt, Brav, & Graham, 2010; Schwert, 1989); thus a rise in correlations caused by excess volatility has only a temporary effect. This feature is in line with the literature on market integration (Bekaert, Hodrick, & Zhang, 2009), which explores the degree of interconnectedness of markets through time, borrowing from Forbes and Rigobon's (2002) analysis the fact that excess interdependence, triggered by volatility, might lead to spurious identification of cases of market integration.

We study comovements amongst financial markets during crises, both in a multi-country and a multi-asset class perspective, contributing to the extant empirical literature on international and intra-asset class shock spillovers. We decompose an average correlation measure into components that are in turn attributed to volatility and exposure. We analyse stock, bond and FX comovements in the United States, Euro Area, United Kingdom, Japan and Emerging Countries, providing an extensive coverage of the global financial markets. Most of the contributions to the literature on comovements entail single asset classes, with the vast majority focusing on stock and bond markets (see *inter alia* Baele, Bekaert, & Inghelbrecht 2010; Bekaert et al., 2009; Driessen, Melenberg, & Nijman, 2003). There is a strand of literature embracing a genuine multi-country and multi-asset-classes approach in the study of shock spillovers. Dungey and Martin (2007) propose an empirical model to measure spillovers from FX to equity markets to investigate the breakdown in correlations observed during the 1997 Asian financial crisis. Ehrmann, Fratzscher, and Rigobon (2011) analyse the financial transmission mechanism across different asset classes

(FX, equities and bonds) in the United States and the Euro Area, using a simultaneous structural model.

The main contribution of this paper is twofold. First, we propose a dynamic factor model that allows to test for the presence of comovements (excess interdependence vs. contagion) in a multi-asset and multi-country framework. Since the seminal works of Ross (1976) and Fama and French (1993), multifactor models for asset returns have been the main tool for studying and characterizing comovements. Moreover, our model is specified with dynamic factor loadings, to accommodate time-dependent exposures of the single assets to the different shocks. This allows us to disentangle the different sources of comovements between financial markets, and to analyse their dynamics during financial crisis periods. Second, we report an empirical application using a sample period which encompasses both the 2007–2009 crisis as well as the current sovereign debt crisis: this is an interesting laboratory to use the proposed framework to explore financial market comovements during crisis periods.

The empirical analysis suggests interesting findings. The global factor is the most pervasive of the considered factors, while the asset class factor is the most persistent and the country factor is negligible in our multiple asset framework. We find evidence of contagion stemming from the US stock market during the 2007–2009 financial crisis and the presence of excess interdependence during the spreading of the European debt crisis from mid-2010 onwards. Any contagion or excess interdependence effect disappears at the overall average level, and because of this, some of the considered assets display diverging repricing dynamics during crisis periods.

The remainder of this paper is organized as follows. In Section 2, we present our dynamic multi-factor model. Section 3 introduces the data. Section 4 reports the relevant empirical results regarding the relevance of global-asset–country factors and the indentification of the situation of contagion and the case of excess interdependence in financial assets. Section 5 concludes.

2. A DYNAMIC MULTI-FACTOR MODEL

In this section, we present the modelling framework we propose. The main novelty of this paper is the formulation and the estimation of a dynamic multi-factor model which allows to test for the presence of *contagion* in the Forbes and Rigobon (2002) sense versus the presence of volatility triggered episodes of *excess interdependence* on financial markets. Contagion is no longer thought as correlation in excess of what is implied by an economic model (as in Bekaert et al., 2005, 2014). It instead corresponds to a specific market situation, that the framework proposed in this paper is able to capture, entailing a persistent change in financial linkages between markets.

Building on the standard latent factor finance literature (Ross, 1976; Fama & French, 1993), let $R_t^{i,j}$ represent the weekly return for the asset belonging to asset class $i = 1, \ldots, I$ and country $j = 1, \ldots, J$ at time t. The general representation of the model is as follows:

$$R_t^{i,j} = \mathbf{E}\left[R_t^{i,j}\right] + F_t^{i,j}\beta_t^{i,j} + \epsilon_t^{i,j} \tag{1}$$

$$\beta_t^{i,j} = diag\left(\mathbf{1} - \phi^{i,j}\right)\beta^{i,j} + diag\left(\phi^{i,j}\right)\beta_{t-1}^{i,j} + \psi^{i,j}Z_{t-1} + u_t^{i,j} \tag{2}$$

where $\mathbf{E}\left[R_t^{i,j}\right]$ is the expected return for asset class i in country j at time t, $\beta_t^{i,j}$ is a vector of dynamic factor loadings, mapping from the zero-mean factors $F_t^{i,j}$ to the single asset returns. We allow the factors $F_t^{i,j}$ to be heteroscedastic, that is, $\mathbf{E}\left[F_t^{i,j\prime}F_t^{i,j}\right] = \Sigma_{F^{i,j},t}$, where $\Sigma_{F^{i,j},t}$ is the time-varying covariance matrix of the factors. The error $\epsilon_t^{i,j}$ is assumed to be white noise and independent of $F_t^{i,j}$, the vector $\beta^{i,j}$ is the long-run value of $\beta_t^{i,j}$, while $\phi^{i,j}$ and $\psi^{i,j}$ are three-dimensional vectors of parameters to be estimated, the errors $\left\{u_t^{i,j}\right\}_{t=1,\ldots,T}$ are independent and normally distributed. We assume $u_t^{i,j}$ to be independent of $\epsilon_t^{i,j}$. Note that $diag(\cdot)$ is the diagonal operator, transforming a vector into a diagonal matrix. Finally, Z_t represents a conditional variable controlling for period of market distress.

Following Dungey and Martin (2007), different sources of shocks are considered, at global, asset class and country level, in a latent factor framework. A first factor, denoted as G_t, is designed to capture the shocks which are common to all financial assets modelled, whereas A_t^i is the asset-class-specific factor for asset class $i = 1, \ldots, I$ and the country factor C_t^j is the country-specific factor for county $j = 1, \ldots, J$ at time t. We denote $F_t^{i,j} \equiv [G_t \ A_t^i \ C_t^j]$ and, correspondingly, for the factor loading we specify $\beta_t^{i,j} \equiv [\gamma_t^{i,j} \ \delta_t^{i,j} \ \lambda_t^{i,j}]'$.

The full model is a multi-factor model with dynamic factor loadings and heteroscedastic factors. This model setting allows us to explore and characterize dynamically the comovements among the considered assets. Time-dependent exposures to different shocks let us disentangle dynamically the different sources of comovement between financial markets, namely, distinguishing among shocks spreading at a global level, at the asset class or rather at the country level. The presence of time-varying exposures to common factors enables us to test for the presence of contagion, controlling at the same time for excess interdependence induced by heteroscedasticity in the factors. In the following sections, we explore the features of the model and use it to characterize financial market comovements during crisis.

In Section 2.1, we describe the estimation of the factors $F_t^{i,j}$, whereas the estimation of Z_{t-1} is presented in Section 2.2.

2.1. Factor Estimation

The factors $F_t^{i,j}$ are estimated by means of principal component analysis (PCA). The choice of PCA is dictated by model simplicity and interpretability, yet providing consistent estimates of the latent factors.[1] The global factor G is extracted using the entire set of variables considered, whereas the other two factors, asset-class- (A) and the country-specific (C), are extracted from the different asset class and country groups, respectively. In this setting, the number of variables from which the factors are extracted, say K, is fixed and small, whilst the number of observations T is large.

2.1.1. Global Factor (G)

Let us first consider the global factor G. In order to estimate it, let $\mathbf{E}\left[R_t^{i,j}\right]$ be the conditional mean by asset class and by country,we define the series of the demeaned returns as $\overline{R}_t^{i,j} \equiv R_t^{i,j} - \mathbf{E}\left[R_t^{i,j}\right]$ and we stack them into the matrix $\mathbf{r}$. We then consistently estimate the variance–covariance matrix of $\mathbf{r}$, say $\Sigma_{\mathbf{r}}$, via maximum likelihood, as

$$\hat{\Sigma}_{\mathbf{r}} \equiv \frac{1}{(T-1)}\mathbf{r}'\mathbf{r} \tag{3}$$

Let $(l_k, \mathbf{w_k})$ be the eigencouples referring to the covariance matrix $\Sigma_{\mathbf{r}}$, with $k = 1, \ldots, K$, such that $l_1 \geq l_2 \geq \cdots \geq l_K$. We estimate $(l_k, \mathbf{w_k})$ by extracting the eigenvalue–eigenvector couples from the estimated covariance matrix of the returns $\hat{\Sigma}_{\mathbf{r}}$, denoted as $(\hat{l}_k, \hat{\mathbf{w}}_{\mathbf{k}})$.

The estimate $\hat{G}$ of the common factor G is given by the principal component extracted using the matrix $\hat{\Sigma}_{\mathbf{r}}$, that is:

$$\hat{G} = \mathbf{r}\hat{\mathbf{w}}_1 \tag{4}$$

$\hat{G}$ is a consistent estimator of the factor G. Indeed, from the standpoint that $\hat{\Sigma}_{\mathbf{r}}$ is a consistent estimator of $\Sigma_{\mathbf{r}}$ as a direct consequence of the invariance property for maximum likelihood estimators, the estimated eigencouples $(\hat{l}_k, \hat{\mathbf{w}}_{\mathbf{k}})$ consistently estimate $(l_k, \mathbf{w}_k)$. See Anderson (2003). Note that $\hat{\Sigma}_{\mathbf{r}}$ is a consistent estimator of $\Sigma_{\mathbf{r}}$ if the number of series is considered as fixed or increases at a slower rate than time.

2.1.2. Asset-Class- (A) and Country-Specific (C) Factors

Following the same procedure used for the estimation of global factor, in order to estimate the asset-class- and the country-specific factors A^i and C^j (with $i = 1, \ldots, I$ and $j = 1, \ldots, J$) respectively, we define $\mathbf{r}^i \equiv [r_t^{i,j}]_{j=1,\ldots,J}$ and $\mathbf{r}^j \equiv [r_t^{i,j}]_{i=1,\ldots,I}$ as the matrices of returns referred to asset class i and country j, respectively. Denote as $\Sigma_{\mathbf{r}^i}$ and $\Sigma_{\mathbf{r}^j}$ the corresponding covariance matrix and let $\hat{w}_1^i$ and $\hat{w}_1^j$ be the eigenvectors corresponding to the largest eigenvalues of the estimates $\hat{\Sigma}_{\mathbf{r}^i}$ and $\hat{\Sigma}_{\mathbf{r}^j}$. The estimates of the asset class and the country specific factors $\hat{A}^i$ and $\hat{C}^j$ are then given by, respectively,

$$\hat{A}^i = \mathbf{r}^i \hat{w}_1^i \tag{5}$$

$$\hat{C}^j = \mathbf{r}^j \hat{w}_1^j \tag{6}$$

As we use demeaned returns, the extracted factors will have zero mean by construction.

For the sake of model interpretability, we orthogonalize the factors, so that the three groups of factors are mutually independent. The preliminary correlation analysis presented in Section 3 suggests that the asset class factors are more pervasive than the country ones. So, we first orthogonalize the asset class factors with respect to the global factor, by regressing the global factor on the asset class factors and using the residuals as the orthogonalised asset class factors. Then, we orthogonalize the country factors with respect to the asset class and the global factors, using the residuals in the regression of the country factors on both the asset class and the global factors. This ensures, for instance, that the US factor is independent of the global factor and of the equity factor. The orthogonalization process, however, is not carried out within the groups of factors, so then the equity factor might have a nonzero correlation with the bond factor, and so the US factor with the EU factor. In the empirical section, we report below, we show that our results are robust to the case in which one orthogonalizes the country factors with the global one and then the asset class factors with respect to the others.

2.2. Factor Loading Specification and Estimation

In our specification (2), Z_{t-1} is a control factor extracted from pure exogenous variables and it is supposed to measure market nervousness and accounts for potential increase in the factor loading during market distress periods. In Section 4, we get an estimate $\hat{Z}_{t-1}$ of Z_{t-1} via the principal component extracted from the VIX, which is widely recognized as indicator of market sentiment, the TED spread and the Libor-OIS spread for Europe, which measure the perceived credit risk in the system. Widening spreads corresponds to a lack of confidence in lending money on the interbank market over short-term maturities, together with a flight to security in the form of overnight deposits at the lender of last resort.

Thus, the specification of Eq. (2) for the factor loadings $\beta_t^{i,j}$ is now

$$\beta_t^{i,j} = diag(\mathbf{1} - \phi^{i,j})\beta^{i,j} + diag(\phi^{i,j})\beta_{t-1}^{i,j} + \psi^{i,j}\hat{Z}_{t-1} + u_t^{i,j} \tag{7}$$

The *conditional time-varying factor loading specification*[2] (7) emphasizes that $\beta_t^{i,j}$ tends to its long-run value $\beta^{i,j}$ while following an autoregressive type of process of order one with a purely exogenous variable Z, with Z a zero-mean variable, $\beta^{i,j}$ can indeed be interpreted as the long-run value for $\beta_t^{i,j}$.

Specification (7) nests two special cases. First, a *static specification* of the form:

$$\beta_t^{i,j} \equiv \beta^{i,j}, \quad \forall i = 1, \ldots, I, \quad \forall j = 1, \ldots, J \tag{8}$$

where we assume that the exposure of all modelled variables to the different groups of factors are kept constant through time.

A second nested case is a *time-varying factor loading specification*

$$\beta_t^{i,j} = diag(\mathbf{1} - \phi^{i,j})\beta^{i,j} + diag(\phi^{i,j})\beta_{t-1}^{i,j} + u_t^{i,j} \tag{9}$$

where it is assumed that no exogenous variables enter in the data generating process of the betas. In Bekaert et al. (2009), the dynamics of the betas is specified using subsamples of fixed length via a rolling window estimation, so that the factor loadings are constant within pools of observations with the factor loadings having the following specification: $\beta_t^{i,j} \equiv \beta^{i,j,s}$, $s = 1, \ldots, S$, where $\beta^{i,j,s}$ is the static factor loading estimate referred to subsample s, while S is the number of subsamples considered. The authors partition the sample in semesters and re-estimate the model every six months. However, the rolling windows estimation is based on changing subsamples of the data and it may not reflect time-variation fairly well especially in small samples as also pointed out, amongst others, by Benerjee, Lumsdaine, and Stock (1992). Thus, in our paper we estimate specification (9) using Kalman filter maximum likelihood estimation to avoid both issues on potential inconsistency of the estimates obtained using sub-samples and any arbitrary choice about the inertia, the subsample length, as to which factor loadings evolve through time.

To summarise, our proposed dynamic multi-factor model is:

$$R_t^{i,j} = \mathbf{E}\left[R_t^{i,j}\right] + \hat{F}_t^{i,j\prime}\beta_t^{i,j} + \epsilon_t^{i,j} \tag{10}$$

$$\beta_t^{i,j} = diag\left(\mathbf{1} - \phi^{i,j}\right)\beta^{i,j} + diag\left(\phi^{i,j}\right)\beta_{t-1}^{i,j} + \psi^{i,j}\hat{Z}_{t-1} + u_t^{i,j} \tag{11}$$

OLS gives consistent estimates of Eq. (10) when using specification (8), corresponding to the static case, which we consider the baseline. When considering the alternative specifications (7) and (9), we allow that the factor loadings show evidence of contagion either in a conditioned way ($\psi^{i,j} \neq \mathbf{0}$) or in an unconditioned way ($\psi^{i,j} = \mathbf{0}$), according to the specified control variable. In these other two cases, estimates are obtained via maximum likelihood by applying the Kalman filter. The models are nested and thus, the standard likelihood ratio test can be employed for model selection.

2.3. Heteroscedastic Factors

In order to distinguish between spikes in comovements due to increasing exposures to common risk factors from the case in which spikes are triggered by excess volatility in the common factors, we allow for heteroscedastic factors. The extend to which the three groups of factors are mutually independent by construction greatly simplifies the estimation. For the case of the global factor G_t, a univariate GARCH(1,1) with normal innovation is employed to estimate time-varying volatility. For the asset class and the country factors, we apply the Engle's (2002) dynamic conditional correlations (DCC) model of order (1,1) with GARCH(1,1) for the marginal conditional volatility processes with normal innovations separately on $\mathbf{A}_t$ and $\mathbf{C}_t$, defined by stacking the factors into matrices as follows: $\mathbf{A}_t \equiv [A_t^i]_{i=1,\ldots,I}$ and $\mathbf{C}_t \equiv [C_t^j]_{j=1,\ldots,J}$. We obtain estimates of the time-varying covariance matrices of the factors, estimating the DCC model via quasi-maximum likelihood estimation.

2.4. Financial Markets Comovements: Contagion versus Excess Interdependence

From the dynamic factor model introduced above, we can derive the time-varying covariance between pairs of financial assets.

To simplifying the notation, let us introduce the one-to-one mapping $n \equiv \aleph(i,j)$. Given the independence between the factors F_t and the error term ϵ_t, from Eq. (1) it follows that the covariance between any pair of assets at time t is given by:

$$\mathrm{cov}_t(R^n, R^m) = \mathbf{E}\left[\beta_t^{n\prime} F_t^{n\prime} F_t^m \beta_t^m\right] + \mathbf{E}\left[\epsilon_t^n \epsilon_t^m\right] \tag{12}$$

for $n = 1, \ldots, N$, $m = 1, \ldots, N$, $n \neq m$. The first term on the right-hand side is what is generally referred to as *model implied*covariance, whereas the second is called *residual* covariance. The empirical counterpart of Eq. (12) is given by:

$$\widehat{\mathrm{cov}}_t(R^n, R^m) = \hat{\beta}_t^{n\prime} \hat{\Sigma}_{F,t}^{n,m} \hat{\beta}_t^m + \hat{\Sigma}_{\epsilon,t}^{n,m} \tag{13}$$

which we rewrite for convenience, as:

$$\widehat{\mathrm{cov}}_{n,m,t} = \widehat{\mathrm{cov}}_{n,m,t}^{F} + \widehat{\mathrm{cov}}_{n,m,t}^{\epsilon} \tag{14}$$

Correspondingly, define the quantities $\widehat{\mathrm{corr}}_{n,m,t}^{\mathrm{F}}$ and $\widehat{\mathrm{corr}}_{n,m,t}^{\epsilon}$ dividing by the appropriate variances. We provide the estimates of $\widehat{\mathrm{corr}}_{n,m,t}^{\epsilon}$ via the DCC framework. We deliberately do not adjust the residuals of the model by heteroscedasticty and/or serial correlation, which are instead treated as genuine features of the data. We denote the *model implied variance* of the n-th market by $\hat{\text{var}}_{n,t}$, which is defined as $\hat{\text{var}}_{n,t} \equiv \hat{\text{cov}}_{n,n,t}$.

During periods of financial distress, soaring empirical covariances are in general observed. Equation (13) shows that the covariance between R^n and R^m can rise through three different channels: an increase in the factor loadings β_t, an increase in the covariance of the factors $\Sigma_{F,t}$, and an increase residual covariance $\Sigma_{\epsilon,t}$. Bekaert et al. (2005) and the related literature identify contagion as the comovement between financial markets in excess of what is implied by an economic model. In this view, contagion is associated with spiking residual covariance between markets, which refers to the second term on the right-hand side of both Eqs. (13) and (14). In our modelling set-up, we take a different stand. Consistently with the case brought by Forbes and Rigobon (2002, pp. 2230–2231), *contagion* is thought as an episode of financial distress characterized by increasing interlinkages between markets. This event finds its model equivalent in a surge in the factor loadings β_t. On the contrary, spiking volatility in the factor conditional covariances is associated with *excess interdependence*. We formalize this notion in Definition 1 (*contagion*) and Definition 2 (*excess interdependence*) further in this paper.

Following Bekaert et al. (2009), we consider the average measure of model implied comovements:

$$\Gamma_t^{\mathrm{F}} \equiv \frac{1}{N(N-1)/2} \sum_{n=1}^{N} \sum_{m>n}^{N} \hat{\text{corr}}_{n,m,t}^{F} \tag{15}$$

and similarly we define Γ_t^{ϵ} as the residual comovement measure.

In order to characterize financial market comovements, we may assume that the residual covariance $\hat{\text{cov}}_{n,m,t}^{\epsilon}$ is negligible and focus our attention on the model implied covariance $\hat{\text{cov}}_{n,m,t}^{\mathrm{F}}$. There are two sources through which the covariance between two markets can surge: an increase in the factor loadings β_t, and/or increase in the factor volatilities $\Sigma_{F,t}$. In other words, assuming that our model fully captures the correlations between assets ($\mathbf{E}\left[\epsilon_t^n \epsilon_t^m\right] = 0$), the possible sources of a surge in the comovements are either soaring factor volatilities or increasing exposures to the factors. We label the former effect as contagion, whereas we call the latter excess interdependence.

We can get further insights into the covariance decomposition outlined in Eq. (12), by recalling that the factors $F_t^{i,j} = \left[G_t \;\; A_t^i \;\; C_t^j\right]$ are by construction mutually independent. Thus, from Eq. (12), denoting $n = \aleph(i_1, j_1)$ and $m = \aleph(i_2, j_2)$, it follows that:

$$\mathrm{cov}_t(R^n, R^m) = \mathbf{E}\left[\gamma_t^{n\prime} G_t^{\prime} G_t \gamma_t^m\right] + \mathbf{E}\left[\delta_t^{n\prime} A_t^{i_1^{\prime}} A_t^{i_2} \delta_t^m\right] + \mathbf{E}\left[\lambda_t^{n\prime} C_t^{j_1^{\prime}} C_t^{j_2} \lambda_t^m\right] + \mathbf{E}\left[\epsilon_t^n \epsilon_t^m\right] \tag{16}$$

with empirical counterpart of the form:

$$\mathrm{cov}_t(R^n, R^m) = \hat{\gamma}_t^{n\prime} \hat{\Sigma}_{G,t} \hat{\gamma}_t^m + \hat{\delta}_t^{n\prime} \hat{\Sigma}_{A,t}^{n,m} \hat{\delta}_t^m + \hat{\lambda}_t^{n\prime} \hat{\Sigma}_{C,t}^{n,m} \hat{\lambda}_t^m + \hat{\Sigma}_{\epsilon,t}^{n,m} \tag{17}$$

which for convenience we write as:

$$\hat{\mathrm{cov}}_{n,m,t} = \hat{\mathrm{cov}}_{n,m,t}^{\mathrm{G}} + \hat{\mathrm{cov}}_{n,m,t}^{\mathrm{A}} + \hat{\mathrm{cov}}_{n,m,t}^{\mathrm{C}} + \hat{\mathrm{cov}}_{n,m,t}^{\epsilon} \tag{18}$$

Our model framework has the advantage that it allows to discriminate among comovements due to global, asset class or country specific shocks. We define a measure of comovement prompted by the global factor as:

$$\Gamma_t^{\mathrm{G}} \equiv \frac{1}{N(N-1)/2} \sum_{n=1}^{N} \sum_{m>n}^{N} \hat{\mathrm{corr}}_{n,m,t}^{\mathrm{G}} \tag{19}$$

where:

$$\hat{\mathrm{corr}}_{n,m,t}^{\mathrm{G}} \equiv \frac{\hat{\mathrm{cov}}_{n,m,t}^{\mathrm{G}}}{\sqrt{\hat{\mathrm{var}}_{n,t}^{\mathrm{F}} \hat{\mathrm{var}}_{m,t}^{\mathrm{F}}}} \tag{20}$$

and can be seen as the part of the correlation between markets n and m, due to the common dependence on the global factor. In the same manner, we define Γ_t^{A} and Γ_t^{C} as the measures of comovements prompted by asset class and country factors, respectively. By construction we have: $\Gamma_t^{\mathrm{F}} \equiv \Gamma_t^{\mathrm{G}} + \Gamma_t^{\mathrm{A}} + \Gamma_t^{\mathrm{C}}$.

Let $\mathcal{I}_i$ be the set of indices from the sequence $n = 1, \ldots, N$ referred to markets belonging to the asset class i, and $\mathcal{J}_j$ be the indices referred to markets in country j, that is:

$$\mathcal{I}_i = \{n | n = \aleph(i,j); \quad j = 1, \ldots, J\} \tag{21}$$

$$\mathcal{J}_j = \{n | n = \aleph(i,j); i = 1, \ldots, I\} \tag{22}$$

The model implied comovement measure for asset class i is given by:

$$\Gamma_t^i \equiv \frac{1}{J(J-1)/2} \sum_{n \in \mathcal{I}_i} \sum_{\substack{m \in \mathcal{I}_i \\ m > n}} \hat{\text{corr}}_{n,m,t}^{\text{F}} \tag{23}$$

and in the same manner for country j, we have:

$$\Gamma_t^j \equiv \frac{1}{I(I-1)/2} \sum_{n \in \mathcal{J}_j} \sum_{\substack{m \in \mathcal{J}_j \\ m > n}} \hat{\text{corr}}_{n,m,t}^{\text{F}} \tag{24}$$

Along with the definition of comovement measures introduced so far, we propose a modification of them, to test for *contagion* versus *excess interdependence*. In the case of Γ_t^{F}, besides the definition in Eq. (15), we consider also:

$$\Gamma_{t,\text{ED}}^{\text{F}} \equiv \frac{1}{N(N-1)/2} \sum_{n=1}^{N} \sum_{m>n}^{N} \hat{\text{corr}}_{n,m,t,\text{ED}}^{\text{F}} \tag{25}$$

$$\Gamma_{t,\text{VD}}^{\text{F}} \equiv \frac{1}{N(N-1)/2} \sum_{n=1}^{N} \sum_{m>n}^{N} \hat{\text{corr}}_{n,m,t,\text{VD}}^{\text{F}} \tag{26}$$

where $\hat{\text{corr}}_{n,m,t,\text{ED}}^{\text{F}}$ and $\hat{\text{corr}}_{n,m,t,\text{VD}}^{\text{F}}$ are the correlation coefficients, respectively, associated with the following covariances:

$$\hat{\text{cov}}_{n,m,t,\text{ED}}^{\text{F}} \equiv \hat{\beta}_t^{n\prime} \hat{\Sigma}_F^{n,m} \hat{\beta}_t^{m} \tag{27}$$

$$\hat{\text{cov}}_{n,m,t,\text{VD}}^{\text{F}} \equiv \hat{\beta}^{n\prime} \hat{\Sigma}_{F,t}^{n,m} \hat{\beta}^{m} \tag{28}$$

$\Gamma_{t,\text{ED}}^{\text{F}}$ differs from Γ_t^{F} in the sense that the correlations used in its definition are computed assuming constant factor volatilities. In this case, the dynamics of the correlation between two markets is triggered by their time-varying exposures to common factors. We call this correlation measure as *exposure driven* (ED). On the contrary, $\Gamma^{F}_{t,\mathrm{VD}}$ is an average measure of comovements triggered by factor volatility only, while the exposures to the factors are kept constant according to their time series average. We call this type of comovements as *volatility driven* (VD). We consider the same two definitions for Γ^{G}_{t}, Γ^{A}_{t} and Γ^{C}_{t}, as well as for Γ^{i}_{t} and Γ^{j}_{t}.

The tools used in the analysis of the resulting time series are based on the Impulse-Indicator Saturation (IIS) technique implemented in Autometrics™, as part of the software PcGive™ (Hendry & Krolzig 2005, Doornik, 2009; Castle, Doornik, & Hendry, 2011). Castle, Doornik, and Hendry (2012) show that Autometrics IIS is able to detect multiple breaks in a time series when the dates of breaks are unknown. Furthermore, the authors demonstrate that the IIS procedure outperforms the standard Bai and Perron (1998) procedure. In particular, IIS is robust in presence of outliers close to the end and the start of the sample.[3]

Following Castle et al. (2012), we look for structural breaks in the generic $\Gamma^{(\cdot)}_{t}$ average comovement measures, by estimating the regression:

$$\Gamma^{(\cdot)}_{t} = \mu + \eta_t \tag{29}$$

where μ is a constant and η_t is assumed to be white noise. We then saturate the above regression using the IIS procedure, which retains into the model individual impulse-indicators in the form of spike dummy variables, signalling the presence of instabilities in the modelled series. These dummies occur in block between the dates of the breaks. In line with the procedure outlined in Castle et al. (2012), we group the dummy variables "with the same sign and similar magnitudes that occur sequentially" to form *segments* of dummies, whereas the impulse-indicators which can not be grouped will be labelled as *outliers*. A segment consists of at least two significant dummies, and at least two consecutive insignificant dummies need to occur to interrupt the segment. We interpret the segments of spike dummies as a step dummy for a particular regime. We can now state the following:

Definition 1 (*Contagion*). A situation of *contagion* is identified when a segment of dummy variables is detected through the IIS procedure for the average comovement measure $\Gamma^{(\cdot)}_{t,\mathrm{ED}}$.

Definition 2 (***Excess Interdependence***). A situation of *excess interdependence* is identified when a segment of dummy variables is detected through the IIS procedure for the average comovement measure $\Gamma^{(\cdot)}_{t,\mathrm{VD}}$.

We set a restrictive significance level of 1%, which leads to a parsimonious specification, as shown in Castle et al. (2012). Section 4.2 gives account of the results of the outlined methodology applied to our data.

3. DATA

We analyse comovements of equity indices, foreign exchange rates, money market instruments, corporate and government bonds in the United States, Euro Area, United Kingdom, Japan and Emerging Countries. Following the literature, to minimise the impact of nonsynchronous trading across different markets, we base our study on end of week data, spanning from 1 January 1999 to 14 March 2012, yielding 690 weekly observations. The starting date coincides with the adoption of the Euro, the Euro Area being one of the key geographical areas considered in the analysis. The sample offers the possibility to explore a variety of different market scenarios. The most notable facts are the speculation-driven market growth of the late 1990s, the financial and economic slowdown of the early 2000s, the burst of the markets during the mid-2000s, the financial turmoil of the period 2007–2009 and the following slow recovery, still pervaded by a high degree of uncertainty, prompted by the sovereign debt crisis in Europe and the United States between 2010 and 2012. This allows us to pick up from an in-sample analysis what the distinctive features of market comovements during crisis periods are.

Details on the time series used in this paper are reported in Table 1. The data sources are Datastream and Bloomberg. We adopt the MSCI definition of Emerging Markets and we select the five most relevant countries in term of size of their economy, according to the ranking based on the real annual GDP provided by the World Bank. Thus, we select China, Brazil, Russia, India, and Turkey as Emerging Countries.[4] We exclude from the analysis money and treasury markets for Japan and Emerging Market, as the series were affected by excess noise caused by measurement errors. We consider the US dollar as the numeraire: all the series are US dollar denominated and the US dollar is the base rate for the

Table 1. List of Variables Used in the Empirical Application.

ID variable	Asset class	Country	Name	Source (Ticker)
CorpBond/US	Corporate Bond	United States	BOFA ML US CORP	Datastream (MLCORPM)
CorpBond/EU	Corporate Bond	Euro Area	BOFA ML EMU CORP	Datastream (MLECEXP)
CorpBond/UK	Corporate Bond	United Kingdom	BOFA ML UK CORP	Datastream (ML£CAU$)
CorpBond/JP	Corporate Bond	Japan	BOFA ML JAP CORP	Datastream (MLJPCP$)
CorpBond/EM	Corporate Bond	Emerging Countries	BOFA ML EMERG CORP	Datastream (MLEMCB$)
EqInd/US	Equity Indices	United States	MSCI USA	Datastream (MSUSAML)
EqInd/EU	Equity Indices	Euro Area	MSCI EMU U$)	Datastream (MSEMUI$)
EqInd/UK	Equity Indices	United Kingdom	MSCI UK U$)	Datastream (MSUTDK$)
EqInd/JP	Equity Indices	Japan	MSCI JAPAN U$)	Datastream (MSJPAN$)
EqInd/EM	Equity Indices	Emerging Countries	MSCI EM U$)	Datastream (MSEMKF$)
FX/EU	Foreign Exchange	Euro Area	FX Spot Rate	Bloomberg (EURUSD Curncy)
FX/UK	Foreign Exchange	United Kingdom	FX Spot Rate	Bloomberg (GBPUSD Curncy)
FX/JP	Foreign Exchange	Japan	FX Spot Rate	Bloomberg (JPYUSD Curncy)
FX/EM	Foreign Exchange	Emerging Countries	FX Spot Rate	Bloomberg (BRLUSD, CNYUSD, INRUSD, RUBUSD, TRYUSD Curncy)
MoneyMkt/US	Money Market	United States	3 month US Libor	Bloomberg (US0003M Index)
MoneyMkt/EU	Money Market	Euro Area	3 month Euribor	Bloomberg (EUR003M Index)
MoneyMkt/UK	Money Market	United Kingdom	3 month UK Libor	Bloomberg (BP0003M Index)
Tr/US	Treasury	United States	US Govt 10 Year Yield	Bloomberg (USGG10YR Index)
Tr/EU	Treasury	Euro Area	EU Govt 10 Year Yield	Bloomberg (GECU10YR Index)
Tr/UK	Treasury	United Kingdom	UK Govt 10 Year Yield	Bloomberg (GUKG10 Index)

Notes: We report the acronyms used to identify each variable (ID variable), the asset class and the country to which they belong, the name of the series, together with the data provider and the ticker for series identification.

Table 2. Descriptive Statistics for the Market Returns.

	Mean (%)	St Dev (%)	Min (%)	Max (%)	Skewness (%)	Kurtosis (%)
CorpBond/US	0.119	0.748	−5.355	3.171	−0.935	8.553
CorpBond/EU	0.103	1.558	−5.815	5.385	−0.194	3.512
CorpBond/UK	0.100	1.612	−13.152	5.628	−1.075	10.651
CorpBond/JP	0.092	1.347	−5.356	8.924	0.572	6.755
CorpBond/EM	0.163	0.826	−9.332	3.724	−3.717	38.973
EqInd/US	0.014	2.747	−20.116	11.526	−0.748	8.850
EqInd/EU	−0.011	3.502	−26.679	12.245	−1.073	9.576
EqInd/UK	−0.009	3.091	−27.618	16.243	−1.249	14.920
EqInd/JP	0.009	2.887	−16.402	11.016	−0.258	4.823
EqInd/EM	0.184	3.380	−22.564	18.538	−0.775	8.889
FX/EU	0.017	1.468	−6.048	4.992	−0.213	3.831
FX/UK	−0.009	1.341	−8.348	5.195	−0.588	6.546
FX/JP	0.050	1.498	−6.027	7.445	0.253	4.304
FX/EM	−0.142	1.517	−17.401	4.786	−2.961	29.634
MoneyMkt/US	−0.350	3.814	−27.877	21.137	−1.850	16.873
MoneyMkt/EU	−0.187	2.087	−11.989	15.021	−0.717	11.723
MoneyMkt/UK	−0.262	2.091	−26.170	8.374	−4.357	43.968
Tr/US	−0.126	3.596	−19.122	12.110	−0.045	5.511
Tr/EU	−0.116	3.056	−17.838	14.018	−0.353	6.476
Tr/UK	−0.105	2.805	−16.758	11.153	−0.362	5.943

Notes: We report summary statistics for the variable used in the empirical application. The number reported refers to the entire sample, which consists of weekly observations from Jan-1999 to Mar-2012.

FX pairs in the dataset. In what follows, we consider simple weekly percentage returns for equity indices, Bond Indices and foreign exchange rates, whereas weekly first differences are considered for money market and goverment rates series. In Table 2, we report some descriptive statistics of the variables.

The most remarkable facts are the extreme values that were recorded in correspondence of the 2008–2009 crisis period. This is particularly evident for stock markets and for short term rates, whereas along the country spectrum, the most hit were Emerging Markets. All series exhibit the typical characteristic of non-normality with high asymmetry and kurtosis. The price series are plotted in Fig. 1. The downturn at the end of the year 2008 is immediately apparent and common to all the considered series.

We propose a dynamic factor model with multiple sources of shocks, at global, asset class and country level. In order to validate this approach, a first preliminary correlation analysis is undertaken. Table 3 reports the in-sample correlation of the modelled variables. We observe high correlation intra-asset class groups. Particularly remarkable are the cases of equity and treasury rates, with correlations in the 70–80% range. We observe substantial correlation even within countries, in particular there is evidence of high interconnection between corporate bonds and FX markets at country level: Euro Area (91.3%), Japan (83.6%) and United Kingdom (83.3%). Hence, there is evidence for the presence of both an asset class and a country effect. However, the asset class effect seems to be systematically more pervasive than the country one. Finally, the correlation is high in three clusters (equity indices, corporate bonds and FX, and treasury rate) and treasury rates.

4. EMPIRICAL RESULTS

In this section, we report the estimates of the dynamic multi-factor model formulated in Section 2. In particular, in Section 4.1 we report the results of the estimation of the factors and the specification of the factor loadings, in Section 4.2 the empirical analysis of market comovements, both the estimates of measures of market comovements (Section 4.2.1) and the regime of contagion versus excess interdependence we identify in market comovements (Section 4.2.2).

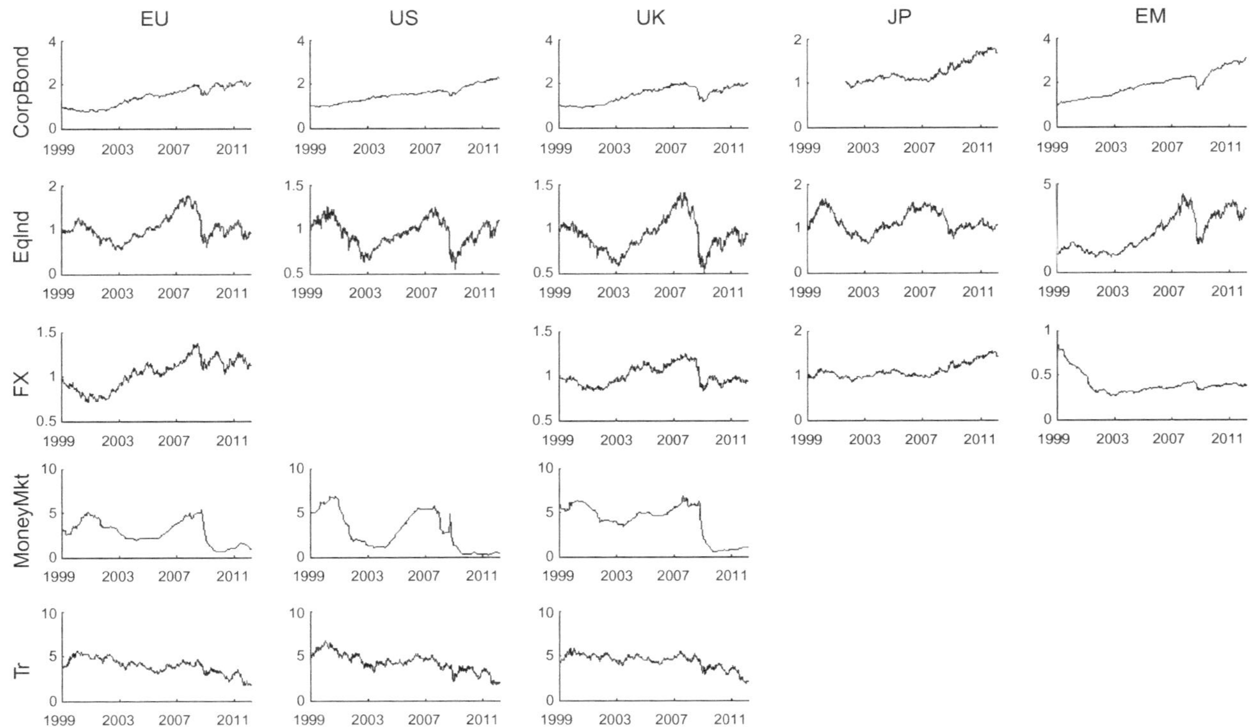

Fig. 1. Price Data Used in the Empirical Application. *Notes*: Asset classes are displayed in the rows, whereas countries are in the columns. We plot the weekly price series for the considered markets. Corporate bond, equity indices and foreign exchange rates (top three rows) are rebased using the first available observation. US foreign exchange is excluded from the analysis because is used as the numeraire. The other missing series are not considered due to lack of data.

Table 3. Sample Correlations among the Market Returns.

	Corp Bond/ US	Corp Bond/ EU	Corp Bond/ UK	Corp Bond/ JP	Corp Bond/ EM	EqInd/ US	EqInd/ EU	EqInd/ UK	EqInd/ JP	EqInd/ EM	FX/ EU	FX/ UK	FX/JP	FX/ EM	Money Mkt/ US	Money Mkt/ EU	Money Mkt/ UK	Tr/ US	Tr/ EU
Corp Bond/EU	0.393																		
Corp Bond/UK	0.462	0.694																	
Corp Bond/JP	0.264	0.312	0.171																
Corp Bond/EM	0.578	0.539	0.516	0.046															
EqInd/US	−0.041	0.087	0.080	−0.260	0.207														
EqInd/EU	0.004	0.411	0.288	−0.161	0.351	0.785													
EqInd/UK	0.028	0.326	0.403	−0.229	0.329	0.764	0.884												
EqInd/JP	0.145	0.238	0.208	0.129	0.272	0.405	0.480	0.444											
EqInd/EM	0.044	0.286	0.272	−0.211	0.448	0.693	0.785	0.755	0.545										
FX/EU	0.178	0.913	0.566	0.232	0.402	0.126	0.435	0.333	0.217	0.291									
FX/UK	0.162	0.616	0.833	0.060	0.358	0.156	0.373	0.502	0.251	0.336	0.642								
FX/JP	0.181	0.295	0.138	0.836	0.054	−0.226	−0.112	−0.177	0.269	−0.140	0.262	0.101							
FX/EM	0.079	0.349	0.296	−0.151	0.356	0.434	0.548	0.522	0.244	0.593	0.338	0.328	−0.094						
Money Mkt/US	−0.342	−0.233	−0.174	−0.133	−0.247	−0.020	−0.094	−0.091	−0.076	−0.088	−0.144	−0.081	−0.102	−0.077					
Money Mkt/EU	−0.177	−0.056	0.001	−0.009	−0.141	0.030	0.053	0.034	0.029	−0.005	0.015	0.036	−0.007	−0.018	0.385				
Money Mkt/UK	−0.227	−0.104	−0.002	−0.077	−0.245	0.061	−0.002	0.033	−0.017	−0.037	−0.023	0.119	−0.083	−0.003	0.536	0.525			
Tr/US	−0.733	−0.246	−0.292	−0.388	−0.230	0.329	0.294	0.262	0.061	0.275	−0.104	−0.037	−0.295	0.152	0.141	0.112	0.090		
Tr/EU	−0.548	−0.171	−0.280	−0.310	−0.152	0.297	0.337	0.276	0.114	0.277	0.045	0.040	−0.219	0.167	0.105	0.152	0.118	0.731	
Tr/UK	−0.531	−0.205	−0.330	−0.313	−0.176	0.266	0.268	0.267	0.121	0.247	−0.031	0.083	−0.208	0.133	0.084	0.105	0.159	0.715	0.798

4.1. Factor Estimates and Factor Loading Selection

We start our empirical analysis by extracting the factors according to the methodology outlined in Section 2.1. We extract the first principal component at a global, asset class and country level from the estimate of the covariance matrix of the demeaned return time series. The factors have by construction zero mean.

The extracted factors account in total for 83.28% of the overall variance, thus explaining a substantial amount of the variation of the considered return series. In particular, the global factor extracts as much as 37.27% of the overall variance, whereas the asset class and the country factors account for a quota in the 50–80% range of the variation in the groups they are extracted from.

We then orthogonalize the extracted factors, so that the system $\hat{F}_t^{i,j} \equiv [\hat{G}_t\ \hat{A}_t^i\ \hat{C}_t^j]$ with $i = 1, \ldots, I$ and $j = 1, \ldots, J$ consists of orthogonal factors. We first orthogonalize each of the asset class factors with respect to the global factor and then orthogonalize the country factors with respect to both the global and the asset class factors. In Section 4.2, we show that all our main results do not depend on the particular way the orthogonalization is carried out.

To validate the interpretations we attached to the factors, we map the contributions of the original variables onto the factors via linear correlation analysis. The result of this analysis is reported in Table 4.

We find that the stock indices are most highly correlated with the global factors, with correlations in the 80–90% range. This characterizes the global factor as the momentum factor. Such an interpretation seems reasonable in view of the fact that the equity asset class can be thought as the most direct indicator of the financial activity among the asset classes considered here.

More generally, when we sort the different markets by the magnitude of their correlation with the global factor, they tend to group by asset class, rather than by country, with the Treasury and the FX market figure in the 30–50% range and the money market and the corporate bond market in the 0–30% range. This again supports the evidence that the asset class effect is more pervasive than the country effect.

To test for excess interdependence prompted by changes in the volatility of the factors, we entertain the possibility that the factor time series might be characterized by volatility clustering. In Table 5, we report the Engle test for residual heteroscedasticity that suggests that at the 1%

Table 4. Correlations between the Market Returns and the Extracted Factors.

	Global	Corp Bond	EqInd	FX	Money Mkt	Tr	US	EU	UK	JP	EM
CorpBond/US	−0.188	0.595	0.684	0.325	−0.337	−0.714	−0.234	0.017	0.032	0.098	0.059
CorpBond/EU	0.237	0.884	0.350	0.822	−0.211	−0.472	0.028	0.150	−0.120	−0.039	−0.044
CorpBond/UK	0.185	0.882	0.425	0.700	−0.131	−0.550	0.038	−0.163	0.145	0.052	−0.022
CorpBond/JP	−0.300	0.450	0.279	0.402	−0.128	−0.245	−0.106	0.107	−0.027	−0.129	0.015
CorpBond/EM	0.281	0.587	0.413	0.377	−0.248	−0.475	−0.007	−0.078	−0.153	0.061	0.278
EqInd/US	0.816	−0.072	0.248	−0.146	0.004	−0.214	0.138	−0.029	−0.035	−0.136	−0.167
EqInd/EU	0.907	0.193	0.275	0.127	−0.066	−0.284	−0.008	0.263	−0.024	−0.129	−0.156
EqInd/UK	0.875	0.206	0.290	0.119	−0.057	−0.307	−0.016	−0.011	0.312	−0.118	−0.188
EqInd/JP	0.541	0.177	0.350	0.096	−0.063	−0.288	−0.143	−0.181	−0.101	0.624	0.036
EqInd/EM	0.854	0.143	0.289	0.078	−0.072	−0.290	0.009	−0.177	−0.167	0.016	0.422
FX/EU	0.313	0.730	0.167	0.830	−0.117	−0.293	−0.012	0.226	−0.114	−0.078	−0.099
FX/UK	0.373	0.683	0.133	0.720	−0.032	−0.260	−0.061	−0.185	0.328	0.027	−0.092
FX/JP	−0.205	0.379	0.226	0.441	−0.104	−0.186	−0.089	0.044	−0.024	−0.026	0.022
FX/EM	0.557	0.225	0.122	0.421	−0.061	−0.222	0.119	−0.131	−0.169	0.088	0.212
MoneyMkt/US	−0.021	−0.241	−0.175	−0.139	0.973	0.187	0.133	−0.019	−0.034	−0.059	0.040
MoneyMkt/EU	0.092	−0.061	−0.138	−0.030	0.551	0.108	−0.360	0.152	−0.021	0.193	−0.096
MoneyMkt/UK	0.065	−0.104	−0.150	−0.010	0.697	0.118	−0.332	−0.031	0.176	0.120	−0.109
Tr/US	0.559	−0.474	−0.716	−0.350	0.153	0.738	0.309	−0.142	−0.123	−0.025	0.028
Tr/EU	0.577	−0.403	−0.700	−0.224	0.130	0.708	−0.229	0.248	−0.045	0.016	−0.025
Tr/UK	0.536	−0.439	−0.695	−0.239	0.118	0.712	−0.250	−0.059	0.266	0.023	−0.017

Notes: We report the correlation between the factors and the market returns from which the factors are extracted. There are 20 series displayed in the rows and 11 factors (one global, five asset class and five country factors), which are displayed in the columns. The numbers reported are in-sample linear correlations.

Table 5. Engle Test for Residual Heteroscedasticity for the Estimated Factors.

FACTOR	STAT
Global	51.982***
CorpBond	7.577***
EqInd	0.458
FX	3.254*
MoneyMkt	59.335***
Tr	0.318
US	31.535***
EU	21.421***
UK	26.668***
JP	3.386*
EM	25.878***

Notes: We report the results of the test for residual heteroscedasticity for the 11 extracted factors (one global, five asset class and five country factors). The first columns reports the name of the factor, the second reports the test statistics in the Engle test for residual heteroscedasticity. In the third column, ***, ** and * indicate rejection of the null of no ARCH effect at the 1%, 5% and 10% significance level, respectively.

Table 6. Likelihood Ratio Test for the Alternative Models.

Null model	Alternative model Time-varing factor loading	Conditional time-varying factor loading
Static factor loading	260,142.36***	261,869.86***
Time-varing factor loading		1,727.50***

Notes: We report the test statistics for the likelihood ratio test comparing the proposed alternative models. The test is employed to evaluate the null hypothesis that the *Null model* provides a better fit than the *Alternative model*. The models refer to the following alternative formulation for the factor loadings: the static factor loading in Eq. (8), the time-varying factor loading in Eq. (9) and the conditional time-varying factor loading in Eq. (7). *** indicates rejection of the null model at the 1% significance level.

confidence level this is indeed the case for 7 out of the 11 estimated factors.

We fit the Engle's DCC model on the series of the estimated factors to get a time-varying estimate of their covariance matrix.

We estimate (10) via OLS when we use the static formulation (8) for the factor loadings, while when the factor loadings are specified as in either the time-varying (9) or the conditional time-varying factor loading (7) model, we estimate (10) via the Kalman filter using maximum likelihood estimation. The models are nested and thus the likelihood ratio test can be employed for model selection. The likelihood ratio statistics are reported in Table 6.

The test strongly rejects the static alternative in favour of the dynamic ones. The conditional time-varying factor loading approach dominates the time-varying factor loading approach. Thus, there is evidence that the fitting of the model improves when we control for market nervousness by means of the control factor Z.

4.2. *Financial Market Comovements Dynamics*

4.2.1. *Measures of Comovements*

We turn now to analyse the average measures of comovements introduced in Section 2.4.

We start with the comparison between Γ_t^F and Γ_t^ϵ. The two measures are plotted in Fig. 2.

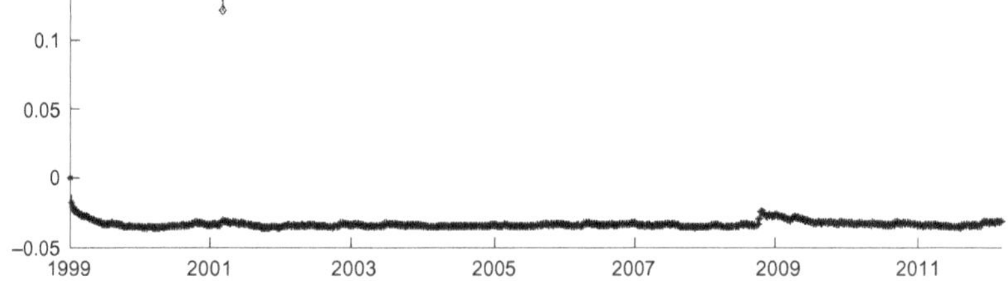

Fig. 2. Model Implied versus Residual Average Correlation Measures. *Notes*: Γ_t^F is the average comovement measure at the overall level, defined as the mean of the model implied correlations between all the couples of asset considered. Γ_t^ϵ is the average residual comovement measure, defined as the mean of the correlations between the error term in the model for all the couples of asset considered.

As it can be clearly seen, the residual component is negligible throughout the sample period and on average does not convey any information about the dynamics of the comovements of the considered markets. We observed only a small jump in the idiosyncratic component in correspondence to late 2008, which has been considered by many the harshest period of the 2007–2009 global financial crisis. The model-implied measure of average comovements Γ_t^F fluctuates around what can be regarded as a constant long-run value of roughly 20%. This erratic behaviour does not allow us to identify any peak in correlation possibly associated to crisis periods. During the period 2007–2009, a slightly lower average correlations seem to be observed instead. We give account of this fact in what follows, by disaggregating the model implied covariation measure Γ_t^F.

We start doing this by considering the decomposition of the overall comovement measure Γ_t^F into Γ_t^G, Γ_t^A and Γ_t^C, which is presented in Fig. 3. The global factor appears to be the most pervasive of all the three factors considered, shaping the dynamics of the average overall measure. The asset class factor is slightly less pervasive, but it is the most persistent of the three, meaning that its contribution is more resilient to change over time. This expresses the fact that the characteristics that are common to the asset class contribute in a constant proportion to the average overall market correlation. The least important factor is the country one, which is almost negligible. Thus, comovements typically propagate through two channels: a

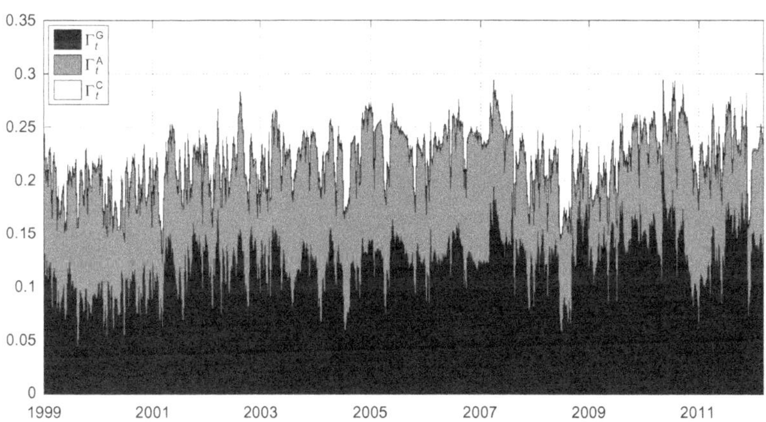

Fig. 3. Decompositions of the Overall Average Comovements by Source of the Shock. *Notes*: Γ_t^G, Γ_t^A Γ_t^C are the average measures of comovement prompted by the global, the asset class and the country factor, respectively.

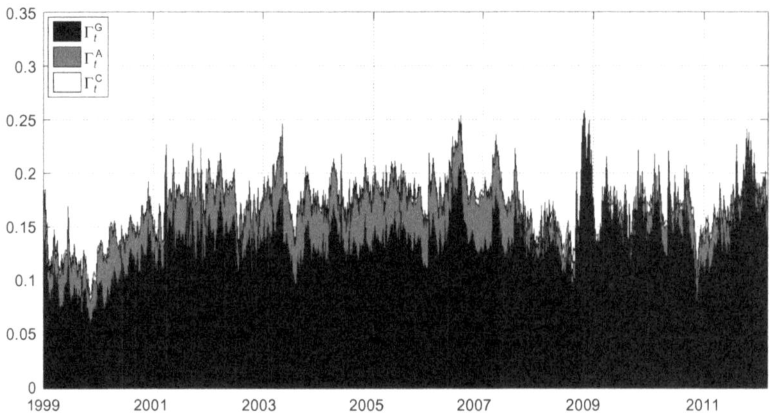

Fig. 4. Robustness Check of the Decomposition by Source. *Notes*: Fig. 3 reports the decompositions of the overall average comovements by source of the shock, for the case in which the asset class factors are first orthogonalized with respect to the global factor and then the country factors are orthogonalized with respect to the asset class and the global factors. Here, we report the same decomposition for the case in which the country factors are orthogonalized with respect to the global factor and then the asset class factors are orthogonalized with respect to the others.

global one, in a time-varying manner, and an asset class channel, according to a constant contribution.

We consider robustness checks of these conclusions, by pursuing an alternative strategy in orthogonalizing the system of factors considered here. We first orthogonalize the country factor against the global and then the asset class one with respect to the other two. Then we re-estimate the model and construct the comovement measures. Fig. 4 shows the results. The dynamics of the comovements are similar. The decomposition changes in favour of the global factor, which is even more pervasive than before. However, the country contribution is almost absent, even when the country factors are extracted and orthogonalized with priority, thus validating our orthogonalization method.

4.2.2. Testing for Contagion versus Excess Interdependence

In this section, we propose an empirical analysis of the comovement measures introduced above by testing for the presence of different regimes in the resulting time series by means of Autometrics. Figs. 5–7 report the

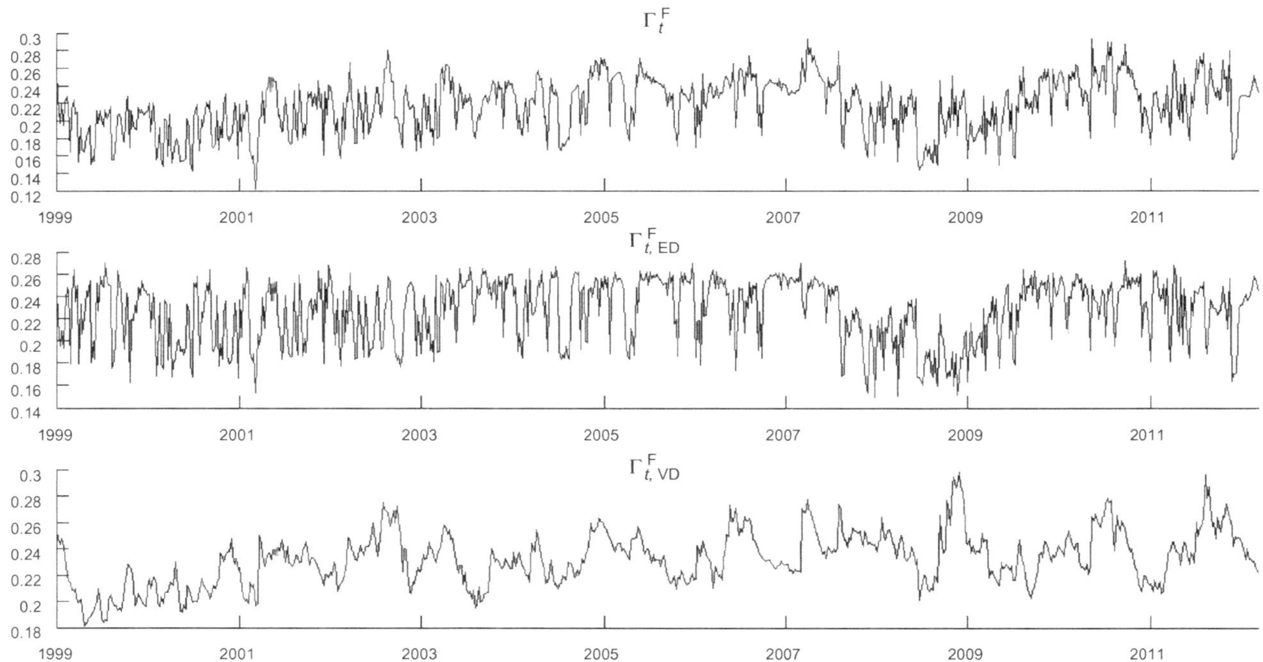

Fig. 5. Average Correlation Measures. *Notes*: Γ_t^F (top panel) is the average comovement measure at the overall level, defined as the mean of the model implied correlations between all the couples of asset considered. $\Gamma_{t,ED}^F$ –mid panel– ($\Gamma_{t,VD}^F$ –bottom panel) considers the correlations for the case in which factor exposures are allowed to vary with time (held at constant) and factor covariances are held at constant (allowed to vary with time).

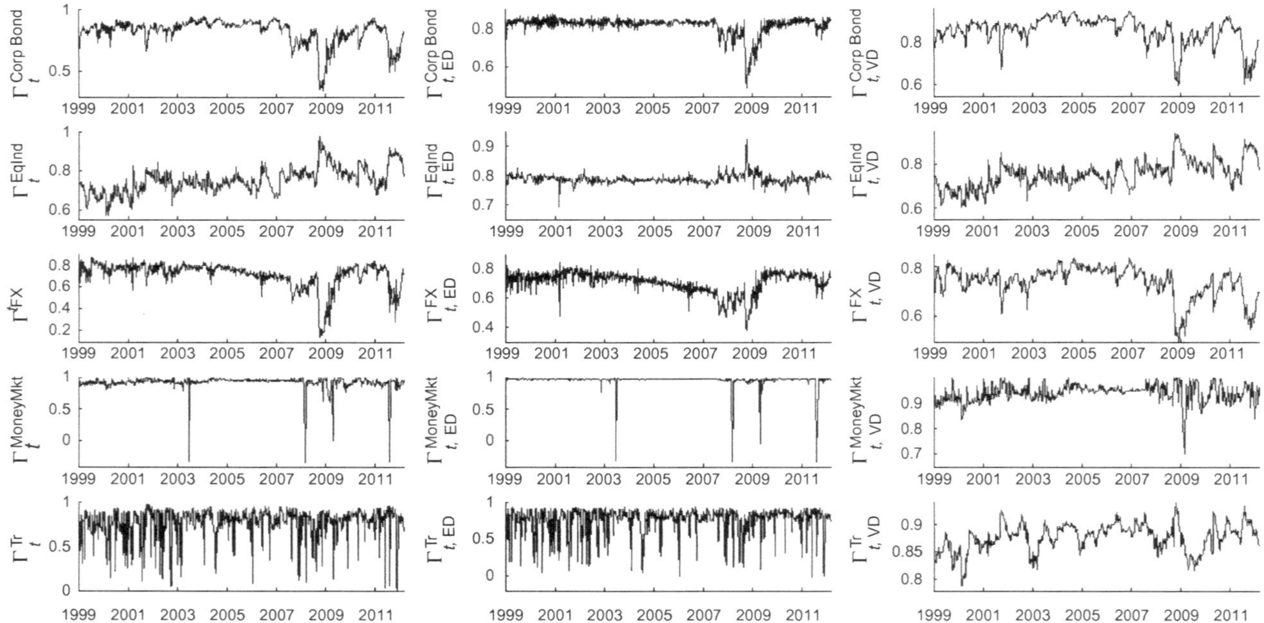

Fig. 6. Average Correlation Measures at the Asset Class Level. *Notes*: $\Gamma_t^{\text{CorpBond}}$ is the average comovement measure within the corporate bond market, defined as the mean of the model implied correlations between all the couples of securities in the corporate bond asset class. $\Gamma_t^{\text{EqInd}}, \Gamma_t^{\text{FX}}, \Gamma_t^{\text{MoneyMkt}}$ and Γ_t^{Tr} are analogously defined for the other asset classes. Exposure-driven (second column) and volatility-driven (third column) comovement measures consider the correlations for the case in which factor exposures are allowed to vary with time (held at constant) and factor covariances are held at constant (allowed to vary with time).

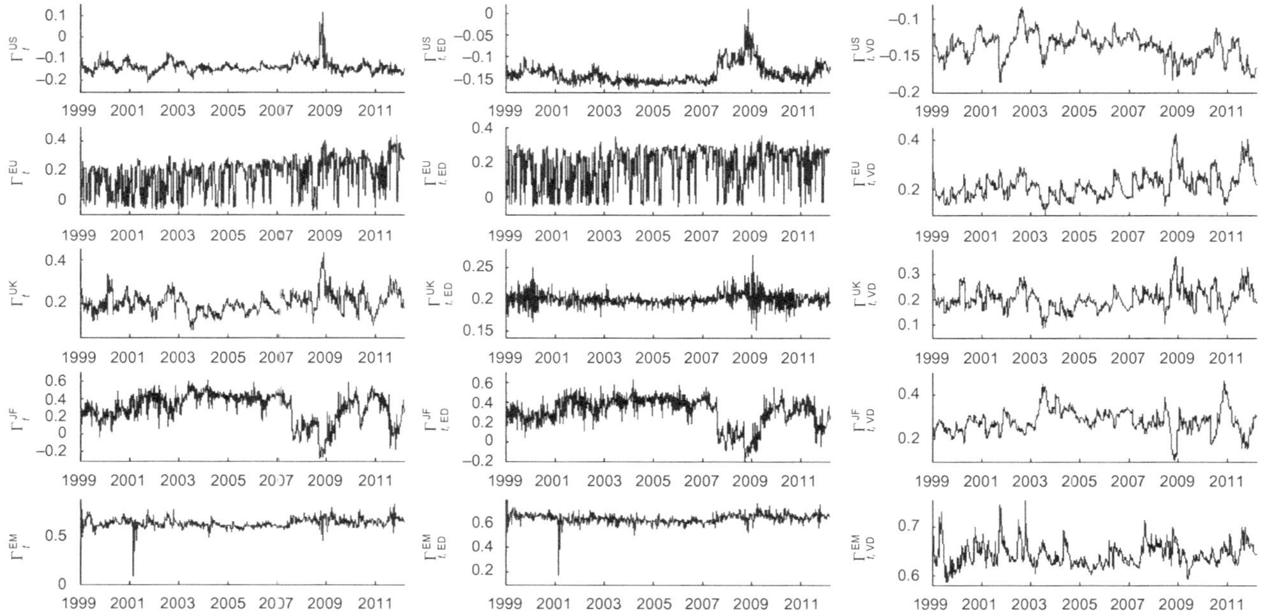

Fig. 7. Average Correlation Measures at the Country Level. *Notes*: Γ_t^{US} is the average comovement measure within the US market, defined as the mean of the model implied correlations between all the couples of securities in the US group. $\Gamma_t^{EU}, \Gamma_t^{UK}, \Gamma_t^{JP}$ and Γ_t^{EM} are analogously defined for the other countries. Exposure-driven (second column) and volatility-driven (third column) comovement measures consider the correlations for the case in which factor exposures are allowed to vary with time (held at constant) and factor covariances are held at constant (allowed to vary with time).

Table 7. IIS Results for the Overall Average Comovement Measures.

Γ_t^F	
Outliers	
26/02/1999	−0.0583**
…	
16/12/2011	−0.0584**
Constant	0.2230***
$\Gamma_{t,\mathrm{ED}}^F$	
Segments	
17/08/2007–21/11/2008	−0.0670***
Outliers	
07/04/2000	−0.0608**
30/06/2000	−0.0607**
09/03/2001	−0.0746***
25/11/2011	−0.0646***
02/12/2011	−0.0583**
Constant	0.2282***
$\Gamma_{t,\mathrm{VD}}^F$	
Segments	
31/10/2008–05/12/2008	0.0564***
12/08/2011–26/08/2011	0.0594***
Outliers	
23/04/1999	−0.0507***
Constant	0.2320***

Notes: Γ_t^F is the average comovement measure at the overall level, defined as the mean of the model implied correlations between all the couples of asset considered. $\Gamma_{t,\mathrm{ED}}^F(\Gamma_{t,\mathrm{VD}}^F)$ considers the correlations for the case in which factor exposures are allowed to vary with time (held at constant) and factor covariances are held at constant (allowed to vary with time). We report the results of the saturation of model in Eq. (29) by means of Autometrics. We report the dates detected via the IIS technique, together with the estimated coefficients. *Segment* refers to group of sequential dummies with the same size and similar magnitude. *Outliers* are dummies which can not be grouped. *Constant* refers to the constant term μ in Eq. (29) (***, ** and * indicate significance of the coefficient at the 1%, 5% and 10% significance level, respectively).

time series analysed. Tables 7–9 show the result of this procedure applied to our data.

Let us start with the analysis of the results for Γ_t^F, $\Gamma_{t,\mathrm{ED}}^F$ and $\Gamma_{t,\mathrm{VD}}^F$ as reported in Table 7. As previously noted for Fig. 2, not surprisingly, we do not find any structural clear pattern in the IIS retained by Autometrics when applied to Γ_t^F. We find outliers only, instead. However, when looking at $\Gamma_{t,\mathrm{VD}}^F$ we find evidence of excess interdependence, that is excess average

Table 8. IIS Results for the Average Comovements Measures at the Asset Class Level.

$\Gamma_t^{CorpBond}$			Γ_t^{EqInd}			Γ_t^{FX}			$\Gamma_t^{MoneyMkt}$			Γ_t^{Tr}		
Segments			*Segments*			*Segments*			*Segments*			*Outliers*		
24/08/2007–26/09/2008	−0.1614	***	10/10/2008–27/03/2009	0.1664	***	26/05/2006–04/08/2006	−0.1623	***	29/01/1999–20/04/2001	−0.0748	***	15/01/1999	−0.4502	***
03/10/2008–23/01/2009	−0.4134	***	12/08/2011–03/02/2012	0.1447	***	16/03/2007–25/07/2008	−0.1681	***	14/03/2008–28/03/2008	−0.9199	***	…		
06/02/2009–04/06/2010	−0.1867	***	*Outliers*			19/09/2008–15/05/2009	−0.3993	***	11/04/2008–19/09/2008	−0.0650	***	16/12/2011	−0.7062	***
12/08/2011–03/02/2012	−0.3610	***	01/01/1999	0.1596	***	12/08/2011–27/01/2012	−0.2528	***	26/09/2008–14/11/2008	0.0521	**	*Constant*	0.8238	***
Outliers			18/02/2000	−0.1695	***	*Outliers*			06/02/2009–22/01/2010	−0.1794	***			
22/01/1999	−0.1643	***	24/03/2000	−0.1718	***	22/01/1999	−0.1107	**	07/05/2010–28/05/2010	0.0525	**			
21/04/2000	−0.1535	***	*Constant*	0.7408	***	07/04/2000	−0.1149	***	15/04/2011–29/07/2011	−0.1010	***			
28/09/2001	−0.1812	***				09/03/2001	−0.1756	***	12/08/2011–19/08/2011	−1.2057	***			
05/10/2001	−0.1859	***				28/09/2001	−0.1433	***	11/11/2011–27/01/2012	−0.1060	***			
12/10/2001	−0.1343	***				21/05/2004	−0.1340	***	*Outliers*					
Constant	0.8557	***				24/07/2009	−0.1781	***	28/09/2001	0.0570	**			
						21/05/2010	−0.1111	**	16/11/2001	0.0550	**			
						Constant	0.7481	***	22/11/2002	−0.1174	***			
									21/02/2003	−0.0593	**			
									28/03/2003	−0.1570	***			
									04/04/2003	−0.0613	**			
									27/06/2003	−1.2632	***			
									01/02/2008	0.0524	**			
									29/02/2008	−0.0606	**			
									05/12/2008	−0.3098	***			
									26/12/2008	0.0502	**			
									02/07/2010	−0.0835	***			
									03/09/2010	0.0522	**			
									26/11/2010	−0.0771	***			
									14/01/2011	−0.0900	***			
									Constant	0.9421	***			

Table 8. *(Continued)*

$\Gamma_{t,\mathrm{ED}}^{CorpBond}$			$\Gamma_{t,\mathrm{ED}}^{EqInd}$			$\Gamma_{t,\mathrm{ED}}^{FX}$			$\Gamma_{t,\mathrm{ED}}^{MoneyMkt}$			$\Gamma_{t,\mathrm{ED}}^{Tr}$		
Segments			*Segments*			*Segments*			*Segments*			*Outliers*		
24/08/2007–05/09/2008	−0.0805	***	24/08/2007–15/05/2009	0.0448	***	19/03/1999–09/07/1999	−0.1037	***	08/11/2002–29/11/2002	−0.0623	***	08/01/1999	−0.1247	**
19/09/2008–03/04/2009	−0.1858	***	*Outliers*			20/05/2005–19/09/2008	−0.1385	***	14/02/2003–20/06/2003	−0.0190	***	…		
10/04/2009–18/09/2009	−0.0686	***	01/01/1999	0.1132	***	26/09/2008–02/01/2009	−0.2525	***	17/08/2007–01/02/2008	−0.0207	***	16/12/2011	−0.7230	***
12/08/2011–23/12/2011	−0.0628	***	26/03/1999	0.0267	***	09/01/2009–15/05/2009	−0.1326	***	14/03/2008–21/03/2008	−1.3015	***	*Constant*	0.8304	***
Outliers			18/06/1999	0.0387	***	12/08/2011–25/11/2011	−0.1176	***	04/04/2008–24/04/2009	−0.0313	***			
22/01/1999	−0.0803	***	02/03/2001	−0.0931	***	*Outliers*			05/06/2009–27/11/2009	−0.0204	***			
22/10/1999	−0.0574	***	28/09/2001	−0.0413	***	22/01/1999	−0.0871	**	14/05/2010–29/07/2011	−0.0311	***			
21/04/2000	−0.0419	***	05/10/2001	−0.0386	***	14/01/2000	−0.1109	***	12/08/2011–19/08/2011	−1.2619	***			
15/09/2000	0.0415	***	15/03/2002	0.0367	***	07/04/2000	−0.1090	***	26/08/2011–16/12/2011	−0.0321	***			
10/11/2000	0.0403	***	09/03/2007	−0.0312	***	09/03/2001	−0.2604	***	*Outliers*					
08/12/2000	0.0493	***	17/07/2009	−0.0435	***	16/03/2001	0.1068	***	18/06/1999	−0.0142	***			
05/01/2001	0.0440	***	14/08/2009	−0.0267	***	*Constant*	0.7349	***	30/07/1999	−0.0134	***			
02/08/2002	−0.0440	***	12/03/2010	−0.0257	***				17/09/1999	0.0157	***			
01/08/2003	0.0403	***	14/05/2010	−0.0460	***				08/10/1999	−0.0175	***			
Constant	0.8276	***	21/05/2010	−0.0336	***				12/01/2001	−0.0233	***			
			20/08/2010	0.0271	***				10/08/2001	−0.0259	***			
			08/04/2011	−0.0345	***				28/12/2001	−0.0165	***			
			15/04/2011	−0.0522	***				08/02/2002	−0.0148	***			
			26/08/2011	0.0293	***				27/06/2003	−1.2770	***			
			23/09/2011	0.0290	***				28/03/2008	−0.2650	***			
			11/11/2011	0.0305	***				01/05/2009	−1.0106	***			
			Constant	0.7860	***				29/05/2009	−0.2165	***			
									Constant	0.9720	***			

$\Gamma_{t,VD}^{CorpBond}$			$\Gamma_{t,VD}^{EqInd}$			$\Gamma_{t,VD}^{FX}$			$\Gamma_{t,VD}^{MoneyMkt}$			$\Gamma_{t,VD}^{Tr}$		
Segments			*Segments*			*Segments*			*Segments*			*Segments*		
28/09/2001–26/10/2001	−0.1392	***	18/02/2000–14/04/2009	−0.1367	***	28/09/2001–02/11/2001	−0.1156	***	29/01/1999–30/04/1999	−0.0511	**	01/10/1999–05/05/2000	−0.0650	***
23/12/2005–27/01/2006	0.0485	*	12/09/2008–13/03/2009	0.1491	***	01/04/2005–15/04/2005	0.0495	*	17/12/1999–10/03/2000	−0.0891	***	06/12/2002–21/03/2003	−0.0536	***
24/08/2007–28/09/2007	−0.1147	***	14/05/2010–28/05/2010	0.1366	***	19/09/2008–04/12/2009	−0.1399	***	26/09/2008–16/01/2009	0.0540	**	31/10/2008–07/11/2008	0.0571	***
17/10/2008–31/07/2009	−0.1823	***	12/08/2011–10/02/2012	0.1347	***	14/05/2010–16/07/2010	−0.0955	***	06/02/2009–20/03/2009	−0.1680	***	17/04/2009–11/09/2009	−0.0502	***
14/05/2010–11/06/2010	−0.1108	***	*Constant*	0.7437	***	15/07/2011–17/02/2012	−0.1452	***	22/05/2009–29/05/2009	0.0534	**	*Outliers*		
08/07/2011–24/02/2012	−0.1627	***				*Outliers*			17/07/2009–11/09/2009	0.0525	**	29/01/1999	−0.0454	**
Outliers						14/05/1999	−0.0794	***	16/10/2009–20/11/2009	−0.0697	***	12/08/2011	0.0545	***
21/04/2000	−0.0857	***				18/10/2002	−0.1335	***	07/05/2010–03/09/2010	0.0541	**	*Constant*	0.8791	***
04/07/2003	0.0503	*				25/10/2002	−0.0899	***	*Outliers*					
12/12/2003	0.0499	*				30/07/2004	0.0523	*	15/10/1999	0.0491	**			
30/09/2005	0.0513	*				*Constant*	0.7604	***	19/09/2008	−0.0521	**			
15/02/2008	−0.0854	***							19/06/2009	−0.0690	***			
Constant	0.8623	***							08/01/2010	−0.0504	**			
									26/11/2010	−0.0522	**			
									11/11/2011	−0.0583	**			
									06/01/2012	−0.0865	***			
									13/01/2012	−0.1019	***			
									Constant	0.9401	***			

Notes: $\Gamma_t^{CorpBond}$ is the average comovement measure within the corporate bond market, defined as the mean of the model implied correlations between all the couples of securities in the corporate bond asset class. $\Gamma_t^{EqInd}, \Gamma_t^{FX}, \Gamma_t^{MoneyMkt}$ and Γ_t^{Tr} are analogously defined for the other asset classes. Exposure-driven (mid-panel) and volatility-driven (bottom panel) comovement measures consider the correlations for the case in which factor exposures are allowed to vary with time (held at constant) and factor covariances are held at constant (allowed to vary with time). Refer to the caption of table 7 for a legend of the results of the estimation.

correlation prompted by the heteroscedasticity of common factors, in correspondence to the most severe period of the 2007–2009 crisis, that is, the last part of 2008, as well as in August 2011, when the sovereign debt crisis spread from the peripheral countries in Europe to the rest of the continent and ultimately to the United States. On the other hand, we detect a significant negative break in the contagion measure $\Gamma^{F}_{t,ED}$ from late 2007 to the end of 2008, which offsets the peak in $\Gamma^{F}_{t,VD}$, so that no peaks are detected in Γ^{F}_{t}, as shown before. When only factor exposures are concerned, we observe an average de-correlation of more than 6%. We further disaggregate the Γ-measures at the asset class and country level. Along with the detected segments, we observe a few outliers. In the case of $\Gamma^{F}_{t,ED}$, we find a couple of outliers in proximity of the Dot-Com bubble burst, witnessing de-correlation on the market. All the other IIS identified by Autometrics are in proximity of the start and the end of the sample, a fact observed also in Castle et al. (2012).

We turn our attention to Table 8 that reports the results referred to the single asset classes. For stock indices, we find evidence of contagion from August 2007 to mid-2009, with correlation significantly up by 5% from the average level of 79%. We also find evidence of excess interdependence for three less extended periods, in correspondence to the most dramatic months of 2008 and 2009, as well as in May 2010 and from August 2011 on, with a surge of 13–15% in the average correlation. We associate the former event to the first EU intervention in the Greece's bailout programme, which marked the triggering of the sovereign debt crisis in Europe. The second identified period has already been epitomized as the moment in which the sovereign debt crisis spread across and outside Europe. At the aggregate level, the 2007–2009 crisis and the debt crisis remain the most relevant episodes in terms of average market correlations.

For the other asset classes, the same periods are detected, but most of them are associated with *decreasing* market correlations. This is particularly evident at the aggregate level for Corporate Bonds (with average slumps in correlation as high as 41.34% in the last part of 2008) and foreign exchange (−39.93% in roughly the same period). This phenomenon is still present when we look for contagion and excess interdependence. The de-correlation observed in the case of foreign exchange rates is due to the contrasting effects of the crisis on the single pairs. Because of the low costs related to a borrowing position in Yen, since the early 2000s, the Japanese currency has been, together with the US dollar, the currency used by investors to finance their positions in risky assets. The massive outflow from the markets experienced in the late 2000s led to the unwinding of these borrowing

Table 9. IIS Results for the Average Comovements Measures at the Country Level.

Γ_t^{US}			Γ_t^{EU}			Γ_t^{UK}			Γ_t^{JP}			Γ_t^{EM}		
Segments			*Outliers*			*Segments*			*Segments*			*Ouliers*		
22/10/1999–29/10/1999	0.0576	***	27/06/2008	−0.2400	**	11/02/2000–21/04/2000	0.1197	***	08/10/1999–12/05/2000	−0.2752	***	22/01/1999	0.1323	***
04/02/2000–18/02/2000	0.0643	***	05/12/2008	0.2282	**	04/07/2003–08/08/2003	−0.1149	***	17/08/2007–24/07/2009	−0.3622	***	…		
24/11/2000–22/12/2000	0.0516	***	19/12/2008	0.2436	**	10/10/2008–13/03/2009	0.1552	***	21/05/2010–25/06/2010	−0.2484	***	09/12/2011	0.0618	**
28/09/2001–02/11/2001	−0.0532	***	10/07/2009	−0.2203	**	12/08/2011–18/11/2011	0.1079	***	12/08/2011–27/01/2012	−0.3288	***	*Constant*	0.6248	***
19/07/2002–27/09/2002	0.0568	***	07/05/2010	−0.2203	**	*Outliers*			*Outliers*					
24/08/2007–26/09/2008	0.0566	***	07/01/2011	−0.2155	**	06/11/2009	0.1012	***	30/05/2003	0.1582	*			
03/10/2008–19/12/2008	0.1524	***	29/07/2011	0.2320	**	16/07/2010	0.1076	***	16/06/2006	−0.2002	**			
02/01/2009–16/07/2010	0.0545	***	12/08/2011	0.2404	**	03/12/2010	−0.0992	***	*Constant*	0.3496	***			
Outliers			26/08/2011	0.2648	**	*Constant*	0.1988	***						
08/08/2003	−0.0515	***	21/10/2011	0.2253	**									
Constant	−0.1399	***	18/11/2011	0.2779	***									
			23/12/2011	0.2197	**									
			06/01/2012	0.2225	**									
			Constant	0.1667	***									
$\Gamma_{t,ED}^{US}$			$\Gamma_{t,ED}^{EU}$			$\Gamma_{t,ED}^{UK}$			$\Gamma_{t,ED}^{JP}$			$\Gamma_{t,ED}^{EM}$		
Segments			*Outliers*			*Outliers*			*Segments*			*Outliers*		
06/08/1999–08/12/2000	0.0331	***	13/08/1999	−0.2253	**	01/01/1999	0.0567	***	27/08/1999–02/06/2000	−0.2321	***	22/01/1999	0.1545	***
19/07/2002–11/10/2002	0.0291	***	04/02/2000	−0.2319	**	…			17/08/2007–10/07/2009	−0.3365	***	…		
23/07/2004–14/04/2006	−0.0147	*	03/03/2000	−0.2266	**	01/10/2010	−0.0249	***	12/08/2011–20/01/2012	−0.2486	***	21/10/2011	0.1071	***

Table 9. *(Continued)*

$\Gamma^{US}_{t,ED}$			$\Gamma^{EU}_{t,ED}$			$\Gamma^{UK}_{t,ED}$			$\Gamma^{JP}_{t,ED}$			$\Gamma^{EM}_{t,ED}$		
17/08/2007–05/09/2008	0.0436	***	14/04/2000	−0.2216	**	*Constant*	0.1989	***	*Outliers*			*Constant*	0.6273	***
12/09/2008–30/01/2009	0.0766	***	23/06/2000	−0.2245	**				12/03/2004	0.2703	***			
06/02/2009–17/07/2009	0.0447	***	23/02/2001	−0.2252	**				04/02/2005	0.1355	*			
12/08/2011–09/03/2012	0.0331	***	07/06/2002	−0.2163	**				27/05/2005	0.1404	*			
Outliers			11/10/2002	−0.2222	**				08/07/2005	0.1449	*			
11/06/2010	0.0252	**	07/03/2003	−0.2263	**				*Constant*	0.3521	***			
03/12/2010	0.0275	***	23/07/2004	−0.2145	**									
Constant	−0.1457	***	20/08/2004	−0.2269	**									
			08/04/2005	−0.2288	**									
			Constant	0.1805	***									

$\Gamma^{US}_{t,VD}$			$\Gamma^{EU}_{t,VD}$			$\Gamma^{UK}_{t,VD}$			$\Gamma^{JP}_{t,VD}$			$\Gamma^{EM}_{t,VD}$		
Segments			*Segments*			*Segments*			*Segments*			*Segments*		
28/09/2001–19/10/2001	−0.0484	***	27/06/2003–08/08/2003	−0.0716	**	04/07/2003–08/08/2003	−0.1111	***	20/06/2003–27/02/2004	0.1288	***	23/04/1999–04/06/1999	0.0683	***
26/07/2002–20/09/2002	0.0451	***	23/01/2004–01/10/2004	−0.0565	*	17/10/2008–20/03/2009	0.1308	***	17/10/2008–19/12/2008	−0.1532	***	30/07/1999–20/08/1999	−0.0542	***
20/06/2008–25/09/2009	−0.0365	***	17/10/2008–26/12/2008	0.1748	***	06/11/2009–13/11/2009	0.0826	**	19/11/2010–14/01/2011	0.1434	***	26/01/2001–02/11/2001	0.0680	***
19/11/2010–17/12/2010	−0.0336	**	02/01/2009–23/07/2010	0.1044	***	02/07/2010–16/07/2010	0.0863	**	12/08/2011–18/11/2011	−0.1162	***	05/07/2002–01/11/2002	0.0660	***
19/08/2011–09/03/2012	−0.0377	***	08/07/2011–27/01/2012	0.1300	***	12/08/2011–11/11/2011	0.1050	***	*Constant*	0.2815	***	07/05/2004–21/05/2004	0.0519	***

Constant	−0.1346	***	*Outliers*			*Outliers*						24/08/2007–14/09/2007	0.0688	***
			31/08/2007	0.0796	**	20/06/2008	−0.0976	***				18/04/2008–31/10/2008	0.0479	***
			Constant	0.2112	***	03/12/2010	−0.0978	***				12/08/2011–11/11/2011	0.0505	***
						Constant	0.1994	***				*Outliers*		
												29/09/2000	0.0593	***
												Constant	0.6394	***

Notes: Γ_t^{US} is the average comovement measure within the US market, defined as the mean of the model implied correlations between all the couples of securities in the US group. $\Gamma_t^{EU}, \Gamma_t^{UK}, \Gamma_t^{JP}$ and Γ_t^{EM} are analogously defined for the other countries. Exposure-driven (mid-panel) and volatility-driven (bottom panel) comovement measures consider the correlations for the case in which factor exposures are allowed to vary with time (held at constant) and factor covariances are held at constant (allowed to vary with time). Refer to the caption of table 7 for a legend of the results of the estimation.

positions, which fuelled a steady appreciation of the Japanese currency. This results in a massive de-correlation of the Yen against the other currencies. As part of the same phenomenon, the Japanese Corporate Bond market, even though it experienced a sharp capital outflow during the first period of the late 2000s financial crisis, continued to grow rapidly (see Shim, 2012), proving to be a safe haven during this period of generalized financial distress. This again triggered de-correlation of the Japan market with the other countries. See Fig. 8 for a graphic comparison of the market dynamics in these periods.

Similarly, the money markets are pervaded by comovements shocks of alternate signs, especially at the aggregate level and when testing for excess interdependence. The series here considered are indicative of the status of the country interbank markets as well as a proxy of the conduct of the monetary policy. The negative breaks in comovements reflect the asymmetries in the shocks on the interbank markets and the differences in the reactions of the monetary policy to the spreading of the crisis.

We detect a positive sign at the aggregate level and at the volatility driven level in correspondence to the joint monetary policy intervention in October 2008 by the FED, the ECB, the Bank of England and the Bank of Japan together with three other central banks of industrialized countries (Canada, Switzerland and Sweden). We find no breaks for Treasury rates at the aggregate level.

We now move on to Table 9 and analyse the same average comovement measures at the country level. We find evidence of a peak in the overall comovements in the United States during the 2007–2009 crisis. In particular, there is strong evidence of contagion at the national level characterized by an escalation in the magnitude of the breaks in correspondence to the worsening of the crisis in the late 2008. Similarly, in the other countries, we observe peaks during financial crises. In particular, in Europe we observe excess interdependence for most of the period between 2008 and 2012. In the United Kingdom, we observe positive breaks in the correlations at the aggregate level and at the volatility driven level both for the 2007–2009 crisis and for the sovereign debt crisis. For Japan, we observe the de-correlation phenomenon described above, with the stock market correlated with the other stock markets, while the national currency was following a steady appreciation path.

The first evidence of contagion during the late 2000s economic and financial crisis was observed for equity markets and the United States, as early as August 2007, anticipating the all-time peak of the S&P500 in October, epitomizing the beginning of the 2007–2009 global financial crisis. This combined

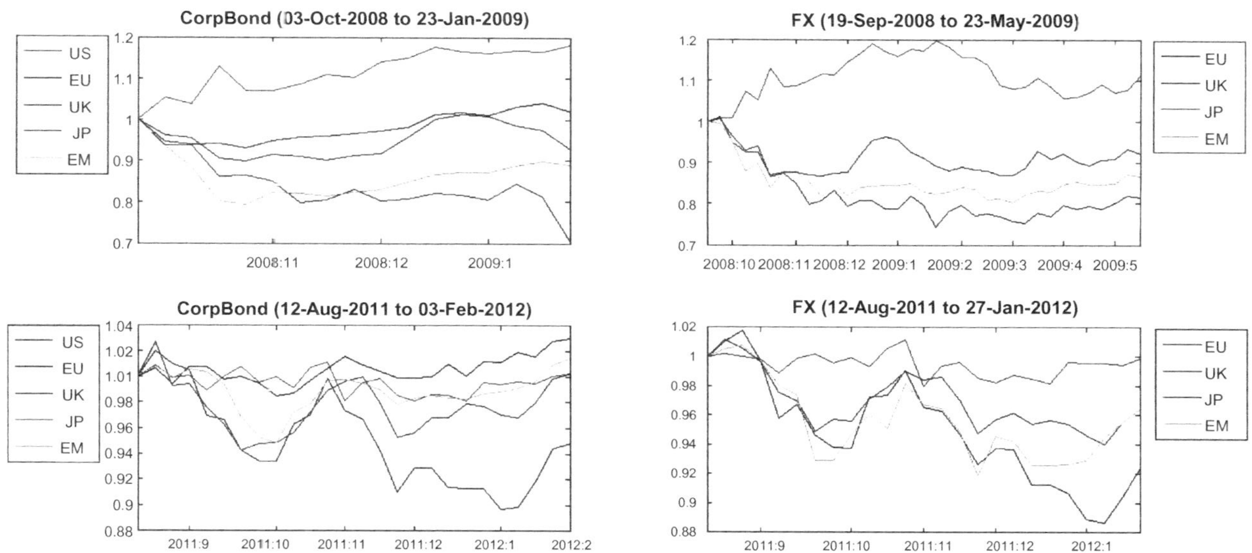

Fig. 8. Comparison among Selected Securities during the Detected Regimes. *Notes*: We report corporate bond and foreign exchange price levels for periods in which decorrelation was detected. The price are rebased using the first observation in each subperiod.

evidence is in line with what has been observed in reality: the crisis originated in the United States, spread across the country and then propagated to the global financial markets, affecting first the global stock markets. On the contrary, there is evidence that the sovereign debt crisis that originated in Europe was characterized by excess interdependence, rather than as an example of contagion. Indeed, in this case the most extended episode of excess interdependence was recorded for equity indices and for Europe.

5. CONCLUSIONS

This paper studied the determinants of the comovements (contagion vs excess interdependence) between different financial markets, both in a multi-country and a multi-asset class perspective. We proposed that a dynamic factor model able to capture multiple sources of shocks, at global, asset class and country level and used it to test for the presence of contagion versus excess interdependence. The model is specified with time-varying factor loadings, to allow for time-dependent exposures of the single assets to the different shocks. We statistically validated the supremacy of this model as compared to a standard static approach and an alternative dynamic approach. The framework is applied to data covering five countries (United States, Euro, United Kingdom, Japan, Emerging Countries), five asset markets (corporate bond yields, equity returns, currency returns relative to the United States, short-term money market yields and long-term Treasury yields) for a total of 20 series. We used weekly data, spanning from 1 January 1999 to 14 March 2012.

The main findings of our empirical analysis can be summarized as follows. First, the global factor is the most pervasive of the considered factors, shaping the dynamics of the comovements of the considered financial markets. On the contrary, the asset class factor is the most persistent through time, suggesting that the structural commonalities of markets belonging to the same asset class systematically contribute in a constant proportion to the average overall comovements. In our multiple asset class framework, the country factor is negligible. In a robustness check, we showed that this result does not depend on the order in which the system of factors is orthogonalized.

Second, we find evidence of contagion stemming from the United States and the stock market jointly in correspondence to the harshest period of the 2007–2009 financial crisis. On the contrary, the currency and sovereign debt crisis, which originated in Europe, are characterized by excess interdependence from mid-2010 onwards. According to the literature on comovements, this lets us characterize the spillover effects during the 2007–2009 financial crisis as persistent, altering the strength of the financial linkages worldwide. On the other hand, the shock transmission experienced during the recent debt crisis has *so far* to be understood as temporary, being prompted by excess factor volatilities, which do not display any trend in the long-term.

Finally, at the overall average level, we do not find any evidence of contagion or excess interdependence. We like to interpret this result as follows. During the crises some of the securities considered in the study, the Japanese currency and corporate bond market in particular, displayed diverging dynamics as result of the unwinding of carry positions, built to finance risky investments.

NOTES

1. In the factor model literature, consistency of the factor estimation is a well-established result for the case in which the factor loading is stable. In this paper, we make use of the limiting theory developed by Stock and Watson (1998, 2002, 2009) and Bates, Plagborg-Møller, Stock, and Watson (2012) for the case of instability of the factor loading, suggesting that factors are consistently estimated using principal components.
2. Specification (7) is within the class of the so-called *conditional time-varying factor loading approach* (see Bekaert et al., 2009), where the factor loadings are assumed to follow a structural dynamic equation (see, for instance, Baele et al. 2010) of the form $\beta_t^{i,j} \equiv \beta(\mathcal{F}_{t-1}, X_t)$ where $\{\mathcal{F}_t\}_{t=1,\ldots,T}$ is the information flow and X_t is a set of conditional variables.
3. The use of the IIS strategy to identify structural breaks using a number of dummy variables has similarities to the contagion test proposed by Favero and Giavazzi (2002).
4. Emerging market weights are the same across different asset classes, are based on GDP and updated annually. The weights for 2012, last year in the sample, are: China 51.3%, Brazil 17.4%, Russia 13.0%, India 12.9% and Turkey 5.4%.

ACKNOWLEDGEMENTS

We wish to thank participants in the Finance Research Workshops at Cass Business School (London, 8 October 2012), in particular A. Beber and K. Phylaktis, in the Fifth Italian Congress of Econometrics and Empirical Economics (Genova, 16–18 January 2013), in the Third Carlo Giannini

PhD Workshop in Econometrics (Bergamo, 15 March 2013), in particular, M. Bertocchi, L. Khalaf and E. Rossi, in the CREATES Seminar (Aarhus, 4 April 2013), in particular, D. Kristensen, N. Haldrup, A. Lunde, and T. Terasvirta, in the Seminari di Dipartimento Banca e Finanza of Università Cattolica del Sacro Cuore (Milan, 13 December 2013), in particular, C. Bellavite Pellegrini, in the 14th OxMetrics User Conference at The George Washington University (Washington, 20–21 March 2014), in the 16th Advances in Econometrics Conference on Dynamic Factor Models at CREATES (Aarhus, 14–15 November 2014), in paricular, J. Breitung, M. Hallin and M. Marcellino, for useful discussions and valuable comments. Special thanks to the Editors, Eric Hillebrand and Siem Jan Koopman, and two anonymous referee for very helpful comments and suggestions that greatly helped to improve this paper. Riccardo Borghi has provided very insightful comments on a previous version of this paper. The usual disclaimer applies. Riccardo Pianeti acknowledges financial support from the Centre for Econometric Analisis at Cass and the EAMOR Doctoral Programme at Bergamo University.

REFERENCES

Anderson, T. W. (2003). *An introduction to multivariate analysis* (Wiley Series in Probability and Statistics). Hoboken, NJ: Wiley.

Bae, K.-H., Karolyi, G. A., & Stulz, R. M. (2003). A new approach to measuring financial contagion. *Review of Financial Studies*, *16*(3), 717–763.

Baele, L., Bekaert, G., & Inghelbrecht, K. (2010). The determinants of stock and bond return comovements. *Review of Financial Studies*, *23*(6), 2374–2428.

Bai, J., & Perron, P. (1998). Estimating and testing linear models with multiple structural changes. *Econometrica*, *66*(1), 47–78.

Bates, B. J., Plagborg-Møller, M., Stock, J. H., & Watson, M. W. (2013). Consistent factor estimation in dynamic factor models with structural instability. *Journal of Econometrics*, *177*(2), 289–304.

Bekaert, G., Ehrmann, M., Fratzscher, M., & Mehl, A. (2014). Global crises and equity market contagion. *Journal of Finance*, *59*(6), 2597–2649.

Bekaert, G., Harvey, C. R., & Ng, A. (2005). Market integration and contagion. *Journal of Business*, *78*(1), 39–69.

Bekaert, G., & Hodrick, R. (1992). Characterizing predictable components in excess returns on equity and foreign exchange markets. *Journal of Finance*, *47*(2), 467–509.

Bekaert, G., Hodrick, R., & Zhang, X. (2009). International stock return comovements. *Journal of Finance*, *64*(6), 2591–2626.

Benerjee, A., Lumsdaine, R. L., & Stock, J. H. (1992). Recursive and sequential tests of the unit-root and trend-break hypotheses: Theory and international evidence. *Journal of Business and Economic Statistics*, *10*(3), 271–287.

Billio, M., & Caporin, M. (2010). Market linkages, variance spillovers, and correlation stability: Empirical evidence of financial contagion. *Computational Statistics and Data Analysis*, *54*(11), 2443–2458.

Brandt, M. W., Brav, A., & Graham, J. R. (2010). The idiosyncratic volatility puzzle: Time trend or speculative episodes? *Review of Financial Studies*, *23*(2), 863–899.

Castle, J. L., Doornik, J. A., & Hendry, D. F. (2011). Evaluating automatic model selection. *Journal of Time Series Econometrics*, *3*(1), Article 8.

Castle, J. L., Doornik, J. A., & Hendry, D. F. (2012). Model selection when there are multiple breaks. *Journal of Econometrics*, *169*(2), 239–246.

Doornik, J. A. (2009). Autometrics. In J. Castle & N. Shephard (Eds.), (2009), *The methodology and practice of econometrics*. Oxford: Oxford University Press.

Driessen, J., Melenberg, B., & Nijman, T. (2003). Common factors in international bond returns. *Journal of International Money and Finance*, *22*(5), 629–656.

Dungey, M., Fry, R. E., González-Hermosillo, B., & Martin, V. L. (2005). Empirical modelling of contagion: A review of methodologies. *Quantitative Finance*, *5*(1), 9–24.

Dungey, M., & Martin, V. (2007). Unravelling financial market linkages during crises. *Journal of Applied Econometrics*, *22*(1), 89–119.

Ehrmann, M., Fratzscher, M., & Rigobon, R. (2011). Stocks, bonds, money markets and exchange rates: Measuring international financial transmission. *Journal of Applied Econometrics*, *26*(6), 948–974.

Engle, R. F. (2002). Dynamic conditional correlation. *Journal of Business and Economic Statistics*, *20*(3), 339–350.

Engle, R. F., Ito, T., & Lin, W. (1990). Meteor showers or heat waves? Heteroscedastic intradaily volatility in the foreign exchange market. *Econometrica*, *58*(3), 525–542.

Fama, E., & French, K. (1993). Common risk factors in the returns on stocks and bonds. *Journal of Financial Economics*, *33*(1), 3–56.

Favero, C. A., & Giavazzi, F. (2002). Is the international propagation of financial shocks non-linear? Evidence from the ERM. *Journal of International Economics*, *57*(1), 231–246.

Forbes, K. J., & Rigobon, R. (2002). No contagion, only interdependence: Measuring stock market comovements. *Journal of Finance*, *57*(5), 2223–2261.

Fry, R., Martin, V. L., & Tang, C. (2010). A new class of tests of contagion with applications. *Journal of Business and Economic Statistics*, *28*(3), 423–437.

Hendry, D. F., & Krolzig, H.-M. (2005). The properties of automatic gets modelling. *Economic Journal*, *115*, C32–C61.

Karolyi, G. A. (2003). Does international financial contagion really exist? *International Finance*, *6*(2), 179–199.

King, M. A., & Wadhwani, S. (1990). Transmission of volatility between stock markets. *Review of Financial Studies*, *3*(1), 5–33.

Longin, F., & Solnik, B. (2001). Extreme correlation of international equity markets. *Journal of Finance*, *56*(2), 649–676.

Pesaran, H., & Pick, A. (2007). Econometric issues in the analysis of contagion. *Journal of Economic Dynamics & Control*, *31*(4), 1245–1277.

Ross, S. M. (1976). The arbitrage theory of capital asset pricing. *Journal of Economic Theory*, *13*(3), 341–360.

Schwert, G. W. (1989). Why does stock market volatility change over time? *Journal of Finance*, *44*(5), 1115–1153.

Shim, I. (2012). *Development of Asia-Pacific corporate bond and securitisation markets.* Working Paper No. 63c. Bank for International Settlements.

Stock, J. H., & Watson, M. W. (1998). *Diffusion indexes.* Manuscript. Cambridge, MA: Harvard University.

Stock, J. H., & Watson, M. W. (2002). Forecasting using principal components from a large number of predictors. *Journal of the American Statistical Association*, *97*, 1167–1179.

Stock, J. H., & Watson, M. W. (2009). Forecasting in dynamic factor models subject to structural instability. In D. F. Hendry, J. Castle, & N. Shephard (Eds.), *The methodology and practice of econometrics: A Festschrift in honour of David F Hendry* (pp. 173–205). Oxford: Oxford University Press.

SPECIFICATION AND ESTIMATION OF BAYESIAN DYNAMIC FACTOR MODELS: A MONTE CARLO ANALYSIS WITH AN APPLICATION TO GLOBAL HOUSE PRICE COMOVEMENT

Laura E. Jackson[a], M. Ayhan Kose[b], Christopher Otrok[c,d] and Michael T. Owyang[d]

[a]*Department of Economics, Bentley University, Waltham, MA, USA*
[b]*World Bank, Washington, DC, USA*
[c]*Department of Economics, University of Missouri, Columbia, MO, USA*
[d]*Federal Reserve Bank of St. Louis, St. Louis, MO, USA*

ABSTRACT

We compare methods to measure comovement in business cycle data using multi-level dynamic factor models. To do so, we employ a Monte Carlo procedure to evaluate model performance for different specifications of factor models across three different estimation procedures. We consider

Dynamic Factor Models
Advances in Econometrics, Volume 35, 361–400

ISSN: 0731-9053/doi:10.1108/S0731-905320150000035009

three general factor model specifications used in applied work. The first is a single-factor model, the second a two-level factor model, and the third a three-level factor model. Our estimation procedures are the Bayesian approach of Otrok and Whiteman (1998), the Bayesian state-space approach of Kim and Nelson (1998) and a frequentist principal components approach. The latter serves as a benchmark to measure any potential gains from the more computationally intensive Bayesian procedures. We then apply the three methods to a novel new dataset on house prices in advanced and emerging markets from Cesa-Bianchi, Cespedes, and Rebucci (2015) and interpret the empirical results in light of the Monte Carlo results.

Keywords: Principal components; Kalman filter; data augmentation; business cycles

JEL: C3; C18; C32; E32

1. INTRODUCTION

Dynamic factor models have gained widespread use in analyzing business cycle comovement. The literature began with the Sargent and Sims (1977) analysis of U.S. business cycles. Since then the dynamic factor framework has been applied to a long list of empirical questions. For example, Engle and Watson (1981) study metropolitan wage rates, Forni and Reichlin (1998) analyze industry level business cycles, Stock and Watson (2002) forecast the U.S. economy, while Kose, Otrok, and Whiteman (2003) study international business cycles. It is clear that dynamic factor models have become a standard tool to measure comovement, a fact that has become increasingly true as methods to deal with large datasets have been developed and the profession has gained interest in the "Big Data" movement.

Estimation of this class of models has evolved significantly since the original frequency domain methods of Geweke (1977) and Sargent and Sims (1977). Stock and Watson (1989) adopted a state-space approach and employed the Kalman filter to estimate the model. Stock and Watson (2002) utilized a two-step procedure whereby the unobserved factors are computed from the principal components of the data. Forni, Hallin, Lippi, and Reichlin (2000) compute the eigenvector–eigenvalue decomposition of the spectral density matrix of the data frequency by frequency, inverse-Fourier transforming the eigenvectors to create polynomials which are then used to construct the factors. This latter approach is essentially a dynamic version of principal components. A large number of refinements to these methods have been developed for frequentist estimation of large-scale factor models since the publication of these papers.

A Bayesian approach to estimating dynamic factor models was developed by Otrok and Whiteman (1998), who employed a Gibbs sampler. The key innovation of their paper was to derive the distribution of the factors conditional on model parameters that is needed for the Gibbs sampler. Kim and Nelson (1998) also developed a Bayesian approach using a state-space procedure that employs the Carter–Kohn approach to filtering the state-space model. The key difference between the two approaches is that the Otrok–Whiteman procedure can be applied to large datasets, while, because of computational constraints, the Kim–Nelson method cannot. The Bayesian approach in both papers is particularly useful when one wants to impose "zero" restrictions on the factor loading matrix to identify group-specific factors. In addition, both approaches, because they are Bayesian, draw inference conditional on the size of the dataset at hand; the classical approaches discussed above generally rely on asymptotics. While this is not a problem when the factors are estimated on large datasets, for smaller datasets – or multi-level factor models where some levels have few time series, it may be problematic. Lastly, the Bayesian approach is the only framework that can handle the case of multi-level factor models when the variables are not assigned to groups a priori (e.g., Francis, Owyang, & Savasçin, 2012).

In this paper, we compare the accuracy of the two Bayesian approaches and a multi-step principal components estimator. In particular, we are interested in the class of multi-level factor models where one imposes various "zero" restrictions to identify group-specific factors (e.g., regional factors). To be concrete, we will label these models as in the international business cycle literature, although the models have natural applications to multi-sector closed economies or to models that mix real and financial variables. We perform Monte Carlo experiments using three different models of increasing complexity. The first model is the ubiquitous single factor model. The second is a two-level factor model that we interpret as a world-country factor model. In this model, one (world) factor affects all of the series; the other factors affect non-overlapping subsets of the series. The third is a three-level factor model that we interpret as world-region-country factor model.

For each model, we first generate a random set of model coefficients. Using the coefficients, we generate the "true" factors and a corresponding set of sample data. We then apply each estimation procedure to the simulated data to extract factors and model coefficients. We then repeat this sequence many times, starting with a new draw for the model parameters each time. The Bayesian estimation approach is a simulation-based Markov chain Monte Carlo (MCMC) estimator, making the estimate of one model non-trivial in terms of time; however, modern computing power makes Monte Carlo study of Bayesian factor models feasible.

In this sense, our paper provides a complementary study to Breitung and Eickmeier (2014), who employ a Monte Carlo analysis of various frequentist estimators of multi-level factor models with their new sequential least-squares estimator. There are three key differences in our Monte Carlo procedures with that of Breitung and Eickmeier (2014). First, they study a fixed and constant set of parameters. As they note in their paper, the accuracy of the factor estimates can depend on the variance of the factors (or more generally the signal-to-noise ratio). To produce a general set of results that abstracts away from any one or two parameter settings, we randomly draw new parameters for each simulation. A second difference is that the number of observations in each of the levels of their factor model is always large enough to expect the asymptotics to hold. In our model specification, we combine levels where the cross-sections are both large and small, which is often the case in applied work. Third, we include in our study measures of uncertainty in factor estimates, while Breitung and Eickmeier (2014) focus on the accuracy of the mean of an estimate. Taken together the two papers provide a comprehensive Monte Carlo analysis of the accuracy of a wide range of the procedures used for a number of different model specifications and sizes.

Our evaluation mainly focuses on the three key features of the results that are important in applied work with factor models. The first is the accuracy of the approaches in estimating the "true" factors as measured by the correlation of the posterior mean factor estimate with the truth. The second is the extent to which the methods characterize the amount of uncertainty in factor estimates. To do so, we measure the width of the posterior coverage interval as well as count how many times the true factor lies in the posterior coverage interval. The third is the correspondence of the estimated variance decomposition with the true variance decomposition implied by the population parameters.[1] In simulation work, we compare two ways to measure the variance decomposition in finite samples. The first takes the estimated factors, orthogonalizes them draw-by-draw, and computes the decomposition based on a regression on the orthogonalized factors (i.e., not the estimated factor loadings).[2]

The second takes each draw of the model parameters and calculates the implied variance decomposition. While the factors are assumed to be orthogonal, this is not imposed in the estimation procedures, which could bias a model where the factors have some correlation in finite samples.

We find that, for the one factor model, the three methods do equally well at estimating a factor that is correlated with the true factor. For models with multiple levels, however, the Kalman-filtered state-space method typically does a better job at identifying the true factor. As the number of levels increases, the Otrok–Whiteman procedure – which redraws the factor at each Gibbs iteration – estimates a factor more highly correlated with the true factor than does PCA, which estimates the factor ex ante. We find that both the state-space and Otrok–Whiteman procedures provide fairly accurate, albeit conservative, estimates of the percentage of the total variance explained by the factors. PCA, on the other hand, tends to overestimate the contribution of the factors.

When we apply the three procedures to house price data in advanced and emerging markets, we find that there does exist a world house price cycle that is both pervasive and quantitatively important. We find less evidence of a widely important additional factor for advanced economies or for emerging markets. Consistent with the Monte Carlo results we find that all three methods deliver the same global factor. We also find that the Kalman Filter and Otrok–Whiteman procedures deliver similar regional factors, which is virtually uncorrelated with the PCA regional factor. The PCA method provides estimates of variance decompositions that are greater than the Bayesian procedures, which is also consistent with the Monte Carlo evidence. Lastly, the parametric variance decompositions are uniformly greater than the factor-based estimates, which is also consistent with the Monte Carlo evidence.

The outline of this paper is as follows: Section 2 describes the empirical model and outlines its estimation using the three techniques – a Bayesian version of principal components analysis, the Bayesian procedure of Otrok and Whiteman, and a Bayesian version of the state-space estimation of the factor – we study. Section 3 outlines the Monte Carlo experiments and describes the methods we use to evaluate the three methods. In this section, we also present the results from the Monte Carlo experiments. Section 4 applies the methods to a dataset on house prices in Advanced and Emerging Market Economies. Section 5 offers some conclusions.

2. SPECIFICATION AND ESTIMATION OF THE DYNAMIC FACTOR MODEL

In the prototypical dynamic factor model, all comovement among variables in the dataset is captured by a set of M latent variables, F_t. Let Y_t denote an $(N \times 1)$ vector of observable data. The dynamic factor model for this set of time series can be written as

$$Y_t = \beta F_t + \Gamma_t \tag{1}$$

$$\Gamma_t = \Psi(L)\Gamma_{t-1} + U_t \tag{2}$$

with $E_t(U_t U_t') = \Omega$,

$$F_t = \Phi(L)F_{t-1} + V_t \tag{3}$$

with $E_t(V_t V_t') = I_M$. Vector Γ_t is an $(N \times 1)$ vector of idiosyncratic shocks which captures movement in each observable series specific to that time series. Each element of Γ_t is assumed to follow an independent AR(q) process, hence $\Psi(L)$ is a block diagonal lag polynomial matrix and Ω is a covariance matrix that is restricted to be diagonal. The latent factors are denoted by the $(M \times 1)$ vector F_t, whose dynamics follow an AR(p) process. The $(N \times M)$ matrix β contains the factor loadings which measure the response (or sensitivity) of each observable variable to each factor. With estimated factors and factor loadings, we are then able to quantify the extent to which the variability in the observable data is common. Our one factor model sets β to a vector of length M, implying all variables respond to this factor.

In multiple factor models, it is often useful to impose zero restrictions on β in order to give an economic interpretation to the factors. The Bayesian approach also allows (but does not require) the imposition of restrictions on the factor loadings such that the model has a multi-level structure as a special case. For example, Kose, Otrok, and Whiteman (2008) impose zero restrictions on β to separate out world and country factors. They use a dataset on output, consumption and investment for G-7 countries to estimate a model with one common (world) factor and seven country-specific factors. Identification of the country factors is obtained by only allowing variables within each country to load on a particular factor, which we then label as the country factor. For the G-7 model, the β matrix (of dimension 21×24 when estimating the model with three dataseries) is:

$$\begin{bmatrix} \beta^{G7}_{US,Y} & 0 & 0 & \beta^{US}_{US,Y} & 0 & 0 & 0 & \cdots & 0 \\ \beta^{G7}_{US,C} & 0 & 0 & \beta^{US}_{US,C} & 0 & 0 & 0 & \cdots & 0 \\ \beta^{G7}_{US,I} & 0 & 0 & \beta^{US}_{US,I} & 0 & 0 & 0 & \cdots & 0 \\ \beta^{G7}_{Fr,Y} & 0 & 0 & 0 & 0 & 0 & \beta^{Fr}_{Fr,Y} & \cdots & 0 \\ \vdots & \vdots & \vdots & \vdots & \vdots & \vdots & \vdots & \vdots & \vdots \\ \beta^{G7}_{UK,Y} & 0 & 0 & 0 & 0 & 0 & 0 & \cdots & \beta^{UK}_{UK,Y} \\ \beta^{G7}_{UK,C} & 0 & 0 & 0 & 0 & 0 & 0 & \cdots & \beta^{UK}_{UK,C} \\ \beta^{G7}_{UK,I} & 0 & 0 & 0 & 0 & 0 & 0 & \cdots & \beta^{UK}_{UK,I} \end{bmatrix}$$

Here, all variables load on the first (world) factor while only U.S. variables load on the second (U.S. country) factor.

The three-level model adds an additional layer to the model to include world, region, and country-level factors. In this setup, all countries within a given region load on the factor specific to that region in addition to the world and country factors. The objective of all three econometric procedures is to estimate the factors and parameters of this class of models as accurately as possible.

2.1. *The Otrok–Whiteman Bayesian Approach*

Estimation of dynamic factor models is difficult when the factors are unobservable. If, contrary to assumption, the dynamic factors were observable, analysis of the system would be straightforward; because they are not, special methods must be employed. Otrok and Whiteman (1998) developed a procedure based on an innovation in the Bayesian literature on missing data problems, that of "data augmentation" (Tanner & Wong, 1987). The essential idea is to determine posterior distributions for all unknown parameters conditional on the latent factor and then determine the conditional distribution of the latent factor given the observables and the other parameters. That is, the observable data are "augmented" by samples from the conditional distribution for the factor given the data and the parameters of the model. Specifically, the joint posterior distribution for the unknown parameters and the unobserved factor can be sampled using an MCMC procedure on the full set of conditional distributions. The Markov chain samples sequentially from the conditional distributions for (parameters | factors) and (factors | parameters) and, at each stage, uses the previous iterate's drawing as the conditioning variable, ultimately yields drawings from the joint distribution for (parameters, factors). Provided samples are readily generated from each conditional distribution, it is possible to sample from otherwise intractable joint distributions. Large cross-sections of data present no special problems for this procedure since natural ancillary assumptions ensure that the conditional distributions for (parameters | factors) can be sampled equation by equation; increasing the number of variables has a small impact on computational time.

When the factors are treated as conditioning variables, the posterior distributions for the rest of the parameters are well known from the multivariate regression model; finding the conditional distribution of the factor given the parameters of the model involves solving a "signal extraction" problem. Otrok and Whiteman (1998) used standard multivariate normal theory to determine the conditional distribution of the entire time series of the factors, $(F_1, \ldots, F_T)$ simultaneously. Details on these distributions are available in Otrok and Whiteman (1998). The extension to multi-level models was developed in Kose et al. (2003). Their procedure samples the factor with a sequence of factors by level. For example, in the world-country model we first sample from the conditional distribution of (world factor | country factors, parameters), then from the conditional distribution of (country factors | world factor, parameters).

It is important to note that in the step where the unobserved factors are treated as data, the Gibbs sampler does in fact take into account the factor estimates' uncertainty when estimating the parameters. This is because we sequentially sample from the conditional posteriors a large number of times. In particular, when the cross-section is small, the procedure will accurately measure uncertainty in factor estimates, which will then affect the uncertainty in the parameters estimates. A second important feature of the Otrok and Whiteman procedure is that it samples from the conditional posteriors of the parameters sequentially by equation; thus, as the number of series increases, the increase in computational time is only linear.

2.2. *The Kim–Nelson Bayesian State-Space Approach*

A second approach to estimation follows Kim and Nelson (1998). As noted by Stock and Watson (1989), the set of Eqs. (1)–(3) comprises a state-space system where Eq. (1) corresponds to the measurement equation and Eqs. (2) and (3) correspond to the state transition equation. One approach to estimating the model is to use the Kalman filter. Kim and Nelson instead combine the state-space structure with a Gibbs sampling procedure to estimate the parameters and factors. To implement this idea, we use the same conditional distribution of parameters given the factors as in Otrok and Whiteman (2008). This allow us to focus on the differences in drawing the factors across the two Bayesian procedures. To draw the factors conditional on parameters, we use the Kim–Nelson state-space approach.

In the state-space setup, the F_t vector contains both contemporaneous values of the factors as well as lags. The lags of the factor enter the state equation (3) to allow for dynamics in each factor. Let M be the number of factors ($M < N$) and p be the order of the autoregressive process each factor follows, then we can define $k = Mp$ as the dimension of the state vector. F_t is then an $(k \times 1)$ vector of unobservable factors (and its lags) and $\Phi(L)$ is a matrix lag polynomial governing the evolution of these factors.

Two issues arise concerning the feasibility of sampling from the implied conditional distribution. The first has to do with the structure of the state space for higher-order autoregressions; the second has to do with the dimension of the state in the presence of idiosyncratic dynamics. To understand the first issue, note that, because the state is Markov, it is advantageous to carry the sequential conditioning argument one step further: Rather than drawing simultaneously from the distribution for $(F_1, \ldots, F_T)$, one samples from the T-conditional distributions $\left(F_j | F_1, \ldots, F_{j-1}, F_{j+1}, \ldots, F_T\right)$ for $j = 1, \ldots, T$. If F_t itself is autoregressive of order 1, then only adjacent values matter in the conditional distribution, which simplifies matters considerably.

When the factor itself is of a higher order, say an autoregression of order $p^{\dagger}$, one defines a new $p^{\dagger}$-dimensional state $X_t = [F_t, F_{t-1}, \ldots, F_{t-p^{\dagger}+1}]$, which in turn has a first-order vector autoregressive representation. The issue arises in the way the sequential conditioning is done in sampling from the distribution for the factor. Note that in $(X_t | X_{t-1}, X_{t+1})$, there is in fact no uncertainty at all about X_t. Samples from this sequence of conditionals actually only involve factors at the ends of the data set. Thus, this "single move" sampling (a version of which was introduced Carlin, Polson, and Stoffer, 1992) does not succeed in sampling from the joint distribution in cases where the state has been expanded to accommodate lags. Fortunately, an ingenious procedure to carry out "multimove" sampling was introduced by Carter and Kohn (1994). Subsequently, more efficient multimove samplers were introduced by de Jong and Shephard (1995) and Durbin and Koopman (2002). We follow Kim and Nelson (1998) in their

Bayesian implementation of a dynamic factor model and use Carter and Kohn (1994). In our analysis of the three econometric procedures, we will not be focusing on computational time.

The second issue arises because, while the multimode samplers solve the "big-T" curse of dimensionality, they potentially reintroduce the "big-N" curse when the cross-section is large. The reason is that the matrix calculations in the algorithm may be of the same dimension as that of the state vector. When the idiosyncratic errors u_t have an autoregressive structure, the natural formulation of the state vector involves augmenting the factor(s) and their lags with contemporaneous and lagged values of the errors (see Kim & Nelson, 1999; chapter 3). For example, if each observable variable is represented using a single factor that is AR(p) and an error that is AR(q), the state vector would be of dimension $p+Nq$, which is problematic for large N.

An alternative formulation of the state due to Quah and Sargent (1993) and Kim and Nelson (1999) avoids the "big-N" problem by isolating the idiosyncratic dynamics in the observation equation. To see this, suppose we have N observable variables, y_n for $n=1,\ldots,N$, and M unobserved dynamic factors, f_m for $m = 1, \ldots, M$, which account for all of the comovement in the observable variables. The observable time series are described by the following version of Eq. (1):

$$y_{n,t} = a_n + b_n f_t + \gamma_{nt} \tag{4}$$

where

$$\gamma_{nt} = \psi_{n,1}\gamma_{n,t-1} + \ldots + \psi_{n,q}\gamma_{n,t-q} + u_{nt} \tag{5}$$

with $u_{nt} \sim \text{iid } N\left(0, \sigma_n^2\right)$. The factors evolve as independent AR(p) processes:

$$f_{mt} = \phi_{m1} f_{m,t-1} + \cdots + \phi_{mp} f_{m,t-p} + v_{mt} \tag{6}$$

where $v_{mt} \sim \text{iid } N(0,1)$. Suppose for illustration that $M = 1$ and $q \geq p$. The "big-N" version of the state-space form for Eqs. (3)–(5) is

$$Y_t = HF_t \tag{7}$$

$$F_t = BF_{t-1} + E_t \tag{8}$$

where $Y_t = (y_{1t}, \ldots, y_{nt})'$,

$$E_t = \left(u_t, 0, \ldots, 0, u_{1,t}, 0, \ldots, 0, u_{2,t}, \ldots, 0\right)'$$

and

$$F_t = \left[f_t, f_{t-1}, \ldots, f_{t-p+1}, \gamma_{1,t}, \gamma_{1,t-1}, \ldots, \gamma_{n,t}, \gamma_{n,t-1}, \ldots, \gamma_{n,t-p+1}\right]'$$

Here, B is the block diagonal with the companion matrix having first row $\phi_1, \phi_2, \ldots, \phi_p$ in the (1, 1) block; the companion matrix with first row $\psi_{11}, \psi_{12}, \ldots, \psi_{1n}$ in the (2, 2) block; …; with the companion matrix having first row $\psi_{n1}, \ldots, \psi_{nq}$ in the southeastern-most block. The matrix H is 0 except for $(b_1, \ldots, b_n)'$ in the first column, and 1's in the columns and rows corresponding to $\gamma_{1,t}$, $\gamma_{2,t}$, etc in F_t.

Alternatively, a system with a lower-dimension state can be obtained by operating on both sides of Eq. (4) by $\left(1 - \psi_{n,1}L^{t-1} - \cdots - \psi_{n,q}L^{t-q}\right)$ to get

$$y^*_{n,t} = a^*_n + b_n\left(1 - \psi_{n,1}L^{t-1} - \cdots - \psi_{n,q}L^{t-q}\right)f_t + u_{nt} \tag{9}$$

where $y^*_{n,t} = y_{n,t} - \psi_{n,1}y_{n,t-1} - \cdots - \psi_{n,q}y_{n,t-q}$, $a^*_n = \left(1 - \psi_{n,1} - \cdots - \psi_{n,q}\right)a_n$. This yields the state-space system

$$Y^*_t = A^*D_t + H^*F_t + U_t \tag{10}$$

$$F_t = BF_{t-1} + E_t \tag{11}$$

where $Y^*_t = \left(y^*_{1t}, \ldots, y^*_{nt}\right)'$, $U_t = (u_{1t}, u_{2t}, \ldots, u_{nt})'$, $F^*_t = \left(f_t, f_{t-1}, \ldots, f_{t-p}\right)'$, $E^*_t = (e_t, 0, \ldots, 0)'$, the nth row of H^* is $\left(b_n, -\psi_{n,1}b_n, \ldots, -\psi_{n,q}b_n\right)$, and B^* has $\phi_1, \phi_2, \ldots, \phi_q, 0, \ldots, 0$ in the first row and 1's on the first subdiagonal. (The extra $q - p + 1$ columns of zeros in B^* accommodate the lags of the factor introduced into the measurement equation by the transformation to serially uncorrelated residuals.) This model has an $l \times m$ dimensional state vector, so N no longer impacts the size of the state vector. Jungbacker, Koopman, and van der Wel (2011) have a discussion comparing the state-space formulation with quasi-differencing (as above) with a formulation that adds the idiosyncratic error terms directly into the state vector. As they note, both formulations will lead to the same answer if the filters are properly normalized.

2.3. *Principal Components*

A third approach to estimating the latent factors employs principal components analysis (hereafter, PCA) which solves an eigenvector–eigenvalue problem to extract the factors before conditioning on said factors. The latter conditioning step treats the extracted factors as observable variables. Thus, PCA identifies the common movements in the cross-sectional data without imposing any additional model structure. The advantage of PCA is that it is simple to use and has been shown, under certain conditions, to produce a consistent approximation of the filtered (i.e., direct) estimate of the factors. An expansive literature has assessed the potential applications of factor modeling techniques and PCA. Bai and Ng (2008) give a detailed survey of the asymptotic properties of static factor models and dynamic factor models expressed in static form. Stock and Watson (2002) deliver key results suggesting that the method of asymptotic PCA consistently estimates the true factor space.

Consider the collection of data at time t, $Y_t = (y_{1t}, \ldots, y_{Nt})'$ to be a random $(N \times 1)$ vector with sample mean and covariance $\bar{y}$ and S, respectively. Normalizing the data to have mean zero, PCA results in the transformation:

$$F_{t,(i)} = (Y_t - \mathbf{1}\bar{y}')g_{(i)}$$

$i = 1, \ldots, N$ and where $\mathbf{1}$ is an $(N \times 1)$ vector of ones. The vector $g_{(i)}$ denotes the standardized eigenvector corresponding to the ith largest eigenvalue of S $(S = GVG')$ where V is a matrix with the eigenvalues of S in descending order along the diagonal. G is an orthogonal matrix of principal component loadings with columns $g_{(i)}$. Thus, $g_{(1)}$ corresponds to the largest eigenvalue associated with the first principal component $F_{t,(1)} = (Y_t - \mathbf{1}\bar{y}')g_{(1)}$.

When extracting static principal components, we find the standardized eigenvalues of the sample covariance matrix and treat the corresponding standardized eigenvectors as the factor loadings relating the static factors to the observable data. In order to impose the structure associated with world, region, and country factors, we extract each factor from the data assumed to load upon each factor. For the multi-level factor models, we extract the first factor, F^W, with its associated eigenvector, g_W, from the entire $(T \times N)$ dataset. Subsequently, letting Y_{n_m} for $n_m = 1, \ldots, m$ correspond to the observable series which load upon the second-level factor m, we adjust the data to remove the first factor: $\tilde{Y}_{n_m} = Y_{n_m} - F^W g_W$. Next, we extract the first principal component from this set of $\tilde{Y}_{n_m}$, $n_m = 1, \ldots, m$.

We perform a similar adjustment for the three-level factor model. Note that uncertainty in estimating F^W is not taken into account in this procedure.

These standardized principal component estimates serve as the latent factor estimates. As in the previous estimation methods, we condition on the latent factor estimates and apply Bayesian estimation to obtain the parameters of the model. However, using the PCA method, we extract the principal components outside of the Gibbs sampler and then treat the unobserved factors as data when we sample from the conditional posterior distributions of the parameters. By construction then, the PCA approach will underestimate the uncertainty in variance decompositions.

3. MONTE CARLO EVALUATION

The three methods presented in the previous section are evaluated using Monte Carlo experiments. For each of the three models (one-factor, two-factor, three-factor), we generate 1,000 sets of true data that includes a set of true factors by simulating U_t and V_t from multivariate normal distributions and then applying Eqs. (1)–(3). The true data consist of 100 time series observations forming a balanced panel. For the one factor case, we generate 21 series of data. For the two factor case, we generate three series of data for each of seven countries. For the three-factor case, we generate a small-region model with three series of data for each of eight countries broken up into two equally sized regions. Additionally, we generate a large-region model with three series of data for each of 16 countries, again broken into two equally sized regions.

In order to assess the methods across a wide range of model parameterizations, we redraw the model parameters at each Monte Carlo iteration. The covariance matrices of each of the innovation processes are fixed. All shocks are normalized to have unit variance and are assumed orthogonal. The AR parameters are drawn from univariate normal distributions with decreasing means for higher lag orders:

$$\Phi_{mi} \sim N\left(0.15(p+1-i), (0.1)^2\right)$$
$$\Psi_{ni} \sim N\left(0.15(q+1-i), (0.1)^2\right)$$

where p and q are the lag orders of factor and innovation AR processes, respectively. The AR parameters for each process are constrained to be stationary; we redraw the parameters if stationarity of the lag polynomial is violated. The factor loadings are also drawn from normal distributions:

$$\beta_{nk} \sim N\left(1, (0.25)^2\right)$$

where the multi-level model zero restrictions on the factor loadings are appropriately applied.

We then estimate the model using the three methods described above. We assume that the number of factors and, in the three factor case, the number and composition of the regions are ex ante known.

3.1. Priors for Estimation

The estimation in our Monte Carlo exercises is Bayesian and requires a prior. The prior for the model parameters are generally weakly informative, with the exception that we impose stationarity on the dynamic components. In the dynamics for the observable data, the prior on the constant, β_{n0}, and each factor's loading, β_{nm}, for $m = 1, \ldots, M$, for each country $n = 1, \ldots, N$ is:

$$\begin{bmatrix}\beta_{n0} & \beta_{n1} & \cdots & \beta_{nM}\end{bmatrix}' \sim N\left(\mathbf{0}_{(M+1)x1}, \operatorname{diag}\left(\left[1, 10 \times \mathbf{1}_{(1\times M)}\right]\right)\right) \tag{12}$$

where $\mathbf{1}_{(1\times M)}$ is a $(1 \times M)$ vector of ones. The prior for the autoregressive parameters of the factors and of the innovations in the measurement equation are truncated normal. In particular, the AR parameters are assumed to be a multivariate standard normal truncated to maintain stationarity. The prior for the innovation variances in the measurement equation is Inverted Gamma, parameterized as $\text{IG}\left(0.05 \times T, 0.25^2\right)$. The factor innovation is normalized to have unit variance and is not estimated. These priors are fairly diffuse. In work on actual datasets (e.g., Kose et al., 2008), we have experimented with different priors and we did not find the results to be sensitive to changes in these priors. Including sensitivity to the prior in our Monte Carlo work would be computationally very expensive.

3.2. Accuracy of Factor Estimates

The first metric for assessing the three estimation methods outlined above is to determine the accuracy of the factor estimates. For each simulation, we have the true values of the factors. In estimation, for each iteration of the Gibbs sampler, we produce a draw from the conditional distribution of the factor. To assess the accuracy of each method, we compare the correlation of the true factor with each draw of the sampler to form a distribution for the correlation. Fig. 1 shows the distribution of this correlation for the world factor in the one-factor case. For the two-factor case, we show the distribution of the correlation for the world factor in Fig. 2. For the country factor, we compute the correlation for each country, then take the average correlation across countries and report this in one PDF in Fig. 3. For the three-factor case with small regions, we show the correlation distribution for the world factor in Fig. 4, the average correlation across regional factors in Fig. 5, and average across countries in Fig. 6. For the three-factor case with large regions, these same plots are shown in Figs. 7–9.

In the one-factor case, all three methods produce similar results with correlations very close to 1. However, when the model is extended to include additional factors, the accuracy of the factor estimates for each of the methods deteriorates, even for the factor estimated across all of the series (the world factor). In addition, for higher level models, the average correlation between the true factor and the estimated factor falls considerably.

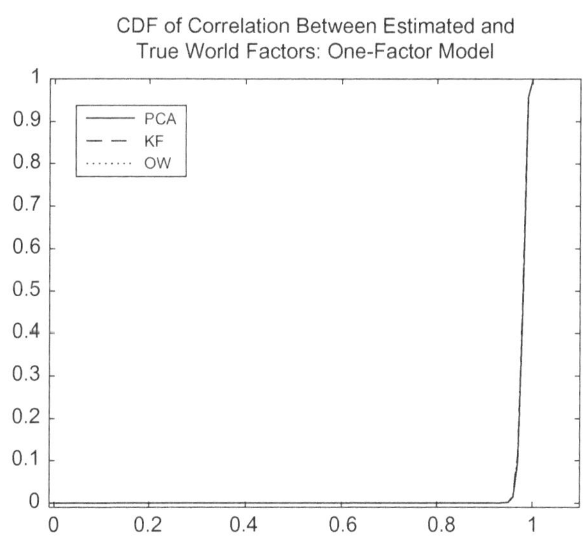

Fig. 1. CDF of the Correlation between the True and Estimated World Factor in the One-Factor Model, over 1,000 MC Simulations.

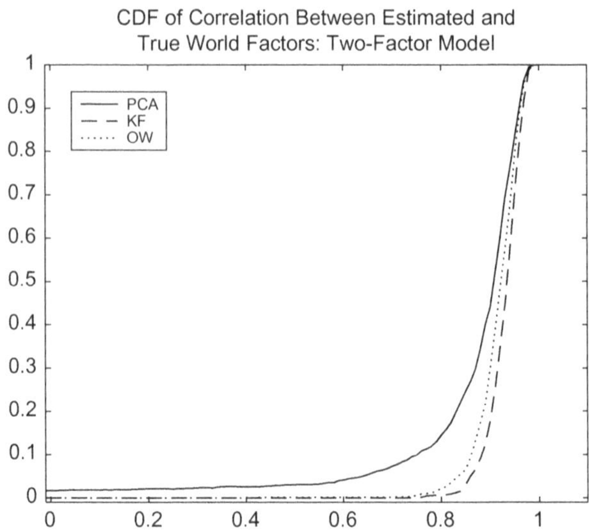

Fig. 2. CDF of the Correlation between the True and Estimated World Factor in the Two-Factor Model, over 1,000 MC Simulations.

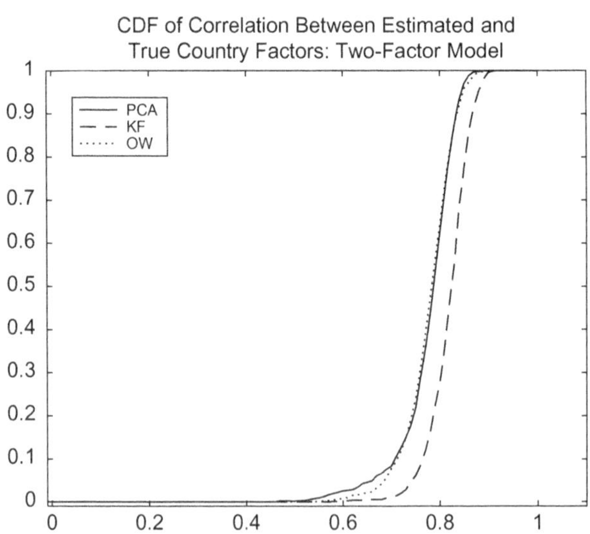

Fig. 3. CDF of the Correlation between the True and Estimated Country Factors in the Two-Factor Model, over 1,000 MC Simulations. The Correlations Are Averaged across Countries.

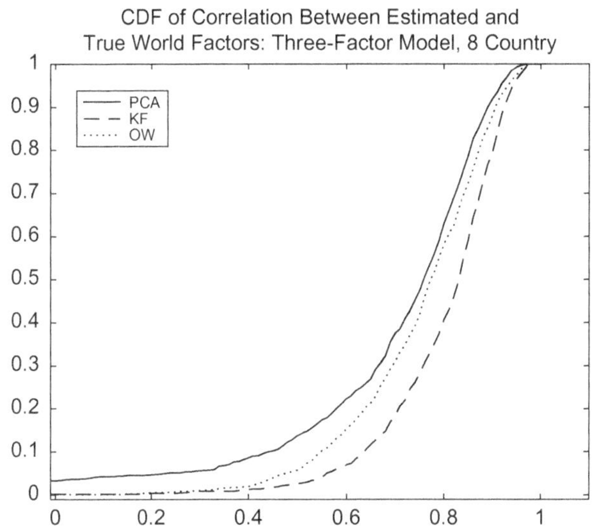

Fig. 4. CDF of the Correlation between the True and Estimated World Factor in the Three-Factor Model with Small Regions, over 1,000 MC Simulations. The Datasets Consist of Eight Countries Broken into Two Equally Sized Regions.

Because the world factor is computed with a larger number of series, we expect the factor to be estimated more accurately.

In each case, the factors that affect smaller number of series are more accurately estimated using the Kalman filter. In most cases, the Otrok–Whiteman procedure performs similar to PCA. One difference occurs when estimating the country factor for the three-factor model. In this case, the country factor is estimated with only three series with two additional layers (factors) contributing to the uncertainty of the estimates. In this case, the Otrok–Whiteman procedure outperforms PCA but continues to be outperformed by the state-space method. This last result may stem from the fact that the state-space imposes orthogonality across the factor estimates, while Otrok–Whiteman procedure assumes it but does not impose it.

3.3. Uncertainty in Factor Estimates

A second method for evaluation of the estimation procedures outlined in the previous section is to determine the uncertainty in the estimates of

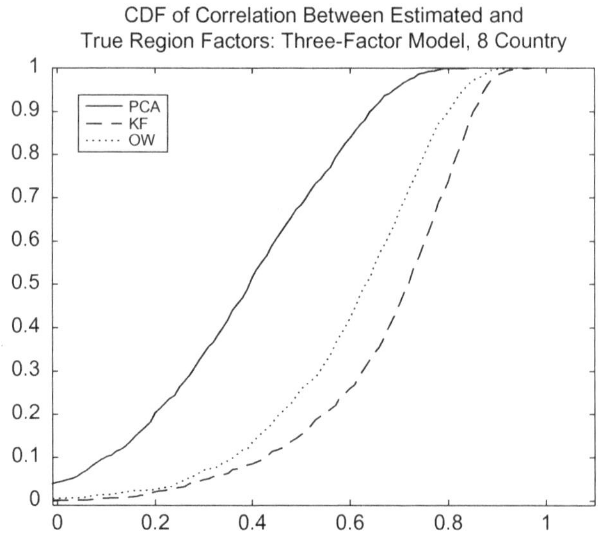

Fig. 5. CDF of the Correlation between the True and Estimated Region Factor in the Three-Factor Model with Small Regions, over 1,000 MC Simulations. *Notes:* The datasets consist of eight countries broken into two equally sized regions. The correlations represent the average across regions.

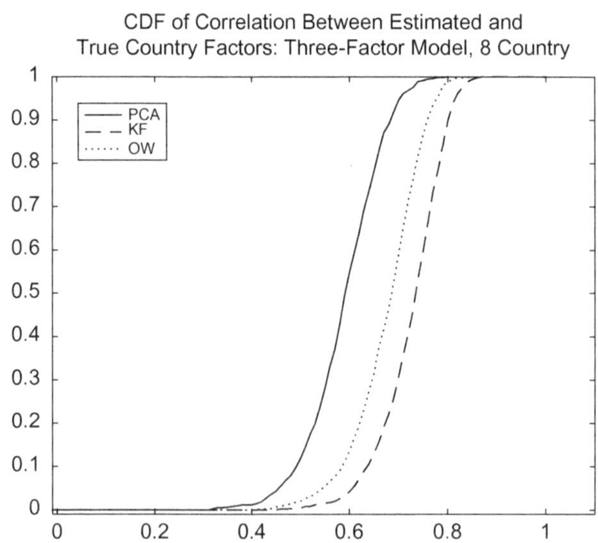

Fig. 6. CDF of the Correlation between the True and Estimated Country Factors in the Three-Factor Model with Small Regions, over 1,000 MC Simulations. *Notes:* The datasets consist of eight countries broken into two equally sized regions. The correlations represent the average across countries.

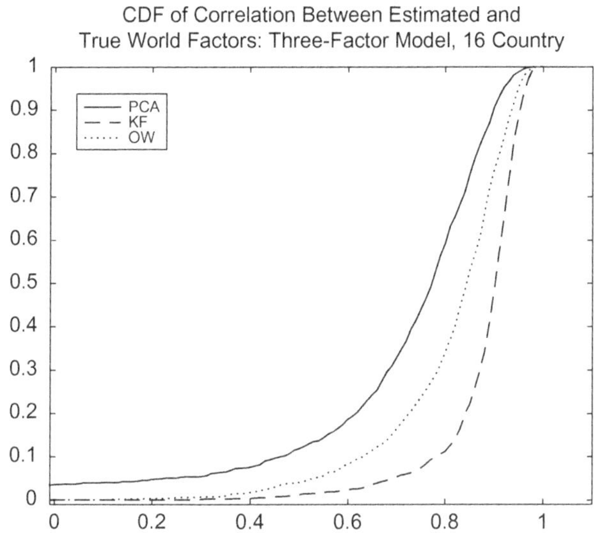

Fig. 7. CDF of the Correlation between the True and Estimated World Factor in the Three-Factor Model with Large Regions, over 1,000 MC Simulations. The Datasets Consist of 16 Countries Broken into Two Equally Sized Regions.

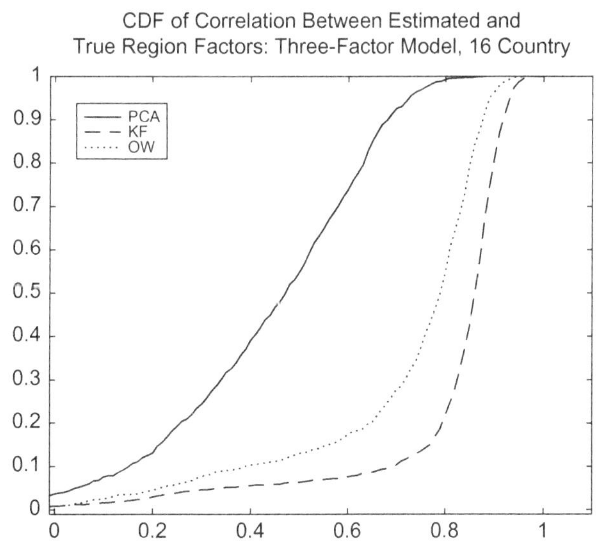

Fig. 8. CDF of the Correlation between the True and Estimated Region Factor in the Three-Factor Model with Large Regions, over 1,000 MC Simulations. *Notes:* The datasets consist of 16 countries broken into two equally sized regions. The correlations represent the average across regions.

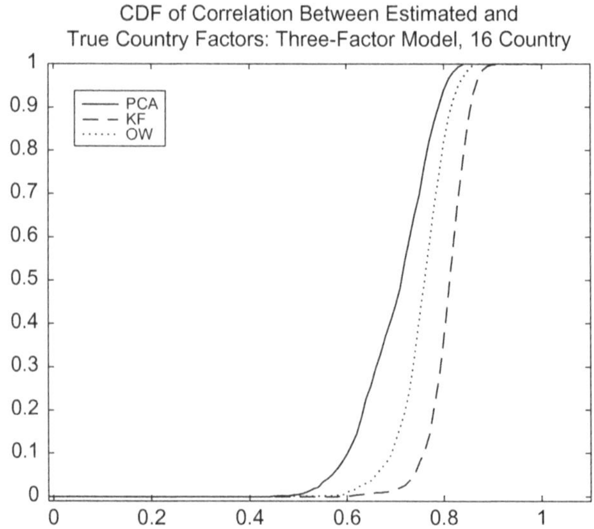

Fig. 9. CDF of the Correlation between the True and Estimated Country Factors in the Three-Factor Model with Large Regions, over 1,000 MC Simulations. *Notes:* The datasets consist of 16 countries broken into two equally sized regions. The correlations represent the average across countries.

the factor. To do this, we construct the average uncertainty in the factor estimates for the procedures. Our measure of average uncertainty is computed by determining the width of the 68% and 90% coverage intervals of the factor for each time period. These intervals are computed for each Monte Carlo iteration across all saved draws of the Gibbs sampler. We then compute the average of these intervals across time and across Monte Carlo iterations and report the numbers in the top panel of Table 1. In the principal components case, the factor is determined outside the Gibbs sampler, and, thus, does not have a small-sample uncertainty measure. We see this as a limitation of this approach as uncertainty in the factor estimate should be part of interpreting the importance of the factor. In each of the three models, the Otrok–Whiteman procedure has, on average, narrower coverages than the state-space method. In particular, the Otrok–Whiteman procedure yields about 20–30% narrower bands than the Kalman filter. Since there is no "true" measure of uncertainty our interpretation is that there are precision gains associated with drawing the factors directly from their distribution as opposed to simulating them in state-space model.

Table 1. Mean Width of Posterior Coverage Intervals.

	PCA	Kalman Filter	Otrok–Whiteman
One-factor model			
World factor: 68% interval	0.000	0.585	0.426
World factor: 90% interval	0.000	0.967	0.706
Two-factor model			
World factor: 68% interval	0.000	0.875	0.622
Country factor: 68% interval	0.000	1.386	0.983
World factor: 90% interval	0.000	1.450	1.004
Country factor: 90% interval	0.000	2.352	1.650
Three-factor model: Eight country			
World factor: 68% interval	0.000	1.007	0.657
Region factor: 68% interval	0.000	1.358	0.976
Country factor: 68% interval	0.000	1.772	0.990
World factor: 90% interval	0.000	1.672	1.055
Region factor: 90% interval	0.000	2.270	1.590
Country factor: 90% interval	0.000	2.998	1.654
Three-factor model: 16 country			
World factor: 68% interval	0.000	0.746	0.494
Region factor: 68% interval	0.000	1.102	0.770
Country factor: 68% interval	0.000	1.425	0.924
World factor: 90% interval	0.000	1.238	0.782
Region factor: 90% interval	0.000	1.836	1.241
Country factor: 90% interval	0.000	2.398	1.547

3.4. Accuracy in Variance Decompositions

A third metric that can be used to evaluate the three estimation procedures is to analyze the accuracy of the variance explained by each of the estimated factors – that is, do the estimated factors explain more or less of the variation explained by the true factors? We measure the variance decomposition in two ways. First, we compute a measure of the variance decomposition using a parameter-based method. This measure uses the draw of the set of model parameters at each Gibbs iteration to calculate the implied variance decomposition. The variance of the factor and idiosyncratic component is constructed using the parameters governing the time series process. These variances, along with the factor loadings, are then used to construct the implied variance of the observable data and the resulting variance decomposition. This procedure relies on the accuracy of the estimates of the full set of the true parameters. In the second procedure, at each iteration of the Gibbs sampler, we orthogonalize the set of estimated factors and then compute the variance decomposition based on a regression of the observable data on the orthogonalized factors. This factor-based method is similar to the approach used in Kose et al. (2003, 2008, 2012). Because the Otrok–Whiteman procedure does not impose orthogonality when estimating the factors themselves, the explained variances could be biased if the factors exhibit some finite sample correlation. The orthogonalization procedure addresses this issue.

To assess accuracy, we first compute the posterior mean of the variance explained by each of the estimated factors across Gibbs iterations. Next, we compute the variance of the data explained by the true factors within each simulated dataset. Finally, we take the difference between the true and estimated variance decompositions for each Monte Carlo iteration. To measure the total size of the bias, we compute the absolute value of the bias. The top panel of Fig. 10 plots the pdfs of the absolute value difference between the true and the estimated factor-based variance decomposition for all Monte Carlo iterations for the three estimation methods applied to the one-factor model. The Kalman filter and Otrok–Whiteman methods produce almost identical results with a nearly identical PDFs. The PCA approach has a distribution that lies to the right (larger absolute error). For parameter-based estimates, the bulk of the distribution is in the same area as the factor based, but has a noticeably larger tail. This larger tail indicates that the parameter-based estimates can occasionally yield large errors in the variance decomposition. Experience tells us that this is driven by nearly non-stationary parameter sets. Small variations in parameters can yield large difference in implied variance when near a unit root. In this sense, the factor-based estimates of variance decompositions are more robust to the underlying model parameters.

Table 2 shows that the mean of the absolute error in the variance decomposition between the true and estimated variance decompositions is 0.092 for the Kalman filter and Otrok–Whiteman methods. The PCA method slightly overestimates the variance attributed to the world factor and as a result, slightly underestimates the variance attributed to the idiosyncratic component. The mean of this distribution, 0.095, is slightly larger in magnitude than that of the other two methods, though the difference is negligible. Average errors, then, appear similar across methods.

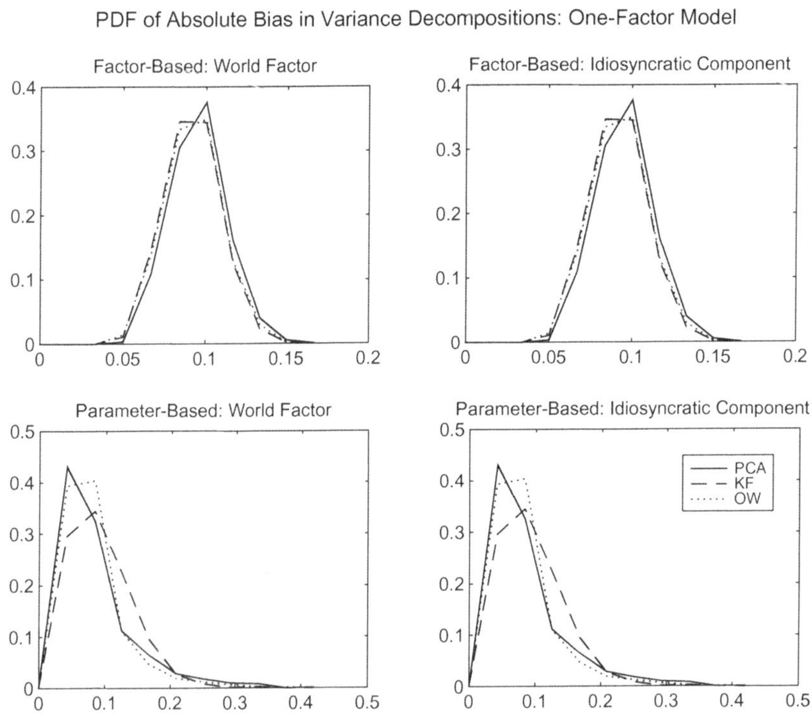

Fig. 10. PDF of the Mean Absolute Bias in the True versus the Estimated Variance Decompositions for the One-Factor Model. The Top Row Shows the Factor-Based Variance Decomposition for the World and Country Factors and the Idiosyncratic Component. The Bottom Row Shows the Parameter-Based Decompositions.

In terms of the sign of the bias, the factor-based estimates do better in all three cases. The Otrok–Whiteman method yields the same number of positive and negative biases regardless of method. For both the PCA and Kalman Filter methods, the number of positive/negative biases goes to a much more asymmetric distribution with a 70/30 split for the parametric case. Taking all these results together, it appears that the factor-based variance decompositions are less likely to lead to significantly wrong answers.

For the two-factor model, the top panel of Fig. 11 illustrates that the Kalman filter and Otrok–Whiteman again produce similar results when using the factor-based variance decomposition. For the country factors, which are the addition relative to Fig. 10, we see that the dispersion is less

Table 2. PDF of Variance Decomposition Biases – One-Factor Model.

	PCA	Kalman Filter	Otrok–Whiteman
Factor-based variance decomposition			
Mean of PDF			
World factor	0.095	0.092	0.092
Idiosyncratic factor	0.095	0.092	0.092
Percent of positive biases			
World factor	0.417	0.486	0.489
Idiosyncratic factor	0.583	0.514	0.511
Percent of negative biases			
World factor	0.583	0.514	0.511
Idiosyncratic factor	0.417	0.486	0.489
Parameter-based variance decomposition			
Mean of PDF			
World factor	0.089	0.094	0.083
Idiosyncratic factor	0.089	0.094	0.083
Percent of positive biases			
World factor	0.674	0.306	0.482
Idiosyncratic factor	0.326	0.694	0.518
Percent of Negative Biases			
World factor	0.326	0.694	0.518
Idiosyncratic factor	0.674	0.306	0.482

diffuse for all three methods with the factor-based approach. With the parameter-based approach, the absolute errors are both larger in mean, more diffuse, and exhibits a larger right tail. This confirms the results of the one-factor model that the factor-based variance decompositions are more robust then then parametric. From Table 3, we can see that the three methods have similar mean absolute errors. However, in this case, the PCA approach has more symmetric errors than either Bayesian which both have more positive errors than negative.

For the three-factor model with small regions, the top panel of Fig. 12 shows the absolute biases resulting from the factor-based variance decomposition. Table 4 gives the mean and percent of positive and negative biases. For the world factor and the idiosyncratic component, the Kalman filter and Otrok–Whiteman methods produce very similar results for all three types of factors. The Bayesian approaches do better than PCA for all three layers of factors. This can all be seen by the means reported in Table 4. The differences in performance for this more complex model are striking. The mean bias for the world factor is 70% higher with PCA than the Kalman filter. The parametric estimates do better for the PCA case than the factor-based estimates. A reversal of previous results. For the Bayesian approach, the factor-based estimates were again superior. Table 5 and Fig. 13 report the same facts for the larger country model. As can be

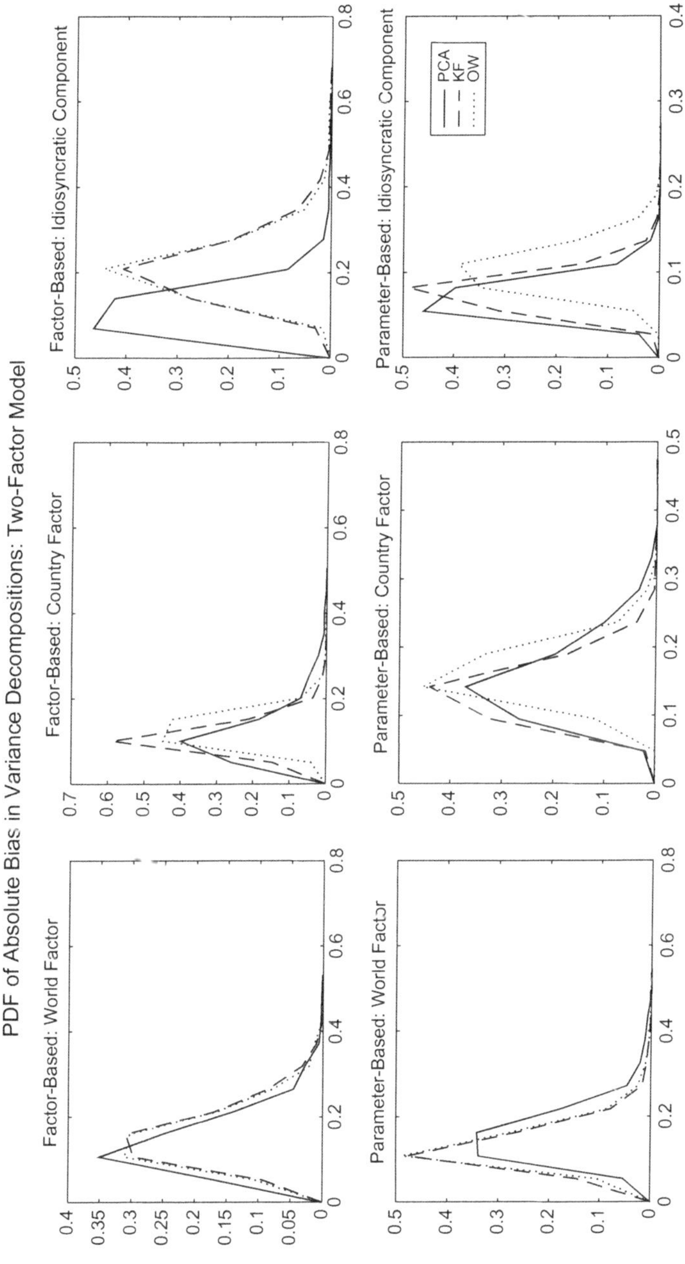

Fig. 11. Model. The Top Row Shows the Factor-Based Variance Decomposition for the World and Country Factors and the Idiosyncratic Component. The Bottom Row Shows the Parameter-Based Decompositions.

Table 3. PDF of Variance Decomposition Biases – Two-Factor Model.

	PCA	Kalman Filter	Otrok–Whiteman
Factor-based variance decomposition			
Mean of PDF			
World factor	0.142	0.159	0.155
Country factor	0.121	0.110	0.130
Idiosyncratic factor	0.120	0.216	0.217
Percent of positive biases			
World factor	0.401	0.838	0.818
Country factor	0.443	0.704	0.679
Idiosyncratic factor	0.752	0.094	0.156
Percent of negative biases			
World factor	0.599	0.162	0.182
Country factor	0.557	0.296	0.321
Idiosyncratic factor	0.248	0.906	0.844
Parameter-based variance decomposition			
Mean of PDF			
World factor	0.161	0.131	0.137
Country factor	0.153	0.138	0.162
Idiosyncratic factor	0.070	0.079	0.104
Percent of positive biases			
World factor	0.457	0.318	0.313
Country factor	0.538	0.653	0.659
Idiosyncratic factor	0.519	0.512	0.492
Percent of negative biases			
World factor	0.543	0.682	0.687
Country factor	0.462	0.347	0.341
Idiosyncratic factor	0.481	0.488	0.508

seen, there is no real difference in the relative performance of the procedures. This indicates that additional cross-section data will likely not change our results, or in the case of the PCA method the smaller model is big enough for asymptotics to apply.

4. AN APPLICATION TO GLOBAL HOUSE PRICES

The recent financial crisis was centered around a global housing price collapse. This has heightened interest in the nature of house price fluctuations. Most work on the issue has been at the national level. One recent exception is Hirata, Kose, Otrok, and Terrones (2013), who use principal-component-based

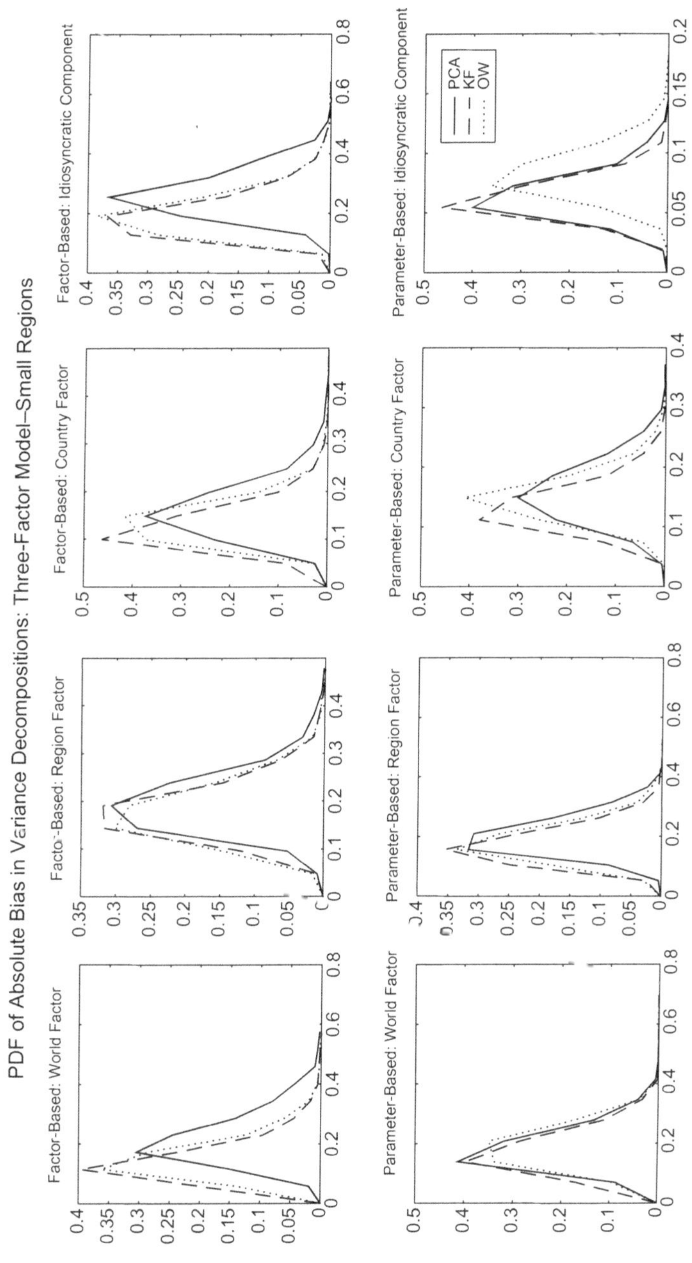

Fig. 12. Component. The Bottom Row Shows the Parameter-Based Decompositions.

Table 4. PDF of Variance Decomposition Biases – Three-Factor Model, Eight Country.

	PCA	Kalman Filter	Otrok–Whiteman
Factor-based variance decomposition			
Mean of PDF			
World factor	0.221	0.144	0.155
Region factor	0.200	0.180	0.179
Country factor	0.162	0.127	0.140
Idiosyncratic factor	0.269	0.198	0.205
Percent of Positive Biases			
World factor	0.264	0.643	0.614
Region factor	0.251	0.199	0.236
Country factor	0.586	0.617	0.604
Idiosyncratic factor	0.860	0.548	0.559
Percent of Negative Biases			
World factor	0.736	0.357	0.386
Region factor	0.749	0.801	0.764
Country factor	0.414	0.383	0.396
Idiosyncratic factor	0.140	0.452	0.441
Parameter-based variance decomposition			
Mean of PDF			
World factor	0.184	0.172	0.191
Region factor	0.205	0.171	0.183
Country factor	0.159	0.134	0.151
Idiosyncratic factor	0.065	0.062	0.083
Percent of positive biases			
World factor	0.408	0.368	0.329
Region factor	0.542	0.551	0.613
Country factor	0.574	0.599	0.604
Idiosyncratic factor	0.438	0.488	0.491
Percent of negative biases			
World factor	0.592	0.632	0.671
Region factor	0.458	0.449	0.387
Country factor	0.426	0.401	0.396
Idiosyncratic factor	0.562	0.513	0.509

factor models to study house price for a panel of advanced economies. A new dataset by Cesa-Bianchi et al. (2015) develops a set of comparable house prices for both emerging markets and advanced economies. Their contribution is to build the dataset, analyze the dataset in terms of moments of house prices and their comovement with the macroeconomy. They then use a panel VAR to

Table 5. PDF of Variance Decomposition Biases – Three-Factor Model, 16 Country.

	PCA	Kalman Filter	Otrok–Whiteman
Factor-based variance decomposition			
Mean of PDF			
World factor	0.208	0.129	0.149
Region factor	0.177	0.175	0.173
Country factor	0.142	0.114	0.129
Idiosyncratic factor	0.273	0.195	0.204
Percent of positive biases			
World factor	0.270	0.690	0.650
Region factor	0.280	0.188	0.226
Country factor	0.480	0.617	0.604
Idiosyncratic factor	0.864	0.533	0.543
Percent of negative biases			
World factor	0.730	0.310	0.350
Region factor	0.720	0.812	0.774
Country factor	0.520	0.383	0.396
Idiosyncratic factor	0.136	0.467	0.457
Parameter-based variance decomposition			
Mean of PDF			
World factor	0.176	0.172	0.180
Region factor	0.201	0.150	0.162
Country factor	0.150	0.126	0.140
Idiosyncratic factor	0.061	0.057	0.079
Percent of positive biases			
World factor	0.411	0.277	0.279
Region factor	0.549	0.538	0.613
Country factor	0.575	0.685	0.634
Idiosyncratic factor	0.447	0.568	0.509
Percent of negative biases			
World factor	0.589	0.723	0.721
Region factor	0.451	0.462	0.387
Country factor	0.425	0.315	0.366
Idiosyncratic factor	0.553	0.432	0.491

understand the differential role of liquidity shocks on house prices across countries in regions defined as advanced and emerging. Here, we extend their work by using multi-layer factor models to measure the importance of world and regional cycles in house prices across advanced and emerging markets. Following the theme of this paper, we will apply all three methods to estimate the factors.

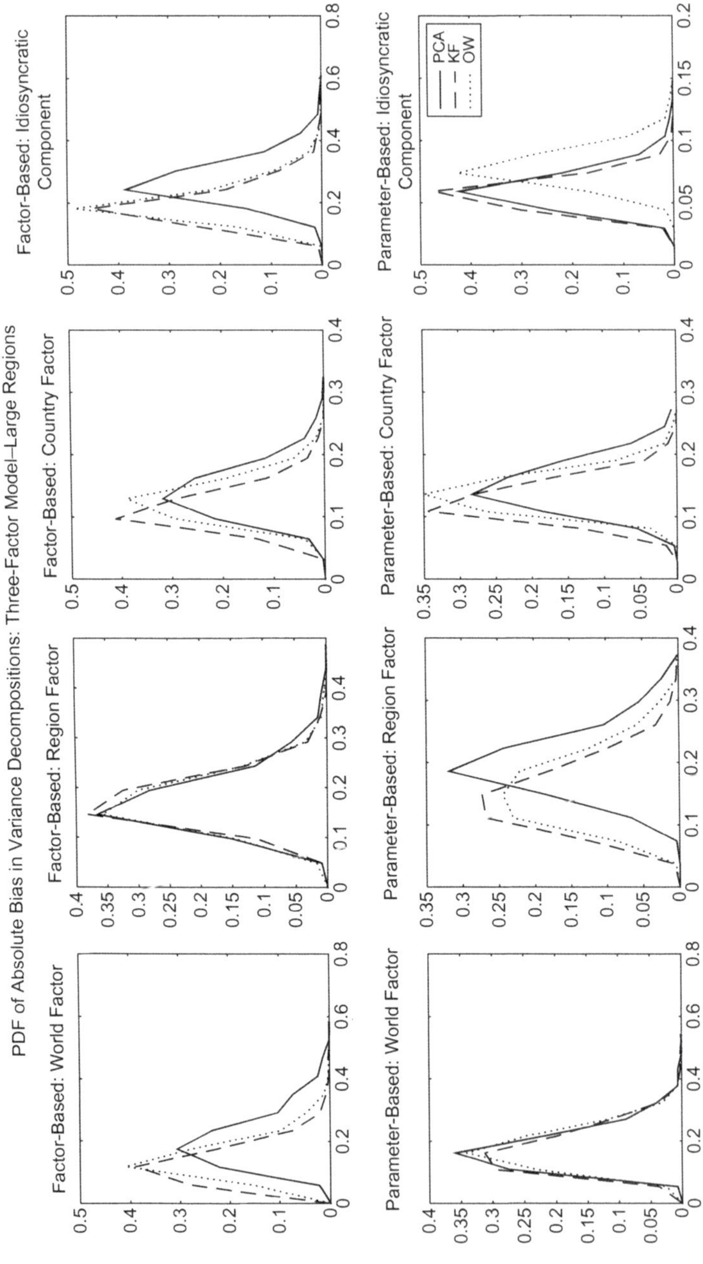

Fig. 13. Component. The Bottom Row Shows the Parameter-Based Decompositions.

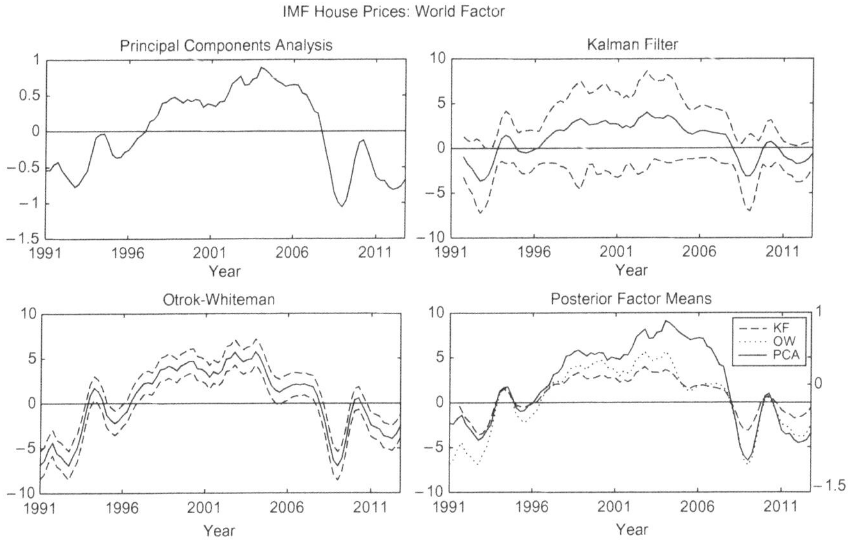

Fig. 14. World Factors Extracted from IMF Real House Price Data in Advanced and Emerging Economies Using Three Estimation Methods: Principal Components Analysis and Bayesian Estimation with Kalman Filtering or the Otrok–Whiteman Method. *Notes:* All plots show the posterior mean factor estimates. The Kalman filter and Otrok–Whiteman plots also include the 68% posterior coverage interval.

In Fig. 14, we plot estimates of the world factor for house prices. The global house price factor exhibits a long and fairly steady growth cycle from 1991 to 2005, when a modest decline begins, followed by a sharp drop in 2007. The global house cycle appears to have little high frequency volatility. The Otrok–Whiteman posterior coverage intervals are tighter than that of the Kalman Filter. This result is consistent with the Monte Carlo results which showed more precise coverage intervals for Otrok–Whiteman. Our Monte Carlo results also showed that the common world factor was generally very similar across procedures. The lower right panel of Fig. 14 plots the means of the factors on the same graph (though PCA is plotted in a different scale). It is clear that they all deliver the exact same message about global house prices.

The factor for advanced economies is plotted in Fig. 15. Posterior coverage intervals for Otrok–Whiteman are again tighter than the Kalman filter. There now appears to be less agreement on the regional factor. The correlation of the PCA factor with Otrok–Whiteman is only 0.19. The correlation

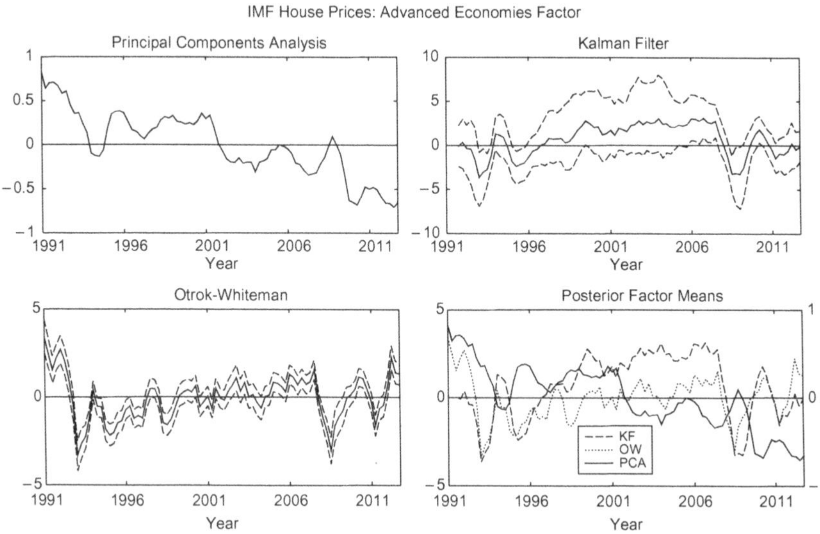

Fig. 15. Regional Factors Extracted from IMF Real House Price Data in Advanced Economies Using Three Estimation Methods: Principal Components Analysis and Bayesian Estimation with Kalman Filtering or the Otrok–Whiteman Method. *Notes:* All plots show the posterior mean factor estimates. The Kalman filter and Otrok–Whiteman plots also include the 68% posterior coverage interval.

between Otrok–Whiteman and the Kalman Filter is 0.62. This is also consistent with our Monte Carlo results in that the two Bayesian procedures deliver similar factors, and we found that in multi-layer models the Bayesian procedures did a better job at finding the true factor. The variance decompositions in Table 6 show that the regional factor is not that important quantitatively, which means there is not a lot of information in the data in the cycle. It is then not surprising that we may get somewhat different estimates of this factor.

The emerging markets factor is plotted in Fig. 16. Consistent with the results for advanced economies, there is little relationship between the PCA factor and the Bayesian factors. The correlation of PCA with the Bayesian estimates is 0, while the correlation between the Bayesian estimates is again 0.6. Taken together, Figs. 14–16 tell an interesting story about the evolution of house prices. In the pre-crisis period, the emerging markets factor is quite volatile, while in the crisis period the factor does not move very much. In contrast, the advanced economies factor is fairly smooth in the

Table 6. Percent of True Observations within Posterior Coverage.

	PCA	Kalman Filter	Otrok–Whiteman
One-factor model			
True world factor in 68% interval	0.000	0.200	0.265
True world factor in 90% interval	0.000	0.333	0.423
Two-factor model			
True world factor in 68% interval	0.000	0.305	0.221
True country factor in 68% interval	0.000	0.468	0.329
True world factor in 90% interval	0.000	0.486	0.349
True country factor in 90% interval	0.000	0.693	0.520
Three-factor model: eight country			
True world factor in 68% interval	0.000	0.337	0.200
True region factor in 68% interval	0.000	0.404	0.284
True country factor in 68% interval	0.000	0.467	0.300
True world factor in 90% interval	0.000	0.523	0.313
True region factor in 90% interval	0.000	0.617	0.441
True country factor in 90% interval	0.000	0.692	0.477
Three-factor model: 16 country			
True world factor in 68% interval	0.000	0.280	0.162
True region factor in 68% interval	0.000	0.405	0.262
True country factor in 68% interval	0.000	0.475	0.304
True world factor in 90% interval	0.000	0.447	0.253
True region factor in 90% interval	0.000	0.614	0.403
True country factor in 90% interval	0.000	0.701	0.484

pre-crisis period while it drops in the crisis period. Because the world factor also drops in this period, the two factors imply that the crash in advanced economies was in fact worse than in emerging markets. The world factor shows that all house prices fell in the crisis, the regional factors indicate a greater drop in advanced economies, while in emerging markets there was relative calm.

The variance decompositions in Table 7 show patterns that would be expected based on our Monte Carlo results. The parametric results are all greater than the factor-based estimates. Our Monte Carlo results suggest the factor based results are more accurate (and conservative) so we will base our discussion on them. Fig. 17 plots the variance decompositions

Table 7. Variance Decompositions.

	PCA						Kalman Filter						Otrok–Whiteman					
	Factor-based			Parameter-based			Factor-based			Parameter-based			Factor-based			Parameter-based		
	World	Region	Idio.	World	Region	Idio.	World	Region	Idio.	World	Region	Idio.	World	Region	Idio.	World	Region	Idio.
Advanced economies																		
United States	0.612	0.023	0.365	0.780	0.064	0.156	0.207	0.130	0.663	0.075	0.110	0.816	0.386	0.048	0.567	0.086	0.059	0.856
Australia	0.370	0.041	0.589	0.691	0.186	0.122	0.186	0.127	0.687	0.363	0.164	0.473	0.360	0.032	0.608	0.548	0.016	0.436
Austria	0.168	0.003	0.829	0.221	0.236	0.543	0.132	0.137	0.730	0.207	0.179	0.614	0.270	0.510	0.220	0.312	0.584	0.104
Belgium	0.523	0.000	0.477	0.842	0.061	0.097	0.120	0.185	0.695	0.191	0.200	0.610	0.241	0.065	0.694	0.244	0.114	0.642
Canada	0.219	0.314	0.467	0.263	0.671	0.066	0.101	0.248	0.651	0.261	0.274	0.465	0.155	0.086	0.759	0.383	0.033	0.584
Denmark	0.487	0.019	0.494	0.858	0.035	0.107	0.161	0.115	0.724	0.355	0.176	0.469	0.301	0.015	0.684	0.536	0.005	0.459
Finland	0.364	0.246	0.390	0.442	0.467	0.090	0.222	0.118	0.660	0.309	0.145	0.547	0.448	0.008	0.543	0.487	0.012	0.501
France	0.657	0.045	0.298	0.735	0.055	0.211	0.206	0.230	0.565	0.126	0.063	0.812	0.414	0.067	0.519	0.179	0.004	0.817
Germany	0.446	0.035	0.519	0.529	0.175	0.296	0.162	0.083	0.755	0.159	0.099	0.742	0.279	0.018	0.703	0.243	0.073	0.684
Greece	0.487	0.300	0.213	0.468	0.499	0.033	0.176	0.078	0.746	0.042	0.023	0.935	0.250	0.009	0.741	0.038	0.005	0.957
Ireland	0.561	0.265	0.174	0.588	0.361	0.051	0.238	0.097	0.665	0.191	0.084	0.725	0.395	0.007	0.597	0.207	0.007	0.786
Italy	0.443	0.017	0.540	0.149	0.148	0.703	0.130	0.158	0.712	0.013	0.008	0.979	0.199	0.119	0.683	0.011	0.001	0.988
Japan	0.062	0.015	0.923	0.293	0.239	0.468	0.013	0.031	0.956	0.074	0.035	0.891	0.038	0.037	0.926	0.132	0.006	0.862
Luxembourg	0.445	0.015	0.540	0.700	0.103	0.197	0.144	0.147	0.709	0.187	0.219	0.594	0.230	0.134	0.635	0.152	0.124	0.724
Malta	0.212	0.039	0.749	0.665	0.091	0.243	0.084	0.062	0.853	0.227	0.109	0.664	0.122	0.004	0.873	0.347	0.012	0.641
Netherlands	0.221	0.358	0.421	0.438	0.241	0.322	0.125	0.058	0.817	0.030	0.044	0.926	0.210	0.058	0.733	0.065	0.012	0.923
New Zealand	0.336	0.070	0.594	0.460	0.335	0.205	0.120	0.128	0.752	0.182	0.113	0.705	0.229	0.013	0.758	0.249	0.008	0.743
Norway	0.267	0.162	0.571	0.496	0.427	0.077	0.166	0.101	0.733	0.344	0.188	0.469	0.362	0.010	0.628	0.572	0.021	0.407
Portugal	0.006	0.329	0.666	0.239	0.462	0.299	0.020	0.035	0.944	0.049	0.048	0.904	0.002	0.089	0.909	0.056	0.043	0.901
Spain	0.749	0.043	0.208	0.759	0.154	0.088	0.263	0.163	0.574	0.159	0.058	0.783	0.418	0.045	0.537	0.196	0.005	0.799
Sweden	0.535	0.073	0.392	0.867	0.052	0.081	0.261	0.171	0.568	0.438	0.234	0.327	0.500	0.045	0.455	0.633	0.028	0.339
Switzerland	0.046	0.635	0.319	0.061	0.875	0.064	0.036	0.073	0.891	0.087	0.047	0.866	0.097	0.017	0.886	0.121	0.007	0.872
United Kingdom	0.733	0.027	0.240	0.928	0.026	0.046	0.373	0.179	0.448	0.583	0.259	0.159	0.738	0.013	0.249	0.870	0.016	0.113
Mean	*0.389*	*0.134*	*0.477*	*0.542*	*0.259*	*0.198*	*0.159*	*0.124*	*0.717*	*0.202*	*0.125*	*0.673*	*0.289*	*0.063*	*0.648*	*0.290*	*0.052*	*0.658*

Median	*0.443*	*0.043*	*0.477*	*0.529*	*0.185*	*0.122*	*0.161*	*0.127*	*0.712*	*0.187*	*0.110*	*0.705*	*0.270*	*0.037*	*0.683*	*0.243*	*0.012*	*0.724*
Emerging economies																		
Hong Kong	0.051	0.566	0.383	0.149	0.723	0.128	0.095	0.036	0.869	0.208	0.019	0.774	0.070	0.014	0.916	0.156	0.008	0.836
Argentina	0.079	0.343	0.578	0.229	0.626	0.146	0.090	0.147	0.763	0.266	0.070	0.664	0.139	0.052	0.809	0.285	0.004	0.711
Chile	0.000	0.010	0.990	0.104	0.200	0.696	0.012	0.553	0.436	0.084	0.467	0.449	0.004	0.734	0.262	0.049	0.776	0.175
Colombia	0.067	0.157	0.776	0.155	0.440	0.405	0.078	0.007	0.914	0.073	0.007	0.921	0.054	0.007	0.938	0.050	0.018	0.932
Croatia	0.006	0.142	0.852	0.175	0.364	0.461	0.009	0.005	0.986	0.082	0.008	0.910	0.026	0.019	0.955	0.144	0.009	0.847
Korea	0.078	0.026	0.897	0.204	0.179	0.616	0.055	0.058	0.887	0.075	0.026	0.899	0.062	0.005	0.933	0.045	0.005	0.950
Malaysia	0.216	0.378	0.406	0.141	0.621	0.239	0.107	0.013	0.880	0.056	0.005	0.940	0.200	0.033	0.767	0.071	0.001	0.928
Singapore	0.033	0.368	0.599	0.117	0.651	0.232	0.067	0.011	0.923	0.177	0.007	0.817	0.023	0.018	0.959	0.089	0.003	0.908
South Africa	0.705	0.098	0.197	0.830	0.120	0.050	0.226	0.006	0.768	0.251	0.003	0.746	0.447	0.002	0.551	0.367	0.001	0.632
Uruguay	0.003	0.072	0.924	0.069	0.463	0.468	0.006	0.055	0.939	0.037	0.045	0.918	0.001	0.009	0.990	0.026	0.007	0.967
Mean	*0.124*	*0.216*	*0.660*	*0.217*	*0.439*	*0.344*	*0.074*	*0.089*	*0.837*	*0.131*	*0.066*	*0.804*	*0.103*	*0.089*	*0.808*	*0.128*	*0.083*	*0.789*
Median	*0.059*	*0.150*	*0.688*	*0.152*	*0.451*	*0.322*	*0.072*	*0.025*	*0.883*	*0.083*	*0.013*	*0.858*	*0.058*	*0.016*	*0.925*	*0.080*	*0.006*	*0.877*

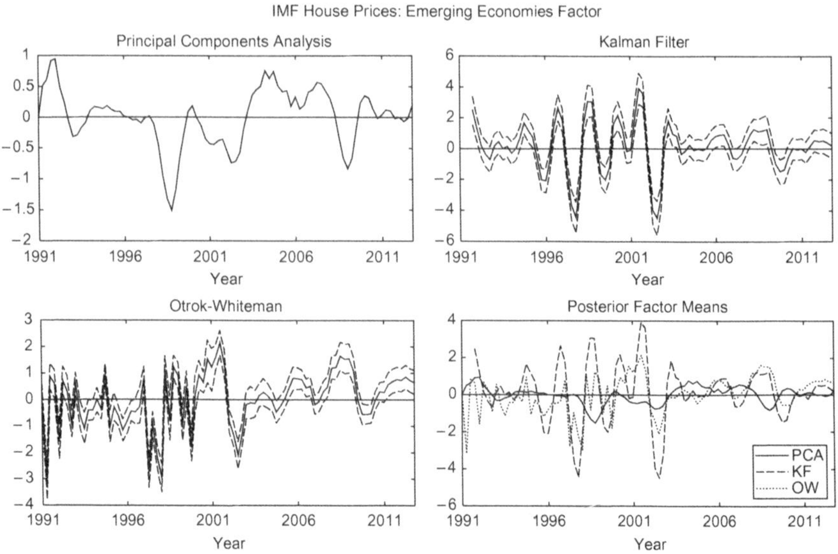

Fig. 16. Regional Factors Extracted from IMF Real House Price Data in Emerging Economies Using Three Estimation Methods: Principal Components Analysis and Bayesian Estimation with Kalman Filtering or the Otrok–Whiteman Method. *Notes:* All plots show the posterior mean factor estimates. The Kalman filter and Otrok–Whiteman plots also include the 68% posterior coverage interval.

from the three models for each country. Its clear that PCA lies above the two Bayesian procedures, consistent with the Monte Carlo results. All three methods show an important role for the world factor, though the PCA perhaps overstates the importance.

For advanced economies their is an important role for the global factor, both PCA and Otrok–Whiteman method attribute about 1/3 of fluctuations to this factor. The advanced regional factor plays a much smaller role, accounting for about 10% of the variation on average. The median variance decompositions are even smaller, suggesting that this factor is only affecting a small number of countries.

Cesa-Bianchi et al. (2015) report average correlations that are greater for advanced economies then for emerging economies. Our results are consistent with these results. The world factor plays a much smaller role for emerging markets as the lower average variance decompositions show. It is also clear that the emerging market comovement is largely through the

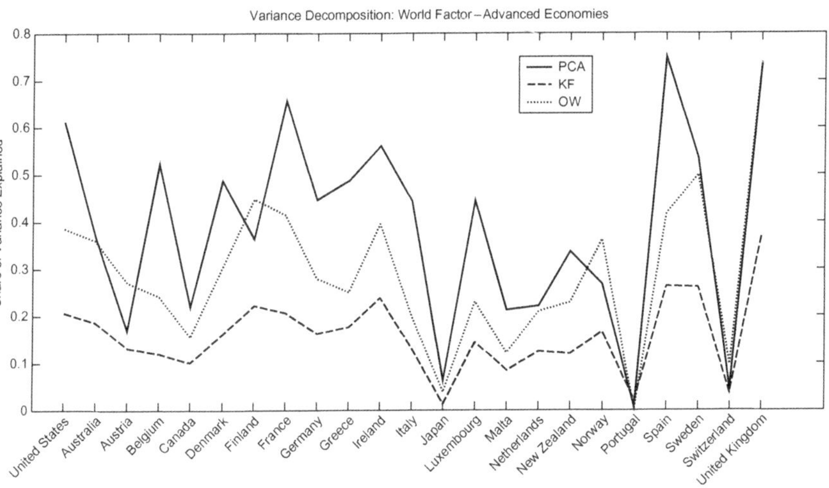

Fig. 17. Variance Decomposition of World Factor for Advanced Economies: The World Factor Is Extracted from IMF Real House Price Data in Advanced and Emerging Economies Using Three Estimation Methods: Principal Components Analysis and Bayesian Estimation with Kalman Filtering or the Otrok–Whiteman Method. *Notes:* The plot shows the share of variation in each country's house prices that is attributable to the world factor.

world factor. The median variance decomposition for the emerging markets factors is only 1%. We conclude then that regional factors are not important in understanding house price movements in either advanced or emerging markets. There is a common global cycle of significant importance.

5. CONCLUSION AND DISCUSSION

In this paper, we use Monte Carlo methods to analyze the accuracy of three methods to estimate dynamic factor models. All three methods worked well in the one factor case. It also is apparent that the factor-based estimates of variance decompositions are better than parametric based estimates. Our experience tells us that this is due to the fact that variances can blow up with parameter estimates near the unit root. In applied work, we have found this to often be true. As model complexity increased the Bayesian approaches yielded more accurate results, which in the case of the three-layer model were substantially different.

This latter results was perhaps unsurprising in sign, though the magnitude of the difference was surprisingly large. The Kalman-filter based method would expected to be most accurate as we are directly drawing from the likelihood of the model. The gains over the Otrok–Whiteman method were surprisingly small. This is goods news in that the Otrok–Whiteman method is more computationally efficient in dealing with large datasets. It is useful to know that its accuracy is a good as the state-space approach. Both the PCA and Otrok–Whiteman approach share an iterative approach to estimation. They differ in that the PCA approach conditions in one direction, while the Otrok–Whiteman conditions (repeatedly) in multiple directions (i.e., world is drawn conditional on country, country then drawn conditional on world). While this is clearly costly in computational terms, it yields more accurate estimates of the importance of factors in complex factor settings.

There are a number of choices to be made in applied work on factor models. The PCA approach will always be best to get quick answers, and in the one factor case there seems to be no accuracy gains from the more complicated methods. On the other hand, the Bayesian methods naturally yield measures of factor uncertainty, which should always accompany applied work in order to establish the statistical legitimacy of the results. In this paper, we have not reported information on computational time, as for any specific application the computation time is not large. Here, for each model type, we have estimated 1,000 different applications (i.e., each model parameter drawn). It follows that estimating one is not a problem for modern computers.

In our application, we did find that PCA yielded factors that appeared to have greater importance then the Bayesian methods. In economic terms though, all methods delivered a similar story in that house price comovement across advanced and emerging markets is through a common factor, and not through group-specific factors.

NOTES

1. One could also consider the accuracy of other model parameters. However, factor analysis has tended to focus on the variance decomposition because it is this output that is most useful in telling an economic story about the data. In addition,

since the scale of a factor model is not identified, the factor loading is not as of as much interest as the scale independent variance decomposition.

2. This is the procedure in Kose et al. (2003), Kose, Otrok, and Whiteman (2008), and Kose, Otrok, and Prasad (2012).

ACKNOWLEDGMENTS

Diana A. Cooke and Hannah G. Shell provided research assistance. We thank two referees and Siem Jan Koopman for helpful comments. Matlab code used in this paper is available at www.runmycode.org. The views expressed herein do not reflect the views of the Federal Reserve Bank of St. Louis, the Federal Reserve System or the World Bank.

REFERENCES

Bai, J., & Ng, S. (2008). Large dimensional factor analysis. *Foundations and Trends in Econometrics*, *3*(2), 89–163.

Breitung, J., & Eickmeier, S. (2014). *Analyzing business and financial cycles using multi-level factor models*, Bundesbank Working Paper No. 11/2014.

Carlin, B. P., Polson, N. G., & Stoffer, D. S. (1992). A Monte Carlo approach to nonnormal and nonlinear state-space modeling. *Journal of the American Statistical Association*, *87*, 493–500.

Carter, C. K., & Kohn, P. (1994). On Gibbs sampling for state space models. *Biometrica*, *81*, 541–553.

Cesa-Bianchi, A., Cespedes, L. F., & Rebucci, A. (2015). Global liquidity, house prices, and the macroeconomy: Evidence from advanced and emerging economies. *Journal of Money, Credit and Banking*, *47*(S1), 301–335.

de Jong, P., & Shephard, N. (1995). The simulation smoother for time series models. *Biometrika*, *82*, 339–350.

Durbin, J., & Koopman, S. J. (2002). A simple and efficient simulation smoother for state space time series analysis. *Biometrika*, *89*(3), 603–615.

Engle, R. F., & Watson, M. (1981). A one-factor multivariate time series model of metropolitan wage rates. *Journal of the American Statistical Association*, *76*(376), 774–781.

Forni, M., Hallin, M., Lippi, M., & Reichlin, L. (2000). The generalized dynamic-factor model: Identification and estimation. *Review of Economics and Statistics*, *82*(4), 540–554.

Forni, M., & Reichlin, L. (1998). Let's get real: A factor analytical approach to disaggregated business cycle dynamics. *Review of Economic Studies*, *65*(3), 453–473.

Francis, N., Owyang, M. T., & Savasçin, Ö. (2012). *An endogenously clustered factor approach to international business cycles*, Working Paper No. 2012-014A, Federal Reserve Bank of St. Louis.

Geweke, J. (1977). The dynamic factor analysis of economic time series. In D. J. Aigner & A. S. Goldberger (Eds.), *Latent variables in socio-economic models*. Amsterdam: North Holland Publishing, Chap. 19.

Hirata, H., Kose, M. A., Otrok, C., & Terrones, M. (2013). Global house price fluctuations: Synchronization and determinants. In *NBER international seminar on macroeconomics 2012* (pp. 119–166). Chicago, IL: University of Chicago Press.

Jungbacker, B., Koopman, S. J., & van der Wel, M. (2011). Maximum likelihood estimation for dynamic factor models with missing data. *Journal of Economic Dynamics and Control*, *35*, 1358–1368.

Kim, C.-J., & Nelson, C. R. (1998). Business cycle turning points, a new coincident index, and tests for duration dependence based on a dynamic factor model with regime switching. *Review of Economic Statistics*, *80*, 188–201.

Kim, C.-J., & Nelson, C. R. (1999). *State space models with regime switching*. Cambridge, MA: MIT Press.

Kose, M. A., Otrok, C., & Prasad, E. S. (2012). Global business cycles: Convergence or Decoupling? *International Economic Review*, *53*(2), 511–538.

Kose, M. A., Otrok, C., & Whiteman, C. H. (2003). International business cycles: World, region, and country specific factors. *American Economic Review*, *93*(4), 1216–1239.

Kose, M. A., Otrok, C., & Whiteman, C. H. (2008). Understanding the evolution of world business cycles. *Journal of International Economics*, *75*(1), 110–130.

Otrok, C., & Whiteman, C. H. (1998). Bayesian leading indicators: Measuring and predicting economic conditions in Iowa. *International Economic Review*, *39*(4), 997–1014.

Quah, D., & Sargent, T. J. (1993). A dynamic index model for large cross sections. In H. S. James & M. W. Watson (Eds.), *Business cycles, indicators, and forecasting* (pp. 285–310). Chicago, IL: University of Chicago Press.

Sargent, T. J., & Sims, C. A. (1977). *Business cycle modeling without pretending to have too much a priori economic theory*, Working Paper No. 55, Federal Reserve Bank of Minneapolis.

Stock, J. H., & Watson, M. (2002). Macroeconomic forecasting using diffusion indexes. *Journal of Business and Economic Statistics*, *20*(2), 147–162.

Stock, J. H., & Watson, M. W. (1989). New indexes of coincident and leading economic indicators. In J. B. Olivier & S. Fischer (Eds.), *NBER Macroeconomics Annual 1989* (Vol. 4, pp. 351–409). Cambridge, MA: The MIT Press.

Tanner, M., & Wong, W. H. (1987). The calculation of posterior distributions by data augmentation. *Journal of the American Statistical Association*, *82*, 84–88.

SMALL- VERSUS BIG-DATA FACTOR EXTRACTION IN DYNAMIC FACTOR MODELS: AN EMPIRICAL ASSESSMENT

Pilar Poncela[a] and Esther Ruiz[b]

[a]*Department of Economic Analysis: Quantitative Economics, Universidad Autónoma de Madrid, Madrid, Spain*

[b]*Department of Statistics, Instituto Flores de Lemus, Universidad Carlos III de Madrid, Madrid, Spain*

ABSTRACT

In the context of Dynamic Factor Models, we compare point and interval estimates of the underlying unobserved factors extracted using small- and big-data procedures. Our paper differs from previous works in the related literature in several ways. First, we focus on factor extraction rather than on prediction of a given variable in the system. Second, the comparisons are carried out by implementing the procedures considered to the same data. Third, we are interested not only on point estimates but also on confidence intervals for the factors. Based on a simulated system and the macroeconomic data set popularized by Stock and Watson (2012), we show that, for a given procedure, factor estimates

Dynamic Factor Models
Advances in Econometrics, Volume 35, 401–434

ISSN: 0731-9053/doi:10.1108/S0731-905320150000035010

based on different cross-sectional dimensions are highly correlated. On the other hand, given the cross-sectional dimension, the maximum likelihood Kalman filter and smoother factor estimates are highly correlated with those obtained using hybrid procedures. The PC estimates are somehow less correlated. Finally, the PC intervals based on asymptotic approximations are unrealistically tiny.

Keywords: Confidence intervals; Kalman filter; principal components; quasi-maximum likelihood

JEL classifications: C18; C38

1. INTRODUCTION

It is often argued that macroeconomic and financial variables are governed by a few underlying unobserved factors. Extracting these factors is becoming a central issue that interests econometricians, practitioners, and policy decision makers.[1] In this context, dynamic factor models (DFMs), originally introduced by Geweke (1977) and Sargent and Sims (1977), are a very popular instrument to deal with multivariate systems of macroeconomic and financial variables.

The availability of large (sometimes huge) systems has generated a debate about whether small- or big-data DFMs should be used to obtain more accurate estimates of the underlying factors. The most popular small-data procedure is based on Kalman filter and smoothing (KFS) algorithms with the parameters estimated by maximum likelihood (ML); see, for example, Engle and Watson (1981) for an early reference. On the other hand, big-data procedures are usually based on principal components (PC) techniques. Allowing for weak cross-correlations between the idiosyncratic noises, the factors are given by the first few PC (ordered by their eigenvalues) of the many variables in the system; see, for example, Stock and Watson (2002) and Forni et al. (2005). Finally, Doz, Giannone, and Reichlin (2011, 2012) propose hybrid methods that combine the PC and KFS (PC-KFS) procedures taking advantage of the best of each of them in such a way that it is possible to deal with big-data systems having efficiency similar to that of KFS. In particular, Doz et al. (2011) propose a two-step Kalman filter (2SKF) procedure which is iterated until convergence in the quasi-maximum likelihood (QML) algorithm of Doz et al. (2012).

Several works compare small- and big-data procedures in the context of forecasting one or several variables of interest; see, for example, Boivin and Ng (2006), Bai and Ng (2008b), Banbura and Runstler (2011), Caggiano, Kapetanios, and Labhard (2011), Alvarez, Camacho, and Perez-Quiros (2012) and Banbura and Modugno (2014). However, few works comparing small- and big-data procedures focus on factor estimates on their own; see, for example, Bai and Ng (2006b) for the importance of an adequate estimation of factors. Diebold (2003), in a short note, implement KFS to small-data and PC to big-data to extract the common factor from an empirical system of macroeconomic variables. After visual inspection of the corresponding plots, he concludes that nearly the same factor is extracted by both procedures. Alvarez et al. (2012) carry out Monte Carlo experiments to compare point factor estimates obtained using small- and big-data procedures. For the big-data case, they implement the QML procedure, while for the small-data, they extract the factors using KFS concluding that factors extracted using the small-scale model have smaller mean-squared errors (MSE) than when they are estimated using the big-data procedure. The differences are more pronounced for high levels of cross-correlation among the idiosyncratic noises and, especially, for high persistence in either the common factors or the idiosyncratic noises. Finally, Doz et al. (2012) also carry out Monte Carlo experiments to compare point estimates obtained using PC and the 2SKF and QML procedures.

In this paper, we compare point and interval factor estimates obtained using the four procedures mentioned above. Our contribution is different from other papers in the literature in several aspects. First, as just mentioned, our focus is on estimating the underlying factors implementing the same procedures to the same data sets; see Aruoba et al. (2009), who suggest that, in order to make a proper empirical comparison among procedures, small- versus big-data approaches should be fitted to the same data set. Furthermore, we compare all the most popular procedures available in the literature, namely, KFS, PC, and the two hybrid procedures. Finally, we compare not only point estimates but also interval estimates; see Bai (2003) and Bai and Ng (2006b) for the importance of measuring the uncertainty when estimating the underlying factors. We carry out this comparison using both simulated data and the real data base of Stock and Watson (2012).

We compare the small- and big-data procedures for different number of variables in the system. Based on asymptotic arguments, several authors argue that the usual methods for factor extraction turn the curse of dimensionality into a blessing.[2] According to Bai (2003), "economists now have the luxury of working with very large data sets." However, one can expect that, when introducing more variables, it is more likely that the weak cross-correlation assumption fails unless the number of factors increases; see Boivin and Ng (2006). Furthermore, when increasing the number of variables, it is very likely that additional sectorial factors may appear; see, for example, Kose, Otrok, and Whiteman (2003), Karadimitropoulou and León-Ledesma (2013), Moench, Ng, and Potter (2013), and Breitung and Eickmeier (2014, 2015), for sectorial factors. Also, by having more variables, the uncertainty associated with the parameter estimation is expected to increase; see Poncela and Ruiz (2015). Therefore, if one wants to estimate a particular factor, for example, the business cycle, it is not obvious that having more variables in the system increases the accuracy. Finally, it is important to mention that several authors conclude that the factors are already observable when the number of variables in the system is around 30; see Bai and Ng (2008b) and Poncela and Ruiz (2015) when extracting the factors using PC and KFS procedures, respectively. We should point out that, in this paper, we do not consider the effect of the parameter uncertainty as the time dimension is considered as fixed.

In this paper, we show that, for a given procedure, factor estimates based on different cross-sectional dimensions are highly correlated. On the other hand, given the cross-sectional dimension, the ML smoothed Kalman filter factor estimates are highly correlated with those obtained using the hybrid PC-KFS procedures. The PC estimates are somehow less correlated. Finally, the PC intervals based on asymptotic approximations are unrealistically tiny. Regardless of the dimension of the system, the two-step procedures are a compromise between the efficiency of KFS and the inefficient but computationally simple and robust PC procedures.

The rest of this paper is organized as follows. In Section 2, we establish notation by briefly describing the DFM and the alternative factor extraction procedures considered. Section 3 reports the results of Monte Carlo experiments carried out to analyze the effect of the number of variables and factors on the properties of the factors extracted using the alternative procedures considered. Section 4 contains the empirical analysis of the Stock and Watson (2012) data base. Section 5 concludes this paper.

2. EXTRACTING COMMON FACTORS

This section establishes notation and briefly describes the DFM and the factor extraction procedures considered, in particular, the PC, KFS, 2SKF, and QML procedures.

2.1. The Dynamic Factor Model

Consider the following DFM in which the factors are given by a VAR(p) model and the idiosyncratic noises are assumed to be a VAR(1) process

$$Y_t = PF_t + \varepsilon_t \tag{1}$$

$$F_t = \Phi_1 F_{t-1} + \cdots + \Phi_p F_{t-p} + \eta_t \tag{2}$$

$$\varepsilon_t = \Gamma \varepsilon_{t-1} + a_t \tag{3}$$

where $Y_t = (y_{t1}, \ldots, y_{tN})'$ is the $N \times 1$ vector of observed variables at time t, $F_t = (f_{t1}, \ldots, f_{tr})'$ is the $r \times 1$ vector of underlying factors and ε_t is the $N \times 1$ vector of idiosyncratic noises. The disturbance, η_t, is Gaussian white noise with finite and positive covariance matrix Σ_η. The idiosyncratic noises, ε_t, are independently distributed of $\eta_{t-\tau}$ for all leads and lags. We also assume that all autoregressive matrices, Φ_i, $i = 1, \ldots, p$, and Γ, are diagonal. Finally, a_t is Gaussian white noise with finite- and positive-definite covariance matrix Σ_a such that the idiosyncratic noises are weakly correlated. The $r \times r$ autoregressive matrices are such that all the roots of the equation $|I_r - \Phi_1 z - \cdots - \Phi_p z^p| = 0$ are strictly larger than one. Therefore, the factors are zero mean and stationary.[3] Similarly, the idiosyncratic noises are assumed to be zero mean and stationary. Consequently, in the remainder of this paper, we assume that, prior to the analysis, all the series in Y_t are demeaned and transformed to stationarity. Since the autoregressive matrices, Φ_i, $i = 1, \ldots, p$, and Γ, are diagonal, the number of parameters is reduced to a manageable size and we avoid to blur the separate identification of the common and idiosyncratic components; see Jungbacker and Koopman (in press) and Pinheiro, Rua, and Dias (2013). The $N \times r$ factor loading matrix is given by $P = [p_{ij}]$ for $i = 1, \ldots, N$ and $j = 1, \ldots, r$; see Bai and Ng (2008a), Breitung and Pigorsch (2013), Stock and Watson (2006, 2011), Breitung and Eickmeier (2006), and Breitung and Choi (2013) for excellent surveys on DFM.

Due to the diagonality of the autoregressive matrices in Eq. (2), the DFM defined in Eqs. (1)–(3) is identifiable as far as the autoregressive parameters in the main diagonal of each of these matrices are different; see Bai and Wang (2014) for a description of identifiability conditions. However, if, for example, the autoregressive matrices are scalar, the rank condition of Bai and Wang (2014) is not satisfied. In this case, the factors and factor loadings are only identified up to a pre-multiplication of an invertible matrix and further restrictions should be imposed for identification. In classical factor analysis, the covariance matrix of the factors, Σ_F, is assumed to be the identity matrix, while $P'P$ is a diagonal matrix; see, for example, Bai (2003). Alternatively, in state-space models, it is rather common to assume that $\Sigma_\eta = I_r$ together with $p_{ij} = 0$ for $j>i$; see Harvey (1989). In both cases, the factors are assumed to be contemporaneously independent which is an appealing property. With any of these restrictions, F and P are uniquely fixed up to a column sign change given the product FP' and the ordering of the factors. These restrictions are arbitrary in the sense that the factors are fixed up to their multiplication by an invertible $r \times r$ matrix. Consequently, the factors obtained may not lead to a particularly useful interpretation. However, once they have been estimated, the factors can be rotated to be appropriately interpreted.

There are several particular cases of the DFM in Eqs. (1)–(3) that have attracted a lot of attention in the related literature. When $\Gamma = 0$ and Σ_a is diagonal, the idiosyncratic noises are contemporaneously and serially independent. In this case, the DFM is known as *strict*. When there is serial correlation with Γ being diagonal, the model is known as *exact*. Chamberlain and Rothschild (1983) introduce the term *approximate factor structure* in static factor models where the idiosyncratic components do not need to have a diagonal covariance matrix.

Next, we briefly describe each of the four procedures considered in this paper to extract the factors in DFM.

2.2. *Principal Components*

In the context of big-data, the factors are usually extracted using procedures based on PC which are attractive because they are computationally simple. Furthermore, they are nonparametric and, consequently, robust to potential misspecifications of the dynamics of factors and idiosyncratic noises. The price to pay for this robustness is that PC extraction looses efficiency with respect to procedures based on well-specified dynamics.

In this subsection, we describe the PC procedure following Bai (2003). PC procedures allow estimating the space spanned by the factors. Consequently, in order to extract the individual factors one needs to know r, the number of factors in the system. The $T \times r$ matrix of PC factor estimates, $\hat{F} = (\hat{F}'_1, \ldots, \hat{F}'_T)$, is given by $\sqrt{T}$ times the r eigenvectors associated with the r largest eigenvalues of YY', where Y is the $T \times N$ matrix given by

$Y = (Y_1', \ldots, Y_T')$, arranged in decreasing order. Then, assuming that $(1/T)\hat{F}'\hat{F} = I_r$, the estimates of the loadings are given by

$$\hat{P} = \frac{Y'\hat{F}}{T}$$

The properties of the PC factor estimator are based on asymptotic arguments when both the cross-sectional, N, and temporal, T, dimensions tend simultaneously to infinity. Stock and Watson (2002) show that, if the cross-correlations of the idiosyncratic noises are weak and the variability of the common factors is not too small, the estimated factors are consistent.[4] Under general conditions that allow for serial and contemporaneous cross-correlation and heteroscedasticity in both dimensions, Bai (2003) shows that the estimated factors can be treated as if they were observed as long as the number of factors is known and fixed as both N and T grow and $(\sqrt{T}/N) \rightarrow 0$ when $N, T \rightarrow \infty$. He also derives the following asymptotic distribution:

$$\sqrt{N}(\hat{F}_t - H'F_t) \xrightarrow{d} N(0, V^{-1}Q\Gamma_t Q'V^{-1}) \tag{4}$$

where $H = \hat{V}^{-1}(\hat{F}'F/T)(P'P/N)$ with $\hat{V}$ being the $r \times r$ diagonal matrix of the first r largest eigenvalues of the matrix $YY'/(TN)$, arranged in decreasing order, and V is its limit in probability, Q being the $r \times r$ limit in probability matrix of $(\hat{F}'F)/T$ and the $r \times r$ matrix Γ_t is defined as follows:

$$\Gamma_t = \lim_{N \rightarrow \infty} \frac{1}{N} \sum_{i=1}^{N} \sum_{j=1}^{N} E(p_{i.}' p_{j.} \varepsilon_{ti} \varepsilon_{tj})$$

with $p_{i.}$ being the ith row of the factor loading matrix P. Given that the factors are estimated according to the normalization $(\hat{F}'\hat{F}/T) = I_r$, an estimate of Q is just the identity matrix. Therefore, the asymptotic variance of $\hat{F}_t$ can be estimated as follows:

$$\mathrm{var}(\hat{F}_t) = \frac{1}{N}\hat{V}^{-1}\hat{\Gamma}_t\hat{V}^{-1} \tag{5}$$

with $\hat{\Gamma}_t$ being a consistent estimate of Γ_t (or more precisely of $H^{-1}\Gamma_t H^{-1}$). Bai and Ng (2006a) propose three different estimators of Γ_t depending on the properties of the idiosyncratic errors. Two of them assume cross-sectionally uncorrelated idiosyncratic errors but do not require stationarity while the third is robust to cross-sectional correlation and heteroscedasticity but requires covariance stationarity. Bai and Ng (2006a) argue that for small cross-correlation in the errors, constraining them to be zero could sometimes be desirable because the sampling variability from estimating them could generate a nontrivial efficiency loss. Consequently, they recommend using the following estimator of Γ_t:

$$\hat{\Gamma}_t = \frac{1}{N}\sum_{i=1}^{N} \hat{p}'_{i.}\hat{p}_{i.}\hat{\varepsilon}^2_{ti} \tag{6}$$

where $\hat{\varepsilon}_t = (\hat{\varepsilon}_{t1}, \ldots, \hat{\varepsilon}_{tN})'$ is obtained as $\hat{\varepsilon}_t = Y_t - \hat{P}\hat{F}_t$.

2.3. Kalman Filter and Smoothing

In the context of small-data, the factors are usually estimated using the KFS algorithms with the parameters estimated by ML. Running the Kalman filter requires knowing the specification and parameters of the DFM in Eqs. (1)–(3). Therefore, the factors extracted using the KFS algorithms can be non-robust in the presence of model misspecification. However, if the model is correctly specified, extracting the factors using the Kalman filter is attractive for several reasons. First, it allows to deal with data irregularities as, for example, systems containing variables with different frequencies and/or missing observations; see Aruoba et al. (2009), Jungbacker, Koopman, and van der Wel (2011), Pinheiro et al. (2013), Banbura and Modugno (2014), and Bräuning and Koopman (2014) for some examples. Second, the KFS procedure is not so affected by outliers as PC procedures that are based on estimated covariance matrices. Third, they provide a framework for incorporating restrictions derived from economic theory; see Bork, Dewachter, and Houssa (2009) and Doz et al. (2012). Fourth, the KFS procedure is more efficient than PC procedures for a flexible range of specifications that include non-stationary DFM and idiosyncratic noises with strong cross-correlations. Finally, it allows obtaining uncertainty measures associated with the estimated factors when the cross-sectional dimension is finite; see Poncela and Ruiz (2015). However, the number of parameters that need to be estimated increases with the cross-sectional dimension in such a way that ML estimation may be unfeasible for large systems.

The DFM in Eqs. (1)–(3) is conditionally Gaussian. Consequently, when the idiosyncratic noises are serially uncorrelated, the KFS algorithms provide Minimum MSE (MMSE) estimates of the underlying factors which are given by the corresponding conditional means. Denoting by $f_{t|\tau}$ the estimate of F_t obtained with the information available up to time τ, and by $V_{t|\tau}$ its corresponding MSE, KFS delivers

$$f_{t|\tau} = E[F_t|Y_1, \ldots, Y_\tau] \tag{7}$$

$$V_{t|\tau} = E\left[\left(F_t - f_{t|\tau}\right)\left(F_t - f_{t|\tau}\right)'|Y_1, \ldots, Y_\tau\right] \tag{8}$$

where $\tau = t-1$, for one-step-ahead predictions, $\tau = t$ for filtered estimates and $\tau = T$ for smoothed factor estimates. It is also important to point out that the KFS procedures deliver out-of-sample forecasts of the factors together with their corresponding mean-squared forecast errors (MSFE). In this paper, our focus is on smoothed estimates so that they can be compared with those obtained from alternative procedures.

When the idiosyncratic noises are serially correlated, the DFM can be reformulated in two alternative ways to preserve the optimal properties of KFS. First, it is possible to express it in state-space form as follows:

$$\begin{gathered}
Y_t = \Gamma Y_{t-1} + \begin{bmatrix} P & -\Gamma P & 0 & \ldots & 0 \end{bmatrix} \begin{bmatrix} F_t \\ F_{t-1} \\ F_{t-2} \\ \vdots \\ F_{t-p+1} \end{bmatrix} + a_t \\
\begin{bmatrix} F_t \\ F_{t-1} \\ F_{t-2} \\ \vdots \\ F_{t-p+1} \end{bmatrix} = \begin{bmatrix} \Phi_1 & \Phi_2 & \ldots & \Phi_p \\ I & 0 & \ldots & 0 \\ 0 & I & \ldots & 0 \\ \vdots & \vdots & \ddots & \\ 0 & 0 & \ldots & I \end{bmatrix} \begin{bmatrix} F_{t-1} \\ F_{t-2} \\ F_{t-3} \\ \vdots \\ F_{t-p} \end{bmatrix} + \begin{bmatrix} \eta_t \\ 0 \\ 0 \\ \vdots \\ 0 \end{bmatrix}
\end{gathered} \tag{9}$$

see Reis and Watson (2010), Jungbacker et al. (2011), and Pinheiro et al. (2013) for implementations of the model in Eq. (9). One can alternatively deal with the autocorrelation of the idiosyncratic noises by augmenting the state vector by ε_t; see, for example, Jungbacker et al. (2011), Banbura and

Modugno (2014) and Jungbacker and Koopman (in press). Both formulations lead to the same results when the initialization issues are properly accounted for. However, note that, in practice, augmenting the state space might be only feasible for relatively small cross-sectional dimensions.

The parameters are usually estimated by ML maximizing the one-step-ahead decomposition of the log-Gaussian likelihood; see Engle and Watson (1981) and Watson and Engle (1983). The maximization of the log-likelihood entails nonlinear optimization which restricts the number of parameters that can be estimated and, consequently, the number of series that can be handled when estimating the underlying factors. Even if the number of factors is considered as fixed, the number of parameters to be estimated increases very quickly with N. Although the EM algorithm allows to maximize the likelihood function of very large DFM, it does not allow the estimation of the parameters in Γ; see Doz et al. (2012). Alternatively, Jungbacker et al. (2011) and Jungbacker and Koopman (in press) propose to transform the observation equation into a lower dimension which leads to a computationally efficient approach to parameter and factor estimation. However, if the cross-sectional dimension is large, this procedure is only feasible if the idiosyncratic noises are serially uncorrelated. Fiorentini, Galesi, and Sentana (2014) also propose an alternative spectral EM algorithm capable of dealing with large systems.

With respect to the uncertainty associated with the KFS estimates, Poncela and Ruiz (2015) obtain expressions of their finite N and T steady-state MSE when the model parameters are known or when they are estimated using a consistent estimator. They show that, in the first case, the MSE are decreasing functions of the cross-sectional dimension regardless of whether the idiosyncratic noises are weakly or strongly correlated. Furthermore, if the idiosyncratic noises are weakly correlated, the minimum MSE are zero for filtered and smoothed estimates while if they are strongly correlated, the minimum MSE are different from zero, so the factor estimates are not consistent. However, it is very important to remark that, in any case, the MSE is very close to the minimum when the number of variables in the system is relatively small, approximately around 30 variables. In the latter case, when the parameters are estimated, if the sample size is fixed, the MSE can even be an increasing function of the cross-sectional dimension. Therefore, in this case, which is the most common when dealing with empirical data, one can have more uncertainty about the underlying factors when the number of series used to estimate them increases.

2.4. Principal Components–Kalman Filter Smoothing

Doz et al. (2011, 2012) propose two procedures to estimate the factors in the presence of big-data systems based on combining the PC and KFS approaches; see Giannone, Reichlin, and Small (2008) for previous empirical applications and Banbura and Modugno (2014) for an extension to systems with missing data.

The 2SKF procedure proposed by Doz et al. (2011) starts extracting the factors by PC, obtaining $\hat{F}_t^{(0)} = \hat{F}_t^{PC}$. Then, the dynamics of the factors are estimated after fitting a VAR(p) model which is estimated by least squares (LS), obtaining $\hat{\Phi}^{(0)}$. The parameters in the covariance matrix of the idiosyncratic noises, Σ_ε, are estimated using the diagonal of the sample covariance matrix of the residuals as follows:

$$\hat{\Sigma}_\varepsilon^{(0)} = \mathrm{diag}\left(\frac{1}{T}\sum_{t=1}^{T}\hat{\varepsilon}_t^{(0)}\hat{\varepsilon}_t^{(0)\prime}\right)$$

where $\hat{\varepsilon}_t^{(0)} = Y_t - \hat{P}^{(0)}\hat{F}_t^{(0)}$ and $\hat{P}^{(0)} = \hat{P}^{PC}$. Setting $\Sigma_\eta = I$ for identification purposes, in the second step, the factors are estimated by running the smoothing algorithm of the Kalman filter implemented in the DFM in Eqs. (1)–(3) with $\Gamma = 0$ and the remaining parameters substituted by $\hat{P}^{(0)}$, $\hat{\Phi}^{(0)}$, and $\hat{\Sigma}_\varepsilon^{(0)}$ and assuming that the idiosyncratic noises are serially and contemporaneously uncorrelated. The MSE of the factors are directly obtained from the Kalman filter.

Doz et al. (2012) propose a QML procedure based on iterating the 2SKF. Actually, this is equivalent to ML estimation implemented through the EM algorithm when the idiosyncratic noises are white noise. Given $\hat{F}_t^{(i)}$, obtained at step i, the procedure is iterated by re-estimating the VAR parameters, the factor loadings and the diagonal covariance matrix of the error term in Eq. (1) as with the EM algorithm. Given this new set of parameters, the common factors are re-estimated at iteration $i+1$ using the Kalman smoother yielding $\hat{F}_t^{(i+1)}$. At each iteration, the algorithm ensures higher values of the log-likelihood. The process converges when the slope between two consecutive log-likelihood values is lower than a given threshold. The MSE of the factors are directly obtained from the Kalman filter in the last step; see Banbura and Runstler (2011) for an application in which they use the MSE to compute the weights for the predictions of a variable of interest.

3. MONTE CARLO EXPERIMENTS

In this section, for each of the four factor extraction procedures considered, we analyze the role of N and r on the finite sample properties of the estimated factors. Furthermore, given N and r, the four procedures are compared in finite samples by carrying out Monte Carlo experiments. The comparisons, made both in terms of point and interval factor estimates, are based on $R = 500$ replicates generated from four different data generating processes (DGP) with $N = 120$ variables, $r = 3$ factors and $T = 200$ observations. The first DGP, denoted as model 1 (M1), is given by

$$Y_t = PF_t + \varepsilon_t \tag{10}$$

$$F_t = \begin{bmatrix} 1.3 & 0 & 0 \\ 0 & 0.9 & 0 \\ 0 & 0 & 0.5 \end{bmatrix} F_{t-1} + \begin{bmatrix} -0.36 & 0 & 0 \\ 0 & 0 & 0 \\ 0 & 0 & 0 \end{bmatrix} F_{t-2} + \eta_t \tag{11}$$

where the weights of the first factor, p_{i1}, for $i = 1, \ldots, N$, are generated by a uniform distribution in $[0, 1]$; see Bai and Ng (2006a), who also carry out simulations generating the weights from a uniform distribution. The weights of the second factor are generated such that $p_{i2} \neq 0$, for $i = 13, \ldots, 60$ and $p_{i2} = 0$ otherwise. When different from zero, the weights are also generated by a uniform distribution. Note that the second factor only affects the variables from $i = 13, \ldots, 60$. Finally, the weights of the third factor, $p_{i3} = 0$ for $i = 1, \ldots, 60$ and are generated from a uniform distribution for $i = 61, \ldots, 120$. Consequently, the third factor affects the last 60 variables in the system. These two latter factors have a block structure as they are specific to subsets of variables. They can be considered as sectorial factors, likely to appear when big-data systems are considered. The idiosyncratic errors are Gaussian white noise with covariance matrix given by $\Sigma_\varepsilon = I_N$ so that they are homoscedastic and temporal and contemporaneously uncorrelated. Note that in Eq. (11), the first factor is given by an AR(2) process with roots 0.9 and 0.4, while the second and third factors are stationary AR(1) processes with roots 0.9 and 0.5, respectively. Finally, the noise in Eq. (11), η_t, is a Gaussian white noise with diagonal covariance matrix such that the variances of the three factors are 1.

Note that, in model M1, the KFS and both hybrid procedures are fitted to the true DGP with uncorrelated idiosyncratic errors. In order to check the effect of misspecification on the results, we also consider a second

DGP, denoted as model 2 (M2), specified as in Eqs. (10) and (11) with temporally correlated idiosyncratic errors which are defined as in Eq. (3) with $\Gamma = 0.5I_N$ and the covariance matrix of a_t defined in such a way that $\Sigma_\varepsilon = I_N$. In this way, comparing the results of models M1 and M2, we can conclude about how the dependence of the idiosyncratic noises affect the conclusions when the ratios of the variances of factors and idiosyncratic noises are kept fixed at 1. The third DGP considered is designed to check how the three factor extraction procedures based on the Kalman filter behave in the presence of misspecification of the factor temporal dependence. Model 3 (M3) is specified as model M1 with the factors given by a VARMA(2,1) model with autoregressive parameters as in Eq. (11) and MA(1) parameters given by 0.7, 0.5, and 0.3 for the first, second, and third factors, respectively. The noises of the factor model are standardized in such a way that $\Sigma_F = I_r$. Finally, the fourth DGP considered, model M4, is designed to take into account the effects on the conclusions of variations in the factor loadings as postulated by, for example, Breitung and Eickmeier (2011). In M4, we introduce a break of magnitude $b = 1$ in the loadings of the first factor, $p_{i1}, i = 1, \ldots, N$, at time $t^* = 100$.

For each DGP and each of the four procedures considered, the factors are extracted with different number of variables and factors in the system. First, we consider $N = 12$ variables (small-data) with the first 12 variables being selected ($r = 1$); second, $N = 12$ variables are selected from the 13th to the 24th so that $r = 2$; third, $N = 12$ variables are selected from the 55th to the 66th so that $r = 3$ and the third factor only has weights on a subset of variables; fourth, $N = 30$ (medium-data) with variables from the 13th to the 42nd so that $r = 2$; fifth, $N = 30$ with variables from the 46th to the 75th being chosen so that $r = 3$; sixth, we consider extracting the factors using all $N = 120$ variables (big-data). Previous to extract the factors, their number is selected by using the procedure proposed by Onatski (2009).[5] For each replicate in which the number of factors is correctly estimated, we estimate P and F by each of the four procedures considered. Consequently, the Monte Carlo results reported later in this paper are conditional on the true number of factors and based on a number of replicates that typically is smaller than 500. The precision of the point factor estimates, $\hat{F}$, obtained by each of the four procedures is measured by the trace R^2 of the regression of the estimated factors on the true ones given by

$$\mathrm{Tr} = \frac{\mathrm{Trace}(F'\hat{F}(\hat{F}'\hat{F})^{-1}\hat{F}'F)}{\mathrm{Trace}(F'F)} \tag{12}$$

The trace measure in Eq. (12), which is smaller than 1 and tends to 1 with the increasing canonical correlations between the estimated and true factors, has been implemented by, for example, Stock and Watson (2002), Boivin and Ng (2006), Doz et al. (2012), and Banbura and Modugno (2014), to assess the precision of factor estimates.

As the objective of this paper is to assess the accuracy of not only point factor estimates but also of interval estimates, we also compute the empirical coverages of the pointwise intervals of the factors extracted by each of the four procedures when the nominal coverage is 95%. The MSE used to construct the PC intervals are the asymptotic MSE in Eq. (5), while the MSE of the other three procedures are obtained from the Kalman smoother when the model parameters are substituted by the corresponding estimated parameters. Note that none of these MSE incorporate the uncertainty associated with the parameter estimation. As mentioned above, the factors are estimated up to a rotation. Consequently, in order to check whether the prediction intervals contain the true factors, the estimated factors are rotated to be in the scale of the true factors as follows:

$$\hat{F}^* = \hat{F}\left[(\hat{P}'\hat{P})^{-1}\hat{P}'P\right]^{-1} \tag{13}$$

where $\hat{P}$ are the matrices of estimated factor loadings obtained after implementing each of the four procedures. Of course, the MSE are also rotated accordingly to Eq. (13). Note that the rotations in Eq. (13) can be carried out when the DGP is one of the models M1, M2, or M3. However, when the DGP is model M4, the true matrix of loadings, P, is not constant and, consequently, it is not obvious which is the adequate rotation to compare the interval estimates of the factors. As a result, in this case, we did not compute the coverage of the interval estimates.

Consider first the results when the DGP is model M1. The first row of Table 1 reports the percentage of failures of the Onatski (2009) test. Observe that the performance is appropriate when the number of series is relatively large with respect to the number of factors. When $N = 12$ and there are 3, 2, and 1 factors in the system, Onatski (2009) procedure detects them correctly in 18%, 55%, and 99.6% of the replicates, respectively. By increasing the number of variables in the system to $N = 30$, the percentage of failures is drastically reduced.

Consider now the Monte Carlo results corresponding to the estimation of the factors in model M1. The top panel of Table 2 reports the Monte Carlo averages and standard deviations of the trace statistic in Eq. (12) computed

Table 1. Percentage of Failures of Onatski (2009) Test to Detect the True Number of Factors through 500 Monte Carlo Replicates for the DFM with White Noise Idiosyncratic Errors (Model 1), VAR(1) Idiosyncratic Noises (Model 2), VARMA(2,1) Factors (Model 3), and Breaks in the Factor Loadings (Model 4).

	$N = 12$ $r = 1$	$N = 12$ $r = 2$	$N = 12$ $r = 3$	$N = 30$ $r = 2$	$N = 30$ $r = 3$	$N = 120$ $r = 3$
Model 1 (%)	0.4	45	82	4.2	20	0.6
Model 2 (%)	3	60	90	12	47	1
Model 3 (%)	0.4	43	81	1.4	15	1
Model 4 (%)	1.2	47.6	87.2	6.2	36	36.6

through those replicates for which the test proposed by Onatski (2009) detects the true number of factors in the system. Note that in model M1, the idiosyncratic errors are white noise and, consequently, the KFS, 2SKF, and QML procedures are based on the true DGP when the parameters are substituted by the corresponding estimates. Table 2 shows that, as expected, regardless of the procedure, if the number of variables is fixed, the trace statistic decreases when the number of factors increases. On the other hand, if r is fixed, the trace statistic increases with the number of variables. Also, it is important to note that the trace statistics of the KFS and QML procedures are very similar in all cases. On the other hand, the trace statistics of PC are clearly smaller while 2SKF is somehow in between. If the DFM is implemented with more than 30 variables, it seems that just one iteration of the Kalman filter is enough to obtain similar factor estimates as with the KFS. Only when $N = 120$ and $r = 3$, the trace statistics of all procedures are similar and over 0.9. Furthermore, note that with $N = 30$, the Kalman-based procedures have statistics over 0.8. Finally, Table 2 shows that, when using the KFS or QML procedures to extract the factors, a remarkably large precision is obtained even with $N = 12$ if there is just one single factor in the system. If by adding more variables, the number of factors increases, the precision is similar. Regardless of the procedure and number of factors, all procedures considered are adequate to estimate the space spanned by the factors if the DFM is implemented with more than approximately 30 variables.

Table 3, which reports the Monte Carlo averages and standard deviations of the coverages of the 95% confidence intervals, shows that the asymptotic MSE used to construct the intervals for the PC factor estimates are clearly small to represent the uncertainty associated with these estimates. Note that

Table 2. Monte Carlo Results for DFM with White Noise Idiosyncratic Errors (Model 1), VAR(1) Idiosyncratic Noises (Model 2), VARMA(2,1) Factors (Model 3), and Breaks in the Factor Loadings (Model 4).

	$N = 12$ $r = 1$	$N = 12$ $r = 2$	$N = 12$ $r = 3$	$N = 30$ $r = 2$	$N = 30$ $r = 3$	$N = 120$ $r = 3$
Model 1						
PC	0.77	0.68	0.63	0.81	0.78	0.93
	(0.08)	(0.06)	(0.04)	(0.04)	(0.04)	(0.01)
KFS	0.91	0.86	0.77	0.91	0.85	0.92
	(0.04)	(0.03)	(0.05)	(0.03)	(0.03)	(0.01)
2SKF	0.86	0.76	0.69	0.87	0.83	0.94
	(0.06)	(0.06)	(0.05)	(0.04)	(0.04)	(0.01)
QML	0.90	0.85	0.76	0.91	0.86	0.94
	(0.04)	(0.04)	(0.05)	(0.03)	(0.03)	(0.01)
Model 2						
PC	0.77	0.69	0.66	0.82	0.79	0.93
	(0.08)	(0.06)	(0.06)	(0.04)	(0.04)	(0.02)
KFS	0.84	0.76	0.66	0.86	0.79	0.89
	(0.07)	(0.06)	(0.08)	(0.04)	(0.04)	(0.02)
2SKF	0.82	0.74	0.68	0.85	0.80	0.93
	(0.08)	(0.06)	(0.07)	(0.04)	(0.04)	(0.02)
QML	0.83	0.76	0.67	0.85	0.80	0.92
	(0.08)	(0.06)	(0.07)	(0.04)	(0.04)	(0.02)
Model 3						
PC	0.78	0.69	0.64	0.81	0.77	0.93
	(0.06)	(0.05)	(0.05)	(0.04)	(0.04)	(0.01)
KFS	0.82	0.76	0.68	0.85	0.80	0.91
	(0.05)	(0.06)	(0.06)	(0.03)	(0.03)	(0.01)
2SKF	0.81	0.72	0.66	0.84	0.79	0.93
	(0.06)	(0.05)	(0.05)	(0.04)	(0.03)	(0.01)
QML	0.82	0.75	0.68	0.82	0.79	0.89
	(0.05)	(0.06)	(0.05)	(0.03)	(0.03)	(0.03)
Model 4						
PC	0.77	0.66	0.61	0.76	0.73	0.88
	(0.06)	(0.06)	(0.06)	(0.04)	(0.04)	(0.02)
KFS	0.80	0.80	0.71	0.84	0.78	0.87
	(0.05)	(0.04)	(0.07)	(0.03)	(0.04)	(0.02)
2SKF	0.79	0.72	0.65	0.81	0.77	0.90
	(0.05)	(0.06)	(0.07)	(0.04)	(0.04)	(0.02)
QML	0.80	0.80	0.71	0.84	0.80	0.87
	(0.05)	(0.04)	(0.07)	(0.03)	(0.03)	(0.02)

Notes: Monte Carlo means and standard deviations (in parenthesis) of the trace statistic are computed through those replicates for which the number of factors is correctly estimated.

Table 3. Monte Carlo Means and Standard Deviations (in Parenthesis) of the Empirical Coverages of 95% Pointwise Factor Intervals Computed through Those Replicates for Which the Number of Factors Is Correctly Estimated in the DFM with White Noise Idiosyncratic Errors (Model 1), VAR(1) Idiosyncratic Noises (Model 2) and VARMA(2,1) Factors (Model 3).

N	12	12		12			30		30			120		
r	1	2		3			2		3			3		
Model 1														
P	0.14	0.13	0.13	0.11	0.14	0.12	0.16	0.17	0.13	0.18	0.15	0.11	0.16	0.12
	(0.07)	(0.07)	(0.07)	(0.05)	(0.06)	(0.05)	(0.10)	(0.08)	(0.07)	(0.07)	(0.06)	(0.07)	(0.08)	(0.07)
K	0.94	0.75	0.74	0.69	0.75	0.71	0.58	0.59	0.65	0.70	0.56	0.39	0.42	0.32
	(0.03)	(0.22)	(0.21)	(0.20)	(0.18)	(0.19)	(0.25)	(0.23)	(0.21)	(0.18)	(0.15)	(0.13)	(0.19)	(0.15)
S	0.96	0.56	0.62	0.62	0.68	0.64	0.44	0.48	0.54	0.60	0.58	0.40	0.42	0.33
	(0.02)	(0.16)	(0.13)	(0.13)	(0.14)	(0.09)	(0.15)	(0.11)	(0.13)	(0.16)	(0.12)	(0.12)	(0.19)	(0.12)
Q	0.97	0.64	0.68	0.70	0.79	0.69	0.48	0.50	0.64	0.68	0.54	0.52	0.43	0.33
	(0.01)	(0.22)	(0.20)	(0.18)	(0.15)	(0.13)	(0.20)	(0.17)	(0.22)	(0.19)	(0.15)	(0.21)	(0.20)	(0.14)
Model 2														
P	0.15	0.13	0.14	0.10	0.14	0.13	0.16	0.17	0.13	0.18	0.16	0.11	0.17	0.12
	(0.07)	(0.07)	(0.06)	(0.04)	(0.05)	(0.05)	(0.10)	(0.09)	(0.07)	(0.07)	(0.07)	(0.07)	(0.08)	(0.07)
K	0.80	0.61	0.58	0.56	0.61	0.64	0.54	0.53	0.60	0.62	0.51	0.38	0.40	0.31
	(0.05)	(0.17)	(0.14)	(0.17)	(0.12)	(0.12)	(0.23)	(0.20)	(0.19)	(0.15)	(0.14)	(0.14)	(0.17)	(0.13)
S	0.88	0.52	0.56	0.57	0.59	0.57	0.42	0.43	0.51	0.57	0.49	0.37	0.40	0.31
	(0.04)	(0.16)	(0.13)	(0.14)	(0.15)	(0.10)	(0.14)	(0.10)	(0.13)	(0.15)	(0.15)	(0.11)	(0.17)	(0.10)
Q	0.86	0.57	0.59	0.63	0.62	0.56	0.46	0.45	0.57	0.60	0.50	0.45	0.39	0.32
	(0.05)	(0.20)	(0.17)	(0.16)	(0.12)	(0.11)	(0.18)	(0.14)	(0.19)	(0.15)	(0.15)	(0.18)	(0.15)	(0.15)

Table 3. *(Continued)*

N	12	12		12			30		30			120		
r	1	2		3			2		3			3		
Model 3														
P	0.14	0.12	0.13	0.10	0.14	0.12	0.17	0.17	0.13	0.18	0.15	0.11	0.16	0.11
	(0.05)	(0.06)	(0.06)	(0.04)	(0.06)	(0.05)	(0.09)	(0.08)	(0.06)	(0.06)	(0.06)	(0.07)	(0.09)	(0.07)
K	0.95	0.80	0.78	0.84	0.71	0.77	0.68	0.69	0.68	0.75	0.66	0.48	0.47	0.41
	(0.02)	(0.17)	(0.18)	(0.16)	(0.14)	(0.19)	(0.24)	(0.24)	(0.17)	(0.15)	(0.14)	(0.18)	(0.19)	(0.20)
S	0.94	0.57	0.64	0.61	0.67	0.66	0.47	0.52	0.55	0.62	0.54	0.42	0.43	0.35
	(0.02)	(0.14)	(0.11)	(0.12)	(0.13)	(0.10)	(0.13)	(0.11)	(0.11)	(0.16)	(0.10)	(0.10)	(0.19)	(0.10)
Q	0.93	0.71	0.76	0.74	0.80	0.73	0.60	0.66	0.66	0.70	0.62	0.57	0.45	0.39
	(0.02)	(0.18)	(0.18)	(0.14)	(0.13)	(0.12)	(0.22)	(0.22)	(0.17)	(0.16)	(0.13)	(0.18)	(0.19)	(0.15)

Note: P stands for PC estimation, K for Kalman filter ML, S for 2SKF and Q for QML.

the coverages are below 0.20 regardless of the cross-sectional dimension and the number of factors. The coverages of the three other procedures considered are appropriate when $N = 12$ and $r = 1$. However, when the number of factors is larger than one, the empirical coverages are well under the nominal. Note that, in each case, the coverages are similar for all the factors in the system. Therefore, if the interest is estimating just one factor (e.g., the business cycle), having more factors in the system could deteriorate the interval estimation. Given the number of factors in the system, the performance of the intervals deteriorates when N increases since the parameter uncertainty also increases and, consequently, the intervals, which do not incorporate this uncertainty, are less reliable. This could explain why the coverages of the intervals can deteriorate as N increases. Therefore, according to the Monte Carlo results reported in Table 3, if one wants to obtain interval estimates of a single factor, it is better to keep the number of variables to be relatively small so that no new additional factors are introduced in the system. This result may have implications for those procedures that treat the estimated factors as if they were the true factors. According to Table 3, the confidence intervals are very tiny. As mentioned above, this could be due to the fact that the MSE used to compute the intervals are too small. However, there is a second cause for the low coverages. If the interest is the estimation of a single original factor common to all variables in the system in a multifactor model, it is not straightforward to find an adequate rotation of the factors that estimates properly this particular original factor. The individual estimated factors could be inappropriate as they involved linear combinations of the original factors and it is not always obvious how to disentangle them. In our experience, when looking at the individual match between each simulated and estimated factor, there are a number of problems that can appear. For instance, the estimated factors are interchanged in a number of replicates, that is, the common factor is estimated as the sectorial factor and vice versa. In practice, although the space spanned by the factors is correctly estimated, each individual factor can be far from the true simulated one. This problem is not important when the factors are used with forecasting purposes as, in this case, the linear combinations of the estimated factors have the predictive power. However, when the objective is the estimation of a single factor of interest, it is not obvious that the factor can be identified individually. Consequently, the individual correlations between each estimated factor and the corresponding true factor could be much smaller than what the trace statistics reported in Table 2 might suggest.

As mentioned above, in model M1, the DGP is the same as that assumed when running the Kalman filter. Consequently, the factors extracted by PC may have a disadvantage. Next, we consider the results for model M2 in which the idiosyncratic noises are a VAR(1). Note that the 2SKF and QML procedures are run assuming that the idiosyncratic noises are white noise. When dealing with the KFS, we also run the filter as if the idiosyncratic noises were white noise. The second row of Table 1, which reports the percentage of failures of the Onatski (2009) test, shows that this percentage is larger than in model M1. Therefore, when the idiosyncratic noises are serially correlated, identifying the correct number of factors and, consequently, a correct estimation of the space spanned by the factors, becomes more complicated. When looking at the results of the trace statistic reported in Table 2, we can observe that the results for PC are very similar to those reported for model M1. This could be expected as the PC factor extraction is non-parametric. When looking at the results for KFS, we can see that the average traces are smaller and the standard deviations are larger. Therefore, by assuming that the idiosyncratic errors are white noise when they are not, the performance of the KFS is worse. However, it is important to point out that when $N = 120$, the results are very similar to those obtained under the true DGP pointing out the possible consistency of the KFS in front of misspecification of the dependencies of the idiosyncratic noises. It is also important to note that the performance of KFS is still better than PC for small N. The results obtained for the hybrid procedures are very similar to KFS. Finally, Table 3 reports the Monte Carlo averages and standard deviations of the empirical coverages of the 95% confidence intervals of the factors. Once more, we can observe that the results for PC are the same as in model M1 while the coverages of the other three procedures deteriorate when the cross-sectional dimension and/or the number of factors increase. In any case, all coverages are well bellow the nominal. As before, the best results are obtained for $N = 12$ and $r = 1$.

In model M3, the dynamics of the common factors are misspecified. The third row of Table 1, which reports the percentage of failures of the Onatski (2009) test, shows that they are very similar to those obtained in model M1. Therefore, if the dynamics of the common factors are misspecified, the test proposed by Onatski (2009) does not deteriorate at all. When looking at the results of the trace statistic reported in Table 2, PC does not deteriorate since this procedure does not contemplate the dynamics in the common factor. However, the results for the remaining three procedures show a deterioration similar to that of model M2. Finally, the coverages given in Table 3 are very similar for PC, while for KFS, 2SKF, and QML, they show similar results to the well specified dynamics. Again, all coverages are well bellow the nominal.

In model M4, we have added breaks to the loadings of the first common factor. However, we have estimated the models as if there were no breaks. Curious enough, the Onatski (2009) procedure does not deteriorate very heavily, but in the case of the large factor model ($N = 120$). The results for the trace statistic, given in Table 2, are very similar to those of the other two misspecifications (models M2 and M3).

4. EMPIRICAL ANALYSIS

In this section, we analyze the monthly series contained in the data base considered by Stock and Watson (2012), which consists of an unbalanced panel of 137 US macroeconomic variables (two of which are deflators and other six are not included in the analysis) observed monthly from January 1959 to December 2011. These variables have been deflated, transformed to stationarity and corrected by outliers as described by Stock and Watson (2012). Furthermore, as it is usual in this literature, they have been standardized to have zero mean and variance one. The variables can be classified into the following 12 economic categories (in parentheses the number of variables in each category): industrial production (13); employment and unemployment (44); housing starts (8); inventories, orders, and sales (8); prices (16); earnings (3); interest rates and spreads (18); money and credit (14); stock prices (3); housing prices (2); exchange rates (6); and other (2). In order to obtain a balanced panel, we select those variables observed without missing values over the whole observation period. The resulting balanced panel has $N = 103$ variables, classified into 11 categories and $T = 628$ observations. The two variables belonging to the *Housing Prices* category disappear from the panel. The objective of this empirical exercise is to answer the following questions: (i) When the interest is to estimate just one factor as, for example, the business cycle, is it worth to use all available variables to extract it? (ii) Are the factor extraction procedure and number of variables used relevant to estimate the factors? (iii) Is the number of factors in the system independent of the number of variables?

We start the analysis by extracting the factors from a system with 11 variables each of them representing one of the categories. Each variable has been chosen as that exhibiting the highest average correlation with respect to the remaining series in the same category; see Alvarez et al. (2012) for this criterion. In this system, we start selecting the number of factors as proposed by Onatski (2009). Note that the number of factors selected is the same when using the procedure proposed by Alessi et al. (2010), who, following Hallin and Liska (2007), introduce a tuning multiplicative constant to improve the performance of the procedure proposed by Bai and Ng (2002). The number of factors selected is one. The factor is then extracted by each of the four procedures described above. When the factor is extracted using PC, it explains 20.34% of the total variability. For the three parametric procedures, the common factor is estimated as an AR(2). Furthermore, full ML is implemented estimating a diagonal VAR(1) for the idiosyncratic noises. Fig. 1 plots the extracted factors together with their corresponding 95% pointwise intervals obtained as explained in the previous section. The factors plotted in Fig. 1 have been rotated as in Eq. (13) substituting the unknown matrix P by $\hat{P}$ estimated using PC. The corresponding root-mean-square errors (RMSE), computed without incorporating the parameter uncertainty, have been reported in the main diagonal of Table 4 for each of the four procedures considered. As already concluded from the simulated system used in the illustration, we can observe that the asymptotic RMSE of the PC procedure are unrealistically small. Fig. 1 illustrates that, regardless of the factor extraction procedure, the point estimates of the factors extracted using the information contained in the 11 variables selected from the original data base are very similar. Table 5, which reports the pairwise correlations between the factor estimates obtained by each of the four procedures, shows that the correlation between PC and 2SKF factor estimates is 0.97 while that between KFS and QML is 0.95. Fig. 1 also plots the periods dated by the National Bureau of Economic Research (NBER) as US business cycle recessions. We can observe that this first common factor detects the periods announced as recessions by the NBER business cycle dating committee for the US economy.

Next, we add into the set of variables used to extract the factors the variables with the second highest average correlation within each category, with $N = 21$ variables. In this case, the number of factors identified using Onatski (2009) is again 1 and the PC factor explains 21.03% of the total variability which is comparable with the percentage explained when the factor is extracted using $N = 11$. Fig. 2 plots the factors extracted by each procedure together with the corresponding pointwise 95% intervals. Comparing Figs. 1 and 2, we can observe that the factors estimated with

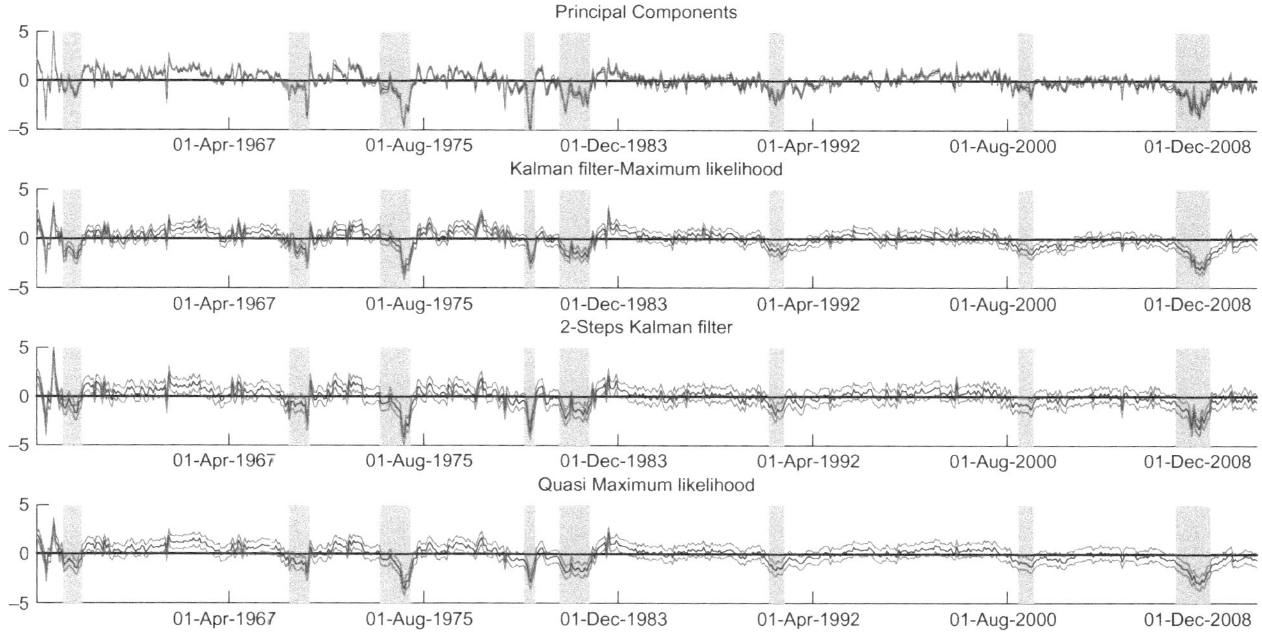

Fig. 1. Factor Extracted by Each of the Four Procedures Using 11 Variables Selected as the Most Correlated in Average within Its Class together with the Corresponding 95% Intervals. *Note*: The gray shadow areas represent the US business cycle recessions as dated by the NBER (National Bureau of Economic Research).

Table 4. Empirical Application of the Stock and Watson (2012) Data Base.

N	PC				KFS				2SKF				QML			
	11	21	91	103	11	21	91	103	11	21	91	103	11	21	91	103
11	0.11	0.96	0.89	0.89	0.27	0.95	0.96	0.96	0.38	0.98	0.94	0.94	0.34	0.91	0.88	0.85
21		0.05	0.94	0.94		0.04	0.96	0.95		0.28	0.98	0.98		0.10	0.93	0.89
91			0.02	1			0.13	0.97			0.13	1			0.15	0.99
103				0.02				0.15				0.13				0.18

Notes: Main diagonal: RMSE of extracted factors. The KFS, 2SKF and QML are computed using the steady-state RMSE obtained from the Kalman filter with estimated parameters. The PC RMSE are obtained using the asymptotic approximation and averaging over time. Off-diagonal elements: correlations between the factors estimated using alternative number of variables.

$N = 21$ are very similar to those estimated with $N = 11$ variables and also very similar among them. Table 5 shows that the correlation between PC and 2SKF is 0.98, while the correlation between KFS and QML is 1.0. Furthermore, for each procedure, Table 4 reports the RMSE together with the correlations between the factors extracted when the cross-sectional dimension changes. We can observe that, in general, the RMSE decrease with N. Also note that the correlation between the factors extracted with 11 and 21 variables is always over 0.91.

We keep adding more variables into the system by adding in each step the variable with the next highest average correlation within its category. In order to save space, we choose reporting the results for $N = 91$ and 103 variables and $\hat{r} = 4$. The first common factor extracted by PC explains 19.60% of the total variability. Once more, this percentage is similar to those obtained with $N = 11$ or 21 variables. The MSE when $N = 91$ and when $N = 103$, reported in Table 4, are very similar. For each procedure, Table 4 also reports the correlations between the factors estimated with different cross-sectional dimensions. The correlations between the factors extracted with 103 and 91 variables are very high. Consequently, Fig. 3 only plots the factors estimated when $N = 91$. Fig. 3 shows that the point estimates are very similar to those plotted in Figs. 1 and 2. Only the intervals are (artificially) tinier than when $N = 11$ or 21.

We also compare the factors extracted using different procedures with the same number of variables. Table 5, which reports the correlations between the estimated factors obtained by the alternative procedures, shows that there is a high correlation between the factor estimates obtained

Table 5. Empirical Application of the Stock and Watson (2012) Data Base.

	$N = 11$			$N = 21$			$N = 91$			$N = 103$		
	KFS	2SKF	QML	KFS	2SKF	QML	KFS	2SKF	QML	KFS	2SKF	QML
PC	0.84	0.97	0.91	0.85	0.98	0.86	0.97	0.99	0.92	0.94	0.99	0.88
KFS		0.92	0.95		0.92	1		0.98	0.95		0.96	0.97
2SKF			0.96			0.93			0.93			0.89

Note: Correlations between the factor estimated by alternative procedures given the number of variables in the system.

using KFS and QML, which is always over 0.95. The same happens with the correlations between the factors extracted using PC and 2SKF which are always over 0.97. These results confirm the conclusions obtained with the Monte Carlo experiments.

Finally, and as a robustness check,[6] we have also carried out the empirical analysis using the sample from 1985 to 2006. As Breitung and Eickmeier (2011) point out, using long samples might lead to parameter instability which, in turn, could inflate the number of required factors. We choose as the starting point 1985 since we do not want to consider data previous to the Great Moderation and end in 2006 to leave out the data from the Great Recession; in a similar data set, Breitung and Eickmeier (2011) find a structural break in the factor loadings around 1984. In this sense, the new data set could be less likely to suffer from parameter instability.

As we are choosing balanced panels, the number of variables corresponding to this new time span is 106. Therefore, following the same criteria to choose the variables as before, the number of variables selected in the second round is 22 instead of 21. Tables 6 and 7 are the replicates for the short sample of Tables 4 and 5, respectively. They report the results obtained for small (11), medium (22), and large (106) systems. For each procedure, Table 6 reports the RMSE and the correlations between the factors extracted when the cross-sectional dimension changes and Table 7 reports the correlations between the estimated factors obtained by the alternative procedures. We can observe that the main conclusions are roughly the same although the correlations between the factor estimates reported in Tables 6 and 7 are slighly smaller than those reported in Tables 4 and 5.

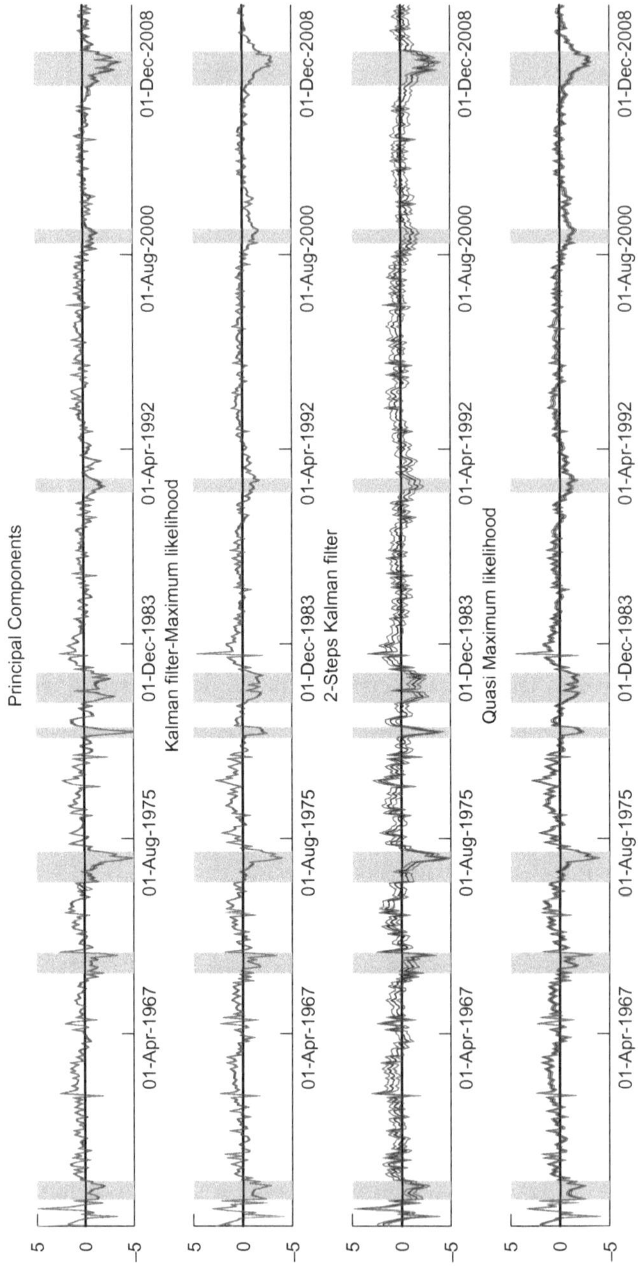

Fig. 2. Factor Extracted by Each of the Four Procedures Using 21 Variables Selected as the Two Most Correlated in Average within Its Class together with the Corresponding 95% Intervals. *Note:* The gray shadow areas represent the US business cycle recessions as dated by the NBER.

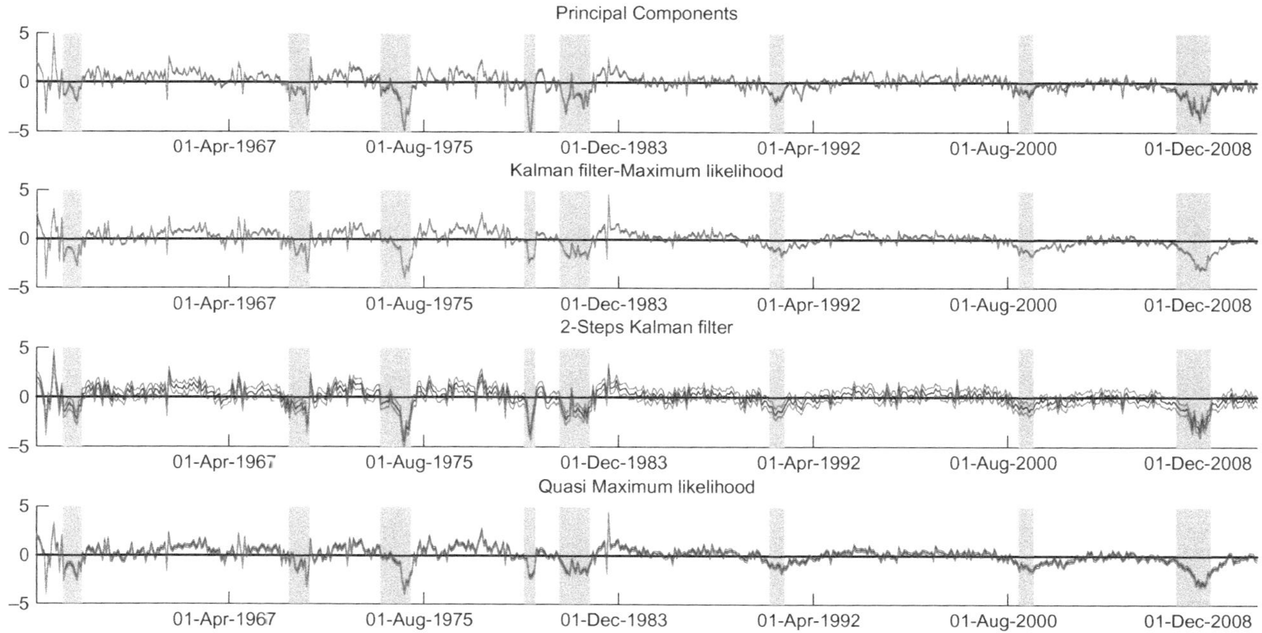

Fig. 3. First Factor Extracted by Each of the Four Procedures Using 91 Variables Selected as the More Correlated in Average within Its Class together with the Corresponding 95% Intervals. *Note*: The gray shadow areas represent the US business cycle recessions as dated by the NBER.

Table 6. Empirical Application of the Stock and Watson (2012) Data Base, Time Span 1985–2006.

N	PC			KFS			2SKF			QML		
	11	22	106	11	22	106	11	22	106	11	22	106
11	0.07	0.97	0.85	0.26	0.79	0.93	0.44	0.98	0.93	0.31	0.95	0.95
22		0.05	0.87		0.06	0.71		0.35	0.95		0.14	0.91
103			0.03			0.13			0.13			0.14

Notes: Main diagonal: RMSE of extracted factors. The KFS, 2SKF and QML are computed using the steady-state RMSE obtained from the Kalman filter with estimated parameters. The PC RMSE are obtained using the asymptotic approximation and averaging over time. Off-diagonal elements: correlations between the factors estimated using alternative number of variables.

Table 7. Empirical Application of the Stock and Watson (2012) Data Base, Time Span 1985–2006.

	$N = 11$			$N = 22$			$N = 106$		
	KFS	2SKF	QML	KFS	2SKF	QML	KFS	2SKF	QML
PC	0.74	0.97	0.80	0.65	0.98	0.75	0.99	0.99	0.95
KFS		0.86	0.98		0.70	0.71		1.00	0.97
2SKF			0.90			0.84			0.96

Correlations between the factor estimated by alternative procedures given the number of variables in the system.

5. CONCLUSIONS

In this paper, we compare small-data and big-data factor extraction procedures implementing the alternative procedures considered to the same data sets. Using simulated and real data, we compare PC, KFS, 2SKF, and QML, given the sample size. We also compare the performance of each procedure for different cross-sectional dimensions. We conclude that, regardless of the procedure implemented and the number of variables used for the factor estimation, the spaces spanned by the extracted factors are very similar. When using simulated data, all procedures extract (conveniently rotated) factors highly correlated with the true unobserved factors. If the objective is estimating a given factor (as, for example, the business cycle) adding more variables into the system may increase the number of factors but the increase in accuracy of point estimates is relatively small. We also show that the asymptotic bounds of PC are too narrow being inadequate to represent the finite sample uncertainty associated with estimated factors. Both ML and QML procedures extract very similar point and interval factors which have, in general, higher correlations with the true factors than PC and 2SKF estimates when the cross-sectional dimension is relatively small.

KFS, 2SKF, and QML are almost identical when they are based on the wrong specification of the dynamics of the idiosyncratic noises and factors. These procedures have a slight advantage over PC if the cross-correlation dimension is small and very similar properties for large N.

In this manuscript, we did not consider the effect of parameter estimation on the construction of intervals for the factors; see Poncela and Ruiz (2015) for this effect in the context of the ML procedure. However, the empirical coverages reported in Table 3 are smaller than the nominal coverages. Furthermore, the interval coverages of all procedures decrease with the number of series, probably as a result of increasing the number of parameters that have to be estimated. Recall that the estimation is only carried out for those cases where the true number of factors is detected. At this regard, the number of factors correctly found by Onatski (2009) test increases with the number of series. On the contrary, the interval coverages decrease with the number of series. Therefore, it seems that incorporating the parameter uncertainty could be important to get more adequate confidence intervals. When dealing with ML or the hybrid procedures, this uncertainty can be incorporated in practice using bootstrap procedures as those proposed by Rodríguez and Ruiz (2009, 2012) in the context of state-space models. However, as far as we know, there are not procedures proposed in the literature to incorporate the parameter uncertainty in the context of PC procedures. Looking at the effects of parameter uncertainty when constructing intervals for estimated factors in empirical applications is within our research agenda. Also, the analysis of real data systems can be extended to consider unbalanced data bases by using, for example, the computationally efficient procedures by Jungbacker et al. (2011) and Jungbacker and Koopman (in press).

Finally, it is important to mention that, in practice, the models fitted could be misspecified. In stationary DFMs, Doz et al. (2011, 2012) show the consistency of the factors estimated using the PC-KFS procedures so that the misspecification of the idiosyncratic noise serial correlation does not jeopardize the consistent estimation of the factors. Considering the effects of misspecification both in the number of factors and/or in the dynamics of factors and idiosyncratic noises is also left for further research.

NOTES

1. Stock et al. (1991), Forni et al. (2000, 2005), Aruoba, Diebold, and Scotti (2009), Altissimo, Cristadoro, Forni, Lippi, and Veronese (2010), Camacho and Perez-Quiros (2010), and Frale, Mazzi, Marcellino, and Proietti (2011) extract factors to estimate the business cycle; Chamberlain and Rothschild (1983), Diebold and Nerlove (1989), Harvey, Ruiz, and Shephard (1994), and Koopman and van der Wel (2013) deal with financial factors; Bernanke, Boivin, and Eliasz (2005), Buch, Eickmeier, and Prieto (2014), Eickmeier, Lemke, and Marcelino (2015), and Han (in press) extract factors to incorporate them in FAVAR models; Banerjee, Marcellino, and Masten (2014), and Bräuning and Koopman (2014) propose incorporating factors in FECM and unobserved component models, respectively.

2. See Stock and Watson (2002) and Forni, Hallin, Lippi, and Reichlin (2000, 2005) for PC consistency results and Doz et al. (2011, 2012) for results on the 2SKF and QML procedures, in stationary systems with weak cross-correlations of the idiosyncratic noises when both the temporal and cross-sectional dimensions tend to infinity.

3. The stationarity assumption is made in order to implement procedures based on PC.

4. Breitung and Choi (2013) have an interesting summary of the conditions for weak cross-correlated idiosyncratic noises and strong factors. Onatski (2012) describes situations in which PC is inconsistent, while Choi (2012) derives the asymptotic distribution of a generalized PC estimator with smaller asymptotic variance.

5. Note that the factors are not uniquely identified which means that even when the objective is the estimation of a unique factor, it is important to know r so that the estimated factors can be rotated to obtain the desired interpretable estimation. There is a large number of alternative proposals for estimating the number of factors, mostly based on the eigenvalues of the sample covariance matrix of Y_t; see, for example, Bai and Ng (2002), Bai and Ng (2007), Amengual and Watson (2007), Alessi, Barigozzi, and Capasso (2010), Kapetanios (2010), and Breitung and Pigorsch (2013), among others. These procedures require that the cumulative effect of the factors grows as fast as N. Alternatively, Onatski (2009, 2010) proposes an estimator of the number of factors that works well when the idiosyncratic terms are substantially serially or cross-sectionally correlated. Onatski (2009) formalizes the widely used empirical method based on the visual inspection of the scree plot introduced by Cattell (1966).

6. We thank an anonymous referee and the editors for suggesting this exercice.

ACKNOWLEDGMENTS

Financial support from the Spanish Government projects ECO2012-32854 and ECO2012-32401 is acknowledged by the first and second authors, respectively. We are very grateful to comments received during the 16th

Advances in Econometrics conference on DFMs held in CREATES, Aarhus university, in November 2014. Also the comments of two referees have been very helpful to obtain a more completed version of this paper. We are indeed grateful to them.

REFERENCES

Alessi, L., Barigozzi, M., & Capasso, M. (2010). Improved penalization for determining the number of factors in approximate factor models. *Statistics and Probability Letters*, *80*, 1806–1813.

Altissimo, F., Cristadoro, R., Forni, M., Lippi, M., & Veronese, G. (2010). New eurocoin: Tracking economic growth in real time. *The Review of Economics and Statistics*, *92*(4), 1024–1034.

Alvarez, R., Camacho, M., & Perez-Quiros G. (2012). *Finite sample performance of small versus large scale dynamic factor models*, WP 1204, Banco de España.

Amengual, D., & Watson, M. W. (2007). Consistent estimation of the number of dynamic factors in a large N and T panel. *Journal of Business and Economic Statistics*, *25*(1), 91–96.

Aruoba, S. B., Diebold, F. X., & Scotti, C. (2009). Real-time measurement of business conditions. *Journal of Business & Economic Statistics*, *27*, 417–427.

Bai, J. (2003). Inferential theory for factor models of large dimensions. *Econometrica*, *71*(1), 135–171.

Bai, J., & Ng, S. (2002). Determining the number of factors in approximate factor models. *Econometrica*, *70*(1), 191–221.

Bai, J., & Ng, S. (2006a). Confidence intervals for diffusion index forecasts and inference for factor-augmented regressions. *Econometrica*, *74*(4), 1133–1150.

Bai, J., & Ng, S. (2006b). Evaluating latent and observed factors in macroeconomics and finance. *Journal of Econometrics*, *131*, 507–537.

Bai, J., & Ng, S. (2007). Determining the number of primitive shocks in factor models. *Journal of Business & Economic Statistics*, *25*(1), 52–60.

Bai, J., & Ng, S. (2008a). Large dimensional factor analysis. *Foundations and Trends in Econometrics*, *3*, 89–163.

Bai, J., & Ng, S. (2008b). Forecasting economic time series using targeted predictors. *Journal of Econometrics*, *146*, 304–317.

Bai, J., & Wang, P. (2014). Identification theory for high dimensional static and dynamic factor and estimation models. *Journal of Econometrics*, *178*(2), 794–804.

Banbura, M., & Modugno, M. (2014). Maximum likelihood estimation of factor models on datasets with arbitrary pattern of missing data. *Journal of Applied Econometrics*, *29*, 133–160.

Banbura, M., & Runstler, G. (2011). A look into the factor model blackbox: Publication lags and the role of hard and soft data in forecasting. *International Journal of Forecasting*, *27*(2), 333–346.

Banerjee, A., Marcellino, M., & Masten, I. (2014). Forecasting with factor-augmented error correction models. *International Journal of Forecasting*, *30*(3), 589–612.

Bernanke, B., Boivin, J., & Eliasz, P. (2005). Factor augmented vector autoregressions (FAVARs) and the analysis of monetary policy. *Quarterly Journal of Economics*, *120*, 387–422.

Boivin, J., & Ng, S. (2006). Are more data always better for factor analysis? *Journal of Econometrics*, *132*, 169–194.

Bork, L., Dewachter, H., & Houssa, R. (2009). *Identification of macroeconomic factors in large panels*. CREATES Research Paper No. 2009-43, School of Economics and Management, University of Aarhus.

Bräuning, F., & Koopman, S. J. (2014). Forecasting macroeconomic variables using collapsed dynamic factor analysis. *International Journal of Forecasting*, *30*(3), 572–584.

Breitung, J., & Choi, I. (2013). Factor models. In N. Hashimzade & M. A. Thornton (Eds.), *Handbook of research methods and applications in empirical macroeconomics*. Cheltenham: Edward Elgar.

Breitung, J., & Eickmeier, S. (2006). Dynamic factor models. In O. Hübler & J. Frohn (Eds.), *Modern econometric analysis*. Berlin: Springer.

Breitung, J., & Eickmeier, S. (2011). Testing for structural breaks in dynamic factor models. *Journal of Econometrics*, *163*, 71–84.

Breitung, J., & Eickmeier, S. (2014). *Analyzing business and financial cycles using multi-level factors*. Discussion Paper No. 11/2014, Deutsche Bundesbank.

Breitung, J., & Eickmeier, S. (2015). Analysing business cycle asymmetries in a multi-level factor model. *Economics Letters*, *127*, 31–34.

Breitung, J., & Pigorsch, U. (2013). A canonical correlation approach for selecting the number of dynamic factors. *Oxford Bulletin of Economics and Statistics*, *75*(1), 23–36.

Buch, C. M., Eickmeier, S., & Prieto, E. (2014). Macroeconomic factors and microlevel bank behavior. *Journal of Money, Credit and Banking*, *46*(4), 715–751.

Caggiano, G., Kapetanios, G., & Labhard, V. (2011). Are more data always better for factor analysis? Results from the euro area, the six largest euro area countries and the UK. *Journal of Forecasting*, *30*, 736–752.

Camacho, M., & Perez-Quiros, G. (2010). Introducing the Euro-STING: Short term indicator of euro area growth. *Journal of Applied Econometrics*, *25*(4), 663–694.

Cattell, R. (1966). The screen test for the number of factors. *Multivariate Behavioral Research*, *1*, 245–276.

Chamberlain, G., & Rothschild, M. (1983). Arbitrage, factor structure and mean-variance analysis in large asset markets. *Econometrica*, *51*, 1305–1324.

Choi, I. (2012). Efficient estimation of factor models. *Econometric Theory*, *28*, 274–308.

Diebold, F. X. (2003). "Big data" dynamic factor models for macroeconomic measurement and forecasting (discussion of Reichlin and Watson papers). In M. Dewatripont, L. P. Hansen, & S. Turnovsky (Eds.), *Advances in economics and econometrics*. Cambridge: Cambridge University Press.

Diebold, F. X., & Nerlove, M. (1989). The dynamics of exchange rate volatility: A multivariate latent factor ARCH model. *Journal of Applied Econometrics*, *4*(1), 1–21.

Doz, C., Giannone, D., & Reichlin, L. (2011). A two-step estimator for large approximate dynamic factor models based on Kalman filtering. *Journal of Econometrics*, *164*, 188–205.

Doz, C., Giannone, D., & Reichlin, L. (2012). A quasi maximum likelihood approach for large approximate dynamic factor models. *Review of Economics and Statistics*, *94*(4), 1014–1024.

Eickmeier, S., Lemke, W., & Marcelino, M. (2015). Classical time-varying FAVAR models – Estimation, forecasting and structural analysis. *Journal of the Royal Statistical Society – Series A*, *178*(3), 493–533.

Engle, R. F., & Watson, M. W. (1981). A one-factor multivariate time series model of metropolitan wage rates. *Journal of the American Statistical Association*, *76*, 774–781.

Fiorentini, G., Galesi, A., & Sentana, E. (2014). A spectral EM algorithm for dynamic factor models, Manuscript.

Forni, M., Hallin, M., Lippi, M., & Reichlin, L. (2000). The generalized dynamic-factor model: Identification and estimation. *Review of Economics and Statistics*, *82*(4), 540–554.

Forni, M., Hallin, M., Lippi, M., & Reichlin, L. (2005). The generalized dynamic factor model: One sided estimation and forecasting. *Journal of the American Statistical Association*, *100*, 830–840.

Frale, C., Mazzi, G., Marcellino, M., & Proietti, T. (2011). EUROMIND: A monthly indicator of the euro area economic conditions. *Journal of the Royal Statistical Society*, *174*, 439–470.

Geweke, J. (1977). The dynamic factor analysis of economic time series. In D. J. Aigner & A. S. Goldberger (Eds.), *Latent variables in socio-economic models*. Amsterdam: North-Holland.

Giannone, D., Reichlin, L., & Small, D. (2008). Nowcasting: The real-time informational content of macroeconomic data. *Journal of Monetary Economics*, *55*, 665–676.

Hallin, M., & Liska, R. (2007). Determining the number of factors in the general dynamic factor model. *Journal of the American Statistical Association*, *102*, 603–617.

Han, X. (in press). Tests for overidentifying restrictions in factor-augmented VAR models. *Journal of Econometrics*, *184*, 394–419.

Harvey, A. C. (1989). *Forecasting structural time series models and the Kalman filter*. Cambridge: Cambridge University Press.

Harvey, A. C., Ruiz, E., & Shephard, N. G. (1994). Multivariate stochastic variance models. *The Review of Economic Studies*, *61*(2), 247–264.

Jungbacker, B., & Koopman, S. J. (in press). Likelihood-based analysis for dynamic factor models. *Econometrics Journal*, *18*, C1–C21.

Jungbacker, B., Koopman, S. J., & van der Wel, M. (2011). Maximum likelihood estimation for dynamic factor models with missing data. *Journal of Economic Dynamics & Control*, *35*, 1358–1368.

Kapetanios, G. (2010). A testing procedure for determining the number of factors in approximate factor models with large data sets. *Journal of Business & Economic Statistics*, *28*, 397–409.

Karadimitropoulou, A., & León-Ledesma, M. (2013). World, country, and sector factors in international business cycles. *Journal of Economic Dynamics and Control*, *37*(12), 2913–2927.

Koopman, S. J., & van der Wel, M. (2013). Forecasting the US term structure of interest rates using a macroeconomic smooth dynamic factor model. *International Journal of Forecasting*, *29*, 676–694.

Kose, M. A., Otrok, C., & Whiteman, C. H. (2003). International business cycles: World, region and country-specific factors. *American Economic Review*, *93*, 1216–1239.

Moench, E., Ng, S., & Potter, S. (2013). Dynamic hierarchical factor models. *Review of Economics & Statistics*, *95*(5), 1811–1817.

Onatski, A. (2009). Testing hypothesis about the number of factors in large factor models. *Econometrica*, *77*(5), 1447–1479.

Onatski, A. (2010). Determining the number of factors from empirical distribution of eigenvalues. *Review of Economics and Statistics*, *92*(4), 1004–1016.

Onatski, A. (2012). Asymptotics of the principal components estimator of large factor models with weakly influential factors. *Journal of Econometrics*, *168*, 244–258.

Pinheiro, M., Rua, A., & Dias, F. (2013). Dynamic factor models with jagged edge panel data: Taking on board the dynamics of the idiosyncratic components. *Oxford Bulletin of Economics and Statistics*, *75*(1), 80–102.

Poncela, P., & Ruiz, E. (2015). More is not always better: Back to the Kalman filter in dynamic factor models. In S. J. Koopman & N. G. Shephard (Eds.), *Unobserved components and time series econometrics*. Oxford: Oxford University Press.

Reis, R., & Watson, M. W. (2010). Relative good's prices, pure inflation, and the Phillips correlation. *American Economic Journal: Macroeconomics*, *2*(3), 128–157.

Rodríguez, A., & Ruiz, E. (2009). Bootstrap prediction intervals in state space models. *Journal of Time Series Analysis*, *30*(2), 167–178.

Rodríguez, A., & Ruiz, E. (2012). Bootstrap prediction mean squared errors of unobserved states based on the Kalman filter with estimated parameters. *Computational Statistics & Data Analysis*, *56*(1), 62–74.

Sargent, T. J., & Sims, C. A. (1977). Business cycle modeling without pretending to have too much a priory economic theory. In C. A. Sims (Ed.), *New methods in business cycle research*, Minneapolis, MN: Federal Reserve Bank of Minneapolis.

Stock, J. H., & Watson, M. W. (1991). A probability model of the coincident economic indicators. In K. Lahiri & G. H. Moore (Eds.), *Leading economic indicators: New approaches and forecasting records*. Cambridge: Cambridge University Press.

Stock, J. H., & Watson, M. W. (2002). Forecasting using principal components from a large number of predictors. *Journal of the American Statistical Association*, *97*(460), 1167–1179.

Stock, J. H., & Watson, M. W. (2006). Forecasting with many predictors. In G. Elliot, C. Granger, & A. Timmermann (Eds.), *Handbook of economic forecasting* (Vol. 1). Amsterdam: Elsevier.

Stock, J. H., & Watson, M. W. (2011). Dynamic factor models. In M. P. Clements & D. F. Hendry (Eds.), *The Oxford handbook of economic forecasting*. Oxford: Oxford University Press.

Stock, J. H., & Watson, M. W. (2012). Disentangling the channels of the 2007–09 recession. *Brooking Papers on Economic Activity*, *Spring*, 81–135.

Watson, M. W., & Engle, R. F. (1983). Alternative algorithms for the estimation of dynamic factor, MIMIC and varying coefficient regression. *Journal of Econometrics*, *23*, 385–400.

PART III
INSTABILITY

REGULARIZED ESTIMATION OF STRUCTURAL INSTABILITY IN FACTOR MODELS: THE US MACROECONOMY AND THE GREAT MODERATION

Laurent Callot[a,b,d] and Johannes Tang Kristensen[c,d]

[a]*Department of Econometrics and OR, VU University Amsterdam, Amsterdam, The Netherlands*
[b]*The Tinbergen Institute, Amsterdam, The Netherlands*
[c]*Department of Business and Economics, University of Southern Denmark, Odense, Denmark*
[d]*CREATES, Aarhus University, Aarhus, Denmark*

ABSTRACT

This paper shows that the parsimoniously time-varying methodology of Callot and Kristensen (2015) can be applied to factor models. We apply this method to study macroeconomic instability in the United States from 1959:1 to 2006:4 with a particular focus on the Great Moderation. Models with parsimoniously time-varying parameters are models with an unknown number of break points at unknown locations. The parameters

Dynamic Factor Models
Advances in Econometrics, Volume 35, 437–479

ISSN: 0731-9053/doi:10.1108/S0731-905320150000035011

are assumed to follow a random walk with a positive probability that an increment is exactly equal to zero so that the parameters do not vary at every point in time. The vector of increments, which is high dimensional by construction and sparse by assumption, is estimated using the Lasso. We apply this method to the estimation of static factor models and factor-augmented autoregressions using a set of 190 quarterly observations of 144 US macroeconomic series from Stock and Watson (2009). We find that the parameters of both models exhibit a higher degree of instability in the period from 1970:1 to 1984:4 relative to the following 15 years. In our setting the Great Moderation appears as the gradual ending of a period of high structural instability that took place in the 1970s and early 1980s.

Keywords: Parsimoniously time-varying parameters; factor models; structural break; Lasso

JEL classifications: C01; C13; C32; C38; E32

1. INTRODUCTION

The Great Moderation is a period of macroeconomic stability in the United States thought to have begun in the 1980s (Kim & Nelson, 1999; McConnell & Perez-Quiros, 2000; Stock & Watson, 2003), or even in the 1950s (Blanchard & Simon, 2001) with an interruption in the 1970s and early 1980s. This period is marked by a decline of inflation and by a relative stabilization of the business cycle and of monetary policy which can be attributed either to a decline in output volatility or to changes in the dynamics of macroeconomic variables (Stock & Watson, 2009). The decline in output volatility is well documented by the authors cited previously, but this does not preclude the possibility of changes in the dynamics of macroeconomic variables. This paper proposes to quantify parameter instability in factor models before and during the Great Moderation using the parsimoniously time varying (ptv) framework of Callot and Kristensen (2015) as it imposes minimal assumptions on the dynamics of the parameters.

The contributions of this paper take both a methodological and an empirical form. From a methodological point of view, we show that the ptv framework proposed by Callot and Kristensen (2015) can be readily used with factor models. In this framework, the parameters are assumed to follow a random walk with a positive probability that an increment is exactly equal to zero, and the resulting parameter paths are estimated using the Lasso. The empirical contribution of this paper consists in estimating a large number of factor models using US macroeconomic data to get a picture of the parameter instability in the last 50 years. We document that the instability is widespread, but that the majority of the breaks occur before the Great Moderation. We also document that allowing for moderate time variation in the parameters can substantially improve the fit of the model in a forecasting context suggesting that improvements in forecasting performance could be possible by taking parameter instability into account.

Factor models have been investigated and applied for more than a decade (for a general review, see Bai & Ng, 2008; Stock & Watson, 2011). The problem of breaks and structural instability in the parameters of factor models remains a field open for exploration though. The seminal work of Stock and Watson (2002) provides insights into the problem, they show that the principal components (PC) factor estimator remains consistent if faced with moderate structural instability in the factor loadings. More recently the literature has seen a number of contributions related to testing for structural breaks in the loadings. The first formal test was proposed by Breitung and Eickmeier (2011), who considered the problem of testing the loadings associated with the individual variables. This has since been followed up by Chen, Dolado, and Gonzalo (2014) and Han and Inoue (2014), who propose tests for breaks in all loadings jointly, and Yamamoto and Tanaka (2013), who proposed a modified version of the Breitung and Eickmeier (2011) test.

More closely related to this present paper, Cheng, Liao, and Schorfheide (2014) use shrinkage methods to determine both the break points and the number of factors, and Corradi and Swanson (2014) propose a test for the joint hypothesis of breaks in the factor loadings and in the parameters of a factor-augmented forecasting model. Although these contributions take different approaches to the issue of structural instability, they do have characteristics in common: They only consider a fixed number of breaks (often a single one) which are typically assumed to be common to all factors and large in magnitude. Breitung and Eickmeier (2011) show how such a set-up will lead to overestimation of the number of factors. This is, in various ways, exploited in the mentioned papers to detect breaks. For example, Cheng et al. (2014) determine the break by minimizing the sum of the numbers of pre- and post-break factors instead of using a traditional sum-of-squared residuals criterion; and Corradi and Swanson (2014) utilize that the information criterion of Bai and Ng (2002) will overestimate the number of factors when the loadings are subject to breaks in order to construct a test statistic.

Our approach to the problem is different from existing papers. Instead of considering only a single or fixed number of breaks, we allow for more general forms of instability in the parameters of the model. The parameters associated with the factors, and those associated with potential autoregressive terms, are allowed to change independently across variables and time. Bates, Plagborg-Møller, Stock, and Watson (2013) show that the PC factor estimator is consistent under general forms of structural instability, including the forms of instability assumed in the ptv framework. The benefit of this is that we can rely on standard results regarding estimation of the factors and that, for example, the information criterion of Bai and Ng (2002) for selecting the number of factors is still valid.

Callot and Kristensen (2015) introduce the idea of ptv parameters for VAR models. This framework allows for an unknown number of break points at unknown locations by estimating the vector of changes in the parameters, which is high dimensional by construction and sparse by assumption, using the Lasso. In this paper, we show that static factor models as well as factor-augmented autoregressive (FAAR) models can be estimated in the ptv framework. The ptv models are well suited for our purpose of investigating the Great Moderation in terms of parameter stability in factor models. The parameters are modelled non-parametrically, the estimation of the number of breaks and of their locations is data driven, and the parameters can remain stable for any duration or even the whole sample in which case our estimator is equivalent to OLS. This allows for the parameters to remain constant, experience a few changes (as in the structural breaks literature), or exhibit much more unstable paths, independently across variables.

We conduct an empirical study of the US macroeconomy using the ptv methodology with both static factor models and FAAR forecasting models. In particular, we follow up on, and expand upon, the empirical investigation of the Great Moderation by Stock and Watson (2009). Stock and Watson (2009) find that the Great Moderation was indeed associated with breaks in both factor loadings and the parameters of FAAR forecasting models. Using the ptv methodology of Callot and Kristensen (2015), we investigate structural instability throughout the sample with a focus on the period of the Great Moderation and investigate whether accounting for structural instability in this fashion is helpful for forecasting.

The remainder of this paper is organized as follows: In Section 2, we present our analytical framework and discuss the estimation of the time-varying parameters and of the factors. Section 3 is dedicated to our empirical investigation of the Great Moderation and is followed by a conclusion summarizing our findings.

2. ANALYTICAL FRAMEWORK

This section presents the theoretical framework for our investigation of the Great Moderation. We begin by discussing the estimation of the static factor model, and FAAR model, with time-varying parameters used in the empirical section. We introduce the process assumed to drive the time-varying parameters, the parsimonious random walk, and present the results established in Callot and Kristensen (2015) for the estimation of models with parsimoniously time-varying parameters. This is followed by a discussion of the estimation of common factors when part of their loadings on the observed variables is assumed to follow a parsimonious random walk, and on the implications of using estimated factors instead of the unobserved factors.

2.1. Models

We make use of two models in this paper. A static factor model is used to estimate the factors and perform structural analysis. An FAAR model is also considered, with the primary purpose of forecasting. In the static factor model, we assume that the variables of interest, X_{it}, $t = 1, \ldots, T$, $i = 1, \ldots, n$, are generated by a factor model with r_F unknown factors, F_t:

$$X_{it} = \lambda_{it}' F_t + e_{it} \tag{1}$$

where $\lambda_{it} \in \mathbb{R}^{r_F}$ is the vector of factor loadings at time t for variable i. We also use this model with a large number of variables n to estimate the factors using PC. We provide consistency results for the estimated factors below, and discuss the data used to estimate the factors in the data section.

Factor models are frequently used for macroeconomic forecasting, in which context the estimated factors are used as predictors in the forecasting models. In the case of linear regression models, this is referred to as FAAR models. Such a model for a h-step ahead direct forecast can be written as

$$X_{it+h}^{(h)} = \mu_i + \beta_{it}' F_t + \sum_{j=0}^{p-1} \gamma_{ijt} X_{it-j} + \varepsilon_{it} \tag{2}$$

To simplify the discussion on the estimation of these models, we write both of them in the same compact form and drop the variable subscript i:

$$y_t = \xi_t' Z_t + \epsilon_t \tag{3}$$

Z_t is a vector of dimension $r \times 1$ containing the factors and, in the case of the FAAR model, the lags of the dependent variable. The $1 \times r$ vector of parameters ξ_t is assumed to be time-varying, and y_t is either X_{it} or $X_{it+h}^{(h)}$.

If the parameters of both these models were assumed to be constant over time, and the factors known or estimated, we could consistently estimate the models by OLS. Instead we let the parameters of the models vary over time and, to be specific, we assume that the parameters follow parsimonious random walks. This process is formalized in the following assumption.

Assumption 1. (Parsimonious Random Walk)
Assume that the parameters follow a parsimonious random walk with ξ_0 *given.*

$$\xi_t = \xi_{t-1} + \zeta_t \odot \eta_t$$

The r-*dimensional vectors* η_t *and* ζ_t *have the following properties:*

$$\begin{aligned}
\alpha_T &= k_\alpha T^{-a}, \quad 0 \leq a \leq 1, \quad k_\alpha > 0 \\
\zeta_{jt} &= \begin{cases} 1, & \text{w.p. } \alpha_T \\ 0, & \text{w.p. } 1-\alpha_T \end{cases} \quad j \in 1, \ldots, r \\
\eta_t &= \mathcal{N}(0, \Omega_\eta) \\
E(\eta_t' \eta_u) &= 0 \quad \text{if} \quad t \neq u \\
E(\eta_t' \zeta_u) &= 0 \quad \forall \quad t, u \in 1, \ldots, T
\end{aligned}$$

The vector of increments to the parameters is given by the element-by-element product of the two vectors of mutually independent random variables ζ_t and η_t. The vector η_t contains a set of i.i.d. random variables with mean zero and bounded variance, and ζ_t is a vector of binary variables in which each element takes value 1 with probability α_T and zero otherwise.

The parsimonious random walk assumption implies that many of the increments to the parameter vectors are equal to zero, the probability of a non-zero increment is controlled by $\alpha_T = k_\alpha T^{-a}$. The constant k_α scales the probability α_T and must be such that $0 \leq \alpha_T \leq 1$. If k_α satisfies this restriction for some T_0, it will satisfy it for any $T \geq T_0$ since $a \geq 0$. Consistency requirements for the Lasso estimator will impose a tighter lower bound on a. Note that we do not set or estimate a (or α_T) but simply assume that a is larger than some quantity which we make explicit later.

2.2. Estimation of the Parsimoniously Time-Varying Parameter Models

Assumption 1 implies that the vector of increments to the parameter vector is sparse, it contains many zeros, but it is also high dimensional since it is at least as large as the sample size, the number of parameters to estimate is of the order of rT.

Recall the compact model,

$$y_t = \xi_t' Z_t + \epsilon_t$$

Define the following matrices:

$$Z^D = \begin{bmatrix} Z_1 & 0 & \cdots & 0 \\ 0 & Z_2 & \cdots & 0 \\ \vdots & \vdots & \ddots & \vdots \\ 0 & 0 & \cdots & Z_T \end{bmatrix}, \quad W = \begin{bmatrix} I_r & 0 & \cdots & 0 \\ I_r & I_r & \cdots & 0 \\ \vdots & \vdots & \ddots & \vdots \\ I_r & I_r & \cdots & I_r \end{bmatrix}, \quad Z^D W = \begin{bmatrix} Z_1 & 0 & \cdot & 0 \\ Z_2 & Z_2 & \cdot & 0 \\ \vdots & \vdots & \ddots & \vdots \\ Z_T & Z_T & \cdots & Z_T \end{bmatrix}$$

We can write the parsimoniously time-varying model (3) as a simple regression model

$$y = Z^D W \theta + \epsilon$$

where the parameter vector $\theta' = \left[\xi_0' + \zeta_1' \odot \eta_1', \zeta_2' \odot \eta_2', \ldots, \zeta_T' \odot \eta_T'\right]$ has length rT, and $y = (y_1, \ldots, y_T)'$, $\epsilon = (\epsilon_1, \ldots, \epsilon_T)'$. The matrix $Z^D W$ contains T observations for rT covariates constructed from the original r variables. The first r elements of θ are the sum of the initial value of the parsimonious random walk ξ_0 and the first increment $\zeta_1 \odot \eta_1$. The subsequent elements of θ are the increments of the parsimonious random walk $\zeta_t \odot \eta_t, t > 1$ so that by cumulating the entries of θ we can recover the full path of the parameters.

The parameter vector we seek to estimate, θ, is sparse and high dimensional, we estimate θ using the Lasso as in Callot and Kristensen (2015). We make extra assumptions below regarding the distribution of the innovations and of the factors, in particular to ensure that all the variances involved are finite.

Assumption 2. (Covariates and Innovations)
Assume that:

(i) $\epsilon_t \sim \mathcal{N}(0, \sigma_\epsilon^2)$ *is a sequence of i.i.d innovation terms,* $\sigma_\epsilon^2 < \infty$.
(ii) $F_t \sim \mathcal{N}(0, \Omega_F^2)$. *For all* $k = 1, \ldots, r_F$, $\mathrm{Var}(F_{kt}) = \sigma_{Fk}^2 < \infty$.
(iii) $E(\epsilon' F) = 0$.
(iv) $\mathrm{Var}(y_t) \leq M < \infty$ *for all* t *and some positive constant* M.

Assumption 2(iv) ensures that the variance of y_t is bounded from above at all points in time. It is a high-level assumption that we give here, a lower level assumption on the dynamics of the model is used instead in Callot and Kristensen (2015) to which the interested reader is referred.

Similarly, we state the following assumption, the restricted eigenvalue condition, which is a standard assumption in the Lasso literature, introduced by Bickel, Ritov, and Tsybakov (2009). Define κ_T^2 as the restricted eigenvalue of the Gram matrix $(1/T)(Z^D W)'(Z^D W)$, we then assume the following restricted eigenvalue condition to hold:

Assumption 3. (Restricted Eigenvalue Condition)
Assume that:

(i) $\kappa_T^2 > 0$.
(ii) $\kappa_T^2 \in \Omega_p(T^{d-1})$ *for some* $d \in (0, 1]$.[1]

The rate of decrease in κ_T^2 stems from technical assumptions on the rate at which the distance between breaks increases asymptotically; the detail of the assumptions and proof of the result can be found in Callot and Kristensen (2015).

We can now state the first theorem from Callot and Kristensen (2015) which provides upper bounds on the prediction and parameter estimation errors. Let $[\sigma_1^2, \ldots, \sigma_{rT}^2] = \mathrm{diag}(\mathrm{Var}(Z^D W))$ and $\sigma_T^2 = \max(\sigma_\epsilon^2, \max_{1 \leq k \leq rT} \sigma_k^2)$, where σ_ϵ^2 is the variance of ϵ and σ_k^2 is the variance of the kth column in $Z^D W$. Define the active set $\mathcal{S}_T$ as the set of indices corresponding to non-zero parameters in θ, $\mathcal{S}_T = \{j \in (1, \ldots, rT) | \theta_j \neq 0\}$, and its cardinality

$|\mathcal{S}_T| = s$. Finally, let $\tilde{\lambda}_T$ be the Lasso penalty parameter. We then have the following result:

Theorem 1. *For* $\tilde{\lambda}_T = \sqrt{8\ln(1+T)^5\ln(1+r)^2\ln(r(T-r+1))\sigma_T^4/T}$ *and some constant* $A>0$, *under Assumptions 1, 2, and 3, and on the set* $\mathcal{B}_T$ *with probability at least equal to* $1-\pi_T^{\mathcal{B}}$ *we have the following inequalities:*

$$\frac{1}{T}\|Z^D W(\hat{\theta}-\theta)\|^2 \leq \frac{16s\tilde{\lambda}_T^2}{\kappa_T^2} \tag{4}$$

$$\|\hat{\theta}-\theta\|_{\ell_1} \leq \frac{16s\tilde{\lambda}_T}{\kappa_T^2} \tag{5}$$

with $\pi_T^{\mathcal{B}} = 2(1+T)^{-1/A} + 2(r(T-r+1))^{1-\ln(1+T)}$.

The bounds given in Theorem 1 are upper bounds on the ℓ_1-norm of the parameter estimation error and on the mean-squared estimation error of the Lasso. These bounds hold on a set that has probability at least $1-\pi_T^{\mathcal{B}}$ for a given value of $\tilde{\lambda}_T$. Nonetheless these bounds are valid for any value of the penalty parameter as long as $\|T^{-1}\varepsilon' Z^D W\|_\infty \leq \tilde{\lambda}_T/2$ is satisfied. Holding everything else constant the probability of this inequality being satisfied decreases with $\tilde{\lambda}_T$ as do the upper bounds in Theorem 1; there is a trade-off between the tightness of the bounds and the probability with which they hold.

Theorem 2 provides an asymptotic counterpart to Theorem 1.

Theorem 2. *Let* a *and* d *be scalars with* $a, d \leq 1, 1-a+d \leq 1$, *and* $\frac{3}{2}-a-d<0$. *Then under Assumptions 1, 2, and 3, and as* $T \to \infty$ *we have:*

$$\frac{1}{T}\|Z^D W(\hat{\theta}-\theta)\|^2 \to^p 0 \tag{6}$$

$$\|\hat{\theta}-\theta\|_{\ell_1} \to^p 0 \tag{7}$$

This theorem establishes the consistency of the Lasso for our models. The parameter estimation error and the prediction error both tend to zero with probability tending to one. Callot and Kristensen (2015) show that, as is usual with the Lasso, the estimator correctly sets many parameters to zero without setting non-zero parameters to zero under the condition that the smallest non-zero parameter is not too small. They also show that the adaptive Lasso, a second-stage estimator using an adaptive penalty constructed using the Lasso estimates, is sign consistent. This means that the adaptive Lasso sets all zero parameters to zero while retaining the nonzero parameters with probability tending to 1. We refer to Callot and Kristensen (2015) for details.

2.3. Factor Estimation with Parsimoniously Time-Varying Loadings

Until this point, we have assumed the factors to be known, in this section we discuss the issue of estimating the factors under the assumption that their loadings can be parsimoniously time varying. We then discuss the effect of using estimated factors instead of the true factors in the model.

If we assume that the factor loadings do not vary over time, then it is well known that the factors can be consistently estimated by means of PC, see, for example, Bai and Ng (2002). Bates et al. (2013) give conditions under which the same holds in the case of time-varying loadings. We use the results of Bates et al. (2013) to show that factors can be consistently estimated when the loadings follow parsimonious random walks as in Assumption 1. We follow the notation of Bates et al. (2013) closely (which also corresponds to the notation of Bai & Ng, 2002). Whereas Bates et al. (2013) give general results in the case of time-varying loadings we will specifically make the following assumption:

Assumption 4. (Time-Varying Factor Loadings)

The loadings in Eq. (1) must satisfy the following:

(i) *With probability $\pi_{n,T}$ the loadings for variable i, λ_{it}, follow a parsimonious random walk as defined in Assumption 1 with fixed initial value λ_{i0}. Alternatively, with probability $1 - \pi_{n,T}$ the loadings are constant.*

(ii) *The probability $\pi_{n,T}$ must satisfy: $\pi_{n,T} = \mathcal{O}\left(1/\min\left(n^{1/2}T^{1-a}, T^{3/2-a}\right)\right)$, where a is the parameter controlling α_T as defined in Assumption 1 with $a \geq 1/2$.*

(iii) *For all (i, j, s, t), e_{it}, the factor model innovations, are independent of $(\zeta_{js} \odot \eta_{js}, F_s)$, the factor loading innovations and the factors. Furthermore, the factor loading innovations, $\zeta_{ip,t}\eta_{ip,t}$, are independent across i, t, and p.*

We should note that Assumption 4(iii) implies that breaks occur independently across variables. Although this does not necessarily correspond well with empirical observations, where series tend to co-break, this assumption is needed to ensure that the factors can be consistently estimated. Note, however, that independence is only needed for the factor estimation, the results of Theorem 1 hold under the more general requirements of Assumption 1.

In addition to this, we make the following standard assumptions (corresponding the Assumptions 1–3 in Bates et al. (2013) or Assumptions A–C in Bai & Ng, 2002):

Assumption 5. (Factors)
$\mathbb{E}\|F_t\|^4 \leq M$ and $T^{-1}\sum_{t=1}^{T} F_t F_t' \rightarrow^p \Sigma_F$ as $T \rightarrow \infty$ for some positive definite matrix Σ_F.

Assumption 6. (Initial Factor Loadings)
$\|\lambda_{i0}\| \leq \bar{\lambda} < \infty$, and $\|\Lambda_0'\Lambda_0/n - D\| \rightarrow 0$ as $n \rightarrow \infty$ for some positive definite matrix $D \in \mathbb{R}^{r_F \times r_F}$.

Assumption 7. (Idiosyncratic Errors)
The following conditions hold for all n and T.

1. $\mathbb{E}(e_{it}) = 0$, $\mathbb{E}|e_{it}|^8 \leq M$.
2. $\gamma_n(s,t) = \mathbb{E}(e_s' e_t/n)$ *exists for all* (s,t). $|\gamma_n(s,s)| \leq M$ *for all* s, *and* $T^{-1}\sum_{s,t=1}^{T} |\gamma_n(s,t)| \leq M$.
3. $\tau_{ij,ts} = \mathbb{E}(e_{it}e_{js})$ *exists for all* (i,j,s,t). $|\tau_{ij,tt}| \leq |\tau_{ij}|$ *for some* τ_{ij} *and for all* t, *while* $n^{-1}\sum_{i,j=1}^{n} |\tau_{ij}| \leq M$. *In addition,* $(nT)^{-1}\sum_{i,j=1}^{n}\sum_{s,t=1}^{T} |\tau_{ij,ts}| \leq M$.
4. *For every* (s,t), $\mathbb{E}\left|n^{-1/2}\sum_{i=1}^{n} [e_{is}e_{it} - \mathbb{E}(e_{is}e_{it})]\right|^4 \leq M$.

Under these assumptions, we have the usual consistency result, the proof of which can be found in the appendix.

Theorem 3. *Let Assumptions 4–7 hold, and $n, T \rightarrow \infty$ with $T^{1-a}/n^{1/2} \rightarrow k$ for some constant $k \geq 0$, then*

$$C_{nT}^2 \left(T^{-1} \sum_{t=1}^{T} \|F_t' - H'F_t\|^2 \right) = \mathcal{O}_p(1) \tag{8}$$

where $C_{nT}^2 = \min(n, T)$ and H is the usual rotation matrix as defined in, for example, Bates et al. (2013).

Note that Assumption 4 puts restrictions on the extent to which the variables are allowed to have time-varying loadings through the probability $\pi_{n,T}$. This is in contrast to the literature on large breaks where it is typically assumed that the break affects all variables. However, the results are comparable to example 3 in Bates et al. (2013) which treats the case of a single large break. The comparable case in terms of our results would be that of a fixed number of breaks which occurs if we have $a = 1$ which implies that $\pi_{n,T} = \mathcal{O}\left(1/\min\left(n^{1/2}, T^{1/2}\right)\right)$. Likewise, example 3 in Bates et al. (2013) requires that at most $\mathcal{O}\left(n^{1/2}\right)$ variables undergo a break. In cases where $a < 1$ we have a trade-off between the (expected) number of breaks in the loadings (which is now increasing in T) and the (expected) number of series which may be associated with breaks as controlled by $\pi_{n,T}$. It is important to note that in order to obtain the result we must restrict the relative growth of n and T. However, by doing so we are able to recover the usual rate of convergence. This is important because it ensures that we can apply the IC_p criterion of Bai and Ng (2002) to determine the number of factors, see Corollary 2 in Bai and Ng (2002).

2.4. Estimation of the ptv Model with Estimated Factors

From Assumption 2 one can see that the covariates are required to be normally distributed with finite variance. Since the estimated factors $\hat{F}$ are a linear combination of the data X, it suffices to ensure that X_t is Gaussian to ensure that the estimated factors satisfy Assumption 2. Bai and Ng (2006) show that using estimated factors does not invalidate consistency and asymptotic normality of the OLS estimator provided $\sqrt{T}/n \to 0$. Given that the objective function used to estimate the ptv models is a penalized least-squares criterion, and that Theorem 3 shows that the factor are estimated with the usual efficiency, we conjecture that the same holds true for the ptv model.

3. EMPIRICAL RESULTS

Stock and Watson (2009) set out to investigate the effect of a structural break on factor models and in particular their ability to forecast. They consider a very specific case, namely, the Great Moderation, which they argue could have caused a break in the mid 1980s. Specifically, they test for a structural break in 1984:1 in both the loadings of the factor model and the parameters of the FAAR forecasting model. We take the analysis, and the data, of Stock and Watson (2009) as our starting point and investigate parameter instability over the entire sample period.

This section contains a description of the data we use and practical details regarding the implementation of the ptv method described above. We then present our empirical results, first focusing on aggregate parameter instability, second investigating more closely some particular variables of interest, and finally looking at the improvements in fit of the ptv model relative to OLS.

3.1. Data

We use the same dataset as Stock and Watson (2009). It consists of 144 quarterly time series for the United States spanning the period 1959:1–2006:4. As is customary in the literature, the series are transformed to be stationary and standardized prior to estimation. Appendix B provides details on the transformations and a complete variable list. Due to the transformations we lose the first two observations, hence the effective sample period is 1959:3–2006:4, that is, a total of $T = 190$ quarterly observations.

A unique feature of this dataset, as discussed in Stock and Watson (2009), is the treatment of disaggregation. The full dataset contains both aggregate and sub-aggregate series, and as argued in their paper the inclusion of series related by identities (being the sum of sub-aggregates) does not add useful information to the estimation problem. For this reason when estimating the factors we only use a subset of the data consisting of 109 series that excludes higher level aggregates related by identities to the lower level sub-aggregates. However, for the analysis of the structural stability of the loadings below, we use all 144 series as they are all related to the factors. Hence, in the first step we estimate the factors using the 109 series, and in the second step we use the methodology described above to estimate the time-varying loadings for all 144 series.

3.2. Estimation

Although we have described the estimation procedure, in general, there are a number of details we must address before the estimation can be carried out in practice. As noted, we estimate the factors by PC, however to do this we must decide on how many factors to include in the model. A number of methods have been proposed for this, and one of the most commonly used is undoubtedly the IC_p information criterion of Bai and Ng

(2002). The authors provide three variants of their criterion, and the results for our dataset differ substantially depending on which is used. Specifically, the criteria choose either 2, 4, or 10 factors. One could argue that the safe choice would then be to use 10 factors, but this could lead to undesirable over-fitting, and hence we follow Stock and Watson (2009) and use four factors throughout this paper.

The four estimated factors we will use throughout this application are plotted in Fig. 1. The first factor appears to decrease before and reaches a minimum during, each recession while the second factor has an almost symmetrical pattern with troughs during recessions and rapid increases at the end of each recession and during the recovery period. These factors could be interpreted as capturing the business cycle. The third and fourth factors are much more volatile than the first two, factor three exhibits a trend while factor four seems to have a higher variance in the middle of the sample than on either end. These last two factors do not lend themselves to an unambiguous economic interpretation. For a possible explanation of this recall that one of the information criteria of Bai and Ng (2002) suggested that there may be only two factors implying that factors 3 and 4 contain little if any relevant information.

For the FAAR forecasting model, we must be specific about the forecast target. We focus on the stability of a four-quarter ahead relationship. For real activity variables the target, $X^{(4)}_{it+4}$, is growth over the next four quarters. For inflation it is average quarterly inflation over the next four quarters minus last quarter's inflation. For variables in levels it is simply the value of the variable four quarters ahead. For details, see Appendix B. Further, in order to fully specify Eq. (2), we must choose the lag length p.

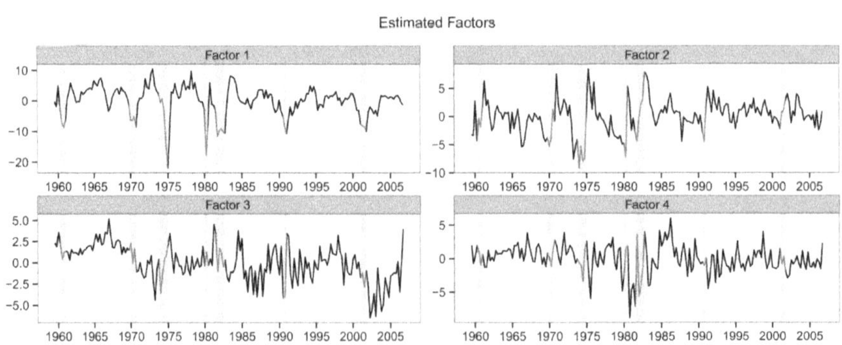

Fig. 1. Estimated Factors, NBER Recessions in Grey.

As we are not concerned with computing actual forecasts, this choice is of less importance, and we simply fix it at $p = 4$.

The actual estimation using the procedure of Callot and Kristensen (2015) is easily implemented using the `parsimonious` package in `R`.[2] As is generally the case the Lasso requires selection of the penalty parameter. This can be done in a number of ways, however, we will follow Callot and Kristensen (2015) and select it using the Bayesian information criterion (BIC). As already noted all variables have been standardized, however, the estimated factors will likely have a much larger variance making it difficult for the Lasso to simultaneous detect breaks in the parameters associated with the factors and the parameters associated with the autoregressive lags in Eq. (2) if they are all penalized equally. To avoid having to introduce different penalty parameters for the factors and the autoregressive lags, we instead also standardize the estimated factors before the Lasso estimation. We will do this in both the cases of the factor model (1) and the FAAR forecasting model (2). The estimation procedure also easily allows for the possibility of having a parsimoniously time-varying intercept by simply including dummies. The need for taking into account instability in the mean of the variables was documented by Stock and Watson (2012). They subtracted a local mean from the variables prior to estimation and found that this mean changed substantially over the sample period. All the presented results are based on models with parsimoniously time-varying intercepts. Our results are, however, robust to this choice and qualitatively similar results are obtained if the intercept is assumed constant. Finally, in the interest of space we only report results for the Lasso and do not consider the adaptive Lasso.

3.3. Results

We begin by examining the stability of the loadings in the factor model (1). Fig. 2 plots the number of breaks in the loadings at every point in time for each factor. Recall that at each point in time for a given factor the vector of loadings is of length $n = 144$, hence if we observe, say, 10 breaks, then that means that 10 out of the 144 variables experience a break.

The first impression given by Fig. 2 is that there is some degree of structural instability over the entire sample period. Closer inspection reveals that in every panel there exist clusters with a large number of breaks, and that outside of these clusters the loadings appear to be relatively stable. This is in particular the case for the parameters associated with the first

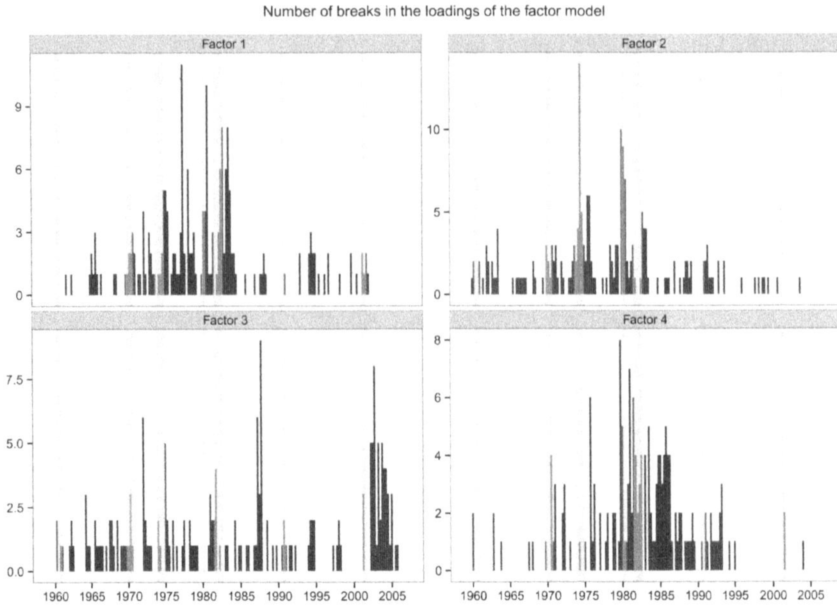

Fig. 2. Breaks in the Factor Model, $n = 144$. The Grey Areas Represent the NBER Recessions.

and second factors where many breaks are detected from the early 1970s to the early 1980s, and much fewer breaks outside this period. From this plot it is not clear whether the instability is greater during recessions.

The parameters associated with the third factor experience breaks relatively uniformly throughout the sample, with a surge in 1987 and after 2000. The breaks in the loadings associated with the fourth factor are concentrated in the 1975 to early 1990s period. Fig. C1 displays the number of breaks in the intercept at every point in time for the factor and forecasting models. For both models the breaks in the intercept appear to be uniformly spread across the sample in contrast to the clusters observed in the factor loadings.

In Fig. 3, we consider mean absolute change in the loadings defined as $n^{-1}\sum_{i=1}^{n}|\lambda_{it}-\lambda_{it-1}|$ for a given point in time t where the mean is taken across all $n = 144$ variables. This illustrates the large differences in the sizes of the breaks. It appears that most of the breaks are (relatively) small, but a few large changes occur from the mid 1970s to the mid 1980s in the

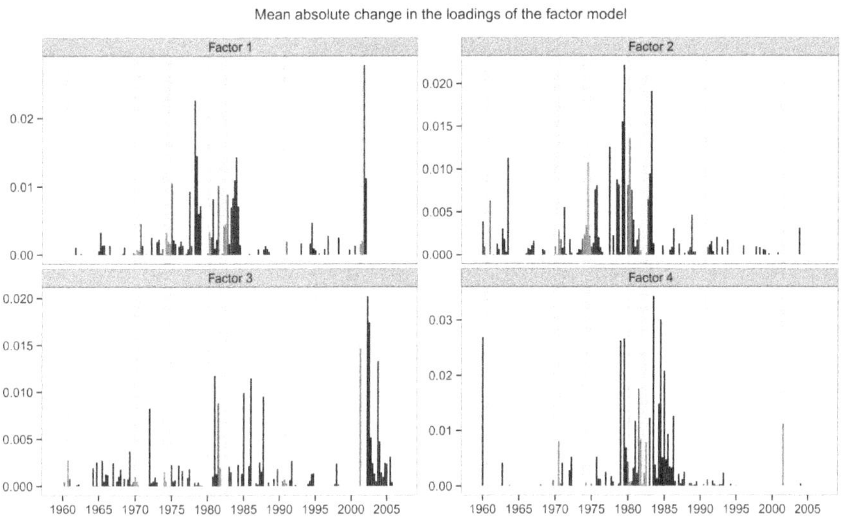

Fig. 3. Size of Breaks in the Factor Model, $n = 144$. The Grey Areas Represent the NBER Recessions.

loadings of factors 1, 2, and 4. This plot provides further evidence in favour of the Great Moderation as a period of stability following a period of structural instability in the 1970s and early 1980s.

In order to illustrate how allowing for breaks affects the fit of the model we turn to the FAAR forecasting model. Fig. 4 shows the cross-sectional root mean-squared error (RMSE), that is, $\sqrt{(1/144)\sum_{i=1}^{144}\hat{\epsilon}_{it}^2}$ for all

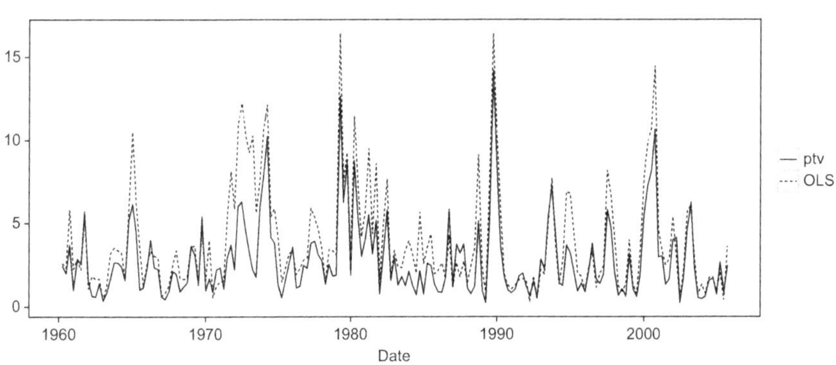

Fig. 4. Cross-Sectional RMSE in the FAAR Model.

$t = 1, \ldots, 190$. The gain in RMSE from allowing parsimoniously time-varying parameters (relative to OLS) is particularly large in the early 1970s, corresponding to the region with more and larger breaks in Figs. 2 and 3 for the factor model. We will return to results on the fit of the FAAR model in more detail later.

First, we now look into the detail of which variables are associated with the breaks in Table 1. In the table we consider both the factor model (1) and the FAAR forecasting model (2). We report the number of breaks over the entire sample (denoted All), the period preceding the Great Moderation from 1970:1 to 1984:4, and the period of the Great Moderation from 1985:1 to 1999:4. For the forecasting model we separately consider the number of breaks in the parameters associated with the factors, β, and the parameters associated with the autoregressive lags, γ. In the interest of space, the table only includes a subset of the variables, most sub-aggregates have been removed, results for the remaining variables can be found in Appendix C.

Regarding the timing of the breaks, Table 1 clearly shows that the majority of the breaks, all of them in some cases, occur in the period preceding the Great Moderation with both models. There is a striking difference in the number of breaks between the two models, though: many more breaks are selected in the forecasting model than in the factor model, and the variables with a large number of breaks are not systematically the same across models. Three possible explanations for the difference between models are, first, the different target variables as explained in Section 3.2; second, the doubling of the number of variables; and third, the potentially high degree of persistence of the extra variables.

Table 1 gives some interesting information as to which variables are subject to instability, despite the differences between the two models. In the static factor model, the breaks are concentrated among monetary and financial variables: MZM (Money Zero Maturity) has 30 breaks and the total amount of reserve 29, interest rates (4, 3, and 2 breaks for the Fed fund rate, the 6 month T-bill, and the 5 year T-bond, respectively), exchange rates (3), and the Dow Jones Industrial Average (1). Many breaks are also found among variables from the real sector of the economy, housing start (28 breaks) and build permits (45), industrial production (5), and the manufacturing indices (NAPM) with 6 and 1 breaks in two out of four variables.

The pattern is different for the FAAR models that have more breaks overall, mostly in the factor loadings, and occurring within different groups of variables. Price variables experience many breaks, up to 56 for CPI Core

Table 1. Number of Breaks in the Loadings of the Factor Model and the Parameters of the Forecasting Model Detailed by Variable.

Series	Factor Model			Forecasting Model					
				Coef. on Factors			Coef. on AR Lags		
	All	1970:1 1984:4	1985:1 1999:4	All	1970:1 1984:4	1985:1 1999:4	All	1970:1 1984:4	1985:1 1999:4
RGDP				1	1				
Cons				4	3		5	4	
GPDInv	1	1		1	1				
Exports									
Imports									
Gov									
IP: total	5	3		1	1				
NAPM prodn	4	3	1						
Capacity Util	18	9	2	11	10	1	8	2	
Emp: total				2	2				
Help wanted indx	1	1							
Emp CPS total				2	2				
U: all				2	2				
HStarts: Total	28	15	4	17	11		11	11	
BuildPermits	45	17	11						
PMI									
NAPM new ordrs									
NAPM vendor del	6	6		9	6		7	4	
NAPM Invent	1	1		2	2		6	6	
PGDP				17	13	3	8	8	
PCED				8	7	1	1	1	
CPI-All				3	3				
PCED-Core				4	4				
CPI-Core				41	29	1	15	15	
PGPDI				24	18	6	5	5	
PFI				23	18	5	4	4	
PEXP				14	12	1	4	4	
PIMP				17	16	1	5	5	
PGOV				3	3				
Com: spot price (real)									
OilPrice (Real)	6	2	4	43	28	13	38		29
NAPM com price	10	5	3	2	2		3	3	
Real AHE: goods	12	7	3	2	2		6	6	
Labor Prod				1		1			
Real Comp/Hour				1	1				
Unit Labor Cost	2	2		7	7		3	3	
FedFunds	4	4		7	5	1	8	8	
6 mo T-bill	3	3		4	4		6	6	

Table 1. (*Continued*)

Series	Factor Model			Forecasting Model					
				Coef. on Factors			Coef. on AR Lags		
	All	1970:1 1984:4	1985:1 1999:4	All	1970:1 1984:4	1985:1 1999:4	All	1970:1 1984:4	1985:1 1999:4
5 yr T-bond	2	2		4	3	1	2	2	
Aaabond	1	1		1	1		3	3	
Baa bond				2	2		6	5	1
M1				2	2				
MZM	30	20	8	12	7	2	4	4	
M2	1	1		9	5	3	4	3	1
MB				2	1	1	2		1
Reserves tot	29	11	7	5	3	1	4		
Bus loans									
Cons credit							1	1	
Ex rate: avg	3	2	1						
S&P 500	9	5	2	12	6	4	4	1	3
DJIA	1		1	2		2			
Consumer expect	2	2		1	1				

and 81 for Oil Price. Breaks are also frequently detected in monetary and financial variables, up to 15 for the 5-year T-bond and around 10 for the other variables of this type.

To illustrate the results discussed above we now take a close look at the estimated loadings of four variables. While we will focus on models in which the estimated parameters vary over time, it should be kept in mind that in a large fraction of loadings no break is found. Furthermore, and as the plots below will make apparent, within models exhibiting breaks the parameters associated with certain variables are still found to be constant.

Fig. 5 shows the estimated loadings of the static factor model for Money Zero Maturity, which is one of the variables with which the largest number of breaks is associated. Most of these breaks, in particular the large ones, occur between the late 1970s and the early 1980s, which is consistent with important changes in monetary policy during that period. The few breaks occurring outside this period are for the most part very small changes to the estimated loading.

Fig. 6 shows the estimated loadings for industrial production, for which five breaks are found. The loading on the first factor has a single

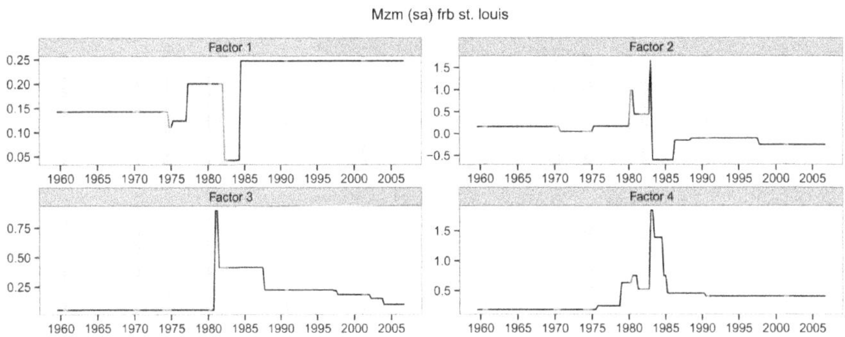

Fig. 5. Time-Varying Loadings of the Factor Model, Money Zero Maturity.

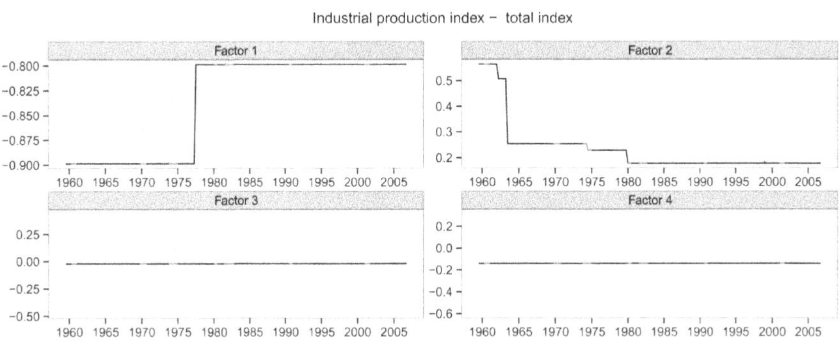

Fig. 6. Time-Varying Loadings of the Factor Model, Industrial Production.

large upward break in 1977, which could be interpreted as indicating that industrial production reacts less to downturns from that point onwards. The second factor has two large downward breaks in the early 1960s and two smaller breaks in the second half of the 1970s, which could be interpreted as indicating that industrial production recovers more slowly. The loadings for the last two factors are found to be constant.

Fig. 7 shows the estimated parameters of a forecasting model for the Capacity Utilization variable, where 11 breaks are found in the factor loadings and 8 in the AR lags. Fig. 7 shows that five out of the eight parameters experience a break and that the instability typically takes the form of few large breaks with very little gradual adjustment.

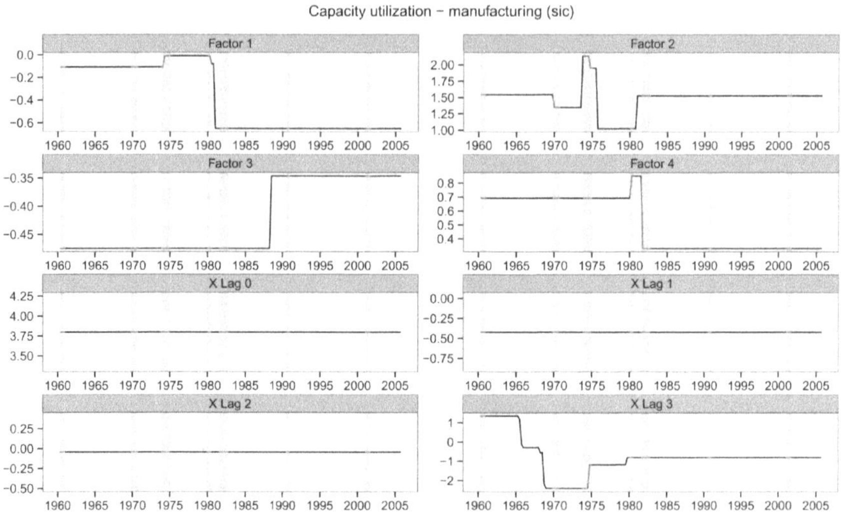

Fig. 7. Time-Varying Loadings of the Forecasting Model, Capacity Utilization.

Fig. 8 is an example of a forecasting model with a large number of breaks with a total of 56, 41 in the factor loadings and 15 in the AR parameters, 44 of these breaks are located in the pre Great Moderation period. In this model, the breaks lead to many substantial and persistent changes in the parameter value. The factor loadings have more breaks than the AR parameters, this results in more jagged paths with several small adjustments in the parameter path. The loadings on the factors could be interpreted as implying that the business cycle has had little effect on CPI Core except for the middle of the 1970s, a period of low economic growth and high inflation in the United States.

In a forecasting context model instability is a concern, and as Table 1 illustrates many forecasting relationships do not appear to be stable over time. We do not view the applied methodology as a means to obtain better forecasts, but as a tool to illustrate that there might be issues when neglecting instabilities or breaks. Nonetheless it is still of interest to see how much the fit of the forecasting model is improved when we allow for a moderate amount of time variation in the parameters. For this purpose, Table 2 reports a number of useful statistics for a subset of variables, the remaining results can be found in Appendix C.

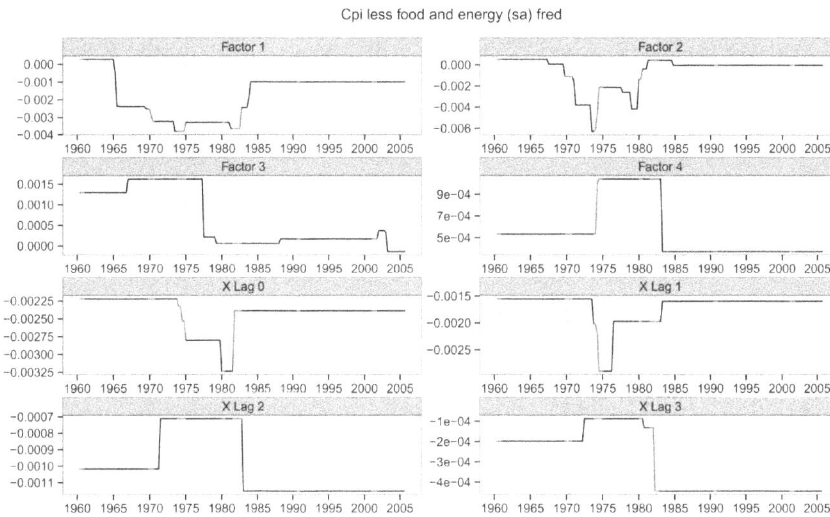

Fig. 8. Time-Varying Loadings of the Forecasting Model, CPI Core.

The first column of the table gives the standard deviation of the series being forecast. The second column gives the RMSE of the residuals of the forecasting model (2) when estimated by the ptv methodology, that is, allowing for breaks. The third column gives the RMSE of the residuals of the forecasting model (2) when simply estimated by OLS, that is, not allowing for breaks. The fourth column gives the relative RMSE of these two approaches and the last column the number of breaks in the corresponding model. The upper bound for the relative RMSE statistic is 1 and corresponds to the case where no breaks were selected, in which case our estimator is equal to OLS. Allowing for breaks must improve the fit to compensate for the penalty in the Lasso estimation, therefore when breaks are estimated the relative RMSE is always below 1. We should stress that this is purely an in-sample comparison of the forecasting models and not a pseudo out-of-sample forecasting experiment.

For certain variables the reduction in RMSE by allowing for breaks is quite substantial, for example a reduction of roughly 70% for CPI-Core (the parameters of this model are plotted in Fig. 8), and 80% for Oil Price. The presence of numerous relatively large breaks (despite the small magnitude of the loadings) explains this reduction in the RMSE. The results in Table 2 show that reduction in RMSE relative to OLS is closely related,

Table 2. Root Mean-Squared Errors (RMSE) of the Residuals of the Forecasting Model when Allowing for Breaks or Not, and the Relative RMSE of These Two Approaches.

Series	Std. Dev. of $X_{it}^{(4)}$	RMSE, ptv	RMSE, OLS	Relative RMSE	# Breaks
RGDP	0.0217	0.0164	0.0174	0.9421	1
Cons	0.0175	0.0118	0.0144	0.8160	9
GPDInv	0.0979	0.0713	0.0736	0.9689	1
Exports	0.0608	0.0517	0.0517	1.0000	0
Imports	0.0694	0.0509	0.0509	1.0000	0
Gov	0.0245	0.0211	0.0211	1.0000	0
IP: total	0.0433	0.0293	0.0317	0.9237	1
NAPM prodn	6.9244	5.7070	6.1090	0.9342	0
Capacity Util	4.5537	1.6163	2.5634	0.6305	19
Emp: total	0.0202	0.0126	0.0134	0.9415	2
Help wanted indx	12.2287	8.2756	8.9425	0.9254	0
Emp CPS total	0.0133	0.0090	0.0098	0.9258	2
U: all	0.9700	0.6328	0.6604	0.9581	2
HStarts: Total	0.2094	0.0862	0.1579	0.5456	28
BuildPermits	0.2467	0.1747	0.1747	1.0000	0
PMI	6.5564	5.1061	5.7118	0.8940	0
NAPM new ordrs	7.3697	5.8167	6.4712	0.8989	0
NAPM vendor del	10.3401	5.2072	8.7598	0.5944	16
NAPM Invent	6.2786	3.6914	5.1907	0.7112	8
PGDP	0.0029	0.0012	0.0022	0.5716	25
PCED	0.0032	0.0022	0.0026	0.8514	9
CPI-All	0.0043	0.0029	0.0031	0.9434	3
PCED-Core	0.0025	0.0019	0.0021	0.9278	4
CPI-Core	0.0038	0.0009	0.0028	0.3157	56
PGPDI	0.0052	0.0021	0.0038	0.5614	29
PFI	0.0053	0.0023	0.0039	0.5984	27
PEXP	0.0102	0.0051	0.0080	0.6420	18
PIMP	0.0190	0.0097	0.0158	0.6167	22
PGOV	0.0051	0.0031	0.0032	0.9479	3
Com: spot price (real)	0.1136	0.0932	0.0932	1.0000	0
OilPrice (Real)	0.2485	0.0464	0.2396	0.1935	81
NAPM com price	14.2807	10.3370	12.1543	0.8505	5
Real AHE: goods	0.0144	0.0082	0.0113	0.7300	8
Labor Prod	0.0165	0.0123	0.0149	0.8216	1
Real Comp/Hour	0.0140	0.0115	0.0130	0.8819	1
Unit Labor Cost	0.0311	0.0152	0.0194	0.7814	10
FedFunds	2.2061	1.3908	1.8203	0.7640	15
6 mo T-bill	1.6417	1.1463	1.4092	0.8135	10
5 yr T-bond	1.3442	1.0934	1.2524	0.8730	6
Aaabond	1.0090	0.7926	0.9385	0.8445	4

Table 2. *(Continued)*

Series	Std. Dev. of $X_{it}^{(4)}$	RMSE, ptv	RMSE, OLS	Relative RMSE	# Breaks
Baa bond	1.1242	0.8188	1.0039	0.8156	8
M1	0.0095	0.0069	0.0072	0.9700	2
MZM	0.0183	0.0076	0.0117	0.6434	16
M2	0.0070	0.0035	0.0050	0.7044	13
MB	0.0066	0.0044	0.0050	0.8809	4
Reserves tot	0.0238	0.0125	0.0150	0.8345	9
Bus loans	0.0148	0.0118	0.0118	1.0000	0
Cons credit	0.0114	0.0091	0.0093	0.9770	1
Ex rate: avg	0.0660	0.0595	0.0595	1.0000	0
S&P 500	0.1435	0.1015	0.1357	0.7481	16
DJIA	0.1400	0.1163	0.1282	0.9067	2
Consumer expect	10.7688	8.5925	9.2908	0.9248	1

but not directly proportional, to the number of breaks. Again this is to be expected as the penalty associated with each extra parameter must be compensated with improvement in fit.

4. CONCLUSION

We have applied the parsimoniously time-varying parameter framework of Callot and Kristensen (2015) to factor models estimated using a set of data from Stock and Watson (2009) containing 144 US macroeconomic variables observed from 1959:1 to 2006:4. The ptv framework allows for an unknown number of breaks at unknown locations to be consistently estimated using the Lasso. We take advantage of this flexibility to study the stability of the parameters of macroeconomic models, and in particular we focus our investigation on the Great Moderation.

We find that for a large share of variables the parameters of either the static factor model or the dynamic forecasting FAAR model are unstable. The number of breaks, their locations, and the resulting parameter paths are very diverse. Nonetheless common patterns emerge, in particular a concentration of parametric instability in the period between 1970 and the middle of the 1980s, and a relative stability of those parameters in the 15 years following. Within our ptv framework, the Great Moderation appears to be a period of stability following a period of instability of the process driving macroeconomic variables. Adequately modelling this instability in time-varying parameter models appears to require more than a single break towards the beginning of the Great Moderation.

Further research could usefully complement our results. On the methodological side, obtaining confidence bands for the estimated parameters is a priority and refining break detection by the introduction of thresholding a possibility. On the empirical side, it would be useful to systematically assess the effect of allowing general forms of parametric instability on the choice of the number of factors and lags, both in terms of model specification and of forecasting performance.

NOTES

1. $f(T) \epsilon \Omega_p(g(T))$ means that there exists a constant $c > 0$ such that $f(T) \geq cg(T)$ for $T \geq T_0$ for a certain T_0 onwards with probability approaching one.
2. Replication files can be found at https://github.com/lcallot/ptv-fac

ACKNOWLEDGEMENTS

The authors would like to thank two anonymous referees and participants at the 2014 Advances in Econometrics conference for their comments and suggestions. Furthermore, support from CREATES, Center for Research in Econometric Analysis of Time Series (DNRF78), funded by the Danish National Research Foundation is gratefully acknowledged.

REFERENCES

Bai, J., & Ng, S. (2002). Determining the number of factors in approximate factor models. *Econometrica*, *70*(1), 191–221.

Bai, J., & Ng, S. (2006). Confidence intervals for diffusion index forecasts and inference for factor-augmented regressions. *Econometrica*, *74*(4), 1133–1150.

Bai, J., & Ng, S. (2008). Large dimensional factor analysis. *Foundations and Trends in Econometrics*, *3*(2), 89–163.

Bates, B. J., Plagborg-Møller, M., Stock, J. H., & Watson, M. W. (2013). Consistent factor estimation in dynamic factor models with structural instability. *Journal of Econometrics*, *177*(2), 289–304.

Bickel, P. J., Ritov, Y., & Tsybakov, A. B. (2009). Simultaneous analysis of lasso and dantzig selector. *The Annals of Statistics*, *37*(4), 1705–1732.

Blanchard, O., & Simon, J. (2001). The long and large decline in US output volatility. *Brookings Papers on Economic Activity*, *2001*(1), 135–174.

Breitung, J., & Eickmeier, S. (2011). Testing for structural breaks in dynamic factor models. *Journal of Econometrics*, *163*(1), 71–84.

Callot, L., & Kristensen., J. T. (2015, April). *Vector autoregressions with parsimoniously time-varying parameters and an application to monetary policy*. CREATES Research Paper No. 2014-41, CREATES, Aarhus University.

Chen, L., Dolado, J. J., & Gonzalo, J. (2014). Detecting big structural breaks in large factor models. *Journal of Econometrics*, *180*(1), 30–48.

Cheng, X., Liao, Z., & Schorfheide, F. (2014, January). *Shrinkage estimation of high-dimensional factor models with structural instabilities*. NBER Working Paper No. 19792, National Bureau of Economic Research.

Corradi, V., & Swanson, N. R. (2014). Testing for structural stability of factor augmented forecasting models. *Journal of Econometrics*, *182*(1), 100–118.

Han, X., & Inoue, A. (2014). Tests for parameter instability in dynamic factor models. *Econometric Theory*, *31*, 1117–1152.

Kim, C.-J., & Nelson, C. R. (1999). Has the US economy become more stable? A bayesian approach based on a Markov-switching model of the business cycle. *Review of Economics and Statistics*, *81*(4), 608–616.

McConnell, M. M., & Perez-Quiros, G. (2000). Output fluctuations in the United States: What has changed since the early 1980's? *American Economic Review*, *90*(5), 1464–1476.

Stock, J., & Watson, M. (2002). Forecasting using principal components from a large number of predictors. *Journal of the American Statistical Association*, *97*, 1167–1179.

Stock, J., & Watson, M. (2003). Has the business cycle changed and why? In *NBER macroeconomics annual 2002* (Vol. 17, pp. 159–230). Cambridge, MA: MIT press.

Stock, J., & Watson, M. (2009). Forecasting in dynamic factor models subject to structural instability. In J. Castle & N. Shephard (Eds.), *The methodology and practice of econometrics: A festschrift in honour of David F. Hendry* (pp. 173–205). Oxford: Oxford University Press.

Stock, J., & Watson, M. (2011). Dynamic factor models. In M. P. Clements & D. F. Hendry (Eds.), *Oxford handbook of economic forecasting* (pp. 35–59). New York, NY: Oxford University Press.

Stock, J., & Watson, M. (2012). Disentangling the channels of the 2007–2009 recession. *Brookings Papers on Economic Activity*, *44*(1), 81–156.

Yamamoto, Y., & Tanaka, S. (2013, December). *Testing for factor loading structural change under common breaks*. Discussion Paper Series No. 2013-17, Graduate School of Economics, Hitotsubashi University.

APPENDIX A. PROOFS

Proof of Theorem 3. The result follows by checking the conditions of corollary 1 in Bates et al. (2013). Define a random variable ν_{ip} that takes on the value 1 with probability $\pi_{n,T}$ and the value 0 otherwise, and is independent from η and ζ and across i, p. Then the loadings for variable i and factor p follow the process:

$$\lambda_{ip,t} = \lambda_{ip,t-1} + \nu_{ip}\zeta_{ip,t}\eta_{ip,t} \tag{A.1}$$

or rewritten:

$$\lambda_{ip,t} - \lambda_{ip,0} = \xi_{ip,t} \tag{A.2}$$

where $\xi_{ip,t} = \nu_{ip}\sum_{s=1}^{t}\zeta_{ip,s}\eta_{ip,s}$ and ζ, η satisfy the conditions in Assumption 1. Note that compared to the expressions in Bates et al. (2013) we have explicitly set $h_{nT} = 1$.

We need to bound two expressions, the first:

$$\sup_{s,t \leq T}\sum_{i,j=1}^{n}|\mathbb{E}(\xi_{ip,s}\xi_{jq,t})| = \sup_{s,t \leq T}\sum_{i,j=1}^{n}\left|\mathbb{E}\left(\nu_{ip}\sum_{u=1}^{s}\zeta_{ip,u}\eta_{ip,u}\nu_{jq}\sum_{v=1}^{t}\zeta_{jq,v}\eta_{jq,v}\right)\right| \tag{A.3}$$

Due to independence across factors, the expression is trivially bounded for $p = q$, hence we only need to consider the case of $p = q$ (and drop the factor index to ease notation). Furthermore, due to independence across variables we get:

$$\begin{aligned} &= \pi_{n,T} n \sup_{s,t \leq T}\left|\mathbb{E}\left(\sum_{u=1}^{\min(s,t)}\zeta_{i,u}\eta_{i,u}^2\right)\right| \\ &= \pi_{n,T} n \alpha_T \sup_{s,t \leq T}\min(s,t)\mathbb{E}\left(\eta_{i,1}^2\right) \\ &= \pi_{n,T} n \alpha_T T \mathbb{E}\left(\eta_{i,1}^2\right) \\ &= \pi_{n,T}\alpha_T \mathcal{O}(nT) = Q_1(n,T) \end{aligned} \tag{A.4}$$

The second expression:

$$\sum_{t,s=1}^{T}\sum_{i,j=1}^{n}|\mathbb{E}(\xi_{ip_1,s}\xi_{jq_1,s}\xi_{ip_2,t}\xi_{jq_2,t}|$$

$$= \sum_{t,s=1}^{T}\sum_{i,j=1}^{n}\left|\mathbb{E}\left(\nu_{ip_1}\sum_{u=1}^{s}\zeta_{ip_1,u}\eta_{ip_1,u}\nu_{jq_1}\sum_{v=1}^{s}\zeta_{jq_1,v}\eta_{jq_1,v}\nu_{ip_2}\right.\right.$$

$$\left.\left.\times\sum_{k=1}^{t}\zeta_{ip_2,k}\eta_{ip_2,k}\nu_{jq_2}\sum_{l=1}^{t}\zeta_{jq_2,l}\eta_{jq_2,l}\right)\right| \tag{A.5}$$

Again, due to independence across factors the expression is trivially bounded if all factor indices differ or one index differs from the rest. The non-trivial cases are thus when all indices are equal, that is, $p_1 = p_2 = q_1 = q_2$, and when they are equal in pairs, say, $p_1 = p_2 \neq q_1 = q_2$. We start with the latter case:

$$= \sum_{t,s=1}^{T}\sum_{i,j=1}^{n}\left|\mathbb{E}\left(\nu_{ip_1}\sum_{u=1}^{s}\zeta_{ip_1,u}\eta_{ip_1,u}\sum_{k=1}^{t}\zeta_{ip_1,k}\eta_{ip_1,k}\right)\right.$$

$$\left.\times\mathbb{E}\left(\nu_{jq_1}\sum_{v=1}^{s}\zeta_{jq_1,v}\eta_{jq_1,v}\sum_{l=1}^{t}\zeta_{jq_1,l}\eta_{jq_1,l}\right)\right|$$

$$= \sum_{t,s=1}^{T}\sum_{i,j=1}^{n}\left|\pi_{n,T}\min(s,t)\alpha_T\mathbb{E}\left(\eta_{ip_1,1}^2\right)\times\pi_{n,T}\min(s,t)\alpha_T\mathbb{E}\left(\eta_{jq_1,1}^2\right)\right| \tag{A.6}$$

$$= \sum_{t,s=1}^{T}\sum_{i,j=1}^{n}\pi_{n,T}^2\alpha_T^2\mathcal{O}\left(T^2\right) = \pi_{n,T}^2\alpha_T^2\mathcal{O}\left(n^2T^4\right)$$

Now consider the case where all factor indices are equal (again omitting the index to ease notation):

$$\sum_{t,s=1}^{T}\sum_{i,j=1}^{n}\left|\mathbb{E}\left(\nu_i\sum_{u=1}^{s}\zeta_{i,u}\eta_{i,u}\nu_j\sum_{v=1}^{s}\zeta_{j,v}\eta_{j,v}\nu_i\sum_{k=1}^{t}\zeta_{i,k}\eta_{i,k}\nu_j\sum_{l=1}^{t}\zeta_{j,l}\eta_{j,l}\right)\right| \tag{A.7}$$

Now, due to independence across variables, whenever $i \neq j$ we can use the same argument as above to show that the summand is $\pi_{n,T}^2 \alpha_T^2 \mathcal{O}(T^2)$. However, when $i = j$ the summand becomes

$$\left| \mathbb{E}\left(\nu_i \sum_{u=1}^{s} \zeta_{i,u}\eta_{i,u} \sum_{v=1}^{s} \zeta_{i,v}\eta_{i,v} \sum_{k=1}^{t} \zeta_{i,k}\eta_{i,k} \sum_{l=1}^{t} \zeta_{i,l}\eta_{i,l} \right) \right| \tag{A.8}$$

which is non-zero whenever all four time indices are equal, or they are equal in pairs. There are $\min(s,t)$ cases where they are all equal and the expression becomes $\mathbb{E}(\nu_i \zeta_{i,1}\eta_{i,1}^4) = \pi_{n,T}\alpha_T \mathcal{O}(1)$. When they are equal in pairs we get: $\mathbb{E}(\nu_i)\mathbb{E}(\zeta_{i,1}\eta_{i,1}^2)\mathbb{E}(\zeta_{i,1}\eta_{i,1}^2) = \pi_{n,T}\alpha_T^2 \mathcal{O}(1)$. In total there are $st + 2\min(s,t)^2$ non-zero summands and we get that Eq. (A.8) is $\pi_{n,T}\alpha_T \mathcal{O}(T) + \pi_{n,T}\alpha_T^2 \mathcal{O}(T^2)$ implying that Eq. (A.7) is $\pi_{n,T}\alpha_T \mathcal{O}(nT^3) + \pi_{n,T}\alpha_T^2 \mathcal{O}(nT^4) + \pi_{n,T}^2 \alpha_T^2 \mathcal{O}(n^2 T^4) = Q_3(n,T)$. According to Bates et al. (2013), we need $Q_1(n,T) = \mathcal{O}(n)$ and $C_{nT}^2 Q_3(n,T) = \mathcal{O}(n^2 T^2)$. From Eq. (A.4), we have

$$\begin{aligned} Q_1(n,T) &= \pi_{n,T}\alpha_T \mathcal{O}(nT) \\ &= \mathcal{O}\left(1/\min\left(n^{1/2}T^{1-a}, T^{3/2-a}\right)\right)\mathcal{O}(T^{-a})\mathcal{O}(nT) \\ &= \mathcal{O}(n) \end{aligned} \tag{A.9}$$

We further have

$$C_{nT}^2 Q_3(n,T) = C_{nT}^2 \left[\pi_{n,T}\alpha_T \mathcal{O}(nT^3) + \pi_{n,T}\alpha_T^2 \mathcal{O}(nT^4) + \pi_{n,T}^2 \alpha_T^2 \mathcal{O}(n^2 T^4) \right] \tag{A.10}$$

$$= \min(n,T)\mathcal{O}\left(1/\min\left(n^{1/2}T^{1-a}, T^{3/2-a}\right)\right)\mathcal{O}(T^{-a})\mathcal{O}(nT^3) \tag{A.11}$$

$$+ \min(n,T)\mathcal{O}\left(1/\min\left(n^{1/2}T^{1-a}, T^{3/2-a}\right)\right)\mathcal{O}(T^{-2a})\mathcal{O}(nT^4) \tag{A.12}$$

$$+ \min(n,T)\mathcal{O}\left(1/\min\left(nT^{2-2a}, T^{3-2a}\right)\right)\mathcal{O}(T^{-2a})\mathcal{O}(n^2 T^4) \tag{A.13}$$

Under the assumption that $a \geq 1/2$ and $T^{1-a}/n^{1/2} \to k \geq 0$, all three terms are $\mathcal{O}(n^2 T^2)$ and the results follow.

APPENDIX B. DATA DESCRIPTION

The dataset used is from Stock and Watson (2009) and can be downloaded from Mark Watson's homepage. The full list of variables along with descriptions from Stock and Watson (2009) has been reproduced in Table B1. The majority of the variables are from the Global Insights Basic Economics Database. The remaining variables are either from The Conference Boards Indicators Database (TCB) or calculated by the authors using Global Insights or TCB data (AC). Transforming the variables to be stationary is done according to the transformation codes (TC), see Table B2 for details as well as details on how the h-quarter ahead version of the variable used in the factor-augmented forecasting regressions is constructed. In addition to this the following abbreviations are used: sa, seasonally adjusted; nsa, not seasonally adjusted; saar, seasonally adjusted at an annual rate. The E.F. column whether the variable was used to estimate the factors (= 1).

Table B1. Data Description.

Short Name	Mnemonic	TC	E.F.	Description
RGDP	GDP251	5	0	Real gross domestic product, quantity index (2,000 = 100), saar
Cons	GDP252	5	0	Real personal consumption expenditures, quantity index (2,000 = 100), saar
Cons-Dur	GDP253	5	1	Real personal consumption expenditures – durable goods, quantity index (2,000 =
Cons-NonDur	GDP254	5	1	Real personal consumption expenditures – nondurable goods, quantity index (200
Cons-Serv	GDP255	5	1	Real personal consumption expenditures – services, quantity index (2,000 = 100),
GPDInv	GDP256	5	0	Real gross private domestic investment, quantity index (2,000 = 100), saar
FixedInv	GDP257	5	0	Real gross private domestic investment – fixed investment, quantity index (200
NonResInv	GDP258	5	0	Real gross private domestic investment – nonresidential, quantity index (2,000
NonResInv-Struct	GDP259	5	1	Real gross private domestic investment – nonresidential – structures, quantity
NonResInv-Bequip	GDP260	5	1	Real gross private domestic investment – nonresidential – equipment & software
Res.Inv	GDP261	5	1	Real gross private domestic investment – residential, quantity index (2,000 = 100
Exports	GDP263	5	1	Real exports, quantity index (2,000 = 100), saar

Table B1. (*Continued*)

Short Name	Mnemonic	TC	E.F.	Description
Imports	GDP264	5	1	Real imports, quantity index (2,000 = 100), saar
Gov	GDP265	5	0	Real government consumption expenditures & gross investment, quantity index (2
Gov Fed	GDP266	5	1	Real government consumption expenditures & gross investment – federal, quantit
Gov State/Loc	GDP267	5	1	Real government consumption expenditures & gross investment – state & local, q
IP: total	IPS10	5	0	Industrial production index – total index
IP: products	IPS11	5	0	Industrial production index – products, total
IP: final prod	IPS299	5	0	Industrial production index – final products
IP: cons gds	IPS12	5	0	Industrial production index – consumer goods
IP: cons dble	IPS13	5	1	Industrial production index – durable consumer goods
IP: cons nondble	IPS18	5	1	Industrial production index – nondurable consumer goods
IP: bus eqpt	IPS25	5	1	Industrial production index – business equipment
IP: matls	IPS32	5	0	Industrial production index – materials
IP: dble mats	IPS34	5	1	Industrial production index – durable goods materials
IP: nondble mats	IPS38	5	1	Industrial production index – nondurable goods materials
IP: mfg	IPS43	5	1	Industrial production index – manufacturing (sic)
IP: fuels	IPS306	5	1	Industrial production index – fuels
NAPM prodn	PMP	1	1	Napm production index (percent)
Capacity Util	UTL11	1	1	Capacity utilization– manufacturing (sic)
Emp: total	CES002	5	0	Employees, nonfarm – total private
Emp: gds prod	CES003	5	0	Employees, nonfarm – goods-producing
Emp: mining	CES006	5	1	Employees, nonfarm – mining
Emp: const	CES011	5	1	Employees, nonfarm – construction
Emp: mfg	CES015	5	0	Employees, nonfarm – mfg
Emp: dble gds	CES017	5	1	Employees, nonfarm – durable goods
Emp: nondbles	CES033	5	1	Employees, nonfarm – nondurable goods
Emp: services	CES046	5	1	Employees, nonfarm – service-providing
Emp: TTU	CES048	5	1	Employees, nonfarm – trade, transport, utilities
Emp: wholesale	CES049	5	1	Employees, nonfarm – wholesale trade
Emp: retail	CES053	5	1	Employees, nonfarm – retail trade
Emp: FIRE	CES088	5	1	Employees, nonfarm – financial activities
Emp: Govt	CES140	5	1	Employees, nonfarm – government
Help wanted indx	LHEL	2	1	Index of help-wanted advertising in newspapers (1967 = 100;sa)
Help wanted/emp	LHELX	2	1	Employment: ratio; help-wanted ads:no. unemployed clf

Table B1. *(Continued)*

Short Name	Mnemonic	TC	E.F.	Description
Emp CPS total	LHEM	5	0	Civilian labor force: employed, total (thous.,sa)
Emp CPS nonag	LHNAG	5	1	Civilian labor force: employed, nonagric. industries (thous.,sa)
Emp. Hours	LBMNU	5	1	Hours of all persons: nonfarm business sec (1982 = 100,sa)
Avg hrs	CES151	1	1	Avg wkly hours, prod wrkrs, nonfarm – goods-producing
Overtime: mfg	CES155	2	1	Avg wkly overtime hours, prod wrkrs, nonfarm – mfg
U: all	LHUR	2	1	Unemployment rate: all workers, 16 years & over (%,sa)
U: mean duration	LHU680	2	1	Unemploy.by duration: average(mean) duration in weeks (sa)
U < 5 weeks	LHU5	5	1	Unemploy.by duration: persons unempl.less than five weeks (thous.,sa)
U 5–14 weeks	LHU14	5	1	Unemploy.by duration: persons unempl.5 to 14 weeks (thous.,sa)
U 15+ weeks	LHU15	5	1	Unemploy.by duration: persons unempl.15 weeks + (thous.,sa)
U 15–26 weeks	LHU26	5	1	Unemploy.by duration: persons unempl.15 to 26 weeks (thous.,sa)
U 27+ weeks	LHU27	5	1	Unemploy.by duration: persons unempl.27 weeks + (thous,sa)
BuildPermits	HSBR	4	0	Housing authorized: total new priv housing units (thous.,saar)
HStarts: Total	HSFR	4	0	Housing starts:nonfarm(1947–1958);total farm&nonfarm(1959-)(thous.,sa
HStarts: NE	HSNE	4	1	Housing starts:northeast (thous.u.)s.a.
HStarts: MW	HSMW	4	1	Housing starts:midwest(thous.u.)s.a.
HStarts: South	HSSOU	4	1	Housing starts:south (thous.u.)s.a.
HStarts: West	HSWST	4	1	Housing starts:west (thous.u.)s.a.
PMI	PMI	1	1	Purchasing managers' index (sa)
NAPM new ordrs	PMNO	1	1	Napm new orders index (percent)
NAPM vendor del	PMDEL	1	1	Napm vendor deliveries index (percent)
NAPM Invent	PMNV	1	1	Napm inventories index (percent)
Orders (ConsGoods)	MOCMQ	5	1	New orders (net) – consumer goods & materials, 1996 dollars (bci)
Orders (NDCapGoods)	MSONDQ	5	1	New orders, nondefense capital goods, in 1996 dollars (bci)
PGDP	GDP272A	6	0	Gross domestic product price index
PCED	GDP273A	6	0	Personal consumption expenditures price index
CPI-All	CPIAUCSL	6	0	Cpi all items (sa) fred
PCED-Core	PCEPILFE	6	0	Pce price index less food and energy (sa) fred
CPI-Core	CPILFESL	6	0	Cpi less food and energy (sa) fred

Table B1. *(Continued)*

Short Name	Mnemonic	TC	E.F.	Description
PCED-Dur	GDP274A	6	0	Durable goods price index
PCED-motorveh	GDP274_1	6	1	Motor vehicles and parts price index
PCED-hhequip	GDP274_2	6	1	Furniture and household equipment price index
PCED-oth dur	GDP274_3	6	1	Other price index
PCED-nondur	GDP275A	6	0	Nondurable goods price index
PCED-food	GDP275_1	6	1	Food price index
PCED-clothing	GDP275_2	6	1	Clothing and shoes price index
PCED-energy	GDP275_3	6	1	Gasoline, fuel oil, and other energy goods price index
PCED-oth nondur	GDP275_4	6	1	Other price index
PCED-services	GDP276A	6	0	Services price index
PCED-housing	GDP276_1	6	1	Housing price index
PCED-hhops	GDP276_2	6	0	Household operation price index
PCED-elect & gas	GDP276_3	6	1	Electricity and gas price index
PCED-oth hhops	GDP276_4	6	1	Other household operation price index
PCED-transport	GDP276_5	6	1	Transportation price index
PCED-medical	GDP276_6	6	1	Medical care price index
PCED-recreation	GDP276_7	6	1	Recreation price index
PCED-oth serv	GDP276_8	6	1	Other price index
PGPDI	GDP277A	6	0	Gross private domestic investment price index
PFI	GDP278A	6	0	Fixed investment price index
PFI-nonres	GDP279A	6	0	Nonresidential price index
PFI-nonres struc Price Index	GDP280A	6	1	Structures
PFI-nonres equip	GDP281A	6	1	Equipment and software price index
PFI-residential	GDP282A	6	1	Residential price index
PEXP	GDP284A	6	1	Exports price index
PIMP	GDP285A	6	1	Imports price index
PGOV	GDP286A	6	0	Government consumption expenditures and gross investment price index
PGOV-Federal	GDP287A	6	1	Federal price index
PGOV-St & loc	GDP288A	6	1	State and local price index
Com: spot price (real)	PSCCOMR	5	1	Real spot market price index:bls & crb: all commodities(1967 = 100) (psccom/pcepilfe)
OilPrice (Real)	PW561R	5	1	Ppi crude (relative to core pce) (pw561/pcepilfe)
NAPM com price	PMCP	1	1	Napm commodity prices index (percent)
Real AHE: goods	CES275R	5	0	Real avg hrly earnings, prod wrkrs, nonfarm – goods-producing (ces275/pi071)
Real AHE: const	CES277R	5	1	Real avg hrly earnings, prod wrkrs, nonfarm – construction (ces277/pi071)
Real AHE: mfg	CES278 R	5	1	Real avg hrly earnings, prod wrkrs, nonfarm – mfg (ces278/pi071)
Labor Prod	LBOUT	5	1	Output per hour all persons: business sec (1982 = 100,sa)

Table B1. *(Continued)*

Short Name	Mnemonic	TC	E.F.	Description
Real Comp/Hour	LBPUR7	5	1	Real compensation per hour,employees: nonfarm business(82 = 100,sa)
Unit Labor Cost	LBLCPU	5	1	Unit labor cost: nonfarm business sec (1982 = 100,sa)
FedFunds	FYFF	2	1	Interest rate: federal funds (effective) (% per annum,nsa)
3 mo T-bill	FYGM3	2	1	Interest rate: u.s.treasury bills,sec mkt,3-mo. (% per ann,nsa)
6 mo T-bill	FYGM6	2	0	Interest rate: u.s.treasury bills,sec mkt,6-mo. (% per ann,nsa)
1 yr T-bond	FYGT1	2	1	Interest rate: u.s.treasury const maturities,1-yr.(% per ann,nsa)
5 yr T-bond	FYGT5	2	0	Interest rate: u.s.treasury const maturities,5-yr.(% per ann,nsa)
10 yr T-bond	FYGT10	2	1	Interest rate: u.s.treasury const maturities,10-yr.(% per ann,nsa)
Aaabond	FYAAAC	2	0	Bond yield: moody's aaa corporate (% per annum)
Baa bond	FYBAAC	2	0	Bond yield: moody's baa corporate (% per annum)
fygm6-fygm3	SFYGM6	1	1	fygm6-fygm3
fygt1-fygm3	SFYGT1	1	1	fygt1-fygm3
fygt10-fygm3	SFYGT10	1	1	fygt10-fygm3
fyaaac-fygt10	SFYAAAC	1	1	fyaaac-fygt10
fybaac-fygt10	SFYBAAC	1	1	fybaac-fygt10
M1	FM1	6	1	Money stock: m1(curr,trav.cks,dem dep,other ck'able dep)(bil $,sa)
MZM	MZMSL	6	1	Mzm (sa) frb st. louis
M2	FM2	6	1	Money stock:m2(m1 + o'nite rps,euro $ g/p&b/d mmmfs&sav&sm time dep(bil (,sa)
MB	FMFBA	6	1	Monetary base, adj for reserve requirement changes(mil $ sa)
Reserves tot	FMRRA	6	1	Depository inst reserves:total,adj for reserve req chgs(mil $,sa)
Reserves nonbor	FMRNBA	6	1	Depository inst reserves:nonborrowed,adj res req chgs(mil $ sa)
Bus loans	BUSLOANS	6	1	Commercial and industrial loans at all commercial banks (fred) billions (sa)
Cons credit	CCINRV	6	1	Consumer credit outstanding – nonrevolving (g19)
Ex rate: avg	EXRUS	5	1	United states;effective exchange rate(merm) (index no.)
Ex rate: Switz	EXRSW	5	1	Foreign exchange rate: switzerland (swiss franc per u.s.)

Table B1. *(Continued)*

Short Name	Mnemonic	TC	E.F.	Description
Ex rate: Japan	EXRJAN	5	1	Foreign exchange rate: japan (yen per u.s.$)
Ex rate: UK	EXRUK	5	1	Foreign exchange rate: united kingdom (cents per pound)
EX rate: Canada	EXRCAN	5	1	Foreign exchange rate: canada (canadian per u.s.$)
S&P 500	FSPCOM	5	1	S&P's common stock price index: composite (1941–43 = 10)
S&P: indust	FSPIN	5	1	S&P's common stock price index: industrials (1941–43 = 10)
S&P div yield	FSDXP	2	1	S&P's composite common stock: dividend yield (% per annum)
S&P PE ratio	FSPXE	2	1	S&P's composite common stock: price-earnings ratio (%,nsa)
DJIA	FSDJ	5	1	Common stock prices: dow jones industrial average
Consumer expect	HHSNTN	2	1	U. of Mich. index of consumer expectations (bcd-83)

Table B2. Variable Transformations.

TC	Transformation (X_{it})	$h-$*Quarter Ahead Variable*$X_{it+h}^{(h)}$
1	$X_{it} = Y_{it}$	$X_{it+h}^{(h)} = Y_{it+h}$
2	$X_{it} = \Delta Y_{it}$	$X_{it+h}^{(h)} = Y_{it+h} - Y_{it}$
3	$X_{it} = \Delta^2 Y_{it}$	$X_{it+h}^{(h)} = h^{-1}\sum_{j=1}^{h} \Delta Y_{it+h-j} - \Delta Y_{it}$
4	$X_{it} = \log Y_{it}$	$X_{it+h}^{(h)} = \log Y_{it+h}$
5	$X_{it} = \Delta \log Y_{it}$	$X_{it+h}^{(h)} = \log Y_{it+h} - \log Y_{it}$
6	$X_{it} = \Delta^2 \log Y_{it}$	$X_{it+h}^{(h)} = h^{-1}\sum_{j=1}^{h} \Delta \log Y_{it+h-j} - \Delta \log Y_{it}$

APPENDIX C. ADDITIONAL RESULTS

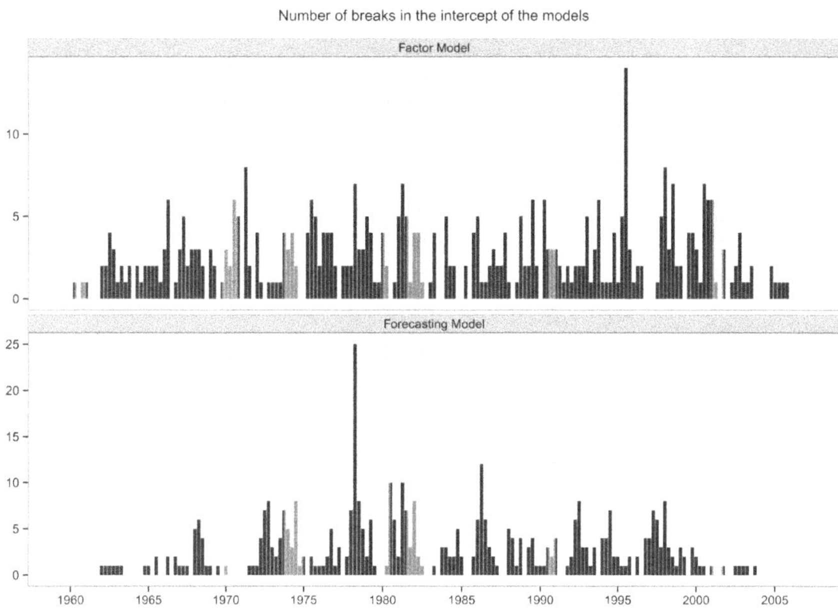

Fig. C1. Breaks in the Intercepts of the Models. The Grey Areas Represent the NBER Recessions.

Table C1. Number of Breaks in the Loadings of the Factor Model and the Parameters of the Forecasting Model Detailed by Variable.

Series	Factor Model			Forecasting Model					
	Coef. on Factors			Coef. on Factors			Coef. on AR Lags		
	All	1970:1 1984:4	1985:1 1999:4	All	1970:1 1984:4	1985:1 1999:4	All	1970:1 1984:4	1985:1 1999:4
Cons-Dur				6	3				
Cons-NonDur									
Cons-Serv				1	1		1	1	
FixedInv				1	1				
NonResInv				4	4		1		1
NonResInv-Struct	2	2		2	2				
NonResInv-Bequip									
Res.Inv	1	1							
Gov Fed				53	17	12	48	15	14
Gov State/Loc				12	5	5	2	1	1
IP: products	1	1		2	2				
IP: final prod	2	2		4	3				
IP: cons gds				4	4				
IP: cons dble				1	1				
IP:cons nondble									
IP:bus eqpt									
IP: matls	42	20	8						
IP: dble mats	11	6							
IP:nondble mats	4	4							
IP: mfg	6	3							
IP: fuels				2	2				
Emp: gds prod	4	3	1	3	3		2	1	1
Emp: mining	51	32	12	48	26	14	15	12	2
Emp: const	4	3							
Emp: mfg	1	1		4	4		1	1	
Emp: dble gds				2	2		1	1	
Emp: nondbles	2	2		15	9	3	5	5	
Emp: services	11	7	1	1	1				
Emp: TTU	2	2		1	1				
Emp: wholesale	2	2							
Emp: retail				1	1				
Emp: FIRE	27	12	11				1		1
Emp: Govt	1	1					2		
Help wanted/emp	18	12							
Emp CPS nonag									
Emp. Hours	1			1	1		1	1	
Avg hrs	9	7							
Overtime: mfg				3	2		2	1	1

Table C1. *(Continued)*

Series	Factor Model			Forecasting Model					
	Coef. on Factors			Coef. on Factors			Coef. on AR Lags		
	All	1970:1 1984:4	1985:1 1999:4	All	1970:1 1984:4	1985:1 1999:4	All	1970:1 1984:4	1985:1 1999:4
U: mean duration				2	2				
U < 5 weeks									
U 5–14 weeks				6	5		2	2	
U 15+ weeks	1	1		3	2				
U 15–26 weeks	1	1		1	1				
U 27+ weeks									
HStarts: NE	4	1	3	4	4				
HStarts: MW	12	8	1	14	6	2	24	24	
HStarts: South	11	10	1						
HStarts: West	38	22	6	31	17	2	22	12	3
Orders (ConsGoods)									
Orders (NDCapGoods)				2	2				
PCED-Dur				20	13	2	3	3	
PCED-motorveh				12	8		2	2	
PCED-hhequip				4	2	2	1	1	
PCED-oth dur				1	1				
PCED-nondur									
PCED-food				5	4		5	5	
PCED-clothing				3	3				
PCED-energy				2	1	1			
PCED-oth nondur				17	12	2	5	5	
PCED-services									
PCED-housing				1	1				
PCED-hhops									
PCED-elect & gas				7	1	4	5		
PCED-oth hhops				1	1				
PCED-transport	47	27	18	20	17	3	7	7	
PCED-medical				10	6	1	7	7	
PCED-recreation									
PCED-oth serv				3	2	1			
PFI-nonres				25	16	6	14	10	3
PFI-nonres struc				21	16	4	5	2	2
PFI-nonres equip	3	2	1	32	17	10	17	12	5
PFI-residential							1	1	
PGOV-FED									
PGOV-SL				3	1	2			
Real AHE: const	3	2	1	6	5		5	5	
Real AHE: mfg	9	5	3	2	2		3	2	1
3 mo T-bill	8	8		6	5	1	6	6	
1 yr T-bond	3	2	1	3	3				

Table C1. (*Continued*)

Series	Factor Model			Forecasting Model					
	Coef. on Factors			Coef. on Factors			Coef. on AR Lags		
	All	1970:1 1984:4	1985:1 1999:4	All	1970:1 1984:4	1985:1 1999:4	All	1970:1 1984:4	1985:1 1999:4
10 yr T-bond	3	3		32	17	9	30	13	16
fygm6-fygm3	24	18	1	14	9	3	7	4	3
fygt1-fygm3	27	19	6	8	7	1	2	1	1
fygt10-fygm3	27	17	8	3	2	1			
fyaaac-fygt10	10	5	3						
fybaac-fygt10	15	7	5				1		
Reserves nonbor	30	19	6	4	3		3	1	2
Ex rate: Switz									
Ex rate: Japan									
Ex rate: UK	4	2	2						
EX rate: Canada	2			4	2	1			
S&P: indust	11	6	3	1		1			
S&P div yield	18	9	7	4	2	2	1	1	
S&P PE ratio	13	2	6	2		2			

Table C2. Root Mean-Squared Errors (RMSE) of the Residuals of the Forecasting Model when Allowing for Breaks or Not, and the Relative RMSE of These Two Approaches.

Series	Std. dev. of $X_{it}^{(4)}$	RMSE, ptv	RMSE, OLS	Relative RMSE	# Breaks
Cons-Dur	0.0627	0.0440	0.0497	0.8856	6
Cons-NonDur	0.0166	0.0148	0.0148	1.0000	0
Cons-Serv	0.0122	0.0085	0.0102	0.8356	2
FixedInv	0.0664	0.0484	0.0506	0.9579	1
NonResInv	0.0686	0.0445	0.0505	0.8821	5
NonResInv-Struct	0.0790	0.0583	0.0634	0.9193	2
NonResInv-Bequip	0.0729	0.0531	0.0531	1.0000	0
Res.Inv	0.1310	0.0978	0.0978	1.0000	0
Gov Fed	0.0445	0.0047	0.0378	0.1250	101
Gov State/Loc	0.0216	0.0124	0.0173	0.7184	14
IP: products	0.0374	0.0242	0.0274	0.8851	2
IP: final prod	0.0368	0.0231	0.0277	0.8325	4
IP: cons gds	0.0318	0.0177	0.0231	0.7666	4
IP: cons dble	0.0756	0.0504	0.0553	0.9120	1
IP:cons nondble	0.0206	0.0162	0.0181	0.8951	0
IP:bus eqpt	0.0720	0.0504	0.0504	1.0000	0
IP: matls	0.0523	0.0379	0.0390	0.9701	0
IP: dble mats	0.0799	0.0573	0.0573	1.0000	0
IP:nondble mats	0.0494	0.0326	0.0387	0.8423	0
IP: mfg	0.0483	0.0342	0.0349	0.9781	0
IP: fuels	0.0449	0.0384	0.0432	0.8899	2
Emp: gds prod	0.0349	0.0210	0.0238	0.8839	5
Emp: mining	0.0706	0.0142	0.0636	0.2227	63
Emp: const	0.0477	0.0338	0.0338	1.0000	0
Emp: mfg	0.0361	0.0224	0.0254	0.8830	5
Emp: dble gds	0.0457	0.0292	0.0314	0.9290	3
Emp: nondbles	0.0239	0.0108	0.0179	0.6060	20
Emp: services	0.0132	0.0070	0.0077	0.9039	1
Emp: TTU	0.0180	0.0103	0.0116	0.8883	1
Emp: wholesale	0.0201	0.0129	0.0136	0.9484	0
Emp: retail	0.0189	0.0102	0.0125	0.8200	1
Emp: FIRE	0.0165	0.0094	0.0102	0.9171	1
Emp: Govt	0.0163	0.0094	0.0104	0.9060	2
Help wanted/emp	0.2971	0.2281	0.2281	1.0000	0
Emp CPS nonag	0.0138	0.0094	0.0100	0.9439	0
Emp. Hours	0.0238	0.0164	0.0174	0.9458	2
Avg hrs	0.5211	0.3246	0.3246	1.0000	0
Overtime: mfg	0.3996	0.2414	0.2873	0.8401	5
U: mean duration	2.0219	0.9627	1.0998	0.8753	2
U < 5 weeks	0.0860	0.0713	0.0713	1.0000	0
U 5–14 weeks	0.1710	0.1125	0.1279	0.8795	8

Table C2. *(Continued)*

Series	Std. dev. of $X_{it}^{(4)}$	RMSE, ptv	RMSE, OLS	Relative RMSE	# Breaks
U 15+ weeks	0.3188	0.1859	0.1968	0.9446	3
U 15–26 weeks	0.2806	0.1892	0.1920	0.9856	1
U 27+ weeks	0.3888	0.2247	0.2247	1.0000	0
HStarts: NE	0.3044	0.1628	0.1843	0.8833	4
HStarts: MW	0.2541	0.0909	0.1915	0.4743	38
HStarts: South	0.2433	0.1628	0.1628	1.0000	0
HStarts: West	0.2800	0.0626	0.1959	0.3198	53
Orders (ConsGoods)	0.0662	0.0467	0.0501	0.9310	0
Orders (NDCapGoods)	0.1274	0.0962	0.0992	0.9699	2
PCED-Dur	0.0050	0.0025	0.0041	0.6201	23
PCED-motorveh	0.0085	0.0041	0.0061	0.6662	14
PCED-hhequip	0.0046	0.0034	0.0038	0.9021	5
PCED-oth dur	0.0062	0.0049	0.0051	0.9779	1
PCED-nondur	0.0069	0.0049	0.0049	1.0000	0
PCED-food	0.0063	0.0038	0.0045	0.8401	10
PCED-clothing	0.0065	0.0038	0.0041	0.9425	3
PCED-energy	0.0540	0.0347	0.0363	0.9564	2
PCED-oth nondur	0.0052	0.0025	0.0042	0.6086	22
PCED-services	0.0026	0.0021	0.0021	1.0000	0
PCED-housing	0.0026	0.0020	0.0021	0.9512	1
PCED-hhops	0.0074	0.0054	0.0054	1.0000	0
PCED-elect & gas	0.0158	0.0083	0.0114	0.7343	12
PCED-oth hhops	0.0061	0.0045	0.0046	0.9836	1
PCED-transport	0.0202	0.0050	0.0084	0.5906	27
PCED-medical	0.0037	0.0023	0.0032	0.7357	17
PCED-recreation	0.0036	0.0025	0.0025	1.0000	0
PCED-oth serv	0.0066	0.0047	0.0051	0.9135	3
PFI-nonres	0.0052	0.0019	0.0042	0.4432	39
PFI-nonres struc	0.0076	0.0035	0.0062	0.5756	26
PFI-nonres equip	0.0056	0.0019	0.0045	0.4241	49
PFI-residential	0.0087	0.0045	0.0046	0.9858	1
PGOV-FED	0.0087	0.0042	0.0042	1.0000	0
PGOV-SL	0.0044	0.0032	0.0035	0.9251	3
Real AHE: const	0.0224	0.0115	0.0155	0.7441	11
Real AHE: mfg	0.0135	0.0092	0.0115	0.7970	5
3 mo T-bill	1.7145	1.1599	1.4675	0.7904	12
1 yr T-bond	1.7022	1.3617	1.4975	0.9093	3
10 yr T-bond	1.1865	0.3193	1.1098	0.2878	62
fygm6-fygm3	0.1847	0.1177	0.1770	0.6650	21
fygt1-fygm3	0.4063	0.2564	0.3702	0.6925	10
fygt10-fygm3	1.2283	0.8486	0.9174	0.9250	3
fyaaac-fygt10	0.4817	0.3101	0.3255	0.9527	0

Table C2. *(Continued)*

Series	Std. dev. of $X_{it}^{(4)}$	RMSE, ptv	RMSE, OLS	Relative RMSE	# Breaks
fybaac-fygt10	0.6662	0.4236	0.4471	0.9476	1
Reserves nonbor	0.0397	0.0178	0.0220	0.8075	7
Ex rate: Switz	0.1122	0.1078	0.1078	1.0000	0
Ex rate: Japan	0.1071	0.0951	0.0992	0.9588	0
Ex rate: UK	0.0954	0.0905	0.0905	1.0000	0
EX rate: Canada	0.0466	0.0367	0.0407	0.9014	4
S&P: indust	0.1491	0.1374	0.1405	0.9779	1
S&P div yield	0.5481	0.4234	0.4616	0.9173	5
S&P PE ratio	3.9634	3.1941	3.2898	0.9709	2

DATING BUSINESS CYCLE TURNING POINTS FOR THE FRENCH ECONOMY: AN MS-DFM APPROACH

Catherine Doz[a] and Anna Petronevich[a,b]

[a]*Paris School of Economics, Université Paris 1 Panthéon-Sorbonne, Paris, France*
[b]*Università Ca'Foscari Venezia, Venice, Italy*

ABSTRACT

Several official institutions (NBER, OECD, CEPR, and others) provide business cycle chronologies with lags ranging from three months to several years. In this paper, we propose a Markov-switching dynamic factor model that allows for a more timely estimation of turning points. We apply one-step and two-step estimation approaches to French data and compare their performance. One-step maximum likelihood estimation is confined to relatively small data sets, whereas two-step approach that uses principal components can accommodate much bigger information sets. We find that both methods give qualitatively similar results and agree with the OECD dating of recessions on a sample of monthly data covering the period 1993–2014. The two-step method is more precise in

Dynamic Factor Models
Advances in Econometrics, Volume 35, 481–538

ISSN: 0731-9053/doi:10.1108/S0731-905320150000035012

determining the beginnings and ends of recessions as given by the OECD. Both methods indicate additional downturns in the French economy that were too short to enter the OECD chronology.

Keywords: Dynamic factor models; Markov-switching models; business cycle turning points

JEL classifications: C32; C34; C55; E32

1. INTRODUCTION

The knowledge of the current state of the economic cycle is essential for policymakers. However, it is not easy to determine. The first problem is that a certain time is to pass before the official institutions announce the state of today. NBER and CEPR produce the reference economic cycle dating for the United States and Europe, respectively, on a basis of a consensus of expert opinions with a lag of several months or years. The OECD dating for Europe also appears with a lag of up to three months as it is based on the quarterly GDP series. Other institutions, such as ECRI,[1] provide dating with at least one year lag. Besides the timing, the second complicated issue is the identification of the list of series which can serve as indicators of the economic cycle. Finally, it is not obvious which method should be used to determine turning points. Several procedures exist, and the results are likely to differ. In this paper, we attempt to tackle these three problems in case of the French economic cycles on the basis of the Markov-switching dynamic factor model.

The dynamic factor model with Markov switching (MS-DFM) was first suggested by Diebold and Rudebusch (1996).[2] This paper relies on the seminal paper by Hamilton (1989) which applies a univariate Markov-switching model to business cycle analysis. It was then formalized for the multivariate case by Kim (1994) and by Kim and Yoo (1995) and used afterwards by Chauvet (1998), Kim and Nelson (1998), and Kaufmann (2000). The model allows to consider two features of an economic cycle as described by Burns and Mitchell (1946), namely, the comovement of individual economic series and the division of an economic cycle into two distinct regimes, recession and expansion. Thus, the common factor of the economic series contains the information on the dynamics of the economic activity, while the two-regime pattern is captured by allowing the parameters of the factor dynamics to follow a Markov-chain process. While the original model assumes switches in mean, other types of non-linearity were proposed by Kholodilin (2002a, 2002b), Dolega (2007), Bessec and Bouabdallah (2015) where the slope of factors or exogenous variables is state dependent; or by Chauvet (1998, 1999), Kholodilin (2002a, 2002b), Kholodilin and Yao (2004), Anas et al. (2007) where the variance of the idiosyncratic component is state dependent; and lastly by Chauvet and Potter (1998) and Carvalho and Lopes (2007) where the authors allow for structural breaks in factor loadings.

The MS-DFM model can be estimated either in one or two steps. The one-step method implies estimation of the parameters of the model and the factor simultaneously, under specific assumptions on the dynamics of the factor. The two-step method consists of (1) extraction of a composite indicator reflecting the economic activity (the factor); (2) estimation of the parameters of the univariate Markov-switching model on the factor series. As usual, each method has its advantages and disadvantages. The one-step approach is given more favor in the literature since, within this method, the extracted factor is designed so that it has Markov-switching dynamics. On the other hand, the one-step approach is subject to convergence problems and is more time-consuming, since the number of parameters to estimate is much larger than in the case of the two-step procedure and increases with the number of series in the database. Thus, it is necessary to choose a set of variables that would reflect the oscillations of the economic activity correctly. The two-step procedure is much easier to implement, it is flexible in the model specification and does not put any restrictions on the number of series by default. This is why it has been used in a number of papers, for example, by Chauvet and Senyuz (2008), Darné and Ferrara (2011), Bessec and Bouabdallah (2015), and others. However, Camacho, Perez-Quiros, and Poncela (2012) argued that this method may face misspecification issues, as the factor extracted in the first step is not supposed to have non-linear dynamics. More precisely, the authors argued that, when estimated with a linear DFM, the factor may give too much weight to the past values of the underlying series, thus being too slow to reflect the most recent changes. In this paper, we analyze and compare the results of these two estimation methods which are used to identify the turning points of the growth rate cycle of the French economy. We estimate the MS-DFM for the period May 1993–March 2014 via the two-step method on a large database containing 151 series and via the one-step method on four series, as suggested by the original paper of Kim and Yoo (1995). We show evidence that, when the factor is estimated by PCA on the first step, and when the number of series is sufficiently large, the two-step estimation method can, in fact, provide satisfactory results.

We determine the key economic indicators that are able to give early and accurate signals on the current state of the growth rate cycle for the one-step method. We then compare the results obtained via the one-step method to the two-step results. This analysis is a contribution to the existing literature on the comparison of the two methods, notably the paper by Camacho et al. (2012), who argued that the one-step method is preferable to the two-step one, although its marginal gains diminish as the quality of the indicators increases and as more indicators are used to identify the nonlinear signal. Their result was illustrated on four series of the Stock–Watson coincident index for the United States, while we perform the comparison on an extensive dataset of 151 French series. Secondly, we decrease the degree of subjectivity regarding the choice of variables for the one-step method by testing all possible combinations of 25 main economic indicators. This is a contribution to existing works on the alternative economic cycle chronologies for France estimated on a small dataset by Kaufmann (2000), Gregoir and Lenglart (2000), Kholodilin (2006), Chen (2007), Chauvet and Yu (2006), Dueker and Sola (2008), and Darné and Ferrara (2011). Finally, we conclude that although both methods provide valid results and outperform the reference dating in timing of the announcement of a current state of the business cycle, the two-step method has the advantage of being easy to implement and able to detect quickly the temporary deterioration in an economy.

The structure of this paper is as follows: in Section 2, we describe the baseline Markov Switching Dynamic Factor model and its two estimation methods. In Section 3, we discuss the dataset and the measures of quality that we use to compare the approaches. Section 4 is devoted to the description of one-step and two-step estimation results and to their comparison. Section 5 concludes.

2. THE MODEL AND THE ESTIMATION METHODS

2.1. The Model

The general framework for Markov-switching factor models has been first proposed by Kim (1994) and was then used by Kim and Yoo (1995) to study the US business cycle. In the present paper, we take the same kind of specification as in Kim and Yoo (1995), and we assume that the growth rate cycle of the economic activity has only two regimes (or states), associated with its low and high levels. The economic activity itself is represented by an unobservable factor, which summarizes the common dynamics of several observable variables. It is assumed that the switch between regimes happens instantaneously, without any transition period (as is considered, e.g., by STAR family models). This assumption can be motivated by the fact that the transition period before deep crises is normally short enough to be omitted. For example, the growth rate of French GDP fell from 0.5% in the first quarter of 2008 to −0.51% in the second quarter of the same year, and further down to −1.59% in the first quarter of 2009.[3]

The model is thus decomposed into two equations, the first one defining the factor model and the second one describing the Markov-switching autoregressive model which is assumed for the common factor. More precisely, in the first equation, each series of the information set is decomposed into the sum of a common component (the common factor loads on each of the observable series with a specific weight) and an idiosyncratic component:

$$y_t = \gamma f_t + z_t \tag{1}$$

where y_t is an $N \times 1$ vector of economic indicators, f_t is a univariate common factor, z_t is an $N \times 1$ vector of idiosyncratic components, which is uncorrelated with f_t at all leads and lags, γ is an $N \times 1$ vector. In this equation, all series are supposed to be stationary, so that some of the components of y_t may be the first differences of an initial non-stationary economic indicator.

The second equation describes the behavior of the factor f_t, which is supposed to follow an autoregressive Markov-switching process with constant transition probabilities.[4] We consider, in most of the paper, that the change in regime affects only the level of the constant with the high level corresponding to the expansion state and the low level to the recession state. Following Kim and Yoo (1995), we also suppose that the lag polynomial $\phi(L)$ is of order 2 so that:

$$f_t = \beta_{S_t} + \phi_1 f_{t-1} + \phi_2 f_{t-2} + \eta_t \tag{2}$$

where $\eta_t \sim \text{i.i.d.}\mathcal{N}(0, 1)$, and ϕ_1 and ϕ_2 are the autoregressive coefficients.

The switching mean is defined as

$$\beta_{S_t} = \beta_0(1 - S_t) + \beta_1 S_t \tag{3}$$

where S_t follows an ergodic Markov chain, that is,

$$P(S_t = j|S_{t-1} = i, S_{t-2} = k, \ldots) = P(S_t = j|S_{t-1} = i) = p_{ij}$$

As it is assumed that there are two states only, S_t switches states according to the transition probabilities matrix $\begin{bmatrix} p_0 & 1-p_0 \\ 1-p_1 & p_1 \end{bmatrix}$, where

$$P(S_t = 0|S_{t-1} = 0) = p_0$$

$$P(S_t = 1|S_{t-1} = 1) = p_1$$

There is no restriction on the duration of each state, and the states are defined pointwise, that is, a recession period may last one month only.

Following Kim and Yoo (1995), we also assume that the idiosyncratic components z_{it} are mutually uncorrelated at all leads and lags, that each of them follows an autoregressive process with a lag polynomial $\psi_i(L)$, and that the degree of this polynomial is 2. Thus,

$$z_t = \psi_1 z_{t-1} + \psi_2 z_{t-2} + \varepsilon_t \tag{4}$$

where ψ_1 and ψ_2 are diagonal matrices of coefficients, $\varepsilon_t \sim \mathcal{N}(0, \Sigma)$, and Σ is a diagonal matrix.

The model can be cast into state-space form:

$$y_t = B\alpha_t \tag{5}$$

$$\alpha_t = T\alpha_{t-1} + \mu_{S_t} + Rw_t \tag{6}$$

where α_t is the state variable,

$$\alpha_t = \left(f_t, f_{t-1}, z_t', z_{t-1}'\right)', \quad \text{with } z_t = (z_{1t}, \ldots, z_{Nt})'$$

$$w_t = \left(\eta_t, \varepsilon_t'\right)', \quad \text{with } \varepsilon_t = (\varepsilon_{1t}, \ldots, \varepsilon_{Nt})'$$

$$E(w_t w_t') = Q = \text{diag}\{1, \sigma_1^2, \ldots, \sigma_N^2\}$$

$$\mu_{S_t} = \left(\beta_{S_t}, 0'_{(2N+1)\times 1}\right)'$$

and B, T, and R are corresponding coefficient matrices.

More explicitly, the state-space representation takes the form:

$$y_t = \begin{pmatrix} \gamma & 0 & I_N & 0 \end{pmatrix} \begin{pmatrix} f_t \\ f_{t-1} \\ z_t \\ z_{t-1} \end{pmatrix} \tag{7}$$

$$\begin{pmatrix} f_t \\ f_{t-1} \\ z_t \\ z_{t-1} \end{pmatrix} = \begin{pmatrix} \phi_1 & \phi_2 & 0 & 0 \\ 1 & 0 & 0 & 0 \\ 0 & 0 & \psi_1 & \psi_2 \\ 0 & 0 & I_N & 0 \end{pmatrix} \begin{pmatrix} f_{t-1} \\ f_{t-2} \\ z_{t-1} \\ z_{t-2} \end{pmatrix} + \begin{pmatrix} \beta_{S_t} \\ 0 \\ 0 \\ 0 \end{pmatrix} + \begin{pmatrix} 1 & 0 \\ 0 & 0 \\ 0 & I_N \\ 0 & 0 \end{pmatrix} \begin{pmatrix} \eta_t \\ \varepsilon_t \end{pmatrix} \tag{8}$$

2.2. One-Step Estimation Method

In this section, we recall the estimation method which has been introduced by Kim (1994) and Kim and Yoo (1995): it is a one-step method, but it can be employed only for a small set of observable series. Using the state-space representation of the model, which is given by Eqs. (7) and (8) the Kalman filter can be written conditionally on the realizations of the state variable at time t and $t-1$. If $X_{t|t-1}^{(j,i)}$ denotes the predicted value of the variable X_t conditional on the information available up to $t-1$ and on the realizations $S_t = j$ and $S_{t-1} = i$, the Kalman filter formulas are the following:

Prediction step:

$$\alpha_{t|t-1}^{(j,i)} = T\alpha_{t-1|t-1}^{(i)} + \mu_{S_t}^{(j)} \tag{9}$$

$$P_{t|t-1}^{(j,i)} = TP_{t-1|t-1}^{(i)}T' + RQR' \tag{10}$$

Error step:

$$v_{t|t-1}^{(j,i)} = y_t - B\alpha_{t|t-1}^{(j,i)} \tag{11}$$

$$\text{Var}\left(v_{t|t-1}^{(j,i)}\right) = H_{t|t-1}^{(j,i)} = BP_{t|t-1}^{(j,i)}B' \tag{12}$$

Updating step:

$$\alpha_{t|t}^{(j,i)} = \alpha_{t|t-1}^{(j,i)} + K_t^{(j,i)}v_{t|t-1}^{(j,i)} \tag{13}$$

$$P_{t|t}^{(j,i)} = \left(I_{(2N+2)} - K_t^{(j,i)}B\right)P_{t|t-1}^{(j,i)} \tag{14}$$

The Kalman gain $K_t^{(j,i)}$ is given by

$$K_t^{(j,i)} = P_{t|t-1}^{(j,i)}B'\left(H_{t|t-1}^{(j,i)}\right)^{-1} \tag{15}$$

As mentioned in Kim (1994) or Kim and Yoo (1995), it is possible to introduce some approximations in order to make the Kalman filter implementable in practice. Instead of producing four sets of values $\alpha_{t|t}^{(j,i)}$ and $P_{t|t}^{(j,i)}$ at each step t, according to the four possible values of (i,j), the idea is to approximate $\alpha_{t|t}$ and $P_{t|t}$ by taking weighted averages over states at $t-1$, which allows to collapse these four sets of values into two. Thus, the following approximations are used:[5]

$$\alpha_{t|t}^{j} = \frac{\sum_{i=0}^{1} P(S=i, S_t=j|I_t,\theta)\alpha_{t|t}^{(j,i)}}{P(S_t=j|I_t,\theta)} \tag{16}$$

$$P_{t|t}^{j} = \frac{\sum_{i=0}^{1} P(S_{t-1}=i, S_t=j|I_t,\theta)\left(P_{t|t}^{(j,i)} + \left(\alpha_{t|t}^{j} - \alpha_{t|t}^{(j,i)}\right)\left(\alpha_{t|t}^{j} - \alpha_{t|t}^{(j,i)}\right)'\right)}{P(S=j|I_t,\theta)} \tag{17}$$

The filtered probability of being in state $j \in \{0; 1\}$ in period t conditional on the information available up to t can then be computed using Hamilton's filter (see Hamilton, 1989) and the previous Kalman filter formulas.

More precisely, if $\theta = (\phi_1, \phi_2, \text{diag}(\psi_1), \text{diag}(\psi_2), \gamma, \sigma_1^2, \ldots, \sigma_N^2, \beta_0, \beta_1, p_0, p_1)'$ is the vector of unknown parameters, if $f(\cdot)$ is the Gaussian density function, and if I_t is the information set available at t, it is possible to compute the filtered probability $P(S_t = j|I_t, \theta)$ through the following equations (based on Bayes' theorem):

$$P(S_t = j|I_t) = \sum_{i=0}^{1} P(S = j, S_{t-1} = i|I_t, \theta) \tag{18}$$

where

$$\begin{aligned} P(S_t = j, S_{t-1} = i|I_t, \theta) &= \frac{f(y_t, S_t = j, S_{t-1} = i|I_{t-1}, \theta)}{f(y_t|I_{t-1}, \theta)} \\ &= \frac{f(y_t|S_t = j, S_{t-1} = i, I_{t-1}, \theta) \times P(S_t = j, S_{t-1} = i|I_{t-1}, \theta)}{f(y_t|I_{t-1}, \theta)} \end{aligned} \tag{19}$$

$$\begin{aligned} f(y_t|S_t = j, S_{t-1} = i, I_{t-1}, \theta) = (2\pi)^{-N/2} |H_{t|t-1}^{(j,i)}|^{-1/2} \\ \times \exp\left\{ -\frac{1}{2}\left(y_t - B\alpha_{t|t-1}^{(j,i)}\right)' \left(H_{t|t-1}^{(j,i)}\right)^{-1} \left(y_t - B\alpha_{t|t-1}^{(j,i)}\right) \right\} \end{aligned} \tag{20}$$

$$P(S = j, S_{t-1} = i|I_{t-1}, \theta) = P(S_t = j|S_{t-1} = i, \theta) \times P(S_{t-1} = i|I_{t-1}, \theta) \tag{21}$$

$$f(y_t|I_{t-1}, \theta) = \sum_{j=0}^{1} \sum_{i=0}^{1} f(y_t, S_t = j, S_{t-1} = i|I_{t-1}, \theta) \tag{22}$$

When $P(S_{t-1} = i|I_{t-1}, \theta)$ is given, every term in Eq. (19) is known, due to the Markovian assumption on S_t. Thus, for any given value of θ, the associated filtered probability $P(S_t = j|I_t)$ can be computed recursively through Eqs. (18)–(22).

The recursion is initialized with the steady state probability of being in state $j \in \{0; 1\}$ at time $t = 0$:

$$P(S_0 = 1|I_0, \theta) = \frac{1 - p_0}{2 - p_0 - p_1} \tag{23}$$

$$P(S_0 = 0|I_0, \theta) = 1 - P(S_0 = 1|I_0, \theta) \tag{24}$$

The previous formulas are also used to compute the log-likelihood function for the whole sample for any given value of θ, since the log-likelihood function for the sample can be written as:

$$\mathcal{L}(y, \theta) = \ln f(y_T, y_{T-1}, \ldots, y_0|I_T, \theta) = \sum_{t=1}^{T} \ln f(y_t|I_{t-1}, \theta) \tag{25}$$

and $f(y_t|I_{t-1}, \theta)$ can be computed using formulas (18)–(22).

The likelihood function can thus be maximized through a numerical optimization algorithm.[6] Then, if $\hat{\theta}$ is the maximum likelihood estimator of θ, the Kalman filter formulas and Hamilton's filter can be used to compute the associated estimated factor and the associated filtered probabilities. In practice, the use of a numerical search algorithms appears to be relatively costly in terms of time and imposes limitations on the number of series included into the model. For instance, the use of four classic series (industrial production index, employment, retail sales, and real income of households) already implies estimation of 22 parameters. Every additional series brings at least four more coefficients to estimate, which extends the estimation time and increases the complexity of the optimal point search. For this reason, we'll mainly apply this method using four series, as it was done by Kim and Yoo (1995) and as it is often done in case of one-step estimation. We also assume that the transition probabilities are time-independent, and in most of the paper we assume that the switch happens in the constant only, as described in Eq. (2).

Within this method, it is thus assumed that the growth rate cycle of the economic activity is described as a common component of just a few series, so the choice of variables is essential and will be discussed in Section 3.

2.3. *Two-Step Estimation Method*

As we just mentioned, the main drawback of the one-step method is that, due to computational constraints, it can only be used with a small set of data. Another possible approach is to proceed in two steps in the following way:

1. The factor f_t is extracted from a large database of economic indicators according to Eq. (1) without taking its Markov-switching dynamics into account. In this paper, we use principal component analysis and we consider that the first principal component $\hat{f}_t$ gives a good approximation of the factor.
2. The parameters of the autoregressive Markov-switching model described by Eqs. (2) and (3) are estimated by maximum likelihood, with f_t replaced by $\hat{f}_t$. This amounts to fit the univariate model of Hamilton (1989) to the estimated factor $\hat{f}_t$, which is taken as if it were an observed variable. The filtered probability of recession $P(S_t = 1|I_t)$ is then calculated as in Eq. (18).

Let us recall that if a Markov-switching model is estimated with Hamilton's method for an observable variable, say z_t, then the log-likelihood

$$\ln f(z_T, z_{T-1}, \ldots, z_0|I_T, \theta) = \sum_{t=1}^{T} \ln f(z_t|I_{t-1}, \theta) \tag{26}$$

is computed along the same lines as in Eqs. (18)–(22) and has to be maximized through a numerical optimization algorithm, too. However, as the number of parameters that have to be estimated is small in this case, the maximization of the loglikelihood through a numerical procedure does not raise any specific problem. This is one of the reasons why the two-step procedure is attractive: in the second step, the number of parameters that have to be estimated is small.

Another attractive feature of the two-step procedure is that it allows to consider a large amount of series, which are used to build the estimated factors in the first step. Here, we take the first principal component of 151 economic indicators concerning the production sector, financial sector, employment, households, banking system, international trade, monetary indicators, major world economic indicators, business surveys, and others. This large set of series is more likely to reflect the business cycle than a small set of series, as it is used in the one-step procedure (as we said before, many authors use only four series when they want to estimate this kind of model).

Finally, as the second step of the two-step procedure is easily tractable, it is possible to introduce additional switching parameters, and to estimate richer models this way. For instance, it is possible to consider a switching variance and to replace σ_η^2 with $\sigma_{\eta_{S_t}}^2$.

The two-step procedure has been employed in several papers (see the non-exhaustive list given in Section 1) but in most of them, the number of series under study is small or moderate. Further, Camacho et al. (2012) argue that this two-step procedure faces misspecification problems, since the Markov-switching dynamics are not taken into account in the first step. We expect that, under standard assumptions, the two-step procedure gives in fact consistent estimators of the parameters. The complete proof of this consistency is addressed in a companion paper (which is still in progress at this time), but the main idea is that, under these standard assumptions, the first principal component consistently estimates the factor. Indeed, as (S_t) is supposed to be a stationary ergodic Markov chain, (f_t) is a stationary process and all the usual sets of assumptions which are commonly used to assert the consistency of PCA for large N and large T (see Bai, 2003; Stock & Watson, 2002, for instance) can be employed in the present setting.

To conclude this section, let us also mention that PCA is not the only way to get a consistent estimator of the factor in the first step. In future work, we intend to extract the factor in the first step either with the two-step estimator based on Kalman filtering, which has been proposed by Doz, Giannone, and Reichlin (2011), or with the QML estimator (Doz, Giannone, & Reichlin, 2012). It seems indeed promising to use these two methods in the first step of the present framework, as they may provide more efficient estimators and as they allow for mixed frequency, missing data, and data with ragged ends.

3. DATA, REFERENCE DATING AND QUALITY INDICATORS

3.1. The Dataset

For the purpose of comparison, one would like to run the two estimation methods on the same dataset. However, due to the different requirements on the number of series in the database for each method (large dataset for the two-step method in order to get the consistency of the PCA factor estimate, small dataset for the one-step method to obtain the convergence of the algorithm), we are unable to perform this kind of analysis. We therefore use a separate dataset for each method.

The database for the two-step procedure is constructed following Stock and Watson (2010) for the United States and Bessec and Doz (2012) for France. It contains 151 monthly series spanning the period May 1993–March 2014.[7] The data cover information on the production sector, financial sector, employment, households, banking system, business surveys, international trade, monetary indicators, major world economic indicators, and other indicators.

For the one-step method, it is crucial to select series properly. The series must be an indicator of the economic cycle and should be available in monthly frequency. We choose 25 series out of the 151 series of the database for the two-step method and to use them in combinations of four, overall $C_{25}^{4} = 12,650$ combinations. The strategy of trying all possible combinations of 4 out of 25 may seem too bulky and inelegant; however, we deliberately avoided any data selection technique in order to minimize its possible impact on the output of the one-step results. The selection was made on the basis of the existing literature on the one-step method applied to business cycle analysis. To the four classical indicators for business cycle dating of the US economy (total personal income, total manufacturing and trade sales, number of employees on nonagricultural payrolls, total industrial production index), we added series used in Kholodilin (2006) (French stock market index CAC40, interest rates on the 3 months and 12 months government bonds, imports, and exports), selected series of business surveys proved to be useful by Bessec and Doz (2012), the components of the OECD Composite Leading Indicator, as well as several series characterizing the dynamics of the major trade partners (Germany, the United States, Asia). Since almost all of these series have already been used in the analysis of the French business cycle (and the others are likely to comove with it), we suppose that the common component of each combination can be considered as an estimate of the business cycle.

All series are seasonally adjusted, tested for the presence of unit roots and transformed to stationarity if necessary, then centered and normalized. Detailed lists of series for both methods are given in Tables A1 and A2.

3.2. Reference Dating

In order to measure the quality of the results of each of the two methods, we need to compare it to some reference business cycle chronology. The choice is not obvious, as the true dating is unknown, whereas the estimates of the true dating provide different sets of turning points. To our knowledge, there are at least three open-source dating chronologies for the European countries: OECD,[8] CEPR,[9] and ECRI.[10] Note that the French National Institute of Statistics and Economic Studies, INSEE, does not publish any official business cycle dating. Fig. 1 shows that these chronologies indeed do not coincide in the starting and the final points of recessions and in the duration of economic cycle phases. Moreover, OECD detects a recession of April 1995–January 1997 which other institutions do not identify.

The difference obviously lies in the methodology and the data taken into consideration. The OECD dating is the output of the Bry–Boschan algorithm (see Bry & Boschan, 1971) applied to the Composite Leading

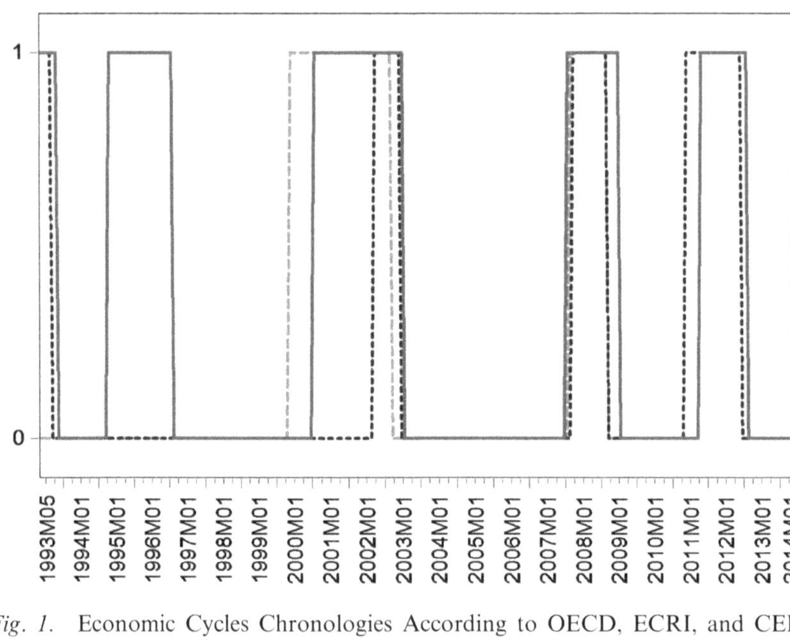

Fig. 1. Economic Cycles Chronologies According to OECD, ECRI, and CEPR. The Recession Phase Corresponds to 1, the Expansion Phase Corresponds to 0. *Note*: The solid line – OECD dating, the dashed line – CEPR dating, the dotted line – ECRI dating.

Indicator (CLI), which is an aggregate of a fixed set of nine series, highly correlated to the reference series (industrial production index or GDP series).[11] The turning point chronologies of CEPR are obtained from the balance of expert opinions on the basis of series selected by the experts involved. The ECRI index is the output of an undisclosed statistical tool on the undisclosed (but probably the most information rich) dataset.

In this paper, we take the OECD dating as a benchmark as it relies on a clear and replicable algorithm. Therefore, the time sample that we consider covers five crises in the French economy as determined by OECD (see Fig. 2):

- March 1992–October 1993: the crisis caused by the oil shock following the first Gulf War, German reunification and tensions in European Monetary system;
- April 1995–January 1997: rather a slowdown in economic growth rates than a real recession, with only one quarter of slightly negative (−0.011) growth rate, caused by the decrease of high public deficit and the consequent strikes throughout the country;
- January 2001–June 2003: the Internet bubble crisis;
- January 2008–June 2009: the Great Recession, the global financial crisis;
- October 2011–January 2013: the sovereign debt crisis.

It can be argued that OECD dating cannot be used as a reference because it represents the chronologies of the growth cycle, whereas we use MS-DFM to identify the growth rate cycle (for most series, in order to achieve stationarity in data, we use differences of logarithms). In our exercise, we avoid cyclical component extraction on purpose as it implies

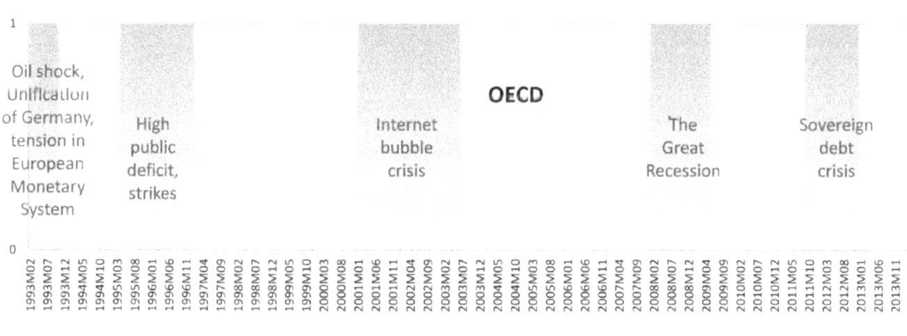

Fig. 2. OECD Reference Turning Points for the French Economy, 1993–2013. *Note*: 1 corresponds to recession, 0 corresponds to expansion.

additional complications inherent to the definition of a trend. However, we support our choice by the fact that the OECD chronology is the closest to the other cyclical indicators calculated for France. In the working paper by Bardaji et al. (2008) (and in a similar paper by Bardaji, Clavel, & Tallet, 2009), the authors propose a reference dating on the basis of the cyclical component of GDP extracted with the Christiano–Fitzgerald filter (see Christiano & Fitzgerald, 2003). We reproduce these estimates on the basis of monthly interpolated GDP growth data. The dating we obtain is indeed very close to OECD results. At the same time, it is rather close to the dating obtained by Anas, Billio, Ferrara, and Do Luca (2007) for Eurostat (see Fig. 3).

Note that the dating on the basis of the Christiano–Fitzgerald filter has two additional recessions (in 1998–1999 and 2004–2005) which are not present in the OECD dating. Interestingly, the Reversal Index[12] (l'Indicateur de Retournement) published by INSEE also detects these

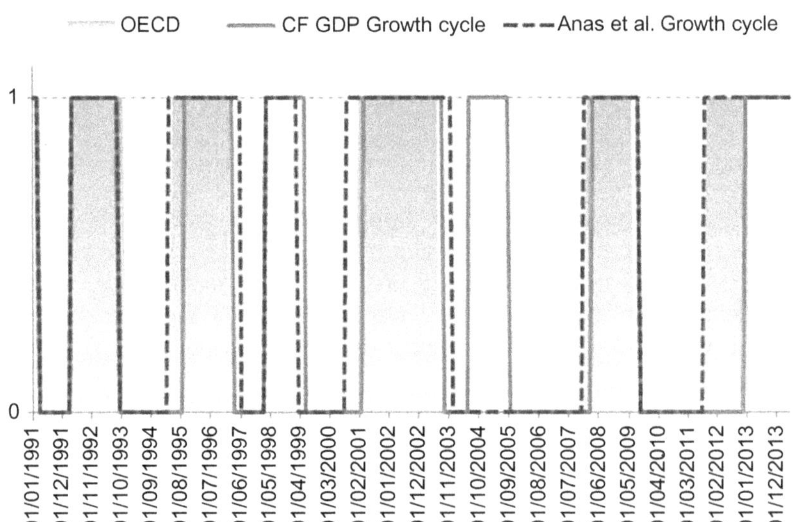

Fig. 3. Turning Point Chronology of the French Economy. *Notes*: Shaded areas correspond to the OECD dating, the dashed line corresponds to the dating for the French economy produced for eurostat by Anas et al. (2007), the solid line corresponds to the GDP growth cycle extracted with Christiano–Fitzgerald filter. 1 to recession, while 0 corresponds to expansion. A Christiano–Fitzgerald filter is applied to the series of French GDP in levels (bandwidth 6–40 quarters), the turning points are considered to take place in the second month of a quarter.

additional recessions, having spikes of high probability of recession in 1998 and 2005 (see Fig. 4). This discrepancy might be due to an important feature of the Bry–Boschan procedure, which is the existence of a lower bound of phase duration (15 months). Consequently, short recessions or expansions do not appear in the OECD chronology.

Indeed, in both cases INSEE detected a temporary deterioration of the economic activity due to different reasons. In 1998–1999 France experienced a significant decline in the net external trade. Undermined by the Asian and Russian crises, the external demand from Japan, China, and Russia, as well as other developing Asian countries and even the United Kingdom, Belgium, and Italy, fell dramatically – from 10% growth rate in 1997 to only 4% in 1998. The depreciation of yen and dollar contributed to the appreciation of the real effective exchange rate of franc. In general, the external balance of France decreased by 7.1%, which resulted into negative contribution to the GDP growth (−0.4 pp).[13] The producers were pessimistic about future activity (also worried about the financial crisis and reducing prices for energy and oil, which threatened to turn into disinflation), decreasing their investment and limiting the inventories.[14] In 2005, the external demand of France decelerated substantially due to uncertainty in the economic situation in the United States and Japan caused by the

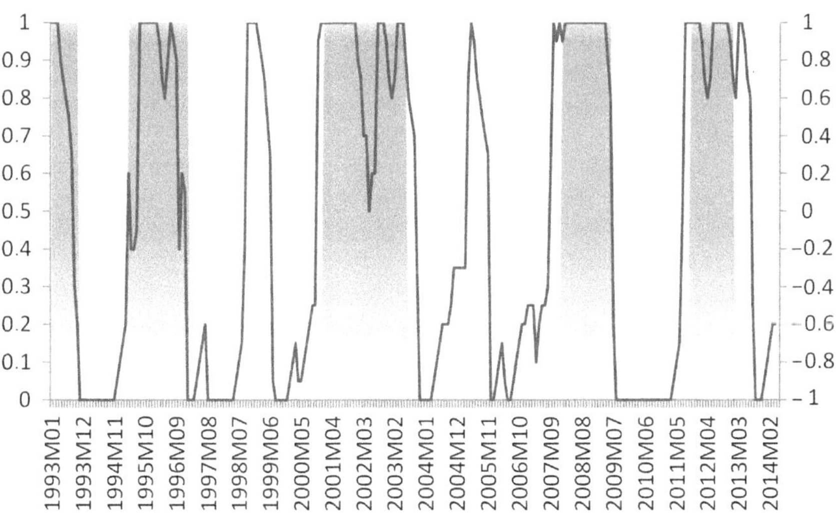

Fig. 4. The Index of Reversal (Solid Line, Right Axis) and OECD Reference Dating (Shaded Areas, Left Axis).

oil price shock. Producers in manufacturing and service acted with caution: the prices for raw materials were rising, and the euro was appreciating in real terms, the saving rate of households fell, the GDP quarterly growth was declining, too.[15,16]

To summarize, the OECD dating largely coincides with the other existing cyclical indicators for the major recessions; however, some other indicators may detect additional shorter recession episodes.

3.3. *Measures of Quality*

To assess the quality of the results of each of the two methods, we use the three following indicators:

- *Quadratic probability score*. This indicator shows the average error of filtered probability as an average quadratic deviation from the reference dating. A high QPS indicates a low quality of the fit of the model.

$$\text{QPS} = \frac{1}{T}\sum_{t=1}^{T}(\text{RD}_t - P(S_t = 1|I_t))^2$$

  where T is the number of periods in the sample, RD_t is the reference dating series of 0 and 1 (1 corresponding to recession, 0 to expansion), and $P(S_t = 1|I_t)$ is the filtered probability of being in a recession in period t.
- *False positives*. This indicator counts the number of wrongly predicted periods. Here, we set the threshold probability on the intuitive level of 0.5.

$$\text{FPS} = \sum_{t=1}^{T}\left(\text{RD}_t - I_{P(S_{t=1}|I_t)>0.5}\right)^2$$

  where $I_{P(S_{t=1}|I_t)>0.5}$ is the indicator function equal to 1 if the estimated filtered probability is higher than 0.5 (determines recession) and 0 otherwise. The lower the FPS, the more qualitatively accurate the model.
- *Correlation*. An accurately estimated filtered probability should have a high correlation with the reference dating. We use a simple sample correlation Corr between the two series.

4. ESTIMATION RESULTS

4.1. One-Step Method

4.1.1. Informative Series

The estimation of 12,650 combinations did not produce 12,650 outputs as for most of the combinations convergence was not achieved, or the series combination produced a factor that does not have a nonlinear structure. Therefore, only 575 combinations achieved convergence, and only 424 of them have interpretable filtered probabilities. Out of this number, we have retained 72 results that are informative in terms of signals of past recessions. Interestingly, the best candidate for the benchmark results – the combination of four series used by Kim and Yoo (1995), Kim (1994), Chauvet (1998), and others for business cycles of the United States (total index of industrial production, employees in nonagricultural payrolls, total personal income less transfer payments, total manufacturing, and trade) did not achieve convergence.

We construct the frequency rating of economic series (given in Table B1) for the integrity of all interpretable results of the one-step estimation. Some series turned out to have weak explanatory content, such as CPI index or CAC-40 financial index, the latter entering none of the successful combinations. Others did much better: the construction confidence indicator, capacity utilization, exports, the retail trade confidence index, and the unemployment rate appear each in 22, 21, 20, and 19 combinations, respectively. This allows us to suggest that the contribution of these indicators is important for the final aggregate factor to follow bi-state dynamics. Interestingly, CPI and the stock market index both enter the OECD CLI, but they do not seem to be very informative for the turning points detection in our framework.

To illustrate the results that we considered as interpretable, we present the output of one of the plausible combinations in Fig. 5. It consists of the four most frequent indicators that we mentioned above: unemployment rate, exports, retail trade, and construction confidence indicator. The resulting filtered probability is one of the best in terms of fit to the OECD official dating.

As we can see from this figure, this combination produces a filtered probability that captures four out of five crises if we consider the economy to be in recession if the filtered probability of a recession is higher than 0.5. One can notice two important features of this example: first, there is an

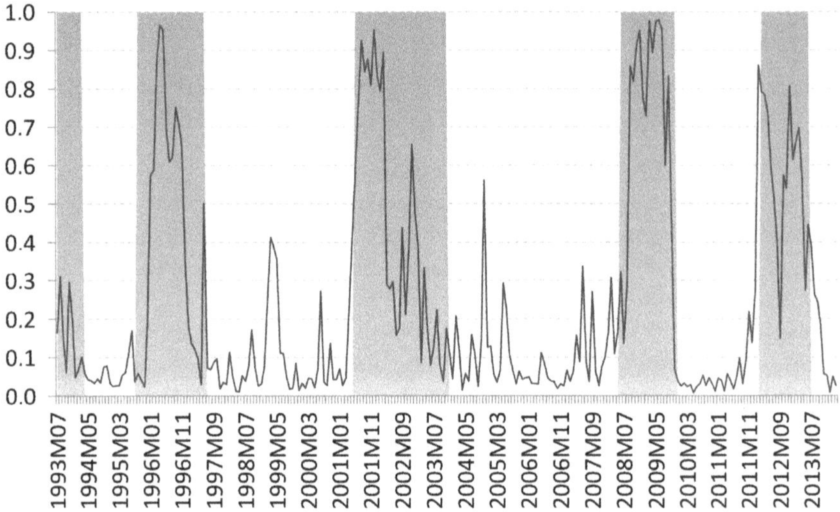

Fig. 5. The Result of One-Step Estimation on Unemployment Rate, Exports, Retail Trade and Construction Confidence Indicator: The Filtered Probability of Recession (Solid Line) and the Reference Dating (Shaded Area, OECD, 1 Corresponds to Recession State).

extra signal in 2004; second, not all the crises are explained equally well. These pitfalls are often present in the other outcomes, so we discuss each of them in detail.

4.1.2. Extra Signals

In general, out of 72 combinations only 27 do not produce any extra signals of recession. The other 45 combinations give an additional alert in the end of 1998, or another one around mid-2005, or both. Among the series with the highest loadings that appear relatively more often in such combinations than in the other ones are manufacturing finished goods stocks level, returns on the FTSE equity index, and the Manufacturing Industrial Confidence Indicator. Indeed, in Fig. 6 we can see that all three series underwent significant downturn in 1998–1999, while in 2005 stocks of manufacturing finished goods and the manufacturing confidence index fell back to the levels of the end of the Internet bubble crisis. However, these events are not captured by the OECD dating.

As we have mentioned above, these signals are not misleading in the sense of producing a false alert of recession when the economy is actually

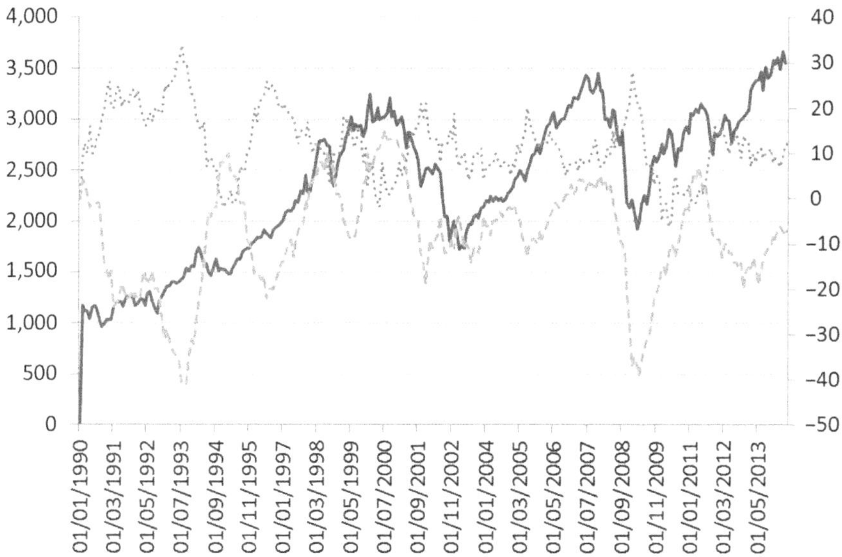

Fig. 6. False Signals Suspects. *Notes*: FTSE 100, All-Share, Index, Price Return, End of Period, GBP (Solid Line, Left Axis), Manufacturing Finished Goods Stocks Level (Dotted Line, Right Axis), Manufacturing Industrial Confidence Indicator (Dashed Line, Right Axis).

growing, and they correspond to a real deterioration of economic conditions. However, for the closest match to the OECD reference, these signals should be avoided. The one-step approach allows to do so, since one can exclude the series that are likely to produce extra signals from the dataset.

4.1.3. Different Set of Series for Different Crises

As for OECD detected recessions, it is important to keep in mind that none of them (at least the five recessions we consider here) had the same origins as any of the other, so it is possible that the determinants of economic activity evolved with time, and so it is likely that the common factor of a particular set of series does not reflect the Great recession as well as it reflected the crisis of 1992–1993. However, in order to construct a universal instrument, it is preferable to find series that would capture the recession in all cases, if possible. For this purpose, we compare the quality indicators of 72 sets of variables for each crisis separately. Table B2 summarizes the information on the best combinations by crisis. Here, FPS shows the proportion of months of each crisis incorrectly determined as expansion, that is, the lower the FPS, the better a crisis is captured.

We can see that:

- the combination consisting of volume of total retail trade, unemployment rate, trade balance, and order books in the building industry, with the highest loadings on unemployment, is the best to detect the first and the last crisis and captures the second crisis well, too;
- the combination consisting of new passenger cars sales and registration, retail trade orders intentions, export, confidence indicator in services, with the highest loadings on retail trade and Confidence Index in services, is leading in case of the second, the third and the fourth crises, being significantly superior to the other combinations for the third and the fourth recessions;
- although good during certain periods of the timeline, unfortunately none of these sets of variables could be used as a "core" set due to their relatively poor performance during the expansion periods and non-detected crises.

The set of data contained in these two combinations appears to be sufficient to identify all five crises with a special role given to unemployment, retail trade orders intentions and confidence index in services.

4.1.4. The Finally Selected Information Set

Considering the observations on the effects of different series on the final filtered probability, we conclude that a good information set (relative to OECD reference) would:

(1) contain the series that determine all five crises,
(2) not contain the series that produce extra signals,
(3) perform well in general in terms of QPS, FPS and Corr.

The top 25 combinations with the lowest QPS and FPS measures and the highest Corr are given in Table B3. The first eight are in the best 10% by all three indicators, so seven of them (we exclude the second combination because of the presence of extra signals) could be candidates for the core sets of economic indicators that enable to match the OECD dating closely. The graphs of corresponding seven filtered probabilities are given in Fig. B1.

It is not surprising that there are several "best" sets of variables, as the restriction of the model to comprise only four series is just a technical limitation, and the factor matching the dynamics of the economic activity is determined by many more series. The analysis of the factor loadings of these seven combinations can give us an idea of the economic indicators that play the most important role. According to our estimations, the heaviest factor loadings belong to (see Table B3):

- France, OECD MEI (Enquete de Conjoncture INSEE), Retail Trade Orders Intentions, SA;
- France, INSEE, Metropolitan, Unemployment, Job Seekers, Men, Total, Categories A, B, & C, Calendar Adjusted, SA;
- France, OECD MEI (Enquete de Conjoncture INSEE), Manufacturing Business Situation Future, SA;
- France, Business Surveys, DG ECFIN, Construction Confidence Indicator, Balance, SA.

The first two of these indicators were also determined as components of the Growth Cycle Coincident indicator by Anas et al. (2007).

Among the other indicators contributing to the factors in the seven selected combinations are:

- France, Consumer Surveys, INSEE, Consumer Confidence Indicator, Synthetic Index, SA,
- France, INSEE, Domestic Trade, Vehicle Sales & Registrations, New, Passenger Cars, Total, Calendar Adjusted, SA,
- France, OECD MEI, INSEE, Total Retail Trade (Volume), SA, Change P/P,
- France, OECD MEI, INSEE, Manufacturing Finished Goods Stocks Level, SA,
- France, INSEE, Foreign Trade, Trade Balance, Calendar Adjusted, SA, EUR,
- France, INSEE, Foreign Trade, Export, Calendar Adjusted, SA, EUR,
- Japan, Economic Sentiment Surveys, ZEW, Financial Market Report, Stock Market, Nikkei 225, Balance,
  United States, Equity Indices, S&P, 500, Index (Shiller), Cyclically Adjusted P/E Ratio (CAPE),
- France, Service Surveys, DG ECFIN, Services Confidence Indicator, Balance, SA.

As an output of this analysis we have thus retained 13 out of 25 series which can be considered as essentially informative of the French business cycle. We tried to use the one-step method on these 13 series simultaneously, in order to take into account all the main information. Unfortunately, the optimization algorithm did not achieve convergence.

Note, however, that when the parameters were set to their initial values (obtained with OLS), the filtered probability did capture the five crises without detecting any extra recessions (see Fig. B2). Therefore, since it seems infeasible to use the information contained in the above listed 13 series simultaneously within the one-step approach, the results of the seven combinations could be used as complements.

4.2. Two-Step Method

4.2.1. First Step: PCA

In the first step of the procedure, we extract the first factor by principal component analysis. The first principal component that we use as a proxy for the factor in the two-step method describes 23.43% of the total variance, which is quite reasonable when considering the size and heterogeneity of the database. The dynamics of the first component and the factor loadings are presented in Figs. 7 and 8. One can note that it is close to the dynamics of GDP growth, so the factor is relevant. Indeed, the correlation on the whole sample is equal to 0.91, while the correlation on the shorter period ending in December 2007 to eliminate the impact of the Great Recession is 0.895.

The three groups with the highest loadings corresponding to the first component belong to: (1) production and consumption series, disaggregated; (2) business surveys; and (3) series concerning the world economy. The first component therefore captures the present behavior of firms and households (including their expectations about the short-term future) and the impact of foreign economies and pays less attention to the banking and financial sector, monetary aggregates, balance of payments and currency indicators.

4.2.2. Second Step: Estimation of a Markov-Switching Model

4.2.2.1. **Basic specification, switch in mean.** In the second step of the two-step estimation procedure, we estimate a Markov-switching model as defined in Eq. (2), with the unobservable factor replaced by the first principal component estimated in the first step.[17] The results are quite satisfactory, with the filtered probabilities capturing all the crises well (but the first one) and without sizable leads and lags (see Fig. 9). As expected, the estimations provide a positive constant for the expansion periods, and a negative one for recessions: $\mu_0 = 1.04$, $\mu_1 = -1.77$, respectively (the estimates are significant at 1% level of significance). For this specification, $QPS = 0.1278$, $FPS/T = 0.1872$, $Corr = 0.75$, and the average lag of the

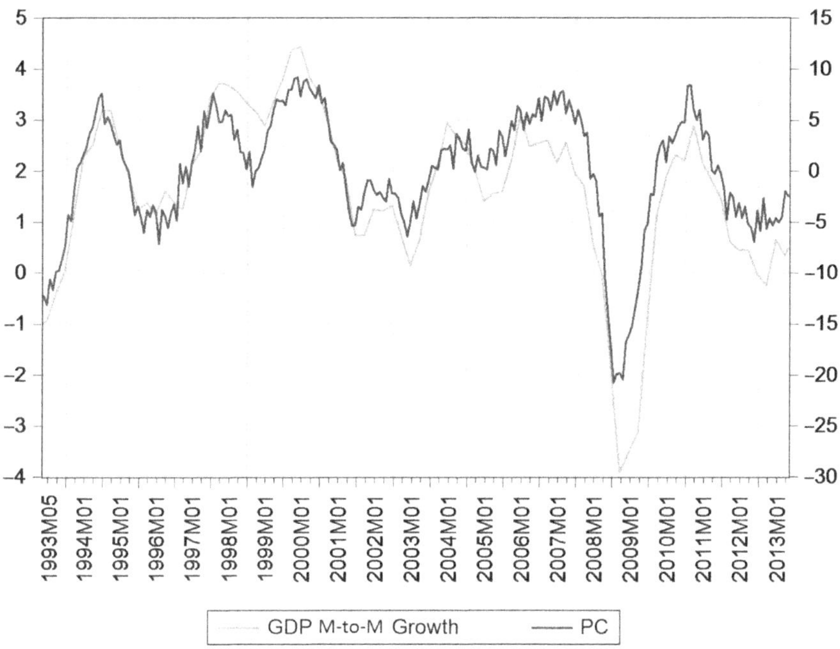

Fig. 7. First Principal Component and Monthly GDP Growth Rate. *Notes*: The black line corresponds to the dynamics of the first principal component of the full dataset (left axis), the gray line corresponds to the French GDP growth series (left axis). The quarterly GDP growth series were converted into monthly series via linear interpolation.

identification of the beginning of a recession is 0.75 months, while the end of a recession is detected one month earlier. Note that this result is comparable to the average result of our one-step estimations (QPS = 0.1346, FPS/T = 0.1779, Corr – 0.69).

The extra signals of 1998–1999 and 2004–2005 are clearly detected by the first component. This may be explained by the fact that the dataset includes the series that induce extra signals for the one-step method, as well as a number of other series that experienced shocks in these periods. We did a simple exercise by trying to eliminate these series from the dataset. It turned out to be impossible to get rid of all extra signals without deteriorating the signals on the OECD recessions. The removal of series undermines the performance of the two-step method and deprives it of its most valuable advantage — the large scale of the dataset.

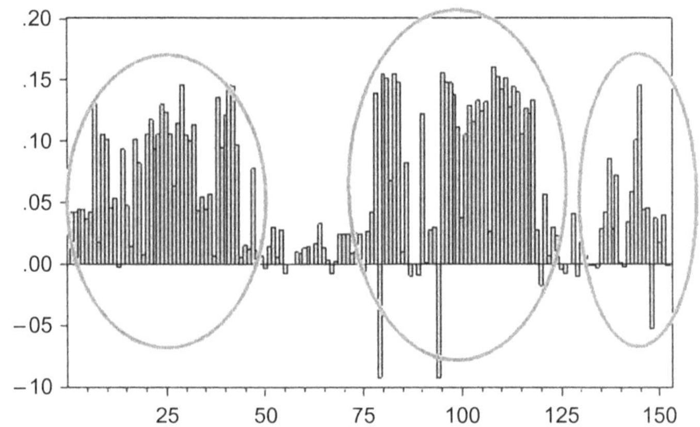

Fig. 8. Factor Loadings Corresponding to the First Principal Component. The three groups of highest loadings of the first component correspond to (in circles, from left to right): (1) production and consumption series, disaggregated; (2) business surveys; and (3) series on the world economy.

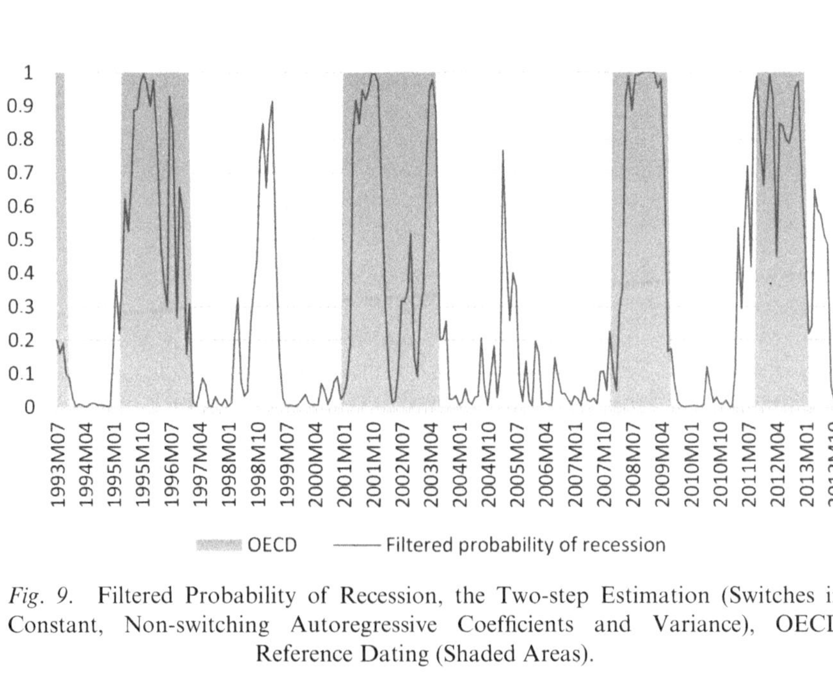

Fig. 9. Filtered Probability of Recession, the Two-step Estimation (Switches in Constant, Non-switching Autoregressive Coefficients and Variance), OECD Reference Dating (Shaded Areas).

Besides extra signals in 1998 and 2005, we can observe a transitory improvement in the middle of the Internet bubble crisis and the earlier detection of the beginning of the sovereign debt crisis, also omitted by the OECD. Similarly, the reasons for this amelioration can be tracked in the INSEE reports.[18]

4.2.2.2. **Alternative specification, switches in mean and variance**. We take advantage of the possibility to introduce switches into other coefficients of the model to check whether this improves the detection of the turning points. Now we allow the variance of the error term in the factor dynamics to be state specific, too, so the model of factor dynamics becomes:

$$f_t = \beta_{S_t} + \phi_1 f_{t-1} + \phi_2 f_{t-2} + \eta_{S_t} \tag{27}$$

where $\eta_{S_t} \sim N(0, \sigma^2_{S_t})$. While on average performing as well as the basic specification (QPS $= 0.1278$, FPS$/T = 0.1885$, Corr $= 0.67$), the alternative specification is slightly better in capturing the beginnings and ends of recessions (the identification lag is 0 and 1 months on average, respectively).[19] As before, the estimations provide a positive constant for the expansion period, and a negative one for recessions $\mu_0 = 1.22$, $\mu_1 = -1.52$. The volatility of the factor dynamics is estimated to be almost two times higher during recessions ($\sigma_0 = 0.4$, $\sigma_1 = 0.75$). The estimates of the other parameters are given in Table C1. Again, the filtered probabilities produced by this specification capture all the crises well (but the first one) and without sizable leads and lags. The dynamics of filtered probabilities for this specification resembles the one for the basic specification, so we do not report the graph here.

4.3. Comparison: One-Step versus Two-Step

We compare the average performance of the one-step method and the two step method in the baseline specification (lines "One-step method, average" and "Two-step method, full dataset" in Table 1). The difference in QPS and FPS is negligible (QPS $= 0.13$, FPS$/T = 0.18$ for the average one-step method vs. QPS $= 0.13$, FPS $= 0.19$ for the two-step method), whereas the correlation with the OECD dating is only slightly higher for the one-step method (Corr $= 0.69$ vs. Corr $= 0.67$). So, on average it is difficult to rank the performance of the methods. However, taking into account that the extra signals are responsible for part of the QPS and FPS/T of the two-step

Table 1. The Comparison of One-Step and Two-Step Estimation Results.

	QPS	FPS/ T	Corr	Start Lag	End Lag	Timing
Benchmark Hamilton (1989) univariate MS-AR model						
Hamilton's AR-MS on IIP (benchmark)	0.3679	0.5231	0.0894	∞	∞	0 M
MS-DFM (Kim & Yoo, 1995)						
One-step method, average	0.1346	0.1779	0.6985	2.5	−2.6	1 M
One-step, combination 1	0.1287	0.1383	0.7155	0.6	0.4	1 M
One-step, combination 2	0.1254	0.1779	0.6899	1.8	−0.2	1 M
One-step, combination 3	0.1328	0.1818	0.7431	3.4	−3.8	0 M
One-step, combination 4	0.1412	0.1818	0.7006	5.2	−4.2	1 M
One-step, combination 5	0.1184	0.1858	0.7082	3.6	−3.6	1 M
One-step, combination 6	0.1493	0.1937	0.6815	3	−8	1 M
One-step, combination 7	0.1492	0.1976	0.5607	1.8	0.6	1 M
Two-step method on 13 series	0.3259	0.3287	0.1649	∞	∞	1 M
Two-step method on 25 series	0.1207	0.2083	0.5703	2.5	−0.5	1 M
Two-step method, full dataset	0.1315	0.1926	0.6712	−0.25	1.4	2 M
Other specifications of MS-DFM						
Two-step method, full dataset, switching σ^2	0.0737	0.1885	0.6724	0	1	2 M
Two-step method, switching μ and σ^2, MS-AR(4)	0.0658	0.1762	0.6751	0	0.8	2 M
Two-step method, switching μ and σ^2+ pc2	0.2495	0.3648	0.3953	∞	∞	2 M
Two-step method, switching μ and + pc4	0.1027	0.1803	0.6602	1.75	−1.75	2 M
Two-step method, switching μ and σ^2+ pc4	0.0699	0.1721	0.7058	2	0.75	2 M
Other results for the French economy						
Kaufmann (2000)	–	0.2151	–	–	–	–
Chauvet and Yu (2006)	–	0.3777	–	–	–	–
Chen (2007)	–	0.2839	–	–	–	–
Kholodilin (2006)	0.152	0.3333	–	–	–	–

Note: For the composition of combination *i* see Tables B3 and A2. Start lag — the number of lags between the estimated beginning of a recession and the OECD determined beginning; End lag – the number of lags between the estimated end of a recession and the OECD determined end; T is the number of periods in the sample.

method, the two-step method is more precise in detecting OECD recessions. In particular, the two-step method is much more accurate with respect to the beginning and the end of recessions, with a tendency to indicate the beginning of a recession on average one quarter of a month earlier; the one-step method dates the beginning 2.5 months late and the end 2.6 months early, on average. In general, both methods produce early estimates: for the one-step method, the data in each of the retained combinations are updated with one month or even zero months lag. This means that the phase of the business cycle in January can be determined either in February or March, with no need to wait for the release of quarterly OECD dating in April. Though the gain in time is not very big, it may still be of a great importance for policy makers. For the two-step method, the estimates are available in two months, which is still less than the timing of the OECD. In this respect, the estimation of the factor on the first step with the help of one of the procedures proposed by Doz et al. (2011, 2012) is very promising since it allows to have the estimator of the factor based on the available information only, without waiting until all series in the database are updated. We leave this exercise for further research.

As for the parameter estimates, the two methods give qualitatively similar results in terms of values of coefficients (see Table C1): there are two distinct regimes, which are characterized by a negative constant in the recession state and a positive constant in the expansion state. The difference between the two constants varies in absolute value as the magnitude of factors is either determined by the underlying economic indicators (for the one-step method) or estimated up to a constant (in case of the two-step method). The estimates of the transition probabilities are similar, too: the phases of the French growth rate cycle are very persistent, with the probability to stay in expansion (on average, $p_0 = 0.96$) a bit higher than the probability to stay in recession (on average, $p_1 = 0.91$). All other estimates of Table C1 cannot be interpreted directly as they refer to different series and are reported for completeness. The estimation of the model with both methods on an expanding sample showed that the estimated coefficients and the resulting filtered probabilities are robust when the sample is up to 50 points shorter, however, the convergence is not always achieved for the one-step method.

The comparison to results in the previous literature by Kaufmann (2000), Chauvet and Yu (2006), Chen (2007), Kholodilin (2006) shows the advantage of the MS-DFM in detecting business cycle turning points, although it should be considered with care since we compare the results on slightly different (although overlapping) time spans.

The final datings for both methods are similar in general, although there are some discrepancies to the OECD dating (see Table 2).

Let us note that although the two methods provide rather similar results, we suggest using the two-step method as it is more robust and easier to estimate and allows to consider big datasets. Furthermore, since the factor is considered as observed, the baseline specification can be extended to include additional autoregressive lags and other explanatory variables. For example, we suggest the following possible extensions of the baseline model (section "Other specifications of MS-DFM" in Table 1): introduction of switching variance, use of two more lags in the autoregressive structure, inclusion of the second and the fourth principal component in the dynamics equation of the factor to take additional information into account. Some of these extensions increase the performance of the baseline specifications, although the improvement is rather minor if not negligible. Nevertheless, this observation leaves room for further research in the direction of multifactor Markov-switching dynamic factor models with a general VAR structure.

As an additional validity check, we make two more comparisons. The first one serves to evaluate the gain from the multivariate analysis. For this purpose we made a comparison with the results of a simple classical Hamilton (1989) model with two autoregressive terms and a switching constant estimated on the growth rate of the index of industrial production (see Table 1). One can see that, contrary to the United States, in the case of France this series contains much less information about the business cycles, at least for the period under consideration. The MS-AR model produces only one signal corresponding to the 2008 recession. This poor performance is reflected in our quantitative indicators as high QPS and FPS and very low correlation with the reference.

Second, to understand the role of the number of series for the two-step method, we analyze its performance on smaller datasets. A number of papers (see, e.g., Boivin & Ng, 2006, 2008) state that using big datasets for factor analysis is not always better than using smaller datasets of appropriately selected series. To evaluate the role of the number of series for the two-step method we estimate the baseline specification on the subset of 25 series which were used for the one-step method as well as on the 13 series which were finally retained ("Two-step method on 25 series" and "Two-step method on 13 series," respectively, in Table 1). As we can see, the use of 13 series does not improve upon the results of the benchmark Hamilton (1989) model. The most likely reason for this is that the PCA

Table 2. Final Dating Produced by One-Step Procedure on Seven Best Sets of Data, Two-Step Procedure and OECD Dating.

		Comb 1	Comb 2	Comb 3	Comb 4	Comb 5	Comb 6	Comb 7	Two-Step	OECD
First crisis	P							1993m02		1992m02
	T							1993m10		1993m10
Second crisis	P	1995m07	1995m06	1995m08	1995m08	1995m09	1995m07	1995m09	1995m01	1995m03
	T	1996m12	1996m12	1996m09	1996m10	1996m10	1997m01	1997m05	1997m01	1997m01
1 false signal	P								1998m09	
	T								1999m04	
Third crisis	P	2001m01	2001m01	2001m02	2001m01	2001m02	2001m01	2001m04	2001m03	2000m12
	T	2003m07	2003m06	2003m04	2002m11	2003m03	2002m07	2003m12	2003m09	2003m06
2 false signal	P								2005m02	
	T								2005m07	
Fourth crisis	P	2007m09	2007m09	2008m04	2008m04	2008m04	2008m04	2008m04	2008m04	2007m12
	T	2009m09	2009m11	2009m05	2009m06	2009m09	2009m04	2009m09	2009m08	2009m06
Fifth crisis	P	2011m09	2011m09	2011m09	2012m07	2012m05	2011m09	2011m06	2011m03	2011m09
	T	2013m07	2013m08	2012m10	2012m11	2013m07	2012m11	2013m07	2013m08	2013m01

Notes: Comb *i* stands for Combination *i*. For the composition of Combination *i* see Tables B3 and A2.

estimate of the factor is not good enough to give meaningful results. However, when the number of series increases to 25, the results become much closer to the results on the full dataset (Two-step method, full dataset): QPS and FPS are almost identical (QPS = 0.12, FPS/T = 0.21 for 25 series, QPS = 0.14, FPS/T = 0.19 for the full dataset), the correlation with the OECD reference is much closer (Corr = 0.57 and Corr = 0.67 for 25 series and the full dataset, respectively), although the beginnings and the ends of recessions are estimated with less precision. To conclude, this exercise shows that the larger the dataset, the more accurate are the estimates of the factor, and therefore the better the quality of the extracted signal, although the marginal gain of a larger number of series decreases.

5. CONCLUSION

This paper focuses on the comparison of the two estimation methods of the MS-DFM model of the business cycle applied to French data. The Maximum Likelihood estimation of the model in one-step can be run only for a very small set of information, whereas the two-step estimation can accommodate much bigger information sets. In this paper, we use an extensive dataset of French series covering the period March 1993–October 2013. We estimate the MS-DFM on 151 series in two steps and on different subsets of four series of main economic indicators in one step. We show that the two-step estimation procedure produces good results in terms of turning points identification. The procedures are transparent and replicable. The model produces turning point estimates up to two months earlier than the reference OECD dating, which is an important gain in timing for economic agents and policymakers.

We find that both estimation methods provide qualitatively similar results: the common factor of several specific economic series (in case of one-step method) and the first principal component of a large set of series (in case of two-step method) can be characterized as having two distinct phases with low and high growth rates, correspondingly. The two-step method also allows to detect the difference in the magnitude of variance in the factor dynamics. For both methods, the periods of high filtered probabilities of recession match the OECD recessions. At the same time, the two-step method and several results of the one-step method identify short recessions in 1998 and 2005 that do not appear in the OECD dating, which is intended to indicate long-lasting phases. We show that these signals are not false, as the worsening of the economic situation was noted in the corresponding short-term INSEE reports, as well as captured by the Index of reversal by INSEE and the datings obtained with the help of the Christiano–Fitzgerald filter.

Both methods largely outperform the results of the univariate Hamilton (1989) model estimated on the index of industrial production, which shows the importance of the multivariate framework for business cycle turning point identification.

The results of the one-step method differ greatly depending on the composition of the four input economic series. We identify series with the highest explicative power (retail trade order intentions, number of job seekers, the survey on manufacturing business situation future and construction confidence index) and the series that produce extra signals (manufacturing finished goods stock level, price return on FTSE equity index and Manufacturing Industrial Confidence Indicator) and determine seven sets of series that perform best in terms of concordance of estimated turning points with the OECD chronology. Since the size of the dataset considered with the one-step method is generally limited to four series, it seems reasonable to use several sets (i.e., several results of the one-step estimation) as complements to overcome the information constraint.

Using a more comprehensive dataset with the two-step method allows us to obtain more accurate estimates of the beginning and the end of recessions. We show that the number of series plays an important role, with larger datasets leading to more accurate identification of the turning points. Introduction of additional autoregressive lags and other principal components further enhances the precision of the two-step results, although the improvements are minor.

We conclude that either method can be used to replicate the OECD dating. Nevertheless, we think the use of the two-step method is very appealing: it allows to get a valid dating of turning points without going through a complicated procedure of series selection, it is much less time-consuming and the numerical convergence problems are not frequent. Another advantage of the two-step method is that it opens the way to different extensions. First, the factor may be estimated within the first step using other methods like the two-step estimator proposed by Doz et al. (2011) or the QML estimator proposed by Doz et al. (2012): this will allow to use data of different frequencies, with missing observations or ragged ends. Second, multifactor Markov-switching models can be estimated. These extensions are left for future research.

NOTES

1. Economic Cycle Research Institute, private organization.
2. The working paper version appeared in 1994 in NBER Working Papers 4643.
3. INSEE, France, Gross Domestic Product, Total, Contribution to Growth, Calendar Adjusted, Constant Prices, SA, Chained, Change P/P.
4. Kim and Yoo (1995) showed that although the assumption of the time dependent probabilities improves the quality of the model, the gain in terms of loglikelihood is not very large.
5. For further details, see Kim (1994) and the references therein.
6. For our estimations, we used Nelder–Mead simplex direct search with maximum function evaluations set to 2000, and tolerance for both function and dependent variables set to 0.001. We set the initial values of the parameters to the estimates of the same state-space model but without switch, that is, the estimates of Stock and Watson (1989a, b) DFM. The latter is, in turn, initialized with the OLS estimates of the system of equations where the first principal component is used as a proxy for the latent common component.
7. The trade-off between the sample size and the number of cross-sections made us to restrict the dataset to just 21 years of observations. A longer period (starting in 1990) would reduce the number of cross-sections to 97, while the full original balanced database (213 series) starts in February 1996.
8. http://stats.oecd.org/mei/default.asp?rev = 2.
9. http://www.cepr.org/content/euro-area-business-cycle-dating-committee.
10. https://www.businesscycle.com/.
11. The CLI components are: (1) New passenger car registrations (number), (2) Consumer confidence indicator (% balance), (3) Production (Manufacturing): future tendency (% balance), (4) SBF 250 share price index (2010 = 100), (5) CPI Harmonized All items (2010 = 100) inverted, (6) Export order books (Manufacturing): level (% balance), (7) Selling prices (Construction): future tendency (% balance), (8) Permits issued for dwellings (2010 = 100), and (9) Expected level of life in France (Consumer Survey) (% balance). All series are detrended, and seasonally, calendar- and noise-adjusted. They are selected so that they have a cycle pattern similar and coincident (or leading) to the one of the reference series. Until April 2012 the industrial production index was taken as a reference series, replaced by monthly estimates of GDP growth afterwards.
12. The Reversal Index is the index comprised between −1 and 1, which shows the difference between the probability to be in expansion in the current period and the probability to be in recession in the current period. The index is based on the business surveys concerning the current, past, and future perceptions of the economic situation.
13. INSEE PREMIERE, NO. 659 – June 1999.
14. INSEE CONJONCTURE, Note de conjoncture, December 1998.
15. INSEE CONJONCTURE, Note de conjoncture, Mars 2005.
16. Interestingly, Bruno and Otranto (2004) also find similar signals of 1998–1999 and 2005 for the chronology of the Italian economic cycle.
17. Following Kim and Yoo (1995), we put two autoregressive lags in the baseline specification. This assumption turned out to be plausible: the correlogram of

the first principal component has high partial autocorrelation for the first two lags. The choice of two lags was also confirmed by Akaike and Schwarz information criteria estimated on the model with one, two, and three autoregressive lags.

18. INSEE observed the improvement of the business climate in 2001 primarily due to the subjective perception that the US had passed the trough of the business cycle; rebound growth in Asia, Germany and the negative oil price shock improved the expectations of investors and entrepreneurs, while the decrease in taxes gave an extra stimulus for household consumption, increasing their purchasing power (INSEE CONJONCTURE, Note de conjoncture, Mars 2002). The reasons for early peaks in 2011 are the deterioration of the business climate in France, the earthquake in Japan, anti-inflation policies in developing countries, as well as budget consolidation politics in the developed countries, positive price shocks for commodities (oil included) increased production costs. All this led to a certain pessimism among French investors (Point de conjuncture October 2011, INSEE).

19. We also tried specifications with switching autoregressive coefficients and different combinations of switching parameters, but none of them were performing as well. To save space, we do not report the results here.

ACKNOWLEDGMENTS

The authors thank the editors and two anonymous referees for useful remarks. All remaining errors are ours. We also acknowledge financial support by the European Commission in the framework of the European Doctorate in Economics – Erasmus Mundus (EDEEM).

REFERENCES

Anas, J., Billio, M., Ferrara, L., & Do Luca, M. (2007). *A turning point chronology for the euro-zone*. Working Papers No. 33, Department of Economics, University of Venice "Ca' Foscari".

Bai, J. (2003). Inferential theory for factor models of large dimensions. *Econometrica, 71*, 135–171.

Bai, J., & Ng, S. (2008). Forecasting economic time series using targeted predictors. *Journal of Econometrics, 146*, 304–317.

Bardaji, J., Clavel, L., & Tallet, F. (2008). Deux nouveau indicateurs pour aider au diagnostic conjoncturel en france, *Dossiers, INSEE*.

Bardaji, J., Clavel, L., & Tallet, F. (2009). Constructing a Markov-switching turning point index using mixed frequencies with an alpplication to French Business Survey Data. *OECD Jounal: Journal of Business Cycle Measurement and Analysis, 2009*(2), 111–132.

Bessec, M., & Bouabdallah, O. (2015). Forecasting GDP over the business cycle in a multi-frequency and data-rich environment. *Oxford Bulletin of Economics and Statistics*, *77*(3), 360–384. Retrieved from http://dx.doi.org/10.1111/obes.12069

Bessec, M., & Doz, C. (2012). Prévision à court terme de la croissance du pib français à l'aide de modèles à facteurs dynamiques. *Economie & Prévision*, *1*, 1–30.

Boivin, J., & Ng, S. (2006). Are more data always better for factor analysis? *Journal of Econometrics*, *132*, 169–194.

Bruno, G., & Otranto, E. (2004). *Dating the italian business cycle: A comparison of procedures*. ISAE Working Papers No. 41, ISTAT, Italian National Institute of Statistics (Rome, ITALY).

Bry, G., & Boschan, C. (1971). Cyclical analysis of time series: Selected procedures and computer programs. National Bureau of Economic Research, Inc.

Burns, A. F., & Mitchell, W. C. (1946). Measuring business cycles. National Bureau of Economic Research, Business & Economics.

Camacho, M., Perez-Quiros, G., & Poncela, P. (2012). *Extracting nonlinear signals from several economic indicators*. Banco de Espaïna Working Papers No. 1202, Banco de Espana. pp. 1–36.

Carvalho, C. M., & Lopes, H. F. (2007). Factor stochastic volatility with time varying loadings and Markov and switching regimes. *Journal of Statistical Planning and Inference*, *6*, 3082–3091.

Chauvet, M. (1998). An econometric characterization of business cycle dynamics with factor structure and regime switching. *International Economic Review*, *39*(4), 969–996.

Chauvet, M. (1999). Stock market Fluctuations and the business cycle. *Journal of Economic and Social Measurement*, *25*(3 and 4), 1–31.

Chauvet, M., & Potter, S. (1998). Nonlinear risk. *Macroeconomic Dynamics* (Vol. 5(04), pp. 621–646). Cambridge: Cambridge University Press.

Chauvet, M., & Senyuz, Z. (2008). *A joint dynamic bi-factor model of the yield curve and the economy as a predictor of business cycles*. MPRA Paper No. 15076, University Library of Munich, Germany. Retrieved from http://ideas.repec.org/p/pra/mprapa/15076.html

Chauvet, M., & Yu, C. (2006). International business cycles: G7 and OECD countrie. *Economic Review, Federal Reserve Bank of Atlanta*, *91*(Q1), 43–54.

Chen, X. (2007). *Evaluating the synchronisation of the eurozone business cycles using multivariate coincident macroeconomic indicators*.

Christiano, L. J., & Fitzgerald, T. J. (2003). The band pass filter. *International Economic Review*, *44*(2), 435–465. Retrieved from http://ideas.repec.org/a/ier/iecrev/v44y2003i2p435-465.html

Darné, O., & Ferrara, L. (2011). Identification of slowdowns and accelerations for the euro area economy. *Oxford Bulletin of Economics and Statistics*, *73*(3), 335–364.

Diebold, F. X., & Rudebusch, G. D. (1996). Measuring business cycles: A modern perspective. *The Review of Economics and Statistics*, *78*(1), 67–77.

Dolega, M. (2007). *Tracking Canadian trend productivity: A dynamic factor model with Markov switching*.

Doz, C., Giannone, D., & Reichlin, L. (2011). A two-step estimator for large approximate dynamic factor models based on Kalman filtering. *Journal of Econometrics*, *164*(1), 188–205.

Doz, C., Giannone, D., & Reichlin, L. (2012). A quasi-maximum likelihood approach for large, approximate dynamic factor models. *The Review of Economics and Statistics*,

94(4), 1014–1024. Retrieved from http://ideas.repec.org/a/tpr/restat/v94y2012i4p1014-1024.html

Dueker, M., & Sola, M. (2008). *Multivariate Markov switching with weighted regime determination: Giving France more weight than Finland.*

Gregoir, S., & Lenglart, F. (2000). Measuring the probability of a business cycle turning point by using a multivariate qualitative hidden Markov model. *Journal of Forecasting*, *102*, 81–102.

Hamilton, J. D. (1989). A new approach to the economic analysis of nonstationary time series and the business cycle. *Econometrica*, *57*, 357–384.

Kaufmann, S. (2000). Measuring business cycles with a dynamic Markov switching factor model: An assessment using Bayesian simulation methods. *Econometrics Journal*, *3*, 39–65.

Kholodilin, K. A. (2002a). Some evidence of decreasing volatility of the US coincident economic indicator. *Economics Bulletin, AccessEcon*, *3*(20), 1–20.

Kholodilin, K. A. (2002b). Two alternative approaches to modelling the nonlinear dynamics of the composite economic indicator. *Economics Bulletin, AccessEcon*, *3*(26), 1–18.

Kholodilin, K. A. (2006). *Using the dynamic bi-factor model with Markov switching to predict the cyclical turns in the large european economies.* DIW Berlin, Discussion paper.

Kholodilin, K. A., & Yao, W. V. (2004). *Business cycle turning points: Mixed-frequency data with structural breaks.*

Kim, C.-J. (1994). Dynamic linear models with Markov-switching. *Journal of Econometrics*, *60*(1–2), 1–22. Retrieved from http://ideas.repec.org/a/eee/econom/v60y1994i1-2p1-22.html

Kim, C.-J., & Nelson, C. R. (1998). Business cycle turning points, a new coincident index, and tests of duration dependence QSED on the dynamic factor model with regime switching. *The Review of Economics and Statistics*, *80*(2), 188–201.

Kim, M.-J., & Yoo, J.-S. (1995). New index of coincident indicators: A multivariate Markov switching factor model approach. *Journal of Monetary Economics*, *36*(3), 607–630.

Stock, J. H., & Watson, M. W. (1989a). New indexes of coincident and leading economic indicators. In *NBER macroeconomics annual 1989* (Vol. 4, pp. 351–409), NBER Chapters. National Bureau of Economic Research, Inc. Retrieved from http://ideas.repec.org/h/nbr/nberch/10968.html

Stock, J. H., & Watson, M. W. (1989b). New indexes of coincident and leading economic indicators. *Macroeconomics Annual*, *4*, 351–401.

Stock, J. H., & Watson, M. W. (2002). Forecasting using principal components from a large number of predictors. *Journal of the American Statistical Association*, *9*(460), 1167–1179.

Stock, J. H., & Watson, M. W. (2010). *Estimating turning points using large data sets.* NBER Working Papers No. 16532.

APPENDIX A. DATASETS

Table A1. Series Used for the Two-Step Estimation

Series Full Name	Source	SA	Lag
Industrial Production by Industry			
General			
France, OECD MEI, Production Of Total Industry, SA, Change P/P	Macrobond	SA	2
France, OECD MEI, Production Of Total Industry, SA, Index	Macrobond	SA	2
France, OECD MEI, Production Of Total Manufactured Intermediate Goods, SA, Index	Macrobond	SA	2
France, OECD MEI, Production In Total Manufacturing, SA, Index	Macrobond	SA	2
France, OECD MEI, Production Of Total Manufactured Investment Goods, SA, Index	Macrobond	SA	2
France, Industrial Production, Total Industry Excluding Construction, Calendar Adjusted, SA, Index	Macrobond	SA	1
France, Capacity Utilization, Total Industry, SA	Macrobond	SA	0
Mining			
France, Eurostat, Industry Production Index, Extraction of Crude Petroleum & Natural Gas, Calendar Adjusted, Change Y/Y	Macrobond		1
France, Eurostat, Industry Production Index, Other Mining & Quarrying, Calendar Adjusted, Change Y/Y	Macrobond		1
France, Eurostat, Industry Production Index, Mining & Quarrying, Calendar Adjusted, Change Y/Y	Macrobond		1
Nondurables			
France, Eurostat, Industry Production Index, Manufacture of Food Products, Calendar Adjusted, Change Y/Y	Macrobond		1
France, Eurostat, Industry Production Index, Manufacture of Beverages, Calendar Adjusted, Change Y/Y	Macrobond		1
France, Eurostat, Industry Production Index, Manufacture of Tobacco Products, Calendar Adjusted, Change Y/Y	Macrobond		1
France, Eurostat, Industry Production Index, Manufacture of Textiles, Calendar Adjusted, Change Y/Y	Macrobond		1
France, Eurostat, Industry Production Index, Manufacture of Wearing Apparel, Calendar Adjusted, Change Y/Y	Macrobond		1
France, Eurostat, Industry Production Index, Manufacture of Leather & Related Products, Calendar Adjusted, Change Y/Y	Macrobond		1
France, Eurostat, Industry Production Index, Manufacture of Paper & Paper Products, Calendar Adjusted, Change Y/Y	Macrobond		1
France, Eurostat, Industry Production Index, Printing & Service Activities Related to Printing, Calendar Adjusted, Change Y/Y	Macrobond		1
	Macrobond		1

France, Eurostat, Industry Production Index, Manufacture of Coke & Refined Petroleum Products, Calendar Adjusted, Change Y/Y		
France, Eurostat, Industry Production Index, Manufacture of Chemicals & Chemical Products, Calendar Adjusted, Change Y/Y	Macrobond	1
France, Eurostat, Industry Production Index, Manufacture of Rubber Products, Calendar Adjusted, Change Y/Y	Macrobond	1
Durables		
France, Eurostat, Industry Production Index, Manufacture of Computer, Electronic & Optical Products, Calendar Adjusted, Change Y/Y	Macrobond	1
France, Eurostat, Industry Production Index, Manufacture of Electric Motors, Generators, Transformers & Electricity Distribution & Control Apparatus, Calendar Adjusted, Change Y/Y	Macrobond	1
France, Eurostat, Industry Production Index, Manufacture of Electrical Equipment, Calendar Adjusted, Change Y/Y	Macrobond	1
France, Eurostat, Industry Production Index, Manufacture of Machinery & Equipment N.E.C., Calendar Adjusted, Change Y/Y	Macrobond	1
France, Eurostat, Industry Production Index, Manufacture of Motor Vehicles, Trailers, Semi-Trailers & of Other Transport Equipment, Calendar Adjusted, Change Y/Y	Macrobond	1
France, Eurostat, Industry Production Index, Building of Ships & Boats, Calendar Adjusted, Change Y/Y	Macrobond	1
France, Eurostat, Industry Production Index, Manufacture of Furniture, Calendar Adjusted, Change Y/Y	Macrobond	1
France, Eurostat, Industry Production Index, Manufacturing, Calendar Adjusted, Change Y/Y	Macrobond	1
France, Eurostat, Construction, Building & Civil Engineering, Construction & Production Index, Buildings, Calendar Adjusted, Change Y/Y	Macrobond	1
France, Eurostat, Construction, Building & Civil Engineering, Construction & Production Index, Civil Engineering Works, Calendar Adjusted, Change Y/Y	Macrobond	1
France, Eurostat, Construction, Building & Civil Engineering, Construction & Production Index, Construction, Calendar Adjusted, Change Y/Y	Macrobond	1
France, Metropolitan, Construction by Status, Number, Permits, Residential Buildings, Total	Macrobond	1
France, Metropolitan, Construction by Status, Number, Housing Starts, Residential Buildings, Total	Macrobond	1
France, Construction by Status, Number, Permits, Residential Buildings, Total	Macrobond	1
France, Construction by Status, Number, Housing Starts, Residential Buildings, Total	Macrobond	1
Utilities		
France, Eurostat, Industry Production Index, Electricity, Gas, Steam & Air Conditioning Supply, Total, Calendar Adjusted, Change Y/Y	Macrobond	1

Table A1. *(Continued)*

Series Full Name	Source	SA	Lag
Industrial production by market			
Durables			
France, Eurostat, Industry Production Index, MIG — Capital Goods, Calendar Adjusted, Change Y/Y	Macrobond		1
France, Eurostat, Industry Production Index, MIG — Consumer Goods (Except Food, Beverages & Tobacco), Calendar Adjusted, Change Y/Y	Macrobond		1
France, Eurostat, Industry Production Index, MIG — Durable Consumer Goods, Calendar Adjusted, Change Y/Y	Macrobond		1
France, Eurostat, Industry Production Index, MIG — Intermediate & Capital Goods, Calendar Adjusted, Change Y/Y	Macrobond		1
France, Eurostat, Industry Production Index, MIG — Intermediate Goods, Calendar Adjusted, Change Y/Y	Macrobond		1
France, Eurostat, Industry Production Index, MIG — Consumer Goods, Calendar Adjusted, Change Y/Y	Macrobond		1
France, Expenditure Approach, Household Consumption Expenditure, Automobiles, Calendar Adjusted, Constant Prices, SA, EUR	Macrobond	SA	1
France, Expenditure Approach, Household Consumption Expenditure, Housing Equipment, Calendar Adjusted, Constant Prices, SA, EUR	Macrobond	SA	1
France, Expenditure Approach, Household Consumption Expenditure, Durable Personal Equipment, Calendar Adjusted, Constant Prices, SA, EUR	Macrobond	SA	1
Nondurables			
France, Eurostat, Industry Production Index, MIG — Non-Durable Consumer Goods, Calendar Adjusted, Change Y/Y	Macrobond		1
France, Eurostat, Industry Production Index, MIG — Energy (Except D & E), Calendar Adjusted, Change Y/Y	Macrobond		1
France, Eurostat, Industry Production Index, MIG — Energy (Except Section E), Calendar Adjusted, Change Y/Y	Macrobond		1
France, Energy Production, Transmission & Distribution, Electric Power Generation, Transmission & Distribution, Calendar Adjusted, SA, Index	Macrobond	SA	1
France, Eurostat, Industry Production Index, Manufacture of Products of Wood, Cork, Straw & Plaiting Materials, Calendar Adjusted, Index	Macrobond		1
France, Eurostat, Industry Production Index, Manufacture of Basic Metals & Fabricated Metal Products, Except Machinery & Equipment, Index	Macrobond		1
France, Expenditure Approach, Household Consumption Expenditure, Textiles & Leather, Calendar Adjusted, Constant Prices, SA, EUR	Macrobond	SA	1

France, Expenditure Approach, Household Consumption Expenditure, Other Manufactured Goods, Calendar Adjusted, Constant Prices, SA, EUR	Macrobond	SA	1
France, Expenditure Approach, Household Consumption Expenditure, Energy, Water & Waste Treatment, Calendar Adjusted, Constant Prices, SA, EUR	Macrobond	SA	1
France, Expenditure Approach, Household Consumption Expenditure, Petroleum Products, Calendar Adjusted, Constant Prices, SA, EUR	Macrobond	SA	1
France, Expenditure Approach, Household Consumption Expenditure, Food, Calendar Adjusted, Constant Prices, SA, EUR	Macrobond	SA	1
France, Expenditure Approach, Household Consumption Expenditure, Goods, Calendar Adjusted, Constant Prices, SA, EUR	Macrobond	SA	1
Equipment			
France, Manufacturing, Computers & Peripheral Equipment, Calendar Adjusted, SA, Index	Macrobond	SA	1
France, Manufacturing, Optical Instruments & Photographic Equipment, Calendar Adjusted, SA, Index	Macrobond	SA	1
France, Manufacturing, Electric Lighting Equipment, Calendar Adjusted, SA, Index	Macrobond	SA	1
France, Manufacturing, Other Electrical Equipment, Calendar Adjusted, SA, Index	Macrobond	SA	1
France, Manufacturing, Repair of Fabricated Metal Products, Machinery & Equipment, Calendar Adjusted, SA, Index	Macrobond	SA	1
France, Manufacturing, Electrical Equipment, Calendar Adjusted, SA, Index	Macrobond	SA	1
Materials			
France, Manufacturing, Clay Building Materials, Calendar Adjusted, SA, Index	Macrobond	SA	1
Employment by skill and gender			
France, Metropolitan, Unemployment, Job Seekers, Men, Total, Categories A, B & C, Calendar Adjusted, SA	Macrobond	SA	1
France, Metropolitan, Unemployment, Job Seekers, Women & Men, Under 25 Years, Categories A, B & C, Calendar Adjusted, SA	Macrobond	SA	1
France, Metropolitan, Unemployment, Job Seekers, Women & Men, Aged 25–49 Years, Categories A, B & C, Calendar Adjusted, SA	Macrobond	SA	1
France, Metropolitan, Unemployment, Job Seekers, Women & Men, Aged 50 & More, Categories A, B & C, Calendar Adjusted, SA	Macrobond	SA	1
France, Unemployment, Job Seekers, Women & Men, Total, Categories A, B & C, Calendar Adjusted, SA	Macrobond	SA	1
France, Metropolitan, Unemployment, Job Seekers, Men, Under 25 Years, Categories A, B & C	Macrobond		1
France, Metropolitan, Unemployment, Job Seekers, Men, Aged 25–49 Years, Categories A, B & C	Macrobond		1

Table A1. (*Continued*)

Series Full Name	Source	SA	Lag
France, Metropolitan, Unemployment, Job Seekers, Men, Aged 50 & More, Categories A, B & C	Macrobond		1
France, Metropolitan, Unemployment, Job Seekers, Men, Total, Categories A, B & C	Macrobond		1
France, Metropolitan, Unemployment, Job Seekers, Women, Under 25 Years, Categories A, B & C	Macrobond		1
France, Metropolitan, Unemployment, Job Seekers, Women, Aged 25–49 Years, Categories A, B & C	Macrobond		1
France, Metropolitan, Unemployment, Job Seekers, Women, Aged 50 & More, Categories A, B & C	Macrobond		1
France, Metropolitan, Unemployment, Job Seekers, Women, Total, Categories A, B & C	Macrobond		1
France, Metropolitan, Unemployment, Job Seekers, Women & Men, Under 25 Years, Categories A, B & C	Macrobond		1
France, Metropolitan, Unemployment, Job Seekers, Women & Men, Aged 25–49 Years, Categories A, B & C	Macrobond		1
France, Metropolitan, Unemployment, Job Seekers, Women & Men, Aged 50 & More, Categories A, B & C	Macrobond		1
France, Metropolitan, Unemployment, Job Seekers, Women & Men, Total, Categories A, B & C	Macrobond		1
France, Unemployment, Job Seekers, Women & Men, Total, Categories A, B & C	Macrobond		1
France, Metropolitan, Unemployment, Job Seekers, Men, Labourers, Categories A, B & C	Macrobond		1
France, Metropolitan, Unemployment, Job Seekers, Women, Labourers, Categories A, B & C	Macrobond		1
France, Metropolitan, Unemployment, Job Seekers, Women & Men, Labourers, Categories A, B & C	Macrobond		1
France, Metropolitan, Unemployment, Job Seekers, Men, Professional Workers, Categories A, B & C	Macrobond		1
France, Metropolitan, Unemployment, Job Seekers, Women, Professional Workers, Categories A, B & C	Macrobond		1
France, Metropolitan, Unemployment, Job Seekers, Women & Men, Professional Workers, Categories A, B & C	Macrobond		1
France, Metropolitan, Unemployment, Job Seekers, Men, Skilled Manual Workers, Categories A, B & C	Macrobond		1
France, Metropolitan, Unemployment, Job Seekers, Women, Skilled Manual Workers, Categories A, B & C	Macrobond		1
France, Metropolitan, Unemployment, Job Seekers, Women & Men, Skilled Manual Workers, Categories A, B & C	Macrobond		1
France, Metropolitan, Unemployment, Job Seekers, Men, Non-Qualified Employed Persons, Categories A, B & C	Macrobond		1
France, Metropolitan, Unemployment, Job Seekers, Women, Non-Qualified Employed Persons, Categories A, B & C	Macrobond		1
France, Metropolitan, Unemployment, Job Seekers, Women & Men, Non-Qualified Employed Persons, Categories A, B & C	Macrobond		1
France, Metropolitan, Unemployment, Job Seekers, Men, Qualified Employed Persons, Categories A, B & C	Macrobond		1

Trade			
Credit			
France, Deposits & Loans, Credit Institutions, Loans, By Entity, to Households & NPISH, Loans for House Purchasing Adjusted for Sales & Securitisation, Total, Flows, EUR	Macrobond		1
France, Deposits & Loans, Credit Institutions, Loans, By Entity, to Households & NPISH, Loans for Other Purposes Adjusted for Sales & Securitisation, Total, Flows, EUR	Macrobond		1
France, Deposits & Loans, Credit Institutions, Loans, By Entity, to Households & NPISH, Loans Adjusted for Sales & Securitisation, Total, Flows, EUR	Macrobond		1
Durables			
France, OECD MEI, CLI New Car Registrations, SA	Macrobond	SA	1
France, OECD MEI, Total Car Registrations, SA	Macrobond	SA	1
France, OECD MEI, Passenger Car Registrations, SA, Index	Macrobond	SA	1
Retail			
France, OECD MEI, Total Retail Trade (Volume), SA, Index	Macrobond	SA	1
France, OECD MEI, Total Retail Trade (Value), SA, Index	Macrobond	SA	1
France, Domestic Trade, Vehicle Sales & Registrations, New, Passenger Cars, Total, Calendar Adjusted, SA	Macrobond	SA	0
France, Eurostat, Retail Trade & Services, Total Market, Retail Sale of Automotive Fuel in Specialized Stores, Calendar Adjusted, Index	Macrobond		2
France, Eurostat, Retail Trade & Services, Total Market, Retail Sale via Mail Order Houses or via Internet, Index	Macrobond		2
France, Eurostat, Retail Trade & Services, Total Market, Retail Sale of Food, Beverages & Tobacco, Trend Adjusted, Index	Macrobond		1
France, Eurostat, Retail Trade & Services, Total Market, Retail Sale of Textiles, Clothing, Footware & Leather Goods in Specialized Stores, Index	Macrobond		2
France, Eurostat, Retail Trade & Services, Total Market, Retail Sale of Textiles, Clothing, Footware & Leather Goods in Specialized Stores, Calendar Adjusted, Index	Macrobond		2
France, Eurostat, Retail Trade & Services, Total Market, Dispensing Chemist, Retail Sale of Medical & Orthopaedic Goods, Cosmetic & Toilet Articles in Specialized Stores, Calendar Adjusted, Index	Macrobond		2
France, Eurostat, Retail Trade & Services, Total Market, Retail Sale of Non-Food Products (Incl. Fuel), Index	Macrobond		1
France, Eurostat, Retail Trade & Services, Total Market, Retail Sale of Non-Food Products (Excl. Fuel), Index	Macrobond		1
Foreign Trade			
France, Foreign Trade, Trade Balance, Calendar Adjusted, SA, EUR	Macrobond	SA	1
France, Foreign Trade, Export, Calendar Adjusted, SA, EUR	Macrobond	SA	1

Table A1. *(Continued)*

Series Full Name	Source	SA	Lag
France, Foreign Trade, Import, Calendar Adjusted, SA, EUR	Macrobond		1
France, OECD MEI, BOP Capital Account Credit, EUR	Macrobond		1
France, OECD MEI, BOP Capital Account Debit, EUR	Macrobond		1
Surveys			
Retail			
France, OECD MEI, Manufacturing Business Situation Future, SA	Macrobond	SA	0
France, OECD MEI, Manufacturing Finished Goods Stocks Level, SA	Macrobond	SA	0
France, OECD MEI, Manufacturing Production Future Tendency, SA	Macrobond	SA	0
France, OECD MEI, Manufacturing Production Tendency, SA	Macrobond	SA	0
France, OECD MEI, Manufacturing Selling Prices Future Tendency, SA	Macrobond	SA	0
France, OECD MEI, Manufacturing Industrial Confidence Indicator, SA	Macrobond	SA	0
France, OECD MEI, Manufacturing Export Order Books Level, SA	Macrobond	SA	0
Consumers			
France, Consumer Surveys, INSEE, Consumer Confidence Indicator, General Economic Situation, Past 12 Months, Balance of Replies, SA	Macrobond	SA	0
France, Consumer Surveys, INSEE, Consumer Confidence Indicator, General Economic Situation, Next 12 Months, Balance of Replies, SA	Macrobond	SA	0
France, Consumer Surveys, INSEE, Consumer Confidence Indicator, Major Purchases Intentions, Next 12 Months, Balance of Replies, SA	Macrobond	SA	0
France, Consumer Surveys, INSEE, Consumer Confidence Indicator, Financial Situation, Last 12 Months, Balance of Replies, SA	Macrobond	SA	0
France, Consumer Surveys, INSEE, Consumer Confidence Indicator, Financial Situation, Next 12 Months, Balance of Replies, SA	Macrobond	SA	0
Industry			
France, Business Surveys, INSEE, Building Industry, Global, Past Activity Tendency	Macrobond		0
France, Business Surveys, INSEE, Building Industry, Global, Expected Activity	Macrobond		0
France, Business Surveys, INSEE, Building Industry, Global, Order Books Level	Macrobond		0
France, Business Surveys, INSEE, Building Industry, Global, Past Workforce Size	Macrobond		0
France, Business Surveys, Bank of France, Industry, Inventories of Final Goods, Manufacturing Industry, SA	Macrobond	SA	0

France, Business Surveys, Bank of France, Industry, Current Order Books, Manufacturing Industry, SA	Macrobond	SA	0
France, Business Surveys, INSEE, Industry, Manufacturing, Personal Production Expectations, Balance of Replies, SA	Macrobond	SA	0
France, Business Surveys, INSEE, Industry, Manufacturing, Demand & Export Order Books, Balance of Replies, SA	Macrobond	SA	0
France, Business Surveys, INSEE, Industry, Manufacturing, General Production Expectations, Balance of Replies, SA	Macrobond	SA	0
France, Business Surveys, DG ECFIN, Retail Trade Confidence Indicator, Balance, SA	Macrobond	SA	0
France, Business Surveys, DG ECFIN, Construction Confidence Indicator, Balance, SA	Macrobond	SA	0
France, Business Surveys, Bank of France, Industry, Current Order Books, Manufacture of Food Products, Beverages & Tobacco Products, SA	Macrobond	SA	0
France, Business Surveys, Bank of France, Industry, Current Order Books, Manufacture of Electrical, Computer & Electronic Equipment, Manufacture of Machinery, SA	Macrobond	SA	0
France, Business Surveys, Bank of France, Industry, Current Order Books, Computer, Electronic & Optical Products, SA	Macrobond	SA	0
France, Business Surveys, Bank of France, Industry, Current Order Books, Machinery & Equipment, SA	Macrobond	SA	0
France, Business Surveys, Bank of France, Industry, Current Order Books, Transport Equipment, SA	Macrobond	SA	0
France, Business Surveys, Bank of France, Industry, Current Order Books, Automotive Industry, SA	Macrobond	SA	0
France, Business Surveys, Bank of France, Industry, Current Order Books, Other Transport Equipment, SA	Macrobond	SA	0
France, Business Surveys, Bank of France, Industry, Current Order Books, Other Manufacturing, SA	Macrobond	SA	0
France, Business Surveys, Bank of France, Industry, Current Order Books, Metal & Metal Products Manufacturing, SA	Macrobond	SA	0
France, Business Surveys, Bank of France, Industry, Current Order Books, Other Manufacturing Industries (Including Repair & Installation of Machinery), SA	Macrobond	SA	0
Services			
France, Service Surveys, DG ECFIN, Services Confidence Indicator, Balance, SA	Macrobond	SA	0
France, Service Surveys, INSEE, Services, Past Trend of Employment, All Non-Temporary Services, Including Transportation, Balance of Replies, SA	Macrobond	SA	0
France, Service Surveys, INSEE, Services, Expected Trend of Activity, All Non-Temporary Services, Including Transportation, Balance cf Replies, SA	Macrobond	SA	0
France, Service Surveys, INSEE, Services, Past Trend of Activity, All Non-Temporary Services, Including Transportation, Balance cf Replies, SA	Macrobond	SA	0

Table A1. *(Continued)*

Series Full Name	Source	SA	Lag
Retail trade			
France, Business Surveys, DG ECFIN, Retail Trade Confidence Indicator, Business Activity (Sales) Development over the Past 3 Months, Balance, SA	Macrobond	SA	0
France, Business Surveys, DG ECFIN, Retail Trade Confidence Indicator, Business Activity Expectations over the Next 3 Months, Balance, SA	Macrobond	SA	0
France, Business Surveys, DG ECFIN, Retail Trade Confidence Indicator, Employment Expectations over the Next 3 Months, Balance, SA	Macrobond	SA	0
France, OECD MEI, Retail Trade Orders Intentions, SA	Macrobond	SA	0
Prices			
France, Consumer Price Index, Total, Index	Macrobond		0
France, Consumer Price Index, Housing, Water, Electricity, Gas & Other Fuels, Rent of Primary Residence, Index	Macrobond		0
France, Eurostat, Producer Prices Index, Domestic Market, Manufacture of Plastics Products, Change P/P	Macrobond		1
Germany, Bundesbank, Price of Gold in London, Afternoon Fixing *, 1 Ounce of Fine Gold = USD ..., USD	Macrobond		0
World, IMF IFS, International Transactions, Export Prices, Linseed Oil (Any Origin)	Macrobond		6
Commodity Indices, UNCTAD, Price Index, End of Period, USD	Macrobond		0
Financial Sector			
Indexes			
NYSE Euronext Paris, cac40 (^ FCHI), price index, beginning of period, EUR	Macrobond		0
United Kingdom, Equity Indices, FTSE, All-Share, Index, Price Return, End of Period, GBP	Macrobond		0
Germany, Bundesbank, Capital Market Statistics, General Survey, Key Figures from the Capital Market Statistics 2, DAX Performance Index, End 1987 = 1000, End of Month, Index	Macrobond		0
Japan, Economic Sentiment Surveys, ZEW, Financial Market Report, Stock Market, Nikkei 225, Balance	Macrobond		0
United States, Equity Indices, S&P, 500, Index (Shiller), Cyclically Adjusted P/E Ratio (CAPE)	Macrobond		0
Exchange rates			
France, FX Indices, BIS, Real Effective Exchange Rate Index, CPI Based, Broad	Macrobond		
France, FX Indices, BIS, Nominal Effective Exchange Rate Index, Broad	Macrobond		0
REER Euro/Chinese yuan, CPI deflated	BCE		0
REER Euro/UK pound, CPI deflated	BCE		0

REER Euro/Japanese yen, CPI deflated	BCE		0
REER Euro/US dollar, CPI deflated	BCE		0
Interest Rates			
France, 3 months treasury bills, reference interest rate – monthly average	BDF		0
France, 12 months treasury bills, reference interest rate – monthly average	BDF		0
France, Government Benchmarks, Eurostat, Government Bond, 10 Year, Yield	Macrobond		0
Loans			
France, Deposits & Loans, Credit Institutions, Loans, By Entity, to Domestic Non-Financial Corporations, Loans Adjusted for Sales & Securitisation, Total, EUR	Macrobond		1
France, Deposits & Loans, Credit Institutions, Loans, By Entity, to Domestic Non-Financial Corporations, Investment Loans Adjusted for Sales & Securitisation, Total, EUR	Macrobond		1
France, Deposits & Loans, Credit Institutions, Loans, By Entity, to Domestic Non-Financial Corporations, Short-Term Loans Adjusted for Sales & Securitisation, Total, EUR	Macrobond		1
France, Deposits & Loans, Credit Institutions, Loans, By Entity, to Domestic Non-Financial Corporations, Other Loans Adjusted for Sales & Securitisation, Total, EUR	Macrobond		1
Monetary Aggregates			
France, Monetary Aggregates, M1, Total, EUR	Macrobond		1
France, Monetary Aggregates, M2, Total, EUR	Macrobond		2
France, Monetary Aggregates, M3, Total, EUR	Macrobond		2
International			
Germany, Economic Sentiment Surveys, ZEW, Financial Market Report, Current Economic Situation, Balance	Macrobond		0
Germany, OECD MEI, Manufacturing Business Situation Present, SA	Macrobond	SA	1
Germany, OECD MEI, Production Of Total Industry, SA, Index	Macrobond	SA	3
United States, Employment, CPS, 16 Years & Over, SA	Macrobond	SA	1
United States, Unemployment, CPS, 16 Years & Over, Rate, SA	Macrobond	SA	1
United States, Industrial Production, Total, SA, Index	Macrobond	SA	1
United States, Domestic Trade, Retail Trade, Retail Sales, Total, Calendar Adjusted, SA, USD	Macrobond	SA	1
United States, Industrial Production, Industry Group, Manufacturing, Total (SIC), SA, Index	Macrobond	SA	1
United States, Equity Indices, S&P, 500, Index, Price Return, End of Period, USD	Macrobond		1

Table A2. List of Series Used for the One-Step Estimation.

No.	Series Name	Publication Lag
1	France, Capacity Utilization, Total Industry, SA	1
2	France, Consumer Surveys, INSEE, Consumer Confidence Indicator, Synthetic Index, SA	0
3	France, Domestic Trade, Vehicle Sales & Registrations, New, Passenger Cars, Total, Calendar Adjusted, SA	0
4	France, OECD MEI, Retail Trade Orders Intentions, SA	0
5	France, OECD MEI, CPI All Items, Change Y/Y	3
6	France, OECD MEI, Production Of Total Industry, SA, Index	3
7	France, OECD MEI, Total Retail Trade (Volume), SA, Change P/P	1
8	France, Metropolitan, Unemployment, Job Seekers, Men, Total, Categories A, B & C, Calendar Adjusted, SA	1
9	France, OECD MEI, Manufacturing Finished Goods Stocks Level, SA	0
10	France, Economic Sentiment Surveys, ZEW, Financial Market Report, Stock Market, CAC-40, Balance	1
11	France, Foreign Trade, Trade Balance, Calendar Adjusted, SA, EUR	1
12	France, Foreign Trade, Export, Calendar Adjusted, SA, EUR	2
13	France, Foreign Trade, Import, Calendar Adjusted, SA, EUR	2
14	France, 3 months treasury bills, reference interest rate – monthly average	3
15	France, 12 months treasury bills, reference interest rate – monthly average	3
16	United Kingdom, Equity Indices, FTSE, All-Share, Index, Price Return, End of Period, GBP	0
17	Japan, Economic Sentiment Surveys, ZEW, Financial Market Report, Stock Market, Nikkei 225, Balance	0
18	United States, Equity Indices, S&P, 500, Index (Shiller), Cyclically Adjusted P/E Ratio (CAPE)	0
19	France, OECD MEI, Manufacturing Business Situation Future, SA	0
20	France, OECD MEI, Manufacturing Industrial Confidence Indicator, SA	3
21	France, Business Surveys, INSEE, Building Industry, Global, Expected Activity	0
22	France, Business Surveys, INSEE, Building Industry, Global, Order Books Level	0
23	France, Business Surveys, DG ECFIN, Retail Trade Confidence Indicator, Balance, SA	0
24	France, Business Surveys, DG ECFIN, Construction Confidence Indicator, Balance, SA	0
25	France, Service Surveys, DG ECFIN, Services Confidence Indicator, Balance, SA	0

APPENDIX B. ONE-STEP ESTIMATION RESULTS

Table B1. Frequency of 25 French Economic Indicators in 72 Successful Combinations for One-Step Estimation.

No.	Frequency	Name of Series
24	22	France, Business Surveys, DG ECFIN, Construction Confidence Indicator, Balance, SA (Business survey)
1	21	France, Capacity Utilization, Total Industry, SA
12	20	France, Foreign Trade, Export, Calendar Adjusted, SA, EUR
8	19	France, Metropolitan, Unemployment, Job Seekers, Men, Total, Categories A, B & C, Calendar Adjusted, SA
23	19	France, Business Surveys, DG ECFIN, Retail Trade Confidence Indicator, Balance, SA (Business survey)
11	18	France, Foreign Trade, Trade Balance, Calendar Adjusted, SA, EUR
4	16	France, OECD MEI, Retail Trade Orders Intentions, SA (Business survey)
7	15	France, OECD MEI, Total Retail Trade (Volume), SA, Change P/P
9	15	France, OECD MEI, Manufacturing Finished Goods Stocks Level, SA
17	13	Japan, Economic Sentiment Surveys, ZEW, Financial Market Report, Stock Market, Nikkei 225, Balance
18	12	United States, Equity Indices, S&P, 500, Index (Shiller), Cyclically Adjusted P/E Ratio (CAPE)
13	11	France, Foreign Trade, Import, Calendar Adjusted, SA, EUR
6	10	France, OECD MEI, Production Of Total Industry, SA, Index
14	10	France, 3 months treasury bills, reference interest rate – monthly average
16	10	United Kingdom, Equity Indices, FTSE, All-Share, Index, Price Return, End of Period, GBP
22	10	France, Business Surveys, INSEE, Building Industry, Global, Order Books Level
3	8	France, Domestic Trade, Vehicle Sales & Registrations, New, Passenger Cars, Total, Calendar Adjusted, SA
25	8	France, Service Surveys, DG ECFIN, Services Confidence Indicator, Balance, SA
2	7	France, Consumer Surveys, INSEE, Consumer Confidence Indicator, Synthetic Index, SA
15	7	France, 12 months treasury bills, reference interest rate monthly average
19	6	France, OECD MEI, Manufacturing Business Situation Future, SA
20	5	France, OECD MEI, Manufacturing Industrial Confidence Indicator, SA
21	4	France, Business Surveys, INSEE, Building Industry, Global, Expected Activity
5	2	France, OECD MEI, CPI All Items, Change Y/Y
10	0	France, Economic Sentiment Surveys, ZEW, Financial Market Report, Stock Market, CAC-40, Balance

Note: The numbers in the last column stand for the length of lag of data updates publication, in months.

Table B2. Crises and Their Most Descriptive Sets of Economic Indicators.

Crisis	Composition				FPS	QPS
March 1992–October 1993	7	8	11	22	0.0000	0.0142
	9	12	18	23	0.0000	0.0331
	9	14	17	23	0.0000	0.0580
April 1995–January 1997	3	4	11	24	0.1828	0.1141
	6	8	12	25	0.1828	0.1171
	7	8	11	22	0.1828	0.1216
January 2001–June 2003	3	9	12	25	0.0000	0.0108
	2	11	16	24	0.0000	0.0460
	4	7	17	24	0.0000	0.0477
January 2008–June 2009	3	9	12	25	0.0000	0.0108
	2	11	16	24	0.0000	0.0460
	4	7	17	24	0.0000	0.0477
October 2011–January 2013	7	8	11	22	0.0000	0.0205
	3	9	12	25	0.0625	0.0828
	7	8	11	23	0.0625	0.0870

Notes: Here, the QPS and FPS are calculated for each recession period only. See Table 4B1 for the series corresponding to the numbers in the column "Combinations."

Table B3. Top 25 Combinations with the Lowest QPS, FPS and the Highest Corr. The First Eight Entries belong to the Best 10% by Three Indicators Simultaneously.

Retained Combinations	Rating	Component Series				FPS	QPS	Corr	Factor Loadings			
Combination 1	1	*4*	7	17	*24*	35	0.1287	0.7155	*0.3665*	0.1006	−0.0026	*0.4463*
–	7	*8*	12	23	24	49	0.1493	0.7554	*−0.1454*	0.0247	0.1301	0.0005
Combination 2	2	2	11	*16*	*24*	39	0.1315	0.6899	0.1775	−0.0584	*0.6137*	*0.3779*
Combination 3	3	3	4	11	*24*	45	0.1254	0.7491	0.0001	0.1807	−0.0461	*0.8876*
Combination 4	4	4	9	*19*	24	46	0.1328	0.7006	0.3058	−0.0799	*0.6722*	0.3337
Combination 5	5	*4*	7	11	*24*	46	0.1412	0.7082	*0.3139*	0.0933	−0.0768	*0.3958*
Combination 6	6	8	18	23	*24*	47	0.1184	0.6815	0.0738	−0.0452	−0.0805	*−0.3292*
Combination 7	8	7	*8*	11	23	50	0.1492	0.5607	0.0216	*−0.1829*	−0.0371	−0.0013
–	9	8	9	15	23	54	0.1674	0.5506	−0.1125	−0.0945	0.0344	0.1510
–	10	1	8	16	22	58	0.1963	0.5929	0.4595	0.0025	0.7876	0.0015
–	11	1	9	13	18	58	0.2135	0.5346	0.0983	−0.1404	0.0905	−0.0010
–	12	4	9	12	24	59	0.1656	0.6488	0.2292	−0.0746	0.1633	0.6130
–	13	1	4	13	15	62	0.2091	0.5255	0.2255	0.1715	0.2888	−0.0014
–	14	4	8	11	23	64	0.1724	0.5033	0.0880	−0.2649	−0.0377	−0.0031
–	15	4	8	19	24	64	0.1795	0.6511	0.3272	−0.0022	0.5487	0.3608
–	16	15	16	17	24	64	0.2092	0.5597	−0.0010	1.0500	−0.0027	0.3222
–	17	9	12	18	23	65	0.1993	0.5321	−0.2665	0.0362	0.0384	0.1614
–	18	18	22	23	24	67	0.2147	0.4249	0.0219	0.0130	0.0738	0.0002
–	19	7	14	24	25	68	0.2233	0.4967	0.0016	−0.0004	0.5101	0.0428
–	20	7	8	11	24	69	0.1702	0.6538	−0.1141	−0.0003	−0.0066	−0.8400
–	21	14	15	18	24	70	0.1799	0.5312	0.0636	0.0063	0.0082	0.7306
–	22	11	16	21	22	71	0.1862	0.5639	−0.0006	0.9812	0.0170	−0.0007
–	23	1	12	13	17	71	0.1863	0.5439	0.8973	0.2825	0.3013	0.0495
–	24	6	8	18	22	71	0.2058	0.4326	0.0262	−0.2072	0.0283	0.0328
	25	12	14	20	23	72	0.1912	0.5221	0.1444	0.0687	0.6750	0.1026

Notes: The series with the highest loadings are in italic. Retained combinations are the combinations retained for the one-step analysis. The second-ranked combination in not included as it produces extra signals. See Table 4B1 for the series corresponding to the numbers in the column "Combinations."

One-step combination 1

4 - France, OECD MEI, Retail Trade Orders Intentions, SA
7 - France, OECD MEI, Total Retail Trade (Volume), SA, Change P/P
17 - Japan, Economic Sentiment Surveys, ZEW, Financial Market Report, Stock Market, Nikkei 225, Balance
24 - France, Business Surveys, DG ECFIN, Construction Confidence Indicator, Balance, SA

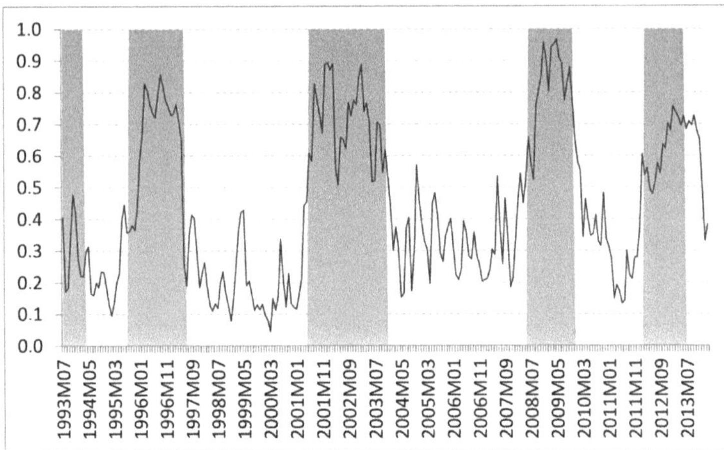

Fig. B1. Results of One-Step Estimation: Filtered Probability to Be in Recession in a Current Period (Solid Line) versus OECD Recession Dating (Shaded Area, 1 Corresponds to Recession, 0 to Expansion).

One-step combination 2

3 - France, Domestic Trade, Vehicle Sales & Registrations, New, Passenger Cars, Total, Calendar Adjusted, SA
4 - France, OECD MEI, Retail Trade Orders Intentions, SA
11 - France, Foreign Trade, Trade Balance, Calendar Adjusted, SA, EUR
24 - France, Business Surveys, DG ECFIN, Construction Confidence Indicator, Balance, SA

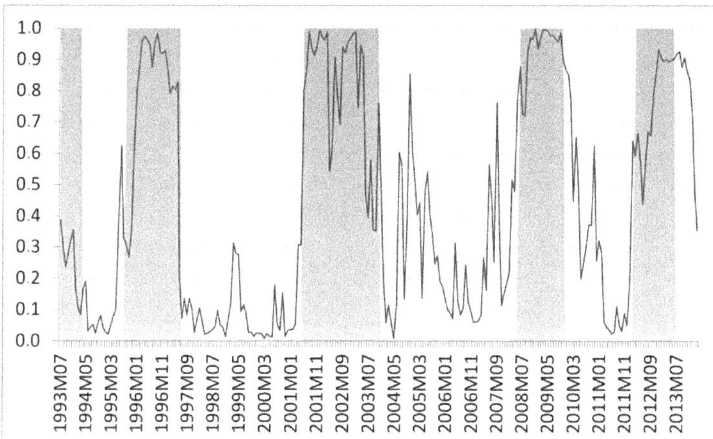

Fig. B1. (*Continued*)

One-step combination 3

4 - France, OECD MEI, Retail Trade Orders Intentions, SA
9 - France, OECD MEI, Manufacturing Finished Goods Stocks Level, SA
19 - France, OECD MEI, Manufacturing Business Situation Future, SA
24 - France, Business Surveys, DG ECFIN, Construction Confidence Indicator, Balance, SA

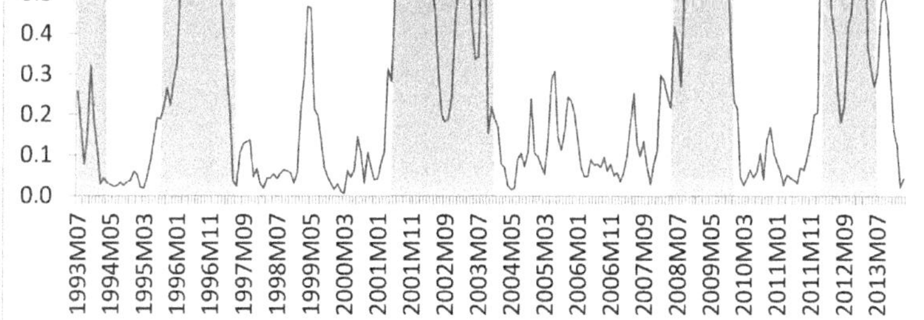

Fig. B1. (*Continued*)

One-step combination 4

4 - France, OECD MEI, Retail Trade Orders Intentions, SA
7 - France, OECD MEI, Total Retail Trade (Volume), SA, Change P/P
11 - France, Foreign Trade, Trade Balance, Calendar Adjusted, SA, EUR
24 - France, Business Surveys, DG ECFIN, Construction Confidence Indicator, Balance, SA

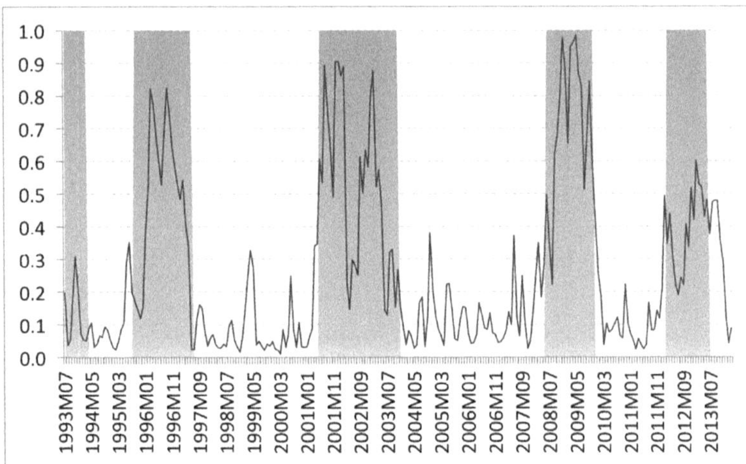

Fig. B1. (*Continued*)

One-step combination 5

8 - Unemployed, total
18 - United States, Equity Indices, S&P, 500, Index (Shiller), Cyclically Adjusted P/E Ratio (CAPE)
23 - France, Business Surveys, DG ECFIN, Retail Trade Confidence Indicator, Balance, SA
24 - France, Business Surveys, DG ECFIN, Construction Confidence Indicator, Balance, SA

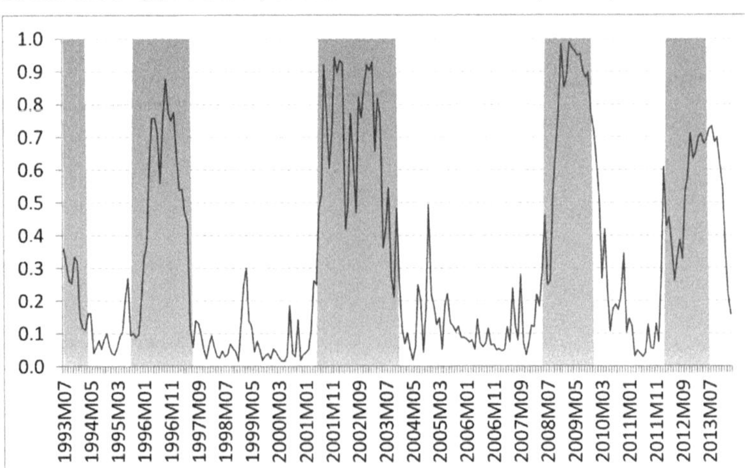

Fig. B1. (*Continued*)

One-step combination 6

8 - Unemployed, total
12 - France, Foreign Trade, Export, Calendar Adjusted, SA, EUR
23 - France, Business Surveys, DG ECFIN, Retail Trade Confidence Indicator, Balance, SA
24 - France, Business Surveys, DG ECFIN, Construction Confidence Indicator, Balance, SA

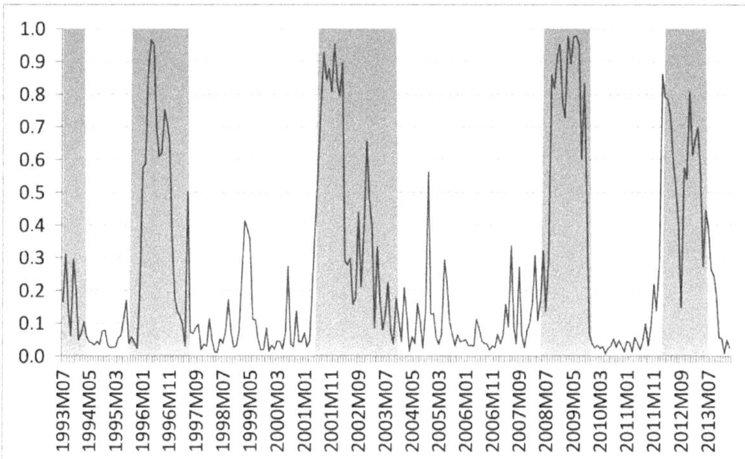

Fig. B1. (Continued)

One-step combination 7

7 - France, OECD MEI, Total Retail Trade (Volume), SA, Change P/P
8 - Unemployed, total
11 - France, Foreign Trade, Trade Balance, Calendar Adjusted, SA, EUR
23 - France, Business Surveys, DG ECFIN, Retail Trade Confidence Indicator, Balance, SA

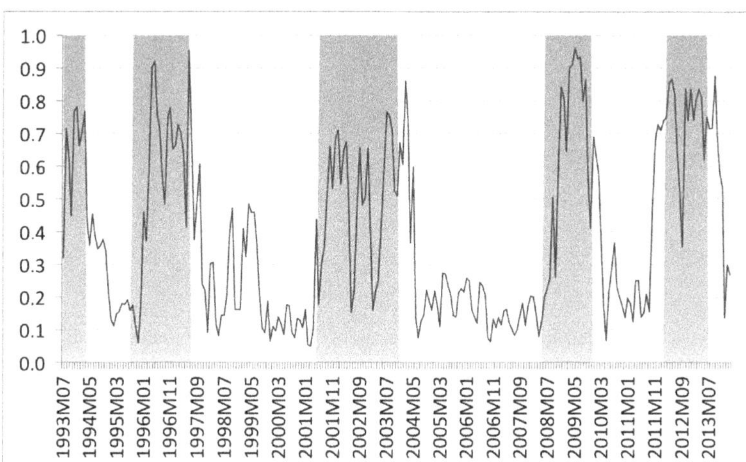

Fig. B1. (Continued)

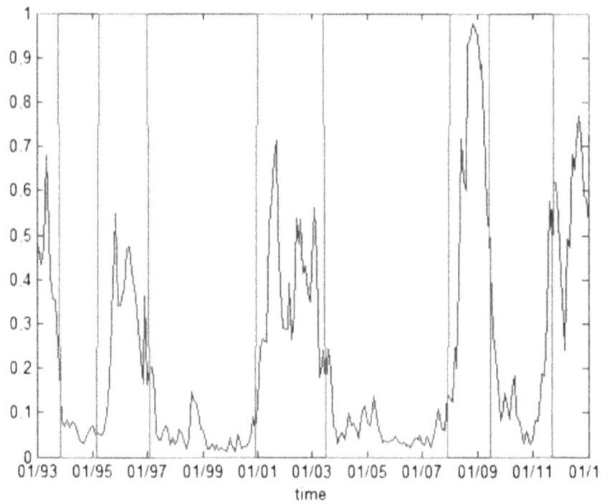

Series:
4 - France, OECD MEI (Enquete de Conjoncture INSEE), Retail Trade Orders Intentions, SA;
8 - Unemployed, total
19 - France, OECD MEI (Enquete de Conjoncture INSEE), Manufacturing Business Situation Future, SA
24 - France, Business Surveys, DG ECFIN, Construction Confidence Indicator, Balance, SA
2 - France, Consumer Surveys, INSEE, Consumer Confidence Indicator, Synthetic Index, SA
3 - France, Domestic Trade, Vehicle Sales & Registrations, New, Passenger Cars, Total, Calendar Adjusted, SA
7 - France, OECD MEI, Total Retail Trade (Volume), SA, Change P/P
9 - France, OECD MEI, Manufacturing Finished Goods Stocks Level, SA
11 - France, Foreign Trade, Trade Balance, Calendar Adjusted, SA, EUR
12 - France, Foreign Trade, Export, Calendar Adjusted, SA, EUR
17 - Japan, Economic Sentiment Surveys, ZEW, Financial Market Report, Stock Market, Nikkei 225, Balance
18 - United States, Equity Indices, S&P, 500, Index (Shiller), Cyclically Adjusted P/E Ratio (CAPE)
23 - France, Business Surveys, DG ECFIN, Retail Trade Confidence Indicator, Balance, SA

Fig. B2. The Filtered Probability of Recession, Estimated with One-Step Method on 13 Series of the Finally Selected Information Set (Solid Line) versus OECD Recession Dating (Vertical Lines).

APPENDIX C. ESTIMATION RESULTS FOR ONE-STEP AND TWO-STEP METHODS

Table C1. Estimated Parameters, One-Step and Two-Step Methods.

Parameters	Two-Step		One-Step						
	(Switch in μ)	(Switca in μ and σ^2)	Comb 1	Comb 2	Comb 3	Comb 4	Comb 5	Comb 6	Comb 7
ϕ_1	0.0010	0.0012	0.0018*	−0.0142*	−0.0031*	−0.0033*	0.8348*	0.1864*	0.0753*
ϕ_2	0.8926*	0.8685*	0.0016	−0.0070	0.0021*	0.0049	−0.2047*	0.7788*	0.6955*
ψ_{11}		–	−0.4524*	−0.4922*	−0.3708*	−0.3659*	0.0027	0.0626*	−0.7269*
ψ_{12}		–	0.0046*	−0.2011*	−0.0063*	−0.0012	−0.0027*	0.1096*	−0.4214*
ψ_{21}		–	−0.7354*	−0.3727*	0.9060*	−0.7475*	0.9889*	−0.5624*	0.2608*
ψ_{22}		–	−0.4208*	0.0016	0.0119	−0.4271*	−0.0003*	−0.2637*	0.3266*
ψ_{31}		–	0.8955*	−0.6637*	0.1708*	−0.6240*	0.5139*	0.4689*	−0.6276*
ψ_{32}		–	−0.0046	−0.2873*	0.1148*	−0.2025*	0.3528*	0.2521*	−0.2046*
ψ_{41}		–	−0.0023	−0.0043*	−0.0014*	−0.0056	0.0007*	0.0040*	0.0007*
ψ_{42}		–	−0.0025*	−0.0075*	−0.0022*	−0.1115*	−0.0003	−0.0043*	−0.0031*
σ_1		–	0.6622	0.6343	0.6470	0.6589	0.6228	0.7256	0.6714
σ_2		–	0.6736	0.6331	0.8150	0.6823	0.9353	0.6539	0.7130
σ_3		–	0.7925	0.6590	0.7136	0.6589	0.7410	0.7500	0.6609
σ_4		–	0.7244	1.0006	0.6676	0.7260	0.6853	0.6905	0.7864
γ_1		–	0.3665*	0.0001*	0.3058*	0.3139*	0.0738*	−0.1454*	0.0216*
γ_2		–	0.1006*	0.1807*	−0.0799*	0.0933*	−0.0452*	0.0247*	−0.1829*
γ_3		–	−0.0026*	−0.0461*	0.6722*	−0.0768*	−0.0805*	0.1301*	−0.0371*
γ_4		–	0.4463*	0.8876*	0.3337*	0.3958*	−0.3292*	0.0005*	−0.0013*
μ_0	1.0452*	1.2251*	0.3089*	0.4710*	0.2107*	0.3075*	0.8431*	0.6056*	0.7906*
μ_1	−1.7789*	−1.5245*	−0.3162*	−0.5072*	−0.7062*	−0.8129*	−0.2725*	−1.6863*	−1.1919*

Table C1. *(Continued)*

Parameters	Two-Step		One-Step						
	(Switch in μ)	(Switch in μ and σ^2)	Comb 1	Comb 2	Comb 3	Comb 4	Comb 5	Comb 6	Comb 7
$\sigma_{\eta 0}$	0.5770*	0.4028*	–	–	–	–	–	–	–
$\sigma_{\eta 1}$	0.5770*	0.7524*	–	–	–	–	–	–	–
p_0	0.9532*	0.9432*	0.9549	0.9585	0.9728	0.9673	0.9636	0.9650	0.9352
p_1	0.9029*	0.9149*	0.9442	0.9525	0.9284	0.9120	0.9385	0.8949	0.9011

Notes: For the composition of Combination *i* see Table B3 and A2. Estimates marked with * are significant on 5% level of confidence probability. $\sigma_{\eta 0}$ and $\sigma_{\eta 1}$ stand for the standard error of η_t (the stochastic term in factor dynamics) in expansion and recession states, respectively.

COMMON FAITH OR PARTING WAYS? A TIME VARYING PARAMETERS FACTOR ANALYSIS OF EURO-AREA INFLATION

Davide Delle Monache[a], Ivan Petrella[b] and Fabrizio Venditti[a]

[a]*Bank of Italy, Rome, Italy*
[b]*Bank of England, Birkbeck University of London and CEPR, London, UK*

ABSTRACT

We analyze the interaction among the common and country-specific components for the inflation rates in 12 euro area countries through a factor model with time-varying parameters. The variation of the model parameters is driven by the score of the predictive likelihood, so that, conditionally on past data, the model is Gaussian and the likelihood function can be evaluated using the Kalman filter. The empirical analysis uncovers significant variation over time in the model parameters. We find that, over an extended time period, inflation persistence has fallen and the

Dynamic Factor Models
Advances in Econometrics, Volume 35, 539–565

ISSN: 0731-9053/doi:10.1108/S0731-905320150000035013

importance of common shocks has increased relatively to that of idiosyncratic disturbances. According to the model, the fall in inflation observed since the sovereign debt crisis is broadly a common phenomenon since no significant cross-country inflation differentials have emerged. Stressed countries, however, have been hit by unusually large shocks.

Keywords: inflation; time-varying parameters; score driven models; state space models; dynamic factor models

JEL classification: E31; C22; C51; C53

1. INTRODUCTION

Inflation has fallen sharply and unexpectedly in most euro area countries following the sovereign debt crisis. At the outset, these developments were perceived as partly temporary, as they were mainly driven by a stabilization of energy prices, and partly idiosyncratic, as the fall in inflation was sharper in countries most hardly hit from the debt crisis (the so-called stressed countries: Greece, Ireland, Italy, Portugal, and Spain). However, inflation weakness persisted long enough to raise fear of a prolonged period of low inflation (*lowflation*), prompting the ECB to deploy a range of unconventional measures, in the attempt to prevent the negative shocks that affected actual inflation to be passed through to inflation expectations. Moreover, the deceleration of consumer prices has spread beyond the stressed countries, as inflation rates touched historically low levels in other countries, like Germany and the Netherlands. The issue goes beyond euro area borders since, depending on how long it takes for euro area inflation to respond to the monetary stimulus, it could imply an asynchronous exit of the monetary policies from the unconventional measures across the world.

In this paper we develop a tool that, starting from national inflation rates, allows us to separate permanent from transitory shocks and common from idiosyncratic components. The model features time variation in the parameters so that it can capture the large number of regime shifts that inflation rates of the countries that form the European Monetary Union (EMU) have experienced since the 1980s. Specifically, the model we propose is a (single) factor model with time-varying volatilities, in which each national inflation rate is separated in a stochastic trend that is common across-countries, and a national component. The latter is less persistent than the common trend, but features a time-varying intercept that captures long-lasting deviations of national inflation from the common trend. The model provides a real-time decomposition of permanent and transitory shocks and as such could be used in real time to monitor imbalances within the euro area, as motivated by the argument developed in Corsetti and Pesaran (2012).

From a methodological point of view, our work is related to two areas of the econometric literature. The first is the work by Creal, Koopman, and Lucas (2013) and by Harvey (2013) on score-driven models where the parameters vary over time as function of the likelihood score. Delle Monache, Petrella, and Venditti (2015) show how the score-driven approach can be used to model parameter variation in a Gaussian state-space model. In this paper, we restrict the general approach therein developed, to a more parsimonious specification. The introduction of time variation in the parameters makes these models highly nonlinear and possibly non-Gaussian, so that computationally intensive simulation-based methods are typically required for estimation. In contrast, when the parameters are driven by the score, the model remains Gaussian, conditional on past data. In this case, Delle Monache et al. (2015) develop a set of recursions that, running in parallel with the standard Kalman filter, allow the evaluation of the score vector and the update of the parameters at each point in time. The likelihood function, which remains Gaussian, can then be evaluated by the means of the Kalman filter and maximized through standard procedures. A second stream of the econometric literature related to our work deals with dynamic factor models, see, for example, Giannone, Reichlin, and Small (2008) and Camacho and Perez-Quiros (2010). Within this branch of the literature our paper is close to the studies that extend traditional dynamic factor models to nonlinear settings, like those by Del Negro and Otrok (2008), Mumtaz and Surico (2012) and Marcellino, Porqueddu, and Venditti (2013). There are a number of differences between our method and those just mentioned. The most important one is that all these papers adopt a Bayesian standpoint and rely on computationally intensive Bayesian methods to estimate the model parameters. In our setup, estimation can be carried out with straightforward maximization of the likelihood function, with some advantages in terms of computational simplicity.

Besides these methodological considerations, our work is relevant for applied economists with an interest in inflation modeling. The topic, which has always attracted much attention in the empirical literature, has received renewed interest in recent years due to the fact that the Great Recession had only a muted impact on consumer prices, an outcome that was at odds with the prediction of existing theoretical models (see Del Negro, Giannoni, & Schorfheide, 2013). In this respect, the inflation rates within the euro area provide an extremely interesting laboratory for nonlinear adaptive models. They have, in fact, undergone a number of breaks and regime shifts in the past 30 years, going from a decade of extreme heterogeneity (the 1980s), through a period of rapid convergence (the 1990s), to a decade of irreversibly fixed exchange rates and centralized monetary policy (the 2000s). In recent years, much like in the United States, the euro area inflation proved to be relatively unresponsive to the Great Recession; it then dropped very rapidly when the Sovereign Debt crisis spread through a large number of countries, bringing the euro area to the brink of deflation. In our model, we capture this complicated narrative through three key ingredients: first, a common driving force that attracts national rates toward a common stochastic trend; second, idiosyncratic cycles that account for potential heterogeneity across countries; third, time-varying coefficients and variances, so that the relative importance of these factors can change over time. Our empirical setup is close in spirit to (but more general than) the models of trend inflation by Cogley (2002), Stock and Watson (2007), and Clark and Doh (2014), where trend/core inflation is modeled as a unit root process that underlies headline inflation. The main difference is that our trend inflation component is extracted from a panel of series, rather than through univariate models. Time variation in the model parameters separates our work from the measures of core inflation proposed by Cristadoro, Forni, Reichlin, and Veronese (2005) and Giannone and Matheson (2007), where core inflation is defined in the frequency domain and extracted from a large dataset, but has time invariant characteristics as opposed to changing variance. The relevance of this latter feature will become clear in the empirical application as we find that the contribution of the persistent common component to individual inflation rates has risen over time. Similar differences arise with respect to the concept of Global inflation proposed by Ciccarelli and Mojon (2010) and Ferroni and Mojon (2014), where a common component is computed from a set of inflation rates as the first principal component or, alternatively, as the cross-sectional mean. In their empirical applications, the common component turns out (but does not need to) be persistent, attractive, and useful for predicting national inflation. We show that both the persistence and the relevance of this common component have changed frequently over time. Finally, heterogeneity across euro area inflation rates is the focus of the paper by Busetti, Forni, Harvey, and Venditti (2007). The main finding of this study is that the cross-country inflation differentials can be very persistent in the EMU. This motivates the introduction in our model of national cycles that account for country-specific idiosyncratic shocks.

Our empirical analysis uncovers significant time variation in the parameters, therefore validating our modeling framework. We find that since the 80s, inflation persistence has gradually fallen, as inflation rates have been disciplined first by the exchange rates agreement underpinning the EMS, then by the ECB monetary policy. At the same time, the importance of common shocks has increased relative to that of the idiosyncratic ones, as a result of inflation convergence and common monetary policy. In more recent years, the stressed countries have been hit by unusually large idiosyncratic shocks, which monetary policy has not been able to neutralize. This, however, has not resulted in significant inflation differentials since, when we take into account filter uncertainty, the idiosyncratic components are not significantly different from zero. We conclude that, despite some short-term volatility in peripheral countries, the recent disinflation observed in the euro area is broadly a common phenomenon.

The rest of this paper is structured as follows. In Section 2, we present the model specification. Section 3 describes the estimation strategy. In particular, we adapt to our setup the algorithm for score-driven time-varying parameters developed in Delle Monache et al. (2015). In Section 4, we discuss the empirical analysis. Section 5 concludes.

2. A MODEL OF INFLATION TREND AND CYCLES

Our empirical setup is meant to capture the features of cross-country inflation differentials in a currency union with national fiscal policies. In this institutional framework, the nominal drift in the economy, which determines steady-state inflation, nominal wage growth and nominal interest rates, is set by the single monetary policy and is common across countries. In our model, this is driven by a random walk plus noise component that is shared by national inflation rates. At the same time, idiosyncratic productivity or fiscal shocks generate fluctuations in relative prices, which are reflected in inflation differentials. However, a persistent deviation from the common trend is equivalent to a persistent real appreciation and leads to competitiveness losses. If too prolonged, inflation differentials may become not sustainable, as current account deficits and debt problems may ensue (see, e.g., De Grauwe & Ji, 2013). In the absence of exchange rate adjustments, a relative price adjustment is bound to occur and national inflation rates can be expected to return toward the common trend. We therefore describe the behavior of the country-specific component of national inflation rates through a time-varying intercept, which reflects longer lasting deviations from the common trend, plus an autoregressive component. We restrict the latter to have non-explosive roots, as explained in details below. The model can therefore be used to assess the *real-time* state of convergence in the inflation rates, which is a key element of a well-functioning monetary union (see, e.g., Corsetti & Pesaran, 2012).

The model is described by the following equations:

$$
\begin{aligned}
\pi_{j,t} &= \mu_t + \psi_{j,t}, & j &= 1, \ldots, N, \quad t = 1, \ldots, n, \\
\mu_t &= \mu_{t-1} + \eta_t, & \eta_t &\sim N(0, \sigma^2_{\eta,t}), \\
\psi_{j,t} &= \gamma_{j,t} + \phi_{j,t}\psi_{j,t-1} + \kappa_{j,t}, & \kappa_{j,t} &\sim N(0, \sigma^2_{j,t})
\end{aligned}
\tag{1}
$$

where $\pi_{j,t}$ are the national inflation rates, μ_t is the common stochastic trend and $\psi_{j,t}$ the idiosyncratic components. We allow for time variation in all the elements of the model, namely the variance of the common stochastic trend $\sigma^2_{\eta,t}$, the intercept of the idiosyncratic process $\gamma_{j,t}$, the autoregressive coefficients of the idiosyncratic component $\phi_{j,t}$, and the variance of the idiosyncratic process $\sigma^2_{j,t}$. A compact state-space representation of the above model is the following:

$$
\begin{aligned}
y_t &= Z\alpha_t, & t &= 1, \ldots, n \\
\alpha_{t+1} &= T_t\alpha_t + \epsilon_t, & \epsilon_t &\sim N(0, Q_t)
\end{aligned}
\tag{2}
$$

where $y_t = \left[\pi_{1,t}, \ldots, \pi_{N,t}\right]'$ is an $N \times 1$ vector of observed inflation rates, α_t is the $m \times 1$ vector of state variables with dimension $m = N + 2$, and Z, T_t and Q_t are the system matrices of appropriate dimension, namely

$$\alpha_t = \begin{bmatrix} \mu_t & 1 & \psi_{1,t} & \cdots & \psi_{N,t}, \end{bmatrix}', \quad Z = \begin{bmatrix} 1_{N\times 1} & 0_{N\times 1} & I_N \end{bmatrix}$$

$$T_t = \begin{bmatrix} I_2 & 0_{2\times N} \\ 0_{N\times 1} & M_t \end{bmatrix}, \quad M_t = \begin{bmatrix} \gamma_{1,t} & \phi_{1,t} & 0 & \cdots & 0 \\ \gamma_{2,t} & 0 & \phi_{2,t} & \ddots & \vdots \\ \vdots & \vdots & \ddots & \ddots & 0 \\ \gamma_{N,t} & 0 & \cdots & 0 & \phi_{N,t} \end{bmatrix}, \tag{3}$$

$$Q_t = \begin{bmatrix} \sigma^2_{\eta,t} & 0 & \cdots & \cdots & 0 \\ 0 & 0 & \ddots & & \vdots \\ \vdots & \ddots & \sigma^2_{1,t} & \ddots & \vdots \\ \vdots & & \ddots & \ddots & 0 \\ 0 & \cdots & \cdots & 0 & \sigma^2_{N,t} \end{bmatrix}$$

Notice that the second element of the diagonal Q_t is set to zero. This means that the breaking intercepts $\gamma_{j,t}$, rather than being driven by a set of random errors like the common level μ_t, will be driven by the score. This point will become clearer below, where we formalize the treatment of the dynamics of the time-varying parameters and discuss the model estimation.

3. ESTIMATION

One way to model the time-varying elements in Eq. (3) is by specifying a law of motion where additional random shocks drive the changes in the parameters. In this case, the Kalman filter looses its optimality and Bayesian simulation methods need to be used, see, for example, Del Negro and Otrok (2008) and Mumtaz and Surico (2012). The alternative approach is to consider an observation-driven model to account for parameters variation as in Koopman, Mallee, and Van der Wel (2010). In this framework, the model remains Gaussian, conditionally on past data, and the likelihood can be computed by means of the Kalman filter and maximized with respect to the parameters of interest.

Recently, a new class of observation-driven models, the so called score-driven models, has been proposed by Creal et al. (2013) and Harvey (2013). The novelty of the approach is represented by the fact that the driver of the time variation is the score of the conditional likelihood. This implies that, at each point in time, the parameters are updated in the direction that maximizes the local fit (i.e., the predictive likelihood). The intuition is that, when the score is zero, the likelihood is at its maximum, so that there is no need to change the parameters. Within this framework, Delle Monache et al. (2015) develop an algorithm that allows to compute the score and to update the parameters for the general class of Gaussian state-space model.

We specialize the algorithm in Delle Monache et al. (2015) to our specific model (2) and (3). Collecting the time-varying parameters in the $k \times 1$ vector f_t, we posit the following law of motion

$$f_{t+1} = f_t + \Theta s_t, \quad t = 1, \ldots, n. \tag{4}$$

The matrix Θ contains the static parameters that govern the speed at which the parameters are updated from one period to the next. The driving mechanism is represented by the scaled score vector of the conditional distribution, $s_t = \mathcal{I}_t^{-1}\nabla_t$, where $\nabla_t = \partial \ell_t / \partial f_t$ is the score, $\mathcal{I}_t$ is a scaling matrix set to be equal to the information matrix $\mathcal{I}_t = -E_t(\partial^2 \ell_t / \partial f_t \partial f_t')$. Finally, ℓ_t is the likelihood function conditional on past observations $Y_{t-1} = \{y_{t-1}, \ldots, y_1\}$, the current value of f_t and the vector of static parameters θ, namely, $\ell_t = \log p(y_t | f_t, Y_{t-1}; \theta)$.

It is important to stress that the time-varying matrices in Eq. (3) are function of past observations only. This implies that the observation and the state vector are still conditionally Gaussian. Specifically, the conditional distribution of the observations and state are $y_t | f_t, Y_{t-1}; \theta \sim N(Za_t, F_t)$ and $\alpha_t | f_t, Y_{t-1}; \theta \sim N(a_t, P_t)$, respectively. Therefore, the log-likelihood function is equal to:

$$\ell_t = -\frac{1}{2}\left[\log(2\pi) + \log|F_t| + v_t' F_t^{-1} v_t\right] \tag{5}$$

and can be evaluated by the Kalman filter:

$$\begin{aligned} v_t &= y_t - Za_t, & F_t &= ZP_tZ', & K_t &= T_tP_tZ'F_t^{-1}, \\ a_{t+1} &= T_ta_t + K_tv_t, & P_{t+1} &= T_tP_tT_t' - K_tF_tK_t' + Q_t, & t &= 1, \ldots, n \end{aligned} \tag{6}$$

The dynamics of the model are completed by adding the recursions for the time-varying parameters. Therefore, at each point in time, the Kalman filter (6) needs to be augmented so that the score s_t can be computed and the vector f_t can be updated as in Eq. (4). Delle Monache et al. (2015) show that the score and the information matrix can be written as:

$$\begin{aligned}\nabla_t &= \frac{1}{2}\left[\dot{F}_t'(F_t^{-1}\otimes F_t^{-1})[\upsilon_t\otimes\upsilon_t - \text{vec}(F_t)] - 2\dot{V}_t'F_t^{-1}\upsilon_t\right],\\ \mathcal{I}_t &= \frac{1}{2}\left[\dot{F}_t'(F_t^{-1}\otimes F_t^{-1})\dot{F}_t + 2\dot{V}_t'F_t^{-1}\dot{V}_t\right]\end{aligned} \tag{7}$$

where "$\otimes$" denotes the Kronecker product, $\dot{V}_t$ and $\dot{F}_t$ denote the derivative of the prediction error, υ_t, and of its variance, F_t, with respect to the vector f_t. Those are computed recursively as follows:

$$\begin{aligned}\dot{V}_t &= -Z\dot{A}_t,\\ \dot{F}_t &= (Z\otimes Z)\dot{P}_t,\\ \dot{K}_t &= \left(F_t^{-1}ZP_t\otimes I_m\right)\dot{T}_t + \left(F_t^{-1}Z\otimes T_t\right)\dot{P}_t - \left(F_t^{-1}\otimes K_t\right)\dot{F}_t,\\ \dot{A}_{t+1} &= \left(a_t'\otimes I_m\right)\dot{T}_t + T_t\dot{A}_t + \left(\upsilon_t'\otimes I_m\right)\dot{K}_t + K_t\dot{V}_t,\\ \dot{P}_{t+1} &= (T_t\otimes T_t)\dot{P}_t - (K_t\otimes K_t)\dot{F}_t + \dot{Q}_t\\ &\quad + 2N_m\left[(T_tP_t\otimes I_m)\dot{T}_t - (K_tF_t\otimes I_m)\dot{K}_t\right]\end{aligned} \tag{8}$$

We have that I_m denotes the identity matrix of order m, and $N_m = \frac{1}{2}(I_{m^2} + C_m)$, where C_m is the commutation matrix.[1] The filter (7) and (8) runs in parallel with the Kalman filter (5) and (6), together with the updating rule (4).

A distinctive feature of this setup is that at each point in time we update simultaneously the time-varying parameters and the state vector of the model. In this respect, we differ from other methods proposed in the literature. In the two-step approach of Koop and Korobilis (2014), for example, given an initial guess of the state vector (usually obtained by Principal Components), the time-varying parameters are computed using the forgetting factor algorithm developed in Koop and Korobilis (2013). Then, conditional on the time-varying parameters, the state vector is estimated through the Kalman filter. This procedure is iterated subject to a stopping rule. It can be shown that such approach is nested as special case of the adaptive state-space model developed in Delle Monache et al. (2015).

The matrix Θ is restricted to be block diagonal, with the diagonal elements depending on the static parameters collected in the vector θ.

We opt for a very parsimonious specification and restrict the number of static parameters to three: one associated with the volatility of the common factor, one with the volatility of the idiosyncratic components and the last one with both the intercept and the autoregressive parameters of the idiosyncratic cycles.[2] The static parameters are estimated by maximum likelihood (ML), namely $\hat{\theta} = \text{argmax}\sum_{t=1}^{n} \ell_t(\theta)$, and the maximization is obtained numerically. Following Harvey (1989, p. 128) we have that $\sqrt{n}\left(\hat{\theta} - \theta\right) \to N(0, \Xi)$, where the asymptotic variance Ξ is evaluated by numerical derivative at the optimum as discussed in Creal et al. (2013, section 2.3).[3] The model is conditional Gaussian and the likelihood can be evaluated by prediction error decomposition (see Harvey, 1989, section 3.7.1). Note that our model requires a diffuse initialization, which is used when the state vector is non-stationary, and it is known to provide an approximation of the likelihood (Harvey, 1989, pp. 120–121). This implies that for Eqs. (5), (7), and (8) $t = d + 1, \ldots, n$, where d is the number of diffuse elements. In principle it would be possible to compute the diffuse likelihood via the augmented KF (see Durbin & Koopman, 2012, section 7.2.3) and therefore amend (5), (6), and (8) for $t = 1, \ldots, d$. However, this is beyond the scope of this paper.

We impose some restrictions on the model parameters. In particular, we let long-run forecasts be bounded by constraining the dynamics of the idiosyncratic components $\psi_{j,t}$ not to be explosive, and this is achieved by restricting the AR coefficients $\phi_{j,t} \in (-1, 1)$. Furthermore, we require the volatilities to be positive. The restrictions are achieved by a transformation of the time-varying parameters through a so-called *link function* that is invariant over time. In particular, we collect the parameters of interest in the vector:

$$\tilde{f}_t = \left[\sigma^2_{\eta,t}, \ \sigma^2_{1,t}, \ \ldots, \ \sigma^2_{N,t}, \ \gamma_{1,t}, \ \ldots, \ \gamma_{N,t}, \ \phi_{1,t}, \ \ldots, \ \phi_{N,t}\right]' \tag{9}$$

we define the link function:

$$\tilde{f}_t = g(f_t), \tag{10}$$

and the elements of $\tilde{f}_t$ fall in the desired region. These restrictions are formalized as follows:

$$\sigma^2_{\eta,t} = \exp(2\cdot), \quad \sigma^2_{j,t} = \exp(2\cdot), \quad \sum_{j=1}^{N} \gamma_{j,t} = 0, \quad \phi_{j,t} = \tan(\cdot), \quad \forall j \tag{11}$$

The first and the second constraints impose positive volatilities, the third one allows to identify the common trend from the idiosyncratic component and the last one leads to AR coefficients with stable roots. In practice, we model the following vector:

$$f_t = \left[\log\sigma_{\eta,t},\ \log\sigma_{1,t},\ \dots\ \log\sigma_{N,t},\ \tilde{\gamma}_{1,t},\ \dots,\ \tilde{\gamma}_{N,t},\ \arctan\phi_{1,t},\ \dots,\ \arctan\phi_{N,t}\right]'$$

where $\gamma_{j,t} = \tilde{\gamma}_{j,t} - (1/N)\sum_{j=1}^{N}\tilde{\gamma}_{j,t}$. Note that, in general, the time-varying parameters collected in f_t enter linearly the system matrices T_t and Q_t, so that $\dot{T}_t = \partial\text{vec}(T_t)/\partial f_t'$ and $\dot{Q}_t = \partial\text{vec}(Q_t)/\partial f_t'$ turn out to be selection matrices. When restrictions are implemented, like those in Eq. (11), the Jacobian of $\tilde{f}_t$ with respect to f_t must be taken into account. In this case the following representation can be used:

$$\dot{T}_t = S_{1T}\Psi_{T,t}S_{2T}, \quad \dot{Q}_t = S_{1Q}\Psi_{Q,t}S_{2Q}$$

where the Jacobian matrices $\Psi_{Q,t}$ and $\Psi_{T,t}$ are equal to

$$\Psi_{T,t} = \begin{bmatrix} \Psi_{\gamma,t} & 0 \\ 0 & \Psi_{\phi,t} \end{bmatrix}, \quad \Psi_{Q,t} = \begin{bmatrix} 2\sigma^2_{\eta t} & 0 & \cdots & 0 \\ 0 & 2\sigma^2_{1,t} & & \vdots \\ \vdots & & \ddots & \vdots \\ 0 & \cdots & \cdots & 2\sigma^2_{N,t} \end{bmatrix},$$
$$\Psi_{\gamma,t} = I_N - \frac{1}{N}1_{N\times N}, \quad \Psi_{\phi,t} = \begin{bmatrix} 1-\phi^2_{1,t} & \cdots & 0 \\ \vdots & \ddots & \vdots \\ 0 & \cdots & 1-\phi^2_{N,t} \end{bmatrix} \tag{12}$$

and S_{1T}, S_{2T}, S_{1Q}, S_{2Q} are selection matrices. Specifically, S_{1T} is constructed starting from the identity matrix of dimension $(N+2)^2$ and selecting only the columns associated with nonzero entries in $\text{vec}(T_t)$, similarly S_{1Q} retains only the columns associated with nonzero entries of $\text{vec}(Q_t)$, whereas S_{2T}, and S_{2Q} identify the positions of the time-varying elements of T_t and Q_t within the vector f_t, namely:

$$S_{2T} = \left[0_{2N\times(k-2N)} \quad I_{2N}\right];$$
$$S_{2Q} = \left[I_{N+1} \quad 0_{(N+1)\times(k-N-1)}\right]$$

Note that the filter for the time-varying parameters in Eq. (4) requires a starting value. We choose the initial values, f_1, by estimating the fixed parameters version of our model on a training sample (10 years of data) that we then discard. Specifically, we approximate the common factor by the cross-sectional average of the data (Pesaran, 2006)[4] and then estimate an AR(1) model on the deviation of each country's inflation from the common component. This allows us to easily compute the initial values $\sigma^2_{\eta,1}$ and $\{\gamma_{j,1}, \phi_{j,1}, \sigma^2_{j,1},\}_{j=1}^N$.

4. EMPIRICAL APPLICATION

We model a panel of 12 inflation rates from a sample of EMU countries (Austria, Belgium, Finland, France, Germany, Greece, Ireland, Italy, Luxembourg, The Netherlands, Portugal and Spain) from 1980:Q1 to 2014:Q3. For each country $\pi_{j,t}$ is the annualized percentage change over the previous quarter of the headline index of consumer prices, $400\times\left(\left(P_{j,t}/P_{j,t-1}\right)-1\right)$. The data source is the OECD Main Economic Indicators database and seasonal adjustment is carried out with Tramo-Seats (Gomez & Maravall, 1996).

A plot of the data is presented in Figs. 1 and 2 reports the cohesion of the inflation series at various frequencies.[5] Three features can be observed. First, there is an evident change in the level of inflation. In the majority of the countries considered here, inflation was in fact at two digits in the early 1980s; it then gradually converged through the 1990s to levels consistent with the Maastricht criteria,[6] and stabilized through the 2000s around the 2% rate targeted by the ECB. This process of convergence toward low and stable rates of inflation is reflected into the high cohesion of the series at low frequencies. In fact, the pronounced disinflationary trend in the euro area is nested within a global tendency toward lower inflation rates, as documented, for example, by Ciccarelli and Monjon (2010) and Mumtaz and Surico (2012). Second, there is a strong decline in the volatility of inflation, which was generally much higher in the first part of the sample, with the temporary exception of the 2008/2009 biennium, when consumer prices were strongly affected by the oil price shock that followed the global financial crisis. A fall in the volatility of inflationary shocks is a common finding in the literature that uses time-varying structural VARs (see, e.g., Benati & Mumtaz, 2007) and it bears important implications for the predictability of inflation, an issue to which we return below. Third, there is a marked

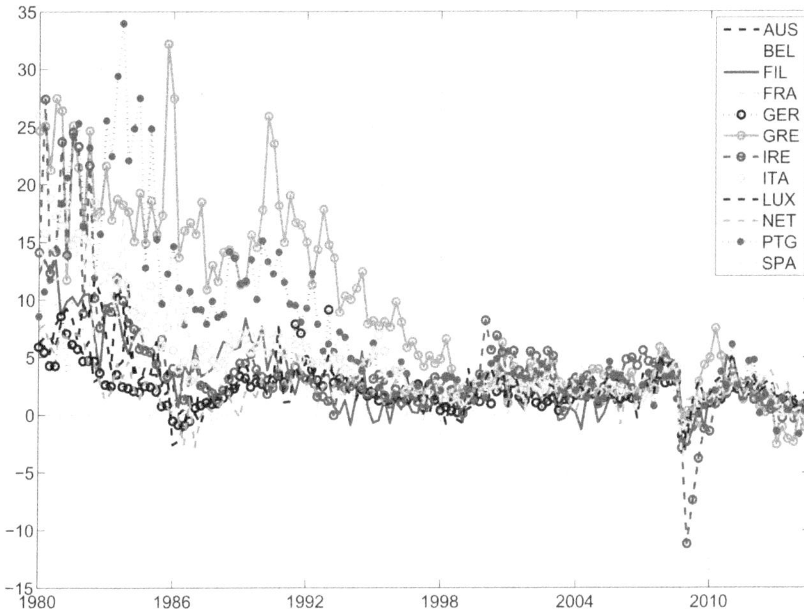

Fig. 1. Data.

increase in the co-movement of consumer prices among countries, as inflation rates become much more synchronized when the third stage of the EMU begins (1999), in particular at business cycle frequencies.[7] We devote the rest of the section to explaining how our model sheds light on these stylized facts.

4.1. The Level of Inflation

In Fig. 3, we present the data together with the estimate of the common stochastic trend μ_t and an alternative measure of co-movement, that is the first principal component of the inflation series. Ciccarelli and Mojon (2010) use principal components (PC) to obtain a measure of "global" inflation, essentially averaging across a relatively large panel of OECD inflation rates. The estimates presented in Fig. 3 can therefore be seen as the euro-area counterpart of their global inflation concept. The first observation is that the

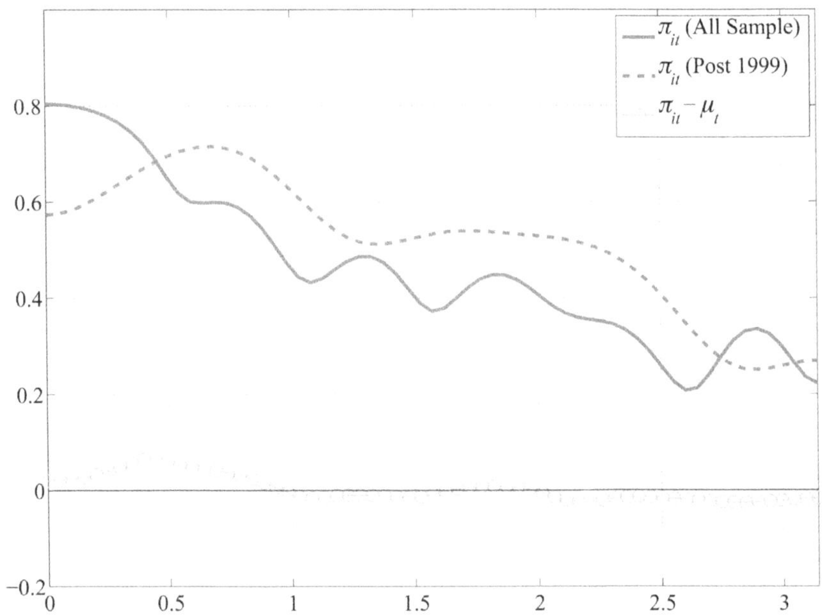

Fig. 2. Cohesion.

common stochastic trend estimated by our model captures very well the downward trend in inflation from the high levels of the 1980s to the moderate/low levels of the 1990s. It is also very smooth as a consequence of the random walk structure that we have imposed on it. Second, the common stochastic trend turns out to be highly correlated with the PC trend. This result, a priori, is not obvious since the latter is obtained through a non-parametric estimator while our common trend obeys a defined law of motion described by the transition equations. A theoretical connection between the two estimators is established by Bai (2004), where it is shown that, if data are generated by a non-stationary factor model, then PC deliver a consistent estimate of the non-stationary factors. This implies that, should a common stochastic trend be present in the data, the first PC would end up capturing it, which is in fact what happens in our case.[8] The common trend captures almost entirely the co-movement of the national inflation rates across all frequencies (see Fig. 2).[9] This highlights how a random walk specification for the common trend does not artificially constrain the common component to capture only the low-frequency variations in the panel.

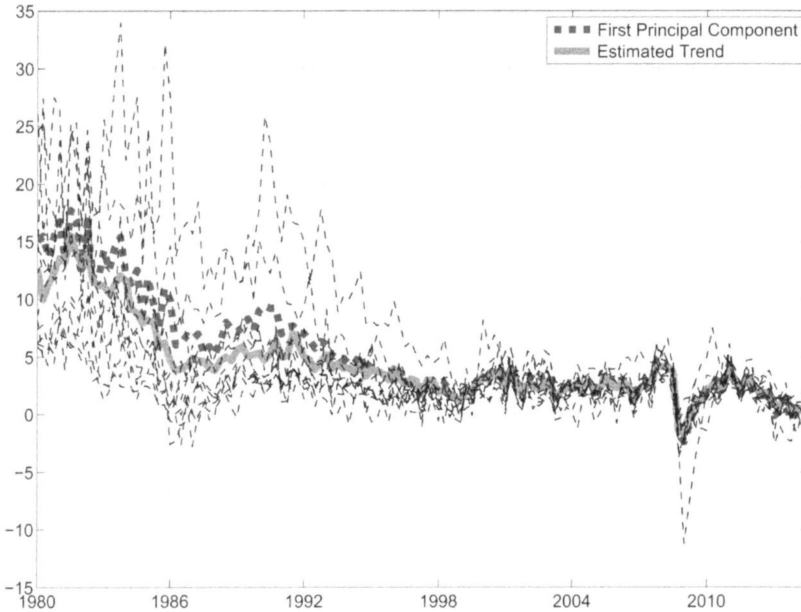

Fig. 3. Common Trend, First Principal Component, and Data.

Fig. 4 compares the estimated common factor with the observed year-on-year aggregate inflation, which is the official target of the European Central Bank. The aggregate factor captures well the persistent movements of aggregate inflation. Interestingly, being estimated in real time from quarterly data, it tends to lead the movements in the year-on-year variation.

In our model, the country-specific level of inflation is captured by the time-varying idiosyncratic cycles ($\psi_{j,t}$), which are shown in Fig. 5. They track quite accurately the various steps that led from the European Monetary System in the late Seventies to the EMU. In particular, three clusters can be identified. The first one is composed of a set of countries whose idiosyncratic elements start out from negative levels in the 1980s, and then slowly converge to zero. These are continental countries that were founding members of the European Monetary System, that is, Belgium, Germany and Luxembourg, together with Austria (whose currency did not take part in the Exchange Rate Mechanism, but was de facto pegged to West Deutsche mark, see Nyberg, Evens, & Horst, 1983). Although the EMS arrangements did not grant a special role to any country, the low

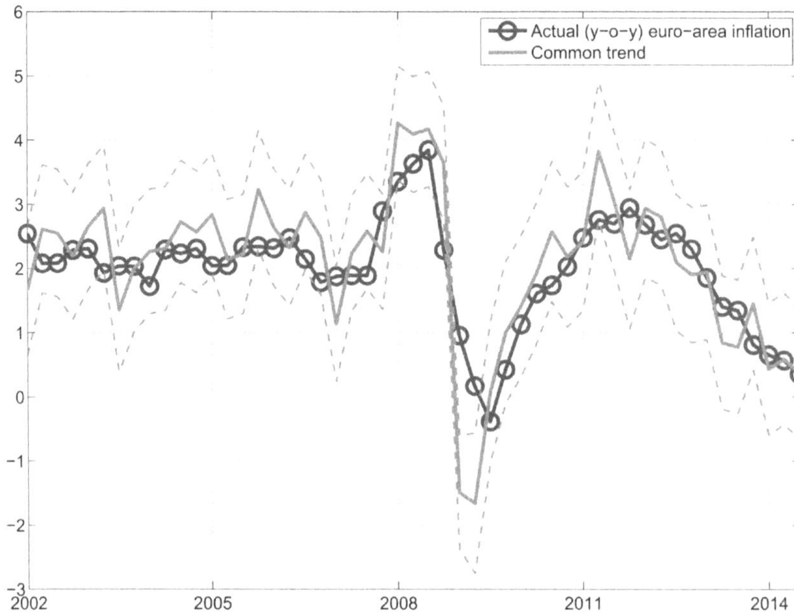

Fig. 4. Actual (y-o-y) Inflation and Common Trend (68% Confidence Interval).

inflation policy that the Bundesbank had pursued since the 1970s (see Benati, 2011) spilled over to the main trading partners and acted as an attractor for the whole EMS, as testified by the fall of the common trend in the first part of the sample. The second block is formed by countries whose idiosyncratic components start positive but converge to zero by the mid-1990s. These are EMS members that managed to sustain temporarily higher inflation rates vis-a-vis Germany through some realignment of their exchange rates but whose price dynamics were eventually attracted to the common trend after the Maastricht criteria imposed a stronger nominal anchor (Italy, Ireland, the Netherlands, Portugal, and, to some extent, Spain, which was not part of the EMS). The third cluster is formed by France and Finland, whose inflation rates roughly fluctuate around the common trend for the whole sample.[10]

Focusing on the period 2002–2014 (i.e., since the introduction of the euro notes in 2002) the national idiosyncratic components fluctuate roughly around zero, see Fig. 6, hence testifying the ability of the centralized monetary policy to overall stabilize inflation rates in the various members of the

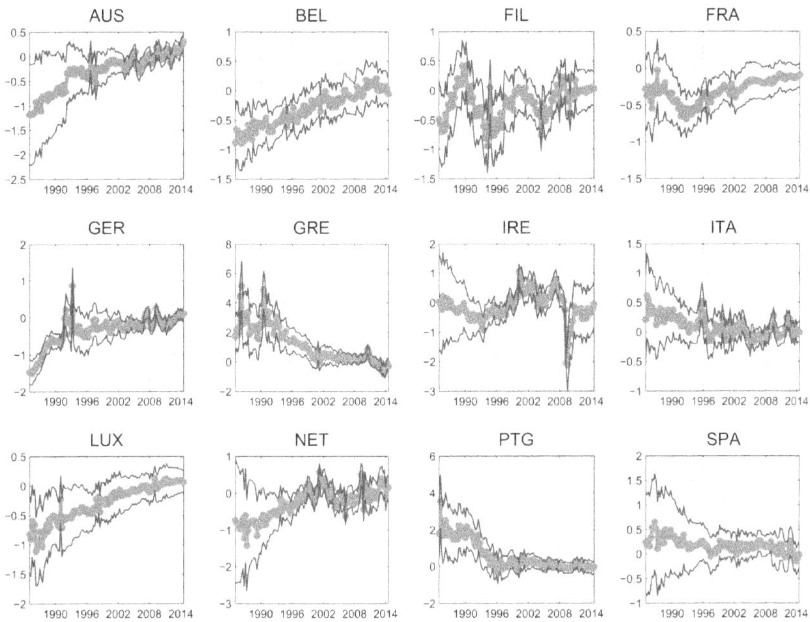

Fig. 5. Idiosyncratic Cycles $\psi_{i,t}$. *Note*: The figure shows the idiosyncratic cycles $\psi_{i,t}$ together with the 68% confidence interval derived from the state covariance estimated with the Kalman filter.

EMU over this period. In the years since the global financial crisis, a slight downward trend in the national-specific component is visible in the so-called vulnerable countries, especially Greece, Portugal, Spain, and Ireland. Corsetti and Pesaran (2012) forcefully argue that real appreciation is the single most important indicator of macroeconomic imbalances (whatever the sources). In a currency union, real appreciation arises through inflation differentials. Our model can therefore be used to shed light on the persistent component of the relative inflation differentials, as captured by the breaking intercepts, $\gamma_{j,t}$. Turning to the post-euro sample Fig. 7 unveils interesting patterns of divergence in the country-specific inflation rates that would be otherwise obscured within broader long-term convergence in relative inflations. Greece, Ireland, Portugal, Spain, and, to a smaller extent, Italy display a markedly persistent positive inflation differential in the run up to the sovereign debt crisis. The model also highlights how different responses to the European debt crisis have contributed to the realignment

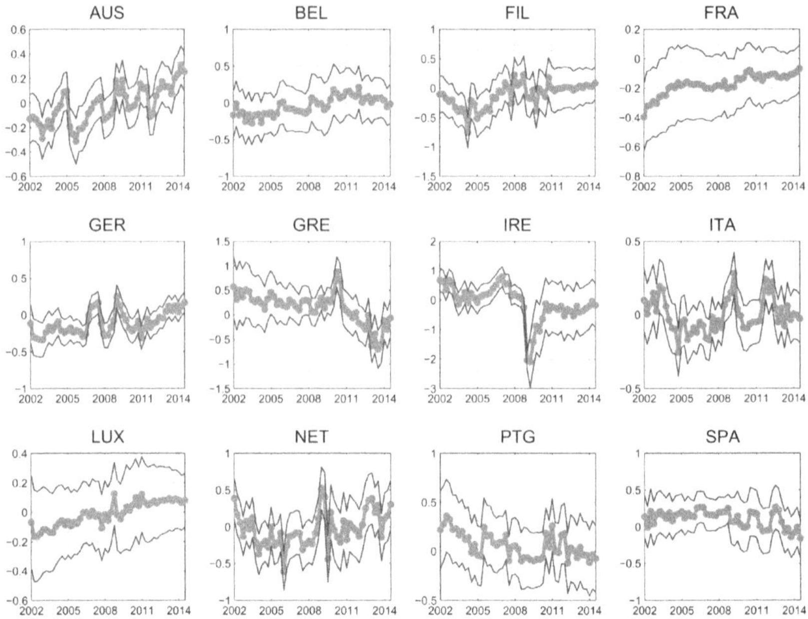

Fig. 6. Idiosyncratic Cycles $\psi_{i,t}$ since the Introduction of the Euro. *Note*: The figure shows the idiosyncratic cycles $\psi_{i,t}$ together with the 68% confidence interval derived from the state covariance estimated with the Kalman filter.

of inflation differentials. For instance, the sharp real depreciations in Ireland and Spain are in stark contrast with the slow correction in Greece.

4.2. *Volatility and Persistence*

Next, we look at the estimated time-varying volatilities, starting from the estimated volatility of the common component, shown in Fig. 8. This variance displays a sharp downward trend since the 1980s up until 2009, when it temporarily increases, only to start falling soon after. Since the common component is a driftless random walk, its variance can be interpreted as a measure of the (common) persistence present in the data. Its falling trajectory is therefore consistent with the decline in inflation persistence in the euro area highlighted by a number of studies within the Eurosystem Inflation Persistence Network (IPN), whose results are summarized in Altissimo, Ehrmann, and Smets (2006). A number of papers within this literature

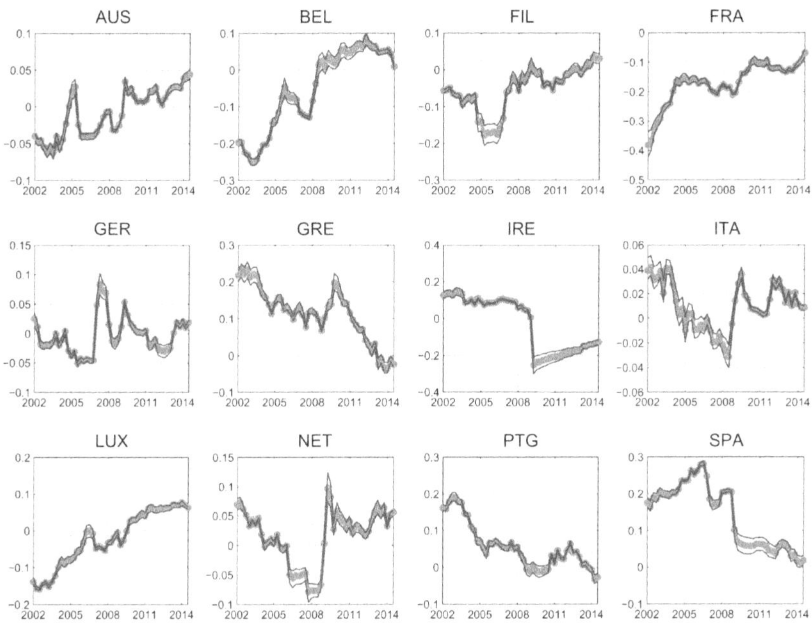

Fig. 7. Idiosyncratic Intercepts $\gamma_{i,t}$. *Notes*: The figure shows the maximum likelihood estimates of the idiosyncratic intercepts together with a 68% confidence interval. To compute the confidence interval, we draw the static parameter from their distribution and run the filter 5,000 times, we then report 16th and 84th percentile of the resulting empirical distribution at every period (see Hamilton, 1986).

(Robalo Marques, 2005; O'Reilly & Whelan, 2005) argue that the evidence in favour of changes in inflation persistence is considerably weaker when the intercept of the inflation models is allowed to change over time. We stress that our model indeed allows for such a break in the country-specific intercepts, yet even accounting for this feature the main result remains. The fall in inflation persistence is not at all specific to the euro area. It is a rather broader phenomenon usually identified either with stronger inflation anchoring by monetary policy, as argued by Benati and Surico (2008), or with more benign inflationary shocks. Stock and Watson (2007), for example, model U.S. inflation using a univariate model featuring a permanent and a transitory component, and allow for changing variances in both components. They find that the variance of the permanent component has fallen significantly over time. Benati (2008) also documents a reduction in inflation persistence in a large number of economies, including the euro area, and points out that

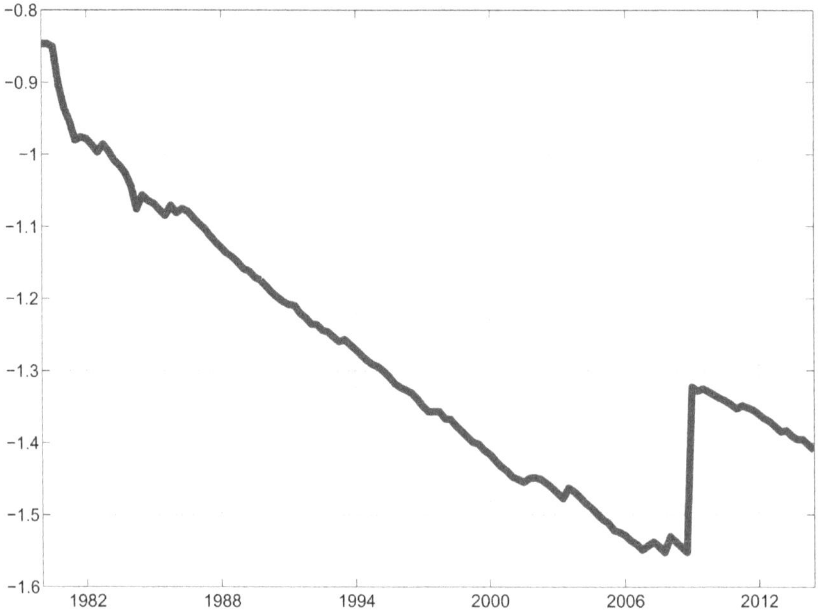

Fig. 8. Common Volatility $\log\sigma_{\eta,t}$.

these changes typically coincide with the adoption of an inflation target. The hypothesis that lower persistence stems from more benign shocks cannot, however, be completely ruled out. Comparing Fig. 8 with Fig. 3 it is in fact clear that the temporary rise in the variance of the common trend is due to the synchronized fall of inflation rates in 2009, which reflected the strong decline in oil prices at the inception of thc financial crisis.

The fall in the volatility of the common permanent component has also strong implications for the predictability of inflation. Indeed it implies that inflation has become easier to forecast by naïve forecasting models, but also that more sophisticated models will have a harder time improving upon simple models, a point made by D'Agostino, Giannone, and Surico (2007).

4.3. Co-movement

Given the fall in the volatility of common shocks discussed in the previous section, it would be tempting to conclude that there has been a fall in the degree of commonality of inflation across euro-area countries. This, however, would be erroneous because co-movement is determined by the relative importance of common and idiosyncratic shocks rather than by the absolute importance of the former. Synchronization of specific inflation rates has, instead, increased over time. We present evidence of this in Figs. 9 and 10. Fig. 9 shows at each point in time the cross sectional standard deviation of the country-specific volatilities, which can be seen as a time-varying measure of dispersion across country-specific inflation shocks (see also Del Negro & Otrok, 2008). Not surprisingly, not only the inflation levels have converged as observed above, but also the amplitude of the shocks has converged considerably over time. Fig. 10 shows the share of the volatility of the common component in the overall forecast error for each of the 12 countries considered.[11] Two main results stand out. First, in all the countries but Belgium, these ratios display an upward trend.[12] Given the decline in the volatility of the common trend documented above, this result implies that the idiosyncratic volatilities not only have diminished over the sample but also that their fall has been relatively faster than

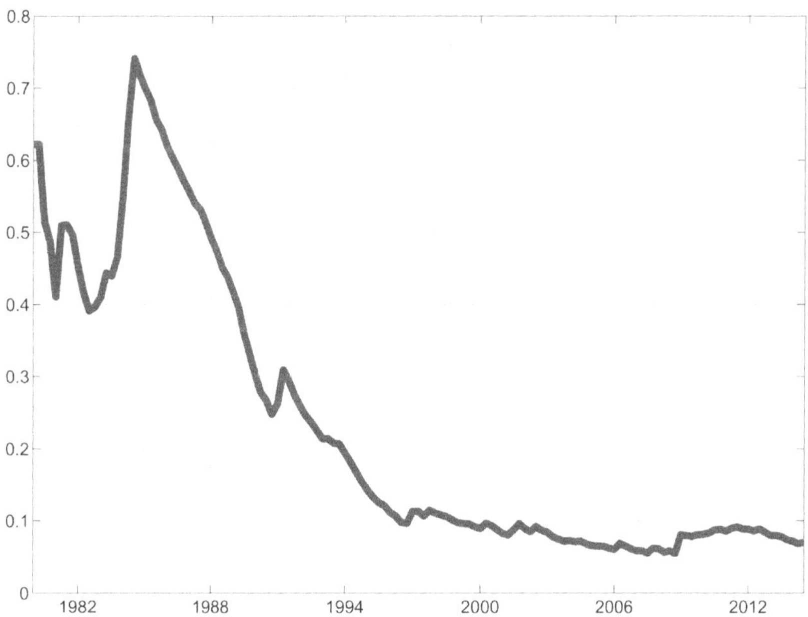

Fig. 9. Cross-Sectional Standard Deviation of Idiosyncratic Volatilities.

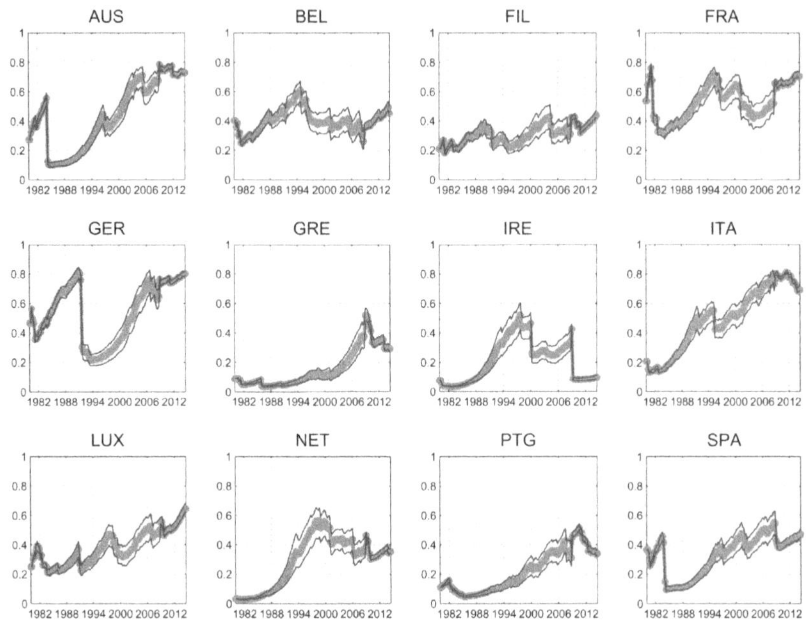

Fig. 10. Share of Common to Overall Volatility.

that of the common trend. In other words, in the context of a general fall in volatility (both common and idiosyncratic), co-movement across inflation rates has actually risen over time. The second result is that in all the stressed countries this trend has partly reversed after 2008. Such development can be better appreciated in Fig. 11, which reports the same ratios of common to idiosyncratic variances over the 2002–2014 period. Over this shorter sample it is more evident that idiosyncratic variances have increased although with a different timing across countries. In Spain and Ireland, whose banking sectors were more directly hit by the 2008 financial crisis, there is a very sharp drop around that year. A similar, although more nuanced, trajectory is visible for Greece and Portugal, where the crisis unfolded in 2010. Finally, in Italy the ratio starts bending in 2011, when the sovereign debt crisis struck the euro area.[13] Comparing these results with the ones on the idiosyncratic cycles discussed above, two conclusions can be drawn. First, peripheral countries in the euro area have been hit by relatively strong idiosyncratic shocks, which the single monetary policy had difficulties in offsetting. Second, this has not resulted in significant

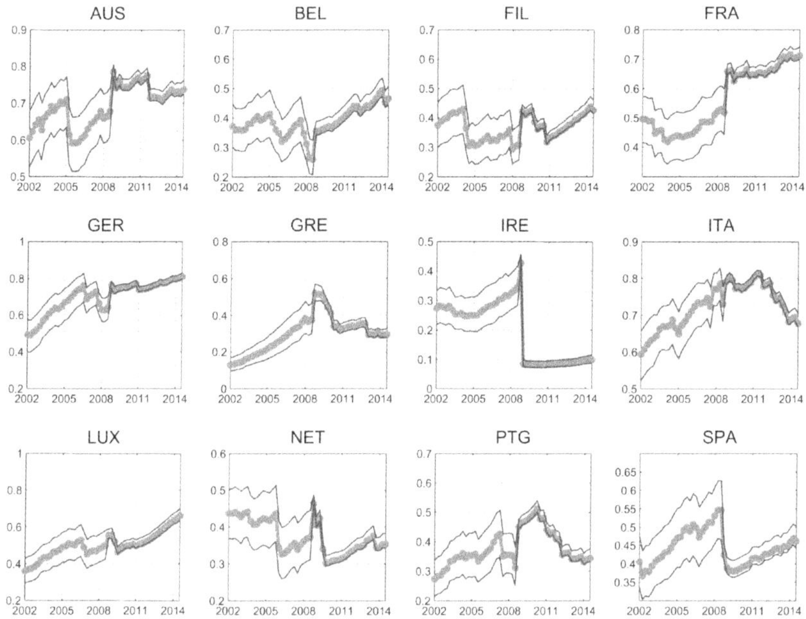

Fig. 11. Share of Common to Overall Volatility since the Introduction of the Euro.

differences across countries in the levels of inflation, so that the disinflation observed since 2013 can largely be interpreted as a common phenomenon (see Fig. 4).

5. CONCLUSIONS

In this paper, we have analyzed inflation developments in the euro area through a factor model with time-varying parameters. The time variation in the model is driven by the score of the predictive likelihood, implying that the estimation can be carried out via maximum likelihood method. The model provides a real-time decomposition of the permanent and transitory shocks to inflations' differentials across countries and could therefore be used in real time to measure the extent of the imbalances within the euro area, as discussed by Corsetti and Pesaran (2012). Our main findings are three. First, the inflation persistence, as measured by the variance of the common component, has decreased over time, in line with the findings obtained in the literature. Importantly, this result is not weakened by the presence of time-varying country-specific intercepts. Second, inflation commonality, estimated as a fraction of the common to idiosyncratic volatility, has risen as a result of the various steps that led to the EMU first, and of the common monetary policy in the last fifteen years. Third, the disinflation experienced since 2011 is largely a common phenomenon, since no significant cross-country inflation differentials have emerged. Since 2008, however, more vulnerable countries have been hit by unusually large shocks, which were only imperfectly offset by the single monetary policy.

We conclude by briefly mentioning avenues for future research that the framework developed in this paper opens. First, the model could be used to analyze the time-varying effects of actual inflation on inflation expectations and possible feedbacks from the latter to the former. Second, it could be used to investigate the relevance of national inflation rates for forecasting the area wide inflation.

NOTES

1. For any square matrix, A, of dimension m, the commutation matrix, C_m, is defined such that $C_m \text{vec}(A) = \text{vec}(A')$.

2. Specifically, denote with θ_1, θ_2 and θ_3 the parameters that govern the law of motion of, respectively, the volatility of the common factor, the volatilities of the idiosyncratic components and the idiosyncratic components (both the constants and the autoregressive coefficients). Then $\Theta = \text{diag}(\theta_1, \theta_2 \iota_{1\times N}, \theta_3 \iota_{1\times 2N})$, where ι denotes a row vector of ones. We also experimented with a specification that allowed for an independent updating coefficient for the constant and the autoregressive parameter in the idiosyncratic component, that is, $\Theta = \text{diag}(\theta_1, \theta_2 \iota_{1\times N}, \theta_3 \iota_{1\times N}, \theta_4 \iota_{1\times N})$, this gives results virtually identical to the ones presented here.

3. Harvey (1989, pp. 182–183) derives the asymptotic normality of non-stationary models with diffuse approximation for non-stationary model but fixed parameters.

4. Initializing the common factor estimates with the first principal component in the data (as in Ciccarelli & Mojon, 2010) gives very similar results.

5. The cohesion measures the average pair-wise dynamic correlations at various frequencies (for more details, see Croux, Forni, & Reichlin, 2001).

6. Formal break tests indeed detect a break in a large number of countries around 1992, see Corvoisier and Mojon (1995).

7. Interestingly, Fig. 2 shows a lower cohesion at low frequencies in the post-euro sample, after the common trend in the inflation series stabilizes. This highlights the presence of persistent deviations from the common trend, confirming the results in Busetti et al. (2007).

8. Ciccarelli and Mojon (2010) find that their measure of global inflation acts as an attractor, that is, they find that deviations of the specific inflation rates from this common force are temporary. Our common trend is an attractor by construction, as a result of the stationarity constraint that we impose on the autoregressive idiosyncratic processes.

9. In addition, it is worth noting that once the common trend is removed from the national inflation rates the average pair-wise correlation in the panel drops from 0.68 to 0.03. Even though the latter is still significant according to Pesaran (2004)'s test for cross-sectional dependence (CSD), the low average pair-wise correlation suggests that the CSD of the deviations from the common trend is likely to be weak.

10. The behaviour of Greek inflation reflects the delay with which Greece met the necessary criteria for joining the EMU.

11. The share is computed at each point in time as $\frac{\sigma^2_{\eta,t}}{\sigma^2_{\eta,t}+\sigma^2_{i,t}}$, that is, as the ratio of the variance of the common trend to the sum of the variance of the common trend and of the idiosyncratic cycle. To compute the confidence interval, we draw the static parameter from their distribution and run the filter 5,000 times, we then report 16th and 84th percentile of the resulting empirical distribution at every period (see Hamilton, 1986).

12. In the case of Germany, the break visible at the beginning of the 1990s is due to the unification of West and East Germany.

13. For a detailed account of the different stages of the euro-zone crisis, see Shambaugh (2012).

ACKNOWLEDGMENTS

The views expressed in this paper are those of the authors and do not necessarily reflect those of the Bank of England or of the Banca d'Italia. While assuming the scientific responsibility for any error in the paper, the authors would like to thank Eric Hillebrand, Siem Jan Koopman, seminar participants at the University of Glasgow, at the Banca d'Italia, at the 2014 Advances in Econometrics – Conference on Dynamic Factor Models and two anonymous referees for useful comments and suggestions.

REFERENCES

Altissimo, F., Ehrmann, M., & Smets, F. (2006). *Inflation persistence and price-setting behavior in the euro area – A summary of the IPN evidence*. Occasional Paper Series No. 46, European Central Bank.

Bai, J. (2004). Estimating cross-section common stochastic trends in non-stationary panel data. *Journal of Econometrics*, *122*(1), 137–183.

Benati, L. (2008). Investigating inflation persistence across monetary regimes. *The Quarterly Journal of Economics*, *123*(3), 1005–1060.

Benati, L. (2011). Would the Bundesbank have prevented the great inflation in the United States? *Journal of Economic Dynamics and Control*, *35*(7), 1106–1125.

Benati, L., & Mumtaz, H. (2007). *U.S. evolving macroeconomic dynamics: A structural investigation*, Working Paper Series No. 0746, European Central Bank.

Benati, L., & Surico, P. (2008). Evolving U.S. monetary policy and the decline of inflation predictability. *Journal of the European Economic Association*, *6*(2–3), 634–646.

Busetti, F., Forni, L., Harvey, A., & Venditti, F. (2007). Inflation convergence and divergence within the European monetary union. *International Journal of Central Banking*, *3*(2), 95–121.

Camacho, M., & Perez-Quiros, G. (2010). Introducing the euro-sting: Short-term indicator of euro area growth. *Journal of Applied Econometrics*, *25*, 663–694.

Ciccarelli, M., & Mojon, B. (2010). Global inflation. *The Review of Economics and Statistics*, *92*(3), 524–535.

Clark, T. E., & Doh, T. (2014). Evaluating alternative models of trend inflation. *International Journal of Forecasting*, *30*(3), 426–448.

Cogley, T. (2002). A simple adaptive measure of core inflation. *Journal of Money, Credit and Banking*, *34*(1), 94–113.

Corsetti, G., & Pesaran, H. M. (2012). *Beyond fiscal federalism: What will it take to save to euro?* VOX, CEPR's Policy Portal.

Creal, D., Koopman, S. J., & Lucas, A. (2013). Generalized autoregressive score models with applications. *Journal of Applied Econometrics*, *28*(5), 777–795.

Cristadoro, R., Forni, M., Reichlin, L., & Veronese, G. (2005). A core inflation indicator for the euro area. *Journal of Money, Credit and Banking*, *37*(3), 539–560.

Croux, C., Forni, M., & Reichlin, L. (2001). A measure of comovement for economic variables: Theory and empirics. *The Review of Economics and Statistics*, *83*(2), 232–241.

D'Agostino, A., Giannone, D., & Surico, P. (2007). *(Un)Predictability and macroeconomic stability*, CEPR Discussion Papers No. 6594.

De Grauwe, P., & Ji, Y. (2013). Self-fulfilling crises in the eurozone: an empirical test. *Journal of International Money and Finance*, *34*, 15–36.

Del Negro, M., Giannoni, M. P., & Schorfheide, F. (2013). *Inflation in the great recession and New Keynesian models*. Staff Reports 618, Federal Reserve Bank of New York.

Del Negro, M., & Otrok, C. (2008). *Dynamic factor models with time-varying parameters: Measuring changes in international business cycles*. Staff Reports 326, Federal Reserve Bank of New York.

Delle Monache, D., Petrella, I., & Venditti, F. (2015). *Adaptive state-space models*. Mimeo.

Ferroni, F., & Mojon, B. (2014). *Domestic and global inflation*. Mimeo.

Giannone, D., & Matheson, T. D. (2007). A new core inflation indicator for New Zealand. *International Journal of Central Banking*, *3*(4), 145–180.

Giannone, D., Reichlin, L., & Small, D. (2008). Nowcasting: The real-time informational content of macroeconomic data. *Journal of Monetary Economics*, *55*, 665–676.

Gomez, V., & Maravall, A. (1996). *Programs tramo (time series regression with arima noise, missing observations, and outliers) and seats (signal extraction in arima time series)*.

Instruction for the User. Working Paper No. 9628 (with updates), Research Department, Bank of Spain.

Hamilton, J. D. (1986). A standard error for the estimated state vector of a state-space model. *Journal of Econometrics*, *33*(3), 387–397.

Harvey, A. C. (1989). *Forecasting, structural time series models and Kalman filter*. Cambridge, MA: Cambridge University Press.

Harvey, A. C. (2013). *Dynamic models for volatility and heavy tails*. With Applications to Financial and Economic Time Series. Cambridge, MA: Cambridge University Press.

Koop, G., & Korobilis, D. (2013). Large time-varying parameter VARs. *Journal of Econometrics*, *177*(2), 185–198.

Koop, G., & Korobilis, D. (2014). A new index of financial conditions. *European Economic Review*, *71*(C), 101–116.

Koopman, S. J., Mallee, M. I. P., & Van der Wel, M. (2010). Analyzing the term structure of interest rates using the dynamic Nelson-Siegel model with time-varying parameters. *Journal of Business and Economic Statistics*, *28*, 329–343.

Marcellino, M., Porqueddu, M., & Venditti, F. (2013). Short-term GDP forecasting with a mixed frequency dynamic factor model with stochastic volatility. *Journal of Business and Economic Statistics*. Forthcoming.

Mumtaz, H., & Surico, P. (2012). Evolving international inflation dynamics: World and country-specific factors. *Journal of the European Economic Association, European Economic Association*, *10*(4), 716–734.

Nyberg, P., Evens, O., & Horst, U. (1983). The European Monetary System: The Experience, 1979–82. International Monetary Fund.

O'Reilly, G., & Whelan, K. (2005). Has euro-area inflation persistence changed over time? *The Review of Economics and Statistics*, *87*(4), 709–720.

Pesaran, M. H. (2004). *General diagnostic tests for cross section dependence in panels*, Cambridge Working Papers in Economics 0435, Faculty of Economics, University of Cambridge.

Pesaran, M. H. (2006). Estimation and inference in large heterogeneous panels with a multifactor error structure. *Econometrica*, *74*(4), 967–1012.

Robalo Marques, C. (2005). *Inflation persistence: Facts or artefacts*? Economic Bulletin and Financial Stability Report Articles, Banco de Portugal, Economics and Research Department.

Shambaugh, J. C. (2012). The Euro's three crises. *Brookings Papers on Economic Activity*, *44*(1), 157–231.

Stock, J. H., & Watson, M. (2007). Why has U.S. inflation become harder to forecast? *Journal of Money, Credit and Banking*, *39*(Suppl. 1), 3–33.

PART IV
NOWCASTING AND FORECASTING

NOWCASTING BUSINESS CYCLES: A BAYESIAN APPROACH TO DYNAMIC HETEROGENEOUS FACTOR MODELS

Antonello D'Agostino[a], Domenico Giannone[b,c,d,e], Michele Lenza[d,f] and Michele Modugno[g]

[a]*European Stability Mechanism, Luxemburg, Luxemburg*
[b]*Federal Reserve Bank of New York, New York, NY, USA*
[c]*CEPR, London, UK*
[d]*ECARES, Brussels, Belgium*
[e]*LUISS, Roma, Italy*
[f]*European Central Bank, Frankfurt, Germany*
[g]*Board of Governors of the Federal Reserve System, Washington, DC, USA*

ABSTRACT

We develop a framework for measuring and monitoring business cycles in real time. Following a long tradition in macroeconometrics, inference is based on a variety of indicators of economic activity, treated as imperfect measures of an underlying index of business cycle conditions. We extend existing approaches by permitting for heterogenous lead–lag

Dynamic Factor Models
Advances in Econometrics, Volume 35, 569–594

ISSN: 0731-9053/doi:10.1108/S0731-905320150000035014

patterns of the various indicators along the business cycles. The framework is well suited for high-frequency monitoring of current economic conditions in real time – nowcasting – since inference can be conducted in the presence of mixed frequency data and irregular patterns of data availability. Our assessment of the underlying index of business cycle conditions is accurate and more timely than popular alternatives, including the Chicago Fed National Activity Index (CFNAI). A formal real-time forecasting evaluation shows that the framework produces well-calibrated probability nowcasts that resemble the consensus assessment of the Survey of Professional Forecasters.

Keywords: Current economic conditions; dynamic factor models; dynamic heterogeneity; business cycles; real time; nowcasting

JEL classifications: C11; C32; C38; E32; E38

1. INTRODUCTION

Macroeconomic and financial variables are characterized by a strong correlation, which is possible only if the bulk of their fluctuations is driven by few common sources. Dynamic factor models (DFM) build on this basic fact to provide a parsimonious and, yet, suitable representation of macroeconomic and financial dynamics. The model assumes that a few unobserved dynamic factors drive the comovement of many observed variables, while the features that are specific to individual series, such as measurement error, are captured by idiosyncratic disturbances.

DFM were initially proposed by Geweke (1977) and Sargent and Sims (1977) as a time series extension of the factor models previously developed for cross-sectional data in psychometrics (see Lawley & Maxwell, 1963, for a comprehensive analysis of factor models for serially uncorrelated data). Over the years, factor models have been successfully used in macroeconometrics for structural analysis and forecasting (see Stock et al., 2011, for a comprehensive survey). DFM have been intensively used in many contexts, ranging from forecasting (Stock & Watson, 2002) and nowcasting (Giannone, Reichlin, & Small, 2008) to the empirical validation of Dynamic Stochastic General Equilibrium models (Boivin & Giannoni, 2006; Giannone, Reichlin, & Sala, 2006), and provide a reliable statistical framework for the estimation of synthetic indexes of business cycle conditions (Stock et al., 1992).

The main feature of business cycle fluctuations is their pervasiveness across the economy.[1] Hence, variables measuring different aspects of the economy can be considered as imperfect measures of a latent common business cycle factor. Formally, the DFM representation for a set of stationary variables, $x_{i,t}$, $i = 1, \dots, n$, is written as follows:

$$x_{i,t} = \sum_{h=0}^{s} \lambda_{i,h} f_{t-h} + e_{i,t} \quad i = 1, \dots, n \tag{1}$$

where f_t is the common factor summarizing the state of the economy and $e_{i,t}, i = 1, \dots, n$ are the idiosyncratic disturbances. The model is identified by assuming that comovement among variables arises only from a single source, the common factor. This amounts at assuming that $e_{1t}, \dots, e_{nt}$ and f_t are orthogonal at all leads and lags.

Following Stock et al. (1992), we will refer to f_t as the *s*ynthetic index of business cycle conditions. Summarizing business cycle conditions using a synthetic index rather than observable measures can enhance timeliness and precision. For example, GDP provides a very comprehensive measure of economic activity and summarizes well the business cycle fluctuations, as shown by the fact that recessions roughly correspond to its decline for two consecutive quarters (see Harding & Pagan, 2002). However, GDP is released with a delay, it is subject to revisions and it is characterized by measurement error. Aggregating the information provided by different variables represents a sort of insurance against the measurement error[2] and, in the assessment of the business cycle conditions, it allows to exploit the different sampling frequency and the different timeliness of macroeconomic data releases.[3]

The DFM is typically specified by assuming $s = 0$ in Eq. (1).[4] We refer to this set of restrictions as *dynamic homogeneity*. This assumption can represent a straightjacket since it imposes that different indicators have the the same lead–lag pattern along the business cycles and proportional impulse response functions to a common shock, that is, to an exogenous shock to the synthetic index. For these reasons, inference is typically performed on pre-selected economic indicators that are judged to be coincident along the business cycle, that is, indicators that "have been tolerably consistent in their timing in relation to business cycle revivals and that at the same time are of sufficiently general interest to warrant some attention by students of current economic conditions" (see Mitchell & Burns, 1938; Moore, 1983).

In this paper, we relax the assumption of dynamic homogeneity and accommodate for heterogeneous dynamics by including a large number of lags ($s \gg 0$) in Eq. (1). The more general structure reduces the risk of model miss-specification, enabling to extract more efficiently the information from economic indicators characterized by a significant degree of dynamic heterogeneity. However, the high level of generality comes at the cost of parameters proliferation. This could increase estimation uncertainty and induce overfitting, which, in turn, could offset the potential benefits of reduced miss-specification, and jeopardize the real-time performance of the model. In order to counterbalance these perverse effects, we combine sample information with a prior belief that lagged effects of the common factor are less important the longer the delay.[5]

We conduct inference using data on real GDP and popular US coincident indices of business cycles. Following a fast growing literature on Bayesian factor models, we estimate the full set of posterior densities for the model's parameters and for the unobserved index of business cycle conditions using Monte Carlo Markov chain (MCMC) techniques.[6]

Our framework encompasses the traditional approaches to the construction of business cycle indicators. In particular, principal components are proportional to the posterior mode of the unobserved factor associated to a static factor model ($s = 0$ and serially uncorrelated factors f_t and idiosyncratic components $e_{i,t}, i = 1, \ldots, n$) with homogenous signal-to-noise ratio and a flat prior. If the factor loadings are also assumed to be the same ($\lambda_i(L) = \bar{\lambda}$), principal components become simple cross-sectional averages. Allowing for serial correlation, but keeping the model dynamically homogenous ($s = 0$), we obtain the Index of Coincident Economic Indicators of Stock et al. (1992).

In-sample inference, based on the most recent available data, shows that the factor provides an accurate characterization of the business cycle dynamics in the United States and suggests that dynamic heterogeneity is an important feature of the data. Indeed, the posterior distribution of the common synthetic index provides a more timely account of peaks and troughs when compared with alternative indicators based on DFM, like the Chicago FED National Activity Index. In addition, the impulse responses of different indicators to a common shock display a relevant degree of heterogeneity.

As stressed above, factor models have proved to be successful not only in the extraction of synthetic indicators, but also for nowcasting in real time. We evaluate the accuracy of our model also along this dimension. This is important also because it reveals whether the in-sample properties we just described are genuine features of the data and not only an artifact due to overfitting. More in details, we study the properties of the model-based predictive distributions for GDP growth and compare them with the consensus probability assessments of the Survey of Professional Forecasters (SPF). In order to meaningfully compare the two sets of nowcasts, we take a fully real-time perspective, that is, we collect the real-time vintages for our variables, which were available at the time the SPF was conducted. Results indicate that the predictive densities are correctly specified – well calibrated – since they cannot be statistically distinguished from the true unconditional data densities. Predictive scores reveal that the predictions of the model are, on average, more accurate than those obtained using a univariate autoregressive benchmark and compare well with the SPF. Overall, the out-of-sample evaluation indicates that dynamic heterogeneity is a genuine and salient feature of the data, and not just the result of overfitting.

The rest of this paper is structured as follows. Section 2 describes the model and the real-time database. Section 3 studies the in-sample properties of our index of business cycle conditions. Section 4 carries out a formal out-of-sample evaluation of the density nowcasts of our model. Section 5 concludes.

2. THE MODEL AND THE DATABASE

2.1. The Dynamic Factor Model

We assume that a set of variables $x_{i,t}$, with $i = 1, \ldots, n$, is characterized by the following equations:[7]

$$x_{i,t} = \lambda_i(L)f_t + e_{i,t}, \quad i = 1, \ldots, n \tag{2}$$

where $\lambda_i(L) = \lambda_{i,0} + \lambda_{i,1}L + \cdots + \lambda_{i,s}L^s$. The process for the common factor f_t and the idiosyncratic components $e_{i,t}$, $i = 1, \ldots, n$ are approximated by finite autoregressive (AR) models:

$$a(L)f_t = u_t, \quad u_t \sim i.i.d.\mathcal{N}(0, 1);$$
$$\phi_i(L)e_{i,t} = v_{i,t}, \quad v_{i,t} \sim i.i.d.\mathcal{N}\left(0, \sigma_i^2\right), \; i = 1, \ldots, n,$$

where $a(L) = 1 - a_1L - \cdots - a_{p_f}L^{p_f}$ and $\phi_i(L) = 1 - \phi_{i,1}L - \cdots - \phi_{i,p_e}L^{p_e}$.

The common shocks u_t are assumed to be orthogonal to the idiosyncratic shocks $v_{i,t}, i = 1\ldots, n$, at all leads and lags. In addition, the idiosyncratic shocks are assumed to be mutually orthogonal at all leads and lags. Under this assumption, the model is known as "exact" since it implies that the cross-correlations among observables are only due to the common factor. Although this assumption may be very restrictive, Doz, Giannone, and Reichlin (2012) have shown that the model is robust to non-Gaussianity and to weak correlation among idiosyncratic components, provided that estimation is carried out with a sufficiently large number of highly collinear variables.

Thanks to the rich dynamics allowed by the polynomials $\lambda_i(L) = \sum_{s=0}^{q} \lambda_{is} L^s$, the model can account for complex heterogeneity in the dynamic effects of the common factors on the observable variables. However, the generality of the model is obtained at the cost of the proliferation in the number of parameters to be estimated. This is the reason why the model is typically estimated with $s = 0$. The most commonly used synthetic indexes are obtained as posterior modes of the following constrained models, when a flat prior is used:

- *Principal components*: Static factor model ($s = p_e = p_f = 0$) with spherical idiosyncratic component ($\sigma_i^2 = \overline{\sigma}^2$);
- *Cross-sectional averages*: Static factor model ($s = p_e = p_f = 0$) with spherical idiosyncratic component ($\sigma_i^2 = \overline{\sigma}^2$) and homogenous loadings: $\lambda_i(L) = \overline{\lambda}_0$;
- *Index of Coincident Economic Indicators of* Stock et al. (1992): Homogenous ($s = 0$) dynamic ($p_e = p_f = 2$) factor model. We will define this model as *DFM*.

The restriction $s = 0$ implies strong homogeneity on the propagation of the common shocks on the variables. In particular, an implication of this assumption is that the fluctuations of all variables are perfectly coincident over the business cycles.[8] We retain the flexibility of the model, relaxing the homogeneity restriction, and we control for the over-fitting due to parameters proliferation by shrinking the model parameters toward those of a simple naïve model, through the imposition of priors. The prior distributions for all the coefficients are centered on zero, with stronger tightness for higher-order lags, so that posterior coefficients of high-order lags of the factors are sufficiently away from zero only if the data strongly favor non-zero values. Formal bayesian inference allows us to combine the information from the data and the prior.

An equivalent representation of the model is obtained by pre-multiplying both sides of Eq. (2) by $\phi_i(L)$:

$$\begin{aligned} \phi_i(L)x_{i,t} &= \theta_i(L)f_t + v_{i,t}, & v_{i,t} &\sim i.i.d.\mathcal{N}\left(0, \sigma_i^2\right), \quad i = 1, \ldots, n, \\ a(L)f_t &= u_t, & u_t &\sim i.i.d.\mathcal{N}(0, 1), \end{aligned}$$

where $\theta_i(L) = \phi_i(L)\lambda_i(L)$. Since $\lambda_i(L)$ and $\phi_i(L)$ are unrestricted, we can estimate the model reparameterized in $\theta_i(L)$ and $\phi_i(L)$.[9] The dynamic effects of the common shocks on $x_{i,t}$ can be retrieved by taking the ratio $\lambda_i(L) = \theta_i(L)/\phi_i(L)$. The priors are specified as follows:

$$\begin{aligned} &\sigma_i^2 \sim IG(1, 3), \\ &\theta_{i,h} \sim N\left(0, \tau\left(1/(h+1)^2\right)\right), \\ &\phi_{i,h} \sim N\left(0, \tau\left(1/h^2\right)\right), \\ &a_h \sim N\left(0, \tau\left(1/h^2\right)\right). \end{aligned}$$

where h indicates the lag of the factor or the variable to which the coefficient is associated. The prior covariance among coefficients associated to different variables and different lags is set to zero. Notice that the variance of the prior is lower for the coefficients associated with more distant lags. The hyperparameter τ controls the scale of all the variances and effectively governs the overall level of shrinkage. We fix this parameter to the conventional value of 0.2.[10] These priors, including the choice of the degree of overall shrinkage, are similar to those proposed by Litterman (1979) in the context of Bayesian Vector Autoregressive models.[11] Our DFM with unrestricted dynamics is shortly defined as heterogenous dynamic factor model (HDFM). In order to be able to capture very general dynamics, we specify the model in order to include 12 lags of the observables, the contemporaneous value and 12 lags of the factors in the equations of the observables and 12 lags of the factors in the equations describing the dynamics of the factors.

As stressed in the introduction, we conduct inference using Gibbs sampling techniques. If all data and also the common factor were observed, drawing from the posterior of the parameters would be simple since the prior is conjugate. Conditionally on the parameters and the observable data, then the common factors and the missing data can be drawn using simulation smoothers (Carter & Kohn, 1994; de Jong & Shephard, 1995; Durbin & Koopman, 2002).[12] In other words, the Gibbs sampler consists in alternating the following two steps:

- given a draw of the parameters, draw the missing data and the latent factor conditional on the observations using the simulation smoother;

- given a draw of the the full data and the latent factors, draw the parameters from their posterior.

The algorithm is initialized by using the parameters associated to principal components computed by fitting missing data by a spline function.

2.2. Data

We study the in-sample properties of the HDFM model and its accuracy in a real-time forecast evaluation by using a relatively small dataset for the US economy, including the most popular coincident indicators: real GDP (GDP), real disposable income (DSPI), employment (EMP), industrial production (IP), and real retail sales (RRS).[13] In addition, we include the purchasing manager index (PMI) because, due to the timeliness of its release, it provides an extremely useful information.[14]

The variables are transformed in order to achieve stationarity. Real GDP enters in the model in terms of quarter-on-quarter growth rates, while real income, employment, industrial production, and real retail sales enter the model in terms of month-on-month growth rates. The PMI is stationary by construction, therefore, it enters the model without being transformed.[15]

This dataset is characterized by mixed frequencies because real GDP is sampled quarterly, while all the other variables are sampled monthly. In order to deal with this issue, we treat real GDP as observable in the last month of the quarter. The first two months of the quarter are treated as missing observations. This approach is convenient since, as explained in Section 2.1, the algorithm used for inference can easily deal with missing data.

The main benchmark in our forecasting evaluation is the SPF. For the sake of comparability, we exactly replicate the information set available to the professional forecasters at the time they produced their own forecasts.[16] Specifically, the forecasts are generated every quarter with the information available on the 14th of the second month in the quarter, which is roughly in line with the deadline for the submission of the SPF questionnaires. The forecasting evaluation is carried out using 11 years of vintages, ranging from Q1-2003 to Q4-2013. For each real-time data vintage the sample starts in January 1993. We start the evaluation in 2003 in order to have a first estimation sample of 10 years.[17]

The real-time exercise introduces an additional source of missing data due to the different availabilities of the data at the time forecasts are generated. In fact, on the 14th of each second month of the quarter, real GDP is available for the previous quarter (e.g., in February real GDP is available up to Q4 of the previous year), employment and PMI are available up to the previous month (e.g., in February they are available up to January), real retail sales, and real disposable income are available up to two months before (e.g., in February they are available up to December). Industrial production is usually released at mid month (between the 13th and the 17th of each month) and, hence, depending on the vintage, it can have either the same availability of employment and PMI or the same availability of disposable income and retail sales.

3. THE SYNTHETIC BUSINESS CYCLE INDICATOR AND DYNAMIC HETEROGENEITY

In this section, we perform in-sample inference using data from the last vintage in our dataset (February 2014). The real-time evaluation is conducted in the next section.

Fig. 1 plots the real GDP growth rate against the other variables included in the database.

Some features stand out. First, all variables tend to comove with GDP, especially during periods of downturn. Second, real disposable income, industrial production and real retail sales display very noisy short-run fluctuations, which tend to hide the lower-frequency fluctuations. Third, the variables exhibit a different lead–lag pattern, which is particularly visible around the great recession. PMI and employment growth tend to lag GDP growth; RRS has a more coincident pattern, whereas DSPI and IP display leading dynamics, providing an early signal of the recession.

Fig. 2 plots the six variables versus the HDFM business cycle indicator (median, 16th and 84th quantiles of the distribution), which is explicitly devised to account for heterogeneous lead–lag structure of the variables.

In general, the HDFM indicator tracks our variables very well. This validates the strategy of estimating a DFM to capture the comovement among the variables. In addition, the indicator is smoother than the individual variables, suggesting that a large part of their high-frequency fluctuations are of idiosyncratic nature. More in details, the indicator is roughly coincident with DSPI, IP and RRS (first three sub-plots) and it clearly leads EMP, PMI and GDP (last three sub-plots), hence it provides an "average" of the variables whose dynamic heterogeneity is properly taken

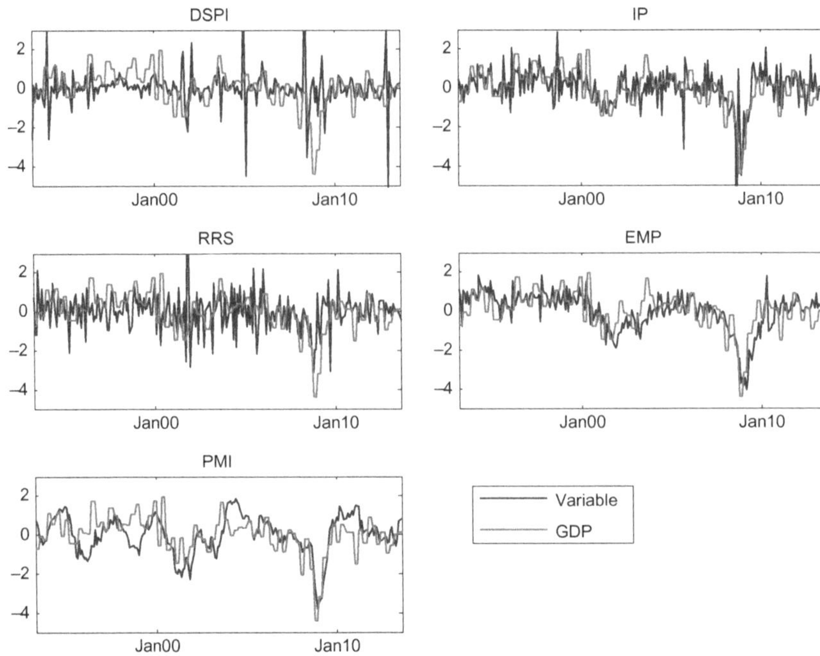

Fig. 1. GDP and Other Variables. *Note*: Gray line: quarter-on-quarter real GDP growth rate; Black line: month-on-month real disposable income growth rate (DSPI), month-on-month industrial production growth rate (IP), month-on-month real retail sales growth rate (RRS), month-on-month employment growth rate (EMP), and level of the purchasing manager index (PMI). Due to the different sampling frequency, GDP growth is reported as constant in the three months of each quarter.

into account. On the other hand, traditional methods for factor extraction would assign most of the weights to the variables with the most persistent dynamics and less volatility (EMP and PMI in our case), and the underlying estimated indicator would be heavily shaped by these series. To illustrate this point, Fig. 3 compares the indicator extracted by employing the HDFM in Section 2.1 with four of the most common methods for factor extraction: the average of the monthly variables included in the panel, the first principal component (PC) of the monthly variables, the Chicago Fed National Activity Index (CFNAI) and the factor extracted from a model that imposes dynamic homogeneity. The latter is the posterior mode of the common factor in our model, estimated under the restriction of complete dynamic homogeneity (DFM).[18]

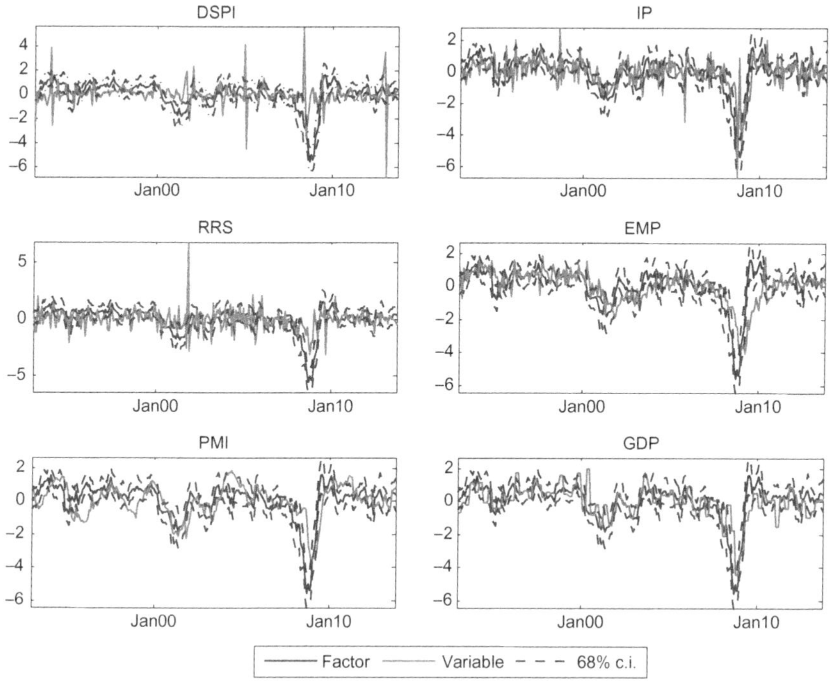

Fig. 2. Common Factor and Variables. *Note*: Black lines: HDFM business cycle indicator (median solid line, 16th and 84th quantiles dashed lines); Gray line: month-on-month real disposable income growth rate (DSPI), month-on-month industrial production growth rate (IP), month-on-month real retail sales growth rate (RRS), month-on-month employment growth rate (EMP), the level of the purchasing manager index (PMI), and the quarter-on-quarter real GDP growth rate, reported as a constant in the three months of each quarter.

The HDFM indicator, which is designed to exploit the dynamic heterogeneity of the variables, leads all the other indicators which do not take into account this important feature of the data. It is worth stressing that this is a pure modeling issue, not related at all with the dimension of the information set; indeed, the CFNAI index, which is extracted from a panel of 85 monthly series, also lags the dynamics of our indicator.[19]

These results have also nontrivial implications on the traditional simulation exercises, which are performed to assess the system dynamics after some exogenous shock. To clarify this point, Fig. 4 reports the impulse response functions (IRFs) of the (log-)levels of the six variables to a

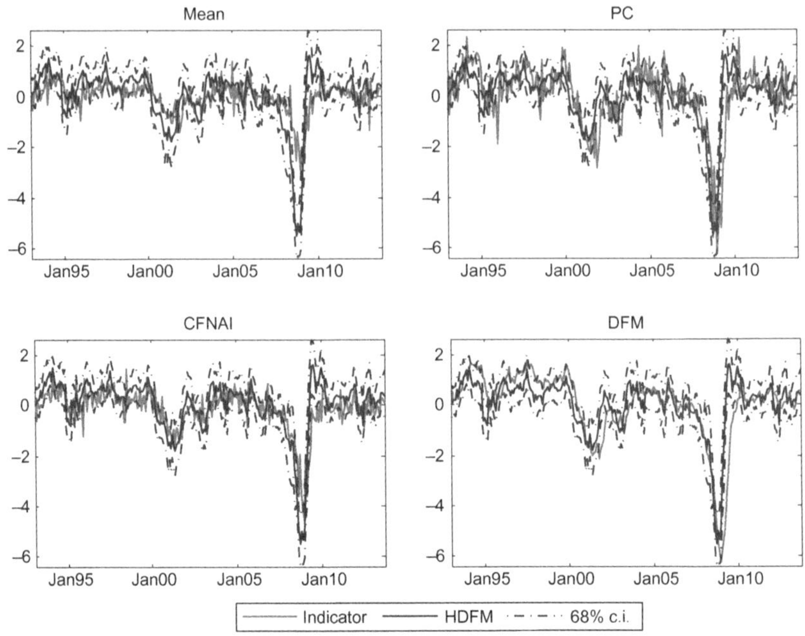

Fig. 3. HDFM and Other Indicators. *Note*: Black lines: HDFM business cycle indicator (median solid line, 16th and 84th quantiles dashed lines); Gray line: simple average of the variables (mean), first principal component of the variables (PC), the Chicago Fed national activity index (CFNAI) and the factor extracted from the homogeneous dynamic factor model (DFM).

common shock, that is an exogenous shock to the synthetic business cycle index. The gray line refers to the median IRF estimated by means of the DFM model, which imposes dynamic homogeneity in the effects of the exogenous shock to the synthetic business cycle indicator. The black lines refer to the IRFs (median, 16th and 84th quantiles of the distribution) of the HDFM. For all variables, the IRFs of the log-levels are obtained by cumulating the IRFs of the growth rates, with the exception of PMI, which is not transformed, and for which the model produces directly the IRFs of the levels.

The most important difference between the two approaches is that, when dynamic heterogeneity is excluded by assumption, the IRFs have essentially the same dynamics, up to a re-scaling factor given by the factor loadings. Indeed, the cumulated IRFs for a model with dynamic homogeneity is

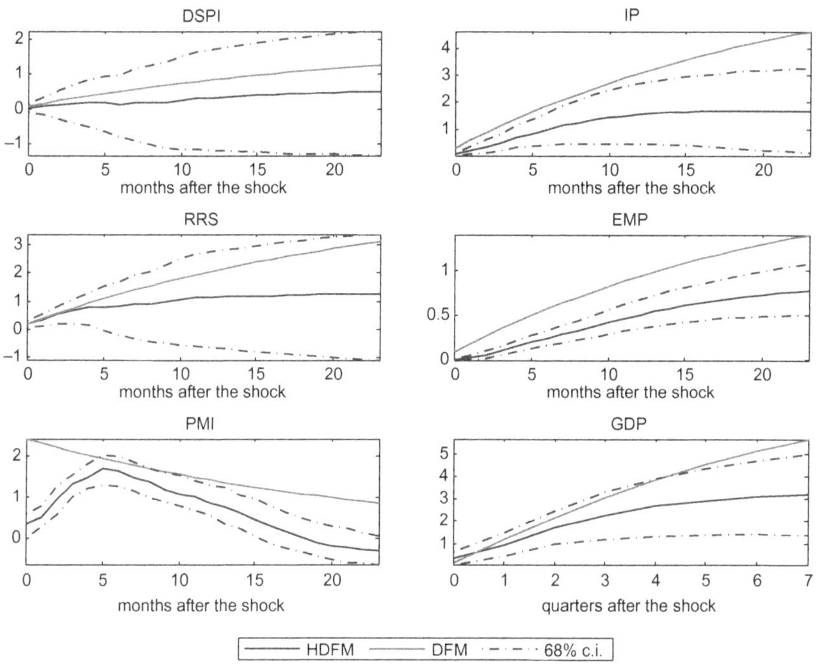

Fig. 4. IRFs of All Variables to a Common Shock. *Note*: Black lines: HDFM IRF of the log-levels of the variables (except for PMI, for which we report levels) to a common shock (median solid line, 16th and 84th quantiles dashed lines); Gray line: DFM IRF of the log-levels of the variables (except for PMI, for which we report levels) of the variables to a common shock (median).

$$\frac{\partial x_{i,t+h}}{\partial u_t} = \lambda_{i,0} \sum_{j=0}^{h} b_j \tag{3}$$

where b_j are the coefficients of the polynomial $b(L) = a(L)^{-1}$, and where $a(L)$ is the polynomial that captures the dynamics of the common factor as described in Section 2.1. As noticed above, the only difference among the IRFs of different variables is their loadings $\lambda_{i,0}$. Instead, accounting for the dynamic heterogeneity, the IRFs are allowed to differ:

$$\frac{\partial x_{i,t+h}}{\partial u_t} = \sum_{j=0}^{h} c_{i,j} \tag{4}$$

where $c_{i,j}$ are the coefficients of the polynomial $c(L) = \theta_i(L)\phi_i(L)^{-1}a(L)^{-1}$. In this case, the IRFs for a specific variable will differ not only by a re-scaling factor, but also by the potentially different importance that lags of the variable itself and of the factors have in explaining the fluctuations. This is evident in Fig. 4, where the dynamics captured by the gray lines are alike among variables. Instead, when a different lead–lag structure among the variables is allowed, the IRFs may have different dynamics. In fact, the black lines in the figures show that the variables have heterogeneous patterns after the shock. The most striking example is PMI that displays a clear hump-shaped reaction to a shock to the common component in the HDFM, while this is not the case for the DFM.

4. EVALUATION OF THE DENSITY NOWCASTS

DFM are known to perform very well as forecasting tools (see Stock et al., 2011, for a survey). However, the specification we advocate in this paper is richly parameterized, hence parameters estimation uncertainty and overfitting are an important concern. For this reason, we evaluate the out-of-sample (real-time) predictive ability of the model.

Bayesian estimation methods allow us to rigorously account for all sources of uncertainty and, hence, we put particular emphasis on density forecast evaluation. We test the density nowcast accuracy of our HDFM against two popular benchmarks: the GDP nowcasts from the SPF and those from a naïve autoregressive model. To our knowledge, this is the first paper that compares probabilistic forecasts of models and institutions in a fully real-time perspective.

For the sake of comparability, our out-of-sample exercise is designed to replicate the features of the SPF. Specifically, we ask what the model would predict if used in "*real time*" to answer the SPF questionnaire for GDP growth. More in details, we collected 44 vintages of data which were available, in real time, to the forecasters in the quarters between 2003-Q1 and 2013-Q4 and, at each point in time, we use the model to derive a nowcast of the GDP growth rate in the current calendar year. For example, by using the data vintage available in the first quarter of 2003, we nowcast GDP growth in that quarter and we forecast GDP growth in the subsequent quarters of 2003 to derive the annual growth rate for 2003.

The annual growth rate g_t for GDP in the calendar year t_y is defined as the growth rate in the average level of GDP over the four quarters in year t_y, compared to the average annual level over the four quarters in year $t_y - 1$:

$$g_{t_y} = 100*\left(\frac{\text{GDP}_{Q1,t_y} + \text{GDP}_{Q2,t_y} + \text{GDP}_{Q3,t_y} + \text{GDP}_{Q4,t_y}}{\text{GDP}_{Q1,t_y-1} + \text{GDP}_{Q2,t_y-1} + \text{GDP}_{Q3,t_y-1} + \text{GDP}_{Q4,t_y-1}} - 1\right)$$

where GDP_{Q_j,t_y} is the level of GDP in the *j*th quarter of year t_y.

Since GDP enters in terms of quarterly growth in the HDFM, the calendar year growth need to be derived starting from the quarterly growth profiles. This is achieved in two steps: first we approximate the year-over-year (yoy) growth rates as a four quarters moving average of annualized quarter-over-quarter (qoq) growth rates; second, we approximate the calendar year growth as the average of the yoy growth rates within the calendar year.[20]

Once again, our choice is driven by the fact that the SPF density forecasts are only publicly available for this definition of the growth rate.[21] The naïve benchmark is an autoregressive model of order two. When looking at data in real time, one issue to address is which data vintage is used to compute the outcome of the target variable. Our choice is to consider the first vintage in the sample in which the data for the full calendar year are made available.

Fig. 5 reports the nowcasts (median, 16th and 84th quantiles, dashed lines) of annual GDP growth for the HDFM and the autoregressive model (AR).[22] The solid line in the charts refers to the realizations of the annual GDP. For both models, we report the nowcasts computed in each quarter of the year.[23]

Fig. 5 shows that the density nowcasts of the HDFM are generally centered around the outcome, already in the first quarter of the year, differently from the AR nowcasts. The uncertainty on the growth rate of GDP in the calendar year decreases and, by consequence, the nowcast densities become narrower.

Next, we evaluate more formally whether the HDFM density nowcasts are a good approximation of the true data densities, by testing the uniformity of the probability integral transforms (PITs).[24] The PITs are the values of the predictive cumulative distribution evaluated at the true realized values of the variables and are widely used to assess the calibration of density forecasts (most recent works include Aastveit, Gerdrup, Jore, & Thorsrud, 2011; Clark, 2011; Geweke & Amisano, 2010; Mitchell & Wallis, 2011). In fact, Diebold, Gunther, and Tay (1998) show that, if the density

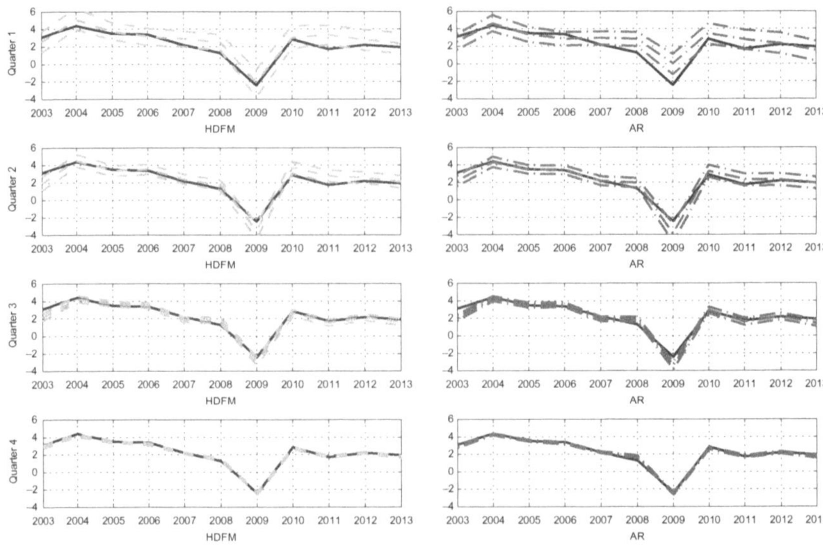

Fig. 5. Nowcasts for the Calendar Year 2003–2013. *Note*: Left panels: HDFM nowcasts (dashed) and out-turns (solid). Right panels: AR nowcasts (dashed) and out-turns (solid). All panels report median, 16th and 84th quantiles of the density nowcasts. From top to bottom, nowcasts produced in quarter 1, 2, 3 and 4.

forecasts approximate well the true density (i.e., are "well calibrated"), then the PITs should be uniformly distributed in the interval $[0-1]$. Assessing the uniformity of the PITs is equivalent to checking whether the inverse normal transformation of the PITs is standard normal. We compare the first four sample moments of the PITs inverse normal transformation with the first four moments of the standard normal distribution (zero, one, zero and three, respectively). Table 1 reports the four sample moments (columns two to five) for each of the nowcasts computed in the four quarters of the year (rows two to five). Following Bai and Ng (2005), we report the heteroskedasticity and autocorrelation consistent (HAC) standard deviation estimator to provide a rough idea of the statistical significance.[25]

Table 1 shows that all the sample moments are close to the theoretical values for the standard normal distribution, indicating that the density nowcasts of the HDFM model are well calibrated. We now turn to the analysis of the "relative" accuracy of the HDFM density nowcasts, comparing their log-scores to those of the alternatives.

Table 1. Tests of Normality, HDFM Nowcasts.

Quarter	First Moment	Second Moment	Third Moment	Fourth Moment
Q1	−0.47	0.52	−0.56	0.72
	(0.58)	(0.70)	(1.02)	(1.42)
Q2	−0.20	0.83	−0.09	1.60
	(0.94)	(0.99)	(1.97)	(2.50)
Q3	0.35	0.92	0.70	2.20
	(0.93)	(1.22)	(2.78)	(5.11)
Q4	0.52	0.77	0.89	1.24
	(0.74)	(0.84)	(1.24)	(1.68)

Note: Sample moments in the four quarters. Standard deviation in parentheses.

Table 2. Evaluation of Density Nowcasts for the Calendar Year, Average Log-Scores.

Quarter	HDFM	AR Minus HDFM	SPF Minus HDFM
Q1	−1.27	−0.16	0.15
		(0.24)	(0.11)
Q2	−1.17	0.22	0.12
		(0.13)	(0.10)
Q3	−0.68	−0.10	−0.03
		(0.12)	(0.18)
Q4	−0.18	−0.01	−0.22
		(0.03)	(0.07)

Note: HDFM (column two), average log-scores. AR (column three) and SPF (column four), average log-scores minus average HDFM log-scores. Standard deviation in parentheses.

In the SPF, the forecasters are asked to report, among other things, a density forecast by allocating probabilities to ranges of possible future outcomes of the annual growth of GDP. The lower bottom interval and the upper interval of the range are open bins, and the interior bins have equal lengths of 1 percentage points.[26] Individual responses are aggregated by computing average probabilities. For the sake of comparability, we organize the output of the model-based nowcasts (HDFM and AR) along the same lines of the SPF questionnaire. In other words, for all models, we compute the percentage (frequency) of the outcomes that fall in the different bins identified in the SPF. Then, we compute the log-scores, for each model and period, defined as the logarithm of the frequency of the bin including the observed annual GDP growth rate. The higher the mean of the log-scores, the higher the accuracy of the density nowcasts. In Table 2, column one indicates the quarter in which the nowcast for the calendar year is produced. In the second column, we report the average HDFM log-scores. For the AR (column three) and SPF (column four), instead, we report the difference of the average log-scores with the HDFM counterpart: positive values indicate that the average log-score of the specific model is higher than the average log-score of the HDFM for that specific quarter, and viceversa. The HAC estimate of its standard deviation are reported in parentheses.[27]

Results in the first column of Table 2 show that, as expected, the accuracy improves as more information becomes available during the year. The

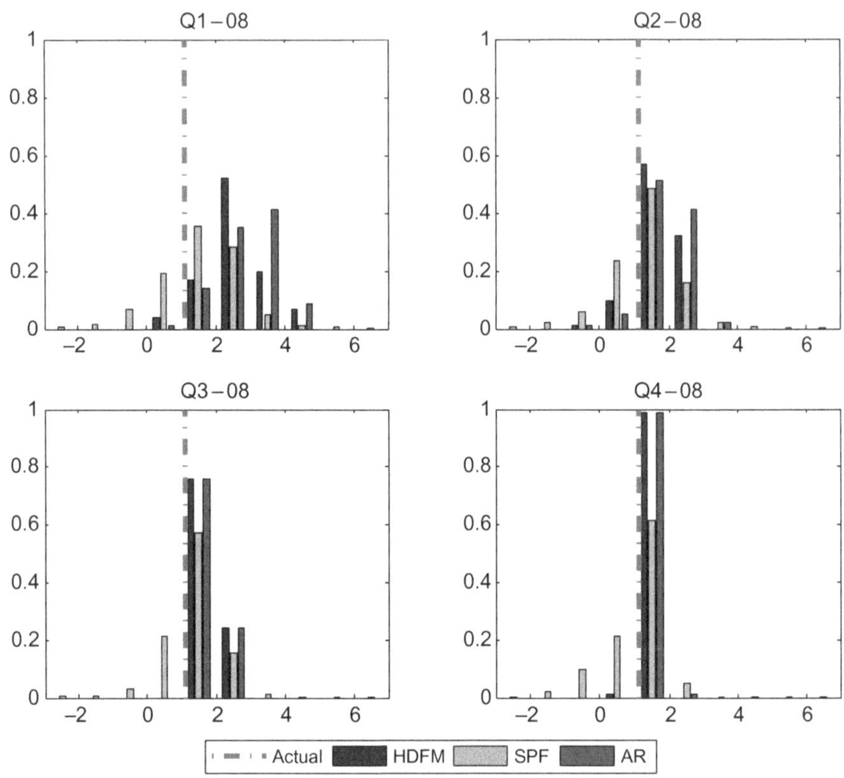

Fig. 6. Case Study – Calendar Year 2008. *Note*: Dashed line: observed calendar year real GDP growth rate; *x*-axis: SPF bins; Black bars: probabilities assigned by the HDFM to the bins; Gray bars: probabilities assigned by the AR to the bins; and Light Gray bars: probabilities assigned by the SPF to the bins.

results in the second column indicate that the HDFM is generally more accurate than the AR model. The third column indicates that, while the SPF density nowcasts are more accurate than those of the HDFM in the first two quarters of the year, the opposite is true in the third and fourth quarter of the year. However, the standard deviations of the sample mean of the difference in log-scores (in parentheses) are quite large compared to the average differences in log-scores, so the differences are unlikely to be statistically significant. Since the evaluation sample is short, the forecasting evaluation should not be seen as a horse race, but rather as an assessment of the validity of the model, aiming to ascertain that the accuracy of the density nowcasts is preserved, in spite of the proliferation of parameters resulting from taking into account general patterns of dynamic heterogeneity.[28]

Figs. 6–8 zoom on three specific calendar years, the 2008, 2009, and 2010. In the three figures, we report the evolution over the four quarters of

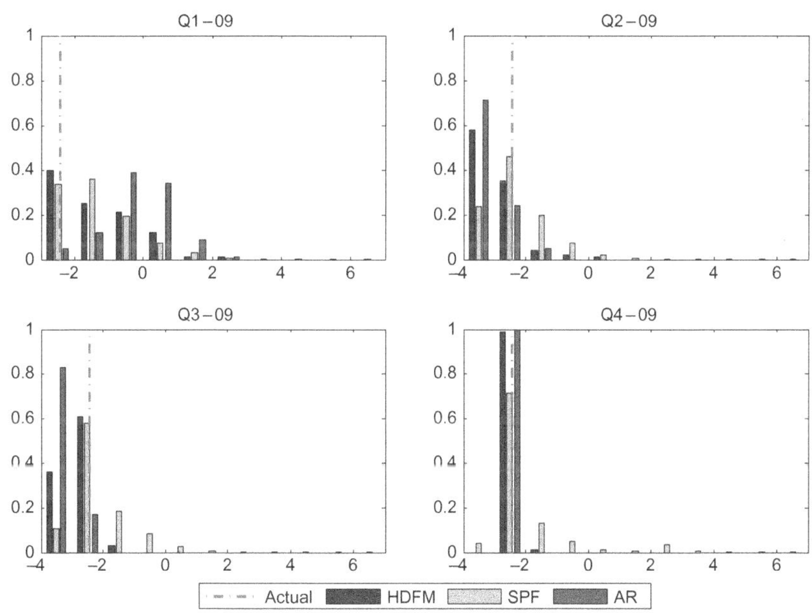

Fig. 7. Case Study – Calendar Year 2009. *Note*: Dashed line: observed calendar year real GDP growth rate; *x*-axis: SPF bins; Black bars: probabilities assigned by the HDFM to the bins; Gray bars: probabilities assigned by the AR to the bins; and Light Gray bars: probabilities assigned by the SPF to the bins.

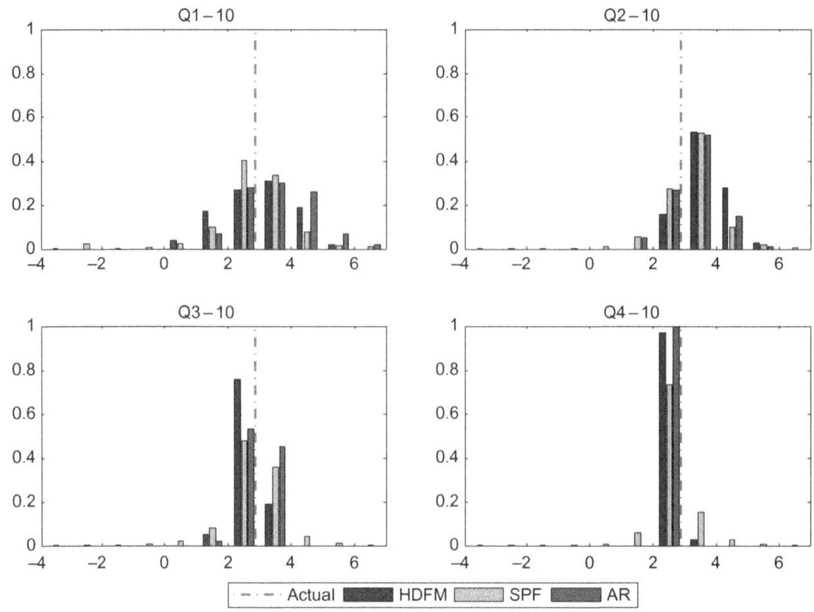

Fig. 8. Case Study – Calendar Year 2010. *Note*: Dashed line: observed calendar year real GDP growth rate; *x*-axis: SPF bins; Black bars: probabilities assigned by the HDFM to the bins; Gray bars: probabilities assigned by the AR to the bins; and Light Gray bars: probabilities assigned by the SPF to the bins.

2008, 2009, and 2010 of the density nowcasts, in form of histograms, of the HDFM, the AR, and the SPF.

Figs. 6–8 show that, particularly in the most acute part of the recession, the HDFM outperforms the AR model and provides very similar outcomes to the SPF. This result shows that accounting for different sources of information is important (e.g., surveys were an important source of information to timely capture the great recession). Moreover, it highlights how the HDFM, in spite of its mechanical nature, is able to replicate the outcomes of the SPF that, presumably, incorporates human judgement. These ability of mechanical models to replicate best practice in nowcasting GDP growth has been extensively documented in the context of point forecasts. The finding above indicate that this stylized fact also holds for density forecasts.

5. CONCLUSIONS

A synthetic indicator of economic activity should condense, in a timely and reliable manner, the information of several alternative observable measures. The indicator proposed in this paper is based on a DFM that explicitly allows for dynamic heterogeneity in the effects of the common factor on the variables. Since the model is richly parameterized, we control for overfitting by combining sample information with a prior belief that the effects of lagged factors on the observed economic indicators are more important the shorter the lag. Empirical results support our modeling strategy and indicate that it is important and feasible to account for general patterns of dynamic heterogeneity in the context of DFM. Indeed, inference based on our framework provides a timely account of the business cycle peaks and throughs and, in real time, it provides accurate and well-calibrated predictive densities.

In this paper, we have focused on a relatively small set of indicators that have been pre-classified as coincident on the basis of a long-established tradition in business cycle analysis. However, the general framework can be used to analyze more general dataset. Indeed, because of the high level of generality, the dynamic heterogenous factor model allows to analyze simultaneously a variety of indicators, without the need of pre-testing or expert judgement for the classification based on lead–lag patterns. This can be particularly important when dealing with datasets characterized by blurred lines of separation between coincident, leading and lagging indicators, as it tends to be the case when considering additional indicators and other countries. Evidence in this direction has been provided by Luciani and Ricci (2013), who have successfully used our methodology to nowcast Norway.

NOTES

1. In their pioneering work, (Burns & Mitchell, 1946, p. 3) define the business cycles as the "type of fluctuation found in the aggregate economic activity of nations that organize their work mainly in business enterprises: a cycle consists of expansions occurring at about the same time in many economic activities, followed by similarly general recessions […]." Pervasiveness is central also in the definition of the NBER dating committee: "During a recession, a significant decline in economic activity spreads across the economy and can last from a few months to more than a year. Similarly, during an expansion, economic activity rises substantially, spreads across the economy, and usually lasts for several years" (www.nber.org/cycles/general_statement.html).

2. See, for example, the statement of the CEPR dating committee: "To reduce the chance that data revisions might lead the Committee to reconsider its choice of turning points in the future, the Committee examines a wide array of economic data in addition to GDP, such as the individual components of output and labor market data."

3. The DFM can be cast in a state space form, which provides a natural environment to deal with missing data and mixed frequencies; it is then a suitable tool for the assessment of economic conditions in real time. See Giannone et al. (2008), Aruoba, Diebold, and Scotti (2009), Camacho and Perez-Quiros (2010), Jungbacker, Koopman, and van der Wel (2011), and Bańbura and Modugno (2014), and for surveys Bańbura, Giannone, and Reichlin (2011), and Bańbura, Giannone, Modugno, and Reichlin (2013).

4. See, for example, Stock et al. (1992), Kim and Nelson (1999), Mariano and Murasawa (2003), Aruoba et al. (2009), Camacho and Perez-Quiros (2010), Bańbura et al. (2013).

5. This is the same logic of the prior beliefs popularized by Litterman (1979) for Bayesian Vector Autoregressions.

6. See Kim and Nelson (1999), Del Negro (2002), Kose, Otrok, and Whiteman (2003), Justiniano (2004), Bernanke, Boivin, and Eliasz (2005), Del Negro and Otrok (2008), Mackowiak, Moench, and Wiederholt (2009), Moench (2012).

7. We assume, without loss of generality, that the variables are demeaned and standardized. In practice, we will estimate the model on demeaned and standardized data and we will re-attribute mean and standard deviation after estimation, as it is common practice in the factor model literature.

8. Dynamic heterogeneity can be taken into account in the context of principal components by including additional new factors, without modeling explicitly that they are lagged versions of each other (see Stock & Watson, 2002). This approach is suitable for forecasting but not for measuring business cycles conditions since it delivers estimated factors that are linear combinations of contemporaneous and lagged values of the index of economic activity.

9. Quah and Sargent (1993) follow the same strategy.

10. We leave for future research the task of conducting inference on the degree of prior tightness, which could be done following the lines of Giannone, Lenza, and Primiceri (2012).

11. We do not need to rescale the variances of the priors to adjust for the different scale of the variables, as it is customary in BVAR applications, since we perform inference using standardized data.

12. For a general discussion about the formulation of the state space in the presence of missing data, see Bańbura, Giannone, and Lenza (2014).

13. These are also the most relevant indicators constantly monitored by the NBER to detect and date peaks and throughs in the business cycle.

14. The importance of survey data for nowcasting has been documented by Giannone et al. (2008), Giannone, Reichlin, and Simonelli (2009), Angelini, Camba-Méndez, Giannone, Reichlin, and Rünstler (2011), and Lahiri and Monokroussos (2013). For a survey, see Bańbura et al. (2011, 2013).

15. Each month, survey respondents are asked to assess their organizations' performance based on a comparison of the current month to the previous month, see http://www.ism.ws/files/ISMReport/ROBBroch08.pdf

16. Real time vintages are downloaded from the Federal Reserve Bank of St. Louis http://alfred.stlouisfed.org/

17. We use a recursive updating scheme in our out-of-sample forecasting evaluation, that is, for every vintage the estimation sample always starts in January 1993.

18. Notice that with flat prior the posterior mode of the model parameters corresponds to the Maximum Likelihood estimates. Following Doz et al. (2012) Maximum Likelihood estimation is performed by using the EM algorithm initialized by principal components. The algorithm is modified to account for arbitrary patterns of missing data following the procedure of Bańbura and Modugno (2014). The algorithm has been shown to be computationally efficient and feasible even with high-dimensional data. Recent results by Jungbacker et al. (2011) and Jungbacker and Koopman (2015) show how computational efficiency can be further improved.

19. See https://www.chicagofed.org/publications/cfnai/index

20. Formally, defining t_q as the last quarter of the calendar year of interest, the computation is equivalent to $(1+L+L^2+L^3)(1-L^4)\log\mathrm{GDP}_{t_q}\times 400$. For a recent discussion, see Crump, Eusepi, Lucca, and Moench (2014).

21. The SPF also targets the GDP growth rate in the current quarter, but the density forecasts for this definition are not available.

22. The order of the autoregressive model is set equal to two, as suggested by the Akaike criteria computed in first estimation sample (1993–2003). Results are similar when using only one lag, when the selection is updated recursively and when the coefficients of the benchmark are restricted to those of the random walk.

23. The density nowcasts of the SPF are not included in the figure because they are available only in terms of histograms.

24. Notice that our evaluation is more demanding that the traditional residual based diagnostics since the predictive densities are computed in real time, hence accounting for parameter estimation uncertainty and overfitting.

25. The implied t-statistics should be taken with caution since the asymptotic distribution is nonstandard due to the recursive estimation of the parameters (see McCracken et al., 2013).

26. Until the first quarter of 2009, the upper bound of the lower bottom interval is 2%. Starting in the second quarter of 2009, it is 3%.

27. See note 25.

28. Results not reported here show that the HDFM model does not significantly outperform the homogenous model in terms of real-time out-of-sample forecasting accuracy. This result indicates that dynamic heterogeneity, although it is a feature of the data, is not so prominent to significantly improve the forecasting performance of the model, at least not in the short evaluation sample considered here.

ACKNOWLEDGMENTS

The views expressed are those of the authors and do not necessarily reflect those of the European Central Bank, the Eurosystem, the European Stability Mechanism, the Federal Reserve Bank of New York, the Board of

Governors of Federal Reserve System. This work was partly supported by the research contracts ARC-AUWB/2010-15/ULB-11 and IAP P7/06 StUDys (DG).

REFERENCES

Aastveit, K. A., Gerdrup, K. R., Jore, A. S., & Thorsrud, L. A. (2011). *Nowcasting GDP in real-time: A density combination approach*. Working Papers No. 0003. Centre for Applied Macro- and Petroleum economics (CAMP), BI Norwegian Business School.

Angelini, E., Camba-Méndez, G., Giannone, D., Reichlin, L., & Rünstler, G. (2011). Short-term forecasts of euro area GDP growth. *Econometrics Journal, 14*, 25–44.

Aruoba, S. B., Diebold, F. X., & Scotti, C. (2009). Real-time measurement of business conditions. *Journal of Business & Economic Statistics, 27*, 417–427.

Bai, J., & Ng, S. (2005). Tests for skewness, kurtosis, and normality for time series data. *Journal of Business & Economic Statistics, 23*, 49–60.

Bańbura, M., Giannone, D., & Lenza, M. (2014). *Conditional forecasts and scenario analysis with vector autoregressions for large cross-sections*. Working Paper Series No. 1733, European Central Bank, International Journal of Forecasting (forthcoming).

Bańbura, M., Giannone, D., Modugno, M., & Reichlin, L. (2013). Nowcasting and the real time data flow. In G. Elliott, & A. Timmermann (Eds.), *Handbook of economic forecasting* (Vol. 2). Amsterdam: Elsevier-North Holland.

Bańbura, M., Giannone, D., & Reichlin, L. (2011). Nowcasting. In M. P. Clements, & D. F. Hendry (Eds.), *Oxford handbook on economic forecasting* (pp. 63–90). Oxford: Oxford University Press

Bańbura, M., & Modugno, M. (2014). Maximum likelihood estimation of factor models on datasets with arbitrary pattern of missing data. *Journal of Applied Econometrics, 29*, 133–160.

Bernanke, B., Boivin, J., & Eliasz, P. S. (2005). Measuring the effects of monetary policy: A factor-augmented vector autoregressive (FAVAR) approach. *The Quarterly Journal of Economics, 120*, 387–422.

Boivin, J., & Giannoni, M. (2006). *DSGE models in a data-rich environment*. NBER Technical Working Papers No. 0332. National Bureau of Economic Research, Inc.

Burns, A. F., & Mitchell, W. C. (1946). *Measuring business cycles. NBER book series studies in business cycles*. National Bureau of Economic Research, Inc. 2011.

Camacho, M., & Perez-Quiros, G. (2010). Introducing the EURO-STING: Short term indicator of euro area growth. *Journal of Applied Econometrics, 25*, 663–694.

Carter, C., & Kohn, P. (1994). On Gibbs sampling for state space models. *Biometrica, 81*, 541–553.

Clark, T. E. (2011). Real-time density forecasts from Bayesian vector autoregressions with stochastic volatility. *Journal of Business & Economic Statistics, 29*, 327–341.

Crump, R., Eusepi, S., Lucca, D., & Moench, E. (2014). *Data insight: Which growth rate? It's a weighty subject*. Liberty Street Economics Blog, Federal Reserve Bank of New York.

de Jong, P., & Shephard, N. (1995). The simulation smoother for time series models. *Biometrika, 2*, 339–350.

Del Negro, M. (2002). Asymmetric shocks among U.S. states. *Journal of International Economics*, *56*, 273–297.

Del Negro, M., & Otrok, C. (2008). *Dynamic factor models with time-varying parameters: Measuring changes in international business cycles*. Discussion paper.

Diebold, F. X., Gunther, T. A., & Tay, A. S. (1998). Evaluating density forecasts with applications to financial risk management. *International Economic Review*, *39*, 863–883.

Doz, C., Giannone, D., & Reichlin, L. (2012). A quasi-maximum likelihood approach for large, approximate dynamic factor models. *The Review of Economics and Statistics*, *94*, 1014–1024.

Durbin, J., & Koopman, S. J. (2002). A simple and efficient simulation smoother for state space time series analysis. *Biometrika*, *89*, 603–616.

Geweke, J. (1977). The dynamic factor analysis of economic time series. In D. Aigner, & A. Goldberger (Eds.), *Latent variables socio-economic models*. Amsterdam: North-Holland.

Geweke, J., & Amisano, G. (2010). Comparing and evaluating Bayesian predictive distributions of asset returns. *International Journal of Forecasting*, *26*, 216–230.

Giannone, D., Lenza, M., & Primiceri, G. (2012). *Priors for vector autoregressions*. Discussion Paper No. 8755, C.E.P.R. Discussion Papers, The Review of Economics and Statistics, (forthcoming).

Giannone, D., Reichlin, L., & Sala, L. (2006). VARs, common factors and the empirical validation of equilibrium business cycle models. *Journal of Econometrics*, *132*, 257–279.

Giannone, D., Reichlin, L., & Simonelli, S. (2009). Nowcasting euro area economic activity in real time: The role of confidence indicators. *National Institute Economic Review*, *210*, 90–97.

Giannone, D., Reichlin, L., & Small, D. (2008). Nowcasting: The real-time informational content of macroeconomic data. *Journal of Monetary Economics*, *55*, 665–676.

Harding, D., & Pagan, A. (2002). Dissecting the cycle: A methodological investigation. *Journal of Monetary Economics*, *49*, 365–381.

Jungbacker, B., & Koopman, S. J. (2015). Likelihood-based dynamic factor analysis for measurement and forecasting. *Econometrics Journal*, *18*, 1–21.

Jungbacker, B., Koopman, S. J., & van der Wel, M. (2011). Maximum likelihood estimation for dynamic factor models with missing data. *Journal of Economic Dynamics and Control*, *35*, 1358–1368.

Justiniano, A. (2004). *Factor models and MCMC methods for the analysis of the sources and transmission of international shocks*. Ph.D. thesis, *Princeton University*.

Kim, C-J., & Nelson, C. R. (1999). *State-space models with regime switching: Classical and Gibbs-sampling approaches with applications* (Vol. 1). Cambridge, MA: The MIT Press.

Kose, M. A., Otrok, C., & Whiteman, C. H. (2003). International business cycles: World, region, and country-specific factors. *American Economic Review*, *93*, 1216–1239.

Lahiri, K., & Monokroussos, G. (2013). Nowcasting US GDP: The role of ISM business surveys. *International Journal of Forecasting*, *29*, 644–658.

Lawley, D. N., & Maxwell, A. E. (1963). *Factor analysis as a statistical method.* London: Butterworths

Litterman, R. (1979). *Techniques of forecasting using vector autoregressions*. Federal Reserve of Minneapolis Working Paper 115

Luciani, M., & Ricci, L. (2013). *Nowcasting Norway, Working Papers ECARES ECARES 2013-10, ULB — Universite Libre de Bruxelles*. International Journal of Central Banking, (forthcoming).

Mackowiak, B., Moench, E., & Wiederholt, M. (2009). Sectoral price data and models of price setting. *Journal of Monetary Economics*, *56*, 78.

Mariano, R. S., & Murasawa, Y. (2003). A new coincident index of business cycles based on monthly and quarterly series. *Journal of Applied Econometrics*, *18*, 427–443.

McCracken, M. W., & Clark, T. E. (2013). Advances in forecast evaluation. In Elliott, G., & Timmermann, A. (Eds.), *Handbook of economic forecasting* (Vol. 2). Amsterdam: Elsevier-North Holland.

Mitchell, J., & Wallis, K. F. (2011). Evaluating density forecasts: Forecast combinations, model mixtures, calibration and sharpness. *Journal of Applied Econometrics*, *26*, 1023–1040.

Mitchell, W. C., & Burns, A. F. (1938). *Statistical indicators of cyclical revivals*. No. mitc38-1 in NBER Books. National Bureau of Economic Research, Inc.

Moench, E. (2012). Term structure surprises: The predictive content of curvature, level, and slope. *Journal of Applied Econometrics*, *27*, 574–602.

Moore, G. H. (1983). The forty-second anniversary of the leading indicators. *Business cycles, inflation, and forecasting* (2nd ed., NBER Chapters, pp. 369–400). National Bureau of Economic Research, Inc.

Quah, D., & Sargent, T. J. (1993). A dynamic index model for large cross sections. *Business cycles, indicators and forecasting* (NBER Chapters, pp. 285–310). National Bureau of Economic Research, Inc.

Sargent, T. J., & Sims, C. A. (1977). Business cycle modeling without pretending to have too much a-priori economic theory. In C. A. Sims (Eds.), *New methods in business cycle research*. Minneapolis, MN: Federal Reserve Bank of Minneapolis.

Stock, J. H., & Watson, M. W. (1992). A probability model of the coincident economic indicators. In G. Moore, & K. Lahiri (Eds.), *The leading economic indicators: New approaches and forecasting record*. Cambridge: Cambrigdge University Press.

Stock, J. H., & Watson, M. W. (2002). Macroeconomic forecasting using diffusion indexes. *Journal of Business & Economic Statistics*, *20*, 147–162.

Stock, J. H., & Watson, M. W. (2011). Dynamic factor models. In M. P. Clements, & D. F. Hendry (Eds.), *Oxford handbook of forecasting*. Oxford: Oxford University Press.

ON THE SELECTION OF COMMON FACTORS FOR MACROECONOMIC FORECASTING

Alessandro Giovannelli[a] and Tommaso Proietti[a,b]

[a]*Department of Economics, University of Rome Tor Vergata, Rome, Italy*
[b]*CREATES, Aarhus, Denmark*

ABSTRACT

We address the problem of selecting the common factors that are relevant for forecasting macroeconomic variables. In economic forecasting using diffusion indexes, the factors are ordered, according to their importance, in terms of relative variability, and are the same for each variable to predict, that is, the process of selecting the factors is not supervised by the predictand. We propose a simple and operational supervised method, based on selecting the factors on the basis of their significance in the regression of the predictand on the predictors. Given a potentially large number of predictors, we consider linear transformations obtained by principal components analysis. The orthogonality of the components implies that the standard t*-statistics for the inclusion of a particular component are independent, and thus applying a selection procedure that takes into account the multiplicity of the hypotheses tests is both correct and computationally feasible. We focus on three main multiple testing*

Dynamic Factor Models
Advances in Econometrics, Volume 35, 595–630

ISSN: 0731-9053/doi:10.1108/S0731-905320150000035015

procedures: Holm's sequential method, controlling the familywise error rate, the Benjamini–Hochberg method, controlling the false discovery rate, and a procedure for incorporating prior information on the ordering of the components, based on weighting the p-values according to the eigenvalues associated to the components. We compare the empirical performances of these methods with the classical diffusion index (DI) approach proposed by Stock and Watson, conducting a pseudo-real-time forecasting exercise, assessing the predictions of eight macroeconomic variables using factors extracted from an U.S. dataset consisting of 121 quarterly time series. The overall conclusion is that nature is tricky, but essentially benign: the information that is relevant for prediction is effectively condensed by the first few factors. However, variable selection, leading to exclude some of the low-order principal components, can lead to a sizable improvement in forecasting in specific cases. Only in one instance, real personal income, we were able to detect a significant contribution from high-order components.

Keywords: Variable selection; multiple testing; *p*-value weighting

JEL classifications: C22; C52; C58

1. INTRODUCTION

The focus of much recent theoretical and applied econometric research has concentrated on the ability to predict key macroeconomic variables, such as output and inflation, using a large number of potential predictors, with little or no a priori guidance over their relevance. This theme, which developed contextually in the statistical and machine-learning literature on data mining and discovery, has received a very distinctive solution, hinging upon the idea that the wealth of information on macroeconomic variables can be distilled by a limited number of common factors.

The common factors capture the comovements among the economic variables and can be consistently estimated by principal components analysis (PCA), as in the static factorial approach proposed by Stock and Watson (2002), or by dynamic PCA, using frequency domain methods, as proposed by Forni, Hallin, Lippi, and Reichlin (2005). Quoting from Stock and Watson (2006), the availability of a factor structure and of closed-form inferences has turned the high dimensionality of the information set from a curse to a blessing.

Once the factors are extracted, they can be used for forecasting the variables of interest, by augmenting an observation-driven model, such as an autoregression, by the estimated factors. This approach, known as the diffusion index (DI), or factor-augmented autoregressive (FAR), forecasting methodology, has become mainstream, owing its success to the ability to incorporate information carried by a large number of potential predictors in a simple and parsimonious way.

Applied economic forecasting has shown that the consideration of the factors as potential predictors has proved successful in macroeconomic forecasting using large datasets; it would be impossible to provide a list of references that could be representative of the research carried out in this field. The reviews in Breitung and Eickmeier (2006), Stock and Watson (2006) and Stock and Watson (2010), as well as Ng (2013), provide ample coverage of the main issues.

As it is well known, the principal components, arising from the spectral decomposition of the sample covariance matrix of the predictors, are ranked according to the size of the corresponding eigenvalue. The current forecasting practice selects the first components according to an information criterion, such as Bai and Ng (2002) and Onatski (2010), and uses them as explanatory variable in the forecasting model in lieu of the original predictors. A potential limitation of this procedure is that the selection of factors is blind to the predictive ability of the principal components, as no consideration is given to their relationship with the predictand by the information criteria commonly used.

The lack of supervision of the principal components in regression has been the matter of an old debate, which is echoed in Cox (1968), Hadi and Ling (1998), Joliffe (1982) and Cook and West (2007), among others. There are essentially two opposite views: the argument of the critics is that there is no logical reason why the predictand should not be related to the least important principal components, and secondly that different predictands, such as output and inflation, cannot depend on the same r principal components. The counter argument, using Mosteller and Tukey quotation of Einstein (Mosteller & Tukey, 1977; pp. 397–398), is that 'nature is tricky, but not downright mean': the first principal components capture the underlying common dimensions of the economy. If this was the case, the leading principal components, those corresponding to the largest eigenvalues, should carry the essential information for predicting economic variables.

The selection of the factors that are relevant for the prediction of macroeconomic variables has attracted a lot of interest and several solutions have been proposed in the literature for supervising the factors, taking into account their ability to predict a specific dependent variable. Bai and Ng (2008) propose distilling the factors, referred to as 'targeted predictors', by performing a PCA on a subset of the original predictors, that are selected according to the strength of the relationship with the variable to be predicted in a marginal regression framework. This is an instance of the method of supervised PCA (Bair, Hastie, Debashis, & Tibshirani, 2006), aiming at finding linear combinations of the predictors that have high correlation with the outcome. Bai and Ng (2009) considered bootstrap aggregation of the predictions arising from an FAR framework, which retains only the significant factors. A comprehensive review of variable selection in predictive regression is Ng (2013).

The research question addressed by this paper is whether many predictors can be replaced by a reduced number of principal components selected according to the strength of the relationship with the predictand, and whether components beyond the first few carry useful information for improving the predictive ability. We propose a simple and operational supervised method based on selecting the factors on the basis of their significance in the regression of the predictand on the predictors. Given a potentially large number of predictors, we consider linear transformations obtained by PCA. The orthogonality of the components implies that the standard t statistics for the inclusion of a particular component in the multiple regression framework are independent, and thus applying a multiple testing procedure to select the components that are significant at a particular level is both correct and computationally feasible.

The selection of the principal components can be seen as a decision problem involving multiple-testing, where a single null hypothesis claims that a specific component ought to be excluded from the model. There are several multiple testing procedures available that focus on controlling some type of error rate, namely, the familywise error rate, such as the Bonferroni–Holm procedure (see Holm, 1979), or the false discovery rate, which is the expected proportion of wrong rejections. Among the procedures controlling for the false discovery rate, we focus on the Benjamini–Hochberg procedure (see Benjamini & Hochberg, 1995) and on a weighted procedure that allows to incorporate prior information about the ordering of the components; see Genovese (2006).

In summary, our methodology has three steps:

1. Orthogonalise the original N predictors by computing the N standardised PCs.

2. Select r principal components on the basis of their correlation with the predictand, taking into account the multiplicity of the testing problem and controlling for the error rate of the selection procedure.
3. Use the selected components in an FAR predictive regression.

Our method can be nested within the shrinkage representation for forecasting using orthogonal predictors proposed by Stock and Watson (2012b) and has analogies with the idea of targeted predictors, although the object of the selection is the principal components, rather than the original predictors: this has the advantage of not having to consider the correlation of the tests statistics for the inclusion of the predictors.

We validate our procedure using a dataset consisting of 121 quarterly U.S. macroeconomic time series observed from 1959-I to 2011-II. A pseudo-real-time rolling forecast experiment is conducted to compare the performance of our selection method to a benchmark autoregressive predictor and to the standard DI forecasts based on the first five components.

This paper is structured as follows. In Section 2, we provide a brief review of the DI methodology. Section 3 considers the issue of estimating supervised factors and reviews the main solutions available in the literature. In Section 4, we present principal components regression as a shrinkage method and discuss the issues posed by the selection of the components and the consequences in terms of forecasting accuracy. Section 5 presents our supervised method using a multiple testing approach to the selection of the principal components in the FAR predictor.

2. FORECASTING USING PRINCIPAL COMPONENTS

Let $X_t = (X_{1t}, \ldots, X_{Nt})'$ denote an $N \times 1$ vector of predictors observed at times $t = 1, \ldots, T$, and y_t a stationary time series to be predicted. We are interested in forecasting the value of the series at a forecast horizon $h>0$, which we denote $y_{t+h}^{(h)}$. The series y_t may arise as a transformation of the original variable Y_t, which depends on its order of integration.

The DI forecasting methodology, originally proposed by Stock and Watson (2002), provides a simple and parsimonious way of incorporating a highly dimensional information set; it is based on the assumption that the predictors have an approximate factor structure, such that the unobserved factors can be estimated consistently by PCA. The factor model is formulated as follows:

$$X_t = \Lambda F_t + \xi_t \tag{1}$$

where $F_t = (F_{1t}, \ldots, F_{rt})$ with $r<N$ are the unobserved common factors, Λ is the $N \times r$ matrix of factor loadings and ξ_t is the idiosyncratic disturbance not explained by the factors. The following normalisation assumptions are made so as to identify the factor model: $E(F_t F_t') = I_r$ and $(1/T)\sum_t F_t F_t' \rightarrow_p I_r$, where I_r is the identity matrix of order r; the elements of Λ are bounded from below and above and $N^{-1}\Lambda'\Lambda \rightarrow \Sigma$, where Σ is an $r \times r$ diagonal matrix with finite positive entries. For the idiosyncratic component, *Assumption M1* in Stock and Watson (2002) holds, allowing ξ_t to be serially and cross-sectionally weakly correlated.

Letting $S = T^{-1}\sum_t X_t X_t'$ and denoting the spectral decomposition of the covariance matrix by $S = VD^2V'$, where $V = (v_1, \ldots, v_N)$ is the $(N \times N)$ matrix of eigenvectors, $V'V = VV' = I_N$, and $D = \text{diag}(d_1, \ldots, d_N)$ is the matrix containing the square root of the ordered eigenvalues, $d_1 \geq d_2 \geq \cdots \geq d_N$, as in Stock and Watson (2002), the common factors F_t are estimated by the first r standardised principal components

$$\hat{F}_t = D_r^{-1} V_r' X_t \tag{2}$$

where $D_r = \text{diag}(d_1, \ldots, d_r)$ and $V_r = (v_1, \ldots, v_r)$.

We assume that we are interested in predicting a variable y_t (which may as well be included in the set X_t) at the horizon $h>0$, by using all the information contained in X_t. For instance, if we are interested in forecasting quarterly industrial production, denoted Y_t, h quarters ahead, we set $y_{t+h}^{(h)} = 400(\ln Y_t - \ln Y_{t-1})$, which assumes that $\ln Y_t$ is difference stationary. In predicting $y_{t+h}^{(h)}$, we include the estimated common factors and the lags of $y_t = (\ln Y_t - \ln Y_{t-1})$, according to the FAR model:

$$y_{t+h}^{(h)} = \mu + \sum_{j=1}^{p} \phi_j^{(h)} y_{t-j+1} + \sum_{k=1}^{r} \beta_k^{(h)} \hat{F}_{kt} + \varepsilon_{t+h} \tag{3}$$

where ε_{t+h} is the forecasting error with variance σ^2. For the forecasting equation (3), *Assumption Y1* in Stock and Watson (2002) holds.

The DI forecasts are obtained according to a two step procedure: in the first step r factors are extracted from the set of predictors by performing a PCA and selecting the number of common factors according to information criteria proposed by Bai and Ng (2002), such as

$$\mathrm{IC}_{p1}(r) = \ln V(r) + r\left(\frac{N+T}{NT}\right)\ln\left(\frac{NT}{N+T}\right)$$

$$\mathrm{IC}_{p2}(r) = \ln V(r) + r\left(\frac{N+T}{NT}\right)\ln\, \min\{N, T\}$$

where $V(r) = (1/NT)\sum_{t=1}^{T}(X_t - \hat{\Lambda}_r \hat{F}_t)'(X_t - \hat{\Lambda}_r \hat{F}_t), \hat{\Lambda} = V_r D_r$. Bai and Ng (2002) show that the value of r that minimizes $\mathrm{IC}_{p1}(r)$ or $\mathrm{IC}_{p2}(r)$ is a consistent estimator, for $N, T \to \infty$, of the number of common factors. In the second step, the estimated factors are used as predictors in Eq. (3). As shown in Bai and Ng (2006), we can treat $\hat{F}_t$ as observed regressors. Alessi, Barigozzi and Capasso (2010) have proposed a modification of the above criteria, introducing a tuning constant which multiplies the penalty term on the right-hand side depending on N and T, as well as a heuristic criterion for determining the number of factors.

Since the factors are selected according to an information criterion that operates on the eigenstructure of X_t, the method is unsupervised. The selection methodology assumes that the factors are ordered according to the size of the corresponding eigenvalue. However, there is no reason why a predictand should not depend on a higher order component or why different predictands, such as output and inflation, should depend on the same factors.

3. APPROACHES TO THE SUPERVISION OF THE FACTORS

Several proposals have been made for supervising the factors, so that the selected factors carry information that is useful for predicting the specific variable under consideration. In this section, we sketch a brief survey of the literature, ignoring the shrinkage and model averaging approaches that are applied directly to the observable predictors, rather than the principal components. For an account of these approaches, see Bai and Ng (2008), De Mol, Giannone, and Reichlin (2008) and Stock and Watson (2006).

In the supervised PC method proposed by Bair et al. (2006), a subset of predictors is first selected on the basis of their correlation with the response variable; more specifically all the predictors for which the estimated regression coefficients are larger than a threshold c are considered

$$\left|\sqrt{T}\frac{\sum_t X_{it} y_{t+h}^{(h)}}{\sum_t X_{it}^2}\right| > c, \quad i = 1, 2, \ldots, N$$

and a PCA is performed on the selected predictors to extract the factors to be used for prediction. The method clearly depends on the threshold c, which is estimated by cross-validation.

Bai and Ng (2008) construct supervised principal components, which they name *targeted predictors*, by pre-selecting a subset of predictors with predictive power for a specific predictand, and conducting a PCA on those. They explore hard thresholding rules constructed on the t-statistics of the regression of $y_{y+h}^{(h)}$ on a single predictor X_{it} (after controlling for a set of predetermined variables, such as the lags of the dependent variables), say t_i^*, selecting those variables for which $|t_i^*| > c$, c alternatively being equal to 1.28, 1.65 and 2.58. Their selection rule does not consider the issue of multiple testing. On page 306, they state however that application of Holm's procedure (see Section 5) did not lead to different results. Other soft thresholding methods, such as the LASSO, least angle regression and the elastic net, are considered and compared. This paper concludes that targeting the predictors to the economic variable to be predicted, they consider inflation in particular, leads to a gain in forecasting accuracy.

Bai and Ng (2009) proposed componentwise and block-wise boosting algorithms for isolating the predictors in FAR models that are most helpful in predicting a variable of interest. The algorithms do not rely on the ordering of the variables (and in the componentwise case do not rely on the ordering of their lags). Starting from the null model including only a constant, the algorithms perform incremental forward stagewise fitting of the mean-square prediction error, by a sequence of Newton–Raphson iterations that iteratively improve the fit. At each step, a single explanatory variable (e.g. a PC), or a block consisting of a regressor and its lags, is fitted by ordinary least-squares regression, and selected according to the reduction in the residual sum of squares. The selected variable contributes to the current predictor with a coefficient which is shrunk towards zero by a fraction known as the learning rate. The algorithm is iterated until a stopping rule is found. Bai and Ng (2009) propose an information criterion for selecting the number of boosting iterations that takes into account the estimation error in the estimation of the factors, which is $O(N^{-1})$.

Inoue and Kilian (2008) present an application of bootstrap aggregation (bagging) of predictors of U.S. inflation using $N = 30$ variables. Among the predictors, they consider selecting by pretesting the PCs among the first K, where K ranges from 1 to 8. Several critical values for selection pretests are considered. Stock and Watson (2012b) and Kim and Swanson (2014) also consider averaging the FAR predictors obtained from independent bootstrap samples with factors selected according to the rule that their (robust) t-statistic must be larger than 1.96 in modulus.

Fuentes, Poncela, and Rodrguez (2014) propose the use of sparse partial least squares to select a small subset of supervised factors, extending to a dynamic setting the static methodology of Chun and Keleş (2010). The candidate factors arise from the spectral decomposition of the matrix $T^{-1}\tilde{X}'yy'\tilde{X}$, where y has generic element $y_{t+h}^{(h)}$ and X is a matrix with rows composed of the elements of X_t', augmented by the lags of the predictand. The loadings are shrunk towards zero by a LASSO-type penalty, aiming at the extraction of sparse supervised components.

Finally, an important class of supervision methods is based on inverse regression. Let (y, X) denote the observable predictand and predictors and let $f(y, X)$ denote their joint density. DI forecasting starts from the factorisation $f(y, X) = f(y|X)f(X)$, assuming a factor structure for X. Obviously, the factors are unsupervised, as only the marginal distribution $f(X)$ is considered. A different approach to the supervision of the factors deals with the factorisation $f(y, X) = f(X|y)f(y)$, using the first conditional density for obtaining a reduced set of predictors incorporating information concerning y, achieving a substantial dimensionality reduction. The reduced set is then used in the prediction of y, according to a maintained model for $f(y|X)$. One such methodology is sliced inverse regression (SIR, Li, 1991): the range of y is partitioned into slices, within which the centroids of the X's are computed; a singular-value decomposition of the matrix of centroids is performed to obtain a few effective dimension-reduction directions. The method of principal fitted components (PFC) analysis, proposed by Cook and West (2007) and Cook and Forzani (2008), is based on inverse regression of X on y to obtain a dimension reduction that preserves the information that is relevant for predicting y; in Cook and West (2007) the conditional mean $E(X|Y = y)$ is estimated by projecting the X's on polynomial terms in y and a PCA is conducted on the conditional mean estimates to obtain the PCFs.

4. PRINCIPAL COMPONENTS REGRESSION AND COMPONENTS SELECTION AS SHRINKAGE METHODS

Our approach is a particular case of the generalised shrinkage model considered in Stock and Watson (2012b). In the sequel we will assume that the forecasting model does not contain lags of the predictand and that the DI predictor results exclusively from principal component regression.

In particular, the data are generated as

$$y^{(h)}_{t+h} = x'_t\delta + \epsilon_{t+h}$$

where ϵ_{t+h} has mean zero and $\mathrm{Var}(\epsilon_{t+h}) = \sigma^2$. We also assume that T observations are available for $y_{t+h}, t = 1, \ldots, T$, and focus on the predictor

$$\hat{y}^{(h)}_{t+h|t} = \sum_{i=1}^{N} \psi_i \hat{\beta}_i \hat{F}_{it} \tag{4}$$

where $\hat{F}_t = D_r^{-1} V'_r X_t$ denotes the $N \times 1$ vector containing the standardised PC scores $\hat{F}_{it}, i = 1, \ldots, N$, such that $(1/T)\sum_t \hat{F}_t \hat{F}'_t = I$, ordered according to the eigenvalues of the matrix S. Moreover, $\hat{\beta}_i = (1/T)\sum_t \hat{F}_{it} y_{t+h}$, is the least-squares estimator of the regression coefficient of y on the i-th PC, and ψ_i is the indicator for the inclusion of the i-th PC. The decision whether to include it or not depends on the strength of the relationship with the predictand and will be discussed shortly.

As $\hat{F}_{it} = x'_t v_i / d_i$, where d_i^2 is the i-th eigenvalue of S and v_i is the corresponding eigenvector,

$$S = VD^2V', \quad V = [v_1, \ldots, v_i, \ldots, v_N], \quad D = \mathrm{diag}(d_1, \ldots, d_N)$$

the predictor (4) can be written $\hat{y}^{(h)}_{t+h|t} = x'_t\hat{\delta}$, for

$$\hat{\delta} = \sum_{i=1}^{N} \psi_i \frac{v_i}{d_i} \hat{\beta}_i$$

The lack of supervision of the ordering of the components, see cautionary note 2 in Hadi and Ling (1998), can be evidenced by a plot of d_i^2, the i-th eigenvalue, versus the increase of the regression residual sum of squares arising from the deletion of the i-th component, measured by $T\hat{\beta}_i^2 = T^{-1}\left(\sum_t \hat{F}_{it} y_{t+h}^{(h)}\right)^2$.

The mean-square error (MSE) of the above predictor, treating the factors as observed variables, is

$$\text{MSE}\left(\hat{y}_{t+h|t}^{(h)}\right) = \left[\text{B}\left(\hat{y}_{t+h|t}^{(h)}\right)\right]^2 + \text{Var}\left(\hat{y}_{t+h|t}^{(h)}\right)$$

where the bias and the variance are given, respectively, by

$$\text{B}\left(\hat{y}_{t+h|t}^{(h)}\right) = \sum_{i=1}^{N} (1-\psi_i)\hat{F}_{it} d_i v_i'\delta, \quad \text{Var}\left(\hat{y}_{t+h|t}^{(h)}\right) = \sigma^2\left(1 + \frac{1}{T}\sum_i \psi_i \hat{F}_{it}^2\right)$$

These simple expressions underlie the usual bias-variance trade-off: removing one factor from the set of predictors (i.e. setting $\psi_i = 0$) reduces the variance, but increases the bias. The bias term features the singular-value d_i, which implies that the bias increase is potentially larger if ceteris paribus a component with high d_i is removed. The bias resulting from the omission of a particular PC further depends on $v_i'\delta$; this term depends on the population relationship between y and the x's and on the loadings of the i-th PC.[1] The main message, conveyed by the above expression for the MSE is that omitting a PC loading heavily on important variables ($(v_i'\delta)^2$ is large) will have more impact if the PC corresponds to a large eigenvalue. Note also that $\text{Var}(\hat{y}_{t+h|t})$ depends solely on σ^2 and the Mahalanobis squared distance of x_T from 0 in the x space, $x_T'\left(\sum_t X_t X_t'\right)^{-1} x_T = \sum_i \hat{F}_{it}^2$; recalling that $\hat{F}_{it} = x_T' v_i / d_i$, the variance will be inflated by the presence of components with small d_i and for which the inner product $x_T' v_i$ is large. The broad conclusion arising from this analysis is that discarding the first PCs is not in general a good idea, and the ordering of the components should be taken seriously. We see this simple result as a possible explanation for the failure of alternative shrinkage methods to outperform the DI approach documented in Stock and Watson (2012b).

In principle, we could determine the optimal set of indicators $\{\psi_i, i = 1, \ldots, N\}$, which minimises the above MSE of prediction, for example, parameterising

$$\psi_i = \psi_i\left(\hat{\beta}_i;\gamma,c\right) = \frac{1}{1+\exp\left[-\gamma\left(|\hat{\beta}_i|-c_i\right)\right]}$$

where, for example, $c_i = c/d_i$ for an unknown positive constant c, and thinking about replacing δ and σ by some estimate, perhaps iteratively. Stock and Watson (2012b) estimate $c_i = c$ by cross-validation and set $\gamma \to \infty$. Hwang and Nettleton (2003) propose a general approach to the problem. We do not pursue this here and consider strategies such that ψ_i is the indicator function that the p-value of the significance test for the i-th regression coefficient is below a given threshold.

Before presenting our methodology, it is perhaps useful to remark that PCR conducted using only the first r principal components, chosen according to an information criterion, poses $\psi_i = I(i \le r)$. Another popular regularisation method, ridge regression, yields the predictor (4) where the shrinkage factor ψ_i varies with i: letting $\rho \ge 0$ denote the penalty parameter in the criterion $S(\rho,\delta) = \sum_t \left(y_{t+h} - x_t'\delta\right)^2 + \rho\delta'\delta$, then

$$\psi_i = \frac{d_i^2}{d_i^2 + \rho}$$

See Hastie, Tibshirani, & Friedman (2009) for a general reference and discussion.

5. THE SELECTION OF THE COMMON FACTORS AS A MULTIPLE TESTING PROBLEM

Consider the set of null hypotheses $H_{0i} : \beta_i = 0, i = 1, \ldots, N$, in the predictive regression model, $y_{t+h}^{(h)} = \sum_{i=1}^{N} \beta_i \hat{F}_{it} + \epsilon_t$, and let p_i denote the two-sided p-value, based on the t-statistic t_i associated to the i-th principal components regression coefficient,

$$t_i = \sqrt{T}\frac{\hat{\beta}_i}{\hat{\sigma}} \qquad (5)$$

An issue is posed by the estimation of the regression standard error σ by $\hat{\sigma}$ in the denominator. The usual estimator, the square root of

$$\hat{\sigma}^2 = \frac{\text{RSS}}{T-N}, \quad \text{RSS} = \sum_t y_{t+h}^{(h)2} - T\sum_i \hat{\beta}_i^2$$

is either infeasible, if $N \geq T$, or severely downward biased due to the overfitting when N is large. We address the issue of estimating σ in Section 5.1.

The testing strategy based on the t_i statistics is multivariate, that is, it treats all the remaining variables as nuisance parameters, when testing for the significance of a particular effect. An alternative strategy, that avoids estimation of σ, is componentwise, being based on the marginal linear regression t_i^* statistics,

$$t_i^* = \sqrt{\frac{T(T-1)}{\text{RSS}_i}}\hat{\beta}_i, \quad \text{RSS}_i = \sum_t y_{t+h}^{(h)2} - T\hat{\beta}_i^2$$

The two t-statistics are related by

$$t_i = t_i^* \sqrt{\frac{\text{RSS}_i/(T-1)}{\text{RSS}/(T-N)}}$$

that is, the univariate t_i^* is multiplied by the root MSE ratio. The componentwise approach is fairly popular in genomics and in hyper-dimensional contexts such that $N > T$. It has been adopted by Bai and Ng (2008) for estimating the targeted predictors.

For T large, the null distribution of the statistic is $t_i \sim \text{N}(0, 1)$, and the p-value is computed as $p_i = 2(1-\Phi(|t_i|))$. Here, in fact, the alternative is two-sided, that is $H_{1i} : \beta_i \neq 0$, $i = 1, \ldots, N$. Let us consider the re-ordering of the components according to the nonincreasing values of $|t_i|$, and let us denote by $\hat{F}_{i,t}$ the i-th component in the new ordering. The corresponding ordered p-values will be denoted by

$$p_1 \leq p_2 \leq \cdots \leq p_i \leq \cdots \leq p_N$$

The inclusion of the i-th PC is based on a multiple testing procedure, which provides a decision rule aiming at controlling overall error rates when performing simultaneous hypotheses tests. The procedures can be distinguished according to the error that is controlled. For this purpose, consider the following confusion matrix:

Actual	Decision		Total
	Accept	Reject	
$H_{0i} : \beta_i = 0$	TN	FP	N_0
$H_{1i} : \beta_i \neq 0$	FN	TP	N_1
Total	A	R	N

$R = FP + TP$ is the total number of hypotheses rejected; FP is the number of false rejections (type I errors, i.e. of falsely rejected hypothesis). There are N_0 true nulls and N_1 false nulls. TN in the number of correctly accepted true nulls, etc. FN is the number of false positive decisions, that is, type II errors and TP (true positives) is the number of correct rejections.

The per comparison error rate (PCER) approach controls for the expected number of true H_{0i} rejected over N, the total number of tests,

$$\text{PCER} = \frac{E(FP)}{N}$$

It amounts to ignoring the multiplicity problem altogether, and uses the critical value corresponding to a preselected significance level α, and thus rejects all hypotheses for which $p_i < \alpha$. This guarantees that α is an upper bound for PCER, as when $N_0 = N$ (all null hypotheses are true) $E(FP) = \alpha N$.

The problem with this approach is that the probability of a false rejection increases rapidly with N; in particular, if $N_0 = N$, $P(R \geq 1) = 1 - (1 - \alpha)^N$. This has led to developing multiple testing strategies requiring that the probability of one or more false rejections does not exceed a given level. Definining the familywise error rate (FWER) as the probability of rejecting any true H_{0i}, $\text{FWER} = P(FP \geq 1)$, the aim is defining decision rules guaranteeing $\text{FWER} \leq \alpha$. The simplest procedure controlling the FWER is the Bonferroni rule, rejecting all H_{0i} for which $p_i < \alpha/N$. However, the power of this rule is typically very low, when $N > N_0$ is large.

A more powerful method for controlling the FWER at level α is due to Holm (1979). Holm's method is a step-down procedure rejecting the (i)-th null hypothesis $H_{0,(i)}$ if

$$p_{(j)} \leq \frac{\alpha}{N - j + 1}, \quad j = 1, 2, \ldots, i$$

Note that $p_{(1)} \leq \alpha/N$ and $p_{(N)} \leq \alpha$, so that at the initial step we apply Bonferroni's rule, while at the final step we get the PCER approach. Thus, if $p_{(1)} > \alpha/N$, all nulls are accepted and the procedure stops. Else, we reject $H_{0,(1)}$ and test the remaining hypotheses at level $\alpha/(N-1)$. In such case, we accept all $H_{0,(i)}, (i) > 1$ if $p_{(2)} > \alpha/(N-1)$, else we reject $H_{0,(2)}$ and test the remaining hypotheses at level $\alpha/(N-2)$, in which case we iterate the procedure, until all the remaining hypotheses are accepted.

Procedures that control for the FWER are unduly conservative when N is very large, despite the improvements offered by step-down procedures such as Holm's. A less conservative approach is offered by procedures that control for the false discovery rate (FDR), which is the expected proportion of falsely rejected nulls:

$$\text{FDR} = E\left(\frac{FP}{R}\right)$$

The main procedure is due to Benjamini and Hochberg (BH, Benjamini & Hochberg, 1995). Let α denote a control rate in the range (0,1). A decision rule that has FDR $= \alpha N_0/N \leq \alpha$ rejects all $H_{o,(i)}, i = 1, \ldots, r$, for which

$$p_{(i)} \leq \alpha \frac{i}{N}, \quad i = 1, \ldots, r, \quad p_{(r+1)} > \alpha \frac{r+1}{N}$$

In high-dimensional settings, BH has been proved to achieve a better balance between multiplicity control and power. Another advantage is that when $N = N_0$ BH controls also the FWER at level α. Adaptive variants and refinements are available in the literature that address the issue of correlation in the tests statistics. See Efron (2010) for a review.

As stated in the introduction, if we can think that nature is benign and that the ordering of the factors carries important information, then the selection should incorporate information on the factor structure. This can be achieved by weighting the p-values, p_i, according to the index i of the factor. A procedure that allows for p-value weighting and achieves control over the FDR is due to Genovese (2006), and it works as follows:

1. Assign weights $w_i > 0$ to each H_{0i} so that $\overline{w} = N^{-1} \sum_i w_i = 1$.
2. Compute $q_i = p_i/w_i$, $i = 1, \ldots, N$.
3. Apply the BH procedure at level α to q_i.

A natural choice for the weights in our context is setting $w_i = d_i^2$; as a matter of fact, if the predictors are standardised, then S is a correlation matrix with eigenvalues $0 \leq d_i^2 \leq N$, and $N^{-1} \sum_i d_i^2 = 1$.

5.1. *Estimation of* σ

Fan, Guo, and Hao (2012) propose a refitted cross-validatory (RCV) estimator of the error variance which is consistent when the number of factors grows at a faster rate than the number of observations. The procedure has two stages: the sample is split into two independent subsamples. In the first stage, variable selection is carried out on the two subsamples by a consistent pretest selection procedure, based on the marginal t statistics $t_i^*, i = 1, \ldots, N$, yielding two sets of selected variables, denoted M_1 and M_2. In the second stage, the two models are estimated on the other subsample (i.e. M_1 is estimated on the second and M_2 on the first subsample), yielding two estimates, s_1^2 and s_2^2, of the error variance. The RCV estimator is the average of the two. Refitting aims at eliminating the influence of variables that have been spuriously selected in the first stage.

5.2. *Controlling for the Lags of the Dependent Variable*

So far we have considered the regression of a predictand on the principal components. If lags of the predictand have to be incorporated, as in the FAR approach, the above variable selection procedures are applied to the regression of the residuals of the regression of $y_{t+h}^{(h)}$ on $\{y_t, y_{t-1}, \ldots, y_{t-p}\}$ on the principal components computed on the set $\{\tilde{X}_{it}, i = 1, \ldots, N\}$, where $\tilde{X}_{it}$ is the residual of the regression of X_{it} on $\{y_t, y_{t-1}, \ldots, y_{t-p}\}$.

6. EMPIRICAL ANALYSIS

6.1. *Data and Methods*

The dataset used in the empirical analysis is derived from that employed by Stock and Watson (2012a), and consists of 211 U.S. macroeconomic time series, available at the quarterly observation frequency from 1959-I to 2011-II. Of the 211 series, 121 were considered for our empirical analysis.[2]

The series are all transformed to induce stationarity by taking first or second differences, logarithms or first or second differences of logarithms. The series are grouped into 12 categories; for a complete list of variables and their transformation see the appendix.

We consider a transformation y_t of the original variable Y_t, depending on the order of integration. Real activity variables are typically integrated of order 1, denoted $Y_t \sim I(1)$; defining $y_t = \Delta \ln Y_t$, the predictand is

$$y_{t+h}^{(h)} = 400 \times \frac{\Delta_h \ln Y_{t+h}}{h}, \quad \Delta_h \ln Y_{t+h} = \ln Y_{t+h} - \ln Y_t$$

the h-period growth at an annual rate. For nominal price and wage series, we assume $Y_t \sim I(2)$, and, in accordance to Stock and Watson (2002b), the series to be predicted is

$$y_{t+h}^{(h)} = 400 \times \left(\frac{\Delta_h \ln Y_{t+h}}{h} - \Delta_1 \ln Y_t \right)$$

The predictors are represented by the 121 standardized principal components obtained from the spectral decomposition of the covariance matrix of the original indicators. A heat map of the squared factor loadings for the 121 quarterly series is provided by Fig. 1. The vertical axis is the series categories reported in the appendix; the horizontal axis shows the order of the principal component. The plot evidences that the first factor loads principally on the growth rates of the indicators of real activity; the second has rather sparse loadings on both real and nominal variables, whereas the third loads eminently on price and wage inflation rates.

The forecasts are obtained using a pseudo-real-time forecasting procedure.[3] The data from 1960:II to 1984:IV are used as a training sample. A PCA on the N standardized indicators is conducted to extract the factors, represented by the standardized principal components. The PCA are selected and the estimated regression coefficients are used at time 1984:IV to forecast $\hat{y}_{T+h}^{(h)}$, $h = 1, 2$, and 4 quarters ahead. Then, the estimation sample is updated by one quarterly observation and downdated by removing the initial one, so that the second set of observations ranges from 1960:III to 1985:I, and so forth. For each rolling window, consisting of $T = 100$ observations, a PCA is conducted to extract the N components, variable selection is performed and the predictor is formed. The process continues until the end of the sample is reached. In our case, the last available data

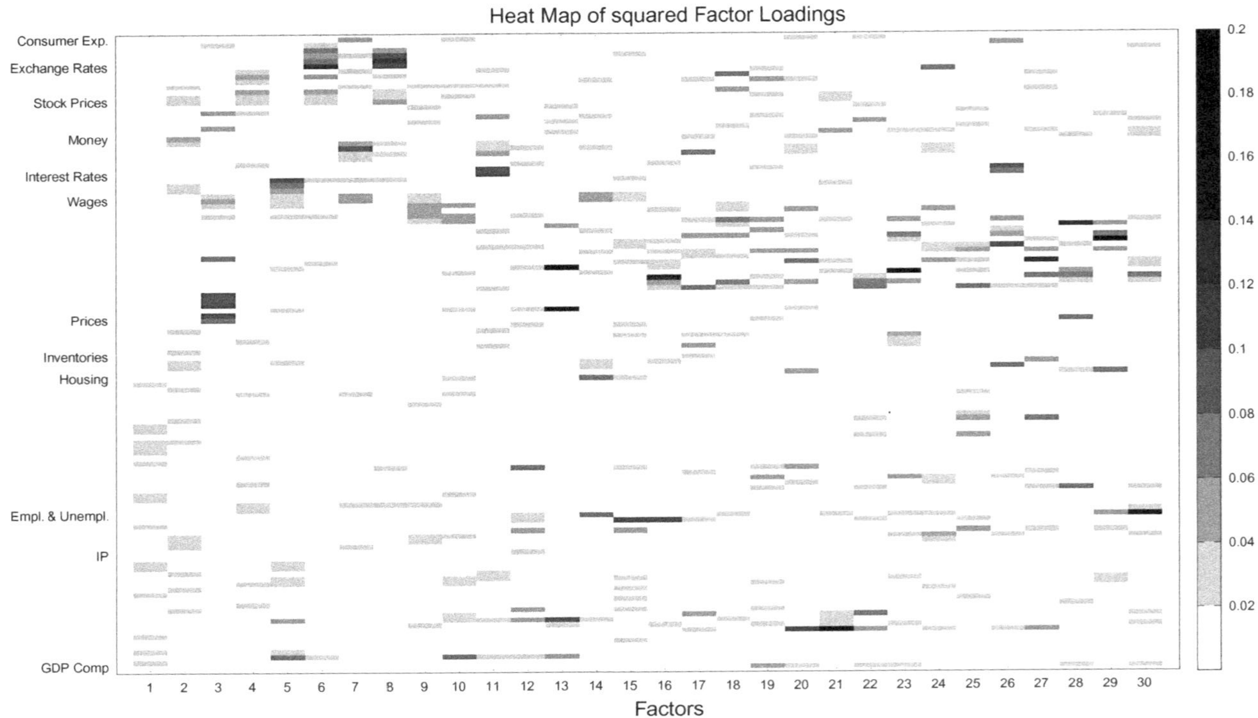

Fig. 1. Heat Map of the Squared Factor Loadings for the 121 Quarterly Series as Given in Table 1. *Notes*: The vertical axis is the series categories as reported in Appendix A; the horizontal axis is the order of the principal component.

are 2010:II when $h = 4$. The experiment delivers around 100 forecasts for each forecasting method, that can be compared with the observed values.

The predictors that are compared are:

- The pure autoregressive (AR) predictor

$$\hat{y}_{t+h}^{(h)} = \hat{\phi}_1^{(h)} y_t + \cdots + \hat{\phi}_p^{(h)} y_{t-p}$$

  where the order is selected according to the Bayesian information criterion (BIC) and the AR coefficients are estimated by ordinary least squares.
- The dynamic factor model predictor (DFM5)

$$\hat{y}_{t+h}^{(h)} = \hat{\beta}_1^{(h)} \hat{F}_{1t} + \cdots + \hat{\beta}_r^{(h)} \hat{F}_{rt}$$

  where $\hat{F}_{it}, i = 1, \ldots, r$, are the first $r = 5$ principal components. This number is selected by the Bai and Ng (2002) criteria and coincides with the factor model benchmark proposed in Stock and Watson (2012b). When lags of the predictand are included, as in Eq. (4), the order p is that selected for the previous case (AR predictor).
- The supervised factor predictor

$$\hat{y}_{t+h}^{(h)} = \hat{\beta}_1^{(h)} \hat{F}_{(1)t} + \cdots + \hat{\beta}_r^{(h)} \hat{F}_{(r)t}$$

  with r factors, ranked according to their p-values, selected according to
  - Holm's multiple comparison procedure, controlling the FWER at the 5% level;
  - the Benjamini–Hochberg procedure, controlling the FDR at the 10% level;[4]
  - the Genovese (2006) procedure with p-values weighted according to the corresponding eigenvalue.

  If the lags of the predictand are considered, as in the FAR approach, then the factors are selected from the principal components computed on the residuals of the projection of the original predictors on the linear space spanned by the first p lags of the dependent variable. We consider two implementations of the variable selection procedure, the first based on the marginal t_i^* statistics and the second based on the t_i statistics computed using the Fan et al. (2012) estimator of the regression error variance.

The performance of the different methods is evaluated using the mean square forecast error (MSFE), defined as follows: let T_0 be the first point in time for out of sample evaluation and T_1 be the last point in time for which we compute the MSFE for $h = 1, 2$, and 4

$$\text{MSFE} = \frac{1}{T_1 - T_0} \sum_{\tau=T_0}^{T_1} \left(\hat{y}_{\tau+h}^{(h)} - y_{\tau+h} \right)^2$$

The results are presented in terms of mean square error (MSFE) relative to the AR (BIC) benchmark

$$R_j = \frac{\text{MSFE}_j(h)}{\text{MSFE}_{\text{AR}}(h)}$$

where $j \in \{\text{DFM5, Holm, BH, GRW}\}$. A value below one indicates that the specified method is superior to the AR (BIC) forecast. We also consider the h steps ahead forecasts obtained from a univariate MA(q) model, with q selected according to BIC. The MA forecasts are expected to perform well for the second differences of the CPI series.

To test the equal predictive ability of factor-based models with respect to the AR (BIC) benchmark, we use the Diebold and Mariano (1995) test for Tables 1 and 3 (the predictive models are not nested), whereas for Tables 2 and 4 we use the Clark and West (2007) test, which is suitable for nested models.

6.2. Empirical Results

The rolling forecast experiment was conducted for the following series:

- Industrial Production Index (IPI)
- Total employment, Non-farm Payroll (NPE)
- Unemployment Rate (UR)
- Housing Starts (HS)
- Consumer Price Index (CPI)
- Treasury Bill 10-years (TB)
- Real Personal Income (RPI)
- Gross National Product (GNP).

Table 1. Relative Mean-Square Forecast Errors of Five Alternative Predictors at Horizons $h = 1, 2$, and $h = 4$. The Selection of the Factors Is Based on the Marginal t_i^* Statistics and Prediction Occurs by Principal Component Regression on the Selected Factors with no Lags of the Predictand.

	IPX	NPE	UR	HS	CPI	TB	RPI	GNP
Panel a: Rolling, $h = 1$								
AR(BIC)	13.164	0.760	0.039	0.005	6.290	*0.172*	12.181	5.573
MA(BIC)	2.106	5.691	2.238	1.013	1.001	1.086	0.962	1.292
DFM5	*0.960*	0.910	*0.693**	0.943	*0.959*	1.207	0.898	0.942
Holm	1.158	*0.771**	0.728	0.914	1.014	1.079	*0.828***	0.900
BH	1.173	0.777	0.728	*0.896*	1.028	1.083	0.897	*0.848**
GRW	1.155	0.771	0.728	0.917	1.022	1.089	0.884	0.876
Panel b: Rolling, $h = 2$								
AR(BIC)	*16.129*	1.103	0.041	0.003	2.068	0.121	5.816	*4.084*
MA(BIC)	1.505	3.746	1.893	1.054	1.010	1.027	0.980	1.346
DFM5	1.026	1.013	*0.680***	1.126	*0.873*	1.246	0.854	1.126
Holm	1.102	0.876	0.757	*0.995*	0.917	*0.985*	0.769	1.000
BH	1.102	*0.873*	0.754	1.041	0.949	1.064	0.774	1.000
GRW	1.102	0.876	0.759	0.995	0.917	0.985	*0.763***	1.000
Panel c: Rolling, $h = 4$								
AR(BIC)	*16.184*	1.694	0.049	0.002	0.553	0.080	3.635	**3.595**
MA(BIC)	1.151	2.184	1.313	0.954	0.979	0.971	0.928	1.173
DFM5	1.124	1.072	*0.713**	1.084	0.869	1.198	0.801	1.112
Holm	1.047	*0.879*	0.759	*0.979*	0.928	0.974	*0.681**	1.008
BH	1.025	0.916	0.735	0.990	*0.859**	*0.963*	0.684	1.028
GRW	1.039	0.917	0.766	0.979	0.954	0.974	0.682	1.005

Notes: Numerical entries are mean-square forecast errors (MSFEs). Forecasts are quarterly, for the period 1985:IV–2010:II for a total of 103 out of sample forecasts. Entries in the first row, corresponding to the AR(BIC) benchmark model, are actual MSFEs, while all other entries are relative MSFEs, such that an entry below one indicates that the specified method is superior to the AR (BIC) forecast. For $*p < 0.1$, $**p < 0.05$, $***\ p < 0.01$, we reject the null hypothesis of equal predictive ability towards AR(BIC).

Tables 1–4 report the relative MSFEs for the five alternative forecasting models under consideration. In particular, Table 1 refers to the case when lags of the predictand are not considered for forecasting using DI and the selection of the PCs is based on the p-values computed on the marginal t_i^* statistics; Table 2 refers to the case when p lags are considered and the

Table 2. Relative Mean-Square Forecast Errors of Five Alternative Predictors at Horizons $h = 1, 2$, and $h = 4$. The Selection of the Factors Is Based on the Marginal t_i^* Statistics and Prediction Occurs by the Principal Component Regression on the Selected Factors Including p Lags of the Predictand.

	IPX	NPE	UR	HS	CPI	TB	RPI	GNP
Panel a: Rolling, $h = 1$								
DFM5	1.195	1.119	0.954	0.939	1.036	1.189	0.996	0.968
Holm	1.005	*0.968**	*0.912***	0.894	1.090	1.123	0.969	0.846
BH	*0.987*	0.968	0.912	*0.873****	1.093	1.115	*0.937**	*0.836**
GRW	1.004	0.971	0.912	0.880	1.075	1.095	0.971	0.847
Panel b: Rolling, $h = 2$								
DFM5	1.230	1.215	1.158	0.984	*0.987**	1.046	0.875	1.067
Holm	1.166	1.204	1.064	*0.923**	1.035	1.033	0.851	1.021
BH	1.161	1.204	1.064	0.932	1.048	1.031	0.871	1.023
GRW	1.168	1.205	1.064	0.939	1.029	1.024	*0.846**	1.020
Panel c: Rolling, $h = 4$								
DFM5	1.084	1.194	1.103	1.033	*0.935**	1.055	0.962	1.101
Holm	1.063	1.189	1.094	1.007	0.970	1.048	*0.912**	1.037
BH	1.064	1.185	1.093	1.009	0.959	1.054	0.928	1.040
GRW	1.077	1.189	1.094	1.009	0.972	1.051	0.919	1.037

Notes: Numerical entries are relative MSFEs, such that an entry below one indicates that the specified method is superior to the AR (BIC) forecast. For $*p<0.1$, $**p<0.05$, $***p<0.01$, we reject the null hypothesis of equal predictive ability towards AR (BIC).

selection is based on the marginal t_i^* statistics. Tables 3 and 4 deal with the selection based on the t_i statistics with σ estimated according to the RCV method by Fan et al. (2012): in Table 3 no lags of the predictands were considered, whereas in Table 4 they were.

There are several conclusions that can be drawn from the empirical evidence summarised in the tables. The first broad consideration is that forecasting methods based on factor models provide accurate forecasts and improve over the AR benchmark in the majority of the cases across all horizons, when the lags of the dependent variable are not considered (which is the case considered in Tables 1 and 3). Notice also that the MA predictor outperforms the AR for the CPI only at yearly forecast horizons and the differences in predictive accuracy are not significantly different.

Table 3. Relative Mean-Square Forecast Errors of Five Alternative Predictors at Horizons $h = 1, 2$, and $h = 4$. The Selection of the Factors Is Based on the t_i Statistics with RCV Estimation of σ and Prediction Occurs by Principal Component Regression on the Selected Factors and no Lags of the Predictand.

	IPX	NPE	UR	HS	CPI	TB	RPI	GNP
Panel a: Rolling, $h = 1$								
DFM5	_0.960_	0.910	_0.693*_	0.943	_0.959_	1.207	0.898	0.942
Holm	1.185	0.744	0.721	0.901	1.040	1.114	_0.828*_	0.816
BH	1.090	0.737	0.749	0.892	1.036	1.181	0.966	0.804
GRW	1.120	_0.728_	0.750	_0.833_	1.051	1.206	0.988	_0.786_
Panel b: Rolling, $h = 2$								
DFM5	1.026	1.013	_0.680*_	1.126	_0.873_	1.246	0.854	1.126
Holm	1.123	_0.889_	0.734	1.065	0.917	1.098	_0.769*_	_0.919_
BH	1.126	0.930	0.778	1.023	0.985	1.178	0.813	0.926
GRW	1.082	0.889	0.796	1.010	0.989	1.149	0.785	0.926
Panel c: Rolling, $h = 4$								
DFM5	1.124	1.072	0.713	1.084	_0.869_	1.198	0.801	1.112
Holm	0.960	1.032	_0.675**_	0.988	0.896	_0.975_	0.686	1.145
BH	0.971	1.003	0.694	_0.963_	0.882	1.026	0.681	1.066
GRW	0.974	_0.993_	0.702	0.963	0.885	1.047	_0.678*_	1.066

Notes: Numerical entries are relative MSFEs, such that an entry below one indicates that the specified method is superior to the AR (BIC) forecast. For $*p < 0.1$, $**p < 0.05$, $***p < 0.01$, we reject the null hypothesis of equal predictive ability towards AR (BIC).

The second general conclusion is that pre-selection of the components by the multiple testing procedures considered leads to several improvements in forecasting accuracy (when no lags of the predictand are considered). The three procedures show the best performances for 52% of the cases across horizons/variables combinations in Tables 1 and 3, whereas DFM5 and AR (BIC) have the best performances only for 21% and 27% of the occurrences, respectively.

Thirdly, when the lags of the target variables are considered in the forecasting model, the predictors based on the factors, regardless of their selection, are more systematically outperformed by the benchmark AR predictor. The combined evidence of Tables 2 and 4 is that the AR predictor is ranked best in 58% of the cases. The last finding has already been

Table 4. Relative Mean-Square Forecast Errors of Five Alternative Predictors at Horizons $h = 1, 2$, and $h = 4$. The Selection of the Factors Is Based on the t_i Statistics with RCV Estimation of σ and Prediction Occurs by Principal Component Regression on the Selected Factors Including p Lags of the Predictand.

	IPX	NPE	UR	HS	CPI	TB	RPI	GNP
Panel a: Rolling, $h = 1$								
DFM5	1.195	1.119	0.954	0.939	1.036	1.189	0.996	0.968
Holm	1.149	0.957	0.921	0.944	1.099	1.192	*0.961**	0.922
BH	1.173	0.963	*0.861***	*0.924****	1.063	1.338	1.018	*0.821****
GRW	1.169	0.967	0.876	0.942	1.070	1.301	1.075	0.821
Panel b: Rolling, $h = 2$								
DFM5	1.230	1.215	1.158	0.984	*0.987**	1.046	0.875	1.067
Holm	1.206	1.184	1.062	*0.951****	1.045	1.059	0.874	1.013
BH	1.236	1.170	1.068	1.012	1.018	1.074	*0.863**	1.019
GRW	1.232	1.172	1.070	0.965	1.021	1.075	0.872	1.019
Panel c: Rolling, $h = 4$								
DFM5	1.084	1.194	1.103	1.033	*0.935**	1.055	0.962	1.101
Holm	1.070	1.178	1.091	1.000	0.959	1.034	*0.915**	1.054
BH	1.073	1.175	1.094	1.014	0.979	1.029	0.942	1.052
GRW	1.073	1.174	1.096	1.016	0.994	1.028	0.933	1.054

Notes: Numerical entries are relative MSFEs, such that an entry below one indicates that the specified method is superior to the AR (BIC) forecast. For $*p<0.1$, $**p<0.05$, $***p<0.01$, we reject the null hypothesis of equal predictive ability towards AR (BIC).

reported and investigated in previous studies, among which we mention D'Agostino and Giannone (2012) and Stock and Watson (2002). A possible explanation is that factor models have the ability to capture efficiently not only the information that is common to other cross-sectional variables, but also the specific dynamic features of each variable to predict. Also, after conditioning for the role of lagged values, the factors computed on the residual variation contribute more to the variability of the forecasts, leading to an increase in the MSFE.

The series for which the multiple testing procedures outperform the DFM5 predictor are Total employment: Non farm Payroll (NPE), Housing Starts (HS), Treasury Bill 10-years (TB), Real Personal Income (RPI). For NPE, HS, TB and RPI, they produce the minimum MSFEe across all horizons (panels a–c of Table 1), with the exception of TB at horizon $h = 1$, for which the AR predictor ranks best. Finally, DFM5 is ranked best for

UR, achieving a 20% reduction in the MSFE over the AR predictor and a 4% reduction over the multiple testing procedures, across all horizons. For the other variables, the results are less sharp and depend basically on the forecast horizon. In Table 1 for IPX and GNP, we observe a slight improvement of 4% of DFM5 and 15% of BH over AR only for $h = 1$, whereas for $h = 2$ and $h = 4$ both multiple testing procedures and DFM5 do not outperform the benchmark.

The choice of the reference test statistics (marginal t_i^* or multiple regression t_i, with RCV estimation of σ) does not seem to affect the results of the multiple testing procedures. The previous results are confirmed examining Table 2. We observe a further improvement only for GNP, where now the best performing predictor is GRW, achieving a 24% MSFE reduction with respect to the benchmark, also for $h = 2$ (Panel b) where the gain in forecasting accuracy amounts to about 10%.

Among the multiple testing procedures, weighting the p-values according to the eigenvalues does not lead to an improvement, with a few exceptions. Holm's sequential procedure clearly outperforms the other predictors in terms of MSFE in 27% of the cases when no lags of the dependent variable are in use, whereas BH and GRW are ranked best in 19% and 6% of the cases, respectively. This result seems to depend exclusively on the conservative nature of the Holm method, compared to the procedures controlling the FDR.

6.3. *Assessment of Real-Time Performance*

Following D'Agostino and Giannone (2012), we evaluate how the forecasting performance of the predictors evolved over time. In Fig. 2, we plot the time series pattern of the MSFEs of the DFM5 predictor (solid line) and the predictor resulting from Holm's selection of the factors (dashed line), relative to the AR benchmark (horizontal line). We consider only three series, namely, NPE, TB and RPI, for which the Holm's selection provided sizable improvements. The relative MSFEs were smoothed over time with a centered moving window spanning two years. The shaded areas are the NBER recessions.

Interestingly, the factor-based methods perform best during the great recession, and present no substantial gain during the great moderation. This empirical finding is consistent with the literature, as during the recession the comovements among economic variables are more prominent and

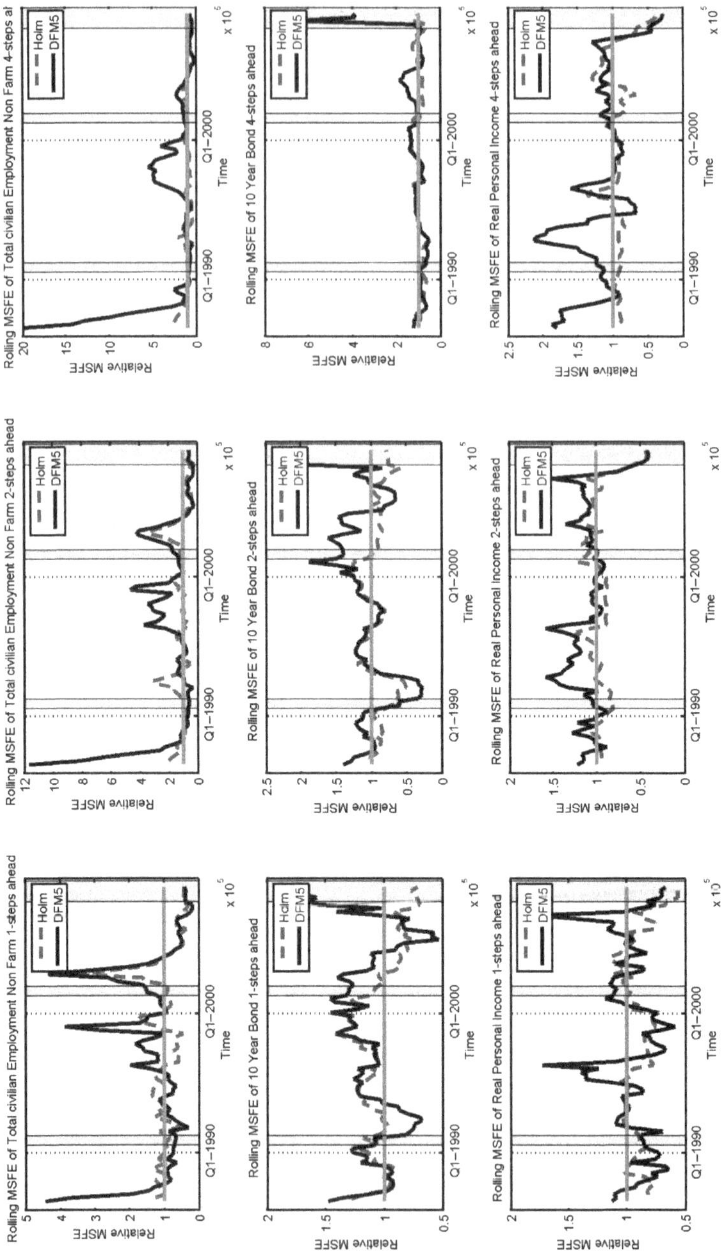

Fig. 2. Time-Varying Performance of Forecasting Methods.

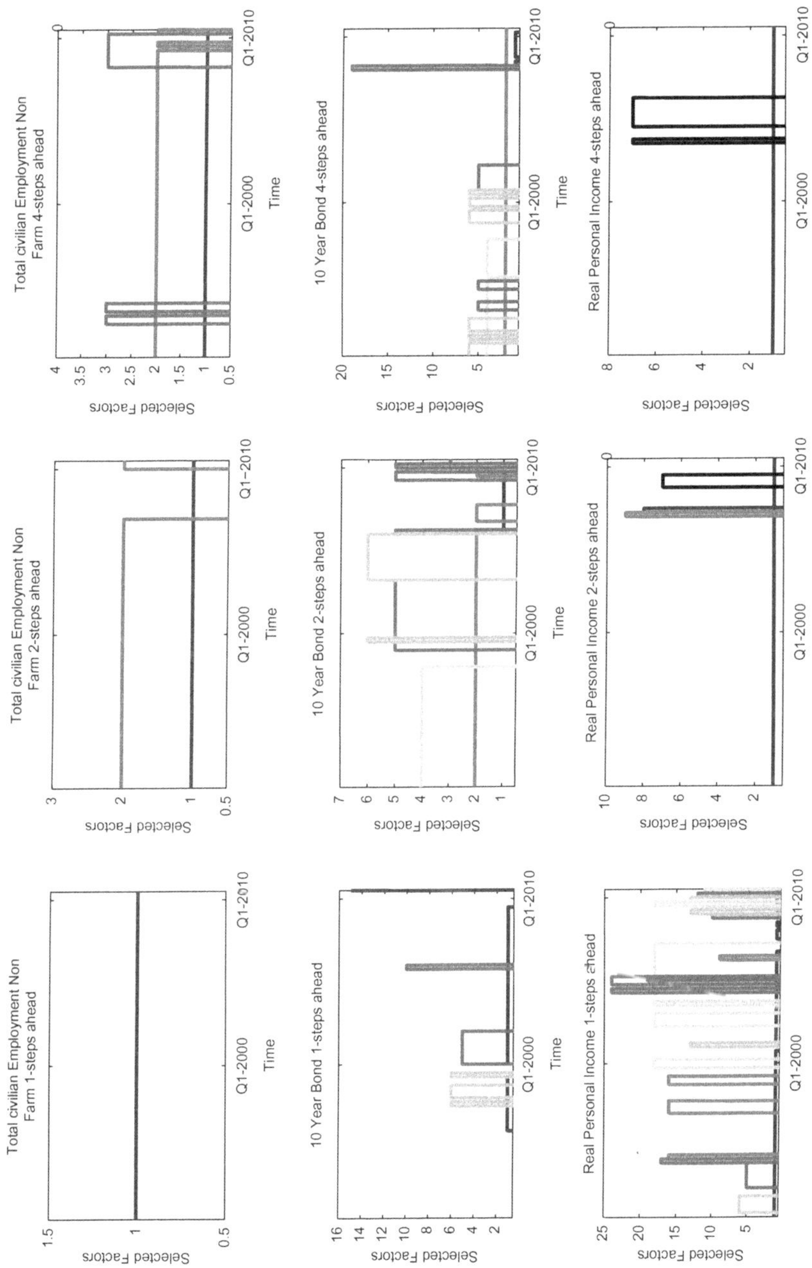

Fig. 3. Selected Factors Using the Holm Procedure.

thus the factors become more useful for forecasting. The selection of the factor leads to greater MSFE reductions in the last five years of the sample, including the great recession.

Further insight into the assessment of the performance of Holm's factor selection method can be gauged from the consideration of which factors are selected by the procedure. Fig. 3 is a plot versus time of the index number of the selected factors arising as a by-product of the rolling forecast experiment. In the case of NPE (first row), the first principal components is always selected, and is the only relevant factor for forecasting one-step-ahead. The second and third factors enter the selection at horizons $h = 2$ and $h = 4$, with the second factor being switched off during the recession and the third emerging during the great recession. Hence, it may be concluded that nature is benign in this case as the information that is essential for forecasting is well represented in the first three factors.

Nature is less benign in the TB case (second row panels). No factor is selected at the beginning and most noticeably at the end of the sample period. High-order components are selected and the intermediate ones receive zero weight. Nature is even more bizarre for the RPI variable (bottom panels); the number of selected factors never exceeds four, but the order of selected factors is surprising. For predicting one-step-ahead, the Holm's procedure, more or less regularly, selects the 16-th, 18-th and 24-th factors. This selection would not be possible if one adopted the usual criteria for determining the number of factors.

7. CONCLUSIONS

This paper has proposed a method for supervising the DI methodology, originally proposed by Stock and Watson (2002), which is based on the simple idea of selecting the relevant factors using a multiple testing procedure, achieving control over either the familywise error rate or the FDR. Prior information about the order of the components may be introduced by weighting the p-values of the test statistics for variable exclusion with weights proportional to the eigenvalues.

Can we conclude that nature is tricky, but essentially benign? The answer is a qualified yes. The information that is needed for forecasting the eight macroeconomic variables considered in this paper is effectively condensed by the first few factors. However, variable selection, leading to exclude some of the low-order principal components, can lead to a sizable improvement in forecasting in specific cases. Only in one instance, real personal income, we were able to detect a significant contribution from high-order components.

NOTES

1. Omitting a component loading on x variables with no effect on the predictand makes a zero contribution to the bias.
2. As in Stock and Watson (2012b), we exclude series at high aggregate level. We decided to exclude also series starting after 1959 or ending before 2011. As a result, our dataset can be considered as an update of that used in Stock and Watson (2012b), which ended in 2009:II.
3. As reported in D'Agostino and Giannone (2012), the exercise is pseudo-real-time because we use the last vintage of data (for this dataset the last vintage is November 2011) and we do not consider the release available at the time of forecasting.
4. This is the target value most often considered in applications; see Efron (2010). Controlling the FDR at the 5% level leads to very similar results.

ACKNOWLEDGMENTS

This paper was presented at the 16th Annual Advances in Econometrics Conference, Aarhus, November 15–16, 2014. The authors gratefully acknowledge financial support by the Italian Ministry of Education, University and Research (MIUR), PRIN Research Project 2010–2011–prot. 2010J3LZEN, Forecasting economic and financial time series. Tommaso Proietti gratefully acknowledges support from CREATES–Center for Research in Econometric Analysis of Time Series (DNRF78), funded by the Danish National Research Foundation.

REFERENCES

Alessi, L., Barigozzi, M., & Capasso, M. (2010). Improved penalization for determining the number of factors in approximate factor models. *Statistics & Probability Letters*, *80*(23), 1806–1813.

Bai, J., & Ng, S. (2002). Determining the number of factors in approximate factor models. *Econometrica*, *70*(1), 191–221.

Bai, J., & Ng, S. (2006). Confidence intervals for diffusion index forecasts and inference with factor-augmented regressions. *Econometrica*, *74*(4), 1133–1150.

Bai, J., & Ng, S. (2008). Forecasting economic time series using targeted predictors. *Journal of Econometrics*, *146*(2), 304–317.

Bai, J., & Ng, S. (2009). Boosting diffusion indices. *Journal of Applied Econometrics*, *24*(4), 607–629.

Bair, E., Hastie, T., Debashis, P., & Tibshirani, R. (2006). Prediction by supervised principal components. *Journal of the American Statistical Association*, *101*(473), 119–137.

Benjamini, Y., & Hochberg, Y. (1995). Controlling the false discovery rate: A practical and powerful approach to multiple testing. *Journal of the Royal Statistical Society. Series B (Methodological)*, *57*, 289–300.

Breitung, J., & Eickmeier, S. (2006). Dynamic factor models. In O. Hübler & J. Frohn (Eds.), *Modern econometric analysis* (pp. 25–40). Heidelberg: Springer.

Chun, H., & Keleş, S. (2010). Sparse partial least squares regression for simultaneous dimension reduction and variable selection. *Journal of the Royal Statistical Society: Series B (Statistical Methodology)*, *72*(1), 3–25.

Clark, T. E., & West, K. D. (2007). Approximately normal tests for equal predictive accuracy in nested models. *Journal of Econometrics*, *138*(1), 291–311.

Cook, R. D. (2007). Fisher lecture: Dimension reduction in regression. *Statistical Science*, *22*, 1–26.

Cook, R. D., & Forzani, L. (2008). Principal fitted components for dimension reduction in regression. *Statistical Science*, *52*, 485–501.

Cox, D. (1968). Notes on some aspects of regression analysis. *Journal of the Royal Statistical Society, Series A*, *131*(3), 265–279.

D'Agostino, A., & Giannone, D. (2012). Comparing alternative predictors based on large-panel factor models. *Oxford Bulletin of Economics and Statistics*, *74*(2), 306–326.

De Mol, C., Giannone, D., & Reichlin, L. (2008). Forecasting using a large number of predictors: Is bayesian shrinkage a valid alternative to principal components? *Journal of Econometrics*, *146*(2), 318–328.

Diebold, F. X., & Mariano, R. S. (1995). Comparing predictive accuracy. *Journal of Business and Economic Statistics*, *13*, 253–265.

Efron, B. (2010). *Large-scale inference: Empirical bayes methods for estimation, testing, and prediction*. Institute of Mathematical Statistics Monographs. New York, NY: Cambridge University Press.

Fan, J., Guo, S., & Hao, N. (2012). Variance estimation using refitted cross-validation in ultrahigh dimensional regression. *Journal of the Royal Statistical Society: Series B (Statistical Methodology)*, *74*(1), 37–65.

Forni, M., Hallin, M., Lippi, M., & Reichlin, L. (2005). The generalized dynamic factor model: One-sided estimation and forecasting. *Journal of the American Statistical Association*, *100*, 830–840.

Fuentes, J., Poncela, P., & Rodrguez, J. (2014). Sparse partial least squares in time series for macroeconomic forecasting. *Journal of Applied Econometrics*, *30*, 576–595.

Genovese, R. (2006). False discovery control with p-value weighting. *Biometrika*, *93*(3), 509–524.

Hadi, A. S., & Ling, R. F. (1998). Some cautionary notes on the use of principal components regression. *The American Statistician*, *52*(1), 15–19.

Hastie, T., Tibshirani, R., & Friedman, J. (2009). *The elements of statistical learning: Data mining, inference, and prediction*. Springer Series in Statistics. New York, NY: Springer.

Holm, S. (1979). A simple sequentially rejective multiple test procedure. *Scandinavian Journal of Statistics*, *6*, 65–70.

Hwang, J. G., & Nettleton, D. (2003). Principal components regression with data chosen components and related methods. *Technometrics*, *45*(1), 70–79.

Inoue, A., & Kilian, L. (2008). How useful is bagging in forecasting economic time series? A case study of U.S. consumer price inflation. *Journal of the American Statistical Association*, *103*, 511–522.

Joliffe, I. T. (1982). A note on the use of principal components in regression. *Applied Statistics*, *31*(3), 300–303.

Kim, H. H., & Swanson, N. R. (2014). Forecasting financial and macroeconomic variables using data reduction methods: New empirical evidence. *Journal of Econometrics*, *178*(2), 352–367.

Li, K.-C. (1991). Sliced inverse regression for dimension reduction. *Journal of the American Statistical Association*, *86*, 316–327.

Mosteller, F., & Tukey, J. W. (1977). *Data analysis and regression: A second course in statistics*. Addison-Wesley Series in Behavioral Science: Quantitative Methods. New York, NY: Addison-Wesley.

Ng, S. (2013). Variable selection in predictive regressions. In G. Elliott & A. Timmermann (Eds.), *Handbook of economic forecasting* (vol. 2, pp. 752–789). Part B, chapter 14. Amsterdam: Elsevier.

Onatski, A. (2010). Determining the number of factors from empirical distribution of eigenvalues. *The Review of Economics and Statistics*, *92*(4), 1004–1016.

Stock, J. H., & Watson, M. W. (2002a). Forecasting using principal components from a large number of predictors. *Journal of the American Statistical Association*, *97*, 1167–1179.

Stock, J. H., & Watson, M. W. (2002b). Macroeconomic forecasting using diffusion indexes. *Journal of Business & Economic Statistics*, *20*(2), 147–162.

Stock, J. H., & Watson, M. W. (2006). Forecasting with many predictors. In G. Elliott, C. Granger, & A. Timmermann (Eds.), *Handbook of economic forecasting* (Vol. 1, pp. 515–554). Amsterdam: Elsevier.

Stock, J. H., & Watson, M. W. (2010). Dynamic factor models. In M. P. Clements & D. F. Hendry (Eds.), *Oxford handbook of economic forecasting* (Vol. 1). New York, NY: Oxford University Press.

Stock, J. H., & Watson, M. W. (2012a). Disentangling the Channels of the 2007–09 Recession. *Brookings Papers on Economic Activity*, *2012*, 81.

Stock, J. H., & Watson, M. W. (2012b). Generalized shrinkage methods for forecasting using many predictors. *Journal of Business & Economic Statistics*, *30*(4), 481–493.

APPENDIX A. LIST OF THE TIME SERIES USED IN THE EMPIRICAL ILLUSTRATION

This appendix reports the time series in the dataset used in the application, the transformation type, the observations frequency (M = monthly and Q = quarterly) and the group to which they belong.

Letting Z_t denote the raw series, the following transformations are adopted:

$$X_t = \begin{cases} Z_t & \text{if Tcode} = 1 \\ \Delta Z_t & \text{if Tcode} = 2 \\ \Delta^2 Z_t & \text{if Tcode} = 3 \\ \ln(Z_t) & \text{if Tcode} = 4 \\ \Delta \ln Z_t & \text{if Tcode} = 5 \\ \Delta^2 \ln Z_t & \text{if Tcode} = 6 \end{cases}$$

Table A1. List of the predictors.

No.	Short Description	Long Description	Tcode	Frequency	Category
NIPA					
1	Disp-Income	Real Disposable Personal Income	5	Q	1
2	FixedInv	Real Private Fixed Investment	5	Q	1
3	Gov.Spending	Real Government Consumption Expenditures & Gross Investment	5	Q	1
4	GDP	Real Gross Domestic Product	5	Q	1
5	Investment	Real Gross Private Domestic Investment	5	Q	1
6	Consumption	Real Personal Consumption Expenditures	5	Q	1
7	Inv·Equip&Software	Real Nonresidential Investment: Equipment & Software	5	Q	1
8	Exports	Real Exports of Goods & Services	5	Q	1
9	Gov Receipts	Government Current Receipts (Nominal)	5	Q	1
10	Gov:Fed	Real Federal Consumption Expenditures & Gross Investment	5	Q	1
11	Imports	Real Imports of Goods & Services	5	Q	1
12	Cons:Dur	Real Personal Consumption Expenditures: Durable Goods	5	Q	1
13	Cons:Svc	Real Personal Consumption Expenditures: Services	5	Q	1
14	Cons:NonDur	Real Personal Consumption Expenditures: Nondurable Goods	5	Q	1
15	FixInv:NonRes	Real Private Nonresidential Fixed Investment	5	Q	1

Table A1. *(Continued)*

No.	Short Description	Long Description	Tcode	Frequency	Category
16	FixedInv:Res	Real Private Residential Fixed Investment	5	Q	1
17	Gov:State&Local	Real State & Local Cons. Exp. & Gross Investment	5	Q	1
18	Inv:Inventories	Real Change in Private Inventories	5	Q	1
19	Inv:Inventories	Ch. Inv/GDP	1	Q	1
20	Output:Bus	Business Sector: Output	5	Q	1
21	Ouput:NFB	Nonfarm Business Sector: Output	5	Q	1
Industrial production					
22	IP: Dur gds materials	Industrial Production: Durable Materials	5	M	2
23	IP: Nondur gds materials	Industrial Production: nondurable Materials	5	M	2
24	Capu Man.	Capu Man. (Fred post 1972, Older series before 1972)	1	M	2
25	IP: Dur Cons. Goods	Industrial Production: Durable Consumer Goods	5	M	2
26	IP: Auto	IP: Automotive products	5	M	2
27	IP:NonDur Cons God	Industrial Production: Nondurable Consumer Goods	5	M	2
28	IP: Bus Equip	Industrial Production: Business Equipment	5	M	2
29	IP: Energy Prds	IP: Consumer Energy Products	5	M	2
Employment and unemployment					
30	Emp: Gov (Fed)	Federal	5	M	3
31	Emp: Gov (State)	State government	5	M	3
32	Emp: Gov (Local)	Local government	5	M	3
33	Emp: DurGoods	All Employees: Durable Goods Manufacturing	5	M	3
34	Emp: Const	All Employees: Construction	5	M	3
35	Emp: Edu&Health	All Employees: Education & Health Services	5	M	3
36	Emp: Finance	All Employees: Financial Activities	5	M	3
37	Emp: Infor	All Employees: Information Services	5	M	3
38	Emp:Leisure	All Employees: Leisure & Hospitality	5	M	3
39	Emp: Mining/NatRes	All Employees: Natural Resources & Mining	5	M	3
40	Emp: Bus Serv	All Employees: Professional & Bus. Services	5	M	3
41	Emp:OtherSvcs	All Employees: Other Services	5	M	3
42	Emp:Trade&Trans	All Employees: Trade, Transp. & Utilities	5	M	3
43	Emp:Retail	All Employees: Retail Trade	5	M	3
44	Emp:Wholesal	All Employees: Wholesale Trade	5	M	3
45	Urate: Age16–19	Unemployment Rate–16–19 years	2	M	3
46	Urate:Age>20 Men	Unemployment Rate–20 years & over, Men	2	M	3
47	Urate: Age>20 Women	Unemployment Rate–20 years & over, Women	2	M	3
48	U: Dur<5wks	Number Unemployed for Less than 5 Weeks	5	M	3
49	U:Dur5–14 weeks	Number Unemployed for 5–14 Weeks	5	M	3
50	U:dur>15–26 weeks	Civilians Unemployed for 15–26 Weeks	5	M	3
51	U: Dur>27 weeks	Number Unemployed for 27 Weeks & over	5	M	3
52	Emp:SlackWk	Employment Level–Part-Time, All Industries	5	M	3

Table A1. *(Continued)*

No.	Short Description	Long Description	Tcode	Frequency	Category
53	AWH Man	Average Weekly Hours: Mfg	1	M	3
54	AWH Overtime	Average Weekly Hours: Overtime: Mfg	2	M	3
55	Emp:nfb	Nonfarm Business Sector: Employment	5	Q	3
Housing starts					
56	Hstarts:MW	Housing Starts in Midwest Census Region	5	M	4
57	Hstarts:NE	Housing Starts in Northeast Census Region	5	M	4
58	Hstarts:S	Housing Starts in South Census Region	5	M	4
59	Hstarts:W	Housing Starts in West Census Region	5	M	4
Inventories, orders and sales					
60	Orders (DurMfg)	Mfrs' new orders durable goods industries (bil. chain 2000 $)	5	M	5
61	Orders(Cons. Goods.	Mfrs' new orders, consumer goods and materials (mil. 1982 $)	5	M	5
62	UnfOrders(DurGds)	Mfrs' unfilled orders durable goods indus. (bil. chain 2000 $)	5	M	5
63	VendPerf	Index of supplier deliveries–vendor performance (pct.)	1	M	5
64	Orders(NonDefCap)	Mfrs' new orders, nondefense capital goods (mil. 1982 $)	5	M	5
65	MT Invent	Manufacturing and trade inventories (bil. Chain 2005 $)	5	M	5
66	Ret. Sale	Sales of retail stores (mil. Chain 2000 $)	5	M	5
Prices					
67	Price:Oil	PPI: Crude Petroleum	5	M	6
68	PPI:FinGds	Producer Price Index: Finished Goods	6	M	6
69	PPI:FinConsGds (Food)	Producer Price Index: Finished Consumer Foods	6	M	6
70	PPI:FinConsGds	Producer Price Index: Finished Consumer Goods	6	M	6
71	PPI:IndCom	Producer Price Index: Industrial Commodities	6	M	6
72	PPI:IntMat	Producer Price Index: Interm. Materials: Supplies & Comp.	6	M	6
73	PCED_MotorVec	Motor vehicles and parts	6	Q	6
74	PCED_DurHousehold	Furnishings and durable household equipment	6	Q	6
75	PCED_Recreation	Recreational goods and vehicles	6	Q	6
76	PCED_OthDurGds	Other durable goods	6	Q	6
77	PCED_Food_Bev	Food and beverages purchased for off-premises cons.	6	Q	6
78	PCED_Clothing	Clothing and footwear	6	Q	6
79	PCED_Gas_Enrgy	Gasoline and other energy goods	6	Q	6
80	PCED_OthNDurGds	Other nondurable goods	6	Q	6
81	PCED_Housing-Utilities	Housing and utilities	6	Q	6
82	PCED_HealthCare	Health care	6	Q	6

Table A1. *(Continued)*

No.	Short Description	Long Description	Tcode	Frequency	Category
83	PCED_TransSvg	Transportation services	6	Q	6
84	PCED_RecServices	Recreation services	6	Q	6
85	PCED_FoodServ_Acc.	Food services and accommodations	6	Q	6
86	PCED_FIRE	Financial services and insurance	6	Q	6
87	GDP Defl	Gross Domestic Product: Chain-type Price Index	6	Q	6
88	GPDI Defl	Gross Private Domestic Investment: Chain-type Price Index	6	Q	6
89	BusSec Defl	Business Sector: Implicit Price Deflator	6	Q	6
Earnings and productivity					
90	CPH:NFB	Nonfarm Business Sector: Real Compensation Per Hour	5	Q	7
91	CPH:Bus	Business Sector: Real Compensation Per Hour	5	Q	7
92	OPH:nfb	Nonfarm Business Sector: Output Per Hour of All Persons	5	Q	7
93	ULC:NFB	Nonfarm Business Sector: Unit Labor Cost	5	Q	7
94	UNLPay:nfb	Nonfarm Business Sector: Unit Nonlabor Payments	5	Q	7
Interest rates					
95	FedFunds	Effective Federal Funds Rate	2	M	8
96	TB-3Mth	3-Month Treasury Bill: section Market Rate	2	M	8
97	AAA_GS10	AAA-GS10 Spread	1	M	8
98	BAA_GS10	BAA-GS10 Spread	1	M	8
99	tb6m_tb3m	tb6m-tb3m	1	M	8
100	GS1_tb3m	GS1_Tb3m	1	M	8
101	GS10_tb3m	GS10_Tb3m	1	M	8
Money and credit					
102	C&Lloand	Commercial and Ind. Loans at All Comm. Banks	5	M	9
103	ConsLoans	Consumer Loans at All Comm. Banks	5	M	9
104	NonBorRes	Non-Borr. Reserves of Dep. Inst. Auction Credit	5	M	9
105	NonRevCredit	Total Nonrevolving Credit Outstanding	5	M	9
106	LoansRealEst	Real Estate Loans at All Comm. Banks	5	M	9
107	TotRes	Total Reserves, Adj. for Chgs in Reserve Reqs.	5	M	9
108	ConsuCred	Total Consumer Credit Outstanding	5	M	9
Stock prices, wealth and household balance sheet					
109	S&P 500	S&P'S COMMON STOCK PRICE INDEX: COMPOSITE	5	M	10
110	DJIA	COMMON STOCK PRICES: DOW JONES INDUSTRIAL AVERAGE	5	M	10
111	HHW:W	Total Net Worth	5	Q	10

Table A1. *(Continued)*

No.	Short Description	Long Description	Tcode	Frequency	Category
112	HHW:TA_RE	TTABSHNO-REANSHNO	5	Q	10
113	HHW:RE	Real Estate–Assets–Households and Nonprofit Orgs	5	Q	10
114	HHW:Fin	Total Financial Assets–Assets–Households and Nonprofits	5	Q	10
115	HHW:Liab	Total Liabilities–Households and Nonprofits	5	Q	10
Echange rates					
116	Ex rate: major	FRB Nominal Major Currencies Dollar Index	5	M	11
117	Ex rate: Switz	FOREIGN EXCHANGE RATE: SWITZERLAND	5	M	11
118	Ex rate: Japan	FOREIGN EXCHANGE RATE: JAPAN	5	M	11
119	Ex rate: UK	FOREIGN EXCHANGE RATE: UNITED KINGDOM	5	M	11
120	EX rate: Canada	FOREIGN EXCHANGE RATE: CANADA	5	M	11
Other					
121	Cons. Expectations	Consumer expectations NSA	1	M	12

ON THE DESIGN OF DATA SETS FOR FORECASTING WITH DYNAMIC FACTOR MODELS

Gerhard Rünstler

European Central Bank, Frankfurt, Germany

ABSTRACT

Forecasts from dynamic factor models potentially benefit from refining the data set by eliminating uninformative series. This paper proposes to use prediction weights as provided by the factor model itself for this purpose. Monte Carlo simulations and an empirical application to short-term forecasts of euro area, German, and French GDP growth from unbalanced monthly data suggest that both prediction weights and least angle regressions result in improved nowcasts. Overall, prediction weights provide yet more robust results.

Keywords: Dynamic factor models; forecasting; variable selection

JEL classifications: E37; C53; C51

Dynamic Factor Models
Advances in Econometrics, Volume 35, 631–664

ISSN: 0731-9053/doi:10.1108/S0731-905320150000035016

1. INTRODUCTION

Dynamic factor models have emerged as a widely used tool for obtaining nowcasts and short-term forecasts of economic activity and inflation (e.g., Giannone, Reichlin, & Small, 2008; Stock & Watson 2002a). From asymptotic considerations, these models are usually applied to large data sets that consist of a wide range of different series. It has been questioned though that increasing the sheer number of series in the data set would necessarily improve forecast performance. Boivin and Ng (2006) have identified conditions, under which enlarging the data set may actually worsen the precision of factor estimates. Bai and Ng (2008) have proposed LARS and related methods to identify efficient sets of predictors in dynamic factor models. These two studies, along with Schumacher (2010), Caggiano, Kapetianos, and Labhard (2011), Alvarez, Camacho, and Perez-Quiros (2012), and Bessec (2013) also present empirical applications that demonstrate gains from using smaller data sets in predictions from dynamic factor models.

In this paper, I propose the use of prediction weights that are obtained from the factor model itself as an alternative method for selecting an efficient set of predictors. As with any linear model, the factor model prediction for a certain target variable can be written as a weighted linear combination of current and lagged values of the predictors. I investigate, whether forecast efficiency can be improved by retaining only predictors with high weights.

Basically, the method parallels stepwise regression, but with the difference that a factor structure is imposed on the data. In its forward selection variant, stepwise regression builds a set of predictors for a certain target variable by an iterative procedure. At each step, it adds the series with the highest marginal predictive gain to the set of series from the previous step. It is well known that this procedure becomes highly inefficient, once the number of series increases. To overcome the dimensionality problem, constrained versions have been proposed, among them LARS and LASSO (Efron, Hastie, Johnstone, & Tibshirani, 2004), the latter being used by Bai and Ng (2008) to select predictors in factor model forecasts of inflation. Another way to deal with high dimensionality is using the factor model itself for estimating the marginal predictive gains of individual series. This amounts to calculating their weights in the factor model prediction.

I provide two pieces of evidence, which suggest that factor model prediction weights are a useful alternative to LARS. First, a Monte-Carlo simulation exercise confirms that both methods are suitable for selecting data sets that result in more efficient predictions. However, factor model prediction weights are more successful than LARS in identifying the appropriate series. Consequently, they also tend to deliver better out-of-sample predictions. LARS, in turn, shows some tendency of overfitting, as pre-sample predictions suggest gains that only partly carry over to the out-of-sample case.

Second, I apply both methods to the now- and forecasting of quarterly GDP growth from large unbalanced monthly data sets. I use the dynamic factor model by Doz, Giannone, and Reichlin (2011), which employs a state-space framework and therefore copes with unbalanced data and mixed frequencies in an efficient way. It has been shown to perform well under these conditions (Angelini et al., 2011; Giannone et al., 2008; Rünstler et al., 2009). As pointed out by Bańbura and Rünstler (2011), prediction weights of individual series can be obtained from an extension of the Kalman filter. LARS is less suited for dealing with unbalancedness and must be applied to quarterly aggregates of monthly data.

I use monthly data sets for the euro area, Germany, and France over the period of 1991–2014. Each data set contains about 70 series. I obtain variable selections from a pre-sample and evaluate their performance from a pseudo-real-time forecast exercise. I find that variable selections of 10–30 series from either method improve nowcasts of euro area GDP. Results for Germany and France are more mixed. Selections from factor model prediction weights provide moderate but consistent gains, while pre-sample LARS selections sometimes suggest gains that revert into losses in out-of-sample predictions. Overall, for nowcasts factor model prediction weights tend to provide more robust variable selections than LARS. Gains for next-quarter forecasts are generally very small.

This paper is organised as follows. Section 2 reviews the basic concepts. Section 3 discusses variable selection in the context of the dynamic factor model by Doz et al. (2011) with unbalanced and mixed-frequency data. Section 4 conducts the Monte Carlo study to investigate the gains from using prediction weights and LARS. Section 5 presents the empirical application. Section 6 concludes this paper.

2. VARIABLE SELECTION IN FACTOR MODELS

Consider the dynamic factor model

$$x_t = \Lambda f_t + \xi_t, \quad \xi_t \sim \mathbb{N}(0, \Sigma_\xi) \tag{1}$$

The model relates the $n \times 1$ vector of series $x_t = (x_{1t}, \ldots, x_{nt})'$ to the $r \times 1$ vector of common factors $f_t = (f_{1t}, \ldots, f_{rt})'$ from the matrix Λ of factor loadings and to the $n \times 1$ vector of idiosyncratic components $\xi_t = (\xi_{1t}, \ldots, \xi_{nt})'$ with covariance matrix Σ_ξ. It holds that $r \ll n$. The common factors f_t and idiosyncratic components ξ_t are assumed to follow certain stochastic processes, which will be specified below.

The purpose of the model is to estimate (and possibly predict) f_t from data x_t, $t = 1, \ldots, T$, and subsequently to predict a scalar target series y_t from the equation

$$y_t = \beta' f_t + \varepsilon_t, \quad \varepsilon_t \sim \mathbb{N}\left(0, \sigma_\varepsilon^2\right) \tag{2}$$

with $r \times 1$ vector $\beta = \left(\beta_1, \ldots, \beta_r\right)'$. The residual ε_t is assumed to be identically independently distributed and to be independent of ξ_t.

As $[n; T] \to \infty$, the factor space of dimension r can be consistently estimated by principal components under various conditions, which include (i) appropriate assumptions on the stationarity and weak time dependence of f_t and ξ_t; (ii) a rank condition on Λ precluding non-trivial factor loadings; (iii) sufficiently weak cross-sectional dependence between f_t and ξ_t; and (iv) sufficiently weak cross-sectional dependence among the elements of vector ξ_t (Bai & Ng, 2002; Stock & Watson, 2002b).

Specifically, under condition (iv), the non-diagonal elements of Σ_ξ should turn sufficiently small as n tends to infinity (e.g., Bai & Ng, 2002). Boivin and Ng (2006) argue that this is likely to be violated in macro-economic data sets, as some correlation among idiosyncratic components would remain. They further show that, in finite samples, forecast precision would not necessarily increase with the number of series if Σ_ξ is non-diagonal. Under certain circumstances, for example with heteroscedasticity in idiosyncratic components or in case that some elements of f_t are irrelevant for predicting y_t, predictions may therefore be improved from removing uninformative series.

Boivin and Ng (2006) and Caggiano et al. (2011) present empirical applications, where forecasts are improved by simple removing the series with the highest cross correlations in idiosyncratic components. In search for more sophisticated selection criteria, Bai and Ng (2008) proposed Least Angle Regressions (LARS) and several variants (LASSO and elastic net algorithms) to select series in factor model forecasts for U.S. inflation. They report considerable gains in precision over a range of specifications. Schumacher (2010) and Bessec (2013) confirm these findings for German and French GDP, respectively. With the exception of Bessec (2013), these studies inspect in-sample forecasts, that is, perform variable selection within the forecast evaluation sample. The studies use diffusion indices (Stock & Watson, 2002a).[1]

LARS is a constrained variant of stepwise forward selection to predict y_t from equation

$$y_t = \beta' x_t^{L,s} + \varepsilon_t^{L,s}, \quad \varepsilon_t^{L,s} \sim \mathbb{N}\left(0, \sigma_{\varepsilon,s}^2\right) \tag{3}$$

where $x_t^{L,s}$ denotes a certain subset of series x_t of size s. Starting with the empty set $x_t^{L,0}$, at each step s, one series is added to $x_t^{L,s-1}$ in order to obtain $x_t^{L,s}$. As with the standard approach, this is the series with the highest marginal predictive gain on top of predictions based on $x_t^{L,s-1}$, that is, the highest correlation with residual $\varepsilon_t^{L,s-1}$. To increase the robustness of forward selection, LARS adjusts coefficients β in Eq. (3) after each step. This is done by increasing the coefficients in their joint least-squares direction until another predictor (not yet contained in $x_t^{L,s}$) displays as much correlation with the residual as the series contained in $x_t^{L,s}$. The process stops at $k = \min(T, n-1)$, and results in a set of selections $\mathcal{L} = \{x_t^{L,s}\}_{s=1}^{k}$.

The purpose of coefficient shrinkage in LARS is to overcome the dimensionality problem that emerges with a high number of predictors and results in highly inefficient selections. Another way to deal with high dimensionality is using the factor model itself for approximating the marginal predictive gains of the individual series. The latter can be obtained from the weights of the individual series in the factor model predictions.

The principle is easily illustrated for a static factor model, with factors being estimated by principal components. The series x_{it} are assumed to be standardized to mean zero and variance one. Consider variable selection $x_t^{W,s}$ and let $(1/T)\sum_{t=1}^{T} x_t^{W,s}(x_t^{W,s})' = V_s D_s V_s'$ be the eigenvalue decomposition of its empirical covariance matrix with eigenvectors V_s. Given the number of factors r, it holds $\hat{f}_t^{(s)} = V_{s,r}' x_t^{W,s}$, where $V_{s,r}$ denotes the matrix containing the first r columns of V_s.

The prediction of y_t is then found with

$$y_{t|t}^{W,s} = \hat{\beta}_s' V_{s,r}' x_t^{W,s} = \left(\omega_0^s\right)' x_t^{W,s} \tag{4}$$

where $\hat{\beta}$ is estimated from a regression of y_t on $\hat{f}_t^{(s)}$ as from Eq. (2) and ω_0^s is the $s \times 1$ vector of prediction weights. These weights represent the marginal predictive gains of the elements of $x_t^{W,s}$ from projecting y_t on f_t.

For a factor model, stepwise backward elimination seems a natural approach. Starting with the entire set of series $x_t^{W,n} = x_t$ at each step $s = n, \ldots, 1$, the factor model is re-estimated based on series $x_t^{W,s}$ and the series with the lowest weight from $x_t^{W,s}$ is removed to obtain selection $x_t^{W,s-1}$. This process results in a set of selections $\mathcal{W} = \{x_t^{W,s}\}_{s=1}^{n}$.

In contrast to most of the earlier literature, I use an out-of-sample forecast design in this paper. I obtain variable selections from a pre-sample and determine the optimal selection size, that is, find those selections in $\mathcal{L}$ and $\mathcal{W}$ that minimize the root-mean-squared error (RMSE) of predictions. This comes closer to application in real time and may reveal issues related to overfitting and spurious selections. Given the heuristic nature of variable selection, the two methods would in general result in different selections and the optimal selection sizes may differ. I therefore determine the optimal selection size separately for each method.

One difference between factor model prediction weights and LARS is that the former would select predictors with high commonality, while LARS would avoid strongly correlated predictors. To see this, consider a group of highly correlated predictors within x_t. From principal components analysis, all elements of the group would attain similar factor loadings and therefore similar model prediction weights. With LARS, by contrast, if one element of the group gets included in set $x_t^{L,s}$, the new residual will have a low correlation with the remaining elements of this group (Bai & Ng, 2008). Hence, the latter would no longer be selected. Overall, LARS is therefore likely to result in a more diverse final set of predictors than prediction weights. Arguably, stepwise regression is therefore not suitable for selecting a subset of variables with high commonality to estimate the principal components in a data set.

3. PREDICTION WEIGHTS FROM A DYNAMIC FACTOR MODEL

This section discusses prediction weights in the context of the dynamic factor model by Doz et al. (2011). The model is given by Eq. (1) together with the law of motion

$$f_{t+1} = \sum_{l=1}^{p} \Psi_l f_{t-l+1} + B\eta_t, \quad \eta_t \sim \mathbb{N}(0, I_q) \tag{5}$$

Common factors f_t are driven by q-dimensional white noise η_t with $r \times q$ matrix B, where $q \leqslant r$. The stochastic process for f_t is assumed to be stationary. Further, the idiosyncratic component ξ_t is modeled from multivariate white noise with diagonal covariance matrix Σ_ξ.[2]

In the empirical application, I will use the factor model to predict quarterly GDP growth from monthly data x_t. To handle these mixed frequencies, I follow Harvey (1989, 309ff) and introduce monthly GDP growth y_t as a latent variable (see also, e.g., Angelini et al., 2011; Mariano & Murasawa, 2010). The dependent variable y_t is assumed to be related to factors f_t by the equation

$$y_t = \mu + \beta' f_t + \varepsilon_t, \quad \varepsilon_t \sim \mathbb{N}\left(0, \sigma_\varepsilon^2\right) \tag{6}$$

This is supplemented with log-linear aggregation rules to relate y_t to observed quarterly GDP growth, y_t^Q. For this purpose, another latent variable Q_t is defined at monthly frequency such that it corresponds to y_t^Q in the third month of the respective quarter, $t = 3k$. Aggregation rules can then be expressed as

$$\begin{aligned} y_t^{(3)} &= y_t + y_{t-1} + y_{t-2} \\ Q_t &= \frac{1}{3}\left(y_t^{(3)} + y_{t-1}^{(3)} + y_{t-2}^{(3)}\right) \\ y_{3k}^Q &= Q_{3k}, \quad k = 1, 2, \ldots, \lfloor T/3 \rfloor \end{aligned} \tag{7}$$

where $y_t^{(3)}$ represents three-month growth rates of monthly GDP, that is, growth rates vis-a-vis the same month of the previous quarter. In the application, y_t^Q is treated as missing in months 1 and 2 of the quarter, but added to observation vector z_t in month 3.

Eqs. (1), (5), (6), and the aggregation rules can be cast in a single state-space form with state vector $\alpha_t = (f_t, \ldots, f_{t-p+1}, y_t, y_{t-1}, y_t^{(3)}, Q_t)$. The state-space form is given in annex 1.

$$\begin{aligned} z_t &= W_t \alpha_t + u_t, \quad u_t \sim \mathbb{N}(0, \Sigma_u) \\ \alpha_{t+1} &= T_t \alpha_t + v_t, \quad v_t \sim \mathbb{N}(0, \Sigma_v) \end{aligned} \tag{8}$$

The Kalman filter and associated smoothing algorithms (see, e.g., Durbin & Koopman, 2001) provide minimum mean square linear (MMSE) estimates $a_{t+h|t} = \mathbb{E}[\alpha_{t+h} | \mathcal{Z}_t]$ of the state vector and their covariance $P_{t+h|h}$ for information set $\mathcal{Z}_t$ and any $h > -t$.

Estimation of the model parameters is described in Giannone et al. (2008). Briefly, estimates of factor loadings Λ and initial estimates of factors f_t are obtained from principal components. The latter are used to estimate Ψ_l in Eq. (5) from OLS. A further application of principal components to the residual covariance matrix of the VAR then gives matrix B. Parameters β and σ_ε^2 are estimated from a quarterly version of Eq. (6), again using the initial estimates of factors f_t with appropriate adjustments (see Angelini et al., 2011).

I will use information criteria to obtain the model specifications. Specifically, r, p, and q are found at the various stages of the estimation process from criterion PCP_2 in Bai and Ng (2002), the Akaike information criterion (*AIC*), and criterion 2 in Bai and Ng (2007), respectively.[3]

As pointed out by Bańbura and Rünstler (2011), prediction weights for y_t^Q can be obtained from an extension of the Kalman filter and smoother due to Harvey and Koopman (2003). For any information set $\mathcal{Z}_t$, the extension provides the weights of individual observations in estimates $a_{t+h|t}$ of the state vector, $h > -t$. As y_t^Q is an element of the state vector, this allows predictions $y_{t+h|t}^Q$ to be expressed as

$$y_{t+h|t}^Q = \sum_{l=0}^{t-1} \omega_{l,t}'(h) z_{t-l} \tag{9}$$

with weights $\omega_{l,t}(h)$. Clearly, weights depend on both the forecast horizon h and the information set $\mathcal{Z}_t$. In recursive forecast evaluation exercises, it is therefore important to define information sets such that the Kalman filter and smoothing algorithms approach their steady state and the time index on weights can be dropped. This holds for pseudo-real-time data sets $\mathcal{Z}_t$ as defined below in Section 5.

Since the Kalman filtering and smoothing algorithms provide MMSE estimates, weights $\omega_{i,l}(h)$ are a measure of the marginal predictive gain in $y_{t+h|t}^Q$ that arises from adding observation $x_{i,t-l}$ to the information set. In the exercises presented in Sections 4 and 5, I will consider cumulative weights $\omega(h) = \sum_{l=0}^{k} \omega_l(h)$ as a measure of the predictive content of series $x_{i,t}$ for $y_{t+h|t}^Q$, where k is chosen sufficiently large.[4]

To obtain selections from LARS, the monthly data must be aggregated to quarterly frequency, x_t^Q. LARS selections for predictions $y_{t+h|t}^Q$ can then be obtained from static regressions of quarterly GDP growth y_t^Q on x_{t-h}^Q as from Eq. (3), $h = 0, 1, \ldots$.

4. A MONTE CARLO SIMULATION

This section conducts a Monte Carlo study to investigate the gains from the two variable selection methods. The simulation design is a variant of simulation 1 in Boivin and Ng (2006). I use the dynamic factor model described in Section 3, but I abstract from mixed frequency issues and assume that y_t is observable.

The data are generated from the equations

$$\begin{aligned} x_t &= \lambda f_t + \xi_t, & \xi_t &\sim \mathbb{N}(0, \Sigma_\xi) \\ f_t &= \psi f_{t-1} + \eta_t, & \eta_t &\sim \mathbb{N}(0, \sigma_\eta^2) \\ y_t &= \beta' f_t + \varepsilon_t, & \varepsilon_t &\sim \mathbb{N}(0, \sigma_\varepsilon^2) \end{aligned}$$

I assume a single latent factor f_t, which is modeled as a first-order autoregressive process with $\sigma_\eta^2 = 1 - \psi^2$ such that $var(f_{i,t}) = 1$. The $n \times 1$ vector of series $x_t = (x_{1t}, \ldots, x_{nt})'$ and the scalar target series y_t are defined as in Eqs. (1) and (2), while the idiosyncratic component ξ_t is assumed to be multivariate white noise with covariance matrix Σ_ξ.

Factor loadings $\lambda = (\lambda_1, \ldots, \lambda_n)'$ are assumed to differ across series. They are drawn from a beta distribution $\mathbb{B}(a, b)$ over support (0, 1). I will consider various values of a and b to inspect the role of dispersion and skewness of factor loadings on the success of variable selection methods. Heterogenous factor loadings translate into heteroscedasticity in idiosyncratic components, as the series x_{it} are standardised to $var(x_{i,t}) = 1$, which implies $var(\xi_{i,t}) = 1 - \lambda_i^2$.

Further, I allow for non-zero cross correlations among idiosyncratic components ξ_{it}. I simply set $corr(\xi_{it}, \xi_{jt}) = \rho$ for all $i, j = 1, \ldots, n, i \neq j$. The elements ij of covariance matrix Σ_ξ are therefore given by

$$\Sigma_{\xi,ij} = \begin{cases} 1 - \lambda_i^2 & \text{for } i = j \\ \rho\sqrt{\left(1 - \lambda_i^2\right)\left(1 - \lambda_j^2\right)} & \text{otherwise} \end{cases}$$

The parameters of forecasting equation (2) for y_t are kept fixed with $\beta = 0.75$ and $\sigma_\varepsilon^2 = 1 - \beta^2$, which implies $var(y_t) = 1$.

As discussed in Section 2, once $\rho > 0$, forecast performance may be improved by using a limited set of variables. With the above simulation design, because the correlations among idiosyncratic components are assumed to be identical across all series, the information content of series $x_{i,t}$ for estimating f_t depends only on its factor loading λ_i. Hence, the best selections would simply consist of the series with highest factor loadings.

The simulations aim at assessing the usefulness of LARS and factor model prediction weights for obtaining predictions $y_{t+h|t}$, $h \geqslant 0$. I use an out-of-sample forecast design. I take 500 draws of of length $T = 180$, which amounts to 15 years of monthly data. The number of series is set to $n = 100$. For each draw $\{f_{t,J}, \xi_{t,J}, \varepsilon_{t,J}, \lambda_J\}$ I obtain $\{x_{t,J}, y_{t,J}\}$ and proceed as follows:

1. I split draw J into two sub-samples 1 and 2 of length $T_1 = T_2 = 90$.
2. From the pre-sample (sub-sample 1), I obtain variable selections of sizes $s = 1, \ldots, n$. Denote with $\mathcal{W}_J = \{x_{t,J}^{W,s}\}_{s=1}^{n}$ and $\mathcal{L}_J = \{x_{t,J}^{L,s}\}_{s=1}^{k}$ the sets of selections according to factor model prediction weights and LARS, respectively, as described in Section 3. For obtaining prediction weights, I either keep the number of factors fixed at the true value of $r = 1$ or estimate r from information criterion PCP_2 in Bai and Ng (2002).

   I further determine the optimal number of series for either selection method from a minimum RMSE criterion. I choose the selection $x_{t,J}^{W,s}$ in $\mathcal{W}_J$, which gives the minimum RMSE of predictions $\hat{y}_{t,J}^{W,s}$ in sub-sample 1. I proceed equivalently for $\mathcal{L}_J$.[5]
3. For all selections $x_{t,J}^{W,s}$ and $x_{t,J}^{L,s}$, I estimate the parameters of the dynamic factor model from sub-sample 1. I then obtain predictions $\hat{y}_{t+h|t,J}^{W,s}$ and $\hat{y}_{t+h|t,J}^{L,s}$ for y_t^J over both sub-samples.

Table 1 shows the findings for the case of a static factor model, $\psi = 0$. The table reports the average RMSE of pre-sample and out-of-sample predictions, the number of series chosen by the in-sample RMSE criterion, and the percentage of correct classification. Denote with $x_{t,J}^{*,s}$ the $s \times 1$ vector of series with the highest s factor loadings among $x_{t,J}$. The percentage of correct classifications is given by the share of elements of $x_{t,J}^{*,s}$ contained in selections $x_{t,J}^{W,s}$ and $x_{t,J}^{L,s}$, respectively.

The following conclusions emerge from Table 1. First, for $\rho = 0$, prediction weights and LARS choose 54 and 20 series, respectively, as opposed to the optimal choice of 100. The smaller selections result in some small losses in out-of-sample predictions, as to be expected.

Second, once $\rho > 0$, both selection methods result in improved out-of-sample predictions. For some of the simulations, these gains are of considerable size. While losses against predictions based on the full set of series may occur, they are always minor. The optimal selections are generally

Table 1. Monte Carlo Simulations Static Factor Model.

		Symmetric Distributions ($r=1$)						
Factor persistence ψ		0.0	0.0	0.0	0.0	0.0	0.0	0.0
Correlation (ρ)		0.0	0.1	0.1	0.1	0.3	0.3	0.3
a		4	2	4	8	2	4	8
b		4	2	4	8	2	4	8
No. of factors (r)		1	1	1	1	1	1	1
No. of series selected	Weights	53.5	16.1	19.7	26.9	7.6	9.9	13.1
	LARS	21.4	11.9	12.6	14.4	6.6	6.7	7.3
Correct classification	Weights	0.88	0.88	0.82	0.75	0.86	0.79	0.69
	LARS	0.35	0.36	0.38	0.37	0.45	0.45	0.41
RMSE pre-sample	All	0.66	0.70	0.72	0.74	0.78	0.80	0.82
	Weights	0.66	0.67	0.69	0.71	0.69	0.72	0.75
	LARS	0.59	0.63	0.64	0.64	0.67	0.70	0.72
RMSE out-of-sample	All	0.68	0.71	0.74	0.75	0.79	0.82	0.83
	Weights	0.68	0.69	0.71	0.74	0.70	0.74	0.78
	LARS	0.69	0.70	0.73	0.75	0.72	0.76	0.80

		Asymmetric Distributions ($r=1$)				Number of Factors Estimated			
Factor persistence ψ		0.0	0.0	0.0	0.0	0.0	0.0	0.0	0.0
Correlation (ρ)		0.1	0.3	0.1	0.3	0.3	0.3	0.3	0.3
a		1	1	3	3	2	4	1	3
b		3	3	1	1	2	4	3	1
No. of factors (r)	All	1	1	1	1	1.99	1.14	1.48	2.00
	Weights	1	1	1	1	1.55	1.01	1.07	1.74
	LARS	1	1	1	1	1.72	1.01	1.09	1.79
No. of series selected	Weights	12.7	6.8	26.3	13.9	33.5	19.1	13.9	29.7
	LARS	10.3	6.2	11.9	8.0	22.7	12.5	10.2	16.3
Correct classification	Weights	0.81	0.77	0.94	0.92	0.70	0.67	0.69	0.75
	LARS	0.49	0.52	0.18	0.24	0.43	0.45	0.51	0.24
RMSE pre-sample	All	0.82	0.92	0.66	0.69	0.67	0.80	0.81	0.66
	Weights	0.72	0.77	0.65	0.66	0.67	0.73	0.78	0.65
	LARS	0.69	0.77	0.62	0.64	0.64	0.70	0.78	0.61
RMSE out-of-sample	All	0.84	0.94	0.67	0.70	0.69	0.82	0.84	0.68
	Weights	0.75	0.79	0.67	0.67	0.70	0.76	0.81	0.68
	LARS	0.76	0.81	0.68	0.69	0.73	0.78	0.84	0.70

Notes: The table shows findings from Monte Carlo simulations for the static factor model ($\psi=0$). ρ is the correlation among idiosyncratic components, while a and b are the parameters of the beta distribution, from which factor loadings λ are drawn (see main text). The table shows four statistics for both LARS and prediction weights: (i) the average number of series in the optimal selections (selections with minimum RMSE pre-sample); (ii) the percentage of correct classifications, that is, of series with highest factor loadings; (iii) the RMSE of the optimal selection in the pre-sample; column "All" refers to the RMSE from all 100 series; and (iv) the out-of-sample RMSE of the optimal selection. The upper panel shows results for symmetric distributions of factor loadings with the number of factors kept fixed at $r=1$. The lower left-hand panel shows results for asymmetric distributions with $r=1$. The lower right-hand panel shows results where the number of factors is estimated from an information criterion (see main text).

small: with one exception, the in-sample RMSE criterion chooses less than 20 series. Optimal selections from prediction weights are somewhat larger than those from LARS.

Third, factor model prediction weights consistently outperform LARS. They provide a lower out-of-sample RMSE, although often only by a small margin, and the percentage of correctly classified series is considerably higher. Whereas the share of correctly classified series amounts to about 0.7 to 0.8 in prediction weight selections, it is less than 0.5 in LARS selections. The overlap among the selections turns out to be moderate. In general, the share of series that are contained in both selections is in between 0.5 and 0.6, depending on the sizes of the selections.[6]

Perhaps more important, LARS shows some tendency of overfitting. For prediction weights, the gains indicated by in-sample predictions largely carry over to the out-of-sample case. The comparatively large pre-sample gains from LARS selections, however, turn out to be spurious, as gains in out-of-sample predictions are much smaller. This is most apparent for the case of $\rho = 0$, where applying either variable selection method results in a slight deterioration of the out-of-sample RMSE, as suggested by asymptotic theory. The pre-sample RMSE suggests however considerable gains from LARS selections.

Fourth, as to the role of ρ and (a, b), higher values of ρ and skewed distributions of factor loadings with a high share of uninformative series give rise to larger gains from variable selection. The beta distribution $\mathbb{B}(a, b)$ is symmetric for $a = b$ with mean 0.5 and its dispersion declines with higher a and b. The case of $a = 1$, $b = 3$ amounts a to left-skewed distribution with mean 0.25, implying a high share of series with low factor loadings, while the opposite case of $a = 3$, $b = 1$ gives a right-skewed distribution with mean 0.75. The gains from variable selection increase with high dispersion and with a left-skewed distribution.

Fifth, for $\rho > 0$, the specific correlation structure of ξ_t in this exercise implies a single high eigenvalue of Σ_ξ and therefore the presence of one principal component in ξ_t. Once the number of factors r is estimated, this is occasionally picked up as a second factor. It turns out, that selection methods then act as an insurance against mis-specification. The right-hand lower panel of Table 1 shows simulation results, where the number of factors r is estimated from information criterion PCP_2 (Bai & Ng, 2002). For $\rho = 0.1$, the criterion chooses $r = 1$ in all cases and gains from variable selection prevail. For $\rho = 0.3$, this still holds for some of the values of (a, b) considered in the simulations. There arise yet two cases where r is estimated predominantly as equal to 2. Predictions from $r = 2$ and the full data set then perform as well as the optimal predictions from $r = 1$, while selection methods do not deliver further gains. In this case, hence, variable selection acts to avoid the losses that would arise from choosing $r = 1$, that is, lower than suggested by the in-sample information criterion. Note that this pattern occurs for precisely those simulations, where selection methods had delivered the largest gains under $r = 1$.

Finally, Table 2 shows that the above conclusions for the case of $\psi = 0$ straightforwardly carry over to $\psi > 0$. Results are reported for both static predictions $y_{t|t}$ and one-step ahead forecasts $y_{t+1|t}$. For static predictions, the above findings remain almost unchanged for both $\psi = 0.5$ and $\psi = 0.8$. For one-step ahead predictions, the gains from variable selection remain in case of highly persistent factor dynamics ($\psi = 0.8$). However, for $\psi = 0.5$, gains decline considerably, as the predictions become generally less informative.

5. FORECASTING GDP GROWTH FROM MONTHLY DATA

This section presents a pseudo-real-time exercise to obtain now- and forecasts of quarterly GDP growth in the euro area, Germany, and France from large unbalanced monthly data sets. I obtain variable selections based on factor model prediction weights and LARS from a pre-sample and evaluate their performance from a pseudo-real-time forecast exercise in the second part of the sample.

The dynamic factor model by Doz et al. (2011) has been applied to predict quarterly GDP growth from unbalanced monthly data for a number of countries, including the U.S. (Giannone et al., 2008), the euro area (Angelini et al., 2011), and several euro area member states, such as Germany and France (Bessec, 2013; Marcellino & Schumacher, 2010; Rünstler et al., 2009). Rünstler et al. (2009) and Marcellino and Schumacher (2010) find the model to perform about as well as other versions of dynamic factor models, while Rünstler et al. (2009) and Angelini et al. (2011) report that it is superior to pooled forecasts from single equations.

I use monthly data sets for the euro area, Germany, and France of about 70 series each. All data start in January 1991. They were downloaded on September 26, 2014. The choice of series is based on Angelini et al. (2011) and includes data on economic activity (such as industrial production, trade, employment), the European Commission business and consumer surveys, financial markets, and the international environment. The series

Table 2. Monte Carlo Simulations Dynamic Factor Model.

		Static Predictions ($h=0$)							
Factor persistence ψ		0.5	0.5	0.5	0.5	0.8	0.8	0.8	0.8
Correlation (ρ)		0.1	0.1	0.3	0.3	0.1	0.1	0.3	0.3
a		4	8	4	8	4	8	4	8
b		4	8	4	8	4	8	4	8
No. of factors (r)		1	1	1	1	1	1	1	1
No. of series selected	Weights	20.2	25.6	9.4	12.4	19.2	24.5	9.5	11.8
	LARS	12.7	14.6	6.6	7.4	12.4	13.9	6.5	7.2
Correct classification	Weights	0.82	0.74	0.79	0.68	0.81	0.71	0.78	0.65
	LARS	0.38	0.36	0.45	0.40	0.37	0.35	0.44	0.40
RMSE pre-sample	All	0.72	0.74	0.81	0.82	0.72	0.74	0.80	0.82
	Weights	0.69	0.71	0.72	0.76	0.69	0.71	0.72	0.76
	LARS	0.64	0.64	0.70	0.72	0.64	0.64	0.70	0.72
RMSE out-of-sample	All	0.74	0.75	0.82	0.83	0.74	0.75	0.82	0.83
	Weights	0.71	0.74	0.74	0.78	0.71	0.74	0.74	0.78
	LARS	0.73	0.75	0.76	0.79	0.72	0.75	0.76	0.79
		One-Step Ahead Predictions ($h=1$)							
Factor persistence ψ		0.5	0.5	0.5	0.5	0.8	0.8	0.8	0.8
Correlation (ρ)		0.1	0.1	0.3	0.3	0.1	0.1	0.3	0.3
a		4	8	4	8	4	8	4	8
b		4	8	4	8	4	8	4	8
No. of factors (r)		1	1	1	1	1	1	1	1
No. of series selected	Weights	26.8	30.2	13.8	15.2	19.3	27.0	9.6	12.8
	LARS	25.1	27.5	11.2	10.7	17.4	20.4	6.7	7.7
Correct classification	Weights	0.83	0.75	0.79	0.69	0.82	0.74	0.78	0.67
	LARS	0.47	0.44	0.47	0.42	0.43	0.41	0.46	0.40
RMSE pre-sample	All	0.93	0.94	0.96	0.96	0.82	0.83	0.89	0.89
	Weights	0.92	0.93	0.93	0.94	0.79	0.80	0.81	0.85
	LARS	0.92	0.93	0.94	0.94	0.79	0.80	0.82	0.84
RMSE out-of-sample	All	0.95	0.95	0.97	0.98	0.84	0.85	0.91	0.91
	Weights	0.94	0.95	0.95	0.96	0.81	0.83	0.84	0.87
	LARS	0.95	0.95	0.96	0.96	0.82	0.84	0.85	0.88

Note: See Table 1 for a description of the contents.

are transformed to monthly rates of change and standardised to mean zero and variance one. Further, they are cleaned from outliers. The series are listed in annex A together with their publication lags and the data transformations used.[7]

5.1. Forecast Design

The forecast design follows Angelini et al. (2011) and aims at replicating the real-time application of the factor model as closely as possible.

First, I account for the timing of data releases. Real-time data sets typically contain missing observations at the end of the sample due to publication lags. Survey and financial market data, for instance, are available right at the end of the respective month, while data on economic activity are usually published with a delay of six to eight weeks. Giannone et al. (2008) and Bańbura and Rünstler (2011) report that differences in the timing of data releases among individual series have large effects on their marginal predictive gains.

I therefore follow those studies in applying the so-called *pseudo-real-time* data sets $\mathcal{Z}_t$, which employ the final data release, but replicate the publication lags from the end of the sample in the earlier periods. Let $z_t' = \left(x_t', y_t^Q\right)$ and denote with $\mathcal{Z}_t$ the information set in period t. Consider the original data set $\mathcal{Z}_T$ as downloaded in period T. Data set $\mathcal{Z}_t$, on which the predictions in period t are based, is obtained by eliminating observation $x_{i,t-l}$, $l \geqslant 0$, if and only if observation $x_{i,T-l}$ is missing in $\mathcal{Z}_T$, $i = 1, \ldots, n$. Quarterly GDP growth is treated in an equivalent way. Kalman filtering and smoothing handles unbalanced data sets in an efficient way. The rows in Eq. (8) corresponding to missing observations in z_t are simply skipped when applying the respective recursions (Durbin & Koopman, 2001:92f).

Second, I inspect six predictions for GDP growth in a certain quarter, which are obtained in consecutive months. I start in the first month of the previous quarter and stop in the third month of the current quarter, 6 weeks before the flash estimate of GDP is released. To predict GDP growth in the second quarter, for instance, the first prediction is run in January and the final (sixth) one in July. Note that predictions 4–6 amount to nowcasting the current quarter.

Third, I inspect out-of sample predictions. I obtain the variable selections and corresponding factor model specifications from a pre-sample and run a forecast exercise with recursive parameter estimation on the remainder. I proceed as follows:

1. I obtain selections $\left\{x_t^{w,s}\right\}_{s=1}^{n}$ and $\left\{x_t^{L,s}\right\}_{s=1}^{k}$ from the pre-sample ranging until 2000 Q4 using stepwise elimination as described in Section 2. I use different selections for now- and next-quarter forecasts. Prediction weight selections are based on mid-quarter weights, that is, weights from prediction 5 for nowcasts (predictions 4–6) and 2 for the next-quarter forecasts (predictions 1–3), respectively.

While factor model prediction weights account for publication lags, a standard application of LARS would ignore them. Bessec (2013) argues that LARS selections can be improved by accounting for publication lags. She proposes to start with unbalanced data and to forecast the missing observations from univariate methods. I follow a proposal by Altissimo et al. (2010) instead and shift the monthly series prior to aggregation.

That is, with series $x_{i,t}$ being subject to a publication lag of l months, I define $x^{\#}_{i,t} = x_{i,t-l}$. I run LARS with quarterly GDP growth being regressed on quarterly aggregates $x^{\#Q}_{i,t}$ at either lag 0 for nowcasts or 1 for next-quarter predictions.

For all selections, model specifications are obtained from the information criteria set out in Section 3. The model is re-specified at each selection step under the restriction that the dimensionality of the model shrinks with the number of series. That is, for example, for prediction weights I obtain specifications $(r^{w,s}, p^{w,s}, q^{w,s})$ related to selection $x^{w,s}_t$ under the restrictions $r^{w,s} \leqslant r^{w,s+1}$, $p^{w,s} \leqslant p^{w,s+1}$, and $q^{w,s} \leqslant q^{w,s+1}$ for $s < n$.

2. I obtain now- and next-quarter forecasts of GDP growth for the period starting with 2001 Q1 based on the variable selections as from step 1. These forecasts employ pseudo-real-time data sets $\mathcal{Z}_t$ and recursive parameter estimates.

   The financial crisis requires some special consideration in the choice of the evaluation sample. It not only implies extremely large forecast errors in 2008 and 2009, but may also constitute a structural break in economic activity in the euro area. The evaluation of variable selection methods may therefore be more safely based on the pre-crisis period. On the other hand, the performance of the models after the crisis is certainly of interest.

   I therefore evaluate the forecasts separately for two samples, a pre-crisis sample from 2001 Q1 to 2007 Q4 and a post-crisis sample ranging from 2010 Q1 to 2014 Q2.

5.2. Results

The factor model specifications chosen by the information criteria are similar across data sets. For the euro area and German full data, the number of factors is estimated as $r = 3$, while p and q are estimated as 2. For the various selections, estimates of r remain at 3, while estimates of p and q shrink as the number of series declines. For France, r and q shrink from 3 to 1, while p stays at 3. For all countries, the average cross correlation among idiosyncratic components is slightly below 0.2. Idiosyncratic components are subject to considerable heteroscedasticity.

Table 3 shows the RMSE of predictions for the sequence of 6 predictions described above. The numbers are averaged over predictions 1–3 (next-quarter) and 4–6 (nowcasts). The RMSE is shown relative to the naive forecast, which is the sample mean of GDP growth.[8] Note that variable selection with LARS stops at $s = 30$, as it is limited by the number of observations.

Table 3. Forecasting Performance of Selections.

	Naive	AR(1)	All	Prediction Weights							LARS			
				60	50	40	30	20	15	10	30	20	15	10
Euro area														
Pre-sample (1991 Q1–2000 Q4)														
4–6	0.488	0.93	0.91	0.79	0.84	0.87	0.82	0.81	0.76	0.77	0.81	0.86	0.88	0.89
1–3	0.488	0.99	0.80	0.80	0.80	0.80	0.81	0.84	0.87	0.90	0.88	0.92	0.93	0.99
1–7	0.488	0.96	0.86	0.80	0.82	0.83	0.81	0.83	0.82	0.83	0.85	0.89	0.90	0.94
Pre-crisis (2001 Q1–2007 Q4)														
4–6	0.339	0.86	0.86	0.85	0.75	0.75	0.75	0.77	0.77	0.78	0.77	0.85	0.82	0.85
1–3	0.344	0.96	0.79	0.79	0.80	0.80	0.81	0.85	0.82	0.84	0.86	0.94	0.94	0.94
1–7	0.342	0.91	0.82	0.82	0.77	0.78	0.78	0.81	0.80	0.81	0.82	0.90	0.88	0.90
Post-crisis (2010 Q1–2014 Q2)														
4–6	0.441	0.76	1.01	0.99	0.88	0.87	0.83	0.79	0.80	0.76	0.82	0.64	0.65	0.72
1–3	0.444	0.85	0.89	0.92	0.91	0.92	0.95	0.95	0.96	0.97	0.83	0.92	0.92	0.93
1–7	0.442	0.80	0.95	0.96	0.89	0.89	0.89	0.87	0.88	0.86	0.83	0.78	0.79	0.82
Germany														
Pre-sample (1991 Q1–2000 Q4)														
4–6	0.712	1.00	0.95	0.95	0.93	0.96	1.10	0.95	0.92	0.92	0.95	0.95	0.94	0.86
1–3	0.712	1.00	0.93	0.91	0.92	0.91	0.92	0.92	0.91	0.92	0.90	0.91	0.91	0.92
1–6	0.712	1.00	0.94	0.93	0.92	0.94	1.01	0.94	0.92	0.92	0.93	0.94	0.94	0.89
Pre-crisis (2001 Q1–2007 Q4)														
4–6	0.565	1.00	0.89	0.89	0.89	0.90	0.89	0.89	0.84	0.84	0.95	0.89	0.90	0.92
1–3	0.569	1.00	0.91	0.91	0.91	0.91	0.92	0.92	0.91	0.91	0.97	0.97	0.97	0.96
1–6	0.567	1.00	0.90	0.90	0.90	0.91	0.90	0.91	0.88	0.88	0.96	0.93	0.94	0.94
Post-crisis (2010 Q1–2014 Q2)														
4–6	0.633	0.96	0.82	0.83	0.77	0.89	0.83	0.83	0.81	0.81	0.86	0.86	0.87	0.89
1–3	0.635	0.97	0.88	0.86	0.86	0.85	0.85	0.85	0.91	0.95	1.00	1.02	1.03	0.99
1–6	0.634	0.97	0.85	0.85	0.81	0.87	0.84	0.84	0.86	0.88	0.93	0.94	0.95	0.94

Table 3. (*Continued*)

	Naive	AR(1)	All	Prediction Weights							LARS			
				60	50	40	30	20	15	10	30	20	15	10
France														
Pre-sample (1991 Q1–2000 Q4)														
4–6	0.446	0.80	0.82	0.80	0.80	0.80	0.83	0.83	0.84	0.83	0.73	0.76	0.76	0.79
1–3	0.446	0.91	0.81	0.82	0.82	0.81	0.82	0.88	0.97	1.05	0.77	0.76	0.76	0.76
1–6	0.446	0.85	0.82	0.81	0.81	0.80	0.83	0.86	0.90	0.94	0.75	0.80	0.80	0.83
Pre-crisis (2001 Q1–2007 Q4)														
4–6	0.339	1.00	0.80	0.75	0.72	0.76	0.76	0.77	0.78	0.80	0.87	0.93	0.94	0.98
1–3	0.342	0.98	0.87	0.83	0.84	0.85	0.86	0.91	0.90	0.97	0.90	0.91	0.93	0.94
1–6	0.340	0.99	0.84	0.79	0.78	0.80	0.81	0.84	0.84	0.89	0.89	0.92	0.94	0.96
Post-crisis (2010 Q1–2014 Q2)														
4–6	0.388	1.00	1.02	0.90	0.89	1.03	1.04	1.07	1.07	1.05	0.83	0.81	0.81	0.77
1–3	0.389	0.95	1.08	1.02	0.98	0.99	0.99	1.37	1.26	0.90	0.96	0.93	0.92	0.91
1–6	0.389	0.98	1.05	0.96	0.94	1.01	1.02	1.22	1.17	0.98	0.90	0.87	0.87	0.84

Notes: Column 1 shows the RMSE of the naive forecast, based on a random walk with drift. The remaining columns show the RMSE relative to the naive forecast for an autoregressive model (AR(1)), the factor model with the full set of series (All), and various selections of different sizes from prediction weights and from LARS. The individual rows show the relative average RMSE over predictions 1–3 (next-quarter forecasts), 4–6 (nowcasts), and 1–6 (overall average), respectively. Results are shown for three separate sub-samples, that is, the pre-sample, used for variable selection, and pre- and post-crisis evaluation samples.

Starting with predictions from the full data set, for the pre-sample and the pre-crisis evaluation sample, the factor model predictions in general improve upon the naive forecast and a first-order autoregression for GDP (AR(1)). The small gains against the AR(1) in the euro area pre-crisis nowcasts fall somewhat short of the findings reported by Angelini et al. (2011), obtained from a shorter sample. The results for Germany and France are largely in line with earlier studies (e.g., Barhoumi et al., 2010; Rünstler et al., 2009; Schumacher, 2010). For the post-crisis sample, the performance of the factor model worsens for the euro area and France, with predictions being outperformed by the AR(1).

The ranking of the series according to the selections from nowcasts are shown in Tables A1–A3 in the annex. Selections from prediction weights are less heterogenous than those from LARS and there is very little overlap among the two. For the euro area, prediction weight selections contain the main items of business surveys (confidence indicators and order books), together with equity price indices, the euro area real effective exchange rate, and raw materials prices. For Germany and France, business, consumer and construction survey items are prominent. Conversely, LARS puts more weight on hard data, such as items of industrial production, and items of construction and retail trade surveys. In line with the discussion in Section 2, LARS also tends to select a more diverse set of series.

I turn to the performance of variable selections. For the euro area, both methods improve predictions 4–6 (nowcasts), but have little effect on predictions 1–3 (next-quarter forecasts). Prediction weight and LARS selections of 10–15 and 20–30 series, respectively, perform best, with gains of about 15% compared to the full data set. Conversely, for predictions 1–3 there are no gains from either method, with small selections of 20 series or less giving rise to sizeable losses. Crucially, these patterns are properly detected in the pre-sample. In real-time application, both methods would therefore have chosen the correct selections.

Results are more mixed for Germany and France. For Germany, the pre-sample indicates moderate gains of less than 10% in nowcasts from small selections of 10–15 series from either method. For factor model prediction weights, these gains carry over to the pre-crisis sample and vanish in the post-crisis one. The application of LARS selections would however result in losses in either sample. The same applies to LARS selections for next-quarter predictions, although the gains indicated in the pre-sample are very small.

For France, the pre-sample indicates gains from both selection methods over the entire horizon. Small gains from factor model prediction weights arise for selections of 40 – 60. Again, these gains carry over to both evaluation samples. The gains from LARS selections are sizeable in the pre-sample. In evaluation samples however, they arise only in the post-crisis samples, whereas in the pre-crisis sample LARS selections give rise to losses at all horizons.

These findings are summarized in Fig. 1, which compares the RMSEs from the full data set with those of the factor model prediction weight and LARS selections that are found to give the smallest RMSE in the pre-sample. Generally, prediction weights appear more robust than LARS selections. They give rise to modest, but stable gains in nowcasts over a range of selection sizes and the pre-sample gives largely correct signals on appropriate selections. While some selections might give rise to losses in next-quarter predictions, these are properly detected in the pre-sample. LARS selections fare equally well for the euro area, but give more mixed results for Germany and France. In particular, the pre-sample gives wrong signals for out-of sample predictions in Germany over the entire horizon, and the pre-crisis sample in France.[9]

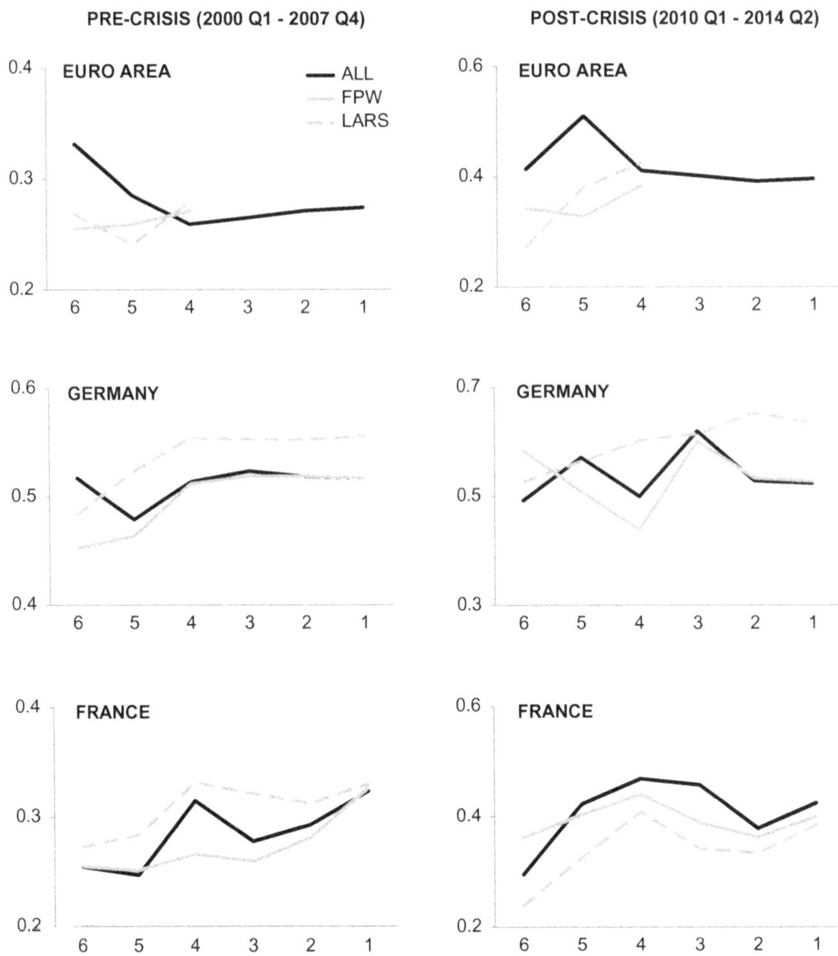

Fig. 1. RMSE from Best Pre-sample Selections. *Notes*: This figure shows the out-of-sample root-mean-squared error (vertical axes) of predictions 1–6 (horizontal axis) from the optimal selections. ALL refers to predictions based on all series, whereas FPW and LARS refer to selections from factor model prediction weights and LARS, respectively. The optimal selections have been determined from the pre-sample. The left-hand and right-hand panels show the RMSE in the pre- and post-crisis evaluation samples, respectively. For the euro area predictions 1–3, the optimal selection is given by the full set of series (ALL).

The above conclusions withstand various robustness checks. First, I used fixed model specifications, that is, applied the specification (r, p, q) as obtained from the full data set to all variable selections. Second, I inspected whether the selections obtained from nowcasts (i.e., prediction 5) may help in improving next-quarter predictions. Third, I used LARS selections derived from the original data $x_{i,t}$ instead of shifted data $x_{i,t}^{\#}$ as described in Section 5.1. These modifications had overall small effects on the results. As one exception, LARS selections for German next-quarter predictions were found to be uninformative in the pre-sample, which avoids the losses in the corresponding predictions in evaluation samples.

6. CONCLUSIONS

The paper has inspected the efficiency gains from variable selection in predictions from a dynamic factor model. I have compared two methods for this purpose, LARS and factor model prediction weights.

In contrast to earlier studies by Bai and Ng (2008), Schumacher (2010) and Caggiano et al. (2011), which performed variable selection in the evaluation sample, this paper inspects the success of variable selection from a pre-sample. The results still confirm the earlier findings that variable selection methods tend to improve the efficiency of predictions. However, gains are moderate and should not be taken for granted. First, both the Monte Carlo simulations and the empirical findings indicate that such gains are small, at best, for one-step ahead forecasts. Second, the Monte Carlo simulations suggest that the relationship between the specification of the factor model and the success of variable selection is not straightforward.

For these reasons, variable selection methods should, first of all, be robust against avoiding potential losses in forecast precision in an out-of-sample context. The evidence presented in this paper suggests that factor model prediction weights perform better than LARS in this respect. In the Monte Carlo simulations they were better in identifying informative series and provided smaller out-of-sample forecast errors. LARS, in turn, showed signs of overfitting: pre-sample forecasts suggested gains that did not necessarily carry over to the out-of-sample case. Similarly, in the empirical application, pre-sample selections from LARS occasionally gave wrong signals that resulted in losses in out-of-sample predictions, whereas prediction weights provided consistent gains.

In the context of a dynamic factor model, factor model prediction weights obviously provide a model-consistent means of variable selection. One question for future research is whether they are useful for the pre-screening of variables also in the context of other forecasting methods.

NOTES

1. Similarly, Barhoumi, Darné and Ferrara (2010) and Alvarez et al. (2012) find that forecasts from small data sets that consist only of aggregate indicators outperform those from larger data sets with a high number of sectoral indicators. Taking a different perspective, Alvarez et al. (2012) and Poncela et al. (2015) show that similar issues arise with the precision of factor estimates.

2. Data x_t load only on current values of factors. However, the representation of Doz et al. (2011) can be derived from a version of a general DFM with q dynamic factors where x_t loads on current and lagged values (see Stock & Watson, 1995).

3. Jungbacker, Koopman, and van der Wel (2011) and Bańbura and Modugno (2014) present maximum likelihood methods to estimate the model, possibly with missing data. They report gains in forecast precision to be limited. Given the high number of estimates in my experiments, with recursive estimation in a variable selection loop, I stick to the less time-consuming two-step estimator.

4. In the application, weights $\omega_l(h)$ decay quickly unless the factors f_t are highly persistent. The choice of k is therefore not critical. Cumulative weights do not measure the predictive gain of a series across all lags precisely. Such a measure could be obtained from $P_{t+h|h}$ to find the loss in forecast precision when eliminating series j from the data (Giannone et al., 2008). However, this becomes computationally very expensive in a stepwise approach as it requires $O(n^2)$ runs of the Kalman filter and smoother.

5. For LARS, I use code by Karl Skoglund (http://www.cad.zju.edu.cn/home/dengcai/Data/code/lars.m).

6. Predictions based on based $x_{t,J}^{w,s}$ fall only marginally short of predictions based on $x_{t,J}^{*,s}$, that is, under the assumption of perfect knowledge about the ranking of series.

7. Outliers are defined as observations that deviate by more than twice the interquintile distance from the median. The interquintile distance is defined as the difference between the 80% and 20% quantiles of the empirical distribution. For principal components, outliers are replaced with the median, for Kalman filtering they are set as missing.

8. The calculation of naive and AR(1) forecasts takes account of the timing of the publication dates of the GDP flash estimates. Forecasts are based on recursive estimates.

9. I do not present tests for forecast accuracy, as they are computationally very costly. The tests give rise to non-standard test distributions, because the individual selections are nested in the full data set, which requires bootstrap techniques. While Hubrich and West (2010) provide a test statistic for nested models that uses standard distributions, the test is not applicable as the (one-sided) alternative hypothesis

goes in the wrong direction: the test examines whether *adding* data to a minimal model would help in reducing forecast errors.

ACKNOWLEDGMENTS

The views expressed in this paper are those of the author and do not necessarily reflect the views of the ECB. The author would like to thank Marta Bańbura, Kirstin Hubrich, Christian Schumacher, Bernd Schwaab, and two anonymous referees for their helpful discussions.

REFERENCES

Altissimo, F., Cristadoro, R., Forni, M., Lippi, M., & Veronese, G. (2010). New EuroCoin: Tracking economic growth in real time. *The Review of Economics and Statistics*, *92*(4), 1024–1034.

Alvarez, R., Camacho, M., & Perez-Quiros, G. (2012). *Finite sample performance of small versus large scale dynamic factor models*, CEPR Discussion Paper No. 8867.

Angelini, E., Camba-Mendez, G., Giannone, D., Reichlin, L., & Rünstler, G. (2011). Short-term forecasts of euro area GDP growth. *Econometrics Journal*, *14*, C25–C44.

Bai, J., & Ng, S. (2002). Determining the number of factors in approximate factor models. *Econometrica*, *70*(1), 191–221.

Bai, J., & Ng, S. (2007). Determining the number of primitive shocks in factor models. *Journal of Business and Economics Statistics*, *25*, 52–60.

Bai, J., & Ng, S. (2008). Forecasting economic series using targeted predictors. *Journal of Econometrics*, *146*, 304–317.

Bańbura, M., & Modugno, M. (2014). Maximum likelihood estimation of factor models on data sets with arbitrary pattern of missing data. *Journal of Applied Econometrics*, *29*(1), 133–160.

Bańbura, M., & Rünstler, G. (2011). A look into the factor model black box: Publication lags and the role of hard and soft data in forecasting GDP. *International Journal of Forecasting*, *27*(2), 333–346.

Barhoumi, K., Darné, O., & Ferrara, L. (2010). Are disaggregate data useful for factor analysis in forecasting French GDP? *Journal of Forecasting*, *29*(1–2), 132–144.

Bessec, M. (2013). Short-term forecasts of French GDP: A dynamic factor model with targeted predictors. *Journal of Forecasting*, *32*, 500–511.

Boivin, J., & Ng, S. (2006). Are more data always better for factor analysis? *Journal of Econometrics*, *132*(1), 169–194.

Caggiano, G., Kapetianos, G., & Labhard, V. (2011). Are more data always better for factor analysis: Results for the euro area, the six largest euro area countries and the UK. *Journal of Forecasting*, *30*, 736–752.

Doz, C., Giannone, D., & Reichlin, L. (2011). A quasi maximum likelihood approach for large approximate dynamic factor models. *Journal of Econometrics*, *164*(1), 188–205.

Durbin, J., & Koopman, S. J. (2001). *Time series analysis by state space methods*. Oxford: Oxford University Press.

Efron, B., Hastie, T., Johnstone, I., & Tibshirani, R. (2004). Least angle regression. *Annals of Statistics*, *32*(2), 407–499.

Giannone, D., Reichlin, L., & Small, D. (2008). Nowcasting: The real-time informational content of macroeconomic data. *Journal of Monetary Economics*, *55*(4), 665–676.

Harvey, A. C. (1989). *Forecasting, structural time series models, and the Kalman filter*. Cambridge: Cambridge University Press.

Harvey, A. C., & Koopman, S. J. (2003). Computing observation weights for signal extraction and filtering. *Journal of Economic Dynamics and Control*, *27*, 1317–1333.

Hubrich, K., & West, K. (2010). Forecast evaluation of small nested model sets. *Journal of Applied Econometrics*, *25*(4), 574–594.

Jungbacker, B., Koopman, S. J., & van der Wel, M. (2011). Dynamic factor analysis in the presence of missing data. *Journal of Economic Dynamics and Control*, *35*(8), 1358–1368.

Marcellino, M., & Schuhmacher, C. (2010). Factor-MIDAS for now- and forecasting with ragged-edge data: A model comparison for German GDP. *Oxford Bulletin of Economics and Statistics*, *72*, 518–550.

Mariano, R., & Murasawa, Y. (2010). A coincident index, common factors, and monthly real GDP. *Oxford Bulletin of Economics and Statistics*, *72*(1), 27–46.

Poncela, P., & Ruiz, E. (2015). More is not always better: Back to the Kalman filter in dynamic factor models. In S. J. Koopman & N. G. Shephard (Eds.), *Unobserved components and time series econometrics*. Oxford: Oxford University Press.

Rünstler, G., Barhoumi, K., Benk, S., Cristadoro, R., den Reijer, A., Jakataine, A., …, van Nieuwenhuyze, C. (2009). Short-term forecasting of GDP using large data sets. *Journal of Forecasting*, *28*(7), 595–611.

Schumacher, C. (2010). Factor forecasting using international targeted predictors: The case of German GDP. *Economics Letters*, *107*(2), 95–98.

Stock, J. H., & Watson, M. W. (1995). *Implications of dynamic factor models for VAR analysis*. Princeton, NJ: Princeton University mimeo.

Stock, J. H., & Watson, M. W. (2002a). Macroeconomic forecasting using diffusion indexes. *Journal of Business and Economics Statistics*, *20*, 147–162.

Stock, J. H., & Watson, M. W. (2002b). Forecasting using principal components from a larger number of predictors. *Journal of the American Statistical Association*, *97*, 1167–1179.

APPENDIX: STATE-SPACE FORM

The transition equation of the model described in Section 3 with $p = 1$ is given by

$$\begin{bmatrix} I_r & 0 & 0 & 0 & 0 \\ -\beta' & 1 & 0 & 0 & 0 \\ 0 & 0 & 1 & 0 & 0 \\ 0 & -1 & -1 & 1 & 0 \\ 0 & 0 & 0 & -\frac{1}{3} & 1 \end{bmatrix} \begin{bmatrix} f_{t+1} \\ y_{t+1} \\ y_t \\ y_{t+1}^{(3)} \\ Q_{t+1} \end{bmatrix} = \begin{bmatrix} 0 \\ \mu \\ 0 \\ 0 \\ 0 \end{bmatrix} + \begin{bmatrix} \Psi_1 & 0 & 0 & 0 & 0 \\ 0 & 0 & 0 & 0 & 0 \\ 0 & 1 & 0 & 0 & 0 \\ 0 & 0 & 1 & 0 & 0 \\ 0 & 0 & 0 & 0 & \Xi_t \end{bmatrix} \begin{bmatrix} f_t \\ y_t \\ y_{t-1} \\ y_t^{(3)} \\ Q_t \end{bmatrix} + \begin{bmatrix} B\eta_t \\ \varepsilon_{t+1} \\ 0 \\ 0 \\ 0 \end{bmatrix}$$

where I_r denotes the $r \times r$ identity matrix. Temporal aggregation rules are implemented in a recursive way from

$$Q_t = \Xi_{t-1} Q_{t-1} + \frac{1}{3} y_t^{(3)}$$

where $\Xi_{t-1} = 0$ in the first month and $\Xi_{t-1} = 1$ otherwise (see Harvey, 1989: 309ff). As a result, the required identities hold in the third month of the quarter, with $y_t^Q = Q_t$.

The equation is to be pre-multiplied by the inverse of the left-hand matrix to achieve the standard state-space form.

The observation equation is given by

$$\begin{bmatrix} x_t \\ y_t^Q \end{bmatrix} = \begin{bmatrix} \Lambda & 0 & 0 & 0 & 0 \\ 0 & 0 & 0 & 0 & 1 \end{bmatrix} \begin{bmatrix} f_t \\ y_t \\ y_{t-1} \\ y_t^{(3)} \\ Q_t \end{bmatrix} + \begin{bmatrix} \xi_t \\ 0 \end{bmatrix}$$

The second row, related to y_t^Q, is skipped in months 1 and 2 of the quarter.

Table A1. Data Euro Area.

No.	Series	Publication Lag (months)	Trans-formation Code	Ranking Prediction Weights	Ranking LARS
1	Index of notional stock – Money M1	1	2	40	
2	Index of notional stock – Money M2	1	2	37	30
3	Index of notional stock – Money M3	1	2	48	
4	Index of Loans	1	2		24
5	ECB Nominal effective exch. rate	0	2	5	17
6	ECB Real effective exch. rate CPI deflated	0	2	4	15
7	ECB Real effective exch. rate producer prices deflated	0	2		
8	Exch. rate: USD/EUR	0	2	60	
9	Exch. rate: GBP/EUR	0	2		
10	Exch. rate: YEN/EUR	0	2	12	
11	World market prices of raw materials in Euro, total, HWWA	2	2	13	
12	World market prices of raw materials in Euro, total, excl energy, HWWA	2	2	6	13
13	World market prices, crude oil, USD, HWWA	0	2	32	
14	Gold price, USD, fine ounce	0	2	35	19
15	Brent Crude, 1 month fwd, USD/BBL converted in euro	0	2	27	
16	Retail trade, except of motor vehicles and motorcycles	2	2		14
17	IP-Total industry	2	2	20	3
18	IP-Total Industry (excl construction)	2	2	18	2
19	IP-Manufacturing	2	2	17	
20	IP-Construction	2	2	57	
21	IP-Total Industry excl construction and MIG Energy	2	2	54	
22	IP-Energy	2	2		
23	IP-MIG Capital Goods Industry	2	2	59	
24	IP-MIG Durable Consumer Goods Industry	2	2	31	
25	IP-MIG Energy	2	2		
26	IP-MIG Intermediate Goods Industry	2	2	16	

Table A1. *(Continued)*

No.	Series	Publication Lag (months)	Trans-formation Code	Ranking Prediction Weights	Ranking LARS
27	IP-MIG Non-durable Consumer Goods Industry	2	2	58	26
28	IP-Manufacture of basic metals	2	2		4
29	IP-Manufacture of chemicals and chemical products	2	2	55	8
30	IP-Manufacture of electrical machinery and apparatus	2	2	21	5
31	IP-Manufacture of machinery and equipment	2	2		
32	IP-Manufacture of pulp, paper and paper products	2	2	28	
33	IP-Manufacture of rubber and plastic products	2	2	19	20
34	Industry Survey: Industrial Confidence Indicator	1	1	3	
35	Industry Survey: Production trend observed in recent months	1	1	14	
36	Industry Survey: Assessment of order-book levels	1	1	2	
37	Industry Survey: Assessment of export order-book levels	1	1	1	
38	Industry Survey: Assessment of stocks of finished products	1	1	10	
39	Industry Survey: Production expectations for the months ahead	1	1	9	
40	Industry Survey: Employment expectations for the months ahead	1	1	11	
41	Industry Survey: Selling price expectations for the months ahead	1	1	15	
42	Consumer Survey: Consumer Confidence Indicator	1	1	24	23
43	Consumer Survey: General economic situation over last 12 months	1	1	22	
44	Consumer Survey: General economic situation over next 12 months	1	1	23	21
45	Consumer Survey: Price trends over last 12 months	1	1	36	11
46	Consumer Survey: Price trends over next 12 months	1	1	53	28

Table A1. (*Continued*)

No.	Series	Publication Lag (months)	Trans-formation Code	Ranking Prediction Weights	Ranking LARS
47	Consumer Survey: Unemployment expectations over next 12 months	1	1	25	22
48	Construction Survey: Construction Confidence Indicator	1	1	39	
49	Construction Survey: Trend of activity compared with preceding months	1	1	44	29
50	Construction Survey: Assessment of order books	1	1	43	27
51	Construction Survey: Employment expectations for the months ahead	1	1	38	9
52	Construction Survey: Selling price expectations for the months ahead	1	1	50	1
53	Retail Trade Survey: Retail Confidence Indicator	1	1	47	
54	Retail Trade Survey: Present business situation	1	1	52	10
55	Retail Trade Survey: Assessment of stocks	1	1		
56	Retail Trade Survey: Expected business situation	1	1	49	
57	Retail Trade Survey: Employment expectations	1	1	41	7
58	New passenger car registrations	1	2	33	12
59	Eurostoxx 500	0	2	8	
60	Eurostoxx 325	0	2	7	
61	US S&P 500 composite index	0	2		16
62	US, Dow Jones, industrial average	0	2	56	
63	US, Treasury Bill rate, three-month	0	1	34	25
64	US Treasury notes & bonds yield, 10 years	0	1	26	
65	Money M2 in the United States	1	2		6
66	US, Unemployment rate	1	1	42	
67	US, IP total excl construction	1	2	45	
68	US, Employment, civilian	1	2	46	
69	US, Production expectations in manufacturing	1	1	29	
70	US, Consumer expectations index	0	1	30	
71	10-year government bond yield	0	1	51	18

Transformation code: 1 = monthly difference, 2 = monthly growth rate; *Rankings*: Ranking of series in stepwise selection (1 = added first/eliminated last).

Table A2. Data Germany.

No.	Series	Publication Lag (months)	Trans-formation Code	Ranking Prediction Weights	Ranking LARS
1	Index of notional stock – Money M1	1	2	36	17
2	Index of notional stock – Money M2	1	2	54	
3	Index of notional stock – Money M3	1	2	31	
4	Index of Loans	1	2	58	11
5	ECB Nominal effective exch. rate	1	2		28
6	ECB Real effective exch. rate CPI deflated	1	2	56	
7	ECB Real effective exch. rate producer prices deflated	1	2	55	
8	Exch. rate: USD/EUR	1	2	50	
9	Exch. rate: GBP/EUR	1	2	24	
10	Exch. rate: YEN/EUR	1	2		
11	World market prices of raw materials in Euro, total, HWWA	2	2	32	
12	World market prices of raw materials in Euro, total, excl energy, HWWA	2	2	30	24
13	World market prices, crude oil, USD, HWWA	1	2	42	
14	Gold price, USD, fine ounce	1	2	28	6
15	Brent Crude, 1 month fwd, USD/BBL converted in euro	0	2	34	27
16	IP-Total industry	2	2	16	
17	IP-Total Industry (excl construction)	2	2	20	
18	IP-Manufacturing	2	2	33	
19	IP-Construction	2	2	46	9
20	IP-Total Industry excl construction and MIG Energy	2	2	17	
21	IP-Energy	2	2		
22	IP-MIG Capital Goods Industry	2	2		15
23	IP-MIG Durable Consumer Goods Industry	2	2	40	13
24	IP-MIG Energy	2	2	60	25
25	IP-MIG Intermediate Goods Industry	2	2	38	
26	IP-MIG Non-durable Consumer Goods Industry	2	2	59	4
27	IP-Manufacture of basic metals	2	2	41	2
28	IP-Manufacture of chemicals and chemical products	2	2	47	21

Table A2. (*Continued*)

No.	Series	Publication Lag (months)	Trans-formation Code	Ranking Prediction Weights	Ranking LARS
29	IP-Manufacture of electrical machinery and apparatus	2	2	45	12
30	IP-Manufacture of machinery and equipment	2	2		
31	IP-Manufacture of pulp, paper, and paper products	2	2		
32	IP-Manufacture of rubber and plastic products	2	2	37	
33	Industry Survey: Industrial Confidence Indicator	1	1	6	
34	Industry Survey: Production trend observed in recent months	1	1	25	
35	Industry Survey: Assessment of order-book levels	1	1	8	
36	Industry Survey: Assessment of export order-book levels	1	1	13	
37	Industry Survey: Assessment of stocks of finished products	1	1	10	26
38	Industry Survey: Production expectations for the months ahead	1	1		18
39	Industry Survey: Employment expectations for the months ahead	1	1	53	3
40	Industry Survey: Selling price expectations for the months ahead	1	1	22	
41	Consumer Survey: Consumer Confidence Indicator	1	1	5	
42	Consumer Survey: General economic situation over last 12 months	1	1	4	
43	Consumer Survey: General economic situation over next 12 months	1	1	3	
44	Consumer Survey: Price trends over last 12 months	1	1	23	
45	Consumer Survey: Price trends over next 12 months	1	1	11	16
46	Consumer Survey: Unemployment expectations over next 12 months	1	1	2	
47	Construction Survey: Construction Confidence Indicator	1	1	7	

Table A2. *(Continued)*

No.	Series	Publication Lag (months)	Trans-formation Code	Ranking Prediction Weights	Ranking LARS
48	Construction Survey: Trend of activity compared with preceding months	1	1	15	
49	Construction Survey: Assessment of order books	1	1	1	20
50	Construction Survey: Employment expectations for the months ahead	1	1	9	7
51	Construction Survey: Selling price expectations for the months ahead	1	1	12	1
52	Retail Trade Survey: Retail Confidence Indicator	1	1		
53	Retail Trade Survey: Present business situation	1	1	35	
54	Retail Trade Survey: Assessment of stocks	1	1	43	
55	Retail Trade Survey: Expected business situation	1	1	26	8
56	Retail Trade Survey: Employment expectations	1	1		10
57	New passenger car registrations	1	2	14	22
58	Index of Employment, Construction	3	2		30
59	Index of Employment, Manufacturing	3	2		
60	Eurostoxx 500	0	2	19	14
61	Eurostoxx 325	0	2	18	
62	US S&P 500 composite index	0	2	39	
63	US, Dow Jones, industrial average	1	2	21	
64	US, Treasury Bill rate, 3-month	1	1	57	
65	US Treasury notes & bonds yield, 10 years	1	1	48	
66	Money M2 in the United States	1	2	44	19
67	US, Unemployment rate	1	1	51	
68	US, IP total excl construction	1	2		
69	US, Employment, civilian	1	2	49	5
70	US, Production expectations in manufacturing	1	1	27	23
71	US, Consumer expectations index	0	1	52	
72	10-year government bond yield	1	1	29	29

Transformation code: 1 = monthly difference, 2 = monthly growth rate; *Rankings*: Ranking of series in stepwise selection (1 = added first/eliminated last).

Table A3. Data France.

No.	Series	Publication Lag (months)	Trans-formation Code	Ranking Prediction Weights	Ranking LARS
1	Index of notional stock – Money M1	1	2	49	
2	Index of notional stock – Money M2	1	2	45	
3	Index of notional stock – Money M3	1	2	34	
4	Index of Loans	1	2		8
5	ECB Nominal effective exch. rate	1	2	30	
6	ECB Real effective exch. rate CPI deflated	1	2	29	
7	ECB Real effective exch. rate producer prices deflated	1	2	52	
8	Exch. rate: USD/EUR	1	2		16
9	Exch. rate: GBP/EUR	1	2	27	19
10	Exch. rate: YEN/EUR	1	2		22
11	World market prices of raw materials in Euro, total, HWWA	2	2		21
12	World market prices of raw materials in Euro, total, excl energy, HWWA	2	2	48	
13	World market prices, crude oil, USD, HWWA	1	2	59	
14	Gold price, USD, fine ounce	1	2	28	12
15	Brent Crude, 1 month fwd, USD/BBL converted in euro	0	2	32	
16	IP-Total industry	2	2	21	6
17	IP-Total Industry (excl construction)	2	2	22	2
18	IP-Manufacturing	2	2	20	
19	IP-Construction	2	2	41	
20	IP-Total Industry excl construction and MIG Energy	2	2	50	
21	IP-Energy	2	2		11
22	IP-MIG Capital Goods Industry	2	2	51	4
23	IP-MIG Durable Consumer Goods Industry	2	2	55	
24	IP-MIG Energy	2	2		
25	IP-MIG Intermediate Goods Industry	2	2	19	
26	IP-MIG Non-durable Consumer Goods Industry	2	2	44	23
27	IP-Manufacture of basic metals	2	2	43	
28	IP-Manufacture of chemicals and chemical products	2	2	42	

Table A3. *(Continued)*

No.	Series	Publication Lag (months)	Trans-formation Code	Ranking Prediction Weights	Ranking LARS
29	IP-Manufacture of electrical machinery and apparatus	2	2	53	26
30	IP-Manufacture of machinery and equipment	2	2	38	17
31	IP-Manufacture of pulp, paper and paper products	2	2	37	
32	IP-Manufacture of rubber and plastic products	2	2	23	7
33	Industry Survey: Industrial Confidence Indicator	1	1	5	
34	Industry Survey: Production trend observed in recent months	1	1	7	
35	Industry Survey: Assessment of order-book levels	1	1	6	
36	Industry Survey: Assessment of export order-book levels	1	1	12	
37	Industry Survey: Assessment of stocks of finished products	1	1	15	28
38	Industry Survey: Production expectations for the months ahead	1	1	14	
39	Industry Survey: Employment expectations for the months ahead	1	1	36	9
40	Industry Survey: Selling price expectations for the months ahead	1	1	16	
41	Consumer Survey: Consumer Confidence Indicator	1	1	9	
42	Consumer Survey: General economic situation over last 12 months	1	1	8	30
43	Consumer Survey: General economic situation over next 12 months	1	1	10	
44	Consumer Survey: Price trends over last 12 months	1	1		14
45	Consumer Survey: Price trends over next 12 months	1	1		20
46	Consumer Survey: Unemployment expectations over next 12 months	1	1	11	
47	Construction Survey: Construction Confidence Indicator	1	1	4	1

Table A3. (*Continued*)

No.	Series	Publication Lag (months)	Trans-formation Code	Ranking Prediction Weights	Ranking LARS
48	Construction Survey: Trend of activity compared with preceding months	1	1	3	
49	Construction Survey: Assessment of order books	1	1	2	3
50	Construction Survey: Employment expectations for the months ahead	1	1	1	
51	Construction Survey: Selling price expectations for the months ahead	1	1	13	13
52	Retail Trade Survey: Retail Confidence Indicator	1	1	18	
53	Retail Trade Survey: Present business situation	1	1	17	
54	Retail Trade Survey: Assessment of stocks	1	1		27
55	Retail Trade Survey: Expected business situation	1	1	24	25
56	Retail Trade Survey: Employment expectations	1	1		
57	New passenger car registrations	1	2	54	10
58	Unemployment rate, total	2	1	33	29
59	US, Dow Jones, industrial average	1	2	58	
60	US, Treasury Bill rate, 3-month	1	1	39	
61	US Treasury notes & bonds yield, 10 years	1	1	31	15
62	Money M2 in the United States	1	2	56	5
63	US, Unemployment rate	1	1	47	24
64	US, IP total excl construction	1	2	40	
65	US, Employment, civilian	1	2	46	
66	US, Production expectations in manufacturing	1	1		
67	US, Consumer expectations index	0	1	25	18
68	Eurostoxx 500	1	2	60	
69	Eurostoxx 325	1	2	26	
70	US S&P 500 composite index	1	2	57	
71	10-year government bond yield	1	1	35	

Transformation code: 1 = monthly difference, 2 = monthly growth rate; *Rankings*: Ranking of series in stepwise selection (1 = added first/eliminated last).